Lecture Notes on Data Engineering and Communications Technologies

Volume 49

Series Editor

Fatos Xhafa, Technical University of Catalonia, Barcelona, Spain

The aim of the book series is to present cutting edge engineering approaches to data technologies and communications. It will publish latest advances on the engineering task of building and deploying distributed, scalable and reliable data infrastructures and communication systems.

The series will have a prominent applied focus on data technologies and communications with aim to promote the bridging from fundamental research on data science and networking to data engineering and communications that lead to industry products, business knowledge and standardisation.

** Indexing: The books of this series are submitted to SCOPUS, ISI Proceedings, MetaPress, Springerlink and DBLP **

More information about this series at http://www.springer.com/series/15362

A. Pasumpon Pandian · Ram Palanisamy ·
Klimis Ntalianis
Editors

Proceeding of the International Conference on Computer Networks, Big Data and IoT (ICCBI - 2019)

 Springer

Editors
A. Pasumpon Pandian
Department of CSE
Vaigai College of Engineering
Melur, Tamil Nadu, India

Ram Palanisamy
Department of Business Administration,
The Gerald Schwartz School of Business
StFX University
Antigonish, NS, Canada

Klimis Ntalianis
Electrical and Computer Engineering
University of Applied Sciences
Egaleo, Attiki, Greece

ISSN 2367-4512   ISSN 2367-4520   (electronic)
Lecture Notes on Data Engineering and Communications Technologies
ISBN 978-3-030-43191-4   ISBN 978-3-030-43192-1   (eBook)
https://doi.org/10.1007/978-3-030-43192-1

This Springer imprint is published by the registered company Springer Nature Switzerland AG
The registered company address is: Gewerbestrasse 11, 6330 Cham, Switzerland

We are honored to dedicate the proceedings of ICCBI 2019 to all the participants and editors of ICCBI 2019.

Foreword

It is with deep satisfaction that I write this Foreword to the Proceedings of the ICCBI 2019 held in, Vaigai College of Engineering, Madurai, Tamil Nadu, December 19–20, 2019.

This conference was bringing together researchers, academics and professionals from all over the world, experts in computer networks, big data and Internet of Things.

This conference particularly encouraged the interaction of research students and developing academics with the more established academic community in an informal setting to present and to discuss new and current work. The papers contributed the most recent scientific knowledge known in the field of computer networks, big data and Internet of Things. Their contributions helped to make the conference as outstanding as it has been. The local organizing committee members and their helpers put much effort into ensuring the success of the day-to-day operation of the meeting.

We hope that this program will further stimulate research in data communication and computer networks, Internet of Things, wireless communication, big data and cloud computing and also provide practitioners with better techniques, algorithms and tools for deployment. We feel honored and privileged to serve the best recent developments to you through this exciting program.

We thank all authors and participants for their contributions.

R. Saravanan

Preface

This conference proceedings volume contains the written versions of most of the contributions presented during the conference of ICCBI 2019. The conference provided a setting for discussing recent developments in a wide variety of topics including computer networks, big data and Internet of Things. The conference has been a good opportunity for participants coming from various destinations to present and discuss topics in their respective research areas.

ICCBI 2019 conference tends to collect the latest research results and applications on computer networks, big data and Internet of Things. It includes a selection of 109 papers from 327 papers submitted to the conference from universities and industries all over the world. All of the accepted papers were subjected to strict peer reviewing by 2–4 expert referees. The papers have been selected for this volume because of quality and the relevance to the conference.

ICCBI 2019 would like to express our sincere appreciation to all the authors for their contributions to this book. We would like to extend our thanks to all the referees for their constructive comments on all papers; especially, we would like to thank organizing committee for their hardworking. Finally, we would like to thank the Springer Publications for producing this volume.

A. Pasumpon Pandian
Ram Palanisamy
Klimis Ntalianis
Guest Editors

Acknowledgments

ICCBI 2019 would like to acknowledge the excellent work of our conference organizing the committee, keynote speakers for their presentation on December 19–20, 2019. The organizers also wish to acknowledge publicly the valuable services provided by the reviewers.

On behalf of the editors, organizers, authors and readers of this conference, we wish to thank the keynote speakers and the reviewers for their time, hard work and dedication to this conference. The organizers wish to acknowledge Dr. R. Saravanan, Dr. R. Thiruchenthuran, Thiru. S. Kamalakannan, Thiru. S. Balasubramanian, Thiru. S. Singaravelan, for the discussion, suggestion and cooperation to organize the keynote speakers of this conference. The organizers also wish to acknowledge for speakers and participants who attend this conference. Many thanks are given for all persons who help and support this conference. ICCBI 2019 would like to acknowledge the contribution made to the organization by its many volunteers. Members contribute their time, energy and knowledge at a local, regional and international level.

We also thank all the chair persons and conference committee members for their support.

Contents

Wireless Sensor Networks Security Issues, Attacks and Challenges: A Survey 1
T. Shanmugapriya, K. Kousalya, J. Rajeshkumar, and M. Nandhini

A Survey on Machine Learning Based Fault Tolerant Mechanisms in Cloud Towards Uncertainty Analysis 13
K. Nivitha and P. Pabitha

IoT Based Attendance Management System Using Google Assistant ... 21
M. Niharika and B. Karuna Sree

A Survey on Privacy Preserving Approaches on Health Care Big Data in Cloud 32
Lipsa Nayak and V. Jayalakshmi

Effective Record Search and Update Using Design Patterns - A Case Study of Blood Bank Mobile Application 39
D. Venkata Subramanian, R. Sugumar, K. Maneesh, V. Manoj, and A. Venkat Kishore

Comprehensive and Comparative Study of Efficient Location Tracking Based on Apriori and Dijkstra Algorithms 48
D. Venkata Subramanian, R. Sugumar, N. Dhipikha, R. Vinothini, S. Kavitha, and A. Harsha Anchaliya

Weather Data Analysis for Jaipur City: Using MongoDB 60
Talha Khan, Zaid Khatri, Shiburaj Pappu, Imtiyaz Khan, and Shakila Shaikh

A Survey on Intrusion Detection System Using Artificial Intelligence 67
Sona Solani and Nilesh Kumar Jadav

**Trusted and Secured E-Voting Election System Based on Block
Chain Technology** 81
A. Anny Leema, Zameer Gulzar, and P. Padmavathy

**Users' Attitude on Perceived Security of Enterprise Systems
Mobility: A Conceptual Model** 89
Ramaraj Palanisamy

Smart Old Age Home Using Zigbee 100
N. Shubha, S. Sahana, K. Pavithra, M. S. Meghana, and K. Panimozhi

**Performance Improvement in 6G Networks Using MC-CDMA
and mMIMO** 112
A. Vijay and K. Umadevi

**An Adaptive and Opportunistic Based Routing Protocol in Flying
Ad Hoc Networks (FANETs): A Survey** 119
O. Aruna and Amit Sharma

**A Comparison Between Robust Image Encryption
and Watermarking Methods for Digital Image Protection** 128
Sreya Vemuri and Rejo Mathew

**Comparing Sentiment Analysis from Social Media
Platforms – Insights and Implications** 139
Lakshmi Prayaga, Chandra Prayaga, and Krishna Devulapalli

A Smart Farm – An Introduction to IoT for Generation Z 145
Lakshmi Prayaga, Chandra Prayaga, Aaron Wade, and Andrew Hart

Algorithms Used for Scene Perception in Driverless Cars 153
Antriksh Tiwari and Rejo Mathew

**Machine Learning Algorithms for Early Prediction
of Heart Disease** 162
Akankasha Sinha and Rejo Mathew

Analysis and Review of Cloud Based Encryption Methods 169
Vicky Yadav and Rejo Mathew

Data Storage Security Issues in Cloud Computing 177
Vedant Paul and Rejo Mathew

Challenges and Solutions in Recommender Systems 188
Abhishek Nair and Rejo Mathews

Review of Machine Learning in Geosciences and Remote Sensing 195
Noel David and Rejo Mathew

Data Storage Security Issues and Solutions in Cloud Computing 205
Tapendra Singh Rathore and Rejo Mathew

**Big Data and Internet of Things for Smart Data Analytics Using
Machine Learning Techniques** 213
J. Betty Jane and E. N. Ganesh

**Comparison of Cloud Computing Security Threats and Their
Counter Measures** 224
Shourya Rohilla and Rejo Mathew

**An Effective Kapur's Segmentation Based Detection
and Classification Model for Citrus Diseases Diagnosis System** 232
C. Senthilkumar and M. Kamarasan

**Anomaly Detection Using Anomalous Behavior at Program
Environment Through Relative Difference Between
Return Addresses** 240
Goverdhan Reddy Jidiga and P. Sammulal

Satellite Image Enhancement and Restoration 252
Vivek Giri and Sudhriti Sen Gupta

Role of AI in Gaming and Simulation 259
Shivam Tyagi and Sudhriti Sengupta

Analysis of MRI Images Using Image Processing Technique 267
Sudhriti Sengupta, Ashutosh Dubey, and Neetu Mittal

Recent Trends of IoT in Smart City Development 275
Disha Kohli and Sudhriti Sen Gupta

Analysis of Biometric Modalities 281
Ayushi Wajhal and Sudhriti Sen Gupta

**Comparative Analysis of Contrast Enhancement Techniques
for MRI Images** 290
Sudhriti Sengupta and Astha Negi

Global User Social Networking Ranking Statistics 297
P. Ajitha, Bana Suresh Kumar Reddy, and Balina Prudhvi

**Detecting Spam Emails/SMS Using Naive Bayes, Support Vector
Machine and Random Forest** 306
Vasudha Goswami, Vijay Malviya, and Pratyush Sharma

Detection of Flooding Attacks Using Multivariate Analysis 314
Priyanka Meel and Tanmay Singh

Human Gait Recognition 325
Vaishnavi Ahuja and Rejo Mathew

Reduction of Traffic on Roads Using Big Data Applications 333
P. Ajitha, A. Sivasangari, G. Lakshmi Mounica,
and Lakshmi Prathyusha

Survey on Utilization of Internet of Things in Health
Monitoring Systems 340
R. Jane Preetha Princy, B. Brenda Jennifer, M. J. Abiya,
B. Thilagavathi, Saravanan Parthasarathy,
and Arun Raj Lakshminarayanan

AgroFarming - An IoT Based Approach for Smart
Hydroponic Farming 348
Aashray Mody and Rejo Mathew

Review on Dimensionality Reduction Techniques 356
Dhruv Chauhan and Rejo Mathews

Security Issues in Cloud Computing 363
Shivani Goyal and Rejo Mathew

A Review on Predicting Cardiovascular Diseases Using Data
Mining Techniques 374
V. Pavithra and V. Jayalakshmi

Adaptive Object Tracking Using Algorithms Employing
Machine Learning 381
Shubham Rai and Rejo Mathew

Emotion Recognition Using Physiological Signals 389
Mrigank Sharma and Rejo Mathew

Prediction of Sudden Cardiac Arrest Due to Diabetes Mellitus
Using Fuzzy Based Classification Approach 397
K. G. Rani Roopha Devi and R. Mahendra Chozhan

Suggesting a System to Enhance Decision Making in Location
Based Social Networks 412
R. Sridevi, G. Bhavani, and R. Meena

Security Challenges in NoSQL and Their Control Methods 420
Mahiraj Parmar and Rejo Mathew

Linux Server Based Automatic Online Ticketing Kiosk 428
Arvind Vishnubhatla

A Reliable Automation of Motorized Berth Climb 440
Arvind Vishnubhatla

Requirement Gathering for Multi-tasking Autonomous Bus
for Smart City Applications 449
Arvind Vishnubhatla

Offline Handwritten Devanagari Character Identification 457
Gita Sinha and Shailja Sharma

Behavior Anomaly Detection in IoT Networks 465
Dominik Soukup, Tomas Cejka, and Karel Hynek

**Review of Digital Data Protection Using the Traditional Methods,
Steganography and Cryptography** 474
Chinmaya M. Dharmadhikari and Rejo Mathew

Waste Management Techniques for Smart Cities 484
Vaibhav Agrawal and Rejo Mathew

**A Light Modulating Therapeutic Wearable Band
for 'Vision Health'** 492
Vijay A. Kanade

"Health Studio" – An Android Application for Health Assessment 500
Suresh Chalumuru, P. Geethika Choudary, Pranav Souri Itabada,
and Vineela Bolla

**Potential Candidate Selection Using Information Extraction
and Skyline Queries** 511
Farzana Yasmin, Mohammad Imtiaz Nur,
and Mohammad Shamsul Arefin

**An Effective Feature Extraction Based Classification Model Using
Canonical Particle Swarm Optimization with Convolutional Neural
Network for Glaucoma Diagnosis System** 523
Narmatha Venugopal and Kamarasan Mari

**Dominant Feature Descriptors with Self Organising Map
for Image Retrieval** 531
S. Sivakumar and S. Sathiamoorthy

**Medical Image Retrieval Using Efficient Texture and Color
Patterns with Neural Network Classifier** 539
C. Ashok Kumar and S. Sathiamoorthy

**Deterministic Type 2 Fuzzy Logic Based Unequal Clustering
Technique for Wireless Sensor Networks** 547
R. Sathiya Priya and K. Arutchelvan

Android Application Based Solid Waste Management 555
Raju A. Nadaf, Fuad A. Katnur, and Susen P. Naik

**An Approach for Detecting Man-In-The-Middle Attack Using DPI
and DFI** 563
Argha Ghosh and A. Senthilrajan

**Smart Irrigation and Crop Disease Detection Using Machine
Learning – A Survey** 575
Anushree Janardhan Rao, Chaithra Bekal, Y. R. Manoj, R. Rakshitha,
and N. Poornima

**Wireless-Sensor-Network with Mobile Sink Using Energy
Efficient Clustering** 582
K. Venkateswara Rao and G. L. Vara Prasad

**Internet of Things and Blockchain Based Distributed Energy
Management of Smart Micro-grids** 590
Leo Raju, V. Balaji, S. Keerthivasan, and C. Keerthivasan

**Bio-inspired Deoxyribonucleic Acid Based Data Obnubilating
Using Enhanced Computational Algorithms** 597
B. Adithya and G. Santhi

**Detection of Type 2 Diabetes Using Clustering Methods – Balanced
and Imbalanced Pima Indian Extended Dataset** 610
S. Nivetha, B. Valarmathi, K. Santhi, and T. Chellatamilan

**Detection of Depression Related Posts in Tweets Using
Classification Methods – A Comparative Analysis** 620
M. Mounika, N. Srinivasa Gupta, and B. Valarmathi

**Aspect Based Sentiment Classification and Contradiction Analysis
of Product Reviews** 631
Md. Shahadat Hossain, Md. Rashadur Rahman,
and Mohammad Shamsul Arefin

**A Novel Approach to Extract and Analyse Trending Cuisines
on Social Media** 645
R. Lokeshkumar, Omkar Vivek Sabnis, and Saikat Bhattacharyya

Alphanumeric Character Recognition on Tiny Dataset 657
Sujit S. Amin and Lata Ragha

**Sentiment Analysis in Movie Reviews Using Document Frequency
Difference, Gain Ratio and Kullback-Leibler Divergence as Feature
Selection Methods and Multi-layer Perceptron Classifier** 668
S. Vigneshwaran

**A Patch - Based Analysis for Retinal Lesion Segmentation
with Deep Neural Networks** 677
A Mary Dayana and W. R. Sam Emmanuel

**Survey of Onion Routing Approaches: Advantages, Limitations
and Future Scopes** 686
Mayank Chauhan, Anuj Kumar Singh, and Komal

A Technical Paper Review on Vehicle Tracking System 698
K. Hemachandran, Shubham Tayal, G. Sai Kumar,
Vamshikrishna Boddu, Swathi Mudigonda,
and Muralikrishna Emudapuram

Role of Wireless Communications in Railway Systems:
A Global Perspective 704
K. Krishna Chaitanya, K. S. Sravan, and B. Seetha Ramanjaneyulu

Protection of Microgrid with Ideal Optimization
Differential Algorithm 712
P. M. Khandare, S. A. Deokar, and A. M. Dixit

An Improved Energy Efficient Scheme for Data Aggregation
in Internet of Things (IoT) 721
Keshvi Sharma and Rakesh Kumar

A Hybrid Approach for Credit Card Fraud Detection Using Naive
Bayes and Voting Classifier 731
Bhagwant Jot Kaur and Rakesh Kumar

Quantum Inspired Evolutionary Algorithm for Web
Document Retrieval 741
Manas Kumar Yogi and Darapu Uma

Signature Recognition and Verification Using Zonewise
Statistical Features 748
Banashankaramma F. Lakkannavar, M. M. Kodabagi,
and Susen P. Naik

Home Security Using Smart Photo Frame 758
Zarinabegam K. Mundargi, Isa Muslu, Susen Naik, Raju A. Nadaf,
and Suvarna Kabadi

Automated System for Detecting Mental Stress of Users in Social
Networks Using Data Mining Techniques 769
Shraddha Sharma, Ila Sharma, and A. K. Sharma

Intelligent Request Grabber: Increases the Vehicle Traffic
Prediction Rate Using Social and Taxi Requests Based on LSTM 778
S. C. Rajkumar and L. Jegatha Deborah

Aggregation in IoT for Prediction of Diabetics with Machine
Learning Techniques 789
P. Punitha Ponmalar and C. R. Vijayalakshmi

EWS: An Efficient Workflow Scheduling Algorithm
for the Minimization of Response Time in Cloud Environment 799
G. Justy Mirobi and L. Arockiam

An Investigation Report on Spotting and Diagnosing Diseases
from the Images of Plant Leaves 811
S. Thenmozhi, Irshadh Ibrahim, Rekha Mohankumar, and Shahla Sohail

Test Case Minimization for Object Oriented Testing Using Random Forest Algorithm 824
Ajmer Singh, Diksha Katyal, and Deepa Gupta

Exploring the Design Considerations for Developing an Interactive Tabletop Learning Tool for Children with Autism Spectrum Disorder 834
Nazmul Hasan and Muhammad Nazrul Islam

A Cloud-Fog Based System Architecture for Enhancing Fault Detection in Electrical Secondary Distribution Network 845
Gilbert M. Gilbert, Shililiandumi Naiman, Honest Kimaro, and Nerey Mvungi

An Autonomous Intelligent Ornithopter 856
Sunita Suralkar, Smit Gangurde, Sanjeevkumar Chintakindi, and Haresh Chawla

Detection of Distributed Denial of Service Attack Using NSL-KDD Dataset - A Survey 866
I. Philo Prasanna and M. Suguna

A Survey on Efficient Storage and Retrieval System for the Implementation of Data Deduplication in Cloud 876
R. Vinoth and L. Jegatha Deborah

Comparison of Decision Tree-Based Learning Algorithms Using Breast Cancer Data 885
M. S. Dawngliani, N. Chandrasekaran, R. Lalmawipuii, and H. Thangkhanhau

Stock and Financial Market Prediction Using Machine Learning 897
Divyanshu Agrawal and Rejo Mathew

Review of Prediction of Chronic Disease Using Different Prediction Methods 903
Aadil Chheda and Rejo Mathew

Review of Software Defined Networking Based Firewall Issues and Solutions 909
Karan Garg and Rejo Mathew

Database Security: Attacks and Solutions 917
Sarvesh Soni and Rejo Mathew

Cloud Based Heterogeneous Big Data Integration and Data Analysis for Business Intelligence 926
T. Jayaraj and J. Abdul Samath

Search Engine Optimization Challenges and Solutions 934
Manashwi Singh and Rejo Mathew

**Performance Assessment of Different Machine Learning
Algorithms for Medical Decision Support Systems** 941
T. Ragupathi and M. Govindarajan

**Implementation of IOT in Multiple Functions Robotic Arm:
A Survey** 948
S. Gowri, Senduru Srinivasulu, U. Joy Blessy,
and K. Mariya Christeena Vinitha

**Real Time Traffic Signal and Speed Violation Control System
of Vehicles Using IOT** 953
S. Gowri, J. S. Vimali, D. U. Karthik, and G. A. John Jeffrey

**Enhanced Financial Module in Intelligent ERP
Using Cryptocurrency** 959
Komal Saxena and Arshdeep Singh Pahwa

Explorative Study of Artificial Intelligence in Digital Marketing 968
Mohd Zeeshan and Komal Saxena

**A Security Model for Enhancement of Social Engineering Process
with Implementation of Multifactor Authentication** 979
Sameer Gupta, Deepa Gupta, and Ruchika Bathla

Implementation of IoT for Trash Monitoring System 988
Vansh Manchanda, Ruchika Bathla, and Deepa Gupta

Author Index 999

Wireless Sensor Networks Security Issues, Attacks and Challenges: A Survey

T. Shanmugapriya[1(✉)], K. Kousalya[2], J. Rajeshkumar[1],
and M. Nandhini[1]

[1] SNS College of Technology, Coimbatore, India
priyamoons@gmail.com, rajeshkumarjmtech@gmail.com,
nandhinim24@gmail.com
[2] Kongu Engineering College, Erode, India
keerthi.kous@gmail.com

Abstract. In recent years, Wireless Sensor Networks [WSN] are developing a real workforce in true condition. WSNs are broadly utilized in the following checking and controlling applications. Inspite of their predominant role in wireless applications, it may also introduce some research complexities. This paper focuses on the survey of various security issues faced by wireless sensor networks such as data security at node and network level and recommends solutions to secure WSN. Despite the difficulties in wireless sensor network, this survey centers around the attacks of different varieties in WSNs such as physical, communication, Base station, and routing protocol attacks and provides some solutions for the attacks. This paper address the issues on versatility, correspondence steering conventions, and security. This study center around security prerequisites, WSN applications, distinctive attacks and safeguards, and various ongoing issues and difficulties. This paper gives a brief overview on challenges of WSN like routing and application based challenges, key management, and new technology based challenges. Moreover, Software Defined Networking (SDN) mechanism is used to support the bendable routing and communication between sensor nodes. This survey discusses the challenges and security issues of SDN in WSN.

Keywords: Wireless Sensor Network (WSN) · Software Defined Networking (SDN) · Routing protocol attacks · Application based challenges · Key management

1 Introduction

Wireless sensor networks are the modern remote systems to screen ecological condition. For example: sound, vibration, temperature, weight movement and to amiably transfer the information through the system to a primary area or sink, where the information can be detected [1]. WSNs are widely utilized in various smart city applications, such as traffic observing, stock control, and planned examination. In particular for military applications WSNs provide outskirt security to anticipate fear mongering and unlawful development of weapons and medications [2]. WSN uses spatially distributed autonomous sensors to monitor the ecological conditions, is called actuator networks. WSN

A. P. Pandian et al. (Eds.): ICCBI 2019, LNDECT 49, pp. 1–12, 2020.
https://doi.org/10.1007/978-3-030-43192-1_1

look into territory has three sorts. For example equipment and programming of sensor nodes, application zone and correspondence & security. WSNs are presented to various sorts of assaults and delivers an extraordinary security test because of the restricted assets of calculation control, and battery correspondence range [3]. Because of openness of scaled down sensor and low control remote correspondence the application spaces of wireless sensor systems are assorted. After the sensor center points are passed on, they are responsible for self-dealing with a reasonable framework structure consistently by enabling multi-bounce correspondence [4] (Fig. 1).

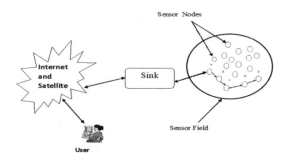

Fig. 1. Model of WSN

2   Wireless Sensor Network Architecture and Its Background

Network manager: A network manager is in charge for structure of the system, the board of directing table, planning correspondence between gadgets, checking and announcing the soundness of the system (Fig. 2).

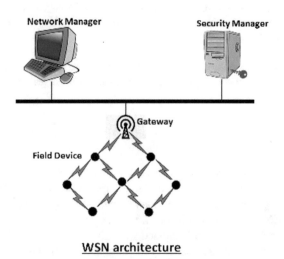

Fig. 2. WSN architecture

Security manager: The key management is done through the security manager.

Field devices: sensor modes or field devices must be fit for steering packets for the benefits of different devices. A router is a remarkable kind of field gadget that doesn't have to process the sensor or control equipment.

Gateway: A Gateway is the path that establishes communication among sensor nodes and end users, in order to forward the data from wireless sensor networks to the server.

Base station is in charge of topology ages and making on spot move against any failing in the system. It gathers intense information: its area ought to be shielded from outside and inside aggressors. On the off chance that a rival thinks about the base station area, that can advance diverse assaults through traffic observing or tracking techniques.

Organization of Wireless sensor node: Sensor nodes are comprised of essential parts, for example, detecting unit, handset unit, handling unit, and control unit appeared in Fig. 3 extra segments are a power generator, an area discovering framework and mobilizer. There are two parts of sensing unit: First one is Analog to Digital Converters (ADCs) and the second one is Sensor. Analog signals created by the sensors are changed to advanced flags through the ADC, and after that it is served into the preparing component. A preparing component is associated with a little stockpiling unit and it can achieve the methods that construct the sensor node to work together with alternate hubs to complete the designated detecting undertakings.

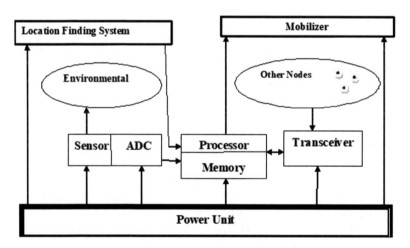

Fig. 3. Sensor network organization

3   Wireless Sensor Network Applications

3.1   Industry Related Applications

The utilization of Wireless Sensor Networks (WSN) for mechanical applications have pulled in a lot of intrigue today. The modern condition for wireless sensor systems remains harsher because of the different varieties in temperature, weight, stickiness, and presence of substantial equipments. Industrial WSN applications with environmental sensors are used to solve the production material pollution. The categories of sensor include humidity, Pressure sensor, temperature sensor, soil moisture, and wind sensor.

3.2 Society

WSNs empower medicinal services and traffic observing frameworks; they give another example of shrewd urban communities where everything will remain associated. Smart parking systems uses WSNs to observe and manage the congestion by monitoring the inward and departing vehicles from the city. A wireless body sensor is employed to monitor the circulatory strain, body temperature, sugar level and senses the feeling of anxiety, in order to remain very helpful for the specialists to analyze any infection with a logical methodology.

3.3 Agriculture

In agriculture, WSNs develop smart fertilization system to reduce the water wastage. The network provides timely information on plant diseases to the farmers. WSNs also help them to identify distinctive types of soil, wind level, temperature, water quality, dampness and daylight power to develop a savvy farming into existence.

3.4 Environmental Protection

WSN in the nuclear compartments needs an unequivocal significance. These systems screen exceedingly emanated zone exclusive for surrendering the life of the reactor's staff. These networks facilitate the governments and states into a case of energy leakage, disastrous situation, normal calamity, and seismic activity. Volcano monitoring ruminate technical, scientific, and social aspects.

4 Requirements and Security Issues in Wireless Sensor Networks

The survey on Wireless sensor network security outlines the security issues, discusses some existing solutions, and suggests possible research directions. The requirements such as Data integrity, Data Confidentiality, Data authentication, Data Freshness are discussed in various levels like data level, node level and network level.

4.1 Security on Data Level

Security of the data confirms the safety from illegal right of use and shields it from modification and exploitation. A huge amount of data is transferred across the WSN, a transferred data can be taken and changed by malevolent user, who can resend or modify the information.

Data Integrity: Integrity of data in sensor systems stands alluring in order to guarantee the unwavering quality of information, which then alludes the capacity toward data support. Integrity refers to accuracy, authenticity, extensiveness and reliability of information resources. Integrity includes the whole life cycle of information upkeep, precision, consistency, and dependability.

Authentication: Authentication guarantees the communication between one node to the other node. Information verification checks the personality of senders. For example: composing a client name and secret key, swiping a savvy card, utilizing advanced testament, utilizing voice acknowledgment and utilizing retina and finger prints.

Confidentiality: Generally, data used in sensor node will lack confidentiality. An authorized person can only access the data and the unauthorized persons cannot access the data. It verifies the security of substance, and shields the substance from unapproved individuals. Secrecy comprises of two sections, one is to get approval and security [1].

4.2 Security in Node Level

In aggressive condition, access to nodes is much troublesome or inconceivable. These circumstances are interested to provide greater defense. Appropriately all nodes must be sited as well as sheltered from vindictive clients. The identification of node, encryption and decryption keys and conventional security are maintained in every node. Legitimate safety on node efforts restrain assailants to get to the node and concentrates majorly on encryption and decryption keys.

Availability: It can be defined as confirmation with the intention of the system that is in-charge for putting away, conveying, keep up and preparing data whenever needed by the authorized one [20]. Network availability ensures strength of the network. A few highlights are: node disappointment, control channel, correspondence interface disappointment, high temperature, and water flood.

Authorization: Authorize the node to modernize data before to take delivery of data. Network administrator assigns different service policies to different users. Diverse arrangements and guidelines of WSNs are allotted to various hubs.

Non-repudiation: It proves the source of packet. The packet sent from the source can be denied by non repudiation. Non-repudiation in WSNs be able to stay away from a few nodes to facilitate that far along periods prevents the security of information. It guarantees the advanced affirmation of that node, which is engaged with the correspondence. In real world example, if the legitimate documents are marked without onlooker, that individual later stages can preclude from claiming his very own mark.

Security on Localization: WSNs have a capability to perform without human intervention will place the sensor nodes within network, this will be more helpful for tracking an object. A sensor network is intended to find blunders and blames that needs definite area data to recognize the fault. In the event that a few opponents distinguish area of a sensor node, the attacker will know the ways to demolish it.

4.3 Security in Network Level

Network security is the term to prevent unauthorized system or data access by compromising various arrangements and standards doled out by the system administrator. It includes the activity related to integrity, versatility unwavering quality and security of system information and hardware.

Self-Organizing: In a WSN, eavesdropper can easily affect the network due to its vulnerable nature. The spectrum analyzer can be used to observe and investigate wireless signals by adversary. Denial of service and man in the middle attack gave a solution in a self-organizing manner. Lying on the off chance that self- organization be inadequate in a sensor organize, the harm coming about because of an attack can prompt inaccessibility of a system [4].

Time Synchronization: Within circulated wireless sensor network, activities in addition to correspondence should be harmonized to do system tasks; the communication entities and logical clock can be synchronized using time synchronization. The information timing surrounded by sensor nodes are necessary with the purpose of identifies the procedures: movement, humidity and temperature. Single hop and multi hop communication different communication protocols are used; [1].

Scalability: A huge number of hubs are sorted out in a system completing spread activities. Versatility alludes to the capacity of a framework to perform valuable work when the size of the framework increments or contribution to framework increase [1]. New node insertion and removal of old node should produce good impact on network tasks. Besides old sensors are fizzled and new sensors are conveyed suggested that forward and in reverse mystery ought to likewise be considered the inward sensor node should not have the capacity just before peruse a few precedent communication and leaving sensor must not have the capacity to peruse any future message [2].

Surrounding Security: WSN environment will be protected as well as secured from external humanity. Malicious activity can be monitored using surveillance cameras mechanism. Arranging and encompassing security may not be a commonsense in certain mission-based applications like checking foe area; regardless, such a system be able to passed on safety structures, in therapeutic fields, common life watching then mechanical presentations [1].

Maximum performance and less Energy Consumption: Insufficient vitality prompts an awful circumstance, like inaccessibility in addition drop a message. The system should utilize conventions or techniques to monitor for every node gave less measure of vitality. High end application like battery control using composite cryptographic technique was designed. The other hand opposite side, monitor towards vitality, short preparing included activities are proposed. For example, dampness checking, object movement and temperature measurement.

5   Attacks on Wireless Sensor Networks

WSN stand predominantly exposed toward attacks on different kinds in light of the assorted arrangement of uses, WSN face numerous sort of attacks: traffic following, traffic investigation, node taking, node altering and information modification [1]. The different kind of attacks are shown in Fig. 4.

5.1   Physical Attack

The goal of physical attack is toward rescind the sensor materially or to take it from the mentioned area [13]. Attacker can consider and break down the inside programme; The attacker be able to separate mystery keys used in cryptographic techniques; attacker know how to transform that one then infuse duplicated node in the system.

Sybil Attack: Redundancy delivers system maintainability and availability. An attacker be able to create a node with many characteristics that feats this functionality. This kind of action describes, one node can counterfeits the characteristics of many nodes in Sybil attack. In this situation very difficult and got confusion to identify the fake and original node. This kind of attack degrades the performance in terms of data integrity, data security and usage of resources [15].

Reverse Engineering Attack: To know the internal functionality order of the process deceived approximately in the real order; To know about working principles and do system analysis to re-produce or copy something is called reverse engineering. The attacker excerpt identical valuable data like software code, cryptography code then knot architecture.

Injection on New Node Attack: The attacker presents the new knot called malicious node, will be injected in to the network to steal the useful information. The malicious node will act as a legitimate node thus it can violate the security of the system.

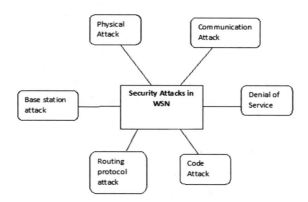

Fig. 4. Types of wireless sensor attacks

5.2 Communication Attack

Collision Attack: In a collision attack [32] attackers damage the legal packets by frequently sending packets in all directions and other packets will be retransmitted. Nodes energy and delay caused by retransmission, service will not be provided to the reasonable nodes.

Jamming Attack: In WSN protocol architecture, jamming attack occur on physical layer. A communication signal has been blocked by the attacker; the attacker uses frequency jammers. The unavailability will be caused by blocking the communication signals.

5.3 Attack on Base Station

Information will be gathered from neighboring nodes and gossip to outer world, where WSN base station uses the internet. Attacks on base station are traffic tracing attack and traffic monitoring.

Attack on Source Location: The attacker trace the data by searching for the signal produced by means of node. The technique called packet tracing, while transmitting the packet an attacker finds one packet, and the source address of the packet should be traced; this technique will repeat until it reaches source node through packet tracing in reverse.

Destination Location Attack: In this attack the attacker searches for the destination node. He aims to find base station or some other aggregator node. Traffic analysis or traffic tracing techniques are used in destination location attack. Traffic density of the node analyzed by the attacker about all surrounding node and anywhere the attacker detects the vast traffic density configuration, he construes the location of destination node.

Tracing traffic Attack: By learning the source and destination field [8]. The packets attacked by attacker, from that information the attacker traces a particular node successfully. He knows how to attack the injured party through forged packets and grounds unreachability. Source and destination gets affected by this kind of attack.

Traffic investigation Attack: An attacker without monitoring the substance of information deducts the area of a base station [3]. The attacker not have an interest to know type of protocol used and what kind of cryptography technique been used. Attackers screens whole system traffic density; he effectively distinguishes the base station by methods for overwhelming traffic and dispatches the assault to crush base station or harm its activity.

5.4 Routing Protocols Attack

A flexible working framework gives security to every one of its activities. An attacker can easily attack the weak operating system:

Black Hole Attack: An attacker makes analyze the everywhere traffic is sent by various nodes, suffocated, what's more, dropped. This attack misuses the creating calculation/ convention's shortcoming; since zero cost node and dark opening node illustrates itself closest to each other. The closest legitimate node considered by encompassing nodes and for all density of traffic it experiences but it does nothing [14]. The infused node act as the genuine node and it is trusted by real genuine nodes.

Hello Flood Attack: A double-cross legitimate node caused by the attacker using high power radio transmission. The limited no of nodes get attacked by the high transmission nodes. In this assault, the genuine hub pursues the assailant's hub, which prompts traffic clog [13, 14] and parcels drop. Every single consequent correspondence are sent to the assailant hub; he screens and peruses the entire system traffic to do pernicious action/tasks.

6   Challenges for Wireless Sensor Network

6.1   Well-Built Routing Protocols

The system commencing malevolent exercises for some solid routing conventions. Conventional developed directing convention can't survive utilized in WSN asset requirement condition. To carryout application necessities WSN ought to contain the most extravagant group of various conventions; a wireless sensor network convention should deal with an aggressive situation. Routing convention must to give a high throughput, and a decline packet loss proportion. Directing calculation should deal with portability and dynamic changing conduct in WSNs [1]. Inconsistent remote media can drop parcels; steering conventions ought to avoid bundle disaster. Planning another routing convention for WSN ought to reflect on security and protection problem.

Mobility on Node: The solid directing convention should deal with mobility on sink node and source node. To gather the information commencing every sensors using sink node versatility. The information gathered by static sink node from every sensor node devoid of altering stable location of the sink node. The versatile target node has its individual consequences for the system, e.g., execution in addition to active modify conduct. The versatile directing plan has been displayed in [1]. Routing conventions must give better network, an effective vitality utilization, a measured overflowing system, effectiveness constructed reasonable use, then the interest group insight [2]. To deal with any disappointment routing resolution should have great topology and the board plan; Node adaption, interface adaption, and over versatility can be maintained by the topology [2].

Trust Relations: The lower rate of trust and higher is usage on cryptographic keys; this kind of technique will provide good performance in network and more preservation on energy. Design of the routing protocols based on application requirement for source limitation procedures: this kind of technology improve the performance of network in terms of memory, balance processing it should balance processing and energy efficiency [15].

6.2 Application Challenges

Interruption Detection Method Incorporation: A unique message will be sent to server by the sensor nodes [14]; the message gets more burden to the system: the situation be able to deplete vitality; likewise uses nodes resources. Additional effort remains finished by [1]; their framework. They have an accepted dominant part of nearby node, which0 is straightforward just before base station—this suspicion isn't valid in down to earth applications.

Programming Wireless Sensor Networks: Programming an extensive system of exceptionally asset requirement gadgets that are self-sorted out and all inclusive steady, with a strong conduct and a progressively evolving condition, is a major test. Programming in a threatening or un-secure condition, to screen the environment, is an overwhelming errand. Programming WSNs must be outfitted with appropriate programming building standards; it must be very much coded, tried, repaired, and should give a defective free plan. A straightforward area explicit language on behalf of WSNs function depiction be exhibited through [14]; it guaranteed that composed language help those student creators who need phenomenal programming limits. As indicated by the best of our insight, a thorough review about WSN programming can be found at [1]. Creators have unmistakably referenced best in class programming ideas, patterns, and difficulties. They have demonstrated distinctive programming approaches, e.g., code depictions [13].

6.3 Key Administration and Challenging Paradigms

Key administration and allocation [12] is a large amount basic test. An incorporated base station having greatest capacity (calculation) can be utilized; in any case, it rises correspondence transparency problem as soon as hubs endeavor in the direction of confirm themselves. The security mechanisms used here are Symmetric key encryption and asymmetric key encryption. These techniques are going out and flow inquires about regions for low power sensors [14].

Wireless Sensor Network and Software Defined Networking: An ongoing test based on wireless sensor network is to regulate the SDN engineering—the abstract design. Software defined network have a central regulator named as SDN controller be in charge of overseeing and controlling the entire system, while actualizing directing, balancing the load, crossing point forming, congestion building, and firewall strategy [11]. Within asset limitation condition, SDN has developed a totally different difficulties. Integrating SDN through WSN facilitates the administration burden, elucidation appropriation issue, series channel problem, inadequate handling, recollection problem, and scheming problem. Innovating a new technology in WSN give numerous favorable circumstances; new issues of security, key administration, load sharing, and existing steering systems ought to be re-designed to work with conventional WSNs.

Cloud computing implementation in Wireless Sensor Network: During this period and the Internet of Things (IoT), the conveyed availability gives distinctive new stages: distributed computing is the new pattern for WSN. These stages are intended for progressively useful equipment where computational limit and memory space isn't constrained. If there should arise an occurrence of WSNs, structuring and coordinating cloud ideas in WSNs must not present novel issue. The cloud sensor structure reconciliation can be explained in [13]. Information gathered as of various applications are prepared here bar/sub intermediary, at that point filled in software application as a service (SaaS). In this structure, end client know how to observe intrigued information lying on a graphical interface; it bolsters wellbeing division, as well as inner-city traffic checking.

7 Conclusion

The applications of WSNs are explicit to build up a verified situation, we should think about the limits of resources (memory, processor, and power supply). Encryption gives security. Test results demonstrate to facilitate the encipher calculations utilizing 64-bit keys used for information protection be capable of smashed down happening four months by means of excellent PCs, which know how to process 1012 code word in a single second. Expanding request in sensor to organize promising applications is increasing in demand. Better methods are required for security, protection, control, computational-ability, and adaptability. Undeniable WSN framework that covers all security necessities, e.g., information protection, information uprightness, information freshness, personality verification, and accessibility is requested application for businesses and associations. The flexible wireless sensor network design be supposed to address the security necessities, Quality of service, assaults, and encryption calculations. Software Defined Networking incorporation of cloud organizations and virtualization advancement needs to be considered. We think, examining and cognizance, this paper gives a total one spot examination designed for WSN then its connected learning.

References

1. Bangash, Y.A., Abid, Q.D., Ali, A.A.A., Al-Salhi, Y.E.A.: Security issues and challenges in wireless sensor networks: a survey. IAENG Int. J. Comput. Sci. **44**(2). IJCS_44_2_02
2. Singh, C., Kaur, R., Kaur, M.: Review of security enhancement techniques for wireless sensor network. Int. J. Electron. Eng. Res. **9**(8), 1185–1196 (2017). ISSN 0975-6450 Research India Publications
3. Mottola, L., Picco, G.P.: Programming wireless sensor networks: fundamental concepts and state of the art. ACM Comput. Surv. (CSUR) **43**(3), 19:1–19:51 (2011)
4. Singh, R., Singh, J., Singh, R.: Security challenges in wireless sensor networks. IRACST Int. J. Comput. Sci. Inf. Technol. Secur. (IJCSITS), **6**(3) (2016). ISSN 2249–9555
5. Pathan, A.S.K., Lee, H.W., Hong, C.S.: Security in wireless sensor networks: issues and challenges. ISBN 89-5519-129-4
6. Rudramurthy, V.C., Aparna, R.: Security issues and challenges in wireless sensor networks: a survey. Int. J. Innov. Res. Comput. Commun. Eng. (An ISO 3297: 2007 Certified Organization) **3**(10), 9648–9656 (2015)

7. Kumar, V., Jain, A., Barwal, P.N.: Wireless sensor networks: security issues, challenges and solutions. Int. J. Inf. Comput. Technol. **4**(8), 859–868 (2014). ISSN 0974-2239
8. Kamat, P., Zhang, Y., Trappe, W., Ozturk, C.: Enhancing source-location privacy in sensor network routing. In: Proceedings of the 25th IEEE International Conference on Distributed Computing Systems (ICDCS 2005), pp. 599–608 (2005)
9. Kumar, A.D., Smys, S.: An energy efficient and secure data forwarding scheme for wireless body sensor network. Int. J. Netw. Virtual Organ. **21**(2), 163–186 (2019)
10. Dener, M.: Security analysis in wireless sensor networks. Int. J. Distrib. Sens. Netw. **10**(10), 303501 (2014)
11. McKeown, N., Anderson, T., Balakrishnan, H., Parulkar, G., Peterson, L., Rexford, J., Shenker, S., Turner, J.: Openflow: enabling innovation in campus networks. ACM SIGCOMM Comput. Commun. Rev. **38**(2), 69–74 (2008)
12. Laouid, A., Messai, M.L., Bounceur, A., Euler, R., Dahmani, A., Tari, A.: A dynamic and distributed key management scheme for wireless sensor networks. In: Proceedings of the International Conference on Internet of Things and Cloud Computing, ICC 2016, New York, NY, USA, pp. 70:1–70:6 (2016)
13. Tripathi, K., Pandey, M., Verma, S.: Comparison of reactive and proactive routing protocols for different mobility conditions in WSN. In: Proceedings of the International Conference on Computing, Communication and Security (ICCCS 2011), pp. 156–161 (2011)
14. Elsts, A., Selavo, L.: A user-centric approach to wireless sensor network programming languages. In: Proceedings of the 3rd International Workshop on Software Engineering for Sensor Network Applications (SESENA), pp. 29–30 (2012)
15. Udhayamoorthi, M., Senthilkumar, C., Karthik, S., Kalaikumaran, T.: An analysis of various attacks in manet. Int. J. Comput. Sci. Mob. Comput. **7**, 883–887 (2014)

A Survey on Machine Learning Based Fault Tolerant Mechanisms in Cloud Towards Uncertainty Analysis

K. Nivitha$^{(\boxtimes)}$ and P. Pabitha

Department of Computer Technology,
Madras Institute of Technology - Anna University, Chennai, India
nivithavimal@gmail.com, pabithap@gmail.com

Abstract. Cloud computing has the tendency to provide on-demand resources. Recently, there has been a large-scale migration of enterprise applications to the cloud. Any unexpected events that occur in cloud due to its dynamic nature is termed as uncertainty. The most cause of uncertainty can be the unexpected fault that arises in cloud environment. Hence the early detection and recovery of fault can abruptly reduce the uncertainty by enhancing the Quality of Service in cloud applications. This paper discusses the types of faults and failures present in cloud environment and it gives an overview on the existing fault handling mechanisms.

Keywords: Cloud computing · Uncertainty · Fault · Quality of Service · Machine learning

1 Introduction

Fault tolerance in cloud can thus be defined as the capability of cloud resources to function seamlessly amidst faults without bringing any hindrance to the overall working of the cloud in order to ensure a good performance of the system. It is the quick repairing and replacement of faulty nodes that has resulted in the failure, usually the term node is used to refer to the virtual machines hosted on the physical machine [1–13]. Various outages have occurred in cloud providers system. For example, most recently the popular cloud application salesforce got blocked out for the entire day [4]. The various services of Apple icloud like email, itunes faced an outage. Due the failure of authentication users were unable to use the services of Apple icloud. The most familiar cloud provider Dropbox faced an outage leading to the unavailable of documents for the Dropbox users [5]. The Big giants like Samsung, Microsoft exchange, Verizon wireless had affected by outages and failures [23]. Thus the reliability of cloud can be enhanced through improved fault tolerance techniques. Due to the uncertainty of undetermined latency and loss of control over computing node are prone to failures in remote cloud systems. It is therefore directly related to the fault tolerance [24]. Therefor fault detection can improve the Quality of Service and reduce the uncertainties in existence.

© Springer Nature Switzerland AG 2020
A. P. Pandian et al. (Eds.): ICCBI 2019, LNDECT 49, pp. 13–20, 2020.
https://doi.org/10.1007/978-3-030-43192-1_2

2 Related Work

The existing system introduces a simple and efficient mechanism that handles different types of failures including crashes, omissions and arbitrary failures. The detailed types of failures and different mechanisms used to identify, manage and recover from failures. A survey on Fault Tolerance and resilience is presented [6]. The presented method discusses the architecture of cloud computing based on the fault tolerance model. It also brief overview of the existing fault tolerance techniques based on the byzantine failure. Proactive and Reactive techniques are presented [7]. The proactive fault tolerance predicts the error before in hand whereas the reactive techniques work after the occurrence of fault in the cloud environment. A fault tolerance model for cloud computing has been developed [12]. The model works based on the principle of reliability on each computing node. In case of high reliability the node is selected else it is rejected. Virtualization technology is implemented [14, 15] to improve availability and reliability of applications that are located in virtual machines in cloud to assess the occurrence of fault.

2.1 Existing Work on Fault Tolerance in Cloud Based on Machine Learning Algorithms

The existing work uses two modules to predict and to detect the failure. The predict module predicts the failure in advance using log messages passed between the components. In case of wrong prediction the occurrence of failure is detected through detection module. Message pattern is extracted through preprocess, Bayes's theorem is used to calculate the probability of failure based on the derived message pattern. Few parameters such as temperature, CPU usage, memory usage of the system, CPU idle time, number of I/O operations are used to build probabilistic model. [16] Fujitsu laboratories have developed failure prediction method which automatically learns patterns of the message and classify it based on their similarity irrespective of their format. Bayesian inference is used to retrieve the relationship between the message patterns and failures. The work was carried out in a real cloud datacenter and the approach has predicted failures with 80% precision and 90% recall [19].

Ensemble of Bayesian submodels are used to represent the multimodal probability distribution [18]. The prior probability is estimated through each submodel with submodel index. To determine the submodel and conditional data probability, Bayesian Expectation Maximization algorithm is performed. The E-step calculates probability of each submodel whereas the M-step updates the probabilities. Failure prediction is also predicted with decision tree which is composed of internal trees with test function. External loss due to hardware failure in data centers are major hassle between cloud providers and cloud users. FailureSim [12], cloudsim based simulator predicts host failure in the cloud 89% accuracy. WEKA library is used for classification; multi-layered perceptron and Elman recurrent neural network is the two neural network algorithm used for classification. On the start of the simulation each host exhibits working behavior, based on the pattern observed the failures are classified into 8 states and analyzed for the prediction accuracy. A detection methodology has been put forth for faults in cloud environment using the unsupervised ML outlier detection method. It makes use of 3 elements, with the Fault Detection System [21], abbreviated as FDS being an integral one. The proposed strategy can be easily implemented on the underlying FDS framework, using the interquartile range of outlier detection method, the various faults are accumulated and sent as a whole to the end use, giving the easiest way to detect fault and attacks.

3 Taxonomy of Faults, Error and Failures

The system lacks the ability to perform the required task which is caused by unusual state or bug located in the parts of a system is termed as Fault. The major types of faults present in cloud environment are categorized as below.

- Network faults: The significant cause of failures in cloud is that they are accessed through network. The faults occur due to congestion, packet loss, corruption, failure of destination node or link etc.
- Physical faults: faults that arise in hardware resources such as CPU, memory, storage, power failure etc.
- Process faults: due to the incompetent processing capabilities, shortage of resource, bugs in software process faults can occur.
- Service expiry fault: Service failures due to the expiry of resource usage time.

Due to the occurrence of any one of the above listed faults, the system may move to an error state. Error is the difference between the true values of the measure and (quantity being measured) and the measurement. The error cannot be completely eradicated but it can be controlled. Proliferation of failure through fault and error is depicted in Fig. 1.

Fig. 1. Relationship between fault, error and failure

The state in which the system does not give the intended result is termed as Failure. A failure is identified through the incorrect output of the system. Failure in a system is due the presence of error which in turn due to the faults.

3.1 Fault Tolerant System

The capability of the system to function without any halt in the ongoing process of a hardware/software failure [8]. A failure is a deviation from agreed result to the actual observed result. Failure arise from fault, the difference is that fault exists without failures. The purpose of fault tolerance is to avoid a fault from being proliferate itself as a failure. Various Fault tolerant metrics are available through which Faults can be detected and rectified thereby reducing the error rate and failure of the system. The parameters or metrics are used to recognize the system capability to perform in presence of failures. Comparison between the various fault approaches is depicted in Table 1 with the respective tools [2, 8, 11], cost [14] associated with the techniques and the type of failures identified.

Table 1. Comparison between fault approaches

Techniques	Policy	Tools	Cost	Failure
Checkpoint	Reactive	AZURE HAProxy	High [9]	Application/host
Replication	Reactive	Hadoop AmazonEC2	Huge [10]	Process/node
Migration	Reactive, Proactive	HAproxy, Hadoop	Varied [20]	Process/node/application
Retry	Reactive	AZURE	None [3]	Host/network
SGuard [3]	Reactive	Hadoop	Low	Node/application
Task Resubmission	Reactive	AmazonEC2	Low [22]	Node/application
Self healing [17]	Proactive	HAproxy, AZURE	High [10]	Process/node/application
Software rejuvenation [17]	Proactive	Eucalyptus, AmazonEC2	High	Process/application
Rescue workflow	Reactive	Hadoop	Low [3]	Node/application

3.2 Approaches for Fault Tolerance in Cloud

The different approaches for Fault tolerance is required as they involve in detection and handling faults that may arise based on the hardware and software faults. The Fault tolerance techniques are broadly into two categories as follows.

3.2.1 Reactive Fault Tolerance

This approach reacts to the system after the occurrence of failure in order to decrease the influence of failure in cloud system. It is further classified as follows.

(a) Check Pointing/Restarting: The tasks are executed again in case of a failure from valid stored checkpoints. The job is restarted from the recent checkpoint to avoid the loss of needy computation.
(b) Retry: This is the simplest approach. The job is submitted to same resource until the successful completion.
(c) Job Migration: The job fails on a specific physical machine due to some fault; it is then moved to another machine to complete the task.
(d) Rescue workflow: Until the rectification of failure the system is allowed to continue working even in presence of failure.
(e) Replication: The tasks continue to execute with the replicas created in case of failure occurrence. This approach provides the redundancy.
(f) S-Guard: Used for the rollback and recovery process.

(g) Task resubmission: In this method either the identical machine or the different machine executes the resubmitted/submitted tasks to it.

(h) User Defined Exception Handling: The user defines certain actions to take place on occurrence of any failure.

3.2.2 Proactive Fault Tolerance

This approach avoids recovery from failure and error by predicting the faults proactively and thereby replacing the suspicious component with the proper running component.

(a) Software rejuvenation: In this approach the system starts with new state after each periodic reboots.

(b) Self-healing: It allows the computing devices to self identify the occurrence of failure independently. This approach follows divide and conquer technique to improve system performance.

(c) Pre-emptive Migration: Based on the feedback-loop control method the application is observed and analyzed.

(d) Load Balancing: when a CPU exceeds the maximum limit the system load is balanced by transferring the load to different CPU which has not reached the maximum limit.

4   Metrics for Occurrence of Fault in Cloud Environment

The occurrence of fault in cloud environment can be categorized into three aspects based on the nature of metrics identified.

4.1   Performance Metrics

An efficient system is the one with high performance features such that the fault tolerant system should be able compete with presence of fault in the ongoing process. In general the response time, throughput falls under the category of performance metrics as depicted in Table 2.

Table 2. Performance metrics for fault tolerant system

Metrics	Definition	Outcome
Response time	The total time taken to respond/reply to a specific client/algorithm	Lesser the response time efficient is the system
Throughput	Successful execution of the tasks per unit time	Throughput should be higher for the highly efficient system

4.2   Economic Metrics

In a fault tolerant system, n resources are needed for the successful completion of any particular task allotted to various virtual machines; hence the price is associated with the number of Workload allocated to number of resources (virtual machines) with respective to the time taken for the successful completion as given in Table 3. Delay time is the difference between actual completion time and expected completion time; it is associated with overhead metrics.

Table 3. Economic metrics for fault tolerant system

Metrics	Definition	Outcome
'Cost	Monetary description of the system	Cost for the execution of workload
Overhead associated	The additional overhead involved during the fault tolerance mechanisms	Reduced overhead associated improves system performance

4.3   General Metrics

General metrics indicate the ability of a fault tolerant system quantitatively. Table 4 gives metrics where the capability of fault tolerant system can be measured. General metrics is comprised of availability of the service, adaptability, usability, efficiency which reflects the performance and reliability that covers the rate of failure. MTBF refers to mean time between failures which is associated with availability parameter and in reliability parameter MTTR conveys mean time to repair the fault occurred in the ongoing process.

Table 4. General metrics for fault tolerant system

Metrics	Definition	Outcome
Availability	Probability in which the system functions after the request	Availability of the service for each request
Adaptability	Based on the current situation and executing the processes automatically on the basis of specific conditions	Continue of service after specific condition (occurrence of faults)
Usability	The usage of the product by the user in order to achieve efficiency and satisfaction	Extent to which the product is being useful to the users
Efficiency	Guarantee the efficiency of the system	Higher the performance higher the efficiency of system
Reliability	Provide accurate output in a particular amount of time	Higher the reliability lowers the failure rate

5 Conclusion

Faults are generally considered as fatal failures in the cloud environment, early detection and mitigation is therefore required. The existing techniques are not good enough to recognize the faults in advance where there is a need for it. In spite of gaining popularity in cloud the occurrence of faults should be detected as early as possible. The uncertainty discussed in this work is mainly focused on fault tolerance while there is a need to address other uncertainties in the cloud environment such as the ones like scalability, resource provisioning, and virtualization. As a future work the Fuzzy rules in combination with various fault tolerance mechanisms (e.g. self-healing, load balancing, replication) can deliver a better enhancement. A machine learning algorithm such as Reinforcement Learning (RL), which learns based on the situation and takes decisions to maximize the reward can also be incorporated with any optimization algorithms in order to provide an optimal solution.

References

1. Amin, Z., Sethi, N., Singh, H.: Review on fault tolerance techniques in cloud computing. Int. J. Comput. Appl. **116**(18), 9–14 (2015)
2. Araujo, J., Matos, R., Maciel, P., Vieira, F., Matias, R., Trivedi, K.S.: Software rejuvenation in eucalyptus cloud computing infrastructure: a method based on time series forecasting and multiple thresholds. In: 2011 IEEE Third International Workshop on Software Aging and Rejuvenation, pp. 38–43. IEEE, November 2011
3. Bodík, P., Menache, I., Chowdhury, M., Mani, P., Maltz, D.A., Stoica, I.: Surviving failures in bandwidth-constrained datacenters. In: Proceedings of the ACM SIGCOMM 2012 Conference on Applications, Technologies, Architectures, and Protocols for Computer Communication, pp. 431–442. ACM (2012)
4. Bort, J.: Salesforce went down or a whole day. Business Insider, May 2016. http://www.businessinsider.com/salesforce-outage-is-an-internet-meme-2016-5
5. Chalermarrewong, T., Achalakul, T., See, S.C.W.: The design of a fault management framework for cloud. In: Proceedings of 9th International Conference on Electrical Engineering/Electronics, Computer, Telecommunications and Information Technology (2012)
6. CRN staff: The 10 biggest cloud outages of 2015 (sofar). CRN, December 2015. http://www.crn.com/slideshows/cloud/300079195/the-10-biggest-cloud-outages-of-2015.htm
7. Davis, N.A., Rezgui, A., Soliman, H., Manzanares, S., Coates, M.: FailureSim: a system for predicting hardware failures in cloud data centers using neural networks. In: IEEE 10th International Conference on Cloud Computing (CLOUD), pp. 544–551 (2017)
8. Garg, A., Bagga, S.: An autonomic approach for fault tolerance using scaling, replication and monitoring in cloud computing. In: 2015 IEEE 3rd International Conference on MOOCs, Innovation and Technology in Education (MITE), pp. 129–134. IEEE, October 2015
9. Goiri, Í., Julia, F., Guitart, J., Torres, J.: Checkpoint-based fault-tolerant infrastructure for virtualized service providers. In: 2010 IEEE Network Operations and Management Symposium-NOMS 2010, pp. 455–462. IEEE, April 2010
10. Hakkarinen, D., Chen, Z.: Multilevel diskless checkpointing. IEEE Trans. Comput. **62**(4), 772–783 (2012)

11. Hasan, T., Imran, A., Sakib, K.: A case-based framework for self-healing paralysed components in distributed software applications. In: The 8th International Conference on Software, Knowledge, Information Management and Applications (SKIMA 2014), pp. 1–7. IEEE (2014)
12. Mugunthan, S.R.: Soft computing based autonomous low rate DDOS attack detection and security for cloud computing. J. Soft Comput. Paradig. (JSCP) 1(02), 80–90 (2019)
13. Jhawar, R., Piuri, V.: Fault tolerance and resilience in cloud computing environments. In: Computer and Information Security Handbook, pp. 165–181 (2017)
14. Jhawar, R., Piuri, V.: Fault tolerance management in IaaS clouds. In: IEEE First AESS European Conference on Satellite Telecommunications (ESTEL), pp. 1–6, October 2012
15. Jhawar, R., Piuri, V., Santambrogio, M.: Fault tolerance management in cloud computing: a system-level perspective. IEEE Syst. J. 7(2), 288–297 (2012)
16. Kaur, J., Kinger, S.: Analysis of different techniques used for fault tolerance. IJCSIT Int. J. Comput. Sci. Inf. Technol. 5(3), 4086–4090 (2014)
17. Kumar, M., Mathur, R.: Outlier detection based fault-detection algorithm for cloud computing. In: International Conference for Convergence for Technology, Pune, pp. 1–4 (2014)
18. Machida, F., Andrade, E., Kim, D.S., Trivedi, K.S.: Candy: component-based availability modeling framework for cloud service management using sysML. In: 2011 IEEE 30th International Symposium on Reliable Distributed Systems, pp. 209–218. IEEE, October 2011
19. Memishi, B., Ibrahim, S., Pérez, M.S., Antoniu, G.: Fault tolerance in MapReduce: a survey. In: Resource Management for Big Data Platforms, pp. 205–240. Springer, Cham (2012)
20. Mohammed, B., Kiran, M., Awan, I.U., Maiyama, K.M.: Optimising fault tolerance in real-time cloud computing IaaS environment. In: 2016 IEEE 4th International Conference on Future Internet of Things and Cloud (FiCloud), pp. 363–370. IEEE, August 2016
21. Myint, J., Naing, T.T.: Management of data replication for PC cluster-based cloud storage system (2011). arXiv preprint arXiv:1112.5917
22. Sharma, S.: Enhance data security in cloud computing using machine learning and hybrid cryptography techniques. Int. J. Adv. Res. Comput. Sci. 8(9), 393–397 (2017)
23. Trivedi, M.: A survey on resource provisioning using machine learning in cloud computing. Int. J. Eng. Dev. Res 4(4), 546–549 (2017)
24. Tsidulko, J.: The 10 biggest cloud outages of 2014. CRN. The Channel Company, December 2014. http://www.crn.com/slideshows/cloud/300075204/the-10-biggest-cloud-outages-of-2014.htm?itc=Refresh. Accessed 5 Apr 2014

IoT Based Attendance Management System Using Google Assistant

M. Niharika$^{(\boxtimes)}$ and B. Karuna Sree

Department of Electronics and Communication, CMR Technical Campus,
Hyderabad, Telangana, India
niharika1997@hotmail.com

Abstract. These days Absenteeism has posed a major problem in Organizations as well as Institutions. Absenteeism affects wages of Employees in organizations, and also affects academics of a Student. Therefore, it is crucial for employees in Organizations and students in Institutions to identify such information on timely basis. In the proposed system, the attendance management has been implemented in real time and monitored using Google Assistant. The attendance was logged using RFID card. As soon as the individual scans the RFID, the Google Assistant, was allowed to access the database and trained to calculate the average attendance. Upon successful calculation, the assistant then responds with Average attendance when asked by the user. With the diverse services of IBM Watson and Google, the system was able to produce accurate results. Portability and Accessibility are the major advantages that have been observed so far. This system can not only be useful for be useful in Educational institutions but also at the same time it can be useful for Payroll Predictions in organizations, on the basis of attendance.

Keywords: Radio Frequency Identification · IBM Watson · Google services · Google Assistant · Institutions · Organizations · Average attendance

1 Introduction

The roots of Voice Assistants can be derived from Electronics back in 1922; with the development of the famous Radio Rex by Elmwood Button Co. Rex was a small brown toy bull dog, which was partially made of metal. The electromagnet under this metal was sensitive to sound patterns. The vowel sound in "Rex" would produce approximately 500 Hz which would then be detected by the electromagnet, and the spring under the dog would expand, making the dog move forward. Later in 1961 William C. Dersch, IBM Shoebox (An IBM Computer), was able to perform Mathematical functions, recognize 16 words and the digits from Zero to Nine.

The study of Natural Language Processing can be traced back in 1980s where it was further classified into two categories. The first Category consisted of a set of complex rules and the other category consisted of Statistical Models. With time, major companies such as IBM, Apple and Google came up with their own digital Virtual

© Springer Nature Switzerland AG 2020
A. P. Pandian et al. (Eds.): ICCBI 2019, LNDECT 49, pp. 21–31, 2020.
https://doi.org/10.1007/978-3-030-43192-1_3

Assistants. The first digital Virtual assistant that was ever released on mobile phone was "Siri". Originally, Siri was released as an Application on iOS, in 2010. The Application was then integrated into IPhone 4S after Siri's Acquisition by Apple in October 4, 2011. Later on May 18, 2016, Google came up with Google Assistant on Android Phones.

Today there are almost above 74% users all over the world [1], and approximately 60% users in India [2], use the voice Assistants only for controlling home appliances. So far, voice Assistants such as Amazon Alexa, Google Home Mini etc. have been useful in controlling home appliances and answering the questions. According to the survey conducted by Comscore [12], among all the Voice Assistants, "Google Assistant" proved out to be the most preferable Voice Assistant, which managed to answer more than 90% of the questions that were asked. According to a recent statistics from "Operating System Market Share 2019" in India [13], there are more number of the Android users. Referring to the above analysis, the platforms that were chosen for the following project were such that, it can be utilized by maximum number of organizations and institutions in future.

It has been predicted that Voice Assistants will also be useful in satisfying the business and other operational needs [10]. This is what the paper aims for. The following paper, elaborates an applied case in an Institution, which will be beneficial for institutions as well as organizations.

2 Project Survey

In order to record the problem statement of the paper, a survey among 198 students was conducted in the institution. The Survey poses several questions regarding the Laboratory management. Since Attendance is one of the crucial things in a laboratory, hence the following survey was conducted [9]. The two main questions, the answers of which have been focused on are:

(i) Have you ever faced attendance issues?
(ii) Would you like a personal assistant on your personal devices that can keep a track of the daily attendance

The results can be observed from the Fig. 1.

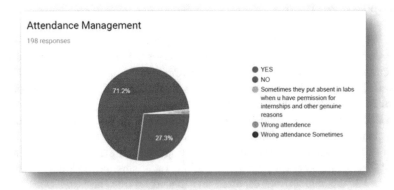

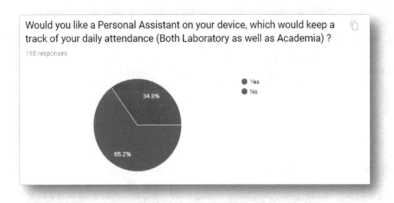

Fig. 1. Survey analysis

Upon the above analysis, Sect. 3 defines a proposed solution for the given case.

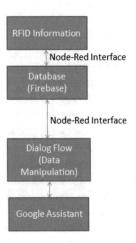

Fig. 2. Flow of attendance management system using Google Asst.

3 Proposed Method

1. The Fig. 2 shows the general flow of the proposed system.
2. The Google Assistant can be used as interface for the user to initiate questions and in return the assistant provides a relevant response.
3. The RFID information can be stored into FIREBASE via a Node-Red Interface. Here Firebase acts "only as a container".
4. The RFID information (which is Attendance Count in this case) can be analyzed and manipulated using a Dialog Flow Editor.
5. At this stage, when the user poses a question through the Google assistant voice input feature, the necessary variables such as for "Name of the Student" and "Roll Number of the student" are captured and looked up into the Database.
6. The above mentioned variables are used to calculate the average attendance, which is done using Dialog Flow.
7. The final percentage i.e. the result, is stored back into the Firebase.
8. The assistant is trained in such a way, that, depending on the relevant attendance percentage belonging to the specific "Name of the student" and "Roll Number of the student", the corresponding result is read by the assistant (Refer to Sect. 4.3) "from Firebase", and the response to the user is given through a user-friendly voice.

4 Implementation

4.1 Platforms Required

A. **IBM Watson™ Internet of Things Platform:** [14] It is a flexible tool kit that provides all the resources needed for accomplishing the tasks for connecting, controlling and monitoring things over the Internet [16]. Watson can collect data from devices through various device gateways, device management, upon which Edge Analytics [4], Machine learning, Visual Recoginition, Speech-Text, Text-Speech, real- time monitoring, Automation [3, 5] including cognitive computing and many more can be performed on real time data or simulation tools. It provides a free Cloud storage of 256 MB for "Standard Lite Plan Users". Further for the premium plans, charges are applied for additional cloud storage.

B. **Node-RED Platform Flow Editor:** [14] It is one of the services in IBM Watson IoT platform. A single Cloud Foundry Application links to a flow editor platform, which can be used to integrate Web services, Hardwares and Application Program Interfaces (API) with the help of inter-connection of various "Nodes". The nodes are formatted in JavaScript Object Notation (JSON).

C. **Google Actions:** The Actions on Google are a way of customizing in-built Google Assistants in Android phones. The customized application can be run on the top of the original Assistant, without altering the algorithms of inbuilt application. Since this platform manipulates the way the application responds to a particular input, the actions are always defined in a tool called "Dialog Flow". A Dialog Flow is thus based on Natural Language processing. As Defined in Sect. 1, the system has incorporated the 1st category of NLP, which is a rule based NLP. A Dialog Flow consists of components such as Fulfillments, Intents, Contexts, Entities, Knowledge, Integrations and Training [7]. Among all the components "Fulfillment" is considered as the logical section of Google Assistant.

D. **Google Firebase:** Firebase is a Real-Time Cloud storage offered by Google in order to perform real-time operations. This Application Program interface provides a way of syncing the data into a personal device [6]. Google Firebase is not only capable of storing but also includes other features such as Authentication, Analytics, Crash Reporting, Hosting, Real-Time Synchronization etc. The dynamic nature of Google Firebase is what has been the major advantage of this project.

E. **Arduino IDE:** [14] An integrated development Environment that consists of text editor and can be used to communicate with a variety of development boards such as Arduino Series, Spark Fun, Intel, WeMos etc. The IDE can be used to compile and upload programs on to the boards.

4.2 Hardware Required

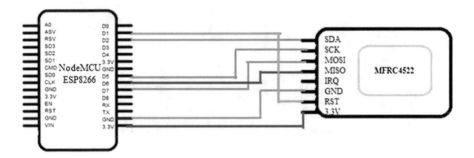

Fig. 3. Connection diagram for attendance logging

A. **NodeMCU (ESP8266):** An ESP8266 Node-MCU [14] has an in-built TCP/IP Protocol Stack, which can be used to connect to various local Wi-Fi networks with ease.

B. **RFID:** A Radio Frequency Identification [8] is a wireless device used for data Transfers. RFID can be classified as Active and Passive Tags. [15] The tags consist of a microchip, antenna and memory, which is capable of managing the data transfers. Passive tags require the power source from external environment whereas Active Tags have its own power source. There are various frequency distributions for the use of RFID tags. Low Frequency RFIDs have been used for the project further, which can detect the radio waves from 30 kHz–300 kHz.

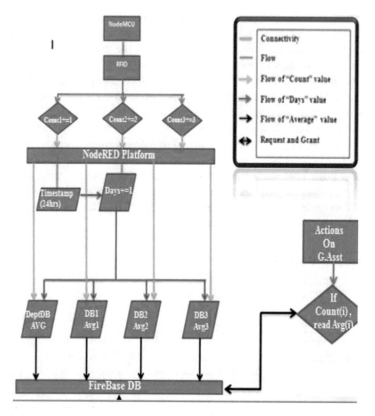

Fig. 4. Block diagram of attendance management system using Google Assistant

4.3   Working

The Fig. 4 can be explained as follows,

1. The connections between NodeMCU and RFID reader is shown in Fig. 3. Two RFID cards of 125 kHz were assumed to be sample Identity cards. The Sample Data of which is shown in Table 1.
2. The UIDs (Unique Identification) of RFID Tags were assigned to Corresponding Names and University Roll Numbers.
3. After the IDs were assigned to the above two parameters, the IDs were scanned again and checked if the Reader was able to detect the presence of Students.
4. Every time the IDs were scanned, the respective "Count Values" got incremented.

Table 1. Sample data

Name	Roll No.	Count	Days	Average attendance (approx.)
Sindhura	424	5	6	100
Radhika	435	6	6	83.33

5. The Count values of Student1 and Student2 were stored in **Count1 and Count2** variables respectively and incremented depending on the number of Logs.
6. The connection of various platforms onto a single model was done using a NodeRED Flow Editor, which is one of the Platform Service provided by IBM Cloud. This editor provides an easy interface among different platforms and embedded devices.
7. IBM IoT Node is built using Node.js in NodeRED Editor and understands JSON scripting from the developer End. This Node is responsible for Collecting IoT Data from various hardware devices. For example the data from NodeMCU (values of Count1 and Count2) was published to NodeRED Editor.
8. The handshaking was possible using MQTT Server which follows a Pub-Sub Model, and hence makes the system more Dynamic.
9. Three Databases in Firebase were created as follows:
   (i)   DeptDB: A database Belonging to Department
   (ii) DB1: Database of Student1
   (iii) DB2: Database of Student2
10. The number of Working days was counted using a "Counter" Node [11] in NodeRED Editor.
11. Considering the variable name as "**days**", the value was incremented for every 24 h.
12. The values of "**Count1**", "**Count2**" and "**days**" were stored in the respective databases. Figure 5 shows a generalized Database representation.

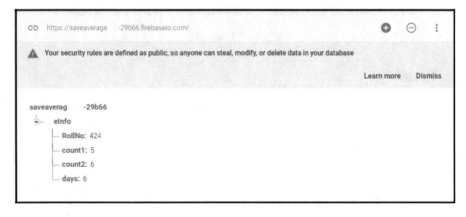

Fig. 5. Data in firebase

13. **Dialog Flow Setup:** Fulfillments [15] were built using "Node.JS" scripting. The three values from Count1, Count2 and days were used to evaluate the Average Attendance of each student.
14. "Google Firebase" was used as a container for device data.
15. The fulfillments in Dialog Flow were used to make a request to the Firebase to retrieve the data.
16. The following basic formulae were used to calculate the average of each student using Node.JS Scripting, as follows,

$$Average1 = Count1/Days$$
$$Average2 = Count2/Days$$

17. The logic defined in fulfillments was used to train the dialog flow agent. The overview of questions-responses can be given as follows:

> **User:** Hi!
> **Agent:** Whose attendance would you like to know?
> **User:** May I Know XYZ attendance?
> **Agent:** Sure, Average attendance of XYZ is 80%.

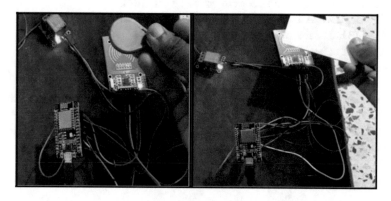

Fig. 6. ID scanning

5 Results

1. Figure 6 depicts the scanning of 2 sample RFID cards. As soon as the ID is scanned, the count is incremented for each of the ID cards depending on the number of logs. This can be monitored using the output window of the NodeRed Flow Editor. As shown in Fig. 7.
2. Further Fig. 5 shows the data which is updated in Google Firebase via the NodeRed Flow editor. The data received in the database is evaluated using Dialog Flow.
3. Figure 8 shows the simulated output of Google Assistant which also works well in real- time. The conversation can be defined as follows,

Fig. 7. NodeRED output

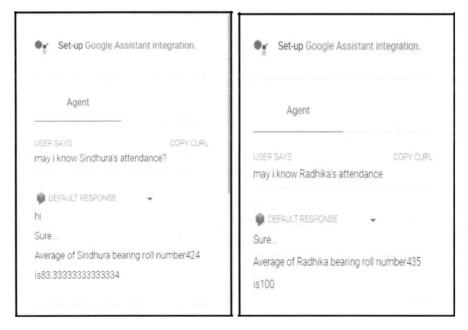

Fig. 8. Google Assistant responses

CASE A: (Student 1)
   User: Hi!
   Agent: Whose attendance would you like to know?
   User: May I Know Sindhura's attendance?
   Agent: Sure, Average attendance of Sindhura is 83.3333333333334%.
CASE B: (Student 2)
   User: Hi!
   Agent: Whose attendance would you like to know?
   User: May I Know Radhika's attendance?
   Agent: Sure, Average attendance of Radhika is 100%.

6 Analysis

The following analysis was made depending on the output observed.

(i) The Google Assistant was able to analyze the students bearing certain Roll Numbers and their corresponding Averages.
(ii) It takes a minimum of 5 s to provide the relevant response.
(iii) The system was able to update the database, dynamically within 10 s.
(iv) According to the Survey conducted on 198 students, 65.2% students demanded and agreed upon a Mobile Application that can keep a track of the daily attendance.

7 Conclusion

So far the Voice Assistants have been mostly useful for regular voice searches and home automation. The proposed idea justifies that Voice Assistants can also have a broader dimension. Within a few years, the Voice Assistant will also be able to satisfy the business and educational needs.

8 Compliance with Ethical Standards

The research or survey doesn't harm the human participants in any aspect, and doesn't reveal any personal information. All Procedures performed in studies involving human participants were in accordance with the ethical standards of the institutional research committee and with ethical comparable standards.

References

1. https://www.pwc.com/us/en/advisory-services/publications/consumer-intelligence-series/pwc-voice-assistants.pdf
2. https://www.thenewsminute.com/article/amazon-echo-google-home-why-voice-based-devices-are-big-hit-india-96529

3. Attitala, S., Chocksi, V., Potdar, M.B.: IBM cloud solutions for home automation. Int. J. Comput. Appl. **164**(4) 2017. ISSN 0975-8887
4. Xu, X., Huang, S., Feagan, L.: Edge analytics as a service (EAaaS). In: 2017 IEEE International Conference on Web Services (ICWS) (June 2017)
5. Shah, N., Sant, N.: Home automation using IBM Watson platform with intrusion detection. IEEE Int. J. Comput. Sci. Mob. Comput. (IJCSMC) **6**(12), 70–77 (2017). ISSN 2320-088X
6. Singh, N.: Study of Google Firebase API for android. Int. J. Innov. Res. Comput. Commun. Eng. **4**(9) (2016)
7. https://dialogflow.com/docs/getting-started
8. https://pdfs.semanticscholar.org/5819/1743afd55aedde6a962b644de2eb72bb80cf.pdf
9. Kaduskar, V.P., Gupta, N., Bhardwaj, Y., Kumar, S.: IoT based lab automation system. Int. J. Curr. Eng. Sci. Res. (IJCESR) **4**(6) 2017. ISSN (PRINT): 2393-8374, (ONLINE): 2394-0697
10. Kar, R., Haldar, R.: Applying chatbots to the Internet of Things opportunities and architectural elements. Int. J. Adv. Comput. Sci. Appl. **7**(11) (2016)
11. Ferrari, P., Flammini, A., Rinaldi, S., Sisinni, E.: Evaluation of Communication Delay in IoT Applications based on OPC-UA. Dept. of Information Engineering, University of Brescia. 978-1-5386-2497-5/18©2018 IEEE
12. https://www.stonetemple.com/digital-personal-assistants-study/
13. http://gs.statcounter.com/os-market-share
14. Niharika, M., Mubeen, S.: Fire detection and tracking without GPS. In: International Congress on Education and Learning (ICEL-18), Institute of Scientific and - Engineering Research, August 25, 2018. ISBN 9788192958093
15. Kaur, M., Sandhu, M., Mohan, N., Sandhu, P.S.: RFID technology principles, advantages, limitations & its applications. Int. J. Comput. Electr. Eng. **3**(1) (2011)
16. Kumar, R.P., Smys, S.: A novel report on architecture, protocols and applications in Internet of Things (IoT). In: 2018 2nd International Conference on Inventive Systems and Control (ICISC), pp. 1156–1161. IEEE (2018)

A Survey on Privacy Preserving Approaches on Health Care Big Data in Cloud

Lipsa Nayak and V. Jayalakshmi$^{(\boxtimes)}$

School of Computing Sciences, Vels Institute of Science,
Technology and Advanced Studies (VISTAS), Chennai, India
info.lipsa@gmail.com, jayasekar1996@yahoo.co.in

Abstract. To reform and enhance healthcare services, the use of information technology is increasing in an unprecedented manner. This technological advancement results in generating substantial medical data from various Healthcare related things. This Medical Big data has a maximum prospective to upgrade patients' treatment and follow up. We anticipate to flare-up some preventable infections, and we can significantly reduce the Healthcare delivery cost, by using the Medical Big data. One specific pattern observed in the medical services industry is serial shifting data to the cloud. By shifting data to the cloud, the accessibility of the patient's medical history in real-time is possible and it can reduce costs. Though we enjoy multiple advantages of cloud computing for Medical Big data, we face specific security issues of our data stored in the cloud. In this paper, a brief analysis on various approaches used to protect the healthcare related information stored in the cloud and further they are secured using cryptographic techniques.

Keywords: Healthcare · Cloud · Security · Medical Big data

1 Introduction

Evolution of digital technology in the Healthcare industry converts the hard copy of documents to electronic documents. These Electronic documents in a Healthcare institution are known as Electronic Medical Records (EMR), Electronic Health Record (EHR), Personal Health Record (PHR). PHR has patients' personal data and taken care by patients or their families [13]. EHR and EMR are health records of patients, Health care professionals like Doctors, and Laboratory experts are taking care of these records [1]. These EHR, EMR, and PHR include a massive variety of information that contains reports of the lab tests, radiology images, billing information, and any additional sensitive patient information, etc. These intricate Data, known as Medical Big Data, contain information of patients that can utilize for analysis and research. This Medical Big Data are responsible for reducing expenses of medical organizations by preventing repetitive health diagnoses like some expensive medical tests or treatment supervision etc. Big data with the cloud will continuously play a significant role in a Healthcare organization [6]. Healthcare cloud is the standard cloud where we store, process, and communicate data between various users through a network controlled by a third-party. Healthcare big data need to store in the cloud. So that all medical service providers and

© Springer Nature Switzerland AG 2020
A. P. Pandian et al. (Eds.): ICCBI 2019, LNDECT 49, pp. 32–38, 2020.
https://doi.org/10.1007/978-3-030-43192-1_4

collaborators can exchange their medical knowledge and can communicate with each other through the cloud server. Regardless of the time and space hurdle, we can store and access Medical Big data in the cloud environment [16, 17]. Apart from these various advantages, there are several security limitations of the healthcare cloud [7]. Security and protection are viewed as essential prerequisites when sharing or getting to persistent information between collaborators.

In this paper, we discussed various benefits of transforming file-based documents of medical institutions to electronic data and shifting those vast data to the cloud. In section two, the security requirement of Medical Big data is described. In section three, some techniques that are already being in use to secure Medical Big data in the cloud and its strengths and weakness are discussed.

2 Security and Privacy Requirement of Medical Big Data in Cloud

In the current era, information multiplication encourages the concept of cloud implementation in Heath care organizations. Patients and medical service providers mutually assisted by transferring Medical Big data to the cloud. Besides various benefits that cloud computing offers us, there are different security and privacy concerns for the sensitive data of the patients in the cloud. Sensitive information leaks, the person in the middle of the cryptographic attack, mocking, and so on are different conceivable assaults [10]. Many organizations suggest many security measures and guidelines. US Congress in 1996 proposes such governmental standards for the US medical organization is the Health Insurance Portability and Accountability Act (HIPAA). Some significant security requirements of Medical Big data in the cloud environment are (1) Information Integrity- It strengthens the inability of any unauthorized entity to change sensitive data. (2) Medical Big Data secrecy- guarantees to keep the delicate information from arriving at unapproved clients. Information encryption is the most generous way to deal with guarantee information confidentiality. (3) The authenticity of Medical Big data - guarantees that only authorized persons can access sensitive data. (4) Accountability of Medical Big data- a commitment to be capable and to legitimize the activities and choices of people or organizations. (5) Audit of Medical Big data- is a prerequisite which guarantees that the data is checked and ensured by monitoring the movement log and guarantees security [8]. As all electronic medical data are stored in third party servers and highly confidential hence, some access control mechanisms are required to protect these data. Some of the leading access restrict techniques in the healthcare organization are Attribute-Based Access Control (ABAC); this technique employs cryptographic and non-cryptographic techniques. Role-Based Access Control (RBAC); this strategy dole out specific roles to the users for accessing information [2]. Identity-Based Access Control (IBAC); this methodology uses client uniqueness for information hiding. Search is a substantial alternate function of a Healthcare cloud [4]. Few Searchable Encryption (SE) methods are there to investigate the scrambled information. Looking encoded information is intense, Searchable Symmetric Encryption (SSE), which allows searches encoded data in the cloud by using keyword. Numerous research examinations led to saving information security in the cloud.

Symmetric Key Encryption, Asymmetric Encryption or Public-key Encryption, are some of the examples of cryptographic mechanisms and RBAC, ABAC, IBAC, etc. are under non-cryptographic arrangements. Cryptography implies concealed writing that analyzes and builds protocols to avoid the reading of secret messages by third parties [14]. The cryptographic mechanism undergoes the entire Encryption and Decryption cycle. It can be symmetric-key cryptography or asymmetric-key cryptography.

3   Overview of Security Techniques for E-Health Cloud

3.1   Proxy Re-Encryption Using Time Limit Model in Cloud

Bhateja et al. [3] describes a framework, which enables clients to find information in some timeframe. Every user can access data by clearing some set of rules within a limited period. Authors presented time as T. Some crucial things of medical information like Client Revocation: to anticipate the activities seen by denied clients; authors discuss the idea of re-keying and re-encryption. To examine the concept of proxy re-encryption as a technique where without any knowledge of unencrypted data servers utilize the ciphertext between clients. It is a feature of the user's request to acquire the appropriate keywords. Conjunctive keyword search and Similarity keyword search, a comparative study was discussed by authors. The objective of the suggested system is to remove unauthorized individuals from accessing and viewing the PHR.

3.2   Attribute-Based Information Retrieval with Semantic Keyword Search

Yang [12] proposes a technique named attribute-based searchable encryption scheme with a semantic keyword search function (SK-ABSE). Concentrate on the multiple sender and user application situations to give an adaptable pursuit approval searchable encryption (SE) technique. The Attribute-Based encryption (ABE) innovation is utilized to help ease of access equivalent word keyword search is empowered in the new conspire. The new technology is named as trait-based available encryption with comparable word keyword search work (SK-ABSE). Attribute-Based Encryption with Keyword Search (ABKS) is a methodology that uses characteristic based encryption for information hiding. This searching technique increases fine-grain access control and promotes flexible keyword searching techniques, proposed by authors.

3.3   Search Mechanism Using Attribute-Based Encryption

Gowda et al. [5] present a keyword search system that maintains the confidentiality of EHR in the healthcare cloud. This system encodes a solitary access system where only trusted specialists could use public and private key sets. The data proprietor sets the entry arrangement and time before re-appropriating the information to the cloud. The authors presented a system model which is having three main entities- such as, Information proprietor (patient), Client (Doctor, Cardiologist, Receptionist, Nurse), and the Cloud service provider. The proprietor of the data extracts the keywords from the records and decodes them into indexes. The owner of information sets the access policy for the data he wants to share with the client. Then the encrypted records are outsourced to the cloud server. It ensures secrecy when conserving control of access and outsourced to the cloud server. It provides confidentiality when conserving the power of access and flexible customer revocation. Authors' experimental findings indicate that the cost of storage and time encryption is better than traditional schemes when comparing with multiple ciphertext files.

3.4 Conjunctive Keyword Search to Secure Healthcare Data

Yang et al. [15] Present a technique named as conjunctive keyword search with assigned analyzer and timing empowered intermediary re-encryption work (Re-dtPECK), to freeze the security and protection of medicinal services information. Authors suggest different forming of public-key encryption with conjunctive keyword search (PECK) over encrypted data. The concept of PEKS with a designated tester (dPEKS) is described to resist the threat. Only a designated examiner, which is usually the server, is capable of carrying on the test algorithm. The authors designed the Re-dtPECK scheme for EHR to achieve goals like Authority delegation: where the data owner could envoy his query without unveiling his private key. Several delegation times for different users: no restriction of time for the data owner is there. Security goals: The privacy concerns of this secure search system are summed up as keyword semantic security, resist KG attacks, a standard model that guarantees a higher security level.

3.5 Cloud-Based Medical Record Access

Rabieh et al. [11] discussed network and risk models. They portrayed a few elements of the system model in network models like SmartPhone (SP): The SP has a place with the patient and records the fundamental exercises by different wearable sensors. It can call the crisis number utilizing cell correspondences to demand a rescue vehicle. If the vehicle is autonomous (AV). The AV is furnished with the on-board unit (OBU) for calculation and correspondence and a Global Positioning System (GPS). The GPS empowers the AVs to find the closest EC in crises. Human Services Provider (HP): It scrambles the patients' therapeutic files, interfaces with the cloud, and transfers them. Cloud Server: The HP transfers the scrambled medicinal records to the cloud server intermittently. Confided in Third Party (TTP): The TTP is in charge of the issuance of cryptographic certifications to the SPs, HPs, APs, and ECs occasionally. The authors discussed the proposed scheme in detail by explaining System Bootstrap, On-vehicle Sensory Data Processing, Declaring Distress Mode, Re-encrypting Medical Records, Medical Records Decryption. The authors present the computational overhead of individual operation in a table and discuss security analysis.

3.6 A Security Model Using a Fog Computing Facility with Pairing-Based Cryptography

Al Hamid et al. [9] state that utilizing fog computing techniques focused on information security in the cloud. Authors present a third party one time validated key agreement rule. It is dependent on the bilinear matching cryptography, which can produce a session key between them so that the members can communicate safely. At last, by executing the Decoy technique, the personal healthcare information stored and used safely. Authors discussed Cloud computing, Fog computing, Decoy technique, Bilinear pairing function, Elliptic Curve Diffie-Hellman, Bilinear Diffie-Hellman Problems. Authors called Medical Big data as MBD, Decoy MBD as DMBD, and original MBD as OMBD. In the proposed system, the Authors presented two picture galleries. The original data is stored covertly, and the decoyed data is utilized as a honey pot and kept in the haze. Consequently, if any unapproved access is found, the attacker will get the decoyed information, which is not real. An effective third-party verified key agreement protocol has been proposed among the client by authors.

The summary of the various Security approaches to protect medical big data in the cloud is presented in Table 1. It provides the analysis of strengths, weaknesses of various security techniques. Security of cloud by identifying and gathering risk based upon multiple techniques are discussed.

Table 1. Comparison of security approaches for health care data in cloud

Author/Year	Techniques	Strength	Weakness	Ref.
Bhateja 2017	Searchable encryption, Proxy re-encryption	Versatile client revocation	Computational overhead	[3]
Gowda 2017	Hierarchy Attribute-Based Encryption	Flexible access control	Linkability among EHR	[5]
Al Hamid 2017	Fog Computing With Pairing-Based Cryptography	Preserves security and privacy	Follow a large number of rules	[9]
Rabieh 2018	proxy re-encryption	Robots and secure access control	Not scalable	[11]
Yang 2015	searchable encryption	Fine-grained access control	Cloud is aware of record access policy	[12]
Ma 2017	Searchable encryption; Conjunctive Keyword Search	Effective user revokement	Obstinate access control	[15]

4 Conclusion

Healthcare big data with cloud computing provide maximum advantages to the health care system. Immeasurable openings are offered for enormous information to make wellbeing research, learning revelation, and clinical health management. The Dominant part of the information in the cloud is exceptionally susceptible to threats and infringements. So it is essential to preserve healthcare cloud from illegal access and all other Security challenges. This analysis is focused on a comprehensive investigation of existing cloud protecting cryptographic and non-cryptographic techniques. In this paper, a brief review of various approaches used to protect the security of the health care information kept in the cloud and securing such data using cryptographic techniques are discussed.

References

1. Li, Z.-R., et al.: A secure electronic medical record sharing mechanism in the cloud computing platform. In: 2011 IEEE 15th International Symposium on Consumer Electronics (2011)
2. Liu, W., Liu, X., Liu, J., Wu, Q., Zhang, J., Li, Y.: Auditing and revocation enabled role-based access control over outsourced private EHRs. In: 2015 IEEE 17th International Conference on High-Performance Computing and Communications (HPCC), New York, NY, USA, pp. 336–341 (2015)
3. Bhateja, R., et al.: Enhanced timing enabled proxy re-encryption model for e-health data in the public cloud. In: 2017 IEEE International Conference on Advances in Computing, Communications, and Informatics (ICACCI) (2017)
4. Attrapadung, N., Furukawa, J., Imai, H.: Forward-secure and searchable broadcast encryption with short ciphertexts and private keys. ASIACRYPT 2006, LNCS 4284, pp. 161–177. Springer, Heidelberg (2006)
5. Gowda, B.K., Sumathi, R.: Hierarchy attribute-based encryption with timing enabled privacy preserving keyword search mechanism for e-health clouds. In: 2017 2nd IEEE International Conference on Recent Trends in Electronics Information & Communication Technology (RTEICT), India, 19–20 May 2017 (2017)
6. Jayalakshmi, V., Nayak, L.: Protecting medical big data in a healthcare cloud using elliptic curve cryptography. J. Adv. Res. Dyn. Control Syst. 10(11-Special Issue), 1054–1059 (2018)
7. Liu, X., Deng, R.H., Choo, K.R., Yang, Y.: Privacy-preserving outsourced clinical decision support system in the cloud. IEEE Trans. Serv. Comput. (2017). https://doi.org/10.1109/tsc.2017.2773604
8. Li, P., Guo, S., Miyazaki, T., Xie, M., Hu, J., Zhuang, W.: Privacy-preserving access to big data in the cloud. IEEE Cloud Comput. 3(5), 34–42 (2016)
9. Al Hamid, H.A., et al.: A security model for preserving the privacy of medical big data in a healthcare cloud using a fog computing facility with pairing-based cryptography, pp. 2169–3536. IEEE (2017)
10. Karthiban, M.K., Raj, J.S.: Big data analytics for developing secure internet of everything. J. ISMAC 1(02), 129–136 (2019)

11. Rabieh, K., Akkaya, K., Karabiyik, U., Qamruddin, J.: A secure and cloud-based medical records access scheme for on-road emergencies. In: 2018 15th IEEE Annual Consumer Communications & Networking Conference (CCNC) (2018)
12. Yang, Y.: Attribute-based data retrieval with semantic keyword search for e-health cloud. J. Cloud Comput. **4**(1), 10 (2015)
13. Chen, Y.Y., Lu, J.C., Jan, J.K.: A secure EHR system based on hybrid clouds. J. Med. Syst. **36**(5), 3375–3384 (2012). https://doi.org/10.1007/s10916-012-9830-6
14. Valanarasu, M.R.: Smart and secure Iot and AI integration framework for hospital environment. J. ISMAC **1**(03), 172–179 (2019)
15. Abbas, A., Khan, S.U.: A review on the state-of-the-art privacy-preserving approaches in the e-health clouds. IEEE J. Biomed. Heal. Inf. **18**(4), 1431–1441 (2014)
16. Yang, Y., Ma, M.: Conjunctive keyword search with designated tester and timing enabled proxy re-encryption function for e-health clouds. IEEE Trans. Inf. Forensics Secur. **11**(4), 746–759 (2016)
17. Kuo, M.H.: Opportunities and challenges of cloud computing to improve health care services. J Med Internet Research **13**(3), e67 (2011)

Effective Record Search and Update Using Design Patterns - A Case Study of Blood Bank Mobile Application

D. Venkata Subramanian$^{(\boxtimes)}$, R. Sugumar, K. Maneesh, V. Manoj, and A. Venkat Kishore

Department of Computer Science and Engineering,
Velammal Institute of Technology, Chennai, India
bostonvenkat@yahoo.com, sugul6@gmail.com,
maneeshvijaykar@gmail.com, manojvelmurugan@gmail.com,
avkish99@gmail.com

Abstract. This is the new era of handling data to develop industry growth by minimizing labor and error, thereby increasing quality as well as extending services to enhance the community. Similarly, there are many technological adjustments brought into the medical sector by managing huge data records and digitizing it. Digitized records provides an efficient data access with great usability. Today, Digitized health records and associated technology plays a vital role in our day to day life. The design patterns like publisher-subscriber, push and query command segregation patterns can be adopted easily across various applications. These advancements enable commercial data services and also assists well-fare groups to serve and contribute to the society. One of the community services addressed by using design pattern is blood donation. This paper briefly discusses about various design patterns and primarily addresses the idea of applying the design patterns to enhance the search and distribution network of blood bank and blood donation. Real-time blood bank needs a dedicated management system with the help of technology [12].

1 Introduction

Blood donation and the associated services are one of the core supporting service for healthcare domain. There are four main types of blood bank administrations maintaining this billion-dollar health care system in India. They are managed by public sectors, Indian Red Cross Society, non-government organizations and corporate or commercial sectors [1]. In support of this, there exists a need to design and develop digitized blood bank application with features like handling of emergency requests based on location constraints and needs. This application maps the nearby donors with requested blood carriers. This application provides the required information on time and also helps in better decision-making. A portable blood bank monitoring system is

© Springer Nature Switzerland AG 2020
A. P. Pandian et al. (Eds.): ICCBI 2019, LNDECT 49, pp. 39–47, 2020.
https://doi.org/10.1007/978-3-030-43192-1_5

required in order to distribute and serve multiple blood donation camps. This application synchronizes the records in blood centres and thereby derives the annual donations and transactions involved in the blood transfusion service and the health care system.

2 Literature Review

There are many design patterns available to solve the problems, but there are few limitations when we follow them, this makes us to approach different design patterns and club them together for our Backend API, Our Blood bank API is based on the Publisher-Subscriber Design Pattern. RESTFul APIs are used for the back end processing in this application. All the database works were carried out using Cloud MongoDB Atlas. The following sub sections briefly discuss about a few important design patterns.

2.1 Publisher Subscriber Pattern

[8] This proposed Blood Bank application follows Publisher Subscriber Pattern where the Users can request the blood as a BroadCast message request only for the users of the same Blood Carrier type (Figs. 1, 2, 3, 4).

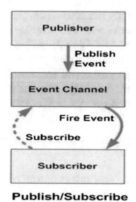

Fig. 1. Publisher-Subscriber Pattern

```
1 admin.post("/donations/:id", (req, res) => {
2   admins
3   .updateOne(
4     { _id: req.params.id },
5     {
6       $push: {
7         donations: {
8           $each: [
9             {
10               user_id: req.body.uid,
11               blood_quantity: req.body.amnt
12             }
13           ],
14           $sort: { blood_quantity: 1 }
15         }
16       }
17     }
18   )
19   .then(val => {
20     res.json({
21       status: "success"
22     });
23   });
24 });
```

Fig. 2. Post a Request for updating the donations on Admin Collections

```
1 users.get("/passchange/:email", (req, res) => {
2   var otp = Math.floor(Math.random() * 1234 + 1000);
3   User.updateOne(
4     { email: req.params.email },
5     { $set:
6       { password_otp: otp }
7     }
8   ).then(val => {
9
10     var mailOptions = {
11       ....
12     };
13     ....
14
15     res.json({
16       status: "success"
17     });
18
19   });
20 });
21
```

Fig. 3. Get Request to Change/Reset the password for the User.

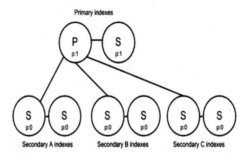

Fig. 4. Query Command Segregation evaluation diagram

3 Implementation

3.1 Interface Security

This is one of the major features of this application by including ONE-STEP VERIFICATION SYSTEM for users through Quick Response (QR) code scanning of government-issued certificates as a unique identity [2, 8]. As a government-issued certificate, this application deployed bases on AADHAR, under the license of the Unique Identification Authority of India. The verification process is the AADHAR QR code scanning of each individual registering in the application. Only legitimate information about the user is collected from the AADHAR. At any cost, only the encapsulated information about the user is publicly operated overall through the application.

The other type of user in the system is the blood bank administrator. They are verified by the license number issued by the Government of India. The registered blood bank details are automatically derived from the Application Programming Interfaces provided by the Open Government Data platform [3, 9]. These features reflect the blood bank-specific functions of this application.

3.2 Request Handling

Every registered user in the application location and blood types are fetched and enclosed in the encapsulated database. During emergency situations, the users can search the required blood samples within the requested location with an intuitive and user friendly graphical map with the blood donor details. This enhances the quality of information sourcing and presentation. In this application, there is a special feature for browsing the blood needs in a radius of approximately 100 km. This is one of the quickest ways of identifying the blood donors available nearby and infamous blood carriers. Using the proposed application portal, anyone can make an appointment with blood banks or hospitals for donating blood near his or her location. Every user is given information about the surrounding active blood bank and campaigns. This application monitors the shortages in order to make online request to active donors in the system.

The publisher-subscriber pattern thrives on its major components to post requests and handle these requests efficiently. The components proposed in this pattern is imposed in the blood request application as donors and requestee. This blood bank application acts as a medium to bridge the gap between the donor and the requester where the donor is a subscriber and the requestor is a publisher. Communication displayed among the donor and the requestor is not user-specific, nor direct but it is managed through the blood bank application which is one of the disciplines of the publisher-subscriber pattern. Websockets are another feature implemented in this blood bank application for instant and quick delivery of request and notification [10, 14].

The users of the blood bank applications are notified in the presence of active and devised blood needs through push notification mechanism using firebase [11]. This notification assignment is channelized by the blood bank application that is the donor and the requestor is not participating in this channel, as the application assigns the donors (Subscribers) for each request received from the requestee (Publishers).

3.3 Data Management and Portability

As far as the users of the application are considered the user's donation track records are maintained in the cloud database and recorded in the user's profile. This is done by generating a unique Quick Response code for every individual in the application. Every time the user donates blood, the unique Quick Response code is scanned and data is updated. With respect to the blood banks, donation records are updated with the feature of Quick Response code scanning after each donation made in the blood bank. Another advantage of Quick Response code scanning is that the blood donations made outside the blood bank can also be recorded. This brings the flexibility for blood donation camps for efficient data tracking.

In large scale data-centric-enterprise applications, the read-write ratio is very high. Hence, the faster read access with different search criteria from the enterprise database is a mandate. To efficiently handle higher data request rates application should be optimally performing using efficient processing steps to reduce the transaction time. With respect to data storage accessibility priorities vary, which induces the effect on query processing situations under the restful application interfaces working by the push pattern.

To solve this problem, The Command Query Responsibility Segregation is very well known for this kind of requirement for multiple reasons [13]. Here this pattern is implemented using MongoDB as primary storage and Synchronous storage in react native as secondary storage.

MongoDB offers strategies to implement Command Query Segregation, operations like create, delete, and update follow these below steps to achieve a better response time.

Every query or read for domain object with search criteria will go through the following stages in Query Service:

1. Try to get the primary keys from the push pattern described Application Interfaces (API) passed as a parameter.
2. Based on the received primary key from the previous step try to get the domain object from the Synchronous storage.
3. If the domain object is not available with cache, then get the domain object from MongoDB. (As retrieval using a primary key is always efficient with MongoDB).
4. Update the domain object into the Synchronous Storage and return the result.

Segregate operations that read data from operations that update data by using separate interfaces. This can maximize the parameters namely performance, scalability, and security. The system should also support the evolution of the system over time

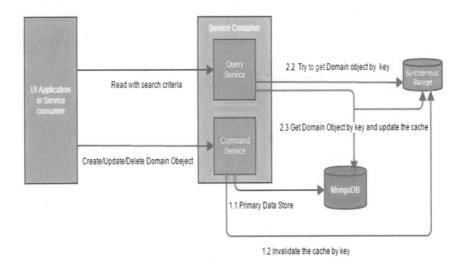

Fig. 5. Query Service Stages under command query segregation pattern.

through higher flexibility, and prevents update commands from causing merge conflicts at the domain level. This is the backend process inbuilt in the blood bank application to solve server requests from users (Fig. 5).

4   Methodology

To make our application robust, the MERN stack is followed (React Native is used instead of React) [15]. A REST API Created for the Blood Bank System to have scalability in our application. Using our REST API Services this application can further be scaled into a web app or a desktop app (for admins). The Endpoints on the REST API are made using the Express.js with the help of Node.js Server as our Backend and we save the user/admin/locations data on the MongoDB Atlas. These endpoints are then called in out React Native Client using a Fetch Library for React called Axios where the get and post methods are called and a JSON is sent either as a request to the server or received as a response from the server (Figs. 6, 7, 8, 9, 10, 11, 12).

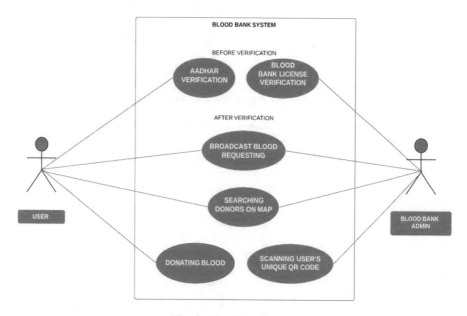

Fig. 6. Use Case diagram

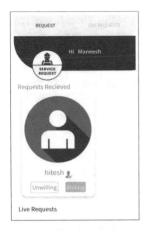

Fig. 7. Requests received to the donor for blood donation without the requestee intervention.

Fig. 8. Unique QR Screen for users which can be scanned only by admin's QR Scanner Screen

Fig. 9. Push notification sent to donors about the need for blood. And this intimation is purely based on the publisher-subscriber pattern.

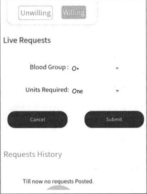

Fig. 10. A QR Scanner Screen for Admins which can only scan the user's unique QR code. Other QR code will not be accepted.

Fig. 11. A publisher screen to make a blood request and publish it online.

Fig. 12. Map Screen Interface to search other Users (under development & deployed for testing)

The requested blood type if it is found on nearby blood banks, then a message along with a notification is sent to the requested user else it maps the other users with same blood on the map with their details and the requested user's nearby blood bank locations and details are shared to the users who express their willingness to donate blood. In a similar manner blood bank admins can also request insufficient blood types and this time it maps only the users with the required blood types on the map with their details, while on the other side the address and other contact details of the blood bank is shared to the user who express their willingness the request (Table 1).

Table 1. Comparison between the existing and proposed system

Authors	Existing system	Proposed system
"Blood donation and lifesaver app" by M R Annish Brislin, Albert Mayan, R Aroul Canessane and M R Anish Hamlin [4]	Request and response method for blood request handling	Broadcasting the required blood carriers over a graphical
"Cloud-Based Information System for Blood Donation" by Dias Fotopoulos, Revekka Palaiologou, Ioannis Kouris, Dimitrios Koutsouris [5]	Blood Donation appointment by volunteered donors is implemented	Blood bank shortages can also make a request to the active donors
"Online Blood Bank Management System using Android" by Ashita Jain, Amit Nirmal, Nitish Sapre, Prof Shubhada Mone [6]	User-specific application with only verification of users	Also, act as a blood bank-specific with registered license verification
"Blood Donation Management System' by KM Akkas Ali, Israt Jahan, Md. Ariful Islam, Md. Shafa-at Parvez [7]	Blood bank records are handled in the static location	Using unique Quick Response code scanning, the blood bank records are updated spontaneously in blood donation camps

5   Conclusion and Future Scope

The mobile application for blood bank is a digitized version of blood banks with automated features. It provides most required details such as location tracking, one-step verification, and structured security. This application enhances the blood bank operations with organized data updates, which can track every blood donation. The blood bank application is clearly a use case scenario of the publisher-subscriber pattern, push and query segregation patterns. Dedicated map interface will monitor and track the blood banks' request and supply needs. The generalization provided by this mobile application can support this growing blood transfusion industry with clean data management. Future scope of this application is to extend this blood request and donation interface to the hospitals nearby, since current scope of the application is limited only to the users and the blood banks. This would also like to improve this application to support both web and mobile environments in the future. Machine learning and deep learning capabilities can be incorporated in order to improve further intelligence and speedy response to handle effective prediction and induce robustness.

References

1. Choudhury, N.: Management in Indian blood banking system: true reality
2. Uzun, V., Biligin, S.: Evaluation and implementation of QR Code Identity Tag system for Healthcare in Turkey
3. Application Programming Interface containing registered blood bank information. https://auth.mygov.in
4. Annish Brislin, M.R., Aroul Canessane, R., Anish Hamlin, M.R., Albert Mayan, J.: Blood donation and lifesaver app
5. Fotopoulos, I., Palaiologou, R., Kouris, I., Koutsouris, D.: Cloud-Based Information System for Blood Donation
6. Jain, A., Nirmal, A., Sapre, N., Mone, S.: Online Blood Bank Management System using Android
7. Ali, A., Jahan, I., Islam, A., Parvez, S.: Blood Donation Management System
8. https://hackernoon.com/observer-vs-pub-sub-pattern-50d3b27f838c
9. https://medium.com/@_JeffPoole/thoughts-on-push-vs-pull-architectures-666f1eab20c2
10. Eugster, P.Th., Felber, P.A., Guerraoui, R., Kermarrec, A.-M.: The Many Faces of Publish/Subscribe
11. Dev Anand, L, Arumugam, P., Harini, J., Chithambarathanu, M: Optimized Service-Oriented Request and Compliance Management Application
12. Kulshreshtha, V., Maheshwari, S.: Benefits of Management Information System in Blood Bank
13. Niltoft, P., Pochill, P.: Evaluating Command Query Responsibility Segregation
14. Ivan, C.: A web based Publish-Subscribe framework for Mobile Computing
15. https://blog.logrocket.com/mern-stack-a-to-z-part-1/

Comprehensive and Comparative Study of Efficient Location Tracking Based on Apriori and Dijkstra Algorithms

D. Venkata Subramanian$^{(\boxtimes)}$, R. Sugumar, N. Dhipikha, R. Vinothini,
S. Kavitha, and A. Harsha Anchaliya

Department of Computer Science and Engineering,
Velammal Institute of Technology, Chennai, India
bostonvenkat@yahoo.com, sugul6@gmail.com,
dhipikha8@hotmail.com, vino14768@gmail.com,
ssikavitha2000@gmail.com, jippuharsha@gmail.com

Abstract. With the advent of Google maps, pointing the precise location and tracking the movement of objects, people and materials have become an integral part of SCM System. The ERP systems include Logistics and Management as core functionalities along with Artificial Intelligent Systems to facilitate positioning and tracking of materials. Similar to material tracking, it is also important to track the location and movement of the person which is a huge demand in present era. However, the designers of such larger ERP systems use patented protocols for material location identification, predicting the shortest distance and tracking the precise location of cargo when they are on the move. The similar techniques neither fully adopted nor applied with humans. This research paper is aimed at finding out the most efficient route by comparing the most popular Apriori and the Dijkstra algorithms. Apriori algorithm involves supervised mining using association rules which can be used for finding right paths whereas the same can be estimated by the Dijkstra's algorithm, taking the connecting nodes in the graph.

Keywords: Supply chain management · ERP · Apriori algorithm · Dijkstra's algorithm · Location tracking · Dataset · Prediction · Location · Tracking · Database · GPS · Notification

1 Introduction

The location dependent systems predict real-time information about an object location using the coordinates through GPS, Wi-Fi or Cell – ID. The data warehouse system supports the management decision making process with its time variant and non-volatile group of multiple datasets [1]. Data mining helps in predicting the hidden data pattern in a warehouse [2]. By discovering the data patterns, it is easy to predict immediate decisions either for business or in our normal life. The modern data mining methods use clustering techniques by its ability to identify similarity in objects. There is a need to identify and distinguish between similar and dissimilar objects in cluster technology [3]. In this paper, Apriori and Dijkstra algorithm are used for identifying a

© Springer Nature Switzerland AG 2020
A. P. Pandian et al. (Eds.): ICCBI 2019, LNDECT 49, pp. 48–59, 2020.
https://doi.org/10.1007/978-3-030-43192-1_6

travel path that is feasible and optimal. The common itemset mining and association rules are predominantly decided by the Apriori algorithm. There is a process in which the individual data set in a database is closely followed up and by expanding them to even larger datasets till those datasets present in the database. This character helps in identifying general trends in the database [4]. In a map, each location is identified as a "node". So multiple locations are linked through these nodes and traceable from any starting point. The Dijkstra algorithm is used for finding the shortest path between the nodes. This uses direction of travel and predicts the path on the map [5]. In this research paper, the Apriori algorithm and the Dijkstra algorithm are compared on key parameters such as efficiency, accuracy, time complexity and space complexity.

2 Literature Review

2.1 GIS in Tracking System

The core functionality of Geographic Information System (GIS) is to create a visualization on the map and 3D scene based on the information collected from the nodes spread across the map. The sophistication of GIS is its ability to formulate patterns and relationship based on the data acquired from time to time, which enables the user to take decisions at the time. The geographic science feeds adequate information for GIS to provide actionable intelligence. GIS is not only used for collecting data, it also analyses and displays a connection to the GIS database. By using the same connection requests from relational database, information exchange becomes spontaneous from the data base to map and vice versa. This technical superiority of GIS enables automatic updates on the maps when database is renewed. GIS Tracking systems are used abundantly in supply chain management system where in the movement of vehicles carrying materials are tracked live. The server which captures the vehicle location uses radio-based transmitters fitted on the vehicle, and a GPS receiver is also used. The locational outline of a moving vehicle on the server is populated by GIS maps. This system holistically helps the vehicles in navigating to their destination [6].

2.2 Tracking System Using GPS

Public transport networks (PTNs) are not so handy to the frequent and infrequent travelers when they travel to unknown areas. Under these circumstances, on-trip navigation systems can be very helpful for such users and the ease of handling such systems can motivate the users to reach the destination. By calculating the location of a user, it would be critical for providing appropriate information. The pertinent outline problems for a user friendly, cost-efficient, on trip navigation service that uses GPS may be considered prior to engaging such systems for an effective comparison. The usage of public transportation systems not only can dramatically bring down the traffic congestion but also contribute on matters related to the environment [7].

3   Adaption of Apriori Algorithm

Association rules are determined by apriori where data mining is used. The authoritativeness of the Apriori algorithm in determining Boolean association rules through mining frequent itemsets must be a definite advantage [8]. The essential property of this algorithm mediates scanning of the database to produce frequent item sets by boolean association rules.

The apriori verifies each method layer by layer, where n - dimensional item sets are used to explore (n + 1) - dimensional item set. Firstly, the frequent set of k-dimensional item sets are denoted by K1, then, K1 is used to find K2, the set of K2 frequent two item sets that are used to find K3, till no more frequent n-dimensional item sets are found [9]. Lastly, after getting the rules from the huge set of data, how Ki−1 is used to find Ki consists of two step process that are prune and join actions [10].

Join step: Join Kn−1 with Kn−1 itself, then combine the same adjunct data that appears to initiate a candidate who has n-dimensional item sets, where Pn denotes the set of candidates, Pn ⊇ Kn.

Pruning step: Calculate the count of each candidate in Pn. If count < minimum count, then delete that count from the candidate itemset 1.

This algorithm utilizes an approach that is interactive which is known as a level wise search [11], where n-item sets traverse (n + 1) itemsets. First, frequent set of k-itemsets are found. This indicates K1. K1 then finds K2, the frequent set of two itemsets that finds K3, there is no more frequent n-item to be found. For discovering each Kn needs to check the whole database. For finding all the repeated item sets, this algorithm selects recursive method as follows [12]:

```
K1 = { large k-itemsets} ;
for (n=2; Kn-1 ≠ Φ ; n++) do {
Pn = Apriori-gen (Kn-1); // new entry of candidates
do {
Pj = subset (Pn, j); // group of j candidates
for each candidates p ∈ Pj
do
p.count++;
}
Kn={ p ∈ Pn | p.count ≥ mincount }
}
Return nKn;
```

For reducing the size of Pn, pruning steps are used. If any (n−1) subset of a candidate n -itemsets not in Kn−1, then the candidate is removed from Pn. The pruning step reduces the calculation cost of the candidate sets by reducing size of candidate sets, thus it improves the performance of discovering the frequent item sets.

4   Adaption of Dijkstra Algorithm

Dijkstra algorithm is known for finding the shortest path. It determines the shortest path from the source to each unseen pinnacle in the graph. In this algorithm, the shortest path is estimated using nodes spread across the map with the starting point is taken as a root. The directed graph with weight produces the shortest path node denoting the starting point. Then all adjacent nodes to 's' are mapped to find the shorted node possible and the arclength is defined as chord-length.

This algorithm estimates the shortest route from the pinnacles of the graph which normally represent 'cities', and the edges constitute 'driving distance between pairs, provided these cities are joined by a direct road.

This implementation of this algorithm is represented in a graph. Here a weighted directed graph G = (V, E) with source vertex n is assigned. 'A' is used for representing the visited vertices. Priority queue P is represented as set of all vertices in the graph.

```
Dijkstra (G, n)
For every vertex v in graph G {
    D[ n] = 0
    D[ n] = ∞
}
Assign visited vertices A in G
A = null
Assign Queue P as set of all nodes in G
P = all vertices V
While P ≠ ∅ {
t = mind (Graph G, distance d)
    A = A+t
for every vertex v in neighbor [ t] {
    If d[ v] > d[ t] + w (t,v)
        Then d[ v] = d[ t] + w (u,v)
    }
Return d
}
```

Here the graph G is initialized with the vertices V and with edges E. Then assign distance for each node and '0' value is set for the current node. After that calculation of distance from the current node to all its neighbor node are marked as a visited node. Then, the node which is already visited cannot be used again. The distance is stored, and the unmarked node is set as less distance.

5   Data Model

Data model in the class diagram consists of classes, attributes and operations. The main class consists of the sub classes. The sub classes are LOCATION, MODE, INTERMEDIATE STATIONS, USER DATA, NOTIFY, DATABASE, RATING (Fig. 1).

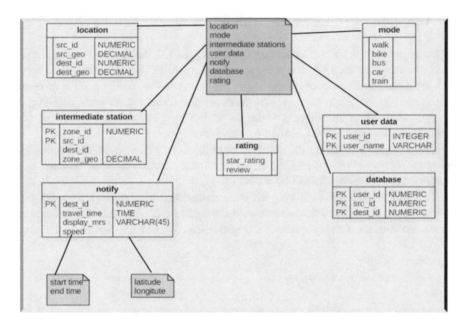

Fig. 1. Data model diagram for location tracking.

Each class contains the attributes and method. Location class has source id, source geometry, destination id and destination geometry. The source id and destination id are declared as number and geometry are declared as decimal variable. The class intermediate stations contains zone_id which is declared as a number, src_id, destination id, all those attributes are declared as the primary key.

MODE class encloses the mode of travel used by the user, the attributes used here are walk, bike, bus, car or train. The class notifies the dest_id which is declared as primary key (unique value), travel_time will be declared as time, display_mrs is declared using varchar with size 20, which is used to get the latitude and longitude of the location and speed which contains the starting and the end time.

The user data class has the user name and user id. Both are assigned as the primary key. Database class will store the user_id, destination_id and the source_id. They are primary keys of unique identification of the user. The class rating will collect the review and rating after the usage of the app.

6   Use Case Diagram

Use case diagram for location tracker system consists of actors, use cases, association and relationship between the use cases and actor. An actor is anything with behavior including the system under discussion itself when it calls upon the services of other system. Here USER, DATABASE, SYSTEM are act as the actors in this diagram. A use case is a list of actions or event steps typically defining the interactions between a role (known in the Unified Modeling Language (UML) as an actor) and a system to achieve a goal (Fig. 2).

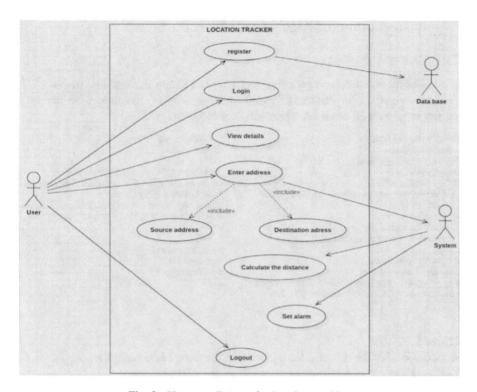

Fig. 2. Use case diagram for location tracking

REGISTER, LOGIN, VIEWDETAILS, ENTER_ADDRESS, SOURCE_AD-DRESS, DESTINATION_ADDRESS, CALCULATION, SET_ALARM, LOGOUT are the use cases used in location tracker system. Register will contain the registration of new user and the login will use the login with the already registered user. The registration details are maintained in the database.

After the login process, the user can view the details present inside the view_details use case. The user will enter the source and the destination address of the location. Once the source and destination address are entered into the system, the system will calculate the distance and set the alarm. The user after entering the source and destination he/she can logout.

7   Implementation of Algorithms Using Datasets

7.1   Apriori Algorithm

In this algorithm, all non-empty subsets of a repeated itemset should be frequent. The main concept of this algorithm is its anti - monotonicity of measures. If an itemset is not frequent, then all other supersets will be infrequent.

```
# Import required libraries
import numpy as n
import matplotlib.pyplot as p
import pandas as d
# Preprocessing of Data
data_set = p.read_csv ('Location.csv', header = None)
transaction = []
for i in range (0, 2):
    transaction.append ([str (data
_set.values [i, j]) for j in range (0, 20)])

# Train the apriori dataset
from apyori import apriori
rule = apriori (transaction, minsupport = 0.003,
minconfidence = 0.2, minlift = 3, minlength = 2)
```

```
# View the result
result = list(rule)
```

Implementing the above code for the data set ('Hariminimahal','karukku','AmbedkarRoad',Saravana Store','Indian oil','PVR','central') ('Hariminimahal','karukku','AmbedkarRoad',Saravana Store','Singaram   pillai   school','New   avadi Road','central')

we get support as
support=0.5714
confidence=0.6667
lift=1.1478

Support
Support specifies the repeated item present in the data. Mathematically, support is the fraction of the no. of total transactions where the itemset is present.

Support_Count(σ)– Frequency of existence of a item set.
Supp = $\sigma(X + Y)\div$ total
It is interpreted as portion of transactions containing X &Y values.

Confidence
The number of times the if then statement occurs and said to be true, in such case it is referred to as confidence. Confidence is conditional probability of existence of consequent presented the antecedent.

conf (X => Y) = supp (X$\cup$Y)$\div$ supp(X)

Lift
The comparison of confidence with the expected conference is called as lift. This says how likely item Y is purchased when item X is purchased while controlling for how popular item Y is. Mathematically,

Lift (X => Y) = conf (X => Y)$\div$supp(Y)

With the million (and more) of rows and columns that can persist in a transactional database, it'll be hard to manually use these mathematical formulas to find relations amongst item sets. In this post, I'll be using the Apriori algorithm from the mixed library. This algorithm is a generally for drawing out frequent item sets.

7.2   Dijkstra's Algorithm

Here in implementing Dijkstra's algorithm, two nodes are taken into account initially to detect the distance between the two nodes after the distance between the first node and the second node is calculated, third node is taken and distance between the second node and the third node is found. This process takes place until the last node is reached.

```
class Graph :

    def _init_ (self) :

    # dictionary that has keys which maps to the
    corresponding vertex

        self.vertex = {}

    def addVertex (self, k) :

        #Add a vertex to the graph corresponding to the
    key k.

        vertex = Vertex (k)

        self.vertex[k] = vertex

    def getVertex (self, k):

        #Return vertex that corresponds to the key.

        return self.vertex[k]

    def _contains_ (self, k) :

        return k in self.vertex

    def addEdge (self, srcKey, destKey, w=1):

        #Add edge from srcKey to destKey with given
    weight

        self.vertex[srcKey].addNeighbour
    (self.vetex[destKey], w)

    def exist(self, srcKey, destKey):

        #Return True if there is an edge from srcKey to
    destKey.

        Return self.vertex [srcKey].pointsto (self.vertex
    [destKey])

    def _iter_ (self) :

        return iter(self.vertex.values())
```

```
class Vertices:

    def _init_ (self, k) :

        self.key = k

        self.pointsto = {}

    def getKey (self) :

        #Return key that corresponds to vertex

        return self.key

    def addNeighbour (self, dest, w):

        # vertex points to desttination with given edge
    weight w

        self.pointsto [dest] = w

    def getNeighbours(self):

        #Return all vertices

        return self.pointsto.keys()

    def getWeight(self, dest):

        return self.pointsto[dest]

    def pointsto(self, dest):

        #If vertex points to destination then return true.

        return dest in self.pointsto

    def dijkstra(g, source):

        """ Return distance where distance[v] is min
    distance from source to v.

    This will return a dictionary distance.

    g is a Graph object.

    source is a Vertex object in g.
```

```python
    """
    unvisited = set(g)
    distance = dict.fromkeys(g, float('inf'))
    distance[source] = 0
    while unvisited != set():
        # find vertex with minimum distance
        closest = min(unvisited, k=lambda v:
    distance[v])
        # mark as visited
        unvisited.remove(closest)
        # update distances
        for neighbour in closest.getNeighbours():
            if neighbour in unvisited:
                new_distance = distance[closest] +
    closest.getWeight(neighbour)
                if distance[neighbour] > new_distance:
                    distance[neighbour] = new_distance
    return distance
g = Graph()
print(" Add Vertex <key>")
print(" Add Edge <src> <dest> <weight> ")
print("shortest <source vertex key> ")
print(" Display ")
print(" Quit ")
while True :
    a = input (" what would you like to do? "). split()
    op = a[0]
    if op == 'add':
        subop = a[1]

            if subop == 'vertex' :
                keys = int (a[2])
                if keys not in g:
                    g.addVertex(k)
                else :
                    print(" vertex is already present ")
            elif subop == 'edge' :
                src = int (a[2])
                dest = int (a[3])
                weight = int (a[4])
                if src not in g:
                    print(" vertex doesn't exist".format(src))
                elif dest not in g:
                    print(" vertex {} does not
    exist.'.format(dest) )
                else :
                    if not g.exist (src, dest ) :
                        g.addEdge (src, dest, weight)
                        g.addEdge(dest, src, weight)
                    else :
                        print(" Edge is already present")
    elif op == 'shortest' :
        k = int (a[1])
        source = g.getVertex (k)
        distance = dijkstra (g, source)
        print(" Distances : ".format(k))
        for v in distance:
            print('Distance to {}: {}'.format(v.getKey(),
    distance[v]))
```

```
print()                                              for dest in v.getNeighbours():

elif op == 'display':                                    w = v.getWeight(dest)

print(" vertices : ", end=" "")                          print(" src={}, dest={}, weight={}
                                                     ".format(v.getKey(), dest.getKey(), w))
for v in g:
                                                     print()
    print(v.getKey(), end=" "")
                                                 elif op == 'quit':
print()
                                                     break
print(" Edge : "")

for v in g:
```

After implementing and distance between the nodes, the accuracy and efficiency of the Dijkstra's algorithm is calculated from the time taken by the code to run.

8   Comparative Analysis of Algorithm's

By using the same data set in both the Apriori and the Dijkstra's algorithm with python language, the efficiency, time complexity, and space complexity are studied and tabulated (Fig. 3).

Algorithm	Time Complexity	Space complexity	Accuracy
Apriori	O(log n)	$O(2^n-1)$	97%
Dijkstra	O(E+VlogV)	O(V)	99%

Fig. 3. Table for comparison of algorithm's

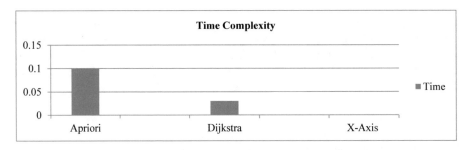

Fig. 4. Graph on time complexity of both algorithm's

The repeated itemsets are identified and populated by apriori algorithm using the previous steps [13]. Notably, Apriori algorithm uses a large no of scans of the dataset [14]. When the apriori algorithm is larger, a large no. of item sets is produced. Algorithm scans database repeatedly for searching frequent item sets. This obviously takes more time and resources while scanning large data sets repetitively, leading to poor efficiency [15]. The Apriori algorithm is found to be not so efficient in case of larger data sets due to these reasons (Fig. 4).

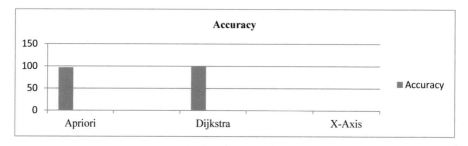

Fig. 5. Graph for accuracy of algorithm's

On the contrary, the Dijkstra's algorithm is found to be better and quicker than apriori algorithm. The apriori algorithm is not recommended for scanning larger networks. The Apriori algorithm also is also known for detecting negative cycles. However when calculating the distance, it will not be in negative form and as a result of this apriori may not be so useful. In reality, there is a need for quick and faster online review of the data in GIS from the data base to map, towards predicting the shorter path. In this circumstance, the Apriori algorithm was not to be ineffective and not recommended (Fig. 5).

9   Conclusion

On establishing the study results, the conclusion is that the dijkstra algorithm is comparatively quick and better than the apriori algorithm and can be extensively used in GIS technology. Furthermore, the Dijkstra's algorithm is found to be quicker in deciding the shortest path even in larger networks. Hence, this uniqueness of the Dijkstra's algorithm is deployed in real-time application of GIS.

9.1   Future Scope

Dijkstra's algorithm can be used for some of the following

1. To find the shortest distance.
2. To calculate the node distance from its source point.
3. To implement the algorithm on different data sets.

Apriori's algorithm can also be used for the following applications with additional automation

1. When Email marketing based on earlier MBA analysis.
2. If any relation of multiple component failures in electric motors similarity in other industry.
3. In Service Engineer recommendation which takes important details to take while fixing an issue.
4. Retails store product recommendation.

Hence, there are additional venues and opportunities available to incorporate machine learning or deep learning functionalities with the data mining algorithms for efficient and proactive prediction without any supervision.

References

1. Ponniah, P.: Data Warehousing Fundamentals—A Comprehensive Guide for IT Professionals, Ist edn. Glorious Printers, New Delhi (2007). ISBN-81-265-0919-8, second reprint
2. An Introduction to Data Mining, Review. http://www.thearling.com/text/dmwhite/dmwhite. htm
3. A Tutorial on Clustering Algorithms, Review. http://home.dei.polimi.it/matteucc/Clustering/ tutorial_html
4. Mining Frequent data sets-Apriori algorithm. http://dwgeek.com/mining-frequent-itemsets-apriori-algorithm.html/
5. https://medium.com/basecs/finding-the-shortest-path-with-a-little-help-from-dijkstra-613149fbdc8e
6. GIS in Tracking system. https://www.roseindia.net/technology/vehicle-tracking/gis-tracking-system.shtml
7. Tracking System Using GPS on Smartphone. https://www.researchgate.net/publication/ 263582815_Analysis_of_Bus_Tracking_System_Using_GPS_on_Smartphones
8. Zhou, Y., Wan, W., Liu, J. Cai, L.: Mining association rules based on an improved Apriori algorithm. IEEE (2010). 978-1-4244-5858-5/10/
9. Fang, L.: The study on the application of data mining based on association rules. In: International Conference on Communication Systems and Network Technologies (IEEE), pp. 477–480, May 2012
10. Yu, C., Ying, X.: Research and improvement of Apriori algorithm for rules. In: 2nd International Workshop on Intelligent Systems and Applications (ISA), pp. 1–4, May 2010
11. Singh, J., Ram, H., Sodhi, J.S.: Improving efficiency of Apriori algorithm using transaction reduction. Proc. Int. J. Sci. Res. Publ. (IJSRP) 3(1), 1–4 (2013). ISSN 2250-3153
12. Waghela, P.P.D.: Comparative study of association rule mining algorithms. In: Proceeding of UNIASCIT, vol. 2, no. 1, pp. 170–172 (2012). ISSN 2250-0987
13. Geetha, K., Mohiddin, S.: An efficient data mining technique for generating frequent item sets. Proc. IJARCSSE 3(4), 571–575 (2013). ISSN 2277-128X
14. Dhanda, M., Guglani, S., Gupta, G.: Mining efficient association rules through Apriori algorithm using attributes. Proc. IJCST 2(3), 342–344 (2011). ISSN 0876-8491
15. Nagpal, S.: Improved Apriori algorithm using logarithmic decoding and pruning. Proc. Int. J. Eng. Res. Appl. 2(3), 2569–2572 (2012). ISSN 2248-9622

Weather Data Analysis for Jaipur City: Using MongoDB

Talha Khan[✉], Zaid Khatri, Shiburaj Pappu, Imtiyaz Khan,
and Shakila Shaikh

Computer Engineering Department,
Rizvi College of Engineering, Mumbai, India
ktalha548@gmail.com, zaidkhatri31@gmail.com,
khanimtiyazkhan786@gmail.com,
shakilashaikh48@gmail.com,
shiburaj@eng.rizvi.edu.in

Abstract. The weather and climate forecast system is often assessed by calculating the available data with the past forecast and their observation. This project aims to forecast the weather by performing analysis in MongoDB, where the proposed system serves as a tool that takes weather data by assuming large amount of data as input and analyse the future weather with the help of min and max temperature, pressure, humidity etc. in an efficient manner. Analysis for this data, visualization and interpretation of weather system over wide geographical areas becomes possible with less effort and errors.

Keywords: MongoDB · Weather · Jaipur · Analysis · Data · Pressure · Temperature · Humidity

1 Introduction

In general, Weather condition is observed to be a highly complex system, which is usually nonlinear in nature [1–5]. Climate change is drawing a lot of attention since a long period of time. Because of uneven changes that occur. Big data is capable of maintaining a huge amount of data that is efficiently processed. To capture, manage and process the data [6–9]. Big data includes data sets with size beyond the ability of commonly used software tools [10–14].

We will refer three papers, which consists of previous year's historical weather data set, we are able to analyse them and find out which one is better and why? Identifying temperature on the dominant abiotic factor which directly affect herbivores insects therefore extremes of temperature may negatively affect insects. In the last decade it is notice that untimely response of nature on flowering and leaf unfolding in spring, similarly many such phonological processes are also not on time. One of the reason is increase in temperature.

Climate change will be experienced million of people worldwide in their daily lives during all the seasons. Rest of the paper is organized as follows. Literature review is discussed in Sect. 2, Methodology in Sect. 3, Design and analysis in Sect. 4. Later, the paper is concluded respectively.

© Springer Nature Switzerland AG 2020
A. P. Pandian et al. (Eds.): ICCBI 2019, LNDECT 49, pp. 60–66, 2020.
https://doi.org/10.1007/978-3-030-43192-1_7

2 Literature Review

In [5] for weather prediction various data mining methodologies are used by many researchers to find out the interesting pattern, K-mean clustering is used which divides the land and other existing area. This data has been collected from the available data (2016–18). The parameter taken are temperature, weather condition and humidity. In the current paper the admin uploads the data like effect of temperature and preventive measures. Later upload this dataset to the system. User has to register to the application at the client side. By doing this user get the predicted current temperature with the help of F-P growth algorithm. It reads the data set and simultaneously maps it. To build a numerical weather prediction model in local weather, system present a first glance on a project. These model are complex and takes more time and resource to accomplish. But complex parameter as input are also accepted and intelligent pattern are generated which successfully predicts the weather.

In [6] there are multiple authors used different data mining methodologies which results into weather predictions. E.G Petre used CART decision tree algorithm and presented a small application. In decision tree algorithm data transformation is required to use WEKA efficiently for prediction of data. Accuracy for this prediction is 83%. Accuracy is good but data transformation and extra computation is required K. Pabreja states that the derivation of sub grid scale weather system is not possible through normal MOS technique, so it uses NWP model. K-mean clustering is applied for two days data of cloud burst. For providing timely and action able information for these events an effort is made by using data mining technique with NWP model in supplement. Accuracy is 100% but it is not suitable for long term prediction. Multiple regression models are used for rainfall prediction

In [7] paper by using Empirical method technique of data mining a short term prediction of rainfall was specific region is made. For grouping, the element clustering technique is used in a particular rainfall region and prediction of rainfall is done in that particular region. Multiple linear regression models are used for rainfall prediction but the result of the prediction have some approximate value it is not 100% accurate. For creating dataset parameters are used such as minimum and maximum temperature, rainfall, humidity, wind pressure. It has a moderate accuracy of 52%. The advantage is data set is also acceptable but on the other hand accuracy rate is low as it provide approximate value in the result.

In this paper of review of Jaipur weather analysis: Jaipur is situated in the eastern part of Rajasthan at an altitude of 431 m. Being a part of the desert state of Rajasthan, Jaipur has a hot semi-arid climate. The city remains quite hot and dry during the summers and is during winters. As per the available data(2016–18) of Jaipur weather minimum temperature in winters results into the most coolest with temperature of 10 °C but in summers the same hypes at 32 °C. Jaipur is a desert area which is known as a hottest place, it's maximum temperature is 46 °C at the time of summers but the same falls up to 18 at the time of winters. Wind pressure has been substituted in three different phases i.e. (Morning, Afternoon and Night). Within two years of data the maximum air pressure calculated is 1026 and minimum is 999. Consistency within the air pressure of two years is high.

3 Methodology

The topic consist of various information and state diagram which measures the weather data and analyse it. It states the working model of the project. It consists of multiple blocks which includes Data extraction, import data, Analyse data. Diagram consists of Structure, behaviour and views of the system. Diagram's description supports structure and behaviour of the system. In this project the data is gathered from governmental websites of last two years from 2016–18 of Jaipur. It is further executed in MongoDB which result into weather prediction. It supports structures and behaviour of the system.

A State Diagram consist of many components, that work's together to implement the overall system. Each block is further explained in detail (shown in Fig. 1).

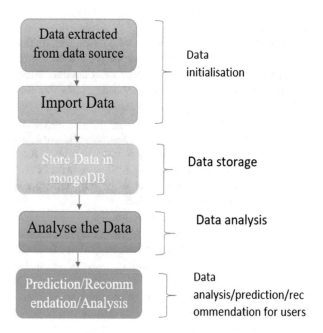

Fig. 1. State diagram

A. *Weather Data*
   This Module contains Weather data which will be used for weather predicting. It contains various parameters such as temperature, pressure, humidity and precipitation. Data set of our Project consist of two years (2016–18) shown below.
B. *MongoDB*
   MongoDB is a language that implements a data store that provide very high level of performance, high availability and automatic scaling is also done. Installation and Implementation of MongoDB is easy and simple.

C. *Import and Export Data*

As MongoDB supports LOC queries, replication, duplication of data as well. It also supports JSON and CSV models. The data handling of importing and exporting become very flexible. MongoDB has the ability to import and export data from collection.

D. *Query and Analysis*

In version 1 the weather analysis using mongoDB is executed successfully. In our project we have used commands like Find, gte (greater than equal), lte (less than equal).

Query:

```
db.JaipurFinalCleanData.find({"mintempm":{$l te:"12"}})
```

As per the available data (2016–18) of Jaipur, the output of the above query is the minimum temperature is noticed 10, the Average temperature counted within this time-period is 19.63. which states that Jaipur is coolest at this phase. In summer's the minimum temperature is around 32, at this phase it is the hottest, but in winters the temperature falls up to 10 and in Rainy it is approximate 20–21 (Fig. 2).

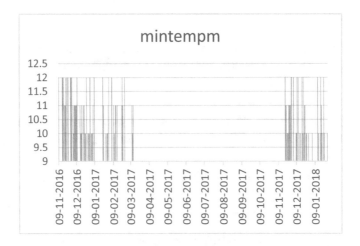

Fig. 2. Analysis of minimum temperature

Query:

```
db.JaipurFinalCleanData.find({"maxtempm":{$g te:"42"}})
```

As per the available data (2016–18) of Jaipur the output of the above query is, the maximum temperature is noticed 46, the Average temperature counted within this time-period is 32.52. which states that Jaipur is hottest at this phase. In summer's the minimum temperature is around 46, at this phase it is the hottest, but in winters the temperature falls up to 18 and in Rainy it is approximate 32–33 (Fig. 3).

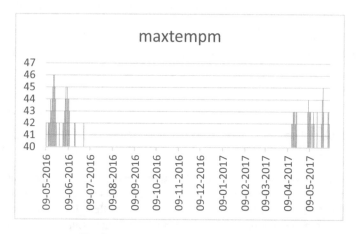

Fig. 3. Analysis of maximum temperature

Query:

```
db.JaipurFinalCleanData.find({},{_id:0,maxpr
essurem_1:1,maxpressurem_2:1,maxpressurem_3:
1,date:1})
```

As per the available data (2016–18) the output of above query is Pressure of Jaipur is further substituted in three different phases i.e. (Morning, Afternoon & Night).

MORNING:
Maximum pressure noticed is '1026' Minimum pressure noticed is '999'
AFTERNOON:
Maximum pressure noticed is '1026' Minimum pressure noticed is '999'
NIGHT:
Maximum pressure noticed is '1026' Minimum pressure noticed is '999'

This states that the consistency is high as because all the three phases are constant and same (Fig. 4).

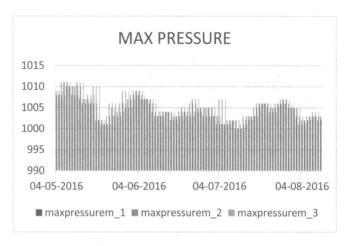

Fig. 4. Analysis of maximum pressure

Query:

```
db.JaipurFinalCleanData.find(
{"meantempm":"34", $or:[{"maxdewptm_1":{$gte:
"$15"}},{"maxhumidity_1":{$gte:"$50"}}]
},{_id:0,meantempm:1,max})
```

Conditonal Query is used (If 'meantemp' = 34, 'maxdewptm' = 15 & 'max-humidity' = 50), whichresults into 35 days in time duration of two years i.e. (2016–18). Which is almost the hottest phase for the concern (Fig. 5).

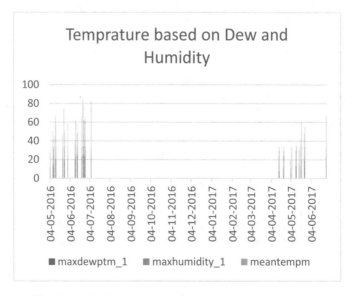

Fig. 5. Analysis of temperature based on Dew and Humidity

4   Conclusion

This paper presents a survey on multiple weather Prediction techniques in which Dhore et al. [5] uses data mining techniques and Chauhan et al. [6] uses machine learning algorithms. These algorithm are used for predicting weather (wind, temperature, rainfall). This concludes that machine learning techniques like decision tree and data mining techniques like K-Mean clustering and K-medoids are not sufficient for the prediction of large datasets. Large data sets can be effectively analysed by using Hadoop framework along with MongoDB database. The approach of MongoDB is better for large datasets and the results are effectively produced.

References

1. Tiwari, A., Sam, R., Shaikh, S.: Analysis and prediction of churn customers for telecommunication industry. In: 2017 International Conference on I-SMAC (IoT in Social, Mobile, Analytics and Cloud) (I-SMAC), Palladam, pp. 218–222 (2017). https://doi.org/10. 1109/i-smac.2017.8058343
2. Navadia, S., Yadav, P., Thomas, J., Shaikh, S.: Weather prediction: a novel approach for measuring and analyzing weather data. In: 2017 International Conference on I-SMAC (IoT in Social, Mobile, Analytics and Cloud) (I-SMAC), Palladam, pp. 414–417 (2017). https://doi.org/10.1109/i-smac.2017.8058382
3. Shaikh, S., Rathi, S., Janrao, P.: IRuSL: image recommendation using semantic link. In: 2016 8th International Conference on Computational Intelligence and Communication Networks (CICN), Tehri, pp. 305–308 (2016). https://doi.org/10.1109/cicn.2016.66
4. Shaikh, S., Rathi, S., Janrao, P.: Recommendation system in e-commerce websites: a graph based approached. In: 2017 IEEE 7th International Advance Computing Conference (IACC), Hyderabad, pp. 931–934 (2017). https://doi.org/10.1109/iacc.2017.0189
5. Dhore, A., Byakude, A., Sonar, B., Waste, M.: Weather prediction using the data mining Techniques. Int. Res. J. Eng. Technol. (IRJET) (2017). www.irjet.net. p-ISSN: 2395–0072
6. Chauhan, D., Thakur, J.: Data mining techniques for weather prediction: a review. Computer Science Himachal Pradesh University (2014). http://www.ijritcc.org
7. Badhiye, S.S., Chatur, P.N., Wakode, B.V.: Temperature and humidity data analysis for future value prediction using clustering technique: an approach. Int. J. Emerg. Technol. Adv. Eng. 2(1), 2250–2459 (2012)
8. Kannan, M., Prabhakaran, S., Ramachandran, P.: Rainfall forecasting using data mining technique. Int. J. Eng. Technol. 2(6), 397–401 (2010)
9. Smys, S., Bestak, R., Chen, J.I.Z.: Special issue on evolutionary computing and intelligent sustainable systems, pp. 1–1 (2019)
10. Smys, S., Bestak, R., Chen, J.I.Z., Kotuliak, I. (eds.) International Conference on Computer Networks and Communication Technologies: ICCNCT 2018, vol. 15. Springer, Heidelberg (2018)
11. Suma, V.: Towards sustainable industrialization using big data and internet of things. J. ISMAC 1(01), 24–37 (2019)
12. Reddy, P.C., Babu, A.S.: Survey on weather prediction using big data analytics. In: 2017 Second International Conference on Electrical, Computer and Communication Technologies (ICECCT), Coimbatore, pp. 1–6 (2017). https://doi.org/10.1109/icecct.2017.8117883
13. www.wikipedia.com
14. www.irjet.net

A Survey on Intrusion Detection System Using Artificial Intelligence

Sona Solani$^{(\boxtimes)}$ and Nilesh Kumar Jadav

Department of Computer Engineering,
Marwadi University, Rajkot, Gujarat, India
ssolani305@gmail.com, nileshjadav991@gmail.com

Abstract. With the increasing demand of the internet in the 21st century, we have the challenges to secure our network and data so, for we demand some software and devices to protect our system. However, these devices want high computational power and have a complex structure. We use firewall and IDS/IPS as an intermediary device and software for protecting our systems. An Intrusion Detection System used to monitor network traffic, and if any suspicious activity is occurring, it will automatically notify the security person to secure future access.

Keywords: Intrusion detection system · Intrusion prevention system · Artificial intelligence · Machine learning · Deep learning

1 Introduction

Nowadays, with the rapid progress in technology, all the information and transactions have been stored inside the internet. Internet use in each domain such as business, agriculture, banking, healthcare etc. [3]. Computing and communications have encountered remarkable changes in recent decades. Computation favoured on the go with a huge demand for mobility assistance in communicating [24, 25]. Due to a large number of users in the wireless environment interface criterion also have shifted to the concept of Cognitive Radio Networks [22, 23] for better utilization of wireless spectrum. Unnecessary to say, the improvement in handheld equipment and the tremendous demand for mobile application leads to the necessity of convenient analysis and security provisioning of communication environment. According to NIST SP 800-39 (CNSSI 4009) "The ability to protect or defend the use of cyberspace from cyber-attacks." Figure 1 illustrate the different areas of cybersecurity in that define methods and rules to protect our cyberspace.

1.1 Cybersecurity and Its Domain

Today's technologies are continuously developed and work with a cloud server, big data, IoT to collect and process the data. In [2] authors have reviewed the basic taxonomies related through Cyber-Physical Systems like smart grid, smart vehicles etc. However, every technology appears with different vulnerabilities and threats. There are some cryptographic algorithms and methods use to protect IoT devices using

© Springer Nature Switzerland AG 2020
A. P. Pandian et al. (Eds.): ICCBI 2019, LNDECT 49, pp. 67–80, 2020.
https://doi.org/10.1007/978-3-030-43192-1_8

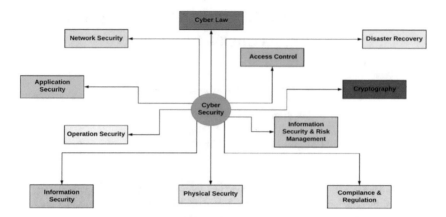

Fig. 1. Different areas of cyber security

Table 1. Basic possible attack in each OSI layer.

Sr. no.	Layer	Attack
1	Physical layer	Eavesdropping, Jamming, Radio interference, Tempering
2	Data link layer	MAC spoofing, Collision, Identity theft, Exhaustion
3	Network layer	Hijacking, Spoofing
4	Transport layer	Flooding, TCP sequence prediction attack
5	Application layer	Malware attack, FTP Bounce, SMTP attack, Data attack

Asymmetric key cryptography for securing end-to-end devices and Lattice-based cryptography for securing Broker Gateway [19]. Using social media every day and store information on the warehouses is the biggest challenge for the security researchers. Most of the attacks are performed using social media and on social networking sites because they store huge amount of confidential data. In [21], authors have been explained the possibilities of various inference attacks performed on social networking sites such as identity theft and communication tracking. Table 1 shows the OSI layers and their possible attacks, while Tables 2 and 3 shows the basic threats and vulnerabilities.

Tables 2 and 3 Mentioned threats and vulnerabilities are common in any technology. However, nowadays, we use an electronic-healthcare product, net banking, online transaction etc. whether they all have their specific threats and vulnerabilities. In [5] they stated that smart medical devices are connected and easy access by everyone. Attackers breach security and may view or alter patient medical records. So, we have to reduce the attacks and protect the systems.

1.2   Solution Related Problem

There are so many security and privacy issues we have, when we are working on the internet however, we can reduce it. Security threats involve three goals confidentiality, integrity and availability. At the primary level of security, we can apply some privacy setting from client-side [4] and, it protects our data at some level. Subsequently, we

Table 2. Basic security threat define in cyber security.

Sr. no.	Basic threats
1	Data restoration from recycle bin
2	Burglary touchy data on media through unapproved access
3	Burglary of gear and touchy media through unapproved access
4	Social engineering of system users
5	System intrusion and unauthorized system access
6	Intentional or accidental denial of service event
7	Framework damage or programming disappointment or breakdown
8	Unchecked data alteration
9	Equipment damage due to natural cause
10	Unrecoverable information because of human or characteristic mistake
11	Failure of network
12	Loss of power
13	Denial of user actions
14	Eavesdropping
15	Purposeful or inadvertent infringement of the framework security strategy
16	Purposeful or unintentional transmission of directed information

Table 3. Basic security vulnerability define in cyber security.

Sr. no.	Basic vulnerability
1	Insufficient patch management in host system
2	Excessive privileges due to lack of a user access review
3	Insufficient authentication mechanism and controls
4	Insufficient backups
5	Insufficient media encryption
6	Insufficient physical controls
7	Absence of anti-virus and malware counteractive action
8	Absence of logging and checking controls
9	Absence of system security
10	Absence of excess power supply
11	Absence of a repetitive framework
12	Absence of transmission encryption
13	Untraceable user actions
14	Possible weak password
15	Critical vulnerabilities on host-system
16	Absence of natural controls
17	Absence of client checking

want some mechanism to hide or to secure our data and, these mechanisms execute on admin or server-side. For instance, in the transmission medium, we implement the cryptographic algorithms for securing the transmission and to add a firewall or IDS/IPS in that define rules that filter out the suspicious activity. In [6] they have given solutions for securing system like metadata removal, spam detection, malware detection etc.

1.3 Major Problem- Malicious Network Traffic

Network traffic creates based on the incoming requests from the clients to the servers [1–4]. At a given point of time, network traffic refers to the range of data moving across the network. Network traffic is directly related to the quality of services. However, in the cyberspace, attackers frequently perform some unwanted or malicious actions to harm the victim's system. Sometimes, it is difficult for servers to handle malicious incoming requests [5–10]. For another type of attacks, we have the software and solution to control it. In network traffic, we cannot predict for incoming requests. Network traffic classifies into the following categories:

- Busy/heavy traffic: Use High bandwidth
- Non-real-time traffic: Use bandwidth throughout working hours
- Interactive traffic: low response when traffic priority is not fix
- Latency-sensitive traffic: lower result for the bandwidth.

1.4 Techniques to Resolve

To overcome this problem, we have mainly two techniques to use, and those are Firewall and IDS/IPS. A firewall is used to prevent the system from unauthorized access and block them [11–13]. But it cannot raise the alert. An Intrusion Detection System consistently monitors the network to look for suspicious activity and, if any suspicious activity happens, it will automatically trigger the alarm [14–18]. It records information about the intruders and gives the ability to configure the rules to prevent further access (Table 4).

Table 4. Different types of possible attack and their solution.

Sr. no.	Types of attack	Solution
1	Malware	Heuristic anti-virus
2	DdoS	Cloud fare or Incapsula- comprehensive protection
3	Man-in-the-middle attack	Intrusion detection system (Monitor your network)
4	Phishing & spear phishing	Two-factor authentication, Web application Firewall
5	Password attack	Create strong password

1.5 IDS

Intrusion Detection System works on both the side, network as well as the host. Figure 2 illustrate the primary classification of an intrusion detection system. In [10],

Table 5. Difference between intrusion detection techniques

Sr. no.	Signature-based	Anomaly-based
1	Effective and simple to detect	Hard to detect
2	Detect attack based on predefine signatures	Detect unknown attacks
3	Good false positive alarm	Higher risk of false positive alarm
4	Alert raised	Alert is not sent to victim
5	Match the pattern of signature	Use statistical measures

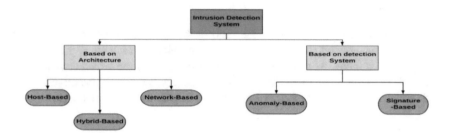

Fig. 2. Classification of intrusion detection system.

Table 6. Difference between intrusion detection architecture

Sr. no.	Host-based	Network-based
1	Requires the software installation on all host system	Requires the establishment of tests on the system
2	Use in centralized system	Use in centralized system
3	Monitor framework exercises, log records and system traffic received	Sniff the network traffic that match with profile and signature
4	OS limitation	Independent from OS

they classify IDS based on architecture and detection techniques. In [1], they have explained the characteristics of all the types of IDS (Tables 5 and 6).

Host-based IDS also monitors OS log files, audit logs, system calls, error messages and also records an attacker performance on the host. The OS version is the most significant part of host-based IDS. However, if intruders remove the IDS tool or the system's network interrelations, this IDS won't generate an alert. While in network-based IDS, sensors use to receive and analyse real network traffic. Probs send an alert, when patterns are detected. NIDS can also detect a port scan and ping sweep in various hosts. Sensor has an interface to monitor the network and a command & control interface. Command & control interface has the ability to communicate with traffic management however, NIDS require packet encryption. Because of hackers try to keep hidden their attack activity from NIDS is to break up in their packets. Network-Based IDS always use with signature-based detection technique [12].

1.6   IDS with Traditional Methods

An attacker can bypass the IDS and may perform their actions. For example, if intruders may turn off the software in the host system or decrypt the packets in transmission media. There are various ways when intruders performing attacks and take the information. An attacker can unauthorized access and steal some victim's data. That time traditional IDS is not working accurately. So, now we want some more powerful techniques to resolve this problem.

1.7   Why AI?

With the use of traditional methods, the ratio of the false alarm rate is increasing. To reduce the false alarm rate now, we are using AI and ML approach rather than using the conventional methods. AI provides a wide range of dataset from which computers can learn with the help of previous data. And, accordingly, to that computers train itself continuously. With the use of AI, we can even mitigate the zero-day attack.

Table 7. Difference between intrusion detection and intrusion prevention system

Sr. no.	Intrusion detection system	Intrusion prevention system
1	Detection and monitor	Control system
2	Do not take their action by own	Accept and reject packets according to the rule database
3	Require human	Require database
4	After the intrusion response	Real time response
5	Passive device	Active & In-line device
6	Reactive (Alerting system)	Proactive (Blocking system)

2   IDS/IPS

In [9] authors stated that, "an intrusion detection system is a set of action that maintain the confidentiality, integrity and availability." Intrusion detection system cannot work alone because it wants some prevention techniques which are used to prevent the system from vulnerability exploits. Intrusion Prevention system (IPS) controls the Intrusion Detection System (Table 7).

2.1   How IDS/IPS Works?

Technical network required so many devices and software for security and protection and, there are so many trusted and untrusted sources, are trying to steal the confidential information. An Intrusion Detection System works as same as the firewall but, it uses to detect anonymous traffic. We can use it in between network or in a particular system also. IDS have so many signatures to define in it. Using this it is filtering out the suspicious websites. IPS is useful to identify potential threats and respond to them. The

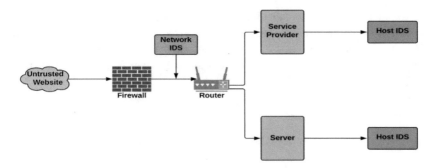

Fig. 3. Architecture of basic network communication

role of IPS is to identify suspicious activity, log information maintaining and block activity for the suspicious website. Figure 3 illustrate the architecture of basic network communication.

2.2 Different Types of IPS

There are mainly four types of intrusion prevention system, which illustrate in Fig. 4 (Table 8).

Table 8. Different types of IPS and their working principles

Sr. no.	Types of IPS	Working
1	Network-based IPS	Analyse protocol activity and untrustworthy traffic
2	Wireless IPS	Analyse protocol activity and untrustworthy traffic in wireless network
3	Host-based IPS	Software package to monitor malicious activity within a host
4	Network behaviour analysis	Examines network flows, pattern matching to detect attack

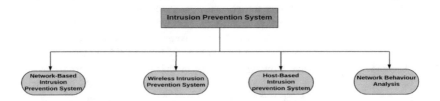

Fig. 4. Intrusion prevention system classification.

2.3 Problem in IDS

Many suspicious websites filtered out through IDS and IDS raised the alarm when it detect the suspicious website or request However, there is an issue to be faced when we are using it because some websites are suspicious but cannot be filtered out in IDS using traditional methods and, that increases the false-negative alarm rate. Wrong alarm or no alarm raised the troubles such as: unauthorized access, data stealing or identity theft issues and when in terms of large organization these are the biggest issues for them. So, now IDS need more upgradation so that it can able to detect suspicious activity with lower rate of false negative alarm rate.

2.4   Different Tools of IDS

Table 11 explain the famous IDS/IPS tools and their working principle.

2.5   How AI Can Improvise the IDS Detection Result

Nowadays, in every field, computers are useful, and they interconnect with each other. AI includes various data science algorithms, statistics and probabilistic methods which are useful in every field. AI has many diverse areas like machine learning, deep learning, Computer Vision etc. in that, we are using dataset, the different feature selection algorithm, classification algorithm and use proper features to get accurate output [11]. AI-based techniques to perform a prominent role in IDS over other methods [7]. Because it gives a variety of algorithms to detect suspicious activity. In [16], authors discussed more than 30 different datasets related to Intrusion detection system and Intrusion Prevention System. Nowadays, most of the application use the combination of mathematics, neural network and some biological event the analyse the pattern [13]. The main advantages of AI include that AI provides flexibility, pattern recognition, fast computing and adaptability. With the use of AI and Machine Learning techniques, in [15] authors customized the snort with Hierarchical Bidirectional Fuzzy Rule Interpolation So, with the use of AI, researchers try to mitigate the false-negative alarm rate. Also in [17], authors proposed the algorithm for Fault detection based on clustering approach (FDBC) which increases the scalability of the system and mitigate

Table 9. Types of artificial intelligence and their working principles

Sr. no.	Types of AI	Working
1	Reactive machines	AI's most basic form in which machine is not able to use the past data to predict current determination. However, it can only deal with current situations. Today, we use the existing application of reactive machine is "Deep Blue". Deep Blue has developed by IBM to play chess
2	Limited memory	It is same as machine learning models which use to derive the knowledge from experience data and build experimental knowledge. An existing form of limited memory is "autonomous vehicle"
3	Theory of mind	It constitutes the ability of decision-making the same as humans. The perfect example of the theory of mind is the virtual voice assistant and SOPHIA. They are not fully humans, but they have the capability of semi-humans
4	Self-awareness	Machine has the ability same as human however, this form is not exist. It will be the most advance form of the AI because machine will do human actions and behaviour

the load of network intrusion detection. However there is another approach of trust-based intrusion detection and clustering to identify malicious nodes and broadcast the energy effective data in wireless sensor network [18].

Table 10. Different artificial intelligence tools and framework

Sr. No.	Types of AI	Working
1	Amazon web service	AI toolkit
2	CNTK (Computational Network Toolkit)	Microsoft open-source AI toolkit
3	DeepLearning4j	Deep learning library for the java virtual machine
4	AI-ONE	API, building agents and document library
5	Apache MAHOUT	Implement on ApacheHadoop
6	OPENNN	Neural network implementation

Table 11. Different intrusion detection tools and their working principles

Sr. no.	Tool Name	Working
1	OSSEC	OSSEC represents Open-Source Security instrument for IDS/IPS. It claims by Trend Micro. In UNIX-like OS, it centres around log and arrangement document, makes a checksum of records and monitors it. In Windows OS it additionally controls unapproved library adjustment. OSSEC is Host-Based IDS. It keeps running on UNIX OS, yet an operator is accessible to ensure Windows OS. At the point when an alarm happens, it will show on reassure and tell through the mail
2	Snort	Snort is Network-Based IDS. It is a blend of more than one tools and packages. It is likewise bundled, sniffer and packet logger. We can customize standards of IDS base on our needs such as a firewall. Standards identify an assortment of events such as stealth port sweeps, cushion flood assaults, CGI assaults and OS fingerprinting. Snort can use with signature-based and anomaly-based
3	Suricata	Suricata is a blend of IDS and IPS. It surrenders the security to the application layer. It parts the bundles and after that monitors it. Like the application layer, it likewise works in the lower-level conventions, for example, TLS, ICMP, TCP and UDP. Suricata circulates its outstanding task over more than a few processor cores and threads for better execution. It likewise utilizes the graphics card for better execution

(continued)

Table 11. (*continued*)

Sr. no.	Tool Name	Working
4	Bro network security monitor	Bro NSM is additionally an open-source tool. It works in two stages, traffic logging and examination. Like Suricata, it likewise surrenders the security to the application layer with better detection. Everything comes in pair with Bro, and its investigation module is comprised of two-component. Firstly, an event engine tracks triggering events, for example, TCP association and HTTP demand. At that point, it broke down by the content to choose whether the alarms raise or not. It tracks HTTP, DNS and FTP action and SNMP traffic. Bro is accessible on UNIX and Linux
5	Open wireless IPS NG	This tool gives security in the wireless network. There are mainly three components. First, a sensor is a device that lone catch remote traffic. Second, a server collects the data, analyse it and then respond. Third, is an interface component which displays the information
6	Samhain	It is a Host-Based IDS tool. It gives file trustworthiness checking and log record examination. It additionally performs rootkit discovery, concealed procedures and port checking. Samhain can keep running on Linux, UNIX and OS X. It can likewise keep running on Windows under Cygwin through agent not server has been tested
7	Fail2Ban	It is a Host-Based IDS and also has some prevention features. It screens log records and auspicious occasions, for example, abuse of and fizzled login endeavours. It naturally refreshes the local firewall principles to obstruct the source IP address of the malware behaviour. The framework accompanies pre-assembled highlights for a portion of the general services such as SSH, FTP, Apache and some more. Prevention is done by changing the host's firewall tables
8	AIDE	AIDE stands for Advanced Intrusion Detection Environment. It is a Host-Based IDS use for rootkit detection and file signature comparison. AIDE use for both signature-based and anomaly-based analysis. AIDE is a data comparison tool

3 Artificial Intelligence and It's Different Area

AI can define as a "Computer or machine that acts the same as a human being". In this new technological generation, AI has the ability in which Computer systems can behave or respond like a human. AI covers its subsets such as Computer Vision, machine learning, augmented reality, NLP, Soft Computing etc. However, every AI technologies use in various area for further development such as soft computing approaches like feature extraction from images, cyber threat detection, cloud computing, deep learning algorithms are using in wireless sensor network and 5G networks [20]. Artificial Intelligence has 4 primary types. Tables 9 and 10 illustrate the types of AI and AI tools respectively.

3.1 Machine Learning

ML can learn and train by itself with its experience without explicitly programmed. It has so many algorithms and statistical models that use to perform the specific task. Machine Learning and Data mining methods used in the intrusion detection system to reduce the ratio of the false alarm rate [8]. ML classified into three major types of learning which is illustrate in Table 12. Reinforcement learning gives the most accurate result because in that learning is with simulated environment, where anomalies are injected in a controlled manner and the rewards are based on correct detection and prediction of anomalies [14]. In ML, we use different datasets to develop machine learning models. Table 14 explained the basic taxonomy of Machine Learning. To develop Machine Learning applications, we generally use Python, R and MATLAB. Python includes a variety of libraries to create ML models. There are so many tools available to develop ML applications. Anaconda Jupiter, Spider, MATLAB use commonly which are illustrate in Table 13.

Table 12. Difference between types of machine learning

Supervised learning	Unsupervised learning	Reinforcement learning
Label data	No label data	Decision process
Direct feedback system	No feedback system	Reward system
Predict next value	Identify cluster	Learn from mistakes
Classification & Regression	Clustering & Dimensionality reduction	Software component learns to react to an environment
Classification: Fraud detection Image classification Diagnostics **Regression:** Forecasting Prediction	**Clustering:** Targeted marketing Recommended system **Dimensionality reduction:** Big-data visualization Structure discovery	AI games Real-time analysis & Decision Learning tasks

3.2 Deep Learning

Deep Learning is a subset of machine learning uses to deal with structured and unstructured data. It includes the network of neurons known as a deep neural network and utilize high-computational raw input data and function to generate the accurate output. In deep learning, learning can be done with supervised learning, semi-supervised learning and unsupervised learning.

Table 13. Machine learning tool and packages

Sr. no.	Tool packages	Platform	Language	Algorithm or features
1	Scikit learn	Linux, Mac OS, Windows	Python, Cython, C, C++	Classification Regression Clustering Preprocessing Model selection Dimensionality Reduction
2	PyTorch	Linux, Mac OS, Windows	Python, C++, CUDA	Auto grad module Optim module NN module
3	TensorFlow	Linux, Mac OS, Windows	Python, C++, CUDA	Dataflow programming
4	Weka	Linux, Mac OS, Windows	Java	Classification Regression Clustering Data preparation
5	Keras.io	Cross-platform	Python	API for neural network
6	Matplotlib	Windows	MATLAB, Python	Data visualization
7	Numpy	Windows	Python	Data analysis
8	Pandas	Windows	Python	Data analysis

Table 14. Machine learning taxonomy and description

Sr. no.	Term	Description
1	Accuracy	Percentage of correct prediction $Accuracy = \frac{TP+TN}{TP+TN+FP+FN}$
2	Precision	Frequency of model for correctly predicting the positive class. $Precision = \frac{TP}{TP+FP}$
3	Re-call	Correctly identify the positive labels from all possible positive labels $Recall = \frac{TP}{TP+FN}$
4	F1-Score or F-measure	Measurement of test accuracy $F-measure = \frac{2*Recall*Precision}{Recall+Precision}$
5	Confusion matrix	Summary of classification model
6	True positive rate or Sensitivity	Correctly predicted the positive class **TPR = True Positive/(True Positive + False Negative)**
7	True negative rate or Specificity	Correctly predicted the negative class **TNR = True Negative/(True Negative + False Positive)**

(continued)

Table 14. (*continued*)

Sr. no.	Term	Description
8	False positive rate	Mistakenly predicted the positive class **FPR = False Positive/(False Positive + True Positive)**
9	False negative rate	Mistakenly predicted the negative class **FNR = False Negative/(False Negative + True Negative)**
10	Overfitting	Create a model which is closely perfect with training dataset, but poor to deal with test dataset
11	Entropy	Measurement of the randomness in the information being processed **One attribute:** $$E(S) = \sum_{i=1}^{c} -p_i \log_2 p_i$$ **Two Attribute:** $$E(T,X) = \sum_{c \in X} P(c)E(c)$$
12	Underfitting	Poor for both training and testing dataset because of over-generalized

4 Conclusion

This paper surveyed the challenges of using an intrusion detection system with conventional methods. We reviewed the basic threats and vulnerabilities that always found in the computer system. An intrusion detection system has the ability to detect the basic threats and vulnerabilities, however every new technology comes with its own threats and vulnerabilities and there is a limitation of using IDS with traditional tools. In this paper, we covered the types of IDS based on detection techniques and architecture designs, traditional tools of IDS/IPS and types of IPS. Those traditional tools are using high-computational power to detect suspicious activity and give high false negative alarm rate with least accuracy. So, now the world is changing because of an artificial intelligence and so that we are applying AI techniques in IDS/IPS and protect our data from hackers and cybercriminals. This paper also covered the types of AI, types of Machine Learning, basic python libraries and packages, AI tools, basic taxonomies of machine learning and brief overview of deep learning technology.

References

1. Liao, H., Lin, C.R., Lin, Y., Tung, K.: Intrusion detection system: a comprehensive review. J. Netw. Comput. Appl. **36**(1), 16–24 (2013)
2. Giraldo, J., Sarkar, E., Cardenas, A., Maniatakos, M., Kantarcioglu, M.: Security and privacy in cyber-physical systems: a survey of surveys. IEEE Des. Test **34**(4), 7–17 (2017)
3. Kabiri, P., Ghorbani, A.: Research on intrusion detection and response: a survey. Int. J. Netw. Secur. **1**(2), 84–102 (2005)
4. Jang-Jaccard, J., Nepal, S.: A survey of emerging threats in cybersecurity. J. Comput. Syst. Sci. **80**(5), 973–993 (2014)

5. Coventry, L., Branley, D.: Cybersecurity in healthcare: a narrative review of trends, threats and ways forward. Maturitas **113**, 48–52 (2018)
6. Rathore, S., Sharma, P., Loia, V., Jeong, Y., Park, J.: Social network security: issues, challenges, threats, and solutions. Inf. Sci. **421**, 43–69 (2017)
7. Kumar, G., Kumar, K., Sachdeva, M.: The use of artificial intelligence based techniques for intrusion detection: a review. Artif. Intell. Rev. **34**(4), 369–387 (2010)
8. Buczak, A., Guven, E.: A survey of data mining and machine learning methods for cyber security intrusion detection. IEEE Commun. Surv. Tutor. **18**(2), 1153–1176 (2016)
9. Kevric, J., Jukic, S., Subasi, A.: An effective combining classifier approach using tree algorithms for network intrusion detection. Neural Comput. Appl. **28**(1), 1051–1058 (2016)
10. Bamakan, S.H., Wang, H., Yingjie, T., Shi, Y.: An effective intrusion detection framework based on MCLP/SVM optimized by time-varying chaos particle swarm optimization. Neurocomputing **199**, 90–102 (2016)
11. Farnaaz, N., Jabbar, M.A.: Random forest modeling for network intrusion detection system. Procedia Comput. Sci. **89**, 213–217 (2016)
12. Shenfield, A., Day, D., Ayesh, A.: Intelligent intrusion detection systems using artificial neural networks. ICT Express **4**(2), 95–99 (2018)
13. Hajisalem, V., Babaie, S.: A hybrid intrusion detection system based on ABC-AFS algorithm for misuse and anomaly detection. Comput. Netw. **136**, 37–50 (2018)
14. Caminero, G., Lopez-Martin, M., Carro, B.: Adversarial environment reinforcement learning algorithm for intrusion detection. Comput. Netw. **159**, 96–109 (2019)
15. Jin, S., Jiang, Y., Peng, J.: Intrusion detection system enhanced by hierarchical bidirectional fuzzy rule interpolation. In: 2018 IEEE International Conference on Systems, Man, and Cybernetics (SMC) (2018)
16. Ring, M., Wunderlich, S., Scheuring, D., Landes, D., Hotho, A.: A survey of network-based intrusion detection data sets. Comput. Secur. **86**, 147–167 (2019)
17. Kumar, D., Smys, S., Smilarubavathy, G., Holzwarth, F.: Fault detection methodology in wireless sensor network. In: 2018 2nd International Conference on I-SMAC (IoT in Social, Mobile, Analytics and Cloud) (I-SMAC), pp. 723–728 (2018)
18. Anguraj, D., Smys, S.: Trust-based intrusion detection and clustering approach for wireless body area networks. Wirel. Pers. Commun. **104**(1), 1–20 (2018)
19. Sridhar, S., Smys, S.: Intelligent security framework for iot devices cryptography based end-to-end security architecture. In: 2017 International Conference on Inventive Systems and Control (ICISC), pp. 1–5 (2017)
20. Smys, S., Bestak, R., Chen, J.: Special issue on evolutionary computing and intelligent sustainable systems. Soft. Comput. **23**(18), 8333 (2019)
21. Praveena, A., Smys, S.: Prevention of inference attacks for private information in social networking sites. In: 2017 International Conference on Inventive Systems and Control (ICISC), pp. 1–7 (2017)
22. Dutta, N., Sarma, H., Polkowski, Z.: Cluster based routing in cognitive radio adhoc networks: reconnoitering SINR and ETT impact on clustering. Comput. Commun. **115**, 10–20 (2018)
23. Dutta, N., Sarma, H.: A probability based stable routing for cognitive radio adhoc networks. Wireless Netw. **23**(1), 65–78 (2015)
24. Dutta, N., Misra, I.: Multilayer Hierarchical model for mobility management in IPv6: a mathematical exploration. Wireless Pers. Commun. **78**(2), 1413–1439 (2014)
25. Dutta, N., Misra, I.: Mathematical modeling of Hierarchical mobile IPv6 based network architecture in search of optimal performance. In: 15th International Conference on Advanced Computing and Communications (ADCOM 2007) (2007)

Trusted and Secured E-Voting Election System Based on Block Chain Technology

A. Anny Leema[1], Zameer Gulzar[2(✉)], and P. Padmavathy[2]

[1] VIT, Vellore, Tamilnadu, India
annyleema@gmail.com
[2] BSA Crescent Institute of Science and Technology, Chennai, India
zamir045@gmail.com, pdma281@gmail.com

Abstract. The blockchain is an open, distributed ledger, which is running under a common software application that must agree before making any change to the network's ledger. The number of blockchain platforms is already available and many companies have already begun applying blockchains to their business. It is not only used in business, but it can also be applied as a service and has the potential to make the voting more accessible. Therefore, NRIs and adult internet users can cast their votes in a matter of seconds despite their location on the day of the election. This will be easy and more secure. In the current system, there's a high risk of false votes and damage to property. People don't want to stand in big queues and wait for a long time. Also, people who conduct the election have to follow a completely different procedure for voting which takes more time and there's no transparency. So we present a novel blockchain technique for voting with a guarantee of a safe and secure election. The election system using Blockchain makes it not only easier for the committee to keep a count, but also for the citizens to vote. The Blockchain system will allow the citizens to cast their votes despite their location. Every year a large group of Indians ends up residing outside the country due to which they are unable to practice their voting rights most of the time. The Blockchain system will be a medium for them to cast their votes even though they are not physically present in India during election time.

Keywords: Blockchain · Models · Perceived usefulness · Recommendation

1 Introduction

Blockchain is a new technology, traversing the past ten years which was first used within Bitcoin. It is a public distributed ledger of all transactions, known for its security, privacy issues which were studied and investigated by many researchers. It's a constantly growing ledger and all the transactions are recorded permanently. The implementation of the blockchain in e-voting reduces the fraud voters and increases voter access. The value of this system is a vote. In this paper, it is discussed how blockchain helps in transferring votes between two peers. No central authority body is required here, it is tamper-proof and cost-effective. The system is very flexible and the voters can cast their votes either from home or from office using a mobile phone, computer or any other electronic gadgets. Some countries have already started implementing blockchain

© Springer Nature Switzerland AG 2020
A. P. Pandian et al. (Eds.): ICCBI 2019, LNDECT 49, pp. 81–88, 2020.
https://doi.org/10.1007/978-3-030-43192-1_9

technologies to reduce fraud voters. The fraudulent election is one of the biggest problems and the democratic countries like India, Japan still suffer from a flawed electoral system. Hacking of the voting machine and illegal way of election manipulation are discussed and to solve these issues new voting model was proposed [1]. Blockchain-based protocol and in the distributed system directed acyclic graph of blocks are used. The pros and cons of using blockchain in e-voting were discussed and how to use this underlying blockchain technology for e-voting complex applications [2–6].

2 Literature Survey

Bitcoin paper (Nakomoto) introduced the first blockchain, focussed on solving the 'double spending problem' for digital currency using a peer-to-peer network. The process of mining works by using a consensus algorithm that aims at having fault-tolerant systems by storing the data in a distributed manner [7]. It is important to note that each block that is added to the blockchain must follow a certain set of consensus rules. The consensus is a self-auditing system to verify whether all the blocks are legitimate and decide on the contributions of various participants in the blockchain. This is called a block reward in Bitcoin's proof-of-work (PoW) algorithm [8]. It is a popular consensus algorithm used by the cryptocurrency networks bitcoin and the mining mechanism consumes high energy and results in longer processing time. This system depends on the miners to prove the accuracy of the transactions in the network.

Proof of stake (POS) is an alternative to the POW which consumes low energy. It is also designed for a public blockchain. It is like a lottery, the more you invest better the chances. In POS, individuals are chosen to generate a block, known as validators based on their economic stake and validate the transaction. Compared to mining in POW, the validation is faster in POS but it results in more vulnerability because the richest stakeholders are permitted to have control over the consensus of the blockchain. Hence the reliable network can be obtained by combining the PoW consensus algorithm and the consensus rules so that agreement has to the shared and the state of the blockchain is achieved. In this paper, we investigated the problems in the election voting systems and proposed the E-voting model to resolve the issues in the existing system. Harsha et al. [9] discussed the potential of blockchain technology and its usefulness in the e-voting scheme but it is not implemented. Kshetri and Voas [10] proposed a blockchain-enabled e-voting system emphasized voters' transparency [11]. Koc and his team [12] proposed a voting system using Ethereum blockchain which creates a transparent environment for e-voting. Sahadevan et al. [13] used the MQTT protocol and proposed an offline-online strategy for IoT applications to tackle real-time corruption at polling booths. Liu and Zhang [14, 15] discussed how to build a secure and practical e-voting system in the industry by using the Weighted voting system.

3 Existing System

Elections are held by using EVMs (Electronic Voting Machine) and at specific locations. These EVMs are taken to different places and then brought back to the main area for counting the votes. A lot of manpower is required and there is a high risk of terrorist attacks or riots because of many movements involved in it. Tampering of data is also possible. In some places, the ballot system is still used. People go to a specific location, stand in long queues, go through a long procedure and then they are finally able to vote. Also, people who conduct the election follow a completely different procedure for voting which takes more time.

3.1 Proposed System

We propose a new election web app that will be made by using Blockchain technology. There's no point of false votes or editing of votes as everything is stored in blocks and there are miners who are always checking the blocks. The objective of our proposed is to provide a safe and secure election that can be done by the people sitting in their homes through their cell phones or laptops so that the security problem that is faced right now is get ridden. The money and labor that is used up for conducting elections in the present scenario will be reduced.

3.2 Benefits of the Proposed System

With the development of the proposed system, people will be able to:

a. Vote from their home and their own devices
b. Safe and secure method
c. Less manpower required
d. Easy counting process as everything will be automated
e. People can host their election.

3.3 Steps Involved in the Blockchain-Based E-Voting System

The e-voting system using blockchain technology is public, distributed and centralized. The steps involved in the design of the web app are as follows:

a. Pre-register for the use of a voting system
b. Option for hosting elections
c. Create the blockchain using python includes previous hash, current hash, and voters details.
d. The voter uses the private key to sign the hash of the vote
e. Submission time of the vote is recorded in the MongoDB with voters details
f. The voter can vote only once and this transaction is made visible to the public.
g. The confidentiality of the voter is maintained but his vote balance is nullified.
h. Voters can cross-check whether his vote is counted or not.
i. Any discrepancies are found then the miner will reject the vote and the block is not created.

3.4   System Overview

The general architecture of the blockchain system is given in Fig. 1. The proposed Web application is based on E-voting using Blockchain. It's a system that needs maximum security as it depends on the trust of the users. It allows the user to vote and also to conduct elections. The user has to pre-register for the use of the voting system and only after registering, he/she can vote. There will be an option for hosting elections that will feature no. of voters and the locations to be covered. Overall the basic idea is that the user will vote to be at home and the new block is generated using the POW algorithm. When the user submits his vote, it is verified by the miner and the new block is added to the existing chain of blocks with vote information. T will be stored in the blockchain and also if a user tries to tamper these votes by voting again or any other methods, he/she cannot change the votes and the hosting committee will get to know about the tampering.

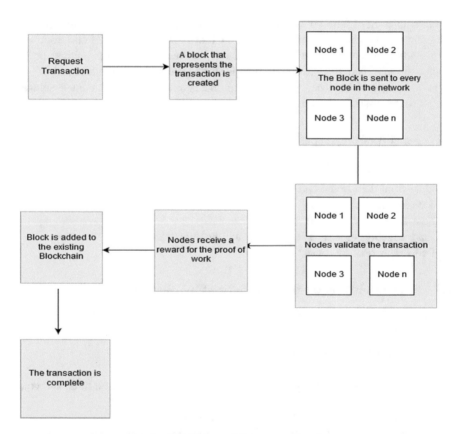

Fig. 1. Architecture of the proposed system

The process involved proposed web application in Fig. 2 using front-end (client-side) in AngularJS and backend server in Node JS (server-side) and MongoDB. AngularJS is a javascript framework to build highly proficient web and mobile applications. The AngularJS framework works by reading the HTML page, which has HTML attributes embedded into it. Angular interprets those attributes and binds the input and output of the page to the model. On the Server side, Node.js will use MongoDB as a database. It provided a callback function to insert, update and delete into a database (a NoSQL Database).

The blockchain part was written in the python script. Basic implementation was that a hash is generated for every block and it is checked. Hash is made using the sha256 algorithm. Every block contains 3 things, i.e., previous hash, Current hash, and transactions. The first block is called genesis block that has a previous hash as 0. All the transactions have the initial value as 0 and when a user enters the system, his value will be 1. When the user votes, his value is decreased to 0 and the party value is increased by 1. The data structure used is the Python dictionary which is an unordered collection of values and holds Key: value pairs. The database contains id, name, username, email, Aadhaar id, voter id and password in hashed value. The primary key is the username and Aadhaar id. As the flask is a framework for Python based, both Mongo DB and Flask part was integrated properly. Python libraries like flask_pymongo and WT-Forms etc. are used for adding more features to the application. Sample Source code is given below for importing flask, connecting it with MongoDB and creation the block.

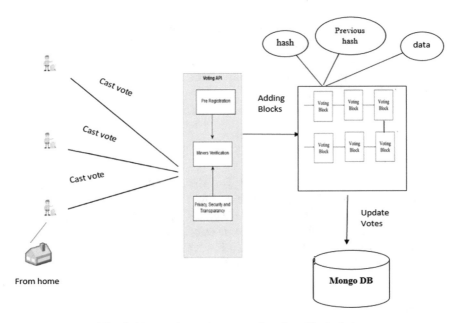

Fig. 2. Proposed e-voting system based on block chain

```
CODE BLOCK.
from flask import Flask, render_template, Response, request, redirect, url_for,
flash, logging, session
from flask_pymongo import Pymongo
from wtforms import Form, StringField, PasswordField, validators from passlib.
hash import sha256_crypt
import pickle
from start import *
app = Flask(_name_)
app.config['MONGO_HOST'] = 'localhost'
app.config['MONGO_USER'] = 'root'
app.config['MONGO_PASSWORD'] = 'notforyou'
app.config['MONGO_DB'] = 'flaskapp'
app.config['MONGO_CURSORCLASS'] = "DictCursor"
mongo=PyMongo(app) userlist=[]
state = {u'p_1':0,u'p_2':0,u'p_3':0} # Define the initial state
genesisBlockTxns = [state]
genesisBlockContents=
{u'blockNumber':0,    u'parentHash':None,u'txnCount':1,u'txns':genesisBlock-
Txns} genesisHash = hashMe( genesisBlockContents)
genesisBlock = {u'hash':genesisHash,u'contents':genesisBlockContents} #gen-
esisBlockStr = json.dumps(genesisBlock, sort_keys=True)
chain = [genesisBlock]
```

3.5 Applications

This system can be used in private sectors, i.e., MNCs for their elections and it can also
be used in Entertainment world for voting. The main advantage of block is that it
provides a new way to vote within a safe system. When more users are using the
website at the same time, there will be some errors regarding the frontend issues like
response time. As the server is running on Flask, GET and POST method takes time
when there are lots of users trying to use the website at the same time.

4   Conclusion

Blockchain is amazing for decentralized applications and security of the transactions.
Implementing a Blockchain in the election process can reduce the cost to a greater
extent and make everyone interestingly participate in voting being in their place.

Forgery of votes is solved because of using blockchain technology in the election process. It is more secure and anonymous compared to the existing e-voting system. There are a lot of things to add on and make this web app as the proper election app. It is planned to add fingerprint security with lots of other security for logging into the system. To show proof of work and for better acceptance, we'll change the server and use Ethereum as the database. It will give our application more mobility and safety. And also planning to make upgrades in this prototype and make it to the market with proper funding as well as proper safety.

Competing Interest. The authors declare that they have no conflict of interest.

Informed Consent. Informed consent was obtained from all individual participants included in the study.

Humans/Animals are not involved in this work.

References

1. Gautam, S., Ashutosh, D., Rajani, S.: Crypto-democracy: a decentralized voting scheme using blockchain technology. In: Proceedings of e-Business and Telecommunications (ICETE 2018), vol. 2, pp. 508–513 (2018)
2. Freya, S., Hardwick, A., Gioulis, R., Naeem, A., Konstantinos, M.: E-voting with blockchain: an e-voting protocol with decentralisation and voter privacy (2018). arXiv:1805. 10258v2
3. Christidis, K., Devetsikiotis, M.: Blockchains and smart contracts for the Internet of Things. IEEE Access **4**, 2292–2303 (2016)
4. Zheng, Z., Xie, S., Dai, H., Chen, X., Wang, H.: An overview of blockchain technology: architecture, consensus, and future trends. In: Proceedings of the 2017 IEEE International Congress on Big Data (BigData Congress), pp. 557–564 (2017)
5. Park, J.H., Park, J.H.: Blockchain security in cloud computing: use cases, challenges, and solutions. Symmetry **9**(8), 164 (2017)
6. Yin, S., Bao, J., Zhang, Y., Huang, X.: M2M security technology of CPS based on blockchains. Symmetry **9**, 193 (2017)
7. Satosh, N.: Bitcoin: a peer-to-peer electronic cash system. J. Cryptol. **6**(7) (2009)
8. Mauro, C., Sushmita, R.A.: A survey on security and privacy issues of bitcoin. IEEE Commun. Surv. Tutor. **20**(4), 3416–3452 (2018)
9. Patil, H.V., Rathi, K.G., Tribhuwan, M.V.: A study on decentralized e-voting system using blockchain technology. Int. Res. J. Eng. Technol. **5**(11), 48–53 (2018)
10. Kshetri, N., Voas, J.: Blockchain-enabled e-voting. IEEE Softw. **35**, 95–99 (2018). https://doi.org/10.1109/MS.2018.2801546
11. Suma, V.: Security and privacy mechanism using blockchain. J. Ubiquit. Comput. Commun. Technol. (UCCT) **1**(01), 45–54 (2019)
12. McCorry, P., Shahandashti, S.F., Hao, F.: A smart contract for boardroom voting with maximum voter privacy. In: Kiayias, A. (ed.) Financial Cryptography and Data Security, pp. 357–375. Springer, Cham (2017)

13. Koc, A.K., Yavuz, E., Cabuk, U.C., Dalkilic, G.: Towards secure e-voting using ethereum blockchain. In: International Symposium on Digital Forensic and Security (ISDFS), Antalya, Turkey, vol. 6, no. 1 (2018)
14. Sahadevan, A., Mathew, D., Mookatana, J., Jose, B.A.: An offline online strategy for IoT using MQTT. In: IEEE 4th International Conference on Cyber Security and Cloud Computing (2017)
15. Liu, Q., Zhang, H.: Weighted voting system with unreliable links. IEEE Trans. Reliab. **66**(2), 339–350 (2017)

Users' Attitude on Perceived Security of Enterprise Systems Mobility: A Conceptual Model

Ramaraj Palanisamy$^{(\boxtimes)}$

Department of Marketing and Enterprise Systems,
Gerald Schwartz School of Business,
St. Francis Xavier University, Antigonish, NS, Canada
rpalanis@stfx.ca

Abstract. Enterprises are implementing mobile technologies for various business applications including Enterprise Systems to rise the flexibility and to gain sustainable competitive advantage. At the same time, the end users are exposed to security issues when using the mobile technologies. The Enterprise Systems have seen breaches and malicious intrusions thereby more sophisticated recreational as well as commercial cybercrimes have been witnessed. This paper examines the impacts of users' attitude towards perceived security of Enterprise systems (ES) mobility. Considering the significance of security in Enterprise Systems mobility, a conceptual model is evolved from the literature and the study proposes to validate the model by empirically collecting data from users of ES mobile systems.

Keywords: Enterprise systems · Mobility · Security · User attitude

1 Introduction

Organizations adopt mobile internet and technologies including tablets, smartphones, and other handheld devices to create value for individual employees and businesses. The mobile technologies have become a strategic concern and have a significant effect on organizational performance and hence organizations evolve an enterprise-wide mobile strategy for competitive advantage [1]. The user base of mobile IT consists of mobile workers who use mobile technologies for convenience, flexibility, job performance, ubiquitous access of relevant data from anywhere and anytime, and for better informed decision making [2].

Despite the benefits of mobile technologies, there are significant security risks in every area of enterprise's operations area [3]. The mobile based Enterprise systems have seen malicious intrusion and breaches in their network systems [4]. With the increased sophistication of cyber crime and rising number of security risks, data protection and authorized access becomes a major necessity. The Enterprise systems have seen commercial and recreational hackings. The research reports more cyber attacks, an increased number of security threats, more number of cyber crime cases in cloud computing services, and development of mobile-platform-related threats [5]. Since the work environment makes use of more mobile devices for accessing ERP systems, cyber

© Springer Nature Switzerland AG 2020
A. P. Pandian et al. (Eds.): ICCBI 2019, LNDECT 49, pp. 89–99, 2020.
https://doi.org/10.1007/978-3-030-43192-1_10

threats tend to focus more on the mobile media. The mobile devices have weaker defence capabilities compared to PCs as they are designed for portability instead of security. The users of ES mobile systems perceive risk in using the system which leads to uncertainties in trusting the system.

The proposed research aims to investigate the influence of users' attitude towards security issues on perceived security of Enterprise systems (ES) mobility. The research objectives are: (i) To find the security issues of Enterprise Systems Mobility (ii) To examine the influence of users' attitude towards the security issues and (iii) To examine the impact of users' attitude towards security issues on perceived security of ES mobility. These objectives are accomplished by empirically testing a security model of ES mobility. A conceptual model is evolved from the literature and the study proposes to validate the model by empirically collecting data from users of ES mobile systems.

2 Literature Review

2.1 Research Model

The research model for users' attitude toward security issues on perceived security of ES mobility is shown in Fig. 1. The section explains the various components of the model.

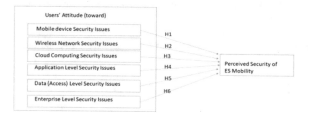

Fig. 1. Theoretical model of users' attitude toward security issues on Perceived Security of ES Mobility

2.2 Enterprise Systems Mobility

Enterprise systems (ES) seamlessly integrate and share organizational information among the various functional systems including finance, accounting, sales, marketing, operations and human resources with external partners. Enterprise System is a suite of integrated software modules based on a central and common database that support various departmental activities [6]. Enterprise systems mobility can be defined as accessing or communicating information using the Enterprise systems through a ubiquitous network technologies on a mobile device anywhere and at anytime [7]. Enterprise mobility provides access to CRM, ERP, SCM, KM and BI systems from anywhere and anytime by using a variety of mobile devices, wireless technologies and applications including smartphones, tablets and other mobile devices/technologies. This is beneficial to employees as they move from place to place the ability to access the enterprise applications also moves freely at all time and at all places.

2.3 Perceived Security of ES Mobility

Perceived value in the marketing context refers to customers' overall perceptions of total losses or gains for buying and consuming a product or a service [8]. Perceived security in the technology context refers to the degree to which users believe that a technological service is secure [9, 10]. Perceived security in the ES mobility context refers to the degree which users believe that the information is secure while using ES mobility. Information security refers to the protection of data from unauthorized access, use, distribution and modification and at the same time the authorized users should be permitted to access and use the data. The nature of security risks remain the same in the case of mobile based ES solutions. Nonetheless, new types and developments were added besides the existing security risks in the ES mobility. The key security properties are integrity, confidentiality, accountability, availability, and non-repudiation through authentication, authorization, and trust management [11]. The security requirements/properties can be defined as follows: Information should not be modified by unauthorized subjects (integrity); Information should be available to authorized subjects when required (availability); Improper disclosure of information should be detected and prevented (confidentiality); the need to ensure that in trading or contracts one cannot afterward deny an action (e.g. signing a contract) that one has carried out (non-repudiation). The goal of security is to ensure that the requirements of confidentiality, integrity, accountability, availability, and non-repudiation of ES mobility are not compromised.

Construct	Authors	Summary
Perceived value	Zeithaml [8]	Customer's overall perceptions of a product or
Perceived security	Rau et al. [9]; Kim et al. [10]	service. Degree to which a technological service is secure
Security goals/properties	Jung et al. [11]	Integrity, confidentiality, accountability, availability, and non-repudiation

2.4 Users' Attitude Towards Mobile Device Security Issues

In the psychology context, the attitude of an individual is defined as the evaluative judgment (positive or negative/favorable or unfavorable), learned predisposition towards an event or idea or an object [12]; and include the psychological behaviours which are affected by the individual beliefs formed by attitudinal learning and doing [13]. In the Enterprise Systems mobility context, users' attitude refers to a psychological state of learned predisposition reflecting the affective or evaluative feelings concerning the security of mobile Enterprise Systems [14] and may influence users to respond favorably or unfavorably toward the security of the ES. Users as a result of using the ES gather information about ES security, understand and associate the information to their current knowledge, where after evaluate the gathered information and form attitudes (favourableness or un-favourableness towards the security of ES) and intentions to use mobile ES.

2.4.1 Mobile Device Security Issues

Ladan [15] defines a mobile device as "any portable device that belongs to a specific user and has computing and storage capabilities". Proliferation of mobile device technologies and increased usage encouraged the hackers to target these mobile device technologies and as a result new types of vulnerabilities started developing [16]. For instance, Android malware attacks install false apps which installs malicious code and contaminate the data and leakage of data. The frequency and sophistication of mobile malware and viruses are rapidly increasing [17] and have caused financial loss and information theft [18]. As many users use their mobile devices to retrieve, access and store vital corporate information, the likelihood of security compromise and new vulnerabilities have also increased. Mobile devices such as iPads, iPhones, and Android smart phones are vulnerable for a new set of scriptless injection attacks which can extract sensitive information through these devices. To deal with malware and other virus attacks on the mobile devices, a key mechanism is to update or install the security software such as firewall or antivirus software on the device on a consistent basis to minimize the risks. IT employees in the enterprise is also responsible to help the users with the regular update and installation of security software.

The employees bring their own device (BYOD) for personal as well as for corporate business use. The BYOD brings many benefits including improving efficiency, productivity and enabling collaboration among employees and increasing potential business opportunities [17]. However, BYOD has a risk of phone loss, untrusted access and can result in data breaches. Despite the convenience, the portable and mobile nature of mobile devices is subject to get lost or stolen and leads to data loss or leak accompanied by new types of security threats. Surprisingly, a majority of smartphone users do not lock their mobile devices by using a password or a pin code; also they save login data on their own mobile devices which is also risky. Many users have a lack of understanding on the need to secure their mobile devices and as a result appropriate security measures are not taken on a regular basis. For instance, Android phone users lack knowledge and have poor comprehension about applications' permissions request [19]. Also, before installing an app, many users hardly pay attention to permission warnings.

2.5 Users' Attitude Towards Wireless Network Security Issues

Mobile technologies use wireless networks which have their own security issues [20, 21]. These issues can fall under the following key security properties: availability, confidentiality and integrity. Confidentiality is where the information privacy of the users can be compromised by a hacker by using the wireless network; integrity is where enterprise data can be modified or interchanged or retransmitted by compromising the wireless network; and availability is where the information or/and service availability could be disrupted [22].

When a location-less hacker intercepts the radio signals of the wireless networks which travel over the air, then tracking the hacker becomes difficult [15]. Besides, majority of the wireless networks depend on other private networks, where the enterprise doesn't have much control on security. These private networks are owned and managed by private service providers. These networks operate on a public-shared infrastructure, where enterprises lack knowledge on security of these infrastructure.

Since the wireless environment is vulnerable for hacking, the hacker gains anonymity without any physical infrastructure [15]. This leads to the Denial of Service attack in which the hacker bombards the communication server with a large number of connection requests and make the server to respond to only the hacker's requests and not for any other authenticated request. Besides, the attacker only requires a mobile device and a wireless card to capture data packets, to read the source and destination fields, to capture the transmission of data, and to gain access to private information (Traffic Analysis). Eavesdropping is another popular security issue in wireless networks. When the wireless network is not adequately secured and when the transmitted information across the network is not encrypted then the hacker closer to the access point can log on to the network. Moreover, in absence of secured wireless network, an attacker can intercept by implanting a malicious host (man-in-the-middle) between the end host and the access point and can change the transmitted information of the session. As a result, all communications will go through the inserted malicious host. Also an attacker who has the full logon details of a legitimate user can capture the transmitted message in the wireless network and can retransmit the (modified) message to same or different destination.

2.6   Users' Attitude Towards Cloud Computing Security Issues

Mobile Cloud computing is the integration of cloud and mobile computing technologies which provides "anytime and anywhere" access and can be defined as "a computing environment where mobile devices communicate with cloud services for data Processing" [23]. Cloud computing provides a number of services including SaaS (software as a service), IaaS (infrastructure as a service) and Paas (platform as a service) [24]. These services benefit the enterprises such as low cost, flexibility, ease of use and convenience. In IaaS, as the service provider owns the storage, hardware, servers, and networking components, the security is in control of the service provider and not with the enterprise. In SaaS environment, as software applications are provided from the cloud, using multiple passwords to access applications is a major security risk [16]. So the basic assumption of carrying out the computing within the organization's firewall is safe is no longer valid because of cloud computing. Also the traditional security technologies (e.g., firewalls, authentication servers, biometrics, cryptography, VPNs and other virus protection measures) are not adequate to address the security concerns in mobile cloud computing [15].

The following are the different types of cloud services: public clouds, private clouds, hybrid cloud, and community cloud. Public cloud allows the users to share the same infrastructure configuration and security protection; Private cloud infrastructure is built for individual enterprise with more data security and control. Comparatively, private clouds have less security and privacy concerns than the public clouds [15]. Hybrid cloud takes the combinations of both the options. Community cloud comprised of shared infrastructure by several organizations within a specific community sector such as healthcare, government, etc. The community cloud has several security benefits as that of a private cloud.

Since the mobile devices lack sufficient computational capabilities, cloud computing enables the user to perform resource intensive computing from the cloud server. So mobile devices are used for transferring the data and visualizing the results. The cloud users are more vulnerable to different and new types of cyberattacks. The hackers could attack a cloud to gain illegal access to the cloud data server.

Among the various vulnerabilities, cloud computing is found to be significantly vulnerable [16]. So security is a critical issue in mobile cloud computing and susceptible for many types of attacks such as man-in-the-middle attack, DDos (distributed denial of services), cloud malware injection attack, information leakage and data theft. In cloud malware injection attack, an attacker damages a cloud application or a service by injecting a SQL command and steal the sensitive information.

2.7   Users' Attitude Towards Application Level Security Issues

Mobile device applications are available in abundance for both Android and Apple iOS platforms. Users download billions of apps and Google celebrated 25 billionth download. With the increase of mobile apps there is an increase of malware attacks especially with Android compared to Apple iOS. These malwares can communicate text messages without the users' knowledge. Fake apps with fake malware versions are the key security issues for both Apple and Android. These fake apps/malware versions are found on third-party application sites where the Android users have easy access compared to Apple users.

These malwares are designed to intercept transactions and redirect them elsewhere; some malware apps are capable of taking videos and pictures of the device locations at any specific time and communicate to the hackers' servers. In absence of upgrading the equipment, many devices may choose to forgo updating the security software. For instance, the latest versions of iOS and Android have apps to check the existence of malware in a device and can remove any malicious behavior of downloaded apps. Attackers gain administrative control to install applications and remove the platform restrictions (jailbreaking) [25]. To defend against malware in case of mobile devices, antivirus software are to be installed.

To manage the security of mobile applications, mobile device management (MAM) software provides features such as data wipe by application, application whitelists and blacklists, enterprise application stores and application security. Similarly, when prompted the device operating systems and applications are to be updated in order to fix known vulnerabilities. For a better protection of security on applications' accounts, the users should avoid storing user names and passwords on the mobile device. To ensure security, applications from untrustworthy sites or from unreliable links within emails should not be download. When downloading or updating the applications, users should be cautious about unnecessary or excessive permission requests to download or update which can increase security risk. The business application systems such as ERP systems need to be updated continuously to avoid any outdated or unsupported security protections. Users should be encouraged to use automatic updater which applies any software application updates when available. Using VPN connection for downloading or updating the applications defend against security threats.

2.8   Users' Attitude Towards Data (Access) Level Security Issues

The increase in the mobility of information provided more benefits to the enterprises including access to data/information anywhere and anytime [26]. Despite the benefits the ES mobility brings a large number of data security threats [27]. Despite the use of intrusion detection software and firewalls in the mobile enterprise systems, extra security measures are required to safeguard sensitive data on the systems [4]. Many of these systems do not have enforcement mechanism to report any suspicious activities on data protection. Due to the loopholes in the authentication scheme, it becomes essential to have data security by preventing the threats from unauthorized users and hackers. The hackers target harvesting data from mobile Enterprise systems [28]. The attackers gather user information and mobile data for creating malicious content. The leaked or stolen data from mobile devices can be used to access mobile ES [17] and can be used to swindle the users for attacks [29].

The data security threats for mobile ES come from external sources as well as from in-house risks [27]. For instance, internally it is paramount to decide who has access to what data (e.g., the system developer need not access salary information of employees). Besides, the permission/authorization to change data should be restricted only to authorized users. Also, audit logs to track any changes in the user/enterprise data need to be maintained. Despite the popularity and benefits of cloud ERP systems, the downside is that organizations are placing 100% of the data security into cloud provider's hands [27]. The reason is the ES data is stored in remotely located cloud servers rather than the local servers. To reduce this risk, organizations need to pay attention to the cloud provider's security processes and their data regulations. As mobile ES have evolved, the capacity to handle a much wider range of functional data as well as sensitive information has increased. Password cracking with the mobile ES is the most common form of hacking. So password alone is not enough to protect confidential business data as the passwords have the risk of stealing or even guessing. As a result for enhanced security, a multi-tier authentication process is needed for accessing the business and sensitive data. For example, along with password authentication, a code can be sent to mobile phone or email address for additional security.

2.9   Users' Attitude Towards Enterprise Level Security Issues

Not having a strategic level security plan for ES mobility is a crucial enterprise level security issue. To get out of the security crisis, the plan should adequately address the rapidly changing security issues in ES mobility. The strategic plan specifies the security goals, policies, and security initiatives. Security programs need to be developed based on rationalistic planning approaches. The security initiatives for ES mobility need to be formulated and implemented in an organization in order to minimize or eliminate security risks arising due to the information handling activities of organizations. Lack of understanding different changes in security context may lead to security crisis. The specific threats are to be identified before planning, designing and developing a secure system. The threats can be characterized as interception, interruption, modification and fabrication which needs to be investigated against the types of security services such as access control, data integrity, authentication, data confidentiality, non-repudiation.

The security policy on sharing access rights is critical to designing secured Mobile ES [30]. The policy on who has to access or modify what needs to be clear in order to prevent any unauthorized access to ES. Hacking is considered to be a game between the firm and a hacker where the hacker aims to compromise the system and the firm tries to minimize loss from a security breach [31]. In this context, user plays a crucial role. Understanding security related behavior of user is paramount for controlling perceived security. User's knowledge and attitude towards Enterprise related security issues such as documentation, data storage, backup and file access practices are significantly related to perceived security-related behavior [32]. The threat of viruses can be addressed by implementing security management policies, anti-virus software, and backup procedures [33]. The organizations need to adopt an appropriate and consistent security policy.

Corporate managers need to be security conscious regarding software acquisition decisions in which user managers with IT security knowledge are to be involved. For example, the acquisition decisions are to be assessed based on the security criteria such as confidentiality, integrity, availability and audit [34]. Besides, organizations need to invest on sophisticated security technologies for information security which is actually a corporate governance responsibility [35]. Doherty and Fulford [36] emphasize on aligning information security management with strategic information systems planning for effective information security policy formulation.

2.10   Research Hypotheses

The below table shows the six hypotheses that are evolved from the literature review:

Number	Hypotheses
H1	Users' Attitude towards Mobile device Security Issues influence Perceived Security of ES Mobility
H2	Users' Attitude towards Wireless Network Security Issues influence Perceived Security of ES Mobility
H3	Users' Attitude towards Cloud Computing Security Issues influence Perceived Security of ES Mobility
H4	Users' Attitude towards Application Level Security Issues influence Perceived Security of ES Mobility
H5	Users' Attitude towards Data (Access) Level Security Issues influence Perceived Security of ES Mobility
H6	Users' Attitude towards Enterprise Level Security Issues influence Perceived Security of ES Mobility

3 Research Methodology

The research objective is to examine the influence of users' attitude towards security issues on perceived security of Enterprise systems (ES) mobility. The research questions are addressed by empirically testing a security model of mobile enterprise systems by collecting data from users of ES mobile systems. For this purpose, a sample of questionnaire survey data will be collected from users of Canadian, American, Indian and Chinese organizations to test the proposed relationships. This is the proposed methodology for collecting data for this ongoing research. In order to obtain relevant data, an online questionnaire will be used and an email request for participation will be sent to potential respondents. The contact details for participants and organizations will be collected from the registry of business companies. Organizations of all sizes from all industries (including manufacturing, service, sales, government, finance, travel, Health, education, and non-profit) will be contacted for possible participation in the questionnaire survey. The user population may include managers, Directors, project managers, and other employees.

The questionnaire instrument has two sections. The first section collects the profile of the organization (e.g., product or services, size in terms of number of employees, and the position of the respondent). The second section has questions to collect the attitude of respondents towards ES mobility security issues and perceived security. This study will be a cross-cultural study and data will be collected from two different culture. For this purpose, the Chinese and English versions of the questionnaire will be prepared. The research methodology explained in this paper is a proposed methodology for this ongoing research. So no research data was collected from any participant. In other words, no participant is involved for this study.

4 Conclusion

The current research has evolved a conceptual model focusing on the user attitude. Six hypotheses were evolved from the literature and the research proposes to collect empirical data based on the questionnaire instrument. The model will be revised after the empirical validation. The revised model will be used for theory development which will be a contribution to the ES mobility and security literature. This research creates an awareness to users for an effective application of security practices in case of ES mobility. The future research can validate the significance of this awareness creation to the adoption of mobile ES. The security model can be validated with different views from the top management and IT staff to gather different perspectives of the ES mobile security. The mobile device designers can make use of this study to incorporate and enhance security features in their devices.

Acknowledgements. This paper is based on conceptual framework and not an empirical study. The research methodology explained in this paper is a proposed methodology for this ongoing research. So no research data was collected from any participant. In other words, no participant is involved for this study. As a result, the ethics committee approval was not obtained.

References

1. Wong, T.Y.T., Peko, G., Sundaram, D., Piramuthu, S.: Mobile environments and innovation co-creation processes & ecosystems. Inf. Manag. **53**, 336–344 (2016)
2. Turnali, K.: The enterprise mobile business intelligence framework. Bus. Intell. J. **23**, 46–58 (2018)
3. Sahd, L.-M.: Significant risks relating to mobile technology. J. Econ. Financ. Sci. **9**, 291–309 (2016)
4. Opara, E.U., Etnyre, V.: Enterprise systems network: SecurID solutions, the authentication to global security systems. J. Int. Technol. Inf. Manag. **19**, 21–35 (2010)
5. Prislan, K.: Efficiency of corporate security systems in managing information threats: an overview of the current situation. Varstvoslovje **16**, 128–136 (2016)
6. Zhaohao, S., Kenneth, S., Sally, F.: Business analytics-based enterprise information systems. J. Comput. Inf. Syst. **57**, 169–178 (2017)
7. Demirkan, H., Delen, D.: Leveraging the capabilities of service-oriented decision support systems: putting analytics and big data in cloud. Decis. Support Syst. **55**, 412–421 (2013)
8. Zeithaml, V.A.: Consumer perceptions of price, quality and value: a means-end model and synthesis of evidence. J. Mark. **52**, 2–22 (1988)
9. Rau, P.L.P., Gao, F., Zhang, Y.: Perceived mobile information security and adoption of mobile payment services in China. Int. J. Mob. Hum. Comput. Interact. **9**, 1179–1198 (2017)
10. Kim, C., Tao, W., Shin, N., Kim, K.S.: An empirical study of customers' perceptions of security and trust in e-payment systems. J. Electron. Commer. Res. Appl. **9**(1), 84–95 (2010)
11. Jung, Y., Kim, M., Masoumzadeh, A., Joshi, J.: A survey of security issue in multi-agent systems. Artif. Intell. Rev. **37**, 239–260 (2012)
12. MacKenzie, S.B., Lutz, R.J.: An empirical examination of the structural antecedents of attitude toward the ad in an advertising pretest context. J. Mark. **53**, 48–65 (1989)
13. Amen, U.: Consumer attitude towards mobile advertising. Interdiscip. J. Contemp. Res. Bus. **2**, 75–87 (2010)
14. Hartwick, J., Barki, H.: Measuring user participation, user involvement, and user attitude. MIS Q. **18**, 59–82 (1994)
15. Ladan, M.I.: A review and a classifications of mobile cloud computing security issues. In: International Conference on Cyber Warfare and Security, p. 214 (2016). https://search.proquest.com/docview/1779927645
16. Kouatli, I.: A comparative study of the evolution of vulnerabilities in IT systems and its relation to the new concept of cloud computing. J. Manag. Hist. **20**, 409–433 (2014)
17. He, W.: A survey of security risks of mobile social media through blog mining and an extensive literature search. Inf. Manag. Comput. Secur. **21**, 381–400 (2013)
18. Chiang, H.S., Tsaur, W.J.: Identifying smartphone malware using data mining technology. In: Proceedings of 20th International Conference on Computer Communications and Networks (ICCCN), pp. 1–6 (2011)
19. Felt, A.P., Ha, E., Egelman, S., Haney, A., Chin, E., Wagner, D.: Android permissions: user attention, comprehension, and behavior. In: Proceedings of the Symposium on Usable Privacy and Security (SOUPS) (2012)
20. Praveena, A., Smys, S.: Efficient cryptographic approach for data security in wireless sensor networks using MES VU. In: 10th International Conference on Intelligent Systems and Control (ISCO), pp. 1–6. IEEE (2016)
21. Smys, S., Josemin, B.G., Jennifer, S.: Mobility management in wireless networks using power aware routing. In: International Conference on Intelligent and Advanced Systems, pp. 1–5. IEEE (2010)

22. Leunga, A., Shengb, Y., Cruickshankb, H.: The security challenges for mobile ubiquitous services. Information Security Group, Royal Holloway, University of London, UK (2007)
23. Dey, S., Sampalli, S., Ye, Q.: MDA: message digest-based authentication for mobile cloud computing. J. Cloud Comput. **5**, 1–13 (2016)
24. Yandong, Z., Yongsheng, Z.: Cloud computing and cloud security challenges. In: International Symposium on Information Technology in Medicine and Education, pp. 1084–1088 (2012)
25. Harris, M., Patten, K.P.: Mobile device security considerations for small- and medium-sized enterprise business mobility. Inf. Manag. Comput. Secur. **22**, 97–114 (2014)
26. Davenport, T.H., Harris, J.G., Morison, R.: Analytics at Work: Smarter Decisions, Better Results. Harvard Business Press, Boston (2010)
27. Adeyelure, T., Kalema, B., Bwalya, K.: A framework for deployment of mobile business intelligence within small and medium enterprises in developing countries. Oper. Res. Int. J. **18**, 825–839 (2018)
28. Markelj, B., Bernik, I.: Mobile devices and corporate data security. J. Educ. Inf. Technol. **1**, 97–104 (2012)
29. Martin, N., Rice, J.: Cybercrime: understanding and addressing the concerns of stakeholders. Comput. Secur. **30**, 803–814 (2011)
30. Yiu, S.M., Yiu, S.W., Lee, L.K., Li, E.K., Yip, M.C.: Sharing and access right delegation for confidential documents: a practical solution. Inf. Manag. **43**, 607–616 (2006)
31. Cavusoglu, H., Mishra, B., Raghunathan, S.: The value for intrusion-detection systems in information technology security architecture. Inf. Syst. Res. **16**, 28–35 (2005)
32. Frank, J., Shamir, B., Briggs, W.: Security-related behavior of PC users in organizations. Inf. Manag. **21**, 127–135 (1991)
33. Post, G., Kagan, A.: Management tradeoffs in anti-virus strategies. Inf. Manag. **37**, 13–24 (2000)
34. Payne, C.: On the security of open source software. Inf. Syst. J. **12**, 61–78 (2002)
35. Gal-Or, E., Ghose, A.: The economic incentives for sharing security information. Inf. Syst. Res. **16**, 186–192 (2005)
36. Doherty, N.F., Fulford, H.: Aligning the information security policy with the strategic information systems plan. Comput. Secur. **25**, 55–63 (2006)

Smart Old Age Home Using Zigbee

N. Shubha, S. Sahana, K. Pavithra, M. S. Meghana,
and K. Panimozhi$^{(\boxtimes)}$

Computer Science and Engineering, BMS College of Engineering,
Bangalore, India
shubha.nagendrakumarm@gmail.com,
sahanaprakash234@gmail.com, kpavithra1998@gmail.com,
msmeghana07@gmail.com, panimozhi.cse@bmsce.ac.in

Abstract. "IoT (Internet of Things) is described as a smart connection, which can track the real-world objects remotely through the internet. The HAS (Home Automation System) is a concept to adopt a safe, smart and automated smart homes into existence. This project mainly focuses on increasing safety and security for the old age people, who are really in need of automation in their day to day life. To achieve reduced human intervention, power consumption, cost and for the communication between the devices we use emerging technologies such as sustainable energy resources (Solar) and Zigbee. The Zigbee is used to create a wireless network, in this project we have implemented two modules which are located inside and outside the home. When there is a movement sensed outside the home the people at home get a notification and take precautions before anything big happens at a very reasonable rate. The Zigbee collects the data systematically at the central location via Wi-Fi and performs the analysis on the data".

Keywords: Wireless Sensor Network · Solar energy · Zigbee · Wi-Fi · Sensors

1 Introduction

IoT (Internet of Things) is an advanced technology which controls the hardware devices through Internet from users. The main propose of using IoT here is to manage the home appliances. Zigbee is a standard communication model that enables us to create the network and communicate among the connected devices. Zigbee uses mesh topology even if one path is down it can use another alternative path for communicating within the nodes connected to the wireless network and when the node is not in use, it goes to sleep mode which saves the power. This technology is very useful to build a Personal Area Network which is cost-effective for the old age home and affordable. Zigbee is best suited for automating old age homes because of its features such as low latency, low data rate, low power consumption, low cost, long battery life and robust which helps in energy management and efficiency. The solar system generates solar energy, which does not have any effects on the environment and is pollution-free. The renewable energy can be accumulated every day, even on cloudy days some amount of power is generated. By combining the advantages of solar energy and Zigbee we are optimizing the usage of power at homes.

© Springer Nature Switzerland AG 2020
A. P. Pandian et al. (Eds.): ICCBI 2019, LNDECT 49, pp. 100–111, 2020.
https://doi.org/10.1007/978-3-030-43192-1_11

2   Literature Overview

The literature overview describes the functionalities of the existing system and taking input from these existing systems, we are overcoming some of the drawbacks of the existing system.

2.1   Fire/Gas Leakage

By Increasing the safety and security of home, we can reduce the outcome of accidental and hazardous events. Early realization helps to overcome many of the accidents.

The gas sensor detects the gas leakage and it gets notified by the GSM module [1–5]. The relay turn's ON the alarm by using ZigBee if there is both fire and gas leakage [6, 7]. Alert, Lightning, Water level and gardening these are the sub-modules in this paper, with the help of ZigBee, DC motor and Relay above data will be sensed and managed [8, 9].

In the proposed system we are overcoming the accidents by sending a message with location to the Fire Engine and turning ON the water pump. Monitoring the gas leakage by turning OFF valve of the cylinder and by turning ON the Exhaust system.

2.2   ZigBee

A network of sensors connected wirelessly is WSN which communicates with other nodes and exchange data wirelessly. Zigbee is more advantageous features compared to other WSN's like delivery rates and utilization of power which is very useful in implementing home automation systems.

Study on various Wireless Technologies with their advantage and disadvantage of using it in the home automation system and power consumption reduced by using solar energy and ZigBee technology [10]. Study on WSN to increase its performance and quality of service while routing the data with the help of Multi stack Architecture [11]. This paper intends to decrease the usage of WSN battery and to increase its performance by prioritizing packets [12]. The problem of bandwidth allocation of increasing traffic has been addressed by using Multi Stack Architecture [13].

In the proposed system we are using Zigbee for making the two individual modules to communicate with one other. Home automation requires less bandwidth and data rate so Zigbee could be the best WSN to use as it as long battery life.

2.3   Smart Door

Nowadays, the growth of crime counts is more so the betterment of home security is required. Illegal Entry of unauthenticated people can cause damage to the house owner as well as a society when the people are not at home.

If an authenticated person tried to enter into the home the door will automatically open and the message will be displayed on Liquid Crystal Display (LCD) [14, 15].

In the proposed system vibration sensor detects and intimates the user that intruders or unauthenticated persons are trying to enter home unethically. Providing control and convenience for homes in terms of safety and security at a very reasonable rate. The keypad is used in our module to achieve it, when a person enters the password correctly, the door automatically gets opened. The user has given an option to change his password and if the incorrect password is entered 3 times the owner is notified.

2.4   Smart Gardens and Water Level Detection

The major aspect of the demolition of plants is people forget to keep track of watering the plants regularly.

Zigbee sends the data to the server which is sensed by the flow and level sensor to the end-users [16, 17]. Sensed data to be sent to the central node from the watered plants in the garden. Gardening system monitor's speed of the wind, moisture content in the soil, temperature, light intensity and humidity level [18]. The amount of fertilization needed for the crops is also sensed by different sensors in the smart irrigation system using Zigbee technology [19]. The irrigation system consists of a network of soil and temperature sensors and the main intention of this design is to minimize water and light [20]. By using Zigbee technology monitoring the crops and water content in the soil can be sensed by using d/f sensors in agriculture [21]. In agriculture, precision farming plays an important role in actuation and decision control based on the data sensed by the various sensors [22]. The moisture parameters are monitored regularly i.e., temperature and humidity to obtain a high-quality environment [23]. With the help of ADC the sensed data is transmitted to Zigbee using UART [24]. Smart Aquaponics system helps the plants to grow naturally for producing food by using the water of fish tank and waste produced by fish is feed as nutrients for plants to grow [25].

In the proposed system soil moisture sensor continuously monitors the moisture level, if barren it automatically turns ON the motor to supply water to the garden on time and need before plants get destroyed due to irregular power supply. For detecting the water level inside the tank, we are using the ultrasonic sensor. The motor automatically gets ON, to lift the water in tanks on time and need by using renewable energy.

2.5   Lighting

There are times when people forget to turn ON/OFF the lights, the lights are automated to provide them a very secure feeling when people walk around.

With the data collected from different sensors, the monitoring station on an individual lamp takes decision independently about lightening the lamp, after this, it will automatically send the response back to the main station. Saving energy is one of the main centers of attention in the world. If in case of any failures Service engineers can get the notification through GSM module. Through GUI Street lights are controlled remotely. It uses a Renewable source of energy which reduces power consumption [26–29]. Without any manual intervention, lights can be switched OFF when not in use and Maintaining the safe level of the temperature of the LED by reducing the current flow through it [30, 31].

In the proposed system we are turning ON the bulbs in garden areas and corridors by detecting the light intensity and presence of human beings in the surroundings. By controlling the intensity of the bulbs, thus reducing energy consumption when someone forgets to OFF the lights.

2.6 Society

As days are passing on, the people at old age homes are increasing, it is very important to provide old age homes with more facilities to make them feel secure and make their work easier, as they forget many things. By implementing the above modules, we are achieving it at very low costs. Old age people are unable to monitor appliances and accidents at home daily. Zigbee helps in monitoring the appliances and takes the action immediately before things get worst and loss of energy.

3 Interface Diagram and Proposed Work

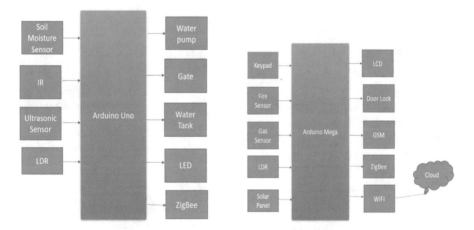

Fig. 1. Inside the home module   **Fig. 2.** Outside the home module

The network comprises multiple sensors as described in Figs. 1 and 2. The sensors forward the sensed information to the Controlling Unit (Arduino) and the required activity is done. Zigbee takes the lead role in establishing the communication between the two modules which are located at home (Fig. 1) and outside the home (Fig. 2) wirelessly. Generally, the power is supplied to the modules with the help of solar, but if there is any insufficiency while providing the power supply through solar power, the main KEB power is supplied to the modules. Taking note of the time the solar power is consumed versus the consumption of the main KEB power is stored in the cloud and validated. The values of these two quantities based on monthly bases are analyzed and depicted in the graph for the generation of a monthly bill.

4 Overall Working of the Module

Figure 3 describes the overall detailed design of the project with a complete overview of each module. The network comprises multiple sensors such as Gas, Flame, Soil moisture, Ultrasonic, and IR sensors. Each of the sensors has a different function to be performed. Firstly, for detecting the presence of fire at homes, a Flame sensor is used. Fire leakage is detected and before it explodes, a message is sent to the owner and fire station with the specific location and simultaneously the water pump is turned ON. For detecting the gas leakage at homes, Gas sensor is used. Gas leakage is detected and the haphazard is overcome by automatically turning OFF the valve of the cylinder, turning ON the exhaust fan and sending a message to the owner. Secure entry to the home is done using a Keypad, where the user enters the password and the password gets evaluated if matched the door gets opened. The user can also change the password and he gets notified when someone enters the wrong password 3 times. When someone tries to enter the home by breaking the door, the vibration sensor detects and notify the owner.

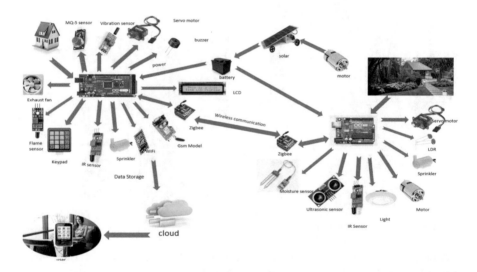

Fig. 3. Shows the detailed system architecture of the overall module

For detecting the distance of an object detected Ultrasonic sensors are used. In this module proposed, the water level is detected using the Ultrasonic sensor inside the water tank, which automatically turns the motor ON when the tank is getting empty. To detect the level of moisture in a particular area, the Soil moisture sensor is used and performs the specific action, in this module we are continuously monitoring the soil in the garden area for detecting the moisture level of it and which automatically turns the motor ON for watering the plants. By using renewable energy, we can do our work on time and on need by not only depending on the normal power supply as we are facing irregular power supply nowadays. To detect the motion in 180°/human presence we make use of IR sensors which is used in multiple areas for different purposes. Firstly near gate, when the human presence is detected and the gate is opened with the help of the servo motor and when people feel that it is not secure to open the gate during night times, the person can switch off the button located inside the home, once the people switch off the button if any person/vehicle presence near gate, Zigbee located outside the home communicates with the other Zigbee which is located inside the home to know the status of switch. If switch is off then it make sure that gate should not open in the presence of human or vehicle and Zigbee sends the notification back to the alarm system located inside the home to indicate the person that someone is there near the gate and Secondly for detecting motion in 180°, when motion is detected lights gets ON automatically based on the light intensity depending on the day/night.

Zigbee with its most attractive advantages makes the home automation system implementation in very reasonable costs and exclusively used to establish the connection in small areas. Here, Zigbee acts a very significant role in making the two modules to communicate with each other wirelessly which is inside and outside the home and also helps to integrate as a portable single unit allowing both the modules to use same alarm system when needed which reduces the number of devices connected and cost. The person who is sitting inside the home can know everything going around his home without putting any effort of going outside and monitoring it every time. She/he can get all the notification easily by using Zigbee.

Renewable resource i.e., solar power that can accumulate power every day including on the cloudy days some energy can accumulate which is much more useful in daily lives. The direction of the solar panel is changed according to the sun's direction for producing maximum power. In our module the main supply of power supplied for every home appliance is supplied from solar power and if any insufficiency in the supply of power for appliances, then main KEB power is used. By taking the note of the time the KEB power is used and renewable solar power is used, we are generating the electricity bill every month by storing the data on the cloud through Wi-Fi. Finally, the graph gets generated based on the analyzed data.

5   Implementation and Results

The project implementation methods and the required outcomes are discussed in this section with the help of the flowchart (Figs. 4 and 5).

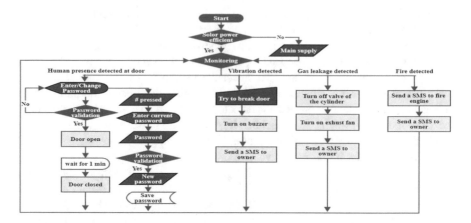

Fig. 4. Flow chart of the module placed inside the home

A fire explosion is detected by using Fire Sensor. If any fire leakage is found, the Arduino will send a message and the location to Fire Engine station and Owner of the house, the Alarm rings and ON's the water sprinkler. Gas leakage is detected using Gas sensor. Our system ON's the exhaust fan which vanishes air, buzzer/alarm rings, cylinder valve is closed and the owner gets notified by sending a message. This will prevent accidents at homes before it turns into a dangerous explosion and improves safety.

When a person enters the password correctly, the door automatically gets opened which provides security for homes and we can also change the password when needed. If someone enters an invalid password for 3 or more times it notifies the owner about the unauthorized entry to the house. When someone tries to enter the home by breaking the door, the vibration sensor detects and notifies the owner. The main supply of power for every home appliance is supplied from solar power and if any shortage in the supply of power for appliances, then main KEB power is supplied.

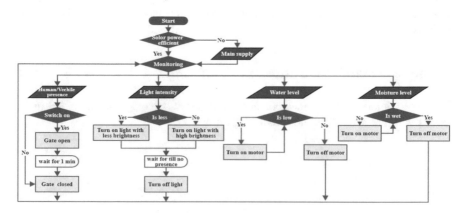

Fig. 5. Flow chart of the module placed outside the home

In Smart gardening, by the help of soil moisture sensor we are detecting the moisture level of the soil in the garden area. It continuously monitors the moisture level, if barren it automatically turns ON the motor to supply water to the garden on time and need before plants get destroyed due to irregular power supply. For detecting the water level inside the water tank we use Ultrasonic sensor, which automatically turns the motor ON when tank is getting empty on time and it needs by using renewable energy and OFF the motor when the tank is full which helps to prevent wastage of water and unreasonable water bills. In smart Lighting we are turning ON the bulbs in garden areas and corridors by detecting the light intensity based on day/night and presence of human beings. By controlling the intensity of the bulbs, that reduces energy consumption and it helps too when someone forgets to turn OFF the light.

6 Snapshots

The following snapshots show the implementation of our modules.

Fig. 6. Module placed outside the house   **Fig. 7.** Module placed inside the house

Figure 6 depicts the module which is placed outside the house which includes gardening, lighting, automatic water level control and smart gate system. Figure 7 depicts the modules which are placed inside the house which includes a smart door lock system, automated gas leakage system, automated fire control, and smart security system.

7 System Testing

Each module is individually tested and checked whether it works completely perfect in all the scenarios (Table 1).

Table 1. System test cases

Sl. No.	Test cases	Expected outcome	Actual outcome
1	Checking whether users can update data to the website	User able to upload data into the website	Data has been updated on the website
2	Checking the sensor's working condition	Sensor's should be working fine and take required actions in the system	Sensors are working fine with required actions performed
3	When the user prefers the solar power, the power is supplied from the battery as same as the main power supply to the system	The power should be supplied from the battery as same as the main power supply to the system	Solar power is supplied to the system
4	When the user prefers the mains as the supply, the main power is supplied from the main KEB power	The power should be supplied from the main power supply to home	The main power is supplied to the system
5	An alert message should be displayed when there is a low battery	When there is low battery an alert message should be shown in the LCD	An alert message displayed on the LCD
6	Zigbee to Zigbee communication should happen between Home model and the garden model	Garden model communicates with the Home model to notify the user by displaying a message in LCD	Communicated Successfully and a message is displayed on LCD

8   Analysis

Solar energy is an inexhaustible energy source that is available every day. By using this we are reducing the home electricity bill and maintenance cost. The user has to pay a bill only when the automation system runs by the main power system. In our system, we are giving preference to the user to select either solar power or main power to get supplied as a main power supply to the home automation system. Using our automation system, the cost savings is more than 30%.

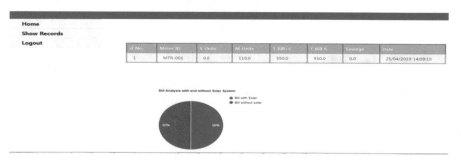

Fig. 8. Bill without solar

By seeing the above Fig. 8 we can analyze that the whole home automation system has been run by using only the main KEB power. i.e. Total Bill with Solar - 550/- and Total Bill without Solar - 550/-. Saving is 0 rupees because solar power is not utilized by the home automation system.

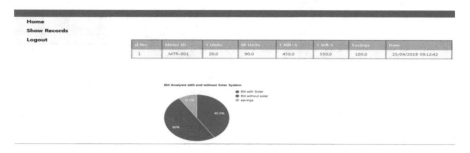

Fig. 9. Bill with solar

By seeing the below Fig. 9 we can analyze that by using both Solar energy and main KEB power as a power supply to the home we are reducing the electricity bill. i.e. Total Bill with Solar - 450/- and Total Bill without Solar - 550/-. We are saving 100/- by using solar energy. So we can say that our home automation is more cost-efficient

For establishing the communication between Home Model and Garden Model we are using this latest technology called ZigBee. Home Automation requires High security, Low power consumption, less bandwidth, and less data rate. So Zigbee could be the best Wireless Sensor Network for home automation whose battery life lives long compared to other Wireless Sensor Network. Our automation system can be adapted with other latest technologies with small rework.

By considering the above scenarios we can say that our home automation system provides higher safety and security and reduce human intervention and cost.

9   Conclusion

Till now the home automation was determined only for switching ON or OFF the appliances by using mobiles, but as technology improves things are operated smartly and easily. Automation merged with WSN and initiated to follow the human instructions. Sensors such as light intensity sensors, humidity sensors, motion sensors, temperature sensors, and smoke detectors are used to develop a home automation system and the data collected from sensors are managed and evaluated by devices.

Our project main aim is to increase security by having a secure door lock system and locking the gate at night time, Increase safety by taking care of accidents that can occur in home before it gets worst such as gas and fire leakage, Turning OFF the appliance when not in use such as Lightening system and watering system helps in reducing power utilization, Need not to worry about many of the day-to-day activities as it is automated which reduces human intervention, The proposed system is remotely controlled by Zigbee and self-charged by solar energy. Using a combination of solar system and ZigBee, our system is optimized by its energy usage and cost.

By analyzing the power consumption and amount charged for it, he/she can reduce the utilization of power consumption based on his budget for that month.

Our project is offered almost all required services for the home, it is used in not only old age homes, but it can also be adaptable in most of the environments like homes, hospitals and other environment based on their requirements. Our project provides flexibility for the latest communication technology, Zigbee can be replaced with any other communication technology because our modules are independent.

Acknowledgment. We would like to thank TEQIP III for their financial support and B.M.S College of Engineering for the encouragement and support.

References

1. Gokula Kaveeya, S., Gomathi, S., Kavipriya, K., Kalai Selvi, A., Sivakumar, S.: Automated unified system for LPG using load sensor. In: International Conference on Power and Embedded Drive Control (ICPEDC) (2017)
2. Fuzi, M.F.M., Ibrahim, A.F., Ismail, M.H., Ab Halim, N.S.: A dedicated fire alert detection system using ZigBee wireless network. In: IEEE 5th Control and System Graduate Research Colloquium, UiTM, Shah Alam (2014)
3. Ransing, R.S., Rajput, M.: Smart home for elderly care, based on wireless sensor network. In: International Conference on Nascent Technologies in the Engineering Field (2015)
4. Nimbal, M.S., Kulkarni, G.A.: Monitoring of home security using GSM and Zig-Bee network. Int. Adv. Res. J. Sci. Eng. Technol. **3**, 129–133 (2016)
5. Ganesh, D., Anilet Bala, A.: Improvement on gas leakage detection and location system based on wireless sensor network. Int. J. Eng. Dev. Res. **3**, 407–411 (2015)
6. Saravanan, R., Vijayaraj, A.: Home security using ZigBee technology. Int. J. Comput. Sci. Inf. Technol. Secur. (IJCSITS) **1** (2011)
7. Chen, D., Wang, M.: A home security ZigBee network for remote monitoring application. In: ICWMMN (2006)
8. Rawate, S.V., Patil, M.D.: Smart sensor network for monitoring and control of society automation. Int. Res. J. Eng. Technol. (IRJET) **3**, 912–915 (2016)
9. Rashid, S., Haider, N., Iqbal, M.: Gas management and disaster system using ZigBee. Int. J. Eng. Res. Technol. (IJERT) **2**, 288–293 (2013)
10. Meghana, M.S., Pavithra, K., Sahana, S., Shubha, N., Panimozhi, K.: Life at ease with technologies-study on smart home technologies. In: Balaji, S., Rocha, Á., Chung, Y.N. (eds.) Intelligent Communication Technologies and Virtual Mobile Networks. Lecture Notes on Data Engineering and Communications Technologies, vol. 33, pp. 511–517. Springer, Cham (2020)
11. Panimozhi, K., Mahadevan, G.: QoS framework for a multi-stack based heterogeneous wireless sensor network. Int. J. Electr. Comput. Eng. (IJECE) **7**, 2713–2720 (2017)
12. Panimozhi, K., Mahadevan, G.: Multi-stack architecture implementation to enhance the QoS in WSN with prioritization of packets. Int. J. Comput. Appl. **120**, 51–56 (2015)
13. Panimozhi, K., Mahadevan, G.: Bandwidth utilization in wireless sensor networks with priority based multi-stack architecture. In: International Conference on Applied and Theoretical Computing and Communication Technology (iCATccT) (2016)

14. Park, Y.T., Sthapit, P., Pyun, J.-Y.: Smart Digital Door Lock for the Home Automation. University Gwangju, Gwangju (2009)
15. Krishna Kanth, B.B.M.: An effective water quality and level monitoring system using wireless sensors through IoT environment. Int. J. Eng. Res. Appl. **7**, 40–44 (2017)
16. Balaji, V., Akshay, A., Jayashree, N., Karthika, T.: Design of ZigBee based wireless sensor network for early flood monitoring and warning system. Easwari Engineering College (2017)
17. Kanagamalliga, S., Vasuki, S., Vishnu Priya, A., Viji, V.: Security monitoring using embedded systems. Int. J. Innov. Res. Sci. Eng. Technol. **3**, 620–623 (2014)
18. Caetano, F., Pitarmaa, R., Reisb, P.: Intelligent management of urban garden irrigation. UBI – University of Beira Interior (2014)
19. Esakki Madura, E., Venkatesa Kumar, V.: Smart agriculture system by using ZigBee technology. Int. J. Curr. Eng. Technol. **6** (2017)
20. Parwatkar, S.A., Bhagat, V.B.: Producing more crops in automated irrigation system using WSN with GPRS and ZigBee. Int. J. Curr. Eng. Technol. **5**, 771–777 (2015)
21. Sahitya, G., Balaji, N., Naidu, C.D.: Wireless sensor network for smart agriculture. In: 2nd International Conference on Applied and Theoretical Computing and Communication Technology (iCATccT) (2015)
22. Kalra, A., Chechi, R., Khanna, R.: Role of ZigBee technology in agriculture sector. In: National Conference on Computational Instrumentation, CSIO Chandigarh (2010)
23. Duraipandian, M., Vinothkanna, R.: Cloud based Internet of Things for smart connected objects. J. ISMAC **1**(02), 111–119 (2019)
24. Sathish Kannan, K., Thilagavathi, G.: Online farming based on embedded systems and wireless sensor network. In: International Conference on Computation of Power, Energy, Information and Communication (TCCPEIC) (2019)
25. Chikankar, P.B., Mehetra, D., Das, S.: An automatic irrigation system using ZigBee in Wireless Sensor Network (2015)
26. Raju, V., Yashaswini, S., Panimozhi, K.: Survey on aqua robotics urban farm system. Int. J. Comput. Sci. Eng. **7**, 614–622 (2019)
27. Srinath, V., Srinivas, S.: Street light automation controller using ZigBee network and sensor with accident alert system. Int. J. Curr. Eng. Technol. **5**, 2819–2823 (2015)
28. Su, M., Su, Y., Lv, P.: Design of the wireless monitoring system of solar lamps based on ZigBee and GPRS. Wuhan University of Technology Huaxia College (2012)
29. Santhosh Kumar, R., Prabu, D., Vijaya Rani, S., Venkatesh, P.: Design and implementation of an automatic solar panel-based led street lighting system using ZigBee and sensors. Middle-East J. Sci. Res. **23**, 573–579 (2015)
30. Mhaske, D.A., Katariya, S.S.: Smart street lighting using a ZigBee & GSM network for high efficiency & reliability. Int. J. Eng. Res. Technol. (IJERT) **3**, 175–179 (2014)
31. Yoon, S., Kim, H.: Development of self-powered LED street light and remote controller utilizing the ZigBee and smart devices. Andong National University (2015)

Performance Improvement in 6G Networks Using MC-CDMA and mMIMO

A. Vijay[1(✉)] and K. Umadevi[2]

[1] Ambal Professional Group of Institutions, Palladam, TN, India
Vijaypgpece@gmail.com
[2] Sengunthar Engineering College, Tiruchengode, TN, India
kindlyuma@gmail.com

Abstract. As several internet-based equipment increases day by day, the need for providing a better quality of service becomes essential which paves the way for evaluation of series of generations in wireless communication. Even though the telecommunications techniques are upgraded and equipment's of transmission and their techniques are changed, there are certain promising conventional methods, which acts as a backbone for 6G and beyond wireless communication Networks. Two major techniques that support future wireless communication are MC-CDMA (Multi carried code division multiple access) and mMIMO (massive multiple input and multiple output). In this article, we have presented the demands and the support of these two technologies concerning upcoming proposed techniques for 6G wireless communication like Underwater Acoustic Channels, Visible light communication, Terahertz Communication, Large Intelligent Surface, Artificial intelligence and Machine learning, Holographic beamforming, Blockchain-Based Spectrum Sharing. Furthermore, this article will be providing possible opportunities in MC-CDMA and mMIMO for the effective optimization of future wireless Communication.

Keywords: MC-CDMA · mMIMO · Deep neural network · Machine learning · Nano-antenna arrays

1 Introduction

As International telecommunications union has released 2030 requirements for wireless communication, creates a demand for research towards 6G wireless communication [14]. Additionally, with the evaluation of new technologies like IoT (Internet of Things), autonomous vehicle and D2D (Device to Device Communication) results in increasing quantity of devices based on internet which will lead to a massive requirement in connectivity of about 1 crore devices per square kilometers. Moreover, the expectation of 6G wireless communication is to provide a data rate of 1 Gb/s for outdoor and 10 Gb/s for indoor environments, it is difficult to achieve the target without invention of new technologies and to integrate them with the promising current technologies [15]. As MC-CDMA can support maximum number of users simultaneously with good bit error rate [16] and mMIMO provides spectral efficiency, energy efficiency and supports user tracking, to optimize the upcoming 6G technology, most

A. P. Pandian et al. (Eds.): ICCBI 2019, LNDECT 49, pp. 112–118, 2020.
https://doi.org/10.1007/978-3-030-43192-1_12

efficient modulation technique like MC-CDMA and transmission methods based on mMIMO will become mandatory.

mMIMO overcomes the limitations of MIMO by providing enhanced throughput, improving energy efficiency and increasing the distance of coverage. Regardless of its features, mMIMO is limited by the uncertainty of channel, variation in channel capacity and reduced reliability. To overcome the drawback of mMIMO proper modulation technique is to be deployed. Out of all the modulation techniques, MC-CDMA supports all the requirements of mMIMO and hence we have concentrated all the possibilities of deploying MC-CDMA along with mMIMO in each one of the upcoming 6G techniques.

In this study paper, initially we discuss about implementing MC-CDMA and mMIMO in under water acoustic channels with the help of machine learning and Artificial intelligence (AI). Furthermore, the role of visible light communication (VLC) in 6G is discussed along with implementation of mMIMO and optical CDMA in indoor environment. Moreover, the implementation of nano antenna arrays along with mMIMO in Teraheartz communication is illustrated. Likewise, our paper presents possibilities of deploying mMIMO in Large Integrated Surface (LIS). Also the need for AI in 6G is also discussed to a little extent. Further, features of holographic beam forming is explored. Finally our article confer about block chain based spectrum sharing, an emerging technology which could replace the role of internet.

2   MC-CDMA and mMIMO Towards Future Wireless Communication Networks

2.1   Underwater Acoustic Channels

As need for underwater communication is increasing in the recent years, under water acoustic channels is used to avoid wired communication under the sea. But it is tough to implement when compared to all other types of communication techniques, hence many underwater applications like site seeing, underwater life research, disaster prevention military, autonomous underwater vehicles transportation and much more applications need a reliable modulation technique. Since CDMA has proven to be a promising modulation technique, many machine learning algorithms like deep neural network is been deployed for studying and increasing the performance of the communication under the sea [1]. in spite of their limitations including limited battery power, limitations in bandwidth usage, reduced sensor life due to corrosion, a large amount of energy absorption by water, it still remains to be a promising technique and has a high demand in the near future as it reduced the cost of laying wires undersea for communication.

AI Controlled Substation. Shortly, there is no doubt that AI (Artificial Intelligence) is going to make a big changeover in the field of communication systems. Implementing AI to control underwater Acoustic channels will improve the efficiency of the system. The architecture of an underwater communication system based on MC-CDMA and mMIMO is as shown in the Fig. 1. Usage of MIMO in underwater communication will pay ways for enhancing performance [2]. Further; it is evident that the MIMO communication under ice water undergoes less distraction and hence when implemented with MC-CDMA will improve the efficiency of the system [3].

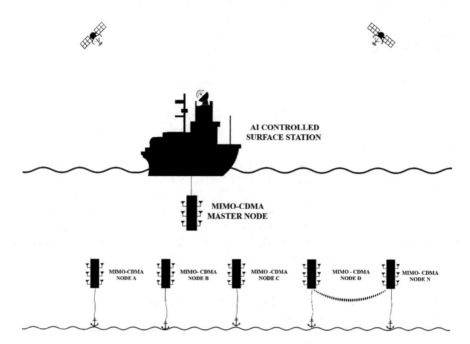

Fig. 1. Architecture of AI controlled underwater acoustic channels based on MC-CDMA and mMIMO.

2.2   Visible Light Communication (VLC)

White LEDs (Light emitting diodes) and Photodiodes are the commonly used transmitters and receivers in the VLC communication. The visible light communication has great attention in the recent days for the development of 6G and beyond wireless communication systems because of its three major properties. Firstly, it can operate in the normal visible spectrum and hence there is no need for spectrum license. Secondly, it can achieve a huge data rate in a transmission that is unimaginable. Finally, it is free from collusion, as electromagnetic waves cannot affect their communication. VLC can be deployed based on a CDMA based system employing two major methods namely unipolar codes and bipolar codes respectively [4]. Additionally, Optical CDMA codes can be used in the implementation of VLC for indoor Localization System as circuit design complexity is less and hence the cost of implementation will cheaper when compared to other modulation techniques [5]. Further, MIMO based VLC communication proves to be a promising technique to fulfil the requirements of 6G and beyond communications techniques, as it can transmit data up to 1 GB/s in indoor communication [6]. Later it can be optimized to work in outdoor employing several AI based protocols.

2.3   Terahertz (THz) Communication

In spite of short-distance communication properties due to propagation loss and restriction to power, THz communication is still a promising technology for 6G and beyond wireless communication as it can provide a high data rates when compared to microwave and millimetre wave bands. To increase the communication distance, mMIMO is been deployed which has proven to increase the performance of the THz band [7]. Developing antennas for providing such huge frequency communication is been a bottleneck and research are going on to design antennas that best support for this massive communication. Recently, nano-antenna arrays antennas are developed and tested for supporting the high-frequency bands [8].

2.4   Large Intelligent Surface (LIS)

LIS contains an array of reflecting elements that can be easily fixed in walls of buildings so that it can reflect the signals to the users from Massive MIMO antennas. Thereby it will be easy for the base station to establish a connection with the users who are in the blind spot region. The gain of mMIMO system increases with the less consumption of power during transmission while implementing LIS [9, 10]. Interference will reduce largely by implementing this technique, it can you used in indoor as well as outdoor environment. Further, it can support technology that is beyond mMIMO [11].

2.5   Artificial Intelligence (AI) and Machine Learning (ML)

There are several solutions to the wireless network demands based on mMIMO and MC-CDMA based techniques but each has its demands and drawbacks. To select the perfect protocols and techniques will need a lot of instant calculations and predictions to pick the right one. For this purpose, a specific method based on AI has been deployed. The AI protocol and techniques where further charted out and a platform was defined for developing AI codes based on network demands. [12]. There is no doubt that AI is going to take over the wireless network shortly (Fig. 2).

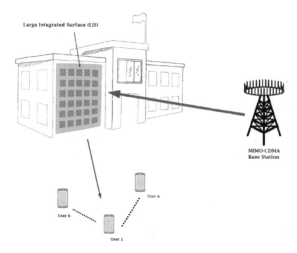

Fig. 2. Large Integrated Surface (LIS) for mMIMO and MC-CDMA based communication networks

2.6   Holographic Beamforming (HBF)

The signal strength from the base station with Software-defined mMIMO antenna is directed towards to a particular user instead of broadcasting in all directions, by means of doing so we can achieve large throughput in the range of gigabits, improving Signal to noise ratio, transmission power is much reduced and even can increase the coverage area. Even though Space division multiple access (SDMA) is been deployed in HBF, MC-CDMA can you deployed in the subcarrier transmission of signals and thereby the overall performance can be optimized. HBF promises to provide cheap manufacturing, less spacious, weightless and works with less power.

2.7   Blockchain-Based Spectrum Sharing

A decentralized form of sharing data among users by spectrum sharing which could perform the same function as the internet. They are categorized to operate under licensed and unlicensed bands. Presence of a nearby node is identified and the spectrum is allocated for effective efficiency improvisation. The primary way of operating is by users finding possible ways based on the mapped information available in the decen-tralized database and choosing the right way to communicate. Before communication, users need to share usage rights among each other for efficient communication [13]. To optimize the payload, applying mMIMO will increase the performance. Further, MC-CDMA can be used as a modulation technique for transmitting the data among users in the decentralized network.

3 Conclusion

In this survey and tutorial paper, we have presented the roles and opportunities of MC-CDMA and mMIMO for 6G wireless communication and even beyond that. There is no doubt that above mentioned seven technologies along with MC-CDMA and mMIMO will improve the efficiency of the 6G networks. Furthermore, most of the technologies discussed above have proven to enhance their performance on merging with MC- CDMA and mMIMO. Finally, our paper provides all the necessary content as well as the outline for those who are willing to start their research in 6G based on MC-CDMA and mMIMO technologies.

References

1. Kim, Y., Lee, H., Ahn, J., Chung, J.: Selection of CDMA and OFDM using DNN in underwater acoustic channels. In: Eleventh International Conference on Ubiquitous and Future Networks (ICUFN), pp. 42–44. IEEE, Crotia (2019). https://doi.org/10.1109/ICUFN. 2019.8805917
2. Bouvet, P.-J., Auffret, Y., Munck, D., Pottier, A., Janvresse, G., Eustache, Y., Tessot, P., Bourdon, R.: Experimentation of MIMO underwater acoustic communication in shallow water channel. In: OCEANS 2017. IEEE, Aberdeen (2017). https://doi.org/10.1109/ OCEANSE.2017.8084719
3. Han, X., Yin, J.-W., Liu, B., Guo, L.-X.: MIMO underwater acoustic communication in shallow water with ice cover. China Ocean Eng. **33**(2), 237–244 (2019). https://doi.org/10. 1007/s13344-019-0023-7
4. Qiu, Y., Chen, S., Chen, H.-H., Meng, W.: Visible light communications based on CDMA technology. IEEE Wirel. Commun. **25**, 1–8 (2017). https://doi.org/10.1109/MWC.2017. 1700051
5. De Lausnay, S., De Strycker, L., Goemaere, J.-P., Stevens, N., Nauwelaers, B.: Optical CDMA codes for an indorr localization system using VLC. In: 3rd International Workshop in Optical Wireless Communications (IWOW). IEEE, Portugal (2014). https:// doi.org/10.1109/IWOW.2014.6950775
6. Azhar, A.H., Ztran, T.-A., Brinen, D.O.: A Gigabit/s indoor wireless transmission using MIM-OFDM visible–light communications. IEEE Photonics Technol. Lett. **25**(2), 171–174 (2013). https://doi.org/10.1109/LPT.2012.2231857
7. Faisal, A., Sareddenn, H., Dahrouj, H., Alnaffouri, T.Y., Alounin, M.-S.: Ultra-massive MIMO systems at terahertz bands: prospects and challenges. Electrical Engineering and Systems Science, pp. 1–7. Cornell University (2019)
8. Han, C., Jornet, J.M., Akyildiz, I.F.: Ultra-massive MIMO channel modeling for graphene-enabled terahertz-band communications. In: IEEE GLOBECOM, Abu Dhabi, UAE (2018)
9. Huang, C., et al.: Energy efficient multi-user MISO communication using low resolution large intelligent surface. In: IEEE GLOBECOM, Abu Dhabi, UAE (2018)
10. Wu, Q., Zhang, R.: Intelligent surface enhanced wireless network: joint active and passive beamforming design. In: IEEE GLOBECOM, Abu Dhabi, UAE (2018)
11. Hu, S., et al.: Beyond massive MIMO: the potential of data transmission with large intelligent surface. IEEE Trans. Sig. Process. **66**(10), 2746–2758 (2018)
12. Vijay, A., Umadevi, K.: Secured AI guided architecture for D2D systems of massive MIMO deployed in 5G Networks. In: Third International Conference on Trends in Electronics and Informatics (ICOEI 2019), pp. 468–472. IEEE, Tirunelveli (2019). https://doi.org/10.1109/ ICOEI.2019.8862712

13. Weiss, M., Werbach, K., Sicker, D., Caicedo, C., Malki, A.: In the application of blockchains to spectrum management. In: 46th Research Conference on Communication, Information and Internet Policy 2018, SSRN (2018)
14. David, K., Berndt, H.: 6G vision and requirements: is there any need for beyond 5G. IEEE Veh. Technol. Mag. **13**(3), 72–80 (2018)
15. Zhang, Z., Xiao, Y., Ma, Z., Xiao, M., Ding, Z., Lei, X., Karagiannidis, G.K., Fan, P.: 6G wireless networks. IEEE Veh. Technol. Mag. 2–15 (2019)
16. Tasneem, A., Majumder, S.P.: BER performance analysis of a MC-CS-CDMA wireless communication system with rake receiver employing MRC under Nakagami m fading. In: 2016 3rd International Conference on Electrical Engineering and Information Communication Technology (ICEEICT). IEEE, Dhaka (2018)

An Adaptive and Opportunistic Based Routing Protocol in Flying Ad Hoc Networks (FANETs): A Survey

O. Aruna[1($\boxtimes$)] and Amit Sharma[2]

[1] Narasaraopeta Engineering College, Narasaraopeta, India
arunasri52@gmail.com
[2] Lovely Professional University, Phagwara, India
amit.25076@lpu.co.in

Abstract. FANET is a special form of ad hoc networks in which UAVs are mobile nodes they can fly in the air autonomously and can be operated remotely. FANETs has many advantages as well as disadvantages compared with MANET and VANET. The existing routing protocols of traditional adhoc networks can't satisfy all the requirements of FANETs. In Multi-UAV systems, during mission operation changes can occur dynamically. Due to unique characteristics of FANETs communication is a big challenging issue. So that it is necessary to develop a new routing protocol that must be able to update routing table dynamically. In this paper, describes functionality of FANETs and collected information from different existing routing protocols for FANETs and used effective routing techniques to increase the efficiency of routing protocol. Communication protocols are also discussed.

Keywords: FANETs · Routing protocol · Communication · UAVs · GCS

1 Introduction

FANET extends from MANET and VANET. It consists of the collection of UAVs that can fly in the air autonomously and can be operated remotely. Compared with Single-UAV System, Multi-UAV system is more advantageous. FANET is only applicable for Multi-UAV Systems. At the same time, all Multi-UAV System do not form a FANET. Recently, FANETs are used in different applications, mostly in military and civilian applications [1]. Compared with ground based networks like MANETs and VANETs, FANETs are more efficient to deliver data communication. But, within the usage of FANETs, Communication between UAVs is a crucial task due to some unique challenges of FANET like the mobility nature of UAVs is very high, continuous changes in network topology, etc.

1.1 Advantages of Multi-UAV Systems

- The cost of small UAVs is very low and more efficient than the large UAVs
- Multi-UAV systems extend scalability of operation using FANET easily compared with large UAVs; it covers limited range of operation.

© Springer Nature Switzerland AG 2020
A. P. Pandian et al. (Eds.): ICCBI 2019, LNDECT 49, pp. 119–127, 2020.
https://doi.org/10.1007/978-3-030-43192-1_13

- If we use single large UAV, it may fail in a mission; there no way to continue the process, the mission can stop. However, if we use multi-UAV system, the operation can continue with other UAVs [1].
- With the help of small UAVs, mission can be completed faster than large UAVs.

1.2   Unique Challenges of Multi-UAV Systems

Compared with single-UAV system, multi-UAV system has some unique challenges. In both single-UAV and multi-UAV systems, communication established between the UAV and infrastructure but the difference is variation of topology. In multi-UAV systems, number of nodes increased, every time the topology will be changed, the distance between nodes is very large and link quality also changed [5]. Depending on these unique challenges, designing the efficient network architecture is a major problem for data communication in multi-UAV system. Completely the multi-UAV system relies on infrastructure based approach. Within the usage of this approach, faced a lot of problems like usage of complicated hardware, range restriction between UAVs and ground station. Another solution for multi-UAV system is FANETs.

1.3   Difference Between MANET, VANET and FANET

FANET extends from Mobile Ad hoc to Vehicular Ad hoc Networks. There are some differences between FANET and the traditional networks [7].

- Mobility: FANET nodes mobility is very high compared with traditional network nodes like MANET and VANET. Mobile Ad hoc and Vehicular Ad hoc nodes are near to ground but FANET nodes fly in the air.
- Topology: FANET network topology changes continuously comparing with MANET and VANET due to high mobility nature UAV nodes
- Distance: The distance between FANET nodes is very high comparing with MANET and VANET.
- Communication Range: In FANET, distance between nodes is very long, depends on that communication range is longer than in MANET and VANET.

The main aim of this paper is to propose a new efficient routing protocol for relaying data between the nodes effectively. Finally, it can satisfy specifications of UAV ad hoc networks.

2   FANET Routing Protocols

The major purpose of routing protocols is to find out proper path for data relaying in a network [1]. In FANET, due to high mobility nature of UAVs the topology will change continuously [4]. The development of efficient routing protocol for FANETs routing is a crucial task and is still under research. In traditional networks, most of the protocols developed for MANET and VANET routing but can't apply directly on FANETs because of UAVs have some unique features [5]. For FANETs routing, some existing protocols modified and some new protocols have been suggested [15]. FANET protocols are divided into 3 categories (Fig. 1).

Fig. 1. Classification of routing protocols

1. Table-Driven or Proactive Routing Protocols

This type of routing protocols stores all the routing information in a routing table. In FANET, there are different table-driven protocols which are not similar to each other. The routing protocol carries the latest information of nodes that is why there is no need to wait and select the path between sender and receiver [14]. When the bandwidth is not used effectively (a lot of traffic between nodes) then this will not be recommended for large communication networks. Other than that the protocol seems to be slow when topology is changed, or a failure occurs [19].

2. On-Demand or Reactive Routing Protocols

Reactive routing protocol can't maintain the topology information. They establish the route from source to destination when it is necessary to relay data packets and it takes more time to find out a path between source to destination. This is the reason; this type of protocols suffers from high delay and packet latency [1].

3. Hybrid Routing Protocols

Hybrid Routing Protocols (HRP) are used to overcome the limitations of proactive and reactive routing protocols as reactive protocols require more time to find routes and proactive protocols have control messages overhead [12]. In Hybrid routing a network is divided into different regions, proactive protocol is used for intra region routing whereas reactive protocol is used for inter region routing [13].

In previous works, used different topology based routing protocols and compared every routing protocol with each other to satisfy the requirements of FANETs and enhance the performance of FANET applications but still no protocol satisfied FANET requirements such as efficient bandwidth, QoS, link connection, end-to-end delay, etc.

Different routing techniques are available in ad hoc networks to relay the data packets strongly [16]. In proposed work, we used greedy forwarding routing technique to reduce delay of transmission and number of hops. We can apply this technique on various topology based routing protocols and compare all the results.

3 Comparison of Routing Protocols

In proactive routing protocols each node maintains complete address to destination but in reactive routing protocol it established a path when it is necessary to relay data [2]. Hybrid routing protocols combines both proactive and reactive routing protocols [13] (Table 1).

Table 1. Comparison of routing protocols

Protocol	Routing structure	Periodic updates	Control overhead	Route possession delay	Bandwidth specification
Proactive	Hierarchical and Flat	Yes, some may use conditional	High	Less delay	Need more bandwidth
Reactive	Mostly flat	Some nodes may require periodic beacons	Less	Delay is very high	Need less bandwidth
Hybrid	Flat	Yes	Average	Lower for Intra-zone; Higher for Inter-zone	Need average bandwidth

4 Functionality of FANET

4.1 Scalability Extension of Multi-UAV Operation

In multi-UAV system, every UAV must be converse with ground control station or satellite for operation otherwise can't operate. But in FANET, no need to communicate all UAVs with ground base. At least one UAV communicates with ground base and remaining all will communicate in ad hoc manner [4]. In single-UAV and multi-UAV systems using UAV-to-Structure-based communication links but in FANETs using UAV-to-UAV communication links and it can extend the scalability of operation. FANET can operate a mission with UAVs even no communication link with infrastructure. For example, In FANET, one UAV can't connect with ground base but no problem, with the help of another UAV we can operate (Fig. 2).

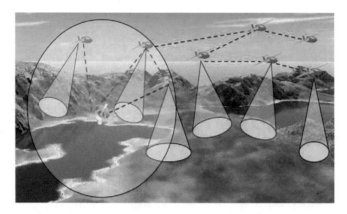

Fig. 2. Scalability extension of multi-UAV systems using FANET.

In Infrastructure-based networks, the communication between nodes also affects due to environment. For example, some obstacles like buildings, mountains or large walls may block the signals to transfer the data between ground base and UAVs especially in urban areas. But in FANET, to operate behind the obstacles also and it can increase the range of UAV operation [16].

4.2 Trustworthy Multi-UAV Communication

In multi-UAV system operations, continuously the topology changes due to high mobility nature of UAVs in a network. During the operation the conditions may change in a mission. In multi-UAV system, all UAVs connected to an infrastructure; there is no chance to establish an ad hoc network, as illustrated in Fig. 3a. In some cases, UAVs may discard their connection in multi-UAV systems due to some problems occurred in the network such as unknown persons entered into the network and take entire control in their hands, suddenly drastic changes occurred in whether, etc. [1]. But in FANET, if one UAV fails, gets no problem with that, we can operate a mission with other UAVs, as it is shown in Fig. 3b. This feature increases the reliability of the multi-UAV systems.

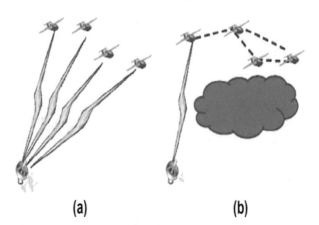

(a)   **(b)**

Fig. 3. Trustworthy communication network using FANET

4.3 UAV Swarms

UAV Swarm is nothing but group of small-UAV nodes co-ordinate together in order to produce a significant result. UAV swarms can handle complex missions with the help of co-ordination for every UAV node in the network. Due to limited payload of small UAVs it's very hard to maintain large and heavy communication hardware [5, 16]. In FANET, using very cheaper and lighter hardware to establish a network connection in small UAVs. UAV swarms can complete complex missions successfully with the help of FANET communication architecture.

4.4 FANET to Decrease Payload and Cost

In small-UAVs, payload is very low because small-UAV systems using lighter and cheaper hardware to establish a communication network. The payload problems occurs not only in small-UAVs, Higher Altitude UAVs also consider payload weights. Multi-UAV systems completely based on UAV-to-Infrastructure links, every UAV must communicate with ground station, it carries higher hardware to data communication [16]. However, FANET uses both UAV-to-Infrastructure link and UAV-to-UAV link, only some of UAVs communicate with ground station and remaining all UAVs can operate with FANET. FANET extends the performance of multi-UAV systems.

5 FANET Design Considerations

5.1 Adaptability

Due to high mobility nature of nodes the parameters can change during operation and always change their location. In FANET, the distance between two nodes is very large, can't be constant because of operational requirements [4]. Because of some reasons like technical problems or attacks operation problems, some UAV's fail in mission operation. That failure decreases number of UAV's. May be that failure mission requires additional UAV injection to maintain network system. Automatically, the injection and failure of UAVs may change FANET parameters. Environment conditions and flight plan updates are also change the FANET specifications. So that we can design FANET with an effective communication architecture and efficient routing protocol to adjust itself in crucial cases against failures.

5.2 Scalability

Comparing with single-UAV system, the swarm of UAVs can improve the performance of the system. FANET extends the scalability of operation [19]. For example, in search and rescue operations the usage of more number of UAVs gives effective performance than the single-UAVs.

5.3 Latency

In every network, latency is a big challenging issue. UAV ad hoc networks also faces this problem in different applications. Especially in military and civilian applications.

In real time applications, the data must be relayed within certain time. The latency is very high in the routing it raises more problems such as collision, collaboration, etc. MANET/VANET routing protocols not satisfied FANET latency requirements. So, we need to develop new routing protocols and algorithms [20, 21].

5.4 UAV Platform Constraints

The weight of hardware and space limitation is platform associated restrictions in FANET design. Weight of hardware means heavy payload. The space limitation is very important for mini UAVs [16].

5.5 Bandwidth Requirement

The main aim of FANET application is to transfer the data from collected location to ground station. For example, in rescue operations, relaying target data with strict delay bound from UAV to command control center, it requires high bandwidth [9].

It's a need to develop an efficient and effective FANET routing protocol to meet the need of bandwidth.

6 FANET Communication Protocols

6.1 Physical Layer

The physical layer is used for data relaying between devices. It contains data within the form of bits. In FANET physical layer, the data bits are modulated to waveforms and relay into the air by using an antenna structure [4]. Finally, the FANET physical layer determined radio propagation model and antenna structure.

The radio propagation model is used to transmit radio signals from source to destination. In FANET, antenna structure plays major role for efficient signal transmission.

6.2 MAC Layer

FANET extends from existing ad hoc networks. Compared with MANET and VANET, FANET has some common design considerations as well as some unique challenges. IEEE802.11 with Omni-directional antenna is most commonly used MAC layer in MANET that MAC layer is used in FANETs first application [17]. This MAC layer overcomes hidden and exposed node problems. FANET unique challenges like high mobility nature of nodes, large distance between nodes, link quality changes, packet latency, etc. directly affects the MAC layer designs. On the other hand, different MAC layers proposed for satisfying FANET requirements but still FANET need a vigorous MAC layer to conquer these challenging tasks.

6.3 Network Layer

The first FANET applications and communication architectures are represents with MANET protocols [4]. Due to FANET unique challenges MANET routing protocols can't satisfy FANET requirements. By the help of MANET routing protocols some routing protocols developed for FANETs. Data centric routing is one type of approach in FANET routing protocols but these algorithms are unexplored [16]. Peer-to-peer communication is a big challenge issue in FANET. For efficient data relay between UAVs in FANETs, it is necessary to develop a new routing protocol.

6.4   Transport Layer

In highly dynamic environments, mainly concentrates on communication architecture and relaying mechanism because these two plays an important role in the success of FANET design [17]. The main liability of a this protocol is as follows:

Consistency: The reliable data relay between source to destination with proper functionalities is a primary responsibility of FANET transport protocol.

Congestion control: If the FANET network is congested, the UAVs can't perform operation properly it increases packet latency and decrease the efficiency. For the avoidance of congestions and reliable FANET design it is necessary to use congestion control mechanism.

Flow control: The receiver overloaded with multiple senders information. In Multi-UAV system, flow of control is a big problem especially in heterogeneous networks.

6.5   Cross-Layer

The cross-layer is used for effective data communication in wired networks but those layered architectures are not effect for these type of networks such as wireless networks [3]. In wireless communication, cross-layered architecture is used to solve the showing problems. The cross-layer maintains information about state of first three layers. This information used to cover the certain requirements of FANET and also enhance performance of FANET especially in link quality.

7   Conclusion

Routing is a big challenging task in FANET due to their unique challenges such as high mobility, frequent changes in network topology, link quality changes, etc. In this survey, we classified different routing protocols and discussed functionality of FANET into different techniques. In continuation of this paper, we used greedy forwarding routing technique to reduce delay of transmission and number of hops. We can apply this technique on various topology based routing protocols and compare all the results.

Our major motivation is to present a problem within the flying ad-hoc network, and encourage researchers working in this area of specialization.

References

1. Yassein, M.B., Alhuda, N.: Flying Ad Hoc Networks: routing protocols, mobility models, issues. Int. J. Adv. Comput. Sci. Appl. (IJACSA) 7(6) (2016)

2. Lalar, S., Yadav, A.K.: Comparative study of routing protocols in MANET. Orient. J. Comput. Sci. Technol. **10**(1), 174–179 (2017). ISSN 0974-6471
3. Alshbatat, A.I., Dong, L.: Cross-layer design for mobile ad hoc unmanned aerial vehicle communication networks. 978-1-4244-6453-1/10. IEEE (2010)
4. Bekmezci, I., Sahingoz, O.K., Temel, S.: Flying Ad Hoc Networks (FANETs): a survey. Ad Hoc Netw. **11**, 1254–1270 (2013)
5. Nadeem, A., Turki Alghamdi, A.Y., Mehmood, A., Siddiqui, M.S.: A review and classification of Flying Ad Hoc Network (FANET) routing strategies. J. Basic Appl. Sci. Res. **8**(3), 1–8 (2018). ISSN 2090-4304
6. Bilal, R., Khan, B.M.: Analysis of mobility models and routing schemes for Flying Ad Hoc Networks (FANETs). Int. J. Appl. Eng. Res. **12**(12), 3263–3269 (2017). ISSN 0973-4562
7. Bekmezci, I., Senturk, E., Turker, T.: Security issues in Flying Ad Hoc Networks (FANETs). J. Aeronaut. Space Technol. **9**(2), 13–21 (2016)
8. Rosati, S., Kruzelecki, K., Heitz, G., Floreano, D., Rimoldi, B.: Dynamic routing for Flying Ad Hoc Networks. IEEE
9. Khan, M.A., Khan, I.U., Safi, A., Quershi, I.M.: Dynamic routing in Flying Ad Hoc Networks using topology-based routing protocols. Drones (2018). https://doi.org/10.3390/drones2030027. www.mdpi.com/Journal/Drones
10. Walia, E., Bhatia, V., Kaur, G.: Detection of malicious nodes in Flying Ad Hoc Networks (FANET). SSRG Int. J. Electron. Commun. Eng. (SSRG-IJECE) **5**(9) (2018)
11. Temel, S., Bekmezci, I.: On the performance of Flying Ad Hoc Networks (FANETs) utilizing near space High Altitude Platforms (HAPs), 978-1-4673-6396-9/13. IEEE (2013)
12. Leonov, A.V.: Modeling of bio-inspired algorithms AntHoc Net and BeeAd Hoc net for Flying Ad Hoc Networks (FANETs), 978-1-5090-4069-8/16. IEEE (2016)
13. Saranya, S., Chezian, R.M.: Comparision of proactive, reactive and hybrid routing protocol in manet. Int. J. Adv. Res. Comput. Commun. Eng. **5**(7) (2016). ISSN (online) 2278-1021, IJARCCE, ISO 3297:2007 Certified
14. Hong, J., Zhang, D.: TARCS: a topology change aware-based routing protocol choosing scheme of FANETs. Electronics **8**, 274 (2019). https://doi.org/10.3390/electronics8030274. www.mdpi.com/journal/electronics
15. Yang, H., Liu, Z.: An optimization routing protocol for FANETs. EURASIP J. Wirel. Commun. Netw. (2019). https://doi.org/10.1186/s13638-019-1442-0
16. Oubbati, O.S., Atiquzzaman, M., Lorenz, P., Tareque, M.H., Hossain, M.S.: Routing in lying Ad Hoc Networks: survey, constraints, and future challenge perspectives. Digit. Object Identifier (2019). https://doi.org/10.1109/access.2923840
17. Chriki, A., Touati, H., Snoussi, H., Kamoun, F.: FANET: communication, mobility models and security issues. Comput. Netw. (2019). https://doi.org/10.1016/j.comnet.2019.106877
18. Jahir, Y., Atiquzzaman, M., Refai, H., Paranjothi, A., LoPresti, P.G.: Routing protocols and architecture for disaster area network: a survey. Ad Hoc Netw. **82**, 1–14 (2018). https://doi.org/10.1016/j.adhoc.2018.08.005
19. Bhalaji, N.: Performance evaluation of flying wireless network with vanet routing protocol. J. ISMAC **1**(01), 56–71 (2019)
20. Yanmaz, E., Yahyanejad, S., Rinner, B., Hellwagner, H., Bettstetter, C.: Drone networks: communications, coordination, and sensing. Elsevier (2017). 1570-8705. www.elsevier.com/locate/adhoc. https://doi.org/10.1016/j.adhoc.2017.09.001
21. Zafar, W., Khan, B.M.: A reliable, delay bounded and less complex communication protocol for multicluster FANETs. Elsevier (2016). 2352-8648. www.elsevier.com/locate/dcan. https://doi.org/10.1016/j.dcan.2016.06.001

A Comparison Between Robust Image Encryption and Watermarking Methods for Digital Image Protection

Sreya Vemuri[(✉)] and Rejo Mathew

Department of Information Technology,
Mukesh Patel School of Technology Management and Engineering,
NMIMS (Deemed-to-be) University, Mumbai, India
sreyavemuri2l@gmail.com, rejo.mathew@nmims.edu

Abstract. Sharing digital image across networks is a common means of multimedia communication. However, these images are subject to all sorts of different attacks, like chosen plaintext attack, noise attack and, geometrical attacks. In order to resist these attacks, encryption and watermarking schemes are designed. This paper compares and analyzes the different encryption and watermarking methods.

Keywords: Image · Encryption · Watermarking · Noise · Robust · Chaotic mapping · Chosen plaintext attack · Geometrical attack · Salt and Pepper noise · Impulse Valued Noise

1 Introduction

The increase in digital image sharing across networks has led to the data becoming susceptible to different kinds of attacks, be it malicious or non-malicious. If the images sent over networks are copyrighted, the images should also be protected from copyright violations. Thus, for secure transmission and reception of digital images, the encryption and watermarking algorithms which are used for the images need to be robust. Robustness is the property of a digital image or its watermark, because of which it becomes resilient against different kinds of attacks. When images are sent through noisy channels, the pixel values change, and the image becomes almost unrecognizable on reception. This noise can be due to the filtering, scaling, cropping, rotation, compression and other processing methods used on the image. Malicious attackers can extract the secret key using the plaintext of the image, and thus compromise the security of the image data.

Using encryption, when images are sent over an insecure network, unauthorized users who do not have the secret key will not be able to recognize the image, and thus protecting the image's data. Many image encryption algorithms use a method called chaotic mapping, which have the characteristics of pseudo-randomness, unpredictability, ergodicity and sensitivity to initial values [1, 2]. Watermarking is the process of directly embedding extra information into the samples of a digital image, audio, or video signal, after which it is transmitted over the channel along with these

© Springer Nature Switzerland AG 2020
A. P. Pandian et al. (Eds.): ICCBI 2019, LNDECT 49, pp. 128–138, 2020.
https://doi.org/10.1007/978-3-030-43192-1_14

signals. This information can be either include block headers or time synchronization markers [3].

Many image encryption and watermarking algorithms are proposed which are robust against attacks like the chosen plaintext attack, noise attack and geometrical attacks. Thus, a need arises to compare these schemes and find which one is not only more robust against these attacks, but also more efficient. This paper compares and analyzes the robustness of different digital image encryption and watermarking algorithms which help in image data protection.

2 Attacks on Digital Images

2.1 Chosen Plaintext Attack

Many image encryption algorithms based on chaotic mapping are vulnerable to an attack known as the chosen-plaintext attack. In this the attacker can randomly choose some plaintexts and their corresponding ciphertexts which are encrypted by the same key. The differences between the plaintexts and the corresponding ciphertexts are also observed by an attacker and used for extracting the secret key. The attack is a cryptanalytic method which studies how differences between plaintext pairs affect the resultant differences [4]. This process is repeated, and the secret key is recovered. Hence it becomes easy to decrypt another ciphertext, which is encrypted by the same key. The chosen plaintext attack is also known as the differential attack. An effective image encryption scheme should be able to withstand the chosen plaintext attack [4].

2.2 Noise Attack

When images are sent over a network, they are subject to noise and interference. One of the types of noise attacks is called Gaussian noise. It is also known as electronic noise because it can come from electronic devices. This noise can be caused by naturally, by the thermal energy generated from vibrating atoms. Generally, Gaussian noise disturbs the grey values in digital images. The next type of noise is called Salt and Pepper noise, or Impulse Valued Noise or Data Drop Noise. The pixel values of the image are replaced by changed pixel values, with either the maximum pixel value which is 255 or minimum pixel value which is 0. The images themselves, or the watermarks embedded within them are prone to noise attacks, and they will appear distorted on reception. Image watermarking schemes need to be robust enough to withstand any kind of noise attacks [5].

2.3 Geometrical Attack

A geometric attack on an image watermark involves the arbitrary displacement of some or all its pixels. Geometric attacks involve distortion of the watermark contents through temporal or spatial transformations [3]. These attacks can arise from image processing techniques, for example scaling, printing, scanning, changing aspect ratio and cropping

[6]. Geometrical attacks can destroy the synchronization in a watermarked bit stream, which is important for most watermarking techniques [7].

3   Image Encryption Methods

Digital images tend to have high capacities of data and also high correlation between adjacent pixels. Therefore, traditional cryptographic methods like RSA, DES and AES are not suitable for encrypting images [8]. Over the years, many digital image encryption algorithms have been proposed. But one of the most popular methods of image encryption algorithms is known as the chaotic image encryption. Chaotic image encryption methods utilize the chaos hypothesis to accomplish cryptography [9]. Chaotic maps are of highly sensitive to their initial conditions and control parameters, the unpredictable in terms of the evolution of their orbits, and the simplicity of their implementation in hardware and software. This results in high rates of encryption [10].

3.1   Add-Image-Feature Algorithm

A solution to the chosen plaintext attack is called the Add-image-feature image encryption algorithm (Chaos-AIF), which is proposed by Deng and Zhong [1] In order to resist the attack, certain information representing the image to be encrypted is extracted. This information is related to the image's pixel value. Certain operations are performed on this information and the secret key, to generate the final initial value of chaos. This value is then utilized to generate the chaotic sequence (Fig. 1).

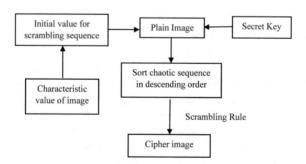

Fig. 1. Image encryption using Chaos-AIF algorithm

3.2   Association of Plaintext Information with Secret Key

This algorithm proposed by Ma et al. [11] is another solution to the chosen plaintext attack. In this algorithm, the information of the plain image is associated with the secret key. This algorithm utilizes chaotic sequences, which are generated by the initial secret key. These sequences are used to scramble the pixels of the image. Then, all the bits of the pixels are divided into different blocks. The number of blocks is set equal to a pixel's bit-width. Parity checking is then performed on each block (this process is called block parity checking). An offset of the initial state values in the secret key, utilizing all of the parity bits. The key after modification is then used for diffusion of pixels in the first round. Therefore, a change in one bit of the plain image will create a completely different secret key, thus avoiding the chosen plaintext attack.

The algorithm also has a method to protect the parameter related to the plaintext from noise and data loss. This method involves the duplication of the parameter to be protected multiple times and then inserted at the end of the ciphertext which is acquired from the first round of encryption. This process is called repetitive coding. Encryption of parameter takes place in the second round of encryption.

For this process, two generators of chaotic sequences are used. The initial state values of the first generator are related to the secret key, and the initial state values of the second generator are related to the plaintext information (Fig. 2).

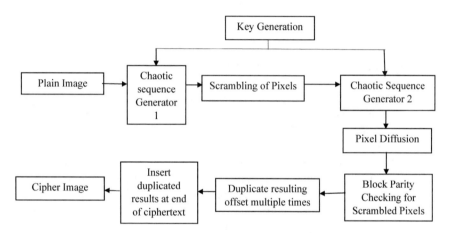

Fig. 2. Image encryption using association of plaintext information with secret key

3.3   Two-Dimensional Logistic Adjusted Sine Map Based Image Encryption Scheme

This algorithm, which is proposed by Hua et al. [12] proposes a two-dimensional logistic adjusted Sine map (2D-LASM). The outputs of this map are more difficult to predict compared to those of Logistic and Sine maps. These values are of random nature. They can change for each encryption round. This ensures that a completely different cipher image is formed, even if the plain image is encrypted multiple times with the same secret key.

Two processes, known as bit manipulation confusion and bit manipulation diffusion are performed on the plain image. Bit manipulation confusion is the process of randomly shuffling the positions of the pixels within the plain image. This shuffling is done according to the chaotic matrix generated by 2D-LASM. The property of confusion shows that the output distribution should depend on the secret key, and the property of diffusion indicates that the ciphertext should show high sensitivity towards a change in the plaintext (Fig. 3).

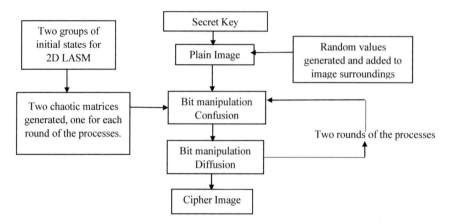

Fig. 3. Image encryption using 2D-LASM and LAS-IES

4   Image Watermarking Methods

Watermarking schemes which are of the transform domain use various transforms for the watermarking process. These transforms are - Discrete Wavelet Transform (DWT), and Singular Value Decomposition (SVD). DWT is a method which breaks down an image into sub-bands.

Using this transform, the image is separated into a lower resolution. The resulting sub-images are then labelled. LL contains the low frequency components and is the overall coarse shape of the image. LH represents the horizontal details of the image, HL represents the vertical details, and HH represents the diagonal details. LH, HL and HH represent the higher frequency information [3].

In SVD based watermarking techniques, the transform usually acts upon the image where watermark is to be embedded. This image is divided into many smaller blocks. Then, SVD is used to decompose the blocks to acquire the Singular Values. Watermark information is embedded using these values [13].

4.1   Optimal DCT - Psychovisual Threshold

This algorithm proposed by Ernawan et al. [15] is a digital image watermark embedding technique based on the psychovisual threshold. Discrete Cosine Transform (DCT), which is utilized in image processing and watermarking applications, is employed for this technique. DCT is utilized to transform the pixel values to the spatial frequencies, which represent the detailed level of information of the image [16].

For embedding the watermark, the image is divided into non-overlapping blocks, which are of 8 × 8 pixels. The selection of blocks to be embedded is based on the modified entropy values. The number of selected blocks is equal to the number of watermark pixels. The watermark bits are then embedded in the frequency coefficients of each selected block of the image. The x and y coordinates which are the location of the watermark embedding are then stored. These stored coordinates are then utilized to find the selected blocks during the process of watermark extraction. In case of an attack, even if the selected blocks can be found by the attacker, it will be difficult to identify the scrambled watermark, and thus protecting the watermark (Figs. 4 and 5).

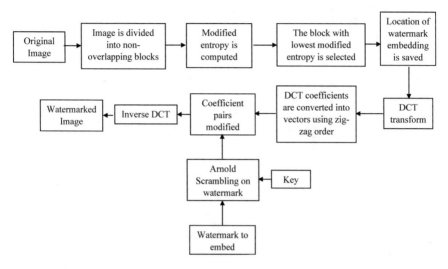

Fig. 4. Watermark insertion using optimal DCT psychovisual threshold

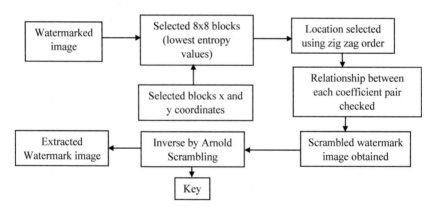

Fig. 5. Watermark extraction using optimal DCT psychovisual threshold

4.2 Logistic Maps and RSA Encryption

This watermarking scheme proposed by Liu et al. [13] uses DWT, SVD, Logistic Maps and RSA, and attempts to lower the consuming time, and improve the watermark's security, also increasing the robustness of the algorithm. Logistic Maps are one-dimensional chaotic maps which are used in security of digital communication, security of multimedia data, and other applications. RSA is an asymmetric public key encryption algorithm, which belongs to the domain of block ciphers (Figs. 6, 7, 8 and 9).

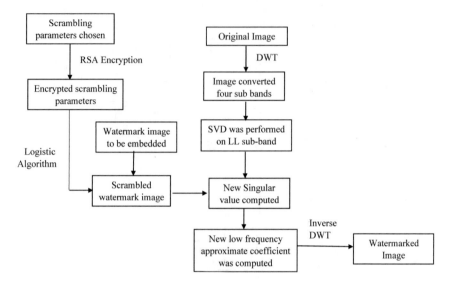

Fig. 6. Watermark insertion using Logistic Map and RSA encryption

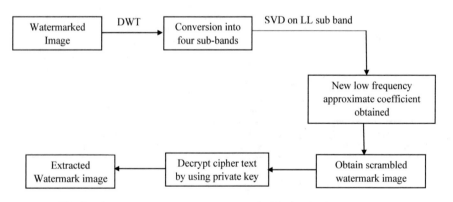

Fig. 7. Watermark extraction using Logistic Map and RSA encryption

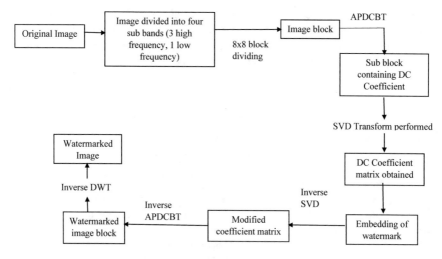

Fig. 8. Watermark insertion using DWT, APDCBT and SVD

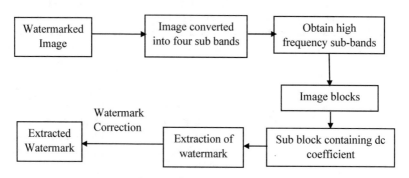

Fig. 9. Watermark extraction using DWT, APDCBT and SVD

5 Comparison and Analysis

5.1 Image Encryption Schemes

To assess the resistance of the image encryption schemes to the chosen plaintext attacks, tests known as the Number of Changing Pixel Rate (NPCR) and Unified Average Changed Intensity (UACI) tests are conducted. These tests are conducted on images where the difference between them is very miniscule, usually, there is only one-pixel difference between the images [17, 18].

All three methods of image encryption pass the NPCR and UACI tests and hence they all are resistant to the chosen plaintext attack to a certain degree. No specific drawbacks seem to be mentioned in the schemes proposed by Deng and Zhong [1] and

Hua et al. [12], but the scheme proposed by Ma et al. [11] needs a more efficient method for hiding plaintext related information in the ciphertext. However, the encryption scheme proposed by Ma et al. [11] is found to perform the best at defending against chosen plaintext attacks, despite its higher complexity compared to the rest of the methods (Table 1).

Table 1. A comparison of image encryption methods

Parameters	Chaos-AIF [1]	Association of plaintext information with secret key [11]	2D-LASM and LAS-IES [12]
Method of resistance to chosen plaintext attack	The characteristic value of the image is added to the plain image in order to resist the attack	No addition of any extra value to the image to resist the attack, but it uses block parity checking as an extra mechanism in the algorithm	Some random values are added to the surroundings of the plain image
Method of resistance to noise and data loss	This method uses asymmetric encryption and decryption to resist noise and data loss	This method uses repetitive coding for protection of the plaintext parameter	This method uses asymmetric confusion-diffusion operations, and it also uses the random values added to the plain image to reduce the impact of noise attacks
Complexity of the method	This method is simple and consists of only one round of encryption and decryption	Two rounds of encryption process are used, so even if security is increased, the complexity of the algorithm also increases	This method uses two rounds of confusion-diffusion operations, thus it slightly more complex
Key space	About $(10^{15})^4$	232 bits	324 bits

5.2   Image Watermarking Schemes

Out of the three watermarking schemes, the scheme by Zhou, Xiao, et al. [16] is found to be the most robust to different kind of attacks, except Gaussian Noise attacks. It also uses APDCBT Transform, which is a better alternative to the DCT transform used in the scheme by Ernawan et al. [14] (Table 2).

Table 2. A comparison of image watermarking methods

Parameters	Optimal DCT - psychovisual threshold [14]	Logistic and RSA encryption [13]	DWT, APDCBT and SVD [16]
Transforms used in the watermarking method	This method uses only DCT transform	Discrete Wavelet Transform (DWT), and Singular Value Decomposition (SVD) were performed	Discrete Wavelet Transform (DWT), Singular Value Decomposition (SVD), and an additional transform, APDCBT was used
Methods used to provide additional security	This method scrambles the watermark using Arnold Chaotic Maps, and, the watermark locations are saved	This method also scrambles the watermark, but it uses Logistic maps, along with additional encryption using RSA	This method does not scramble the watermark like the other two, but it uses hybrid transforms instead. It also uses APDCBT which provides more protection to the watermark
Drawbacks	The resistance of this scheme to rotational attacks is not satisfactory	This algorithm uses RSA as an encryption method, thus the time taken is more. Also, since it uses DWT, it has higher computational complexity	This scheme suffers from the false positive problem, and also has a higher computational complexity because it uses DWT

6   Conclusion

In this paper, three image encryption schemes and three image watermarking schemes were compared and analyzed. On comparing the image encryption schemes, since all of them passed the NPCR and UACI tests, they can resist the chosen plaintext attack. The method proposed by Ma et al. [11] was found to be the most effective in defending against the chosen plaintext attack. On comparing the image watermarking schemes, the method proposed by Zhou, Xiao et al. [16] was found to be the most effective in resisting noise and geometrical attacks. The protection of digital images when they are sent over networks is of deep concern, and further research can help in finding a more efficient method of image data protection.

References

1. Deng, Z., Zhong, S.: A digital image encryption algorithm based on chaotic mapping. J. Algorithms Comput. Technol. (2019)
2. Alvarez, G., Li, S.: Some basic cryptographic requirements for chaos-based cryptosystems. Int. J. Bifurcat. Chaos **16**(8), 2129–2151 (2006)

3. Tao, H., et al.: Robust image watermarking theories and techniques: a review. J. Appl. Res. Technology **12**(1), 122–138 (2014)
4. Li, C., et al.: Cryptanalysis of a chaotic image encryption algorithm based on information entropy. IEEE Access **6**, 75834–75842 (2018)
5. Boyat, A.K., Joshi, B.K.: A review paper: noise models in digital image processing. arXiv preprint arXiv:1505.03489 (2015)
6. Licks, V., Jordan, R.: Geometric attacks on image watermarking systems. IEEE Multimed. **3**, 68–78 (2005)
7. Dong, P., et al.: Digital watermarking robust to geometric distortions. IEEE Trans. Image Process. **14**(12), 2140–2150 (2005)
8. Luo, Y., et al.: A novel chaotic image encryption algorithm based on improved baker map and logistic map. Multimed. Tools Appl. 1–21 (2019)
9. Zahmoul, R., Ejbali, R., Zaied, M.: Image encryption based on new Beta chaotic maps. Opt. Lasers Eng. **96**, 39–49 (2017)
10. Li, C., et al.: An image encryption scheme based on chaotic tent map. Nonlinear Dyn. **87**(1), 127–133 (2017)
11. Ma, S., Zhang, Y., Yang, Z., Hu, J., Lei, X.: A new plaintext-related image encryption scheme based on chaotic sequence. IEEE Access **7**, 30344–30360 (2019)
12. Hua, Z., Zhou, Y.: Image encryption using 2D logistic-adjusted-sine map. Inf. Sci. **339**, 237–253 (2016)
13. Liu, Y., et al.: Secure and robust digital image watermarking scheme using logistic and RSA encryption. Expert Syst. Appl. **97**, 95–105 (2018)
14. Praveena, A., Smys, S.: Efficient cryptographic approach for data security in wireless sensor networks using MES VU. In: 2016 10th International Conference on Intelligent Systems and Control (ISCO), pp. 1–6. IEEE (2016)
15. Ernawan, F., Kabir, M.N.: A robust image watermarking technique with an optimal DCT-psychovisual threshold. IEEE Access **6**, 20464–20480 (2018)
16. Abu, N.A., Ernawan, F.: A novel psychovisual threshold on large DCT for image compression. Sci. World J. **2015**, 11 (2015)
17. Zhou, X., Zhang, H., Wang, C.: A robust image watermarking technique based on DWT, APDCBT, and SVD. Symmetry **10**(3), 77 (2018)
18. Wu, Y., Noonan, J.P., Agaian, S.: NPCR and UACI randomness tests for image encryption. Cyber J. Multidiscip. J. Sci. Technol. J. Sel. Areas Telecommun. (JSAT) **1**(2), 31–38 (2011)

Comparing Sentiment Analysis from Social Media Platforms – Insights and Implications

Lakshmi Prayaga[1(✉)], Chandra Prayaga[1], and Krishna Devulapalli[2]

[1] University of West Florida, Pensacola, USA
{lprayaga, cprayaga}@uwf.edu
[2] Indian Institute of Chemical Technology, Hyderabad, India
krishnad2@gmail.com

Abstract. The ubiquitous presence of social media is exerting its influence on several institutions in society by both positive and negative ways. In this context it behooves researchers to understand the validity of the influence of social media and what it means to the functionality of the institutions in society which include healthcare, politics, education, marriage, etc. This paper presents results and insights obtained from comparing sentiment analysis applied to Twitter and YouTube data on a set of topics. The focus of this study was to observe differences among sentiments expressed on different social media platforms. In other words, was there any influence generated by the social media platform on the individual's expression of sentiments. Additionally, we also developed an app to encourage citizen data scientists to search for a topic relevant to their area of interest and obtain sentiment analysis for that topic.

Keyword: Social media · Sentiment analysis

1 Introduction

Social media data generated from various platforms is creating a rich minefield of data. The existing live data collected using a plethora of tools offer an invaluable source for researchers, educational institutions, marketing companies, pharmaceutical companies, political parties, philanthropic organizations, and many other institutions. These institutions use the data to fine tune their message and direct it towards specific groups of audience to get the best ROI from their messages. Research has found that social media outlets have resulted in improving student learning outcomes in their ability to formulate cogent arguments, improve communication skills and improve their skills in collaborative activities [1, 2]. Citizen Science communities are found to thrive by communicating through social media [3]. Similarly, studies have indicated that social media platforms have helped patients to connect with other patients with similar medical conditions and get emotional support during stressful times, such as battling cancer and other health related problems [4–8]. Social media also provides a forum for consumers to post their likes and dislikes on specific products, which can have an impact on the marketability of those products.

Social media can also be used in a negative way that could again affect individuals and organizations. Several studies have indicated that social media can also have

© Springer Nature Switzerland AG 2020
A. P. Pandian et al. (Eds.): ICCBI 2019, LNDECT 49, pp. 139–144, 2020.
https://doi.org/10.1007/978-3-030-43192-1_15

adverse effects, a classic example is cyberbullying and its associated outcomes [9, 10]. Social media may also be used to invade individual privacy and use it as a leverage to access secure information of individuals and organizations [11–15].

Given the pros and cons of social media, it is interesting and helpful to analyze the posts on different social media and their associated sentiment values. This paper presents sentiment analysis results from using two social media platforms Twitter and YouTube. These results offer an opportunity to delve deeper and gain insights on what these tweets or sentiments mean and how they could help in shaping policy and decision making in multiple domains.

2 Methodology

This study used data from two popular social media platforms, Twitter and YouTube, on four broad topics across multiple domains. These topics are:

(1) Medicare
(2) Gun control
(3) Fitbit
(4) Apple watch

These topics were chosen specifically to get a broader representation of the population and a perspective of the insights gained in contrast to focusing on a single topic that could be limited in scope.

Research question
The primary research question for this study was to observe if the sentiment analysis results on each of these topics were similar for both platforms or were the numbers different for each platform.

Hypothesis
The hypothesis for this research was that a larger percentage of tweets exhibited a negative sentiment compared to sentiments expressed via comments on YouTube.

Following the norms used in selecting an optimal sample size, 1000 tweets and 1000 YouTube individual comments on each of the four topics were extracted. The software packages used were, RStudio to extract the Twitter and YouTube data, and Python to perform the Sentiment Analysis. Also, a shiny app was developed to encourage citizen data scientists to obtain data from Twitter and perform the analysis on the topics relevant to their interests.

3 Analysis

Sentiment analysis was performed on the text files containing Twitter and YouTube data, using R and Python packages. The following are the steps in the analysis:

1. The R package extracts the data into .CSV files. Shown below is the portion of the R script which accomplishes this, and a sample tweet from the extracted .CSV file:

```
RScript:
hc_tweets <- searchTwitter("health care", n = 1000, lang = "en")
hc_tweets
tweetsdf <- twListToDF(hc_tweets)
tweetsdf
write.csv (tweetsdf, file=
'c:/users/cprayaga/documents/healthcaretweets/hctweets.csv',
row.names = F)
    hctweets <-
read.csv("c:/users/cprayaga/documents/healthcaretweets/hctweets.cs
v")
    head(hctweets)
```

Sample Tweet:

"RT @DrJenGunter: And here is a good article on anti vaccine views of many Canadian chiropractors @CaulfieldTim knows more about this than I..."

2. The Python package provides an API for natural language processing (NLP) tasks such as sentiment analysis, noun phrase extraction, etc. (16). It reads the .CSV files and performs the sentiment analysis and groups the Twitter data into seven sentiment categories, strongly positive (SP), positive (P), weakly positive (WP), neutral (NEU), weakly negative (WN), negative (N), and strongly negative (SN), given with percentage numbers for each category. On the other hand, in the case of the YouTube data, the Python sentiment analysis package groups the comments into only two categories, positive (P) and negative (N), as shown in the sample data Table 1 given below.

Table 1. Sentiment analysis data

	Gun Control	Medicare	Fitbit	Applewatch
Twitter				
SP	1.4	1.0	5.4	5.0
P	4.7	2.9	12.3	12.8
WP	31.0	22.7	15.6	15.9
NEU	38.2	12.2	55.5	53.5
WN	13.9	60.1	2.4	7.0
N	7.2	1.1	7.3	4.2
SN	3.1	0.0	1.4	1.2
Twitter data converted				
P	56.2	32.7	61.1	60.5
N	43.3	67.3	38.9	39.2
YouTube				
P	28.6	23.9	21.8	40.3
N	71.4	76.1	78.2	59.7

3. In order to compare the data from the two sources, the Twitter data was further grouped, into the same two categories, positive and negative. This further grouping of the Twitter data was done manually, according to:

Positive + weakly positive + strongly positive = positive;

Negative + weakly negative + strongly negative = negative

Further, the percentage number for the neutral category in the Twitter data on each topic was equally divided into the positive and the negative categories. This is also shown in Table 1 below.

Figures 1, 2, 3 and 4 below show the comparative sentiment analysis results of the Twitter and YouTube data. In each figure, twitter and YouTube data are plotted in as stacked bars, each stacked bar showing positive and negative data in blue and orange respectively.

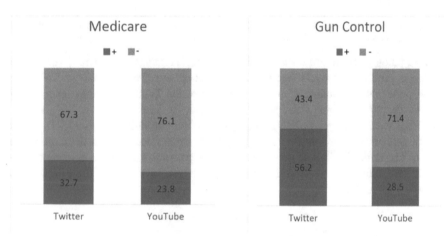

Fig. 1. Medicare   **Fig. 2.** Gun Control

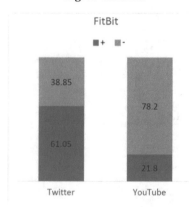
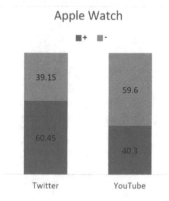

Fig. 3. Fitbit   **Fig. 4.** Apple Watch

4   Results

The comparative percentage numbers for positive (blue) and negative (orange) categories in Twitter and YouTube data, for each of the four topics listed above, are shown above, graphically, in Figs. 1, 2, 3 and 4.

In the case of Medicare, the data from Twitter and from YouTube seem to be in reasonable agreement, with more people giving a negative response than those giving a positive one. But in the other three topics, Gun Control, Apple Watch and Fitbit, there is a reversal of opinion between the two media. The YouTube responses are significantly more negative than the Twitter responses.

4.1   Insights and Implications from This Study

It was interesting to find that the sentiment analysis from Twitter and YouTube had almost flipped between the positive and negative scores. This raises some questions which are summarized below:

1. Is there a difference between the type of audience for each of the social media platforms? Does it mean that folks choosing Twitter are different from those choosing YouTube?
2. Does it also imply that folks using Twitter have a more negative perspective on any given topic? Are folks using Twitter more often to vent their emotions compared to other social media platforms?
3. Is there a validity problem when analyzing Tweets or YouTube posts? If the same people participated in formal polls would they express the same sentiments or would they give in to pollster bias?
4. Is it more evident that, given the pervasiveness of social media, pollsters, and policy makers, marketing agencies must take these informal media more seriously and maybe even weight them heavily in their decision - making process. Lack of such conscious focus on the relevance of social media might contribute to completely disregarding an entire segment of population.

Example screen shots from the shiny app are attached at the end of the paper.

5   Conclusion

While social media has a significant positive impact on several aspects related to daily life, it should also be observed that the implications of validity of social media data, security concerns, negative impacts of social media must be addressed to use this medium of communication effectively. Future work includes further refinement of the analysis based on geographic locations and other demographics where applicable, to obtain more focused insights of the population being represented.

References

1. Akgunduz, D., Akinoglu, O.: The impact of blended learning and social media-supported learning on the academic success and motivation of the students in science education. Educ. Sci. **42**, 69–90 (2017). https://doi.org/10.15390/EB.2017.6444
2. Ferna´ndez-Ferrer, M., Cano, E.: Int. J. Educ. Technol. High Educ. **13**, 22 (2016). https://doi.org/10.1186/s41239-016-0021-2
3. Daume, S., Galaz, V.: "Anyone know what species this Is?" – Twitter conversations as embryonic citizen science communities. PLOS ONE **11**, e0151387 (2016). https://doi.org/10.1371/journal.pone.0151387
4. Jones, E.C., Storksdieck, M., Rangel, M.L.: How social networks may influence cancer patients' situated identity and illness-related behaviors. Front. Public Health **6**, 240 (2018). https://doi.org/10.3389/fpubh.2018.00240
5. Mohammadzadeh, Z., Davoodi, S., Ghazisaeidi, M.: Online social networks - opportunities for empowering cancer patients. Asian Pac. J. Cancer Prev. **17**(3), 933–936 (2016)
6. Petit, A., Cambon, L.: BMC Public Health **16**, 552 (2016). https://doi.org/10.1186/s12889-016-3225-4
7. Kim, J.Y., Wineinger, N.E., Steinhubl, S.R.: J. Med. Internet Res. **18**(6), e116 (2016). https://doi.org/10.2196/jmir.5429
8. Agha Zade, H., Habibi, L., Arabtani, T.R., Sarani, E.M., Farpour, H.R.: Functions of social networks in a community of cancer patients: the case of Instagram. Int. J. Netw. Commun. **7**(4), 71–78 (2017). https://doi.org/10.5923/j.ijnc.20170704.01
9. Garett, R., Lord, L.R., Young, S.D.: Associations between social media and cyberbullying: a review of the literature. mHealth **2**, 46 (2016). https://doi.org/10.21037/mhealth.2016.12.01
10. Dennehy, R., Cronin, M., Arensman, E.: Involving young people in cyberbullying research: the implementation and evaluation of a rights - based approach. Health Expect. Int. J. Public Particip. Health Care Health Policy **22**(1), 54–64 (2018)
11. Gupta, A., Dhami, A.: J. Direct Data Digit. Mark. Pract. **17**, 43 (2015). https://doi.org/10.1057/dddmp.2015.32
12. Roy, R., Gupta, N.: Digital capitalism and surveillance on social networking sites: a study of digital labour, security and privacy for social media users. In: Kar, A., Sinha, S., Gupta, M. (eds.) Digital India (2018)
13. Advances in theory and practice of emerging markets. Springer, Cham [6]. Kandias, M., Galbogini, K., Mitrou, L., Gritzalis, D.: Insiders trapped in the mirror reveal themselves in social media. In: Lopez (2013)
14. Lopez, J., Huang, X., Sandhu, R.(eds.): Network and System Security. NSS 2013. LNCS, vol. 7873. Springer, Heidelberg (2013)
15. Sarikakis, K., Winter, L.: Social media users' legal consciousness about privacy. Soc. Media + Soc. (2017). https://doi.org/10.1177/2056305117695325

A Smart Farm – An Introduction to IoT for Generation Z

Lakshmi Prayaga(✉), Chandra Prayaga, Aaron Wade, and Andrew Hart

University of West Florida, Pensacola, USA
{lprayaga, cprayaga, awadel}@uwf.edu,
ach64@students.uwf.edu

Abstract. This paper describes an Internet of Things (IoT) project within the context of an inexpensive table top model of a smart farm. The purpose of the project is to use advances in technology including Bluetooth, the Internet and micro sensors to design an inexpensive model of a smart farm. The project brings ideas related to IoT and smart cities within the reach of a common man. Advances in technology and simplicity of high level programming languages make it also possible for anyone who is interested to easily build a smart connected system that can be communicated with, using inexpensive components and a simple program. This project demonstrates one such novel application of a smart farm with the use of a microbit, a humidity sensor and WiFi to monitor moisture levels of plants and depending on a set threshold the program sends an appropriate message to the owner of the plants. The use of a very simple microcontroller (the microbit) makes the project accessible to students at college and even school level. This technology is not only scalable, but it is also expandable to other domains, such as inventory management, power systems, etc. This project has broad applicability and fits in with the recent trends in resource planning and allocations.

Keywords: Internet of Things · Automation · Smart device · Smart farm

1 Introduction

The idea of smart cities is no longer an abstract concept, especially for the current generation Z, who live in a connected world. [1–4] and value a personalized immersive learning environment [5]. The researchers in this project used the notion of smart cities [6] as a context to design projects related to Internet of Things (IoT). IoT is defined [7] as "A global infrastructure for the information society, enabling advanced services by interconnecting physical and virtual things based on existing and evolving interoperable information and communication technologies." IoT devices can take data or information and use it for the purpose of creating a smarter system. This can be done by combining individual technologies into a single device or system, or, as the definition indicates, into a larger interconnected system. The purpose of both is to utilize the information gathered by a system, for a multitude of purposes. Advances in technology [8] including Bluetooth and relatively inexpensive microsensors make it possible to

© Springer Nature Switzerland AG 2020
A. P. Pandian et al. (Eds.): ICCBI 2019, LNDECT 49, pp. 145–152, 2020.
https://doi.org/10.1007/978-3-030-43192-1_16

make the esoteric ideas of IoT and smart cities common place. Particularly projects such as model smart farms described here are designed specifically to appeal to generation z students who can build the projects with a minimal learning curve.

The objective of this project was to create a simple and cost-effective plant monitoring system that sends data to a server for a user to read the data and make decisions on the use of resources. The hardware and software used were chosen to be cost effective and simple to monitor a single or multiple plant and receive notifications when the plants needed attention. An additional outcome of this project is that it can be used as a tool to educate students on the concepts of IoT and the efficient use of resources such as water and electricity to design a sustainable environment [9, 10].

2 Technologies Used

2.1 Hardware

The microcontroller used is the microbit, which was chosen for simplicity and on board features. When it is connected to a breadboard, through an adapter for its edge connector, it provides the user with inputs to connect sensors. This allows us to use a soil sensor to measure humidity. The microbit also has built in Bluetooth, which allows it to be connected to other nearby Bluetooth devices. By using a microcontroller with these functionalities, the system we create can be easier to understand, not requiring knowledge of multiple devices and microcontrollers.

The sensor used is from Adafruit, and it returns the ambient temperature and the humidity of the soil over I2C [11]. An all in one sensor was chosen to keep the cost low and make the system easy to wire; only four wires are required to connect the sensor to the microbit. The sensor works on capacitance instead of resistance which increases the resolution and longevity of the sensor.

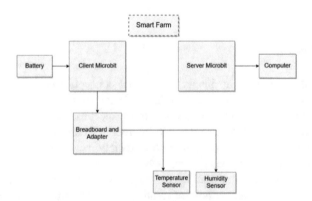

Fig. 1. Block diagram representing the hardware used. A client Microbit, powered by a battery, is connected by an adapter to a breadboard. The humidity and temperature sensor is then connected to the breadboard to collect data. A server Microbit is connected to a computer to receive the information sent by the client.

In order to send data to an online server, a second, "server," microbit is used. The data is sent from the first, or "client," microbit, through the built-in proprietary Bluetooth. The temperature and humidity are then monitored constantly and the data is sent to a server on the internet [12] (Fig. 1).

2.2   Software

The microbit uses a proprietary website (makecode.microbit.org) for creating code that uses code blocks to represent real code. This allows students to understand the code they write without requiring instruction on languages and syntax. The actual code is written in JavaScript, and this can be accessed in the online Microbit IDE. If added

```
1   let humidity = 0
2   let id = ""
3   let i2cAddress = 0
4   function ReadHumidity() {
5       pins.i2cWriteNumber(
6           i2cAddress,
7           3856,
8           NumberFormat.UInt16BE,
9           false
10      )
11      basic.pause(5)
12      humidity = pins.i2cReadNumber(i2cAddress, NumberFormat.UInt16BE, fals
13      basic.pause(100)
14  }
15  function ReadTemperature2() {
16      pins.i2cWriteNumber(
17          i2cAddress,
18          4,
19          NumberFormat.UInt16BE,
20          false
21      )
22      basic.pause(5)
23      temp = pins.i2cReadNumber(i2cAddress, NumberFormat.UInt32BE, false)
24      temp = temp & 0x3FFFFFFF
25      temp = temp * 0.00001525878
26      basic.pause(100)
27  }
28  let temp = 0
29  i2cAddress = 54
30  id = "12349876"
31  radio.setGroup(1)
32  humidity = 0
33  temp = 0
34  basic.forever(function () {
35      ReadHumidity2()
36      ReadTemperature2()
37      radio.sendString("" + id + "_" + "t" + "_" + temp)
38      radio.sendString("" + id + "_" + "h" + "_" + humidity)
39      if ((temp > 37.7 || temp < 0) || (humidity < 500)) {
40          basic.showIcon(IconNames.Sad)
41      } else {
42          basic.showIcon(IconNames.Happy)
43      }
44      basic.pause(2000)
45  })
```

Fig. 2. Code used on the client Microbit. This code declares two functions to control the temperature and humidity functions of the sensor. Then, it declares variables tlo use in the forever function. The forever function then works on a loop to call the sensor functions, send the input as a string, and display a message to show if it is above or below a threshold.

functionality is needed, such as with a third-party sensor, it is possible to code directly in JavaScript and create a unique extension.

In order to use the sensor with the Microbit, a custom function was created to program the Microbit to read humidity and temperature. These are separate functions, written in JavaScript. The code programs the Microbit to look for input from the sensor in a specific location, one that is mapped to a pin on the adapter, which is connected to the sensor through the breadboard. The value of humidity and temperature are then returned by the sensor and stored in variables. The value of temperature requires a bit mask and value conversion to be stored in *Celsius/Fahrenheit/Kelvin*. Once the values are stored, they can be used by the Microbit program (Fig. 2).

The software on the microbit first declares variables for humidity and temperature, establishes where data will be read from, and sets a radio group and id to send the information to the server microbit. The previously explained humidity and temperature functions are also declared. A single if else statement is used to read the temperature and the humidity levels. Once the humidity gets too low or the temperature is below freezing or above one hundred degrees, an alert is displayed on the microbit. All of the data readings from the sensor are sent by the client microbit to the server microbit [13].

2.3 Server

In order to display the data, it is sent from the client microbit to the server micro bit over Bluetooth. The server microbit uses code that specifies both what data is being received and what client microbit it is connecting to. Once the data is received as a string, it is sent through a computer's serial ports (Fig. 3).

```
 1  radio.onReceivedString(function (data) {
 2      serial.writeLine("" + data)
 3  })
 4  let id = 0
 5  let group = 1
 6  radio.setGroup(group)
 7  basic.forever(function () {
 8      basic.showNumber(group)
 9  })
10
```

Fig. 3. Code used on the server MicroBit. A radio function is declared that when data is passed to the server, it will write the data to the serial port of the computer. The bluetooth group of the server is then set. In the forver function, it displays the group of microbitis it receives data from, to receive data from only the correct clients.

After sending the code through the serial port, it is sent to Node RED flows [14]. Node RED is a flow tool that helps us visually program hardware to a web based UI. The flows that were used first decode and separate the data that is received. Once there, it is then sent to a web based UI. This UI will display the data received by the Microbits so that it can be read by a user [15].

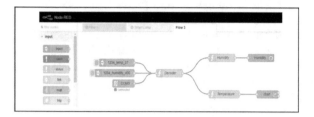

Fig. 4. Node Red Drag and Drop Interface for Hardware

3 Materials and Methods

Summer Camp Testing. This activity was tested in one-week summer camps with fifty middle school students. Students built their devices with Microbits, sensors, jumpers and breadboards. Once they built their devices, they then build the code in MakeCode.

The goal was to a. introduce students to the idea of Iot, and b. develop computational thinking skills including algorithmic and logical thinking practices. To achieve these outcomes the smart farm activity was done in stages, a. introduce the idea of IoT and elicit everyday examples of an IoT application, b. introduce conditional if-then-else and else if statements, use built in function calls to connect the devices, and finally add server commands, to send data as a string.

A smart farm application was designed for this project. The design included testing moisture and humidity levels in the soil of a plant. Depending on the levels of humidity appropriate messages were sent to the owner of the plant, in this case it was the student. Conditional statements such as

If moisture < .30
message = "I am thirsty, I need water" else
message = "I am happy"

This component addressed the systematic algorithmic and computational thinking aspects of the project. Students used the appropriate sensors and the code segments to execute the logic of the program.

4 Results and Discussion

Few students (approximately 25%) knew what the exact meaning of the word IoT. However, when the discussion on what IoT stands for was introduced several (almost 80%) students were able to relate to it and offered examples of IoT applications such as smart parking lots, smart refrigerators, smart video bells, smart cars etc.

In terms of evaluating comprehension of algorithmic practices and computational thinking, more than 90% of the students were able to use conditional statements in their projects and used them to test what happens if the test condition for humidity level increases or decreases in the humidity. They played with those values and observed the responses from the program. Students also exchanged their code with other students and observed the different results. Students also were interested in observing what happens when you add other sensors to detect attributes such as light, nutrients etc. This demonstrates that given a relevant context it is evident that the younger generation can become contributors to the growth of technology vs. simply being consumers of available technology.

Platform and Software Discussion. The microbit platform has benefits and drawback. In terms of hardware, they are extremely simple Microcontrollers that lend themselves to teaching the basic uses and applications of a microcontroller. Connecting the microbits to other devices is also simple with a breadboard adapter, and allows for a wide range of applications. The built-in features with the microbit are also incredibly useful, allowing communication between microbits, and other devices such as a computer. The MakeCode environment is intuitive and easy to use when teaching the concepts of computer science and information technology to new students. Rather than focusing on a language, it allows students to focus on algorithm design which is a very important computational skill.

However, the next part of integrating the sensors and microbit is not a simple exercise. The sensor used is built for Arduino, and the Arduino language. There are no specific sensors for the microbit. In order for the sensor to be used with microbit, the library functions have to be converted to Javascript. It also requires an entire function to be built and added to the microbit extension list, which must be done manually on every computer used to program the microbits. This is a complex task to expect middle and high school students to perform. These steps were automated for students in this age group.

The Node RED application was used to configure the server. This part of the project was also a complex task for younger audience. For this reason, all server-side work was also preconfigured for students to work in a plug and play interface (Fig. 5).

Fig. 5. A dashboard visualization of the smart farm

The device itself worked well. Data was read from our input device, and sent to the server, where a user could see the results. Multiple pages could be brought up to display the results from multiple devices as shown in Fig. 4.

5 Conclusions

This system works well if it is taught in a step by step manner using some guidelines discussed in the paper, specifically abstracting some of the concepts related to the server. The MakeCode and microbit platforms provides a simple medium of instruction to the elementary and middle school students.

Future work will include displaying results on a single page, rather than displaying it on multiple pages so as to mimic a dashboard. Also, the use of multiple input devices with a single controller could increase the amount of data received about an environment.

Future development. We plan to develop a complete miniaturized smart city with several functional units such as smart parking lots, smart healthcare facilities, smart parks etc. using the notion of IoT. The main purposes of this development is a. to build model cities and b. provide a learning environment for students to be prepared to function in and contribute to a connected world.

Compliance with Ethical Standards
All author states that there is no conflict of interest.
Humans/Animals are not involved in this work
We used our own data.
Informed consent was obtained from all individual participants included in the study.

References

1. Siddiqui, N., Mishra, S.: Exploring the newbies: a comparative study of Gen Y and Gen Z. In: People Management in 21st Century Practices and Challenges, pp. 205–225, August 2018
2. Schwieger, D., Ladwig, C.: Reaching and retaining the next generation: adapting to the expectations of Gen Z in the classroom. Inf. Syst. Educ. J. **16**(3), 45 (2018). https://files.eric. ed.gov/fulltext/EJ1179303.pdf
3. Adobe: Gen Z in the Classroom: Creating the Future. Adobe Education Creativity Study (2016). http://www.adobeeducate.com/genz. Accessed 10 June 2017
4. Merriman, M., Valerio, D.: One tough customer: how Gen Z is challenging the competitive landscape and redefining omnichannel. Ernst & Young Report (2016). http://www.ey.com/ Publication/vwLUAssets/EY-one-tough-customer/$FILE/EY-onetough-customer.pdf. Accessed 28 Apr 2017
5. Videnovik, M., Zdravevski, E., Lameski, P., Trajkovik, V.: The BBC micro:bit in the international classroom: learning experiences and first impressions. https://ieeexplore.ieee. org/document/8424786
6. Ismagilova, E., Hughes, L., Dwivedi, Y.K., et al.: Smart cities: advances in research - an information systems perspective. Int. J. Inf. Manage. **47**, 88–100 (2019)

 7. ITU-Telecom. https://www.itu.int/ITU-T/recommendations/rec.aspx?rec=y.2060
 8. Kumar, R.P., Smys, S.: A novel report on architecture, protocols and applications in Internet of Things (IoT). In: 2018 2nd International Conference on Inventive Systems and Control (ICISC), pp. 1156–1161 (2018)
 9. Walter, A., Finger, R., Huber, R., Buchmann, N.: Opinion: smart farming is key to developing sustainable agriculture. Proc. Nat. Acad. Sci. USA **114**(24), 6148–6150 (2017). https://www.ncbi.nlm.nih.gov/pmc/articles/PMC5474773/
10. Kodali, R., Nimmanapalli, K., Jyothirmay, S.: Micro:bit Based Irrigation mintoring. https://ieeexplore.ieee.org/abstract/document/8777721
11. Trivedi, D., Khade, A., Jain, K., Jadhav, R.: SPI to I2C protocol conversion using verilog. https://ieeexplore.ieee.org/document/8697415
12. Jindarat, S., Wuttidittachotti, P.: Smart farm monitoring using Raspberry Pi and Arduino (April 2015). https://ieeexplore.ieee.org/document/7219582
13. Sharma, P., Padole, D.V.: Design and implementation soil analyser using IoT (March 2017). https://ieeexplore.ieee.org/document/8697415
14. Lekic, M., Garasevic, G., IoT sensor integration to Node-RED platform. https://ieeexplore.ieee.org/document/8345544
15. Nerukar, U.: Web User Interface Design, Forgotten Lessons. https://ieeexplore.ieee.org/document/8345544

Algorithms Used for Scene Perception in Driverless Cars

Antriksh Tiwari[✉] and Rejo Mathew

Mukesh Patel School of Technology Management and Engineering,
Mumbai, India
antrikshtiwari157@gmail.com, rejo.mathew@nmims.edu

Abstract. Research on developing Driverless/Autonomous cars have been going on from a long time now, still there are many functionalities on which a human cannot be completely replaced, for example recognizing a scene and reacting to it properly, handling different real-time situations, going according to other cars with drivers inside them, etc. This paper will highlight the issues and their solutions related to the domain of scene perception. The issues related to the same were gridding, speed efficiency, and complex scene recognition. Solutions that should be implemented to solve the related problems are Pyramid Scene Parsing Network (PSP Net), Convolutional Neural Network (CNN), Hybrid Dilated Convolution (HDC) and SegNET and Efficient Neural network. Each algorithm has its own ownership and unique way of identifying the scene, therefore they can be compared along the parameters of scene perception. These solutions have been evolved with time, implementing these technical aspects into the current cars can help replace a human with a driverless car with the highest possible precision.

Keywords: Scene segmentation · Max pooling · Driverless cars · Convolution · Datasets

1 Introduction

Scene Perception is the visible understanding of a surrounding as considered by using an observer at any given time. It consists of not only the appreciation of the object but adds some meaning to it with respect to the location and environment. To encounter the problems faced in scene perception, the first step is to define a dataset that includes different categories of objects and scenes. The new ADE20K [1, 7] is used for the same. So, a large number of labels and vast distributions of scenes come into existence. Before relying on any database, it should be properly captured for all the possible conditions. A flaw in such a dataset which is so widely used can lead to the failure of many technologies. For scene perception, the first part is scene segmentation Object identification [10].

Currently, the scenario is that to fill in the city information into the car via the LIDAR [3, 4] sensor normally mounted on the top of the car, a person has to initially drive it through the roads of the city and only then the car will be able to succeed in after operations. It is not possible or feasible to drive a car on every road to fill in the

© Springer Nature Switzerland AG 2020
A. P. Pandian et al. (Eds.): ICCBI 2019, LNDECT 49, pp. 153–161, 2020.
https://doi.org/10.1007/978-3-030-43192-1_17

information about the surroundings so that the next time the car goes on the same road it recognizes it. Therefore, Scene Perception can play a vital role to develop the technology of driverless cars to its ultimate evolution. Studies and researches have been going on to accomplish the goal of truly driverless cars. Completely Driverless cars are on the edge to enter the global market. It's better for researchers to solve these problems as soon as possible so that they won't create a hindrance in the coming years once they are launched [9].

2   Problems Encountered in Scene Segmentation

2.1   Mixed Matched Relationship in Images

Mixed Matched relationship was a major problem, initially noticed in FCN [1]. In scene perception it is important to understand complex scenes and that it is capable of differentiating and categorizing between complex scene samples. Most of the Technologies use the nearby environment to predict and identify an object, but this was not the optimal way for the same. Nowadays the most widely used dataset is ADE20K [1] which tests any algorithm for scene segmentation. Even humans get confused between so closely placed data items so machine technology could easily be diverted from their goal, these datasets are made so closely linked and confusing to test the technology to its extreme limits. Also, the problems which are encountered in scene perception is that several important objects are of different sizes [1] and if they are small and aren't recognized, it could create a problem. Anything like a signboard means very important for a car could be ignored because of its small size so algorithms should be planned accordingly to encounter such problems but FCN was not at par in every aspect of the scene segmentation.

The goal of inventing FCN was to produce semantic segmentation. FCN outputs the image of a similar size and resembles the original input but each pixel is categorized according to the identification algorithm. Fully convolutional means that each layer is a convolutional layer. There is no connected layer in between. So, only the convoluted layers are used for segmentation and identification purposes [1].

Upsampling is done once all the layers are convoluted. Deconvolution is done for upsampling purpose to get better knowledge of each pixel. In upsampling the categorization takes place because it is reverse of convolution, therefore all the layers are separated in the output [15].

Downsampling is the next step in which all the layers are then downsampled but the problem which researchers were facing in this step was that a lot of spatial information was getting lost in the end result [14].

At last, all the layers which were divided are combined to get an output that resembles the original image and is of the same size but due to mixed match relationship, it needs to be upgraded (Fig. 1).

Fig. 1. Explaining the process of FCN briefly

2.2 Gridding in Samples

Gridding is a problem which is faced when using dilated convolution. The output image ends up as a grid-like pattern in the generated feature maps. The root cause of this is when the input image has a higher frequency than the dilation rate of the system. These grid patterns hinder the process of scene segmentation because of the information loss in the image.

Dilated convolution is a scene segmentation technique in which the input image is first passed through CNN for getting the layer-wise information and then dilated convolution is applied onto it. Dilated Convolution is just a normal convolution applied to the input image with predefined gaps. With this definition, given the input is a 2D image, dilation rate $k = 1$ is normal convolution and $k = 2$ means skipping one pixel per input and $k = 4$ means skipping 3 pixels. As the value of k changes, the convolution output becomes more scattered [2].

But due to the limitation of non-participation of all the pixels, It was not successful.

2.3 Speed Efficiency of Perception

In Driverless cars the efficiency of recognizing a scene is important as it is the actual basis of Driving the car autonomously. The decisions further taken will revolve around how fast is the scene recognized. The correctness should not be compromised when taking speed into account as it will lead to errors that are extremely sensitive in this case. It is the point of life and death. Therefore Speed matters but only when correctness is not compromised.

3 Counter-Measure Algorithms

3.1 Pyramid Based Pooling Method

The first step is to pass the input through Convolutional Neural Network (CNN) to get a featured map of the scene. This supports PSPNet as the output produced can be used for further dilated convolution [1].

Dilated convolution is used to widen the receptive field without reducing the feature map resolution which was the output of CNN. PSPNet uses Pre-trained ResNet models with dilated convolution which produces an output which is 1/8 of the size of the original input image.

Pyramid Pooling Module comes into the process once the image has gone through dilated convolution because the pooling module includes upsampling which will increase the size of the image to counter that dilated convolution is necessary [13].

Later the convoluted layers are concatenated to give an output where each pixel of the image belongs to an identified category (Fig. 2).

Fig. 2. PSPNET process

The problem which was first encountered by FCN is taken care of by the Pyramid Pooling Module, it is a mechanism followed by Pyramid Scene Parsing Net (PSPNet). It was Introduce to give a wider perspective to scene perception theory. One thing to note is that PSPNet is based on the ResNet framework which is more flexible [1]. The neural network helps to increase real-time object understanding.

In the working of PSPNet, DUC (Dilated Up-sampled Convolution) boosts the performance by upsampling the layers which are segmented. It divides the given sample into maximum possible layers and then starts to identify the scene, layer by layer and pixel by pixel.

Now in Driverless cars, the processing of information should be very fast because delay is not at all acceptable in such vehicles, for that the algorithm which are used should be light and easily processable. The average prediction time for PSPNet is very less. It is better than FCN in categorizing and dividing the object in complex datasets and overcomes the problems faced in it.

Since PSPNet uses the property of layering with upsampling it is easier to identify a scene as every layer gets identified separately. It breaks the sample by the property of the individual objects. Small objects also get into the account when layers are being formed to overcome the challenges faced by the FCN algorithm. The same is the case with larger objects like a mountain or a hill. The pixel accuracy of PSPNet is higher in comparison with FCN [1].

3.2   Hybrid Dilated Convolution (HDC)

Dilated convolution is explained as normal convolution with defined gaps known as dilation rates. Different dilation rate is used for every layer in HDC [2].

For 2-D dilated convolution, holes/zeros are inserted between each pixel in the convolutional function. The issue of gridding is not solved by dilated convolution; therefore, the research went on and came up with Hybrid Dilated Convolution (HDC). This solves the problem of the information loss we were having previously with dilated convolution [2].

In HDC the sample is scaled up and down but only square samples are considered so that the information loss does not occur for the sample. Basically, instead of using the same dilation rate for all layers after the downsampling occurs, it uses a different dilation rate for each layer.

One thing to note here is the top layer is not disturbed in any of the steps, only the layers below it gets dilated so that the top layer can get a broader range of pixels. HDC is also based on CNN for the further recognition and categorization of the scene as well as objects. It feeds information to CNN in the form of a dilated layer which makes the job easier for CNN to segment the scene. The information sustenance capacity of this algorithm is very large due to the fact that it takes different dilation rates (Fig. 3).

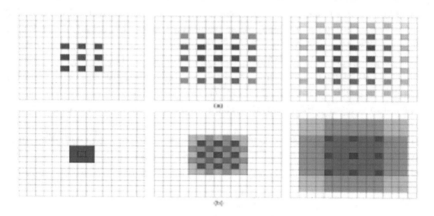

Fig. 3. Working of HDC

3.3 SegNet

It constitutes 2 major networks, namely Encoder network, Decoder network and a final pixel-wise classification layer [5].

The encoder network is used for object classification purposes. The encoder network also reduces the number of parameters used for identification. By using convolution with a filter bank, the layers get fully independent to retain high-resolution feature maps. After this step, max-pooling [5] is performed and the resultant output is sub-sampled by a factor of 2 [11].

Later the input layer is batch normalized by adjusting and scaling the parameters. The increasing boundary details lead to loss of boundary-information; therefore, it is necessary to capture the boundary information before sub-sampling is performed. Max-pooling indices [5] are used to store the information in an optimal way.

The decoder network upsample the input using the memorized max-pooling indices and the output feature map from the encoder network. Output is then convoluted with a trainable decoder filter bank. An output feature map is created by a decoder network.

This map is then fed to a soft-max classifier to identify each pixel independently. These three steps are combined together to construct a SegNet for scene perception purposes. SegNet is used to focus on boundary delineation, it works very well with moving objects.

3.4   E-Net (Efficient Neural Network)

The process of ENet is divided into several stages, with convolution, each layer is separated and then merged back with an element-wise addition. A max-pooling layer is added to the main branch in the case when the process is downsampling with batch normalization. The type of convolution used can either be normal, dilated or full convolution. To regularize we use Spatial dropout [7].

Between each convolutional layer, Batch Normalization is added to normalize the feature map. While going through the decoder the max-pooling which occurs in case of encoding is replaced by max un-pooling [7].

The softmax layer at the end of the process is used for segmentation of each pixel in a layer.

The whole idea of encoder-decoder network is to ease out the upsampling and the downsampling process for convolution. The functions are divided into encoder and decoder respectively for avoiding any confusion or failiure due to load on one function [11] (Table 1).

E-Net is also efficient in terms of speed because it counts as a major characteristic if it is implemented in driverless cars (Fig. 4).

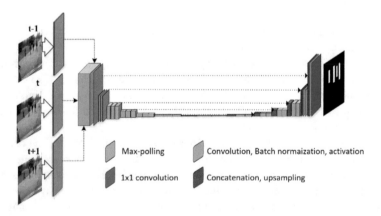

Fig. 4. Encoder decoder network [7]

Table 1. Comparison of algorithms for scene perception in driverless cars

Parameters	Solution			
	PSPNet	HDC	SegNet	ENet
Framework	Based on Resnet framework which is flexible	Based on Dilated convolution	Based on encoder-decoder framework	Based on encoder-decoder framework
Architecture	Upsampling and Downsampling	Upsampling and Downsampling with dilation rates and square samples	Encoder-Decoder	Cov-net Architecture with Encoder-Decoder
Process between convoluted layers	No connected layer	Value of next dilation constant	Filter banks which consist of predefined categories	Layer of Batch Normalization
Identification process	Layer wise Identification	Layer wise Identification	Pixel wise identification	Element wise identification
Output Size	1/8 of the size of original input	Same as the input image	Depends on the number of filters used	Less than the original image
Focus	Simple implementation with better results	Different dilation rate leading to low information loss	Improve Boundary delineation	Faster storage and efficiency in speed
Contribution	Skip-connections to merge lower-level finer features with higher-level coarse features	Increased Information sustenance.	Reduces computational time by storing max-pooling indices Reduce inference time while upsampling	Early Downsampling along with Dilated and Asymmetric Convolutions Reduces large number of floating points
Drawbacks	Concatenating is helpful but brings more noise and redundancy	Large number of floating points required	Large number of floating points required	———
Problems tackled	Mixed matched relationship Confusing categories	Gridding	Low level image to high level	Efficiency, in sense of speed and storage capacity

4 Conclusion and Future Work

Technology for Driverless cars have gone too far in scene perception and segmentation, object identification. Even though all the techniques of scene perception have their unique features, like HDC changes its dilation rate, SegNET with a encoder decoder network but the best researchers have found that ENet has high accuracy and faster processing speed, which are basic things required for the scene segmentation in a driverless car. It contains a batch normalization layer which normalizes the output feature map which makes it advanced from others. With faster speed it also has an efficient storage capacity. It solves all the problems which were faced in the domain of scene perception. ENet is till now the most advanced technology for scene perception and segmentation. Research is still going on every aspect and technology which is until now been implemented to improve the decision making and reducing the cost.

References

1. Zhao, H., Shi, J., Qi, X., Wang, X., Jia, J.: Pyramid scene parsing network. In: IEEE Conference on Computer Vision and Pattern Recognition (CVPR) (2017)
2. Wang, P., Chen, P., Yuan, Y., Liu, D., Huang, Z., Hou, X., Cottrell, G.: Understanding convolution for semantic segmentation. In: IEEE Winter Conference on Applications of Computer Vision (WACV) (2018)
3. Fridman, L., Brown, D.E., Glazer, M., Angell, W., Dodd, S., Jenik, B., Terwilliger, J., Patsekin, A., Kindelsberger, J., Ding, L., Seaman, S., Mehler, A., Sipperley, A., Pettinato, A., Seppelt, B., Angell, L., Mehler, B., Reimer, B.: Large-scale naturalistic driving study of driver behavior and interaction with automation. IEEE Access (2019)
4. Naveen, A., Bandaru, N.: Obstacle detection using stereo vision for self-driving cars. In: IEEE 17th International Conference (2012)
5. Badrinarayanan, V., Kendall, A., Cipolla, R: SegNet: a deep convolutional encoder-decoder architecture for image segmentation. In: IEEE Transactions on Pattern Analysis and Machine Intelligence Year (2016)
6. Chaurasia, A., Kim, S., Culurciello, E.: ENet: A Deep Neural Network Architecture for Real-Time Semantic Segmentation published on 7 June 2016
7. Tompson, J., Goroshin, R., Jain, A., LeCun, Y., Bregler, C.: Efficient object localization using convolutional networks. In: IEEE Conference on Computer Vision and Pattern Recognition (2015)
8. Zhan, X., Liu, Z., Luo, P., Tang, X., Loy, C.C.: Mix-and-match tuning for self-supervised semantic segmentation. In: AAAI Conference on Artificial Intelligence (AAAI) 2018 (2018)
9. Chen, S., Zheng, N.-N., Chen, Y., Zhang, S.: A novel integrated simulation and testing platform for self-driving cars with hardware in the loop. IEEE Trans. Intell. Veh. (2019)
10. Li, X., Wang, Y., Yan, L., Wang, K., Deng, F., Wang, F.-Y.: ParallelEye-CS: a new dataset of synthetic images for testing the visual intelligence of intelligent vehicles. IEEE Trans. Veh. Technol. (2019)
11. Yasrab, R.: ECRU: An Encoder-Decoder Based Convolution Neural Network (CNN) for Road-Scene Understanding (2018)
12. Tian, J., Hu, J.: Image Target Detection Based on Deep Convolutional Neural Network (2019)

13. Xu, L., Ren, J.S., Liu, C., Jia, J.: Deep Convolutional Neural Network for Image Deconvolution (2014)
14. Dumitrescu, D., Boiangiu, C.-A.: A Study of Image Upsampling and Downsampling Filters (2019)
15. Long, J., Shelhamer, E., Darrell, T.: Fully Convolutional Networks for Semantic Segmentation (2015)

Machine Learning Algorithms for Early Prediction of Heart Disease

Akankasha Sinha$^{(\boxtimes)}$ and Rejo Mathew

MPSTME, NMIMS (deemed-to-be) University, Mumbai, India
akankasha.sinha72@nmims.edu.in, rejo.mathew@nmims.edu

Abstract. Machine learning has been effective in assisting the decision making and analysis of huge data produced by healthcare sector. By using machine learning model for prediction and prognosis of heart disease is now becoming an important method for treatment. This study discusses about various machine learning techniques used in predicting heart disease. This paper aims to provide a comparative analysis on various machine learning algorithms used for the prediction of heart disease.

Keywords: Machine learning · Heart disease · Cardiovascular disease · Artificial Neural Network · Decision Tree · Support Vector Machine · Naïve Bayes

1 Introduction

Machine Learning focuses on the development of computer programs that can access data and use it to learn by themselves. It combines computer science with statistics. Computer Science verifies that whether the problems are solvable, and statistics helps in data modeling [10]. Machine learning helps in establishing a relationship between the characteristics and creates a model to describe the data [11]. Machine Learning integrates computer system with the medical field for efficient diagnosis and quality treatment by medical experts [1]. Machine learning models are being used in collection, prediction and analysis of diseases. Classification is one of the most important and essential tasks in machine learning [6].

Medical diagnosis is a very crucial step in treatment. It is proved that survival rate is 88% after five year of diagnosis, and 80% after 10 years from diagnosis [8], thus early prediction is important and one of the most significant steps in the investigation process. Accurate diagnosis helps in deciding the correct and suitable therapy at an early stage, thus increasing the survival rate of patients. Identifying how to use healthcare data will help in saving human beings and early detection of abnormalities in heart disease. Statistics show that early detection of disease has a huge influence on the survival rate of the patient. Prediction of disease is challenging but can hugely impact the mortality rate if it is detected at early stage and proper medication is provided. Machine learning helps to analyze large, complex datasets and delivers accurate analysis at a faster rate [3].

© Springer Nature Switzerland AG 2020
A. P. Pandian et al. (Eds.): ICCBI 2019, LNDECT 49, pp. 162–168, 2020.
https://doi.org/10.1007/978-3-030-43192-1_18

Heart disease is one of the most significant causes of mortality today [5]. Various types of cardiovascular diseases are heat attack, congenital heart disease, coronary artery disease. Indian Heart Association states that 50% of heart strokes happen under the age of 50 years and 25% heart strokes under the age of 40 years [4]. Statistics from WHO shows that one third of population across the globe die due to heart disease [17–20]

Section 2 briefly discusses about the various symptoms and causes of heart disease it also highlights the difference between the traditional method of diagnosing and how is ML helping in diagnosis. Section 3 discusses about the various machine learning classifiers and how are they implemented for prediction. Section 4 gives a brief elaboration on how the input attributes are applied in various algorithms to determine the prediction. Section 5 gives a brief comparison of supervised machine learning algorithms. Section 6 is conclusion of the study.

2 Challenges in Conventional Method of Diagnosis

Clinical decisions are often based on doctor's knowledge and experience and not on account of hidden knowledge of data [18]. Patient must undergo numerous blood tests, ECG, MRI scan, blood pressure, heart rate and other lab tests to know the presence of disease. Predicting heart disease is a tough job, because of the complexity of organ. This process is time consuming and at times, because of this long process the detection of disease happens late. Machine Learning classifiers take in consideration of various factors such as age, sex, type of chest pain and others to train their models. These models help in quick and accurate prediction for presence of disease. These models help clinicians to provide proper early and proper medication to the patients. Various algorithms of machine learning, such as Decision Tree, Artificial Neural Network are being integrated to analyze these data.

3 Classifiers

There are various types of machine learning strategies which help in training and testing of dataset, these algorithms take in the input value perform the function and give the output value. The output value calculated is compared with the expected output value to determine the accuracy of the algorithm. 70% of dataset is used for training the model and 30% of data set is used for testing the model. Classifiers take in account the features like age, sex, blood pressure, type of pain and accordingly classify the patient with respect to the condition [18] (Fig. 1).

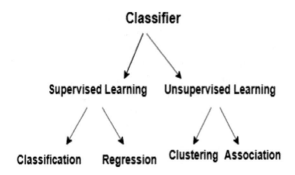

Fig. 1. Machine learning classifiers

3.1 Supervised Learning Algorithm

In supervised learning the objective is to predict the output value for given input value. Using this algorithm, the dataset is trained according to the desired result, both the input and expected output is given to train the dataset [15]. Classification and regression are types of learning classifiers. Classification helps in identifying where the data belongs [10]. The output can either have a continuous value or a discrete value, if the output value is continuous regression is used and if the output value is discrete classification is used. Naïve Bayes, Support Vector Machine, Decision Tree are supervised learning algorithms.

3.2 Unsupervised Learning Algorithm

In unsupervised learning the output is predicted based on the similarities between the input vectors. This method is used when the data is not labeled [12]. Clustering and association are two learning categories of unsupervised learning algorithm. In clustering method, similar type of data is grouped based on their distance; it follows the rule that intra cluster similarity should be high and inter cluster similarity should be low. Association helps to determine relationship between the data values or objects which are repeated together [12]. This algorithm requires much larger training set than supervised learning algorithm [5].

4 Methodology

For prediction of heart disease various attributes are considered like age, gender, cholesterol level, color of blood, type of chest pain, blood sugar level, maximum heart rate, resting blood pressure [13].

4.1 Support Vector Machine

Using this algorithm, each data item is plotted in an n-dimensional space, where n indicates the number of attributes considered for prediction. Classification is performed on these data items by finding the hyper plane. Hyper plane is a line which separates the classes. Coordinates of each observation are called support vectors. Support vector works by maximizing the margin, margin is the distance between the hyperplane and support vector [10]. The hyperplane acts as a decision boundary between two clusters, existence of a decision boundary. The resulting classifier achieves considerable generalization that can be used to classify samples [13].

4.2 Artificial Neural Network

It is a mathematical model that uses the characteristic of neurons to pass message or information. it constitutes of three elements the input, the output and the transfer function [1]. Input is extraordinary values and weights (path between neurons); they are modified in the training process of the data set. Output of the artificial neural network is calculated for the known class; the weight is recomputed using the error margin between the output of predicted and actual class.

Multilayer Perceptron is a neural network model which follows feed forward architecture and supervised training [14]. Supervised training is a process, in which the inputs are given to the network and the outcome is compared with the expected output. Perceptron comprises of three layers; input layer which includes the data of patient like age, sex, type of chest pain, cholesterol, blood pressure, etc.; hidden layer which includes the weight and the output calculated and the output layer. It follows back propagation method as training algorithm. Back propagation means that the algorithm repeats the presentation of input data values to neural network. Every time when the algorithm is repeated, output derived is compared with the expected output, if any error it is calculated and the feedback is back propagated to the network. This method modifies the weights and the required output is calculated through iterations.

4.3 Decision Tree

Decision tree is simply a nested if else loop [20] for which two steps are required, building a tree and applying the tree to data [19]. Decision Tree is like tree graph, initiating from a single node which keeps on branching to represent all possible outcomes [9] (Fig. 2).

Root node is the topmost decision node, which checks for the presence or absence of the attributes and then splits. For example, let, root node be chest pain, if the patient suffers chest pain it will be concluded that the patient has high risk of suffering a heart disease, if the patient has not been detected with chest pain, next condition is checked for high blood pressure and so on so forth the conditions are checked. Thus, depending upon the if, else condition, the node splits and a conclusion is made.

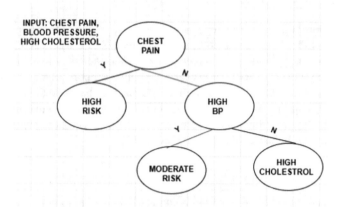

Fig. 2. Decision Tree

4.4   Naïve Bayes

In Naïve Bayes algorithm, the dataset used for training helps to find out the conditional probability value of vectors for a given class. It considers that all the variables are independent of each other. It is highly scalable [10]. Naïve Bayes is based on Bayesian Theorem [22], which determines of how likely an event can occur given that another event has already occurred.

$$P(A|B) = \frac{P(B|A) * P(A)}{P(B)} \qquad (1)$$

P(A) = probability that patient has history of heart attack
P(B) = probability of patient being male
P(A|B) = probability that the patient will suffer a heart attack given that the patient is male
P(B|A) = probability of a male patient having history of heart attack.

5   Comparison of Algorithms

In this section supervised machine learning algorithms are compared on listed parameters (Table 1).

Table 1. Comparison table

Classifier condition	Support vector machine	Decision tree	Artificial neural network	Naïve Bayes
Multiclass prediction	Cannot be extended to multiclass	Can be used for multiclass classification	Can be used for multiclass classification	Can be used for multiclass classification
Non linear problems	Non linear classifier	Non linear classifier	Non linear classifier	Linear classifier
Overfitting	Less overfitting	Prone to overfitting	Prone to overfitting	Prone to overfitting
Size of dataset	Can work on all types of dataset, but more time is required with large dataset	Can perform on large dataset	Requires large amount of dataset	Can perform on large dataset with high accuracy
Value	Can handle both continuous and binary values	Can work for continuous values	Can work for continuous values	Works better for discrete values
Clustering	Uses clustering technique	Cannot perform clustering	Uses clustering technique	Can perform clustering
Parallel processing	Cannot be easily implemented in parallel	Have strength to perform parallel processing	Have strength to perform parallel processing	Have strength to perform parallel processing
Cost	Is not expensive computationally	Computationally expensive	Computationally expensive	Computationally very expensive

6   Conclusion and Future Scope

Machine learning has vast scope of application in healthcare sector which is being explored and studied. There is potential future development of the presented work. Early detection of heart disease plays significant role in the medication process, the survival rate increases because of early diagnosis. There have been lot of recent advancement in machine learning technology which is helping the medical sector to treat patients. Rather going through a series, a test, using few attributes will be of great help for the medical diagnosis. Other machine learning classifiers, hybrid classifiers and algorithms can be explored and testes on medical dataset not only for early prediction of heart disease but also for early diagnosis and prediction of other ailments like diabetes, Alzheimer's.

References

1. Haq, A.U., Li, J.P., Memom, M.H., Nazir, S., Sun, R.: A hybrid intelligent system framework for the prediction of heart disease using machine learning algorithm. Mob. Inf. Syst. (2018)
2. Maity, N.G., Das, S.: Machine learning for improved diagnosis and prognosis healthcare (2017)

3. Gavhane, A., Pandya, I., Kokkula, G., Devadkar, K.: Prediction of heart disease using machine learning. In: Second International Conference on Electronics, Communication and Aerospace Technology. IEEE (2018)
4. Mohan, S., Thirumalai, C., Srivastava, G.: Effective heart disease prediction using hybrid machine learning techniques. IEEE Access **7**, 81542–81554 (2019)
5. Sinawane, J.S., Patil, D.R., Thakare, V.S.: Survey on decision support system for heart disease. Int. J. Adv. Technol. **4**(1), 89–96 (2013)
6. Ahmad, L.G., Eshlaghy, A.T., Poorebrahimi, A., Ebrahimi, M., Razavi, A.R.: Using three machine learning techniques for predicting breast cancer recurrence. J. Health Med. Inform. **4**(2), 124 (2013)
7. Kohli, P.S., Arora, S.: Application of machine learning in disease prediction. In: 4th International Conference on Computing Communication and Automation IEEE (2018)
8. Pavithra, D., Jayanthi, A.N.: A study on machine learning algorithm in medical diagnosis. Int. J. Adv. Res. Comput. Sci. **9**(4) (2018)
9. Bayrak, E.A., Kirci, P., Ensari, T.: Comparison of machine learning methods for breast cancer diagnosis. In: Scientific Meeting on Electrical-Electronics & Biomedicl Engineering and Computer Science. IEEE (2019)
10. Umasankar, P., Thiagarasu, V.: Machine learning and deep learning methods in Heart Disease (HD) research for the prediction of heart disease: a literature survey. Asian J. Comput. Sci. Technol. **8**(1), 1–6 (2019). ISSN 2249-0701
11. Bharat, A., Pooja, N., Reddy, R.A.: Using machine learning algorithm for breast cancer rosk prediction and diagnosis. In: IEEE Third International Conference on Circuits, Control, Communication and Computing (2018)
12. Abushariah, M.A.M., Alqudah, A.A.M., Adwan, O.Y., Yousef, R.M.M.: Automatic heart disease diagnosis system based on Artificial Neuron Network (ANN) and Adaptive Neuro-Fuzzy Inference Systems (ANFIS) approaches. J. Softw. Eng. Appl. **7**, 1055–1064F (2014)
13. Kusuma, S., Divya Udayan, J.: Machine learning and deep learning methods in Heart Disease (HD) research. Int. J. Pure Appl. Math. **119**(18), 1483 (2018)
14. Kirmani, M.M., Ansarullah, S.I.: Prediction of heart disease using decision tree a data mining technique. Int. J. Comput. Sci. Netw. **5**(6) (2016)
15. Sahran, S., Qasem, A., Omar, K., Albashih, D., Adam, A., Norul, S., Abdullah, H.S., Abdullah, A., Hussain, R.I., Ismail, F., Abdullah, N., Hayati, S., Pauzi, M., Shukor, N.A.: Machine learning methods for breast cancer diagnostic. Breast Cancer Surg. IntechOpen (2018)
16. Joseph, S., Thanakumar, I.: Survey of data mining algorithm's for intelligent computing system. J. Trends Comput. Sci. Smart Technol. (TCSST) **1**(1), 14–24 (2019)
17. Pattekari, S.A., Parveen, A.: Prediction system for heart disease using Naïve Bayes. Int. J. Adv. Comput. Math. Sci. **3**(3), 290–294 (2012). ISSN 2230-9642
18. Karthiga, A.S., Mary, M.S., Yogasini, M.: Early prediction of heart disease using decision tree algorithm. Int. J. Adv. Res. Basic Eng. Sci. Technol. **3**(3) (2017)
19. Chaturvedi, S., Patil, S.: Oblique decision tree learning approaches – critical review. Int. J. Comput. Appl. **82**(13), 0975–8887 (2013)
20. Arun, R., Deepa, N.: Heart disease prediction using Naïve Bayes. Int. J. Pure Appl. Math. **119**(16), 3053–3065 (2018)

Analysis and Review of Cloud Based Encryption Methods

Vicky Yadav$^{(\boxtimes)}$ and Rejo Mathew

Department of Information Technology, NMIMS (Deemed-to-be) University,
Mumbai, India
yadavvicky2d@gmail.com, rejo.mathew@nmims.edu

Abstract. Data security plays a crucial role in cloud computing. Cloud storages are secured and managed centrally by cloud service providers. Data stored in cloud is secured by encrypting them with the help of various encryption techniques and models. Encryption is very important from a data confidentiality point of view and should match the system and security requirements. Complex encryption methods increase the computational time of the cloud system whenever any data query is being processed. The encryption method has to produce a cipher text which cannot be easily predicted or decrypted without the encryption key. This paper discusses various encryption methods being developed and provides an analysis and comparative study on them.

Keywords: Encryption · Cloud computing · Attribute based encryption · Security · Data privacy

1 Introduction

Cloud computing has been classified into different service models named as Software as a Service (SaaS), Platform as a Service (PaaS), and Infrastructure as a Service (IaaS) [13]. Depending on the deployment models the cloud can be categorized into private cloud, Community cloud, Public cloud, Hybrid cloud [13]. The data stored in clouds are critical for the user and thus the data should be protected from unauthorized access, providing confidentiality and integrity to the cloud data. The cloud storages are managed by Cloud service providers (CSPs) and the security is managed by them, centrally. The data is being encrypted by the user before being uploaded to the cloud. The CSPs being semi trusted agents it's risky to upload plain text data as it hinders the data confidentiality and privacy. Even though encryption has various drawbacks, it is the most widely used data securing method. Encryption increases the need for computational power of the system every time an authorized user tries to access the data. A great number of methods and algorithms are developed for encryption data in order to ensure data confidentiality. Public key encryption with equality test was initiated by Yang et al. [4], which enabled the equality testing of messages, further to provide authorization, a novel PKE-ET was suggested by Tang [5]. To provide a scalable authorization in [5], a scheme was introduced with delegated equality testing by Tang [6] and Ma et al. [10]. Attribute based encryption (ABE) technique is a public key encryption mechanism which produces the secret key by using the user attributes like

© Springer Nature Switzerland AG 2020
A. P. Pandian et al. (Eds.): ICCBI 2019, LNDECT 49, pp. 169–176, 2020.
https://doi.org/10.1007/978-3-030-43192-1_19

the designation, country, name etc. In ABE, the message can be decrypted only by the user having the same set of attributes as that of the cipher text. ABE implements access control with fine granularity on encrypted data, it is very flexible. There are various ways to implement ABE in cloud. Section 2 of this paper discusses the issues in cloud computing, Sect. 3 describes the various solutions for the issues mentioned, Sect. 4 provides a comparison between the solutions mentioned in Sects.3 and 5 provides a conclusion and future scope of these solutions.

2   Privacy Issues in Cloud Computing

2.1   Data Encryption Is Required as the Service Providers Are Semi-trusted

Data security is a major issue in cloud computing, it has to protected from eternal attackers and snoopers (employees of the cloud provider) [15]. It is reported that users' data leakage incidents happened on the Google's cloud storage system [1]. Cloud service providers are semi-trustable agents which can lead to confidentiality and integrity issues with the sensitive data. Plain-text cannot be directly uploaded to the cloud instead cipher-text is stored in cloud and at the receiving end the cipher text is converted back to original message. A secure encryption process with proper key management such that only users with the encryption key can access data along with access control mechanism. Miao Zhou et al. [12] proposes a flexible access control mechanism. Complicated and lengthy encryption methods and access control methods, hinder the computational power of the system hence degrading the performance over all and making the system slow. The encryption process desires to be secure meanwhile not affect the system computational power and to achieve this; a number of encryption systems are proposed and developed every day. The main aspects of privacy are the lack of user control, potential unauthorized secondary use, data proliferation and dynamic provisioning [14].

3   Encryption Methods

3.1   Key-Policy Attribute-Based Encryption [7]

This technique provides a fine-grained authorization. It comprises of public key encryption and equality testing [8] and key-policy Attribute based encryption.
   The steps in KP-ABE are:

  i. The encrypted data is finely grained in this scheme which helps in reducing the computational time of the system. In testing algorithm it is changed from one-many from one-one [7].
 ii. This scheme tests if the cipher tests are same information encrypted by two different public keys [7].
iii. It achieves one-way against chosen-cipher text attack [7].

iv. This proves that the Twin-decision bilinear Diffie-Hellman problem is equally tough as the DBDH problem [7].
v. It has the security model of authorization unlike other methods mentioned in Sect. 3
vi. It proves the security of authorization based on the tDBDH assumption [7].

This provides a more practical approach for storing data in form of cipher text. The equality test however is performed by using trapdoor techniques.

If the user wants the CSPs to perform the equality tests, they need to submit their attribute and the cipher text to the server depending on which the server responds. It protects the user's identity, which can be risky as the CSPs are not trustworthy [7].

3.2   Multi-authority Proxy Re-encryption [2]

The plain idea of this technique is to provide more flexible and fine-granted access control policies. It is associated with the access rights of the cipher text. Then, a client's private key is related with the attribute set of the client. The private key is generated by a combination of private attribute keys that associated to each attribute of the client. If a client's attribute set fulfills the attributes in the structure of tree associated to the cipher text, they have enough information in their private attribute keys to generate the private key required to access the cipher text so that user is authorized to decrypt it [2]. MPRE-CPABE consists of six phases: initialization, key generation, partition, encryption, decryption, and proxy re-encryption [2].

The steps in MPRE-CPABE:

i. A partition scheme is presented. This scheme splits data into two different-size blocks that is one big and one small block. The small block is kept at client side, used as the symmetric key to encrypt the big block that is stored in the cloud. When a malicious client tries to decrypt the cipher text, they cannot get the complete file easily [2].
ii. A multi-authority structure is introduced into the attribute-based proxy re-encryption system. Each authority distributes private attribute keys to a disjoint set of clients which composes its domain. This domain shares risks of authority failure and improves the efficiency of the generation of private attribute key components. The access tree defined by data owner contains the set of attributes required for access. This access tree is mapped to a weighted access structure list, which enables the multi-authority system to perform fine-grained access control operations [2].
iii. The proxy authority is used to re-encrypt both cipher text and private key when a client's access is revoked. This improvement shifts some CPU-intensive tasks from the client-side to the cloud data center, especially suitable for battery-powered devices of clients [2].
iv. This technique protects cloud storage systems against force attacks, such as bypassing keys. The small data block is extracted from the file, and the rest of the file is incomplete. Then, the two blocks will be encrypted further. Thus, if an illegal client tries to bypass the symmetric key cipher text decryption and adopts

the brute force to get plaintext, he/she can only get the incomplete large data block of the file [2].

The MPRE-CPABE architecture has four entities:

A. K-Attribute authorities: The whole set of attributes is separated into k separate sets, controlled and achieved by k attribute-authorities. They need powerful computing capabilities to generate their own main key parameters during the initial phase, generate private attribute keys for clients during the file sharing phase. Send these private attribute keys to the central authority and clients according to the results of proxy re-encryption [2].
B. Central authority: The central-authority has private keys for clients based on private attribute key components received from each attribute-authority. It is not responsible for encryption [2].
C. Data storing & processing module: The data storing and processing module receives files and related information, stores them, verifies signatures and implements proxy re-encryptions [2].
D. Clients: The cloud storage clients include data owners and data consumers. Each user can be either data owner and consumer or both. They use the client-side components for automatically splitting/assembling files and performing encryption/decryption processes [2].

This method has a long workflow when it comes to accessing or uploading any file. The system working when the client wants to upload a file [2]:

i. The client splits the file into one big and one small block, and encrypts the bigger one with the small block as symmetric key.
ii. The client uses public key from any attribute authority to encrypt the small block (symmetric key).
iii. The client uploads blocks to the cloud.

The system working when the file has to be accessed [2]:

i. The client downloads both file blocks.
ii. Client asks the central-authority and all attribute-authorities about attribute associated to the file.
iii. Generate the file private key locally with the above information receive
iv. Decrypt the small block and the big block
v. Assemble these two blocks to obtain the complete file.

This scheme has certain assumptions in this method where the success of the method is based on

(a) Trustworthiness of the central attribute as it receives and stores the private keys.
(b) Trustworthiness of the clients as they encrypt the files locally before uploading them.
(c) Semi-trustworthiness of the attribute-authorities and data storing, processing modules.

3.3 Dynamic Encrypted Data Sharing Scheme [3]

This technique openly re-encrypts the data in cloud storage without revealing the data to anyone, not even proxy. The method is experimentally and theoretically tested. [3] This scheme has the support of user dynamics from Delerablee et al. scheme [11–15]. The CPBRE-DS architecture has four entities [3]:

A. Client: The user who stores and shares data in the cloud with other clients.
B. Broadcast Centre: They allocate the secret key and re-encryption key to the client and also set the security parameters for the entire system. They send the generated keys to the respective client.
C. CSP: They provide the data sharing and storage services.

The workflow of this method is:

i. Client sends its ID to the broadcasting server for the purpose of generation of secret key [3].
ii. Client encrypts the data with this secret key [3].
iii. This data is sent over open channel to the servers [3].
iv. If client desires to share any data, the data has to be re-encrypted with the re-encryption key (provided by broadcast centers) and then send the key to the servers through a secure channel [3].
v. The server re-encrypts the data after receiving the re-encryption key provided by objective clients [3].
vi. The objective clients transfer the encrypted data from the cloud storage and decrypt it with their personal keys [3].

Note: There is no need to send the re-encryption key to the server to share new data encrypted under similar conditions [3].

3.4 Cipher Text-Policy Attribute-Based Encryption [9]

It provides data security by enabling the one-many encryption which increase the flexibility of the system.
Steps in CP-ABE [9]:

i. This scheme introduces the idea of Public key encryption assisting equality test into the CP-ABE-based. A semi-trusted Cloud service provider in attribute based encryption and equality testing can be substituted to execute a similarity test on this technique type cipher-text which is encrypted by diverse access policies. The substituted individual has no knowledge about the original plain text [9].
ii. This scheme is using the bilinear pairing and Vi' ete's formulas technique. The size of cipher-texts is constant. Comparing to the solution presented in [7] which has more excessive access policies which include wildcard attributes, which is proved to be selective security against a chosen plain-text attack in the typical model under Decisional n-Bilinear Diffie-Hellman Exponent assumption [9] (Table 1).

4 Comparison and Analysis

4.1 Encryption Schemes

Table 1. A comparison of encryption methods

Parameters	Key-Policy attribute-based encryption	Multi-authority proxy re-encryption	Dynamic encrypted data sharing	Cipher text-Policy attribute-based encryption
Approach	Fine-grained authorization	Fine-granted access control policies	Directly re-encrypting the data	Equality test of data under different access policies
Proof of authorization	Yes	No	No	No
Complexity factors	Requirement of total attributes	Number of attributes required	Increases with re-encryption	Less as no re-encryption
Selective security	No	No	No	Yes
Backdoor access	Trapdoor	No	No	Trapdoor
Equality testing	Yes	No	No	Yes
Time	Equality testing reduces the throughput	Depends on the access tree branch	Faster if data encrypted under similar conditions	Equality testing is time consuming
Re-encrypting the data	No	Yes	Yes	No
Key generation	Based on attributes	Small block is symmetric key, Attribute authority generates public key	Secret key is based on the client ID, provided by broadcasting server	Based on attributes
Assumption	(1) Bilinear Diffie-hellman (2) Twin-decision Diffie-Hellman	Trustworthiness of the users and central authority	None	Decisional n-Bilinear Diffie-Hellman Exponent
Drawbacks	(1) The user has to provide the attribute and cipher text to server (2) CSPs have access to the servers where the attribute and cipher text is Provided	(1) Not suitable if the domain of the users varies a lot. (2) More the number of attributes, more is the number of authorities required to manage	(1) No need to send the re-encryption key to the server to share new data encrypted under same conditions (2) Broadcasting servers are responsible for the entire process which is accessible by CSPs	(1) Depends on the CSPs for the equality testing by similarity testing (2) Constant sized cipher text

5 Conclusion

Cloud computing still has a long way to go when it comes to data confidentiality and privacy, there are a lot of encryption methods at present and being developed. Protecting data from misuse by semi-trusted CSPs and other unauthorized user access is necessary. In the techniques discussed in this paper the data is encrypted differently and tested accordingly and in order to maintain data confidentiality and privacy and also preserving the system performance. The nature and size of data being stored is changing with time and the techniques used to encrypt the data have to evolve likewise. Choosing the correct scheme for encryption depends on the needs and nature of data of the organization or cloud service providers. Data privacy is critical to the user as well as the efficiency of the cloud storage and both of them depend on the encryption scheme used thus selecting the appropriate encryption method is very important. This paper attempts to offer a clear assessment between several encryption methods being proposed and assist the readers in the same.

References

1. Feng, D., Zhang, M., Zhang, Y., et al.: Study on cloud computing security. Chin. J. Softw. **22**(1), 71–83 (2011)
2. Xu, X., Zhou, J., Wang, X., Zhang, Y.: Multi-authority proxy re-encryption based on CPABE for cloud storage systems. J. Syst. Eng. Electron. **27**(1), 211–223 (2016)
3. Jiang, L., Guo, D.: Dynamic encrypted data sharing scheme based on conditional proxy broadcast re-encryption for cloud storage. IEEE Access **5**, 13336–13345 (2017)
4. Yang, G., Tan, C.H., Huang, Q., Wong, D.S.: Probabilistic public key encryption with equality test. In: Proceedings of the International Conference Topics in Cryptology (CT-RSA), vol. 5985, pp. 119–131 (2010)
5. Tang, Q.: Towards public key encryption scheme supporting equality test with fine-grained authorization. In: Proceedings of the Australasian Conference on Information Security and Privacy, vol. 6812, pp. 389–406 (2011)
6. Tang, Q.: Public key encryption supporting plaintext equality test and user-specified authorization. Secur. Commun. Netw. **5**(12), 1351–1362 (2012)
7. Zhu, H., Wang, L., Ahmad, H., Niu, X.: Key-policy attribute-based encryption with equality test in cloud computing. IEEE Access **5**, 20428–20439 (2017)
8. Yang, G., Tan, C.H., Huang, Q., Wong, D.S.: Probabilistic public key encryption with equality test. In: Proceedings of the Cryptographers' Track at the RSA Conference, pp. 119–131 (2017)
9. Wang, Q., Peng, L., Xiong, H., Sun, J., Qin, Z.: Ciphertext-policy attribute-based encryption with delegated equality test in cloud computing. IEEE Access **6**, 760–771 (2018)
10. Ma, S., Zhang, M., Huang, Q., Yang, B.: Public key encryption with delegated equality test in a multi-user setting. Comput. J. **58**(4), 986–1002 (2014)
11. Delerablée, C., Paillier, P., Pointcheval, D.: Fully collusion secure dynamic broadcast encryption with constant-size ciphertexts or decryption keys. In: Proceedings of the International Conference on Pairing-Based Cryptography, pp. 39–59 (2007)
12. Wang, B., Li, B., et al.: Privacy-preserving public auditing for secure cloud storage. IEEE Trans. Comput. **62**(2), 362–375 (2012)

13. Sridhar, S., Smys, S.: A survey on cloud security issues and challenges with possible measures. In: International Conference on Inventive Research in Engineering and Technology, vol. 4 (2016)
14. Karthiban, K., Smys, S.: Privacy preserving approaches in cloud computing. In: 2018 2nd International Conference on Inventive Systems and Control (ICISC), pp. 462–467. IEEE (2018)
15. Praveena, A., Smys, S.: Ensuring data security in cloud based social networks. In: 2017 International Conference of Electronics, Communication and Aerospace Technology (ICECA), vol. 2, pp. 289–295. IEEE (2017)

Data Storage Security Issues in Cloud Computing

Vedant Paul$^{(\boxtimes)}$ and Rejo Mathew

Mukesh Patel School of Technology Management and Education,
NMIMS, Mumbai, India
{vedant.paul44, rejo.mathew}@nmims.edu.in

Abstract. Cloud computing allow users to store information in data centers, it also enables on-demand computation of resources. Remote servers can access the information inside the cloud. Cloud computing serves as an internet-accessible resource allocation business model and offers pay-per-use computing infrastructure. This paper primarily discusses about the main data storage security issues- in cloud computing and certain solutions which could be adopted as a remedy to such issues. It also includes an analysis review which helps in picking out the best fitting solution.

Keywords: Client · Confidentiality · Data · Integrity · Issues · Storage · Server · Technology

1 Introduction

Cloud computing provides applications and resources which are delivered on demand over the internet as services. Cloud computing is very convenient for data storage, data is stored in data centers across the world and can be easily accessed whenever and wherever needed. It is one of the most emerging technology which is being used by almost everyone due to its leading benefits like scalability, easy deployment, reduced costs and flexibility. Due to so many benefits, many companies are either using this technology or providing this technology to consumers but none of the companies have perfected the data security aspect of this technology. The reliability of this technology reduces due to data security loopholes. This paper explores data confidentiality issues like data breaches and Dos attacks and data integrity issues which reduces the reliability of cloud computing. The solutions to these problems are also discussed and their comparison analysis is conducted at the end so that one could choose the right solution according to their needs. It is important to get rid of such issues so that more companies adopt this technology and this technology becomes more reliable.

2 Issues and Challenges

2.1 Data Loss Issues

Enterprises are giving their whole data to cloud service providers (CSP) which includes the companies extremely sensitive data [1]. There is a high possibility that this data

A. P. Pandian et al. (Eds.): ICCBI 2019, LNDECT 49, pp. 177–187, 2020.
https://doi.org/10.1007/978-3-030-43192-1_20

could be lost due to malicious attacks or the server would crash or get deleted unintentionally by the CSP without keeping backups. Even natural disasters, fire and earthquakes would damage data centers hence resulting in data loss. Even events leading to harming the encryption keys can finally result in data losses [2].

2.2 Data Breaches Issues

Cloud environment generally consists of various consumers and enterprises whose data are stored in the same place. If the cloud gets breached then all of the consumers and enterprises data would get exposed. Also due to multiple users using the same location, customers using different apps on the same VM would be sharing the same database, any event of corruption that would occur due to one app would affect others data and app too [2].

2.3 Malicious Insiders

Malicious insiders are generally those individuals who are appointed to control data such as server admins and CSP staff or contractors accessing data or associates. These people could be involved in data theft and sell it to other companies or could also do it due to disliking the company they work for due to underpayment. Sometimes even the CSP's aren't aware of such people [2].

2.4 Data Location Issues

Data centers are widespread across the world. Data location becomes an issue as the client is generally unaware about the location of the data center where his data is stored, this leads to less control in the hands of the client. Some nations, according to rules and regulations need their enterprises to store the data locally. The data location also matters as data should not be kept in locations which are prone to wars and natural disasters.

2.5 Data Integrity Issues

Clients give their data to cloud service providers but this data is actually computed at a random remote server hence the data should constantly be checked to see whether the integrity is maintained or altered. During computation, the data could be altered due to software bugs or configuration errors.

2.6 Data Confidentiality Issues

As data is directly outsourced by the client to the CSP, data has to be given a lot of priority as its only managed by the service provider and not by the client. When many entities are sharing data, there is high chance they could access unauthorized data.

3 Solutions

3.1 Methods to Prevent Data Loss

3.1.1 Strong API
A Strong API can be used for access control [3].

3.1.2 Protection of Data
Encryption and protection of integrity of data should be conducted while it is still in transit. Data protection should be analyzed at run and design time [3].

3.1.3 Backup, Clearing of Data and Key Maintenance
Backup and retention policies should be pre-specified. Service provider should erase the retained data before letting it go to the cloud. Powerful key generation should be implemented and storage, destruction and management practices should be used [3].

3.2 Methods for Avoiding Malicious Insiders

3.2.1 Creating Processes and Transparency
Processes should be created which would notify the service provider whenever data breaches occur. All information security and cloud services practices should be made transparent [3].

3.2.2 Human Resources Requirements and Use of IDs
Human resources requirements should be defined in the legal contract. Supply chain management ID should be made stricter and a comprehensive supplier assessment should be conducted [3].

3.3 Solutions to Data Location Issues

3.3.1 Informed About Location
The client should be informed about the location of the data center where his data would be stored as this would enable the client to know the risks.

3.3.2 Backup
A backup of the client data should be stored in another location. This would help the client in recovering his data from the other location in case there is loss of data at the first location.

3.4 Protecting Data Integrity

3.4.1 Appointing Third Party
A third party can be appointed for this who first downloads the file and hash value is checked. In this method a message authentication code algorithm (MAC) is used which takes two values, secret key and a random length of data that produces MAC tag. Using this method, the algorithm is used by the third party. Posterior to receiving the MAC using the algorithm, the third party uploads the data. Then later to check the integrity,

he downloads the data which was uploaded earlier and uses MAC to calculate the tag and compares it with the earlier one, if it matches, the integrity is maintained [4].

3.4.2   Using a Hash Tree

A hash value could also be computed using a hash tree [4].

3.4.3   Using Provable Data Possession (PDP)

PDP is basically a remote server method for information integrity assurance. In Provable Data Possession, a user that has saved data on an unfaithful server can check that without retrieving it, the server has the data that is original. 'Ateniese' was the earliest to examine government auditing capacity to ensure the ownership of documents on untrusted storage in their specified "provable information ownership" model. The working principle includes two stages

(a) Pre-process and store: By using probabilistic key generation algorithms the client produces matching public and secret keys. Then the public key and file are sent to the server by the client for storing it.

(b) Using server stage to verify the file proprietorship: For subdivisions of blocks in the file, the user challenges the server to check if it has a proof of proprietorship and then he checks the response from the server [9].

3.4.4   Using Basic PDP Scheme Based on MAC

Message Authentication Code based PDP is uses a simple way to ensure data integrity of file X (X is an arbitrary file) stored on cloud storage. The data owner calculates and locally stores the entire Message Authentication Code of the file with a set of secret keys before it is outsourced to the Cloud Service Provider. It only maintains the computed Message Authentication Code in its local storage and gives the file to the CSP. If a person who verifies, needs the integrity of file X to monitor, then a request is sent to recover the file from CSP, it reveals a secret key to the cloud server and asks for re-computed Message authentication of the entire file [10].

3.4.5   Using Scalable PDP

The Scalable PDP is an enhanced PDP version. The main distinction is the use of symmetrical encryption by Scalable PDP while the initial PDP utilizes a public key to decrease overhead computations. Dynamic operation for remote information can be carried out by Scalable PDP. All problems for Scalable PDP are pre-computerized and a restricted amount of updates are provided. No bulk encryption is necessary for Scalable PDP. It is based on a more effective symmetric key than the encryption of public key. It therefore does not provide public verification [11].

3.4.6   Using Dynamic PDP

Dynamic PDP is a collection of 7 algorithms "(Perform Update DPDP, Prepare Update DPDP, GenChallenge DPDP, Prove DPDP, KeyGen DPDP, Verify DPDP, Prepare Update DPDP)". Full dynamic operations such as insert, update, edit, delete, etc. are supported. In this method the standard authenticated directories and a skip list are used to insert and delete functions. It has a DPDP computational complexity and remains efficient. In order to verify a 500 MB test file, for instance, DPDP generates only 208 KB of evidence data and an overhead calculation of 15 ms. This technology offers a dynamic operation, such as modification, deletion, insertion, etc. With its full dynamic operation, computer, communication and overhead storage are relatively higher. All the challenges and responses are generated dynamically [12].

3.4.7   By Using Proof of Reliability (POR)

Proof of retrievability is a cryptographic means of remotely testing the completeness of cloud files without storing a local copy of the original user files. In a scheme, users back up their data file to a potentially dishonest cloud storage server together with some authentication data. With the authentication key, the user can inspect for the integrity of the data that is stored with the CSP without recovering data from the cloud. There are two phases in POR, a setup stage and a sequence of verification stage.

(a) Setup stage: In this stage, to generate some authentication code, the user pre-processes his data file using his private key. He then sends the data file to the cloud storage server in accordance with authentication code and removes it from his local disk. As a result, the user has his private key on his local disk at the end of this phase, and Cloud Service Provider has both the data file and the appropriate authentication code.

(b) Sequence of Verification stages: The user generates a random challenge query in each verification stage sequence and the Cloud Service Provider is expected to generate a brief response or proof based on the user's data file and the respective authentication information on the received challenge request. Users will check the reaction of CSP using their private key at the end of a verification stage and decide to recognize or dismiss this reaction from CSP [13].

3.4.8   Using POR Based on Keyed Hash Function hk(F)

POR is a method for big files using sentinels. Using this technique it is possible to use only one single key regardless of the file size or the amount of documents that the user requires to access only a tiny part of file X. Indeed, this tiny part of file X is independent of X's duration. Special sentinels blocks hidden in the information file X are randomly embedded in the information blocks in this technique. During the verification phase, the user challenges the CSP by specifying positions of a sentinel collection to ensure the integrity of the data file X and requests the CSP to return related sentinel values [14].

3.4.9   Using Hail

HAIL is a high-availability and integrity layer for cloud storage, where it enables users to save their information to various servers which makes the information redundant. The main principle of this technique is to guarantee the information integrity of the file through information redundancy. HAIL uses MAC, pseudorandom function, and a universal hash function to guarantee that the integrity of the process. The evidence is produced by this technique, which is not dependent on the magnitude of the information and small in magnitude [16–18].

3.5   Encryption Techniques to Prevent Data Confidentiality Issues

There are several encryption techniques which could be adopted to prevent data confidentiality issues in cloud computing.

Primarily 2 techniques of encryption exist-

(a) Symmetric key algorithms: This requires a single key for both encrypting and decrypting the message. E.g.: DES, AES, Blowfish Algorithm and IDEA.

(b) Asymmetric key algorithms: This utilizes 2 types of keys, public and private key which are used for encryption and decryption respectively. E.g.: RSA Algorithm and Homomorphic Encryption [5] (Fig. 1).

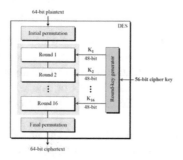

Fig. 1. DES

3.5.1   Advanced Encryption Standard (AES)

AES or Rijndael is a symmetric key algorithm which was founded by NIST in 2001. "In AES each cipher has a 128-bit block size and key sizes of 128, 192 and 256 bits respectively. AES algorithm makes sure that the hash code is also encrypted in secure manner."

The application procedure is such that users decide to use cloud services and migrate their cloud information. Then User submits to the Cloud Service Provider (CSP) its service requirements and selects the best-specified services provided by the provider. When data migration to the selected CSP occurs and when any data is uploaded to the cloud by an application in the future, the data will initially be encrypted using the AES algorithm and then sent to the CSP. Later it will be uploaded to the cloud, any request to read the data will occur after it is decrypted at the end of the users and then the user can read plain text data. The information in plain text will not be stored on the cloud anywhere. This involves all data types. This encryption method is applications transparent and could be swiftly integrated without any application modifications. Next to the encrypted information, the key is never stored as it can also compromise the key. A physical key management server should be included in the premises of the user to store the keys. This encryption protects data and keys and ensures that they stay under the control of the user and are never released in storage or transit. As DES' 56 bit keys were no longer deemed secure, AES has substituted the DES.

Algorithm

1. First Key Expansion occurs
2. In the Initial Round, the round keys are added
3. Rounds, Sub bytes- A non-uniform substitution stage where each byte is replaced by a table-based one.
4. Rows are shifted- Transposition stage where a certain number of steps is cyclically moved from each row of the system.
5. Columns are mixed- A state column mixing process combining the four bytes in each column.
6. Add Round key- Each byte of that particular state is combined with a round key and a key schedule is used to derive each round key from the given cipher.
7. Final Round followed by sub bytes then shift rows and add round key [6].

3.5.2 Blowfish Algorithm

Blowfish Algorithm is basically a symmetric key algorithm which was created in 1993 by Schneiser. It is an unpatented and license free algorithm which is made available to all users for free. It is a variable length key and a 64-bit block cipher. This algorithm has one of the best processing time [5].

3.5.3 IDEA Algorithm

IDEA algorithm is a symmetric key algorithm and was created by James Massey and Xuejia Lai in 1991. It accepts 64-bit plain text and the key size is 128-bit. The 64-bit of data is divided into 4 blocks, each having a size of 16-bits. Basic operations like addition, multiplication and XOR are applied on sub blocks. There are 8 and half rounds in IDEA algorithm and each round consists of different sub keys. There are total 52 keys used in all the rounds [7].

3.5.4 RSA Algorithm

RSA or Rivest Shamir Adleman Algorithm is an asymmetric key algorithm and was developed in 1977. In this a public key is circulated to all to encrypt the message and then a private key is used to decrypt the message, this private key is not shared with everyone. There are three steps involved in RSA: Key generation, Encryption and Decryption. This algorithm is used in cloud computing to guarantee data security. With the assistance of the RSA algorithm, we encrypted our information to provide safety. The sole aim of obtaining information is to make it accessible only to authorized users. The information being encrypted is stored in the cloud. Upon request, you can place a request with the cloud provider to access the information. The CSP authenticates the client and provides the client with the information. As RSA is a Block Cipher that maps each message to an integer. Public key is made aware to all in the proposed cloud environment, while Private

Key is made aware only to clients who initially own the data. Thus, the cloud service provider will encrypt and the client will decrypt. Once the data is encrypted with the Public Key, only the corresponding Private Key will decrypt it [7].

Algorithm

1. Choose 2 prime nos p and q
2. N = p*q then φ(n) = (p–1) * (q–1) [where φ is Euler's totient function]
3. 1 < e < φ(p*q) e is the public key exponent
4. d = e^{-1} mod(φ) d is the private key exponent
5. public key = (e, n) and private key = (d, n)
6. Encipher: c = m^e mod n
7. Decipher: m = c^d mod n

3.5.5   Homomorphic Encryption
Homomorphic Encryption uses unequal key algorithm, in this asymmetric key algorithm, a public and private key is used for encryption and decryption. Homomorphic is basically the conversion of one data set to another without losing relation between them. In this, complex functions are used to encode the data and related reverse functions are used to decode it [7].

3.5.6   Diffie Hellman Key Exchange
Diffie and Hellman introduced this key exchange protocol in 1976 with the help of discrete log problem. This is a asymmetric key algorithm.

Algorithm

1. Person A and B finalize a prime number 'p', a base 'g'
2. Person A finalizes a secret number a, sends it to Person B '(g^a mod p)'
3. Person B finalizes a secret number b, sends it to Person A '(g^b mod p)'
4. Person A computes '((g^b mod p) ^{a} mod p)'
5. Person B computes '((g^a mod p) ^{b} mod p)'
6. Person A and B can use this number as key

4   Analysis of the Methods Used

See Tables 1 and 2.

Table 1. Comparison tables of the different encryption techniques used to overcome data confidentiality issues

Parameters	DES	AES	Blowfish	IDEA	RSA	Homomorphic	Diffie Hellman key exchange
Security from the perspective of user and provider	Secure for both provider and user	Secure for both provider and user	Secure for both provider and user	Secure for user only	Secure for user only	Secure for both provider and user	Secure for both provider and user
Scalability	Scalable	Scalable	Scalable	Not Scalable	Not Scalable	Scalable	Scalable
Authentication Type	Less authentic than AES	Best authenticity provider	Best authenticity provider	Less authentic	Robust authentication	Less authentic than AES	Robust authentication
Data Encryption Capacity	Less than AES	Huge amount of data	Less than AES	Less amount of data	Less amount of data	Less than AES	Less amount of data
Memory Usage	More than AES	Low RAM needed	Executes in less than 5 Kb	Highest usage of RAM	Highest usage of RAM	More than AES	Highest usage of RAM
Execution Time	Fast	Fast	Lesser time to execute	Requires max time	Requires max time	Requires max time	Requires max time

Table 2. Comparison tables of the different techniques used to overcome data integrity issues

Techniques	Methods used for data integrity	Advantages	Disadvantages
PDP	Key generation algorithm	• This technique provides a strong proof of data integrity • Protection from low levels of corruption	•Failure to correct codes to solve corruption issues • There is no privacy preservation • Lack of dynamic support • Unbound no of queries
PDP Scheme based on MAC	Message authentication Code	• Gives powerful evidence of data integrity • It is a simple and very secure technique	• To compute new MACs, which is impossible for huge file, the data owner must retrieve the entire X file from the server • Limited number of verifications with a restricted amount of secret keys
Scalable PDP	Cryptographic hash function and key symmetric key encryption	• It offers safe encryption of PDP • Dynamic operations on outsourced data blocks is supported • It is a lightweight PDP	• Does not insert blocks where only insertions of append-type are feasible • Limited no of challenges and updates • Problems with large files

(*continued*)

Table 2. (*continued*)

Techniques	Methods used for data integrity	Advantages	Disadvantages
Dynamic PDP	Rank-based authentication skip list	• Efficient verification of integrity is performed through querying and updating the DPDP situation.	• Not used for thin clients • Extra computation is performed by client • Doesn't include provisions for robustness
POR	Encryption	• It minimizes the size of data integrity evidence as it decreases the bandwidth of the network • Reduces both clients and CSP's computational and storage overhead	• Can be only implemented on static data
POR based on Keyed hash function hk	Key Hash function	• Simple and easy to implement	• Computation cost is high • There is computational burden on the server and client • More no of keys are needed for each check
Hail	MAC, Hash function and Pseudorandom function	• Users can store data on multiple cloud	• Cannot be implemented on thin client • Can be only implemented on static data

5 Conclusion and Future Work

One could say that their cloud computing infrastructures data is secure only if it could achieve three aspects of data security, i.e. Confidentiality, Integrity and Availability.

If an encryption method is selected and the right access control mechanism is set up then it would be very difficult to lose data to unauthorized users. One has to choose the right encryption method wisely to ensure data confidentiality. There are also several solutions provided to prevent data integrity issues which could be adopted to overcome such issues.

References

1. Subashini, S., Kavitha, V.: A survey on security issues in service delivery models of cloud computing. J. Netw. Comput. Appl. **34**(1), 1–11 (2011)
2. Group, T.T.W., et al.: The Notorious Nine: Cloud Computing Top Threats in 2013. Cloud Security Alliance (2013)
3. Hubbard, D., Sutton, M.: Top Threats to Cloud Computing. Cloud Security Alliance (2010)
4. Aldossary, S., Allen, W.: Data security, privacy, availability and integrity in cloud computing: issues and current solutions. (IJACSA) Int. J. Adv. Comput. Sci. Appl. **7**(4), 485–498 (2016)

5. Chatterjee, R., Roy, S.: Cryptography in cloud computing: a basic approach to ensure security in cloud. IJESC (2017)
6. Jayant, B., Swapnaja, U., Subhash, P., Kailash, K., Sulabha, A.: Developing secure cloud storage system by applying AES and RSA cryptography algorithms with role bases access control model. Int. J. Comput. Appl. (2015)
7. Mohanaprakash, T.A., Vinod, A.I., Raja, S., Kalyan, A.P., Babu, C.B., Vivek, G.: A study of securing cloud data using encryption algorithms. IJSRCSEIT (2018)
8. Thakur, N., Sharma, A.K.: Data integrity techniques in cloud computing: an analysis. Int. J. Adv. Res. Comput. Sci. Softw. Eng. (2017)
9. Ateniese, G., Burns, R., Curtmola, R., Herring, J., Kissner, L., Peterson, Z., Song, D.: Provable data possession at untrusted stores. In: Proceedings of 14th ACM Conference on Computer and Communication Security (2007)
10. Kumar, R.S., Saxena, A.: Data integrity proofs in cloud storage. In: Third International Conference on Communication Systems and Networks (COMSNETS) (2011)
11. Chandran, S., Angepat, M.: Cloud computing: analyzing the risks involved in cloud computing environments. In: Proceedings of Natural Sciences and Engineering (2010)
12. Bowers, K.D., Juels, A., Oprea, A.: Proofs of retrievability: theory and implementation. Proc. J. Syst. Softw. (2012)
13. Curtmola, R., Khan, O., Burns, R., Ateniese, G.: MR- PDP: multiple-replica provable data possession. In: Proceedings of 28th IEEE ICDCS (2008)
14. Bowers, K.D., Juels, A., Oprea, A.: HAIL: a high availability and integrity layer for cloud storage. In: Proceedings of 16th ACM conference on Computer and Communications Security (2009)
15. Aguiar, E., Zhang, Y., Blanton, M.: An overview of issues and recent developments in cloud computing and storage security in high performance cloud auditing and applications. Springer (2014)
16. Gul, I., Islam, M.: Cloud computing security auditing. In: International Conference on Next Generation Information Technology (2011)
17. Shah, M.A., Baker, M., Mogul, J.C., Swaminathan, R.: Auditing to keep online storage services honest. In: Proceedings of the 11th USENIX Workshop on Hot Topics in Operating Systems (2007)
18. Neelaveni, R.: Performance enhancement and security assistance for vanet using cloud computing. J. Trends Comput. Sci. Smart Technol. (TCSST) 1(01), 39–50 (2019)

Challenges and Solutions in Recommender Systems

Abhishek Nair$^{(\boxtimes)}$ and Rejo Mathews

Mukesh Patel School of Technology Management and Education, NMIMS,
Vile Parle West, 400056 Mumbai, India
abhishek.nair38@nmims.edu.in, rejo.mathew@nmims.edu

Abstract. Recommender Systems play a huge part in most of our lives today. A large portion of today's digital customers rely on such programs to shape their usage of online markets. The main objective of such systems is to build relationships between the products and its users and to help them make the best decisions depending on their needs. There are 4 main types of Recommender Systems that follow different methods in order to satisfy user preferences by filtering through data in an efficient manner.

Content-based filtering systems suggest items based on attributes they share with other similar items. Collaborative filtering systems analyze the past behaviour of the customer and recommend items they might find interesting. Demographic filtering uses pre-existing and already compiled data about the behaviour of the customer based on population statistics and uses it to create a list of recommendations and the Hybrid filtering system is a combination of all these systems. Essentially, recommender systems filter through large amounts of data to give the user personalized results. This paper explains and analyzes the Filtering Systems in depth and delves into the challenges these systems face in order to produce accurate suggestions.

Keywords: Recommender systems · Content based filtering systems · Collaborative filtering systems · Hybrid filtering · User profiles

1 Introduction

Ever since the improvements in the world wide web, the way people communicate with each other has changed. People depend on the web for a lot of daily objectives now but these objectives come at the price of information and cognitive overload for the users. Thus, recommender systems help in finding relevant and related items by making relevant suggestions to the user. Such programs traverse the internet to provide appropriate recommendations to users on the basis of their specific interests or objective behaviour [1].

Recommender systems are of 4 types.

(I) Content based: A content based recommender system indicates the products that the customer has already liked in the past that are comparable in features. Profiles are then updated implicitly or explicitly and utilities allocated to objects

© Springer Nature Switzerland AG 2020
A. P. Pandian et al. (Eds.): ICCBI 2019, LNDECT 49, pp. 188–194, 2020.
https://doi.org/10.1007/978-3-030-43192-1_21

on the basis of utilities previously assigned to items observed. These kinds of filtering systems have 3 main components [2, 3].
These are:

(1) Content Analyser: Preprocessing steps are required in situations where text is present to extract the relevant information. The content analyser thus analyses various kinds of data to get the most relevant and important information.

(2) Profile learner: By collecting user preference data, this creates a user profile and attempts to generalize this data. Normally, the generalization strategy is implemented through machine-learning techniques that are capable of understanding a good example of user interests, starting with items that have been ordered in the past.

(3) Filtering component: This uses user data profile to find matching items related to the list of items, but new products and present this new feature to the use The client will be shown a recommended list of potentially interesting products.

(II) Collaborative filtering: These systems create different user groups and suggest objects that users of this group likes. This essentially pairs users with other users with similar interests. Based on this information, it then exploits correlations amongst multiple users or objects to predict missing ratings and therefore makes correct recommendations [3].

(III) Demographic: This type of recommender system utilizes profile data such as gender age, demographic region, etc. to identify prevalent consumers with comparable scores and interests dividing users by age group and living area [4]

(IV) Hybrid: These systems combine content based and collaborative filtering systems to get the best results. Thus it helps gain better results and reduce the issues and challenges of these applications. A few examples of hybrid systems are:

- Weighted: We combined numerically each recommended element with a different system rating.
- Switching: The program has multiple user suggestion choices and selects the chosen one by user preferences.
- Mixed: Program simultaneously suggests multiple different things to the user.
- Feature Combination: Different sources of information are combined to establish features of the recommendation system.
- Feature Augmentation: One of the important parts of the next method is the functionality increase used to measure a range of recommender device features.
- Cascade: Recommenders list has a weighted priority since the high rate first appears, then products with an increasing slow ranking.
- Meta-level: Is one of the to generate input techniques used and produces some kind of template to the nextstep of the algorithm of the recommender method.

2   Issues and Challenges

2.1   Cold Start Problem: This issue occurs with new customers entering the system or adding fresh products to the system. Therefore, neither the taste of the fresh customers nor the fresh products can be rated in such cases. This means that suggestions are less accurate and the recommender system is less efficient.

2.2   Synonymy: This problem occurs when two or more distinct names represent the same item. The scheme here is therefore unable to comprehend whether these are distinct objects or the same object [5].

2.3   Shilling attacks: This problem occurs when a malicious consumer or competitor joins the system and starts to give false ratings on certain items either to increase the popularity of the product or to decrease its popularity [6].

2.4   Privacy: The problem of data privacy and security is a major issue with the design of recommender systems. This makes users feel hesitant to feed their information into the process.

2.5   Grey Sheep: This issue occurs when a new user enters the system and his/her opinions do not match with any user groups and thus is unable to get the benefit of recommendation.

2.6   Sparsity: This is due to the big quantity of products in the scheme and users' disinclination to rate products. This results in a distributed matrix of the profile and consequently leading to less precise recommendations [5].

2.7   Scalability: The rate of growth of data in different systems leads to an issue in the processing of data by recommender systems.

2.8   Latency Problem: This problem occurs in certain recommender systems where items are added very frequently. In such systems, only the already rated items are suggested and new items are not shown.

2.9   Evaluation and availability of online data sets: This issue has mainly to do with recommender systems. It becomes very difficult to evaluate a recommender system and its quality. Thus finding the right criteria to rate a recommender system is very important.

2.10   Context Awareness: The absence of context awareness is a significant problem with primitive recommender systems. Properties such as present place, present activity and time can assist the system get better customer suggestions.

3   Solutions

3.1   Increasing user input: Increasing user output can help solve a number of problems such as the cold start problem and the grey sheep problem. This can be done by asking the users to state their taste as well as rate their items well in advance [1, 2].

3.2   Single Value Decomposition (SVD): Single Value decomposition is a method of linear algebra in which a real or complex matrix is factorized. Its eigen-decomposition of a positive semi definite normal matrix is then taken via an extension of polar decomposition. This is used to solve a lot of problems in Recommender Systems such as Synonymy, Scalability and Sparsity issues.

3.3 Incorporating detection methods: This method is mainly used to solve the issue of Shilling Attacks. These kinds of attacks are categorized into bandwagon, random, average and reverse bandwagon attacks. These different kinds of attacks can be detected by methods such as model specific attributes, generic attributes, prediction shift and hit ratio.

3.4 Cryptographic Mechanisms: Cryptographic mechanisms are used to protect against the intentional or accidental alteration of data. Thus, it helps solve the problem of privacy in such sytems. The primary privacy problem in recommendation systems is that user details are stored in a central repository that can be damaged and therefore lead to information misuse. Cryptographic mechanisms or other techniques such as randomized perturbation techniques can be used to allow user publishing without divulging personal details.

3.5 Filtering/Clustering: The grey sheep problem is a very common problem found in CF Systems. It can easily be resolved using CB filtering or hybrid filtering methods. Another solution is using offline clustering techniques like K-mean clustering in which data is divided into small clusters [7].

3.6 Better categorization: Latency is a common issue observed in recommender systems in which some items are excessively selected. In tandem with the consumer definition, a category-based approach can be used.

3.7 Evaluation and availability of online datasets: The absence of assessment metrics in the Recommender Systems sector contributes to a major issue with how distinct recommender systems can be evaluated.

3.8 Improved user analysis: Three major techniques are used to overcome this issue of Context awareness. These are: detecting facial expressions, recording speech interpretation, and physiological signals analysis [11–15].

4 Analysis/Review

See Table 1.

Table 1. Comparison table

Solutions/types of recommender Systems	Content based recommender systems	Collaborative recommender systems	Demographic recommendation systems	Hybrid recommendation systems
Increasing user input	Solves cold start problem	Creates bigger and better user groups and thus suggest better items, solving issues like the grey sheep problem	Better user groups depending on the demographic features is formed and thus the recommendation system works better	Increasing user input helps create better user groups and thus helps in all kinds of hybrid recommender systems

(*continued*)

Table 1. (*continued*)

Solutions/types of recommender Systems	Content based recommender systems	Collaborative recommender systems	Demographic recommendation systems	Hybrid recommendation systems
Single value decomposition	Problems such as synonymy, scalability and sparsity are solved in these systems	Solve the issue of sparsity in these systems	Solves the issue of sparsity and scalability in these systems	Problems such as synonymy, scalability and sparsity are solved in these systems
Incorporating detection methods	Solves the issue of shilling attacks in these systems as it stops malicious users from entering the system	Solves the issue of shilling attacks in these systems as it stops malicious users from entering the system	Is not of much use in these systems because shilling attacks aren't very common in these systems	Solves the issue of shilling attacks in these systems as it stops malicious users from entering the system
Cryptographic mechanism	Protects the data of the users in such systems. These mostly include the products the user has liked in the past	Protects the personal data of the user such as age, likes, other users with similar likes etc	Protect the personal data of the user such as age, height, address etc from malicious users	Protects all the personal data of the user in such systems
Filtering/clustering	Isn't used in such systems because the grey sheep problem isn't observed here	Solves the issue of Grey sheep in such systems	Solves the issue of Grey sheep in such systems	Isn't used in such systems because the grey sheep problem isn't observed here
Better categorization	Helps solve the issue of Latency in such Systems	Helps solve the issue of Latency in such Systems	Helps solve the issue of Latency in such Systems	Helps solve the issue of Latency in such Systems
Improved user analysis	Helps better predict what the user likes to make better and accurate predictions	Detecting facial expressions, recording speech interpretation and physiological signals analysis leads to better creation of user groups	Such Systems are heavily based on user analysis and thus will lead to better recommendations for users	Detecting facial expressions, recording speech interpretation and physiological signals analysis leads to better creation of user groups for such systems

5 Conclusion

Recommendation Systems are an important research area today. Such programs are used to recommend items to the user, based on their tastes and the previous purchases and/or preferences. The various types of suggested systems and the advantages and disadvantages of these systems are discussed in this article. It also discusses solutions to these problems and highlights the need for further research into this up and coming area of product management and consumer relations. We have concluded from this research paper that hybrid recommender systems are the best recommendation system currently present in the market place due to its ability to combine the best features of content-based and collaborative recommender systems. These systems have widespread use in the tourism domain such as the personalized context-aware hybrid travel recommender system (PCAHTRS).

References

1. Pu, P., Chen, L., Hu, R.: Re-evaluating recommender systems from the user's perspective: survey of the state of the art. User Model. User-Adap. Inter. 22(4–5), 317–355 (2012)
2. Adomavicius, G., Tuzhilin, A.: Toward the next generation of recommender systems: a survey of the state-of-the-art and possible extensions. IEEE Trans. Knowl. Data Eng. 17, 734–749 (2005)
3. Montaner, M., López, B., de la Rosa, J.L.: A taxonomy of recommender agents on the Internet. Artif. Intell. Rev. 19(4), 285–330 (2003)
4. Jain, S., Grover, A., Thakur, P.S., Choudhary, S.K.: Trends, problems and solutions of recommender system. In: International Conference on Computing, Communication and Automation (2015)
5. Su, X., Khoshgoftaar, T.M.: A survey of collaborative filtering techniques. Adv. Artif. Intell. (2009)
6. Schafer, J.B., Frankowski, D., Herlocker, J., Sen, S.: Collaborative filtering recommender systems. In: The Adaptive Web (2007)
7. Ghazanfar, M.A., Prügel-Bennett, A.: Leveraging clustering approaches to solve the gray-sheep users problem in recommender systems. Expert Syst. Appl. 41(7), 3261–3275 (2014)
8. Adomavicius, G., Sankaranarayanan, R., Sen, S., Tuzhilin, A.: Incorporating contextual information in recommender systems using a multidimensional approach. ACM Trans. Inf. Syst. (TOIS) 23(1), 103–145 (2005)
9. Melville, P., Mooney, R.J., Nagarajan, R.: Content-boosted collaborative filtering for improved recommendations. In: AAAI/IAAI (2002)
10. Shahabi, C., Banaei-Kashani, F., Chen, Y.S., Yoda, M.D.: An accurate and scalable web-based recommendation system. In: Cooperative Information Systems (2001)
11. Xu, S., Jiang, H., Lau, F.: Personalized online document, image and video recommendation via commodity eye-tracking. In: Proceedings of the 2008 ACM Conference on Recommender Systems (2008)
12. Shah, K., Ali, Z., Ullah, I.: Recommender Systems: Issues, Challenges, and Research Opportunities. Springer, Singapore (2016)
13. Mohamed, M.H., Khafagy, M., Ibrahim, M.H.: Recommender systems challenges and solutions survey. In: 2019 International Conference on Innovative Trends in Computer Engineering (ITCE 2019), Aswan, Egypt, 2–4 February 2019

14. Batet, M., Moreno, A., Isern, D.: Agent-based personalised recommendation of touristic activities. Expert Syst. Appl. **39**(8), 7319–7329 (2012)
15. Niraki, A.S., Kim, K.: Ontology based personalized route planning system using a multi-criteria decision. Expert Syst. Appl. **36**(2), 2250–2259 (2012)

Review of Machine Learning in Geosciences and Remote Sensing

Noel David$^{(\boxtimes)}$ and Rejo Mathew

Mukesh Patel School of Technology Management and Engineering,
NMIMS University (Deemed-to-be), Mumbai, India
noel.david69@nmims.edu.in, rejo.mathew@nmims.edu

Abstract. It is known that the machine learning algorithms are able to process data without the need of human intervention. A lot of the research and analysis involved in Geosciences and Remote Sensing is labor intensive and demanded high amount of resources. The problems in Geosciences are usually different to what is encountered in other applications, requiring unique techniques and formulations. A more time and cost effective method is needed to help classify, identify, and collect the required data. Machine learning is used to solve various problems in geosciences and remote sensing effectively. Machine learning is a collection of various algorithms such as support vector machines, gradient boosting machines, trees, etc. which provide us with different options of mapping our results, including classification, identification and prediction. The main objective is to use Machine learning as a tool to learn from the given data and effectively solve problems in remote sensing and geosciences.

Keywords: Geosciences · Genetic Programming · Machine learning · Remote sensing · RSI images · Regression · Support Vector Machine

1 Introduction

Geosciences are the study about the Earth, especially how it all functions, the atmosphere, oceans, lakes, soil and so much more. An important part of geosciences is acquiring data; to make better predictions and simulations [10]. This paper uses Machine learning to analyze and answer the problem(s) more efficiently [2]. Machine learning is part of Artificial Intelligence where the algorithm improves itself overtime without the need of any human intervention. Genetic Programming is used to characterize and classify various rock types with accurate computer program simulations [2]. With the addition of sensors, algorithms can be used to find the composition of aerosol particulate, which plays a major role in finding the impact of these aerosols on the human body and in case of availability of a large amount of information [2]. Machine learning is also used to detect changes in satellite images over a period of time from data collected since a decade, 1998 to 2011 to detect changes made in the water bodies, topography and if any urban development as taken place [3]. Using a hybrid environment, satellite images are classified with the help of three different machine learning algorithms [7]. This paper consists of the various problems related to geosciences and remote sensing,

© Springer Nature Switzerland AG 2020
A. P. Pandian et al. (Eds.): ICCBI 2019, LNDECT 49, pp. 195–204, 2020.
https://doi.org/10.1007/978-3-030-43192-1_22

including their solutions. These solutions are also compared and analyzed and followed with a conclusion and references.

2   Related Work

Different problems active in geosciences and remote sensing are shown below, different applications of Genetic Programming over Neural and Artificial Networks are given [9], GP is also used to solve engineering in different applications [5, 15], PM 2.5 levels vary throughout the world, and a method is required to measure its impact on the world and on the human body. Dust Sources are linked directly to climate change and they affect the energy budget of the impacting cloud cover, if the dust sources present all around the world could be classified, the data would be very useful, helping to derive patterns and predictions in changes in climate. Satellite Data is traditionally mapped and identified manually, which is very time consuming, automation of this process is needed for more accurate results [3].

3   Problems in Geosciences and Remote Sensing

3.1   Choosing the Right Machine Learning Model

Machine learning being a newer aspect in the field of geosciences and remote sensing, there is lot of scope of improvement and growth but it is difficult to choose the correct algorithm/models to use with so many available options such as Artificial Neural Network, Support Vector Machine, and Decision Trees etc. [2]. The algorithm being employed must satisfy the requirements of the problem, but many of the above mentioned models are not capable of generating practical prediction equations or some are downright incompatible. This makes them black box models, choosing the correct model is of utmost importance - One that can handle the data and produce the necessary results in remote sensing and geosciences applications [2].

3.2   Classifying and Characterizing Airborne Particulate Matter Using Machine Learning

Climate change has become a increasingly big problem that cannot be ignored any more, by a survey held by the WHO [6], 7 million deaths across the world by 2012 was because of air pollution, aerosols included [2]. These aerosols have a diameter of less than 10 um can enter the lungs and ones with a diameter of less than 2.5 um can enter the gas exchange region of the lungs hence directly affecting the human health [2], long or short term [2]. Another problem is that the PM 2.5 levels cannot be accurately measured if the different regions are not covered by the sensors properly, which can lead to unstable results. The current PM2.5 levels in Delhi is 193 which is considered unhealty [12]. The average PM2.5 levels are rising throughout India. A solution needs to be found to effectively measure the PM 2.5 levels and its impacts using Machine learning and Remote Sensing [12].

3.3   Change Detection in Satellite Images in Delhi, India

Change detection in satellite images is used for many purposes, land/forest cover, change in buildings and settlements, change in water bodies, etc. [3]. Change detection is usually done manually but the process is very time consuming and labour intensive. The outcome of manual analysis is also heavily affected by the opinions of the analyst doing the analysis. Some may agree while some may disagree – there is no concrete answer. Furthermore, mechanical automation of the process is difficult and complex [3]. A method needs to be found to efficiently analyze these images and provide concrete results that keep true with time.

3.4   Classification of Satellite Images in a Hybrid Environment

In the recent times, with the advent of the exploration age, the popularity of acquiring data through remote sensing has increased exponentially. Researchers have access to large amounts of data through which they can analyze and study the behavioral patterns of the earth, but acquiring the data is not the challenge here [7]. The original analysis method was text based but this method failed to accurately describe or provide data for the images. Efficiently analyzing and classifying is the current drawback being faced by the analysts. Deep learning became a viable solution with the advent of higher resolution equipment but didn't gain much traction due it requirements, .i.e. a very large data set; it becomes incredibly difficult to process and analyze the computations and the results [7]. A technique was needed which could reduce the dimension of the images for faster computations and provide adequate amount of information [7].

4   Solutions and Countermeasures for Geosciences and Remote Sensing

4.1   Using Machine Learning Techniques as an Efficient and Accurate Solution to Geo-Scientific Problems of Classification and Characterization

4.1.1   Genetic Programming

The fairly new Genetic Programming Model (GP) [2, 9] is used to solve machine learning problems. Genetic Programming is a special case of Genetic Algorithm (GA) where each different individual (raw data) is a made into a computer program. These "computer programs" are then evolved over time to perform the needed task. Genetic algorithms on the other hand are search algorithm that uses the process of natural selection. A helpful feature of the Genetic programming is that it can produce the required equations without the need to enter the type of the existing relationship [2]. MEP (Multi Expression Programming), is very accurate in predicting and classifying the different types of soil [4]. Genetic Programming is especially great in analyzing and characterizing different types of rocks such as limestone, granite rock(s) among many others such as rock mass modulation as stated in [11].

4.1.2   Support Vector Machine

Support Vector Machine is a very commonly used technique in solving machine learning problems especially when dealing with limited amount of data or data with a lot of parameters. In the paper [1], Support Vector Machines have been used to classify potato diseases with an accuracy of 95%. This Support Vector Machine technique analyzes plant images and detects diseases based on different patterns. The data set is publicly available consisting of over 50,000 different images of health and diseased plants [1]. This paper [1] uses a multiclass Support Vector Machine model to classify images. The areas of interest is isolated, the disease regions have a big difference in colour compared to the healthy areas of the leaves. Image background, healthy areas of the leaves are removed from the images and leaving only the disease regions. This paper [1] extracts up to 10 different features and the SVM model is trained based on the isolated information. SVM provided high accuracy with comparatively lower computational costs.

4.2   Using Machine Learning to Analyze Aerosol and Particulate Matter

Using Machine learning, the data from the study can contribute to the Health Impact Assessment (HIA) and geographical coverage which can contribute to revealing any hidden impacts of PM 2.5 on health [2]. The premise of the solution is to use about 50 sensors to collect data [2]. Machine learning is also to predict the aerosol particulate accurately, especially at the lower atmosphere levels where PM 2.5 can most commonly come in contact with humans. The gaps in coverage due to the weak sensor framework can be taken care of by using AODs (Aerosol Optical Depth).

4.2.1   Using a Gradient Boosting Machine Learning Model

This paper [14] uses a Geographically Weighted Gradient Boosting Model (GW-GBM) to predict daily concentration levels of PM 2.5 particulate matter. The significant advantage provided by this model is that it can provide accurate predications despite of insufficient sensor coverage [14]. The GW-GBM is made up of multiple Gradient Boosting Models. Each Gradient Boosting machine was used to cover a 0.5×0.5 area [14]. Insufficient data due to limited coverage over the region is also compensated by replacing the missing data with the most behaviourally similar data for better results.

4.2.2   Linear Support Vector Machine

Linear SVM is the simplest form of Support vector Machines; it is a regression and classification algorithm which uses a hyperplane in an Nth Dimension to classify data [15]. A hyperplane is a fancy way of representing a boundary; points on either side of this boundary represent a different class of data. In a way, L-SVM breaks down the plane into different parts. The main objective of a Support Vector Machines too specialize data and isolate the classes to the greatest extent. As the name suggests, L-SVM uses a linear kernel, i.e. the data exists in a linear format and can be represented by the algorithm in a linear fashion [15].

4.3   Using Machine Learning Algorithms to Analyses Satellite Images

The paper [3] uses an three different Machine learning Algorithm to detect change and analyze the input satellite images; three different types of machine learning algorithms are used [3]:

4.3.1   K Means (Partition Based - Clustering)

The Clustering as the name suggests means classifying similar or related data into groups. In k-means clustering, we divide the data in predefined classes. So K Means cluster method divides 'n' number observations into different 'k' predefined clusters and the observation follows the cluster with the closest mean, and acts as its prototype. K-Means is also knows as Lloyd's Algorithm [3, 13], and helps to provide a basic structure to the data [8, 13].

4.3.2   FCM (Fuzzy Network – Fuzzy C Means)

Fuzzy C Means is also a clustering technique, but as an exception allows a piece of data to exist in 1 or more clusters. Fuzzy C Means produces the correct number of partitions by minimizing the error of the function [3]. Fuzzy C Means is also efficient in recognizing patterns in images which is especially useful in analyzing the input satellite images. Fuzzy C Means assumes that every data point belongs to very cluster rather than it belonging to just one cluster [8].

4.3.3   EM (Probability Based – Expectation Maximization)

Expectation Maximization involves adding variable to unknown values by imputation [3]. In ML a lot of values are observable at some point of time then unobservable for the rest. Expectation Maximization is a probability based Machine Learning algorithm which helps to predict values of the data when it is not observable. Expectation Maximization is an unsupervised learning algorithm. With each iteration of the algorithm, the accuracy of the prediction increases [3].

The use of spectral indices helps classification. Any bright portion in the images signifies high density in that portion. K Means being the most accurate. The study accurately predicted there has been a 20% increase in development areas, while vegetation has decreased by 18% [3]. All the 3 algorithms showed that Machine learning can be used to effectively show changes over a period of time in satellite images [3].

4.4   Classification and Analysis of Images Using Machine Learning

This paper [7] uses 2 different feature extraction techniques to extract information from the images, Binarization and Morphological Operator. Feature extraction is the process of identifying and extracting important parts of images which will later be used for classification. 1000 images [7] (256 × 256) are taken from an open source website and classified under different areas of interest. This paper uses a hybrid solution with two different fusion techniques, early and late fusion [7]. The 2 feature extraction techniques are standardized to remove any bias and fused to form the training data for these 3 algorithms. K-Nearest Neighbour, Support Vector Machine and Random Forest algorithms are different in the way they process data, therefore covering a broad spectrum of analysis results:

4.4.1   K Nearest Neighbour

K Nearest Neighbour is one of the most basic classification Machine Learning algorithms. K Nearest Neighbour is a non parametric algorithm, because it has no knowledge or makes no presumptions of the distribute data. We can also call this a 'Lazy' algorithm. Using K Nearest Vector, an element A can find the 'k' number of elements closest to it. A will then be classified to the majority winning surrounding elements' class [7].

4.4.2   Support Vector Machine

Support vector machine is another classification and regression algorithm. It classifies data by drawing a hyperplane between the two sets of data [7]. Every Support Vector Machine uses a kernel to draw the hyperplane. This kernel turns the problem into an algebraic equation. A linear kernel deals with only simpler classifications while the polynomial kernel is used for more complex distribution patterns [7]. Maintaining a good margin is also very important. The hyperplane should be equidistant from both the sets to ensure good accuracy [7].

4.4.3   Random Forrest

Random Forrest algorithm helps make predictions for classification and regression. Random forest uses a collection of data trees to classify data [7]. When an element needs to be classified each forest provides data to classify that element. This method is considered to a 'black box model' because we have little to no control over the inner workings of the model. Random Forests consists of decision trees. These trees can fit even complex data [7]. Any element which needs to be classified starts form the root node of multiple decision trees and proceeds slowly working its way to the bottom, till it can't be split anymore. This last leaf node counts as 1 vote for a set. Each tree will cast such as vote and the majority of the vote will be the set in which the element will be assigned to [7].

5   Analysis of the Methods Used

See Tables 1, 2 and 3.

Table 1. Comparison tables of the different machine learning techniques

Parameters	Optimal ML Model for Geosciences and Remote Sensing Problems		Analysis of Particulate Matter using Machine Learning	
	Genetic Programming (GP) [2]	Support Vector Machine (SVM) [1, 2]	Gradient Boosting Machine (GBM) [14]	Linear Support Vector Machine (L-SVM) [14, 15]
Training Dataset	Large dataset required due to its high search rates	Limited data is sufficient. SVM only generates a classifier based on the training data	Large dataset required as multiple data trees required to analyze data	Fewer amounts of data is also sufficient as linear kernel works on single polynomial equations
Type of Algorithm	Characterization	Clustering	Prediction Models	Clustering Classification
Resource Requirements	Very resource intensive as it simulates entire computer programs	Less resource intensive, as the classifier depends on the input data	Very resource intensive uses trees to analyze training data	Less resource intensive
Time Requirements	Quick outputs compared to SVM	Comparatively slower than GP	Faster	Slower, due to amount of clusters
Learning Model	Supervised/Unsupervised	Supervised	Supervised	Unsupervised - Clustering
Accuracy	Depends on fitness of dataset, generally more accurate than SVM as in [2]	Classification accuracy of 95% as mentioned in [1]	Better Accuracy when compared to SVM. 85% as mentioned in [14]	GBM outperforms SVM in different conditions
Complexity of problems	Can handle complex problems with ease	Difficulty in execution increases with complexity of problems	Can work with complex problems	Difficulty in execution increases with complexity of problems
Output	Computer Programs	Binary, Numeric Format	Converts weak dataset to strong dataset	Outputs in Numeric, Variable Format

Table 2. Comparison tables of the different machine learning techniques

Parameters	Machine learning algorithms to classify images		
	K-Means [3]	Fuzzy C Means (FCM) [3]	Expectation Maximization (EM) [3]
Training Dataset	Land Satellite images taken from 1998 to 2011		
Type of Algorithm	Partition based clustering algorithm [3]	Fuzzy based clustering algorithm (Soft K-Means)	Probability Based
Time Requirements	Faster than FCM [3], fastest method as in [3]	Slower than K-Means as in [3]	Slowest of all methods as in [3]
Silhouette Value	Highest compared to FCM and EM as mentioned in [3]	Lower than K Means, higher than EM	Lowest
Accuracy	Highest Accuracy	Lower than K-Means	Lowest accuracy
Dataset Complexity	Struggles with larger clusters, such clusters need to be generalized	Deals with larger more efficiently compared to K-Means and EM	Efficiency decreases with increase in cluster size
Output	Centroids of the predefined clusters and the labeled data	Outputs Labeled data	Parameters Conditional probability of the data points in the clusters

Table 3. Comparison tables of the different machine learning techniques

Parameters	Image Classification/Feature Extraction using Machine Learning		
	KNN (K-Nearest Neighbor) [7]	SVM (Support Vector Machine) [7]	RF (Random Forest) [7]
Dataset	Requires more data than SVM as it is a slow learner	Least amount of data required	Most amount of data required, uses multiple trees to decide on a common vote
Type of Algorithm	Classification and Regression	Clustering	Classification
Result Significance in [7]	Significant as closest to theoretical values	Least Significant as not matching with the theoretical values	Most Significant, closest to theoretical values
Computational Ability	Cannot handle large data sets, may lead to inaccurate results	Can manage larger dataset with polynomial kernel	Worse with increasing size of dataset
Error Rate	Highest error rate when dealing with multiple parameters	Low error rate	Least Error Rate due to large data set
Learning Model	Supervised/Unsupervised	Supervised	Supervised
Accuracy	Second Best accuracy, not friendly to large training dataset	Least Accurate as in [7]	Highest accuracy due to its majority voting process
Output	Outputs a discreet value	Labels unlabeled data	Outputs tree like structure for classification

6   Conclusion

Machine learning Algorithms give us the results without the need of human intervention. Genetic Programming, an extension of the Evolution Model paved the way to identify various properties of rocks and remote sensing problems such as liquefaction phenomenon, ground movement etc. [3, 10, 15]. Accurate Analysis and identification of PM2.5 particulate matter was possible through various supervised and unsupervised learning methods where Gradient Boosting was more efficient and accurate than its competition [2]. In remote sensing, K-Means was the most accurate unsupervised model to classify images. For image classification through hybrid fusion and feature extraction, K Nearest Neighbor was the most accurate in correctly classifying images with the new hybrid method as mentioned in [7]. Although manmade object detection is still a challenge, using CNN (Region Based CNN), [17, 18] and transfer learning, airplanes can be detected easily. To gather hyper spectral data for remote sensing, [16] suggests using Hyper2Lidar network and Parameter learning to gather hyper spectral data. This hyper spectral data compared to RGB images are more precise and are capable of improving the current models in machine learning. Having better data sources will increase the scope of machine learning in remote sensing and geosciences as the models will be able to perform efficiently and with increased accuracy.

References

1. Monzurul, I., Anh, D., Khan, W., Pankaj, B.: Detection of potato diseases using image segmentation and multiclass support vector machine. In: CCECE (2017)
2. Lary, D.J., Alavi, A.H., Gandomi, A.H., Walker, A.L.: Machine learning in geosciences and remote sensing. Geosci. Front. 7(1), 3–10 (2015). China University of Geosciences
3. Bhatt, A., Ghosh, S.K., Kumar, A.: Automated change detection in satellite images using machine learning algorithms for Delhi, India. In: IEEE International Symposium on Geoscience and Remote Sensing (2015)
4. Alavi, A.H., Gandomi, A.H., Sahab, M.G., Gandomi, M.: Multi expression programming: a new approach to formulation of soil classification. Eng. Comput. 26(2), 111–118 (2010)
5. Javadi, A.A., Rezania, M., Nezhad, M.M.: Evaluation of liquefaction induced lateral displacements using genetic programming. Comput. Geotech. 33, 222–233 (2006)
6. WHO: World Health Organization Press Release (2014). http://www.who.int/mediacentre/news/releases/2014/air-pollution/en/
7. Das, R., De, S., Thepade, S.: Machine learning in hybrid environment for information identification with remotely sensed image data. In: Gavrilova, M., Tan, C. (eds.) Transactions on Computational Science XXXIV. LNCS, vol. 11820. Springer, Heidelberg (2019)
8. Khatami, R., Mountrakis, G., Stehma, S.V.: A meta-analysis of remote sensing research on supervised pixel-based land-cover image classification processes: general guidelines for practitioners and future research. Remote Sens. Environ. 177, 89–100 (2016)
9. Joseph, S.I.T.: Survey of data mining algorithm's for intelligent computing system. J. Trends Comput. Sci. Smart Technol. (TCSST) 1(1), 14–24 (2019)
10. Koza, J.: Programming of Computers by Means of Natural Selection. MIT Press, Cambridge (1992)

11. Karpatne, A., Ebert-Uphoff, I., Ravela, S., Babaie, H.A., Kumar, V.: Machine learning for the geosciences: challenges and opportunities. IEEE Trans. Knowl. Data Eng. **31**(8), 1544–1554 (2019)

12. Ravandi, E.G., Rahmannejad, R., FeiliMonfared, A.E., Ravandid, E.G.: Application of numerical modeling and genetic programming to estimate rock mass modulus of deformation. Int. J. Min. Sci. Technol. **23**(5), 733–737 (2013)

13. Lary, D.J., Faruque, F., Malakar, N., Moore, A., Roscoe, B., Adams, Z., Eggelston, Y.: Estimating the global abundance of ground level presence of microscopic particulate matter. Geosp. Health **8**(3), S611–S630 (2014)

14. Kanungo, T., Mount, D.M., Netanyahu, N.S., Piatko, C.D., Silverman, R., Wu, A.Y.: An efficient k-means clustering algorithm: analysis and implementation. IEEE Trans. Pattern Anal. Mach. Intell. **24**, 881–892 (2002)

15. Zhan, Y., Luo, Y., Deng, X., Chen, H., Grieneisen, M.L., Shen, X., Zhu, L., Zhang, M.: Spatiotemporal prediction of continuous daily PM2.5 concentrations across China using a spatially explicit machine learning algorithm. Atmos. Environ. **155**, 129–139 (2017)

16. Kleine Deters, J., Zalakeviciute, R., Gonzalez, M., Rybarczyk, Y.: Modeling PM2.5 urban pollution using machine learning and selected meteorological parameters. J. Electr. Comput. Eng. **2017**, 14 (2017)

17. Ozkan, S., Bozdagi Akar, G.: Hyperspectral data to relative lidar depth: an inverse problem for remote sensing. In: Proceedings of the IEEE Conference on Computer Vision and Pattern Recognition Workshops (2019)

18. Hassan, A., Hussein, W.M., Said, E., Hanafy, M.E.: A deep learning framework for automatic airplane detection in remote sensing satellite images. In: 2019 IEEE Aerospace Conference, pp. 1–10. IEEE (2019)

Data Storage Security Issues and Solutions in Cloud Computing

Tapendra Singh Rathore$^{(\boxtimes)}$ and Rejo Mathew

Department of Information Technology,
NMIMS (Deemed-to-be) University, Mumbai, India
tapendrarathore15@gmail.com, rejo.mathew@nmims.edu

Abstract. Cloud computing is the backbone of most of the modern technological functioning employed across the globe. It's primary concern is to store
data in an efficient and secured manner. In cloud computing, users can access
their data from server on which their data is stored. User access data from
servers when required via the cloud service providers. The server is most vulnerable to threats as it is the main hub where the data is stored, not only server is
vulnerable to attacks but also the channels via which the user access data hence
everywhere precautions has to be taken. In order to provide a secure access of
data to the users various measures have to be made. Cloud computing is vulnerable to various threats as the data is accessed from remote server that are
away from users, this also make users doubtful to prefer cloud over traditional
storage. This paper reviews various security issues being faced with data storage
in cloud computing.

Keywords: Cloud storage · Coprocessors cloud storage · Searchable
encryption · Kerberos · Data obfuscation · Steganography

1 Introduction

Cloud computing is the model for providing on demand services to the users on the
internet via remote servers rather than their local server. Cloud computing provides
various services like IaaS which stands for Infrastructure as Service, Software as a
Service and Platform as a service. Such services could help in providing users an
environment in which they can perform their task efficiently. Many industries are
shifting towards cloud computing as it is an on demand service. Many industries shave
to deal with a large amount of vital data which could be easily stored on cloud and can
be accessed when required.

1.1 Cloud Storage

Cloud computing provides users a storage platform in which users can store and access
their data according to their convenience. Cloud storage consists of a large amount of
data that is provided from various kinds of users. Data in cloud storage could be highly
confidential and important to the users. In such cases a high level of security has to be
provided to the data. Not only providing a secure storage is the concern but also

© Springer Nature Switzerland AG 2020
A. P. Pandian et al. (Eds.): ICCBI 2019, LNDECT 49, pp. 205–212, 2020.
https://doi.org/10.1007/978-3-030-43192-1_23

providing a secure access to the storage is an important aspect in cloud computing, due to the presence of a large amount of data, availability of the user's data has to be taken care of by the host.

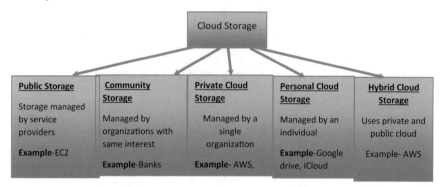

1.1.1 Personal Cloud Storage [1]
In this type of storage an individual stores their data on cloud, it usually consists of small files like photos, videos, documents etc. In such cloud storage user have more administrative power [1].

1.1.2 Public Cloud Storage
In this type of cloud storage the storage provider manages the user's cloud storage. It requires less administrative controls. The user authorized can access the service [1].

1.1.3 Private Cloud Storage
In this type of cloud storage access is provided to a single organization. It uses the resources with its own internal platform. It does not share its access or authorities with any other organization [1].

1.1.4 Hybrid Cloud Storage
It is the combination of public and private cloud storage. In this type of cloud storage a user can store their important data on private cloud and data that is not usually used or the data that is not of high importance in public cloud [1].

1.1.5 Community Cloud Storage
In this type of cloud storage many organizations with similar interests uses a common cloud platform.

Community Clouds are a subset of Public Clouds tweaked to a particular vertical industry, for example, medicinal services or account and government organizations that all offer an assortment of administrations including Software as a Service (SaaS) or Platform as a Service (PaaS) [2].

2 Issues and Attacks in Cloud Computing Data Storage

2.1 Weak Cryptography

In order to make the process of cloud computing much faster the major cloud storage providing hosts have diverted their primary focus to provide a faster and simpler data storage rather than providing a secure data storage to the users. Encryption uses an addition bandwidth to encrypt data before it is sent to the cloud, hence many service provider usually avoid a secure encryption [3].

2.2 Lack of Secure Client Confirmation Access to the Cloud Storage

Providing clients an access to the cloud storage is difficult in the current world situation where highly advanced attacks are made on a user's active internet sessions. Key logging and other highly advanced attacks are made on a regular basis to capture the password of a user, in such a situation the confirmation of client's identity without revealing their credentials like password is very difficult [4, 5].

2.3 Attack on Storage Server

In cloud computing the data is stored in various host server, hence it becomes the hub hotspot for the attacks in order to access data or cause harm to the data stored [6].

2.4 Types of Attacks

2.4.1 Side Channel Attack

In side channel attacks a malicious virtual machine is placed on the same server as the main virtual machines and it is used to extract the cryptography algorithm's implementation. Side channel consists of two major steps Placement and Extraction. Placement is placing the malicious virtual machine near the main virtual machine. Extraction is the process of extraction of vital file and information from the main virtual machine [7, 8].

2.4.2 DoS (Denial of Service)

DoS is the most common attack on the server, in cloud computing various users use servers to access and store data. Due to already a large amount of users trying to use server it becomes easier to make a DoS attack on cloud storage server. DoS attack usually intent to reduce the resources of cloud in order to block the supply of cloud services. DoS attack takes the advantage of various loopholes in the cloud system in order to achieve their goal [7, 9, 15].

2.4.3 Malware Injection

In malware injection the attacker uploads a modified copy of victim's service instance initiating the execution of malicious code. In such scenario a hacker's queries are assumed to be a victim's query and hence when the request is uploaded the malware injection takes place successfully [7, 10].

3 Countermeasures for Storage Issues in Cloud

3.1 Integrating Transposition and Substitution for Encryption [3]

Cryptography's idea is mainly divided into two ground level concept transposition and substitution [11].

In substitution ciphers, letters are supplanted by different letters in request to create the cipher text, while transposition cipher includes scrambling the letters of a plaintext with the end goal that they result in an alternate position [12].

The given algorithm uses transposition and substitution cipher for improving the classical encryption. Plain text is converted into the corresponding ASCII value of the alphabets. Its key range is between 1 to 127.

3.1.1 Encryption Algorithm for Integrated Transposition and Substitution Encryption [3]

The following flow chart describes how Transposition and Substitution Encryption can be used for Encrypting data in order to make it hard for attackers to decode it. It also display how the Decryption process is used for decrypting the ciphered text (Fig. 1).

3.1.2 Decryption Algorithm for Integrated Transposition and Substitution Encryption [3]

3.2 Kerberos Protocol for Storage Access Authentication [5]

To secure delicate information of a user Kerberos protocol is proposed. Kerberos is network authentication protocol in which the server grants a token to the user when a user applies to get a token. Kerberos server consists of two different parts Authentication Server and Ticket Granting Server. AS authenticates the user and TGS grants the token to user. It works in three different [5] (Fig. 2).

3.3 Encrypted Data Storage Coprocessor for Separate Encryption and Storage [1]

Using Co-Processor can solve the major issue of data protection on the server, using coprocessor it is possible to store extremely sensitive data inside coprocessor itself. Coprocessor consists of memory which can store a limited amount of information, coprocessor has an encrypting system which can encrypt data. It has to be installed on server. If this coprocessor is tempered by attackers then it will clear all of its information. It is not possible to store all information in this coprocessor as it will end up heating itself hence it is possible for user that they could store the encrypt decrypt keys or other set of vital keys or information [1].

IBM 4758 Cryptographic Coprocessor (IBM) is a cryptoprocessor, it consists of a programmable PCI board and a CPU and it also has memory storage that can delete itself when an attempt of tampering is detected.

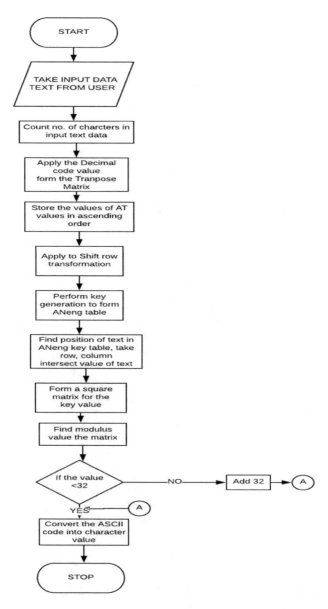

Fig. 1. Encryption algorithm for integrated transposition and substitution encryption

3.4   Searchable Encryption [1]

In searchable encryption an index is built for the files carrying sensitive or important information. Generally while encrypting the entire data is encrypted and then if a user needs to access anything in data entire data has to be decrypted. In this method index file is encrypted and stored along with the file so while searching for a particular set of

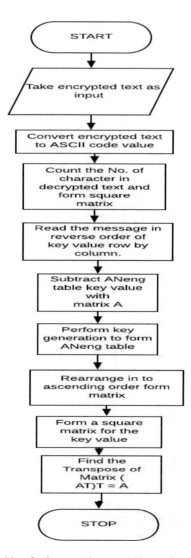

Fig. 2. Decryption algorithm for integrated transposition and substitution encryption [3]

information only the keywords are required to be searched, this will save a large amount of time when implemented [1].

3.5 Data Obfuscation and Steganography for Enhanced Public Cloud Storage [13]

The following technique uses a combination of data obfuscation and steganography for preventing unauthorized access. In this technique data is hidden in images, in order to prevent an attacker to get the data in image first the data is obfuscated format using MRADO Technique and then the obfuscated data in implanted in the image using LSB embedding method.

3.5.1   MRADO Technique [14]

MRADO Technique [14] enhances the classical obfuscation technique by integrating ASCII values, Transposition and Substitution methods.

Methodology:

1. At first the line (L is read from the text)
2. Then the characters in the line are converted to ASCII values which is later multiplied with the position of character

3.5.2   Pixel Processing Using Least Significant Bit (LSB) [15]

Pictures are made using three shades red, green and blue i.e. RGB. It is understood that in 8 bits the essential piece is Most Significant Bit (MSB) and the last piece Least-Significant Bit (LSB) [15]. Here LSB bit is used for stowing ceaselessly scrambled data inside the image. Consequently, simply last piece of pixels is changed as the quality of the picture is just change by 1 or 0 in the wake of concealing the data, the adjustment in power is either 0 or 1 on the grounds that the change last bit. For example 11111000 11111001. The change is just one bit with the goal that the quality of picture isn't influenced excessively and information can be effectively moved.

4   Analysis/Review

Comparison table

	Integrating Transposition and Substitution [3.1]	Kerberos Protocol [3.2]	Encrypted Data Storage [3.3]	Searchable Encryption [3.4]	Data obfuscation and steganography [3.5]
Method used	Encryption and Decryption	Token Authentication	Encryption	Encryption	Obfuscation and Steganography
Hardware based	NO	Yes	Yes	No	No
Cost effectiveness	Not Expensive	Expensive	Expensive	Not Expensive	Not Expensive
Advantages	This is much more secure and faster	It maintain authenticity by secure token authentication	If coprocessor is attacked the data is erased that maintain confidentiality	It makes it easier to find certain keywords	It is highly secure
Disadvantages	It has complex mechanism	If the server is down no authentication is possible	It has small storage space	Only the keywords are encrypted which increase vulnerabilities	It is complex and can reduce picture quality in case of large data

5 Conclusion

Cloud computing provides a storage platform to the users. Due to presence of large amount of data, storage in cloud computing requires a high end security. Security to the data could be provided in various forms, for example by using secure encryption methods and by using secure data access methods to access cloud storage data confidentiality and integrity of data could be maintained. The threats to the cloud storage has increased as the future of data storage lies in it. The advancement in technology has also made it easier for attackers to search for possible loopholes in the secured network hence users and providers both need to be proactive and take precautions for a secured usage.

References

1. Venkatesh, A., Eastaff, M.S.: A study of data storage security issues in cloud computing. IJSRCSEIT **3**(1) (2018)
2. Patel, F.: Community cloud computing with association of cloud computing paradigm: a study. IJARCSSE **4**(5) (2014)
3. Sugumar, R., Raja, K.: Enhanced data security methodology for cloud computing environment. IJSRCSEIT (2018)
4. Shinde, S., Wanaskar, U.H.: Keylogging: a malicious attack. IJARCCE **5**(6) (2016)
5. Babu, V.S., Kumar, M.M.V.M.: An efficient and secure data storage operations in mobile cloud computing. IJSRSET **4**(1) (2018)
6. Purushothaman, D., Abburu, S.: An approach for data storage security in cloud computing. IJCSI **9**(2) (2012)
7. Singh, A., Shrivastava, M.: Overview of attacks on cloud computing. IJEIT **1**(4) (2012)
8. Sevak, B.: Security against side channel attack in cloud computing. IJEAT **2**(2) (2012)
9. Gairola, T., Singh, K.: A review on DOS and DDOS attacks in cloud environment & security solutions. IJCSMC **5**(7) (2016)
10. Rao, R.K., Vasudha, S.B.: A review on malware injection in cloud computing. IJIRCCE **6**(4) (2018)
11. Vaudenay, S.: A Classical Introduction to Cryptography Applications for Communications Security. Springer, Boston (2006)
12. Stallings, W.: Cryptography and Network Security - Principles and Practice, 5th edn. Pearson Education, Inc., London (2011)
13. Amalarethinam, D.I.G., Mary, B.F.: Data security enhancement in public cloud storage using data obfuscation and steganography. In: WCCCT (2017)
14. Amalarethinam, D.I.G., Mary, B.F.: Data security enhancement in public cloud storage using data obfuscation. Perspect. Sci. (2016). (communicated)
15. Smys, S.: DDOS attack detection in telecommunication network using machine learning. J. Ubiquitous Comput. Commun. Technol. (UCCT) **1**(01), 33–44 (2019)
16. https://www.lia.deis.unibo.it/Courses/RetiDiCalcolatori/Progetti98/Fortini/lsb.html

Big Data and Internet of Things for Smart Data Analytics Using Machine Learning Techniques

J. Betty Jane$^{(\boxtimes)}$ and E. N. Ganesh

Department of Computer Science and Engineering,
Vels Institute of Science and Technology, Chennai, India
blessyjane@gmail.com, enganesh50@gmail.com

Abstract. With the increase in growth of technologies and communications all over the world there is a rise in IOT based internet connected sensor devices. With the rise of IOT devices, all the applications work smarter and they are time efficient. Since the IOT sensor devices produce large amount of data per day they generate big data in the form of volume, velocity and variance. The processing and the analysis of this big data intelligently, help in developing smart applications. In this paper, we discuss about the smarter big data analysis with the use case of smart parking system using machine learning algorithms and IOT. The CNN machine learning algorithms is used for the smart occupancy of parking slots.

Keywords: IOT · Big data · Machine learning · Convolution neural networks · Arduino · Raspberry

1 Introduction

The sensor, mobile applications have become smarter over the past few decades, and hence enable the communication and complex execution tasks work smarter. The internet connected sensor devices crossed over 25 billion over the population in global rate. With the increase in the number of IOT devices such as smart phones and sensors all the electronic devices generate large data which leads to the IOT era [1]. since the IOT devices generates data in a large amount, the data science helps in making the IoT applications work smarter. The process of finding and understanding new innovations from the large amount of data with the combination of different fields such as machine learning and data science and mining [2]. All these devices connected to the internet are associated with the life of the human closely. The technology has been perfectly fit in the human routine life. The process of complicated platform is facilitated and is provided by the digital space. Hence the computing devices and sensors have been deployed. The internet produces more information and they consumed by machines where the communication becomes easier and also helps the improvement in human life. IOT has been involved in various fields such as healthcare, disaster management [3]. In healthcare management a Sensor or a chip is being implemented into every individual in the hospitals for finding their signs of health to predict whether it is serious or normal. In the past few decades the machine learning techniques helps in providing the hidden data and make decisions accordingly in the fields such as

© Springer Nature Switzerland AG 2020
A. P. Pandian et al. (Eds.): ICCBI 2019, LNDECT 49, pp. 213–223, 2020.
https://doi.org/10.1007/978-3-030-43192-1_24

healthcare, weather forecasting and for all smart data analysis. since the data production is large the traditional learning methods is difficult to have deeper and hidden insights. Basically, the traditional base machine learning algorithms load the data in the memory which do not have the big data context [4]. Different technologies are developed by various organizations and use IOT for its real-world applications. In order to find the IOT applications with design issues the researchers are focusing on the machine learning. The limitations in the work of research and defining how IOT system performs applications with machine learning algorithms [5] (Fig. 1).

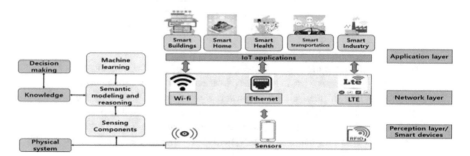

Fig. 1.   IOT and machine learning architecture

2   Related Works

Since there is a rise in the electronic devices and gadgets the IOT technology growth is more and it gains more attention. People are surrounded by the gadgets at all time. In this paper they have introduced the human device for safety called HAR that is achieved by the IOT devices and sensors. These sensors help in finding the changes in the pressure human heart beat, temperature etc., and they perform the analysis of data by machine learning algorithms to analyze whether the recorded reading is correct or unusual. On these readings to check whether the readings are usual or unusual. The sensors that are considered are pulse sensor, temperature sensor [6]. The tremendous growth in the IOT devices has resulted in a need of security concern and privacy. The next issue is the scalability that is the size of the data in order to analyze it in the cloud. The cloud model gives the less reaction time to security issues and they lead to congestion of network. For the purpose of storing and moving computation the fog computing gives a perfect solution for these issues. In this paper a clear picture of the fog computing, ML techniques, and its growth are all discussed. The abnormalities are discussed by ML techniques and also illustrating the solutions for the IOT data growth and security issues of the fog computing [7]. The internet of things is connected to the sensing devices which deliver large amount of data all over time at a huge amount of volume and velocity and build applications. The nature of the application will deliver real time and fast data applications. Hence by applying analytics techniques such data will discover new understanding future decision controls and make IOT, a well worthy technique for businesses and for day to day life. This paper describes a complete overview on the machine learning techniques and how the deep learning helps in analysis of data and for IOT area. Two different data characteristics are used for data analytics, one is, and IOT data analytics and streaming data analytics as a possible approach for analytics is also discussed. The capability of DL techniques and its challenges are also discussed [8]. Big data analytics has been gaining lots of interest in the academic and in the industry side due to the need in understanding the deeper insights from large datasets. Due to the enormous growth of the sensors and networks and IOT has become a common for all devices the collection of data increases in a large area in the field of smart cities, health care, educational institutions and various other fields. Though the data is collected they are uncertain due to the noisiness, inconsistency and incompleteness. So, in order to analyze such a large amount of data they are using the predicting techniques for reviewing and for making decisions with advanced strategies [9]. In the parking areas the sensors in the grounds are kept to predict the spaces of the parking slots. Due to this the sensors need to be installed which is expensive [10]. Since there is a increase in the number of owned private cars there is a problem in locating the cars in the respective slots and this has become an issue nowadays. In this paper they have used parking guidance and information systems for finding the vacant places in the congestion areas [11]. This paper helps to give a real time parking occupancy by using a combination of structural algorithms and deep learning. At the stage of preprocessing a decision module on structural similarity is added as preprocessing stage to find the changes of status in the occupancy. Some changes in the images that are given as inputs are done and given to the binary classifier to finalize the status of parking occupancy [12]. Streets that are crowded create larger problems in cities. Major problems occur in the street parking. The empty spots are been spread out in an area rather than being in a clustered area for maintaining a target rate in some cities such as San Francisco and Los Angeles. In this paper, the use case given is San Francisco's SF park system [13]. The parking that is illegal is becoming a serious and vital problem. Background segmentation is used for detecting vehicle that are illegally parked. The method is environment sensitive and weakly robust. This paper proposes deep learning for illegal parking of vehicles. The vehicles that are captured by camera is classified by Single Shot Multi Box Detector (SSD) algorithm [14]. The costly parking spaces to build, enforcing parking payments and drivers excessive time searching for vacant occupancy. The developers and municipalities will give the accurate quantification for space and design, while drivers can save time with real time measurements. In this paper, a video system of real time for internet of things and smart applications for accurate measurements [15, 16].

3   System Overview

The machine learning algorithm helps in analyzing large amount of data and finally gives a précised data which helps in improving the decision making for the companies and industries. Hence the process of the analyzing the data is connected with three different processes. They are

  i. IOT applications connected with sensors.
 ii. Analyzing the data, generating and storing big data.
iii. Applying machine learning techniques on the big data, the stored data.

3.1   IOT Applications Connected with Sensors

A huge number of devices are connected to sensors called sensor devices which collect large amount of data from various fields of applications such as agriculture, education, marketing and in various other fields. Based on the type of the application the devices work accordingly and give the real time data in a fast manner. IOT helps in developing smarter devices and the IOT is connected to sensors processing networks and analyze

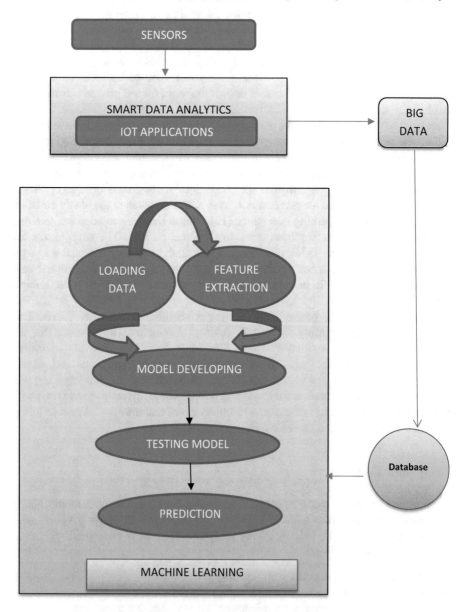

Fig. 2. Operation Of analyzing using IOT data

data and also monitors the data. The process of IOT connected with sensors is they collect they information from the given environment (Fig. 2).

3.2   Analyzing the Data, Generating and Storing Big Data

The data is collected from the sensors and they are tested for finding insights and some understanding from the raw data. Then the data will be in state to get transferred to the servers connected to the internet.

3.2.1   Storing the Big Data

Big data has a storage infrastructure where we can store the data and can be accessed.

There is another important feature known as computing framework in IOT. The computing framework is used for processing and analyzing the data. There are two types of data processing; one is they process data after generation and the second one is they process the data instantly. The instant processing is known as fog computing. There are different types of computing they are

- **Fog computing-** analyzes the data at the network edge instead of moving to the cloud
- **Edge computing-**data processing near the edge of the data when it is generated.
- **Cloud computing-**accessing data at any system with internet
- **Distributed computing-**A distributed computing in which the components are present in different networks and gets coordinated by message passing.

3.3   Applying Machine Learning Techniques on the Big Data, the Stored Data

The machine learning algorithm takes a group of samples and it is known as training set. There are three different machine learning techniques they are (Fig. 3)

- Supervised learning
- Unsupervised learning
- Reinforcement learning

Machine learning algorithm	Data processing technique
K-nearest neighbours	classification
Naïve bayes	classification
Support vector machine	classification
Linear regression	Regression
Random forest	Both regression and classification
k-means	clustering

Fig. 3.   Machine learning algorithms for smart data analysis

3.3.1 Classification

Classification is the process of organizing data in to small categories in order to have an effective use of those data.

3.3.2 Regression

The regression performs the prediction of an event with the relationship between the data set and the variables.

3.3.3 Clustering

The clustering is used for the statistical data analysis purpose to predict the data to which group they get assigned by analyzing it through the deeper understanding.

4   IOT Applications and Algorithms for Smart Data Analysis

4.1   Smart Traffic and Environment

In the smart analysis of traffic and environment the algorithms to be used are the classification and clustering algorithms. The metrics that are needed to find are traffic prediction and data abbreviation.

4.2   Smart Weather Prediction

In the smart weather prediction, the support vector regression algorithm is used. It helps in forecasting the weather conditions and for predicting.

4.3   Smart Market Analysis

In this analysis the linear regression is used. The metrics needed to optimize is reducing the amount of data and for real time prediction. This algorithm is also used in smart energy usage.

4.4   Smart Citizen and Agriculture

In this analysis, the K-nearest algorithm and Naïve bayes algorithm is used. The optimizing metrics are finding the passengers travel pattern. The naïve bayes classifier is used for the assumption of a particular feature that is not related to the other feature that is present.

4.5   Analysis on Smart Home and Smart City

In the smart home and smart city analysis, the algorithms used are k-means algorithm. The K-MEANS clustering is the simple way of classification of data through selected clusters.

4.6   Smart Health Analysis

In the smart health analysis, the neural network algorithm is used. The neural network algorithm is used for recognizing the patterns as such of human brain. The neural networks help in clustering and classification.

4.7   Smart Monitoring

In the smart monitoring analysis, the principal component analysis (PCA) algorithm is used. The metrics to optimize is the fault detection. Principle component analysis is an process extracting selected variables from the large variables in the data set.

5   Use Case for Smart Parking System

The smart parking system that is to be used helps in getting the data of the parking slot from the sensors and the sensors send the data through the microcontroller and the data is processed and the parking status will be given as user display for efficient and easy parking.

This parking system can be used in covered parking, street side parking also.

5.1   Hardware

- Raspberry Pi 3 Model B (2pcs)
- Arduino UNO (2pcs)

5.2   Software Used

- Arduino IDK.

5.3   Programming Language

- Python, JavaScript

5.4   Project Implementation

The smart parking system will use cloud storage for storing the data of the parking slots. The centralized server will be there to store the number of parking slots in the area.

The components in the automated parking system are:

- Centralized server-The server gives the slots of parking and the availability information.
- Arduino-This board helps in inputs reading like sensor light etc.
- Camera and Sensors-camera and sensors are used in photo taking and finding the car parking occupancy.

- Navigation System-This helps in navigating the driver to the vacant parking slot.
- Display device-The display device helps in giving the messages and images to the vehicle real time operator.
- User device-The data can be accessed by the users using the mobile devices (Fig. 4).

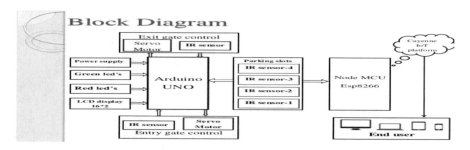

Fig. 4. Smart Parking System with IOT Devices

6   CNN Algorithm (Convolution Neural Networks) of Machine Learning for Smart Parking Analysis

The algorithm used is the convolutional neural networks. The CNN algorithm has a flexible infinite function and these algorithms are scalable and fast. The convolution networks have a filter that helps to find the features from the given object and then each filter gets activated to those pixels that give a match to the filter shape (Fig. 5).

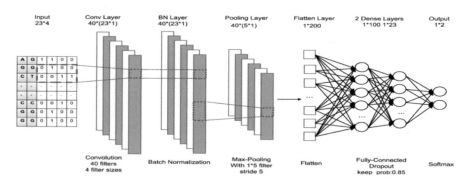

Fig. 5. Layers of CNN architecture

The steps involved in the CNN algorithm is:

- **Input Layer**-The input of the raw image is taken with its depth, height, width.
- **Convolution layer**-The main building blocks of CNN is the convolution layer. The convolution is applied to the input that gives a result that are highly featured which is detected in any place from the given input image.

- **Pooling Layer:** This layer is inserted periodically. The function of the Pooling layer is to reduce the volume size and hence it makes the fast computation and overfitting is reduced.

There are two types of pooling:
Max pooling and Average pooling-Max pooling helps in reducing the image dimensionality and average pooling helps in down sampling by division of input in to rectangular regions and for computing.

- **Fully connected Layer-**This layer helps in the representations learning and helps in making decisions.

The Smart parking system process involves as follows:

- Collecting the data
- Data preprocessing
- Test case creation
- Model implementation
- Evaluation of the model (Fig. 6)

Fig. 6. Smart parking system

6.1   Collecting the Data

The data that are collected from the sensors are stored and is been processed.

6.2   Data Preprocessing

The data preprocessing involves filtering the data, labeling and segmenting the data.

6.3   Test Case Creation

Creating testing and training the data sets for each and every case.

6.4   Model Implementation

The models implemented are VGG 19, resnet 50.

6.5   Evaluation of the Model

The evaluation model involves the running the tests on the models and capturing the performance and evaluating is involved. This is the way the CNN helps in the prediction of the parking slots that are not occupied and that helps in the smart parking analysis.

7   Conclusion

This paper gives the detailed approach on the techniques that are involved in the Big data analytics for IOT with machine learning. Smart data analytics is performed on big data that are generated from the sensors and they are applied to an analytic process by using relevant machine learning algorithms. A use case on the smart parking system has been explained in a detailed way for detecting the occupancy of vacant parking slots. The CNN algorithm helps in robustness and gives good accuracy in the new parking spaces. The future research of the paper is introducing the detection of automated parking slots in new locations by applying machine learning algorithms since it will help in reducing the process of system installation in new locations.

References

1. Zantali, F., Koulouras, G., Karabetso, S., Kandris, D.: A review of machine learning and IoT in smart transportation. Future Internet **11**(4), 94 (2019)
2. Mahdavinejad, M.S., Rezvan, M., Barekatain, M., Adibi, P., Barnaghi, P., Sheth, A.P.: Machine learning for Internet of Things data analysis: a survey. J. Digit. Commun. Netw. **4**(3), 161–175 (2018)
3. Tulasi. B., Vemulkar, G.J.: Blending IOT and big data analytics. IJESRT, April 2016. ISSN 2277-9655
4. Qiu, J., Wu, Q., Ding, G., Xu, Y., Feng, S.: A survey of machine learning for big data processing. EURASIP J. Adv. Signal Process. **2016**(67), 1–16 (2016)
5. Desai, A.M., Jhaveri, R.H.: The role of machine learning in Internet-of-Things (IoT) research: a review. Int. J. Comput. Appl. **179**(27), 0975–8887 (2018)
6. Sharma, K., Londhe, D.: IOT based sensor data analysis using machine learning. Int. J. Comput. Eng. Appl. **XII** (2018). Special Issue, ISSN 2321-3469
7. Moh, M., Raju, R.: Machine learning techniques for security of Internet of Things (IoT) and fog computing systems. In: International Conference on High Performance Computing & Simulation (2018)
8. Mohammadi, M., Sorour, S.: Deep learning for IoT big data and streaming analytics: survey. IEEE Commun. Surv. Tutor. **20**(4), 2923–2960 (2018)
9. Hariri, R.H., Fredericks, E.M., Bowers, K.M.: Uncertainty in big data analytics: survey, opportunities, and challenges. J. Big Data **6**(1), 44 (2019)
10. Amato, G., Carrara, F., Falchi, F., Gennaro, C., Vairo, C.: Car parking occupancy detection using smart camera networks and deep learning. In: IEEE Symposium on Computers and Communication (ISCC) (2016)
11. Acharya, D., Yan, W.: Real-time image-based parking occupancy detection using deep learning. In: Proceedings of the 5th Annual Conference@locate (2018)

12. Ng, C.-K., Cheong, S.-N., Foo, Y.-L.: Low latency deep learning based parking occupancy detection by exploiting structural similarity. In: Alfred, R., Lim, Y., Haviluddin, H., On, C. (eds.) Computational Science and Technology. Lecture Notes in Electrical Engineering, vol. 603, pp. 247–256. Springer, Singapore (2020)
13. Pasumpon, P.A.: Artificial intelligence application in smart warehousing environment for automated logistics. J. Artif. Intell. 1(02), 63–72 (2019)
14. Simhon, E., Liao, C., Starobinski, D.: Smart parking pricing: a machine learning approach. In: 6th Workshop on Smart Data Pricing IEEE Conference on Computer Communications Workshops (INFOCOM WKSHPS): SDP 2017 (2017)
15. Xie, X., Wang, C., Chen, S., Shi, G., Zhao, Z.: Real-time illegal parking detection system based on deep learning. Association for Computing Machinery (2017). ISBN 978-1-4503-5232-1
16. Cai, Y., Alvarez, R., Sit, M., Duarte, F., Ratti, C.: Deep learning based video system for accurate and real-time parking measurement bill. IEEE Internet Things J. 6(5), 7693–7701 (2019). Student Member. IEEE: Special issue on enabling a smart city: Internet of things meet AI

Comparison of Cloud Computing Security Threats and Their Counter Measures

Shourya Rohilla$^{(\boxtimes)}$ and Rejo Mathew

Department of Information Technology, NMIMS (Deemed-to-be) University,
Mumbai, India
shouryarohilla10@gmail.com, rejo.mathew@nmims.edu

Abstract. One of the major driving forces in the current technological world, is the concept of cloud computing. Many companies and organizations have their precious data stored on various cloud computing platforms. Organizations depend on cloud computing will effectively manage the excessive data quantity and handle them. In such a scenario, a breach of that data or any major threat can cause huge damage to the image and financial position of the company. Due to a security breach, if all of those data is leaked to an unauthorized intruder, it can lead to severe manipulations of that data and cause huge losses to the assets of the company. Hence, various counter measures to tackle those security issues are discussed in this paper. All the possible data issues are listed out and the various methods to counter them are discussed.

Keywords: Cloud computing · Confidentiality · Integrity · Availability · Authentication · Intrusion Detection Systems

1 Introduction

Cloud computing comes out as one of the fastest growing technologies in the world of data storage and management [1]. Even after being highly successful and effective in its purpose, cloud computing technology has various security threats and issues. These threats include data loss, data leakage to a third party, network threats [2], etc. Various methods and solutions are discussed to tackle those issues in this paper.

Cloud computing challenges are usually related to the protection of the CIA triad [1]. Any harm to any of the component of the CIA triad can lead to hampering of its business. Moreover, maintaining CIA is more difficult in cloud computing because of its architectural pattern that allows customers to share computing resources in a private or public cloud and also the nature of the infrastructure which is usually distributed [1].

The paper is organized in the following sections. Section 1 gives the brief introduction about the paper and the issues and solutions it is going to tackle. Section 2 discusses about the problems or methods that are described in the papers that pose a challenge to cloud computing. Section 3 discusses about the various solutions/counter measures to those problems. Section 4 is the analysis/review table that discusses or compares the features of the six solutions. Section 5 concludes the paper with the conclusion write up and the corresponding future work.

© Springer Nature Switzerland AG 2020
A. P. Pandian et al. (Eds.): ICCBI 2019, LNDECT 49, pp. 224–231, 2020.
https://doi.org/10.1007/978-3-030-43192-1_25

2 Security Issues in Cloud Computing

2.1 Data Threats

Data is the most fundamental asset of any organization or body [2]. Here, we discuss data loss and data breaches as they are defined by CSA as major threats to data security [2, 7]. They are discussed below:

2.1.1 Data Breaches

If any untrusted user gains access to either of customer data, or organization data in an unauthorized way, then we term it as a data breach [3]. Results of such a breach include impact on company's financial assets, loss of confidential data that may lead to disclosing of some of company's private financial statements, etc. Major reasons for such incidents to happen are infrastructural flaws, application designing, operational issues, lack of authentication, and audit controls [8]. Malicious users that have a common virtual machine can attack the system.

2.1.2 Data Loss

Like data breaches, data loss is one of the major issues tackling many organizations. Data loss can happen due to many factors like natural disasters, loss of data encryption keys, etc. Attacks like malware attack, which perform malicious activities on the cloud platform without the customer having any knowledge about it, also occur which thereby results in data destruction [2].

2.2 Network Threats

Many organizations do not pay major attention to network security problems. Most severe threats occurring in cloud computing world related to network security are account or service hijacking and denial of service attacks (DoS) [2, 5].

2.2.1 Account or Service Hijacking

This type of attack occurs when the credentials of users are stolen, which leads the attacker to the access of the user's data, account or his personal data. This can lead to loss of privacy of the user as the attacker can now spy, monitor and analyze the user data, which may result in breach of data confidentiality and integrity [2]. These types of attacks also include phishing, fraud, cross site shifting, botnets, etc.

2.2.2 Denial of Service Attack

When the actual users of cloud services are prevented from accessing the cloud resources, cloud network, etc. we say that a Denial of service attack is taking place [2]. According to a recent study 81% users consider DoS attacks as a major threat in security world [7, 8]. The results of such attacks include delayed cloud operations, which makes it difficult for the cloud to interact with other users and services.

2.3 Cloud Environment Specific Threats

Cloud service providers control most of the cloud environment. Albeit, a study done by Alert Logic [9] reveals that 50% of people evaluate issues occurring due to service provider as a serious threat in the world of cloud computing. Some threats under this section are providing insecure interfaces, malicious cloud users, etc. They are discussed below:

2.3.1 Insecure Interfaces and APIs

Software applications communicate with each other through internet. This communication is defined by a set of standards and protocols called Application Programming Interface i.e. API [2]. Communication with other services also use cloud API's. In order to give more and widespread services to customers, API's of the cloud providers are offered to various third parties by them. In such a case, API's that are weak may result in the third parties having all the vital information as well as the security keys. APIs should be able to encrypt the data and transfer through the interfaces.

2.3.2 Malicious Insiders

Any individual who is a cloud based worker or a company partner who can make an entrance to cloud network, apps, services, information and misuse that entrance for unprivileged operations is termed as a malicious insider [3]. Other types of malicious insiders include hackers who do it for hobby purposes, who are administrators who want to obtain unlawful delicate data just for pleasure, and corporate spying that includes carrying off confidential company data for corporate reasons that domestic governments could sponsor [2].

2.3.3 Insufficient Due Diligence

The word due diligence relates to people/clients who have full information before using their services to assess the hazards associated with a company [4]. Thus, this threat occurs because organizations adopt cloud to save costs, with insufficient knowledge of other threats. Cloud clients have no concept of inner safety processes, auditing, information storage, information access, resulting in the creation of unknown cloud risk profiles.

2.4 Abuse and Nefarious Use

This is one of the top threats identified by Cloud Security Alliance. The reason for the arrival of this threat is due to comparatively fragile registration systems present in the cloud computing environment [3]. The multi-tenancy feature of cloud can be misused by intruders to hack into other organizational data [15].

2.5 Issues Due to Shared Technology

Cloud consists of IaaS functionality [6]. It enables users to access any equipment which was common amongst them. Hypervisor helps in connecting the guest operating system to other resources. This puts the system at danger as the operating system gains entry to even unwanted concentrations that affect other systems on the mesh [6].

2.6   Security Challenges in the Authentication and Access Control

In order to confirm and verify the identity of the individual, to connect, access and use cloud resources, Authentication and Access control is performed [1]. An authentication process that involves verification based on passwords does not give the public cloud the benefits of complete safety. Using many approaches like a brute force attack, dictionary attack, phishing, passwords may be revealed. It is therefore very essential for the cloud service provider to include extremely secured techniques of authentication in a public cloud [1].

3   Solutions to Security Issues in Cloud Computing

The safety techniques for avoiding the exploitation of threats listed in Sect. 2 are discussed in this section. The application at distinct levels of these safety methods to protect the cloud against threats has been portrayed.

3.1   Data Security

The solutions to the problems listed in the Sect. 2, are discussed below:

3.1.1   Protection from Data Breaches

Numerous security measures have been proposed to counter data breaching losses. One of them, is to encrypt the data to a cipher text and then store them in the network, or the cloud. The Attribute-based encryption can be performed [15]. Here, the encrypted data can only be accessed by those whose keys match the attributes. The access control policies are defined by the attributes.

3.1.2   Protection from Data Losses

To counter this issue, the most vital of measures is to have a substitute of all the data on cloud network, which in situation of any data damage, can be accessed. The method of trusted computing to provide data protection is proposed by Chow [10]. A trusted server can monitor cloud server information features and provide information proprietor with the full audit report. Hence, the owner will be sure that the data access protocols have not been disrupted [10].

3.2   Network Security

The solutions to the problems listed in the Sect. 2, are discussed below:

3.2.1   Protection from Account or Service Hijacking

This attack can be countered with variety of measures. These include using cloud-based intrusion detection systems (IDS) to audit network traffic and nodes for malicious activity disclosure. Cloud efficiency, compatibility and virtualization based context [12], must be considered before IDS and other network systems must be designed. Authentication by multi-factor for distant access using two credentials can be used to prevent account hijacking threats. Also, to prevent these threats, auditing all of the user's privileged operations along with information security incidents produced from it should be performed [12].

3.2.2 Protection from Denial of Service
In order to avoid attacks or threats related to services being denied to users, it is essential to analyze and achieve basic security requirements. After developing to confirm that they have no loop holes that the attackers can exploit, applications should be screened [2]. Jin et al. suggested a hop count filtering method that can be used to remove fabricated IP packets and helps to reduce such attacks by 90% [14].

3.3 Cloud Environment Security

The solutions to the problems listed in the Sect. 2, are discussed below:

3.3.1 Protection from Insecure Interfaces and APIs
It is pivotal for the designers to develop these APIs by tracing conventions of trusted computing. Cloud suppliers must also guarantee that all cloud-based APIs are safely designed and checked for possible faults before deployment [2]. In order to protect information and services from insecure interfaces and APIs, powerful authentication methods and access conducts must also be introduced. In addition, clients are responsible for analyzing cloud provider interfaces and APIs before shifting their information to cloud [11].

3.3.2 Protection from Malicious Insiders
Protection from such attacks can only be accomplished by restricting access to the approved staff to the hardware infrastructure. Strong access control must be carried out by the service provider. Employee audits should also be carried out to verify their suspect behavior [2]. It is also possible to implement encryption in storage and government networks to avoid information from malicious insiders.

3.3.3 Protection from Insufficient Due Diligence
Organizations should know the possibilities of risk associated with cloud before they shift their work or other official work on it. After certain intervals, cloud providers should conduct risk assessments using various methods to verify data storage, flow, and processing [9]. The service suppliers must reveal to customers the relevant records, infrastructure such as the firewall to take steps to secure their apps and information.

3.4 Protection from Abuse and Nefarious Use

Negative usage of cloud resources can be avoided by implementing a process that performs registration and acceptance before entering the cloud system [3]. Many value-added services can be used to evaluate clients using their own services without influencing the general process. Stricter registration and validation process can also be implemented. Also, detailed introspection of user's network traffic can be worked upon [9].

3.5 Protection from Issues Due to Shared Technology

These issues can be countered by following solutions. One way is to get the best safety measures for installation or configuration purposes. Also, the need of strengthened authorization processes for administrative and other operations can be boosted up [11]. Promotion of service level agreement for installing vulnerability assessments can be done [11, 13]. Scanning of vulnerabilities from time to time should take place.

3.6 Protection from Threats in Authentication and Access Control

It is possible to adopt cryptosystem of RSA that can accept various authentication examples such as two-factor authentication, and adaptive authentication [1]. Wherever feasible, administer single-sign-on policy. Biometric Authentication an also be used as a single sign-on policy. Third-party alternatives such as Microsoft Azure Active Directory, Cloud Identity Manager McAfee, can also be used [1, 8].

4 Analysis/Review

Comparison Table

Problems	Data Threats - Data Losses and breaches	Network Threats - Account or service hijacking, Denial Of Service Attacks	Cloud environment specific threats - Insecure interfaces and APIs, Malicious Insiders, Insufficient Due Diligence	Threats arising from Abuse and Nefarious use of cloud computing	Issues arising due to shared technology	Security Challenges in the Authentication and Access Control
Solutions	Data Security	Network security	Cloud environment security	Protection from Abuse and Nefarious Use	Protection from issues due to shared technology	Protection from issues in Authentication and Access Control
Methods used as solutions	Attribute - based encryption [15]	Intrusion Detection Systems (IDS) [3, 13]	Service Level Agreement definition language [7, 9]	Original registration and acceptance process [3]	Promotion of service level agreement for installing vulnerability assessments [11, 13]	Security Assertion Markup Language (SAML) [1]
Principles used to protect data from various threats	Trusted computing [2]	Multi-factor authentication [12]	Principles of trusted computing [6]	Detailed introspection of user's network traffic [9]	Strengthened authorization processes for administrative and other operations [11]	A Single-sign-on policy Multifactor authentication can also be employed [1, 8]

(*continued*)

(continued)

Problems	Data Threats - Data Losses and breaches	Network Threats - Account or service hijacking, Denial Of Service Attacks	Cloud environment specific threats - Insecure interfaces and APIs, Malicious Insiders, Insufficient Due Diligence	Threats arising from Abuse and Nefarious use of cloud computing	Issues arising due to shared technology	Security Challenges in the Authentication and Access Control
Data CIA protection	Data Confidentiality and data Integrity [5, 6]	Data Availability and Confidentiality [2, 5]	Data Integrity and Availability [5, 9]	Data confidentiality and integrity [9]	Protects data availability and confidentiality [11]	Data confidentiality, availability [1, 8]
Advantages	Provides solutions to prevent data losses and data breaches preserving the integrity and confidentiality of the data [5]	Provides solutions to prevent account or service hijacking and DoS attacks keeping the data safe from attackers [2]	Provides solutions to prevent the data from Malicious Insiders, from insufficient Due Diligence and protection from Insecure Interfaces [7]	Provides solutions like detailed introspection of user's network traffic to detect and mitigate nefarious use of cloud computing [3]	Provides solutions to mitigate the impact of issues due to sharing of technology and resources [12]	Risk decisions can be made for access management [1]
Disadvantages	The Attribute-based encryption is complex to implement [2, 12]	The Intrusion detection System has a drawback of false alarms [3, 13]	Policies must be made part of the Service Level Agreement (SLA) between user and service provider, which is not feasible [2]	Implementation of an original registration and validation process [3] is a complex process and requires careful and minute attention	Scanning of vulnerabilities from time to time is tough. The check for vulnerability requires correct implementation of the checking program [12]	Requires cloud applications to use open standards where applicable, like SAML for exchanging authentication [1, 8]

5 Conclusion

As seen, various threats and issues clog the successful running of cloud computing services. The six major issues damage the data confidentiality, integrity and availability of the data. The solutions like Intrusion Detection systems and Attribute - based encryption provide extensive mechanisms of preventing any cause of harm to the data. The IDS specially improves the data security and helps IT staff in the organizations to quickly detect if any invader or attacker is trying to break in. But IDS also gives many false alarms, which causes unnecessary confusion. So, in a situation where the numbers of users in the system are less, IDS works very effectively in preserving the data CIA. The Nefarious use of cloud services can be avoided by implementing methods to introspect detailed analysis of user's network traffic. IDaaS solutions are used in corporate infrastructure to tackle the challenges arising from Authentication and Access Control. The future work is concerned with more solutions to tackle the emerging cloud computing threats and preserve the data CIA of many organizations who have their data stored and managed on various cloud computing platforms.

References

1. Ravi Kumar, P., Herbert Raj, P., Jelciana, P.: Exploring data security issues and solutions in cloud computing. In: 6th International Conference on Smart Computing and Communications, ICSCC 2017, 7–8 December (2017)
2. Kazim, M., Zhu, S.Y.: A survey on top security threats in cloud computing. Int. J. Adv. Comput. Sci. Appl. (IJACSA) **6**(3) (2015)
3. Kumar, S.V.K., Padmapriya, S.: A survey on cloud computing security threats and vulnerabilities. Int. J. Innov. Res. Electr. Electron. instrum. Control Eng. **2**(1), 622–625 (2014)
4. Ahmat, K.A.: Emerging Cloud Computing Security Threats, vol. 2, Department of Information Technology City University of New York (2016)
5. Worlanyo, E.: A survey of cloud computing security: Issues, challenges and solutions, 20 August 2017
6. Jakimoski, K.: Security techniques for data protection in cloud computing. Int. J. Grid Distrib. Comput. **9**(1), 49–56 (2016)
7. Group, T.T.W., et al.: The Notorious Nine: Cloud Computing Top Threats, Cloud Security Alliance (2013)
8. Alliance, C.S.: Top Threats to Cloud Computing v1. 0, Cloud Security Alliance (2010)
9. Cloud Security Report Spring: JSAI M International Conference of Smart Computing. Accessed 08 Nov 2014
10. Chow, R., Golle, P., Jakobsson, M., Shi, E., Staddon, J., Masuoka, R., Molina, J.: Controlling data in the cloud: outsourcing computation without outsourcing control. In: Proceedings of the 2009 ACM Workshop on Cloud Computing Security. ACM (2009)
11. Takebayashi, T., Tsuda, H., Hasebe, T., Masuoka, R.: Data loss prevention technologies. Fujitsu Sci. Tech. J. **46**(1), 47–55 (2010)
12. Modi, C., Patel, D., Borisaniya, B., Patel, H., Patel, A., Rajarajan, M.: A survey of intrusion detection techniques in cloud. J. Netw. Comput. Appl. **36**(1), 42–57 (2013)
13. Roschke, S., Cheng, F., Meinel, C.: Intrusion detection in the cloud. In: Eighth IEEE International Conference on Dependable, Autonomic and Secure Computing, 2009. DASC 2009. IEEE (2009)
14. Jin, C., Wang, H., Shin, K.G.: Hop-count filtering: an effective defense against spoofed DDoS traffic. In: Proceedings of the 10th ACM Conference on Computer and Communications Security. ACM (2003)
15. Sridhar, S., Smys, S.: A Survey on cloud security issues and challenges with possible measures. In: International Conference on Inventive Research in Engineering and Technology, vol. 4 (2016)

An Effective Kapur's Segmentation Based Detection and Classification Model for Citrus Diseases Diagnosis System

C. Senthilkumar[(⊠)] and M. Kamarasan

Department of Computer and Information Science, Annamalai University,
Chidambaram, India
senthilkumar_c@hotmail.com, smkrasan@yahoo.com

Abstract. Agriculture remains as an important occupation, the decrease in its production will lead to huge economical loss. Under the class of plants, citrus is employed as major nutrient resources such as vitamin C over the globe. But, citrus disease severely affects the growth as well as quality of citrus fruits. From the past ten years, computer vision applications are commonly employed to detect and classify the plant diseases effectively. This paper introduces a new segmentation based classification model to identify the presence of citrus disease. In addition, it classifies the different types of citrus diseases in a significant way. The presented model involves a two stage process namely Kapur's based segmentation and particle swarm optimization with support vector machine (PSO-SVM) based classification. The presented KPS model is evaluated using Citrus Disease Image Gallery Dataset and the experimentation section validated the superior nature of the KPS model in terms of classification accuracy.

Keywords: Citrus · Classification · Segmentation · Kapur's method

1 Introduction

Fruit plants play a vital role in any part of the agro-economic civilization. From the available numerous fruit plants, citrus plants is highly beneficial which offers Vitamin-C to human body and also it can be utilized as a unprocessed matter in several agricultural based industries. But, presently, the growth of citrus is highly damaged due to the presence of citrus diseases namely black spot, canker and so on as shown in Fig. 1. Keeping this aspect, diverse automation models has been developed to detect the diseases present in citrus plants and significant outcome is noticed. So, presently, several studies begun to identify an automated computer vision based method to identify the existence of the citrus disease. Since the process of identifying and classifying citrus lesion spots involves a set of 4 main steps namely Preprocessing, extracting features, segmentation and Classification, this study concentrated on the process of segmentation and classification.

© Springer Nature Switzerland AG 2020
A. P. Pandian et al. (Eds.): ICCBI 2019, LNDECT 49, pp. 232–239, 2020.
https://doi.org/10.1007/978-3-030-43192-1_26

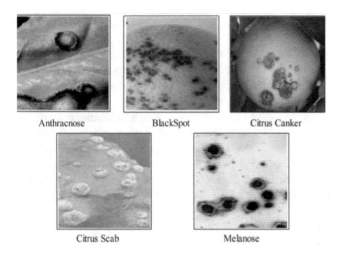

Anthracnose   BlackSpot   Citrus Canker

Citrus Scab   Melanose

Fig. 1. Types of citrus disease

A new model to detect, segment and measure the disease present in citrus plants is developed in [1]. It includes a set of sub-processes namely separating objects, reducing shadows, K-means clustering and blob identification stages. [2] developed a model which enhances the opposing effect generated by the spherical shaped objects at image gathering process. [3] presented a model to determine the mass as well as volume of 4 different citrus fruits namely lemon, orange, lime and tangerine. The method utilizes fundamental frustum for evaluating the weight of the fruit. Then, [4] devised a model to sort and grade the damaged citrus fruits. The presented model effectively recognizes the kind of citrus fruits from the images of the mixed ones. This method categorizes the fruits into diverse groups depending upon the variables of Gray-Level Co-Occurrence Matrix (GLCM) variables. [5] devised an efficient method to detect and classify the presence of citrus diseases by the use of ΔE color variation model to detect the affected region in the image. [6] introduced a color co-occurrence technique to classify the citrus diseases. Based on the experimentation, an enhanced classification is attained with the accuracy of 95%. [7] employed an efficient model to detect greening disease in citrus depending upon huanglongbing (HLB) and cost. Based on the performance, the employed model attains an accuracy of 91.93% with minimum cost as well as computation time over other classifier models. [8] introduced a way to detect and classify citrus diseases using a 2-stage processes. At the initial stage, selection of features takes place by the use of feature ranking and threshold method. [9] introduced a novel way of automatically detecting the citrus canker from leaves through the integration of global as well as local characteristics. At the end, extraction of color as well as local texture features takes place. [10] utilizes the process of inspecting leaves to identify the earlier identification of citrus diseases. It makes use of GLCM model for feature extraction and SVM based data classification for identifying the affected leaves. Some of the commonly utilized segmentation models are K-means thresholding, clustering, region growing and morphological based models. But, they suffer from the limitations that they fails to perform well in case of complicated images and low contrast images. In addition, the GLCM characteristic does not offer better classifier results. In addition, some of the methods face other issue of degraded accuracy.

To resolve these issues, in this paper, a new segmentation based classification model to identify the presence of citrus disease. In addition, it classifies the different types of citrus diseases in a significant way. The presented model involves a two stage process namely Ostu based segmentation and particle swarm optimization with support vector machine (PSO-SVM) based classification. The presented KPS model is evaluated using Citrus Disease Image Gallery Dataset and the experimentation section validated the superior nature of the KPS model in terms of classification accuracy.

2 Proposed Work

The new segmentation based classification model to detect the citrus diseases involves three sub processes namely preprocessing, segmentation and classification. The entire working procedure is shown in Fig. 2. Initially, the input citrus plant image is provided. Then, the pre-processing stages perform several filtering operations. Next, the process of image enhancement takes place to improve the quality of the image. Then, Kapur method for threshold based segmentation is applied. Finally, PSO-SVM based classification process is executed which detects as well as classifies the type of citrus disease. Here, the PSO algorithm is applied for optimizing the variables of SVM to achieve proper classifier accuracy.

2.1 Pre-processing

In this stage, the quality of the input image will be enhanced by the elimination of diverse issues like brightness effect, illumination, and problems because of low contrast. The preprocessing acts as a significant part in the domain of image processing due to the low contrast images affects the lesion segmentation accuracy. This method utilizes the contrast stretching method depending upon the Top-hat filter and Gaussian function. At the beginning stage, Top-hat filter is executed on the input image and then the lesion contrast is enhanced by the inclusion of Tophat filter and difference-Gaussian images.

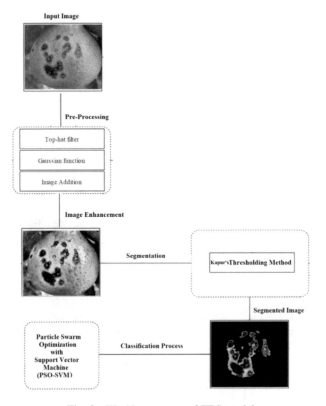

Fig. 2. Working process of KPS model

2.2 Kapur's Method for Segmentation

A new entropy dependent technique is presented for the maximization of the entropy of the segmented histogram. It makes each individual region has a central circulation. In the beginning level, bi-level thresholding approach is applied for determining the threshold values present in the histogram for extracting the objects from the background image. The bi-level thresholding is represented in Eq. (1):

$$Maximization \ of \ f(t) = HI_0 + HI_1 \qquad (1)$$

Where $HI_0 = -\sum_{n=0}^{t-1} \frac{P_n}{\omega_0} ln \frac{P_n}{\omega_0}$ and $\omega_0 = \sum_{i=0}^{t-1} P_n$; $HI_1 = -\sum_{n=t}^{L-1} \frac{P_n}{\omega_1} ln \frac{P_n}{\omega_1}$ and $\omega_1 = \sum_{i=0}^{L-1} P_n$.

The best possible threshold values in the gray scale values are maximized using Eq. (1). The entropy approach is extended to solve the multi-level thresholding problem. It is assumed as an n-dimensional optimization issue for determining a total of k optimum thresholds for a provided input image (t1, t2, …, tn), where the main intention is the maximization of the objective efunction.

$$f[(t_1, t_2, .., t_k] = HI_0 + HI_1 + .. + HI_n \qquad (2)$$

Where $HI_0 = -\sum_{n=0}^{t_1-1} \frac{P_n}{\omega_0} ln \frac{P_n}{\omega_0}$ and $\omega_0 = \sum_{n=0}^{t_1-1} P_n$; $HI_1 = -\sum_{n=0}^{t_2-1} \frac{P_n}{\omega_0} ln \frac{P_n}{\omega_0}$ and $\omega_1 = \sum_{n=0}^{t_2-1} P_n$; $HI_2 = -\sum_{n=0}^{t_3-1} \frac{P_n}{\omega_0} ln \frac{P_n}{\omega_0}$ and $\omega_2 = \sum_{n=0}^{t_3-1} P_n, \ldots$

2.3   PSO-SVM Based Classification

SVM is an important classification model which holds the advantages of global optimization and maximum generalization capability. Additionally, it avoids the overfitting issue and offered a spare solution over the classification models like NN. The formulation of SVM classifier is given below. Under a training set $T = \{(x_1, y_1), .., (x_n, y_n)\}$, with $x_i \in \mathbb{R}^m$ and $y_i \in \{\pm 1\}$ assuming n examples from m real time variables, learning a hyperplane $<w, x> + b = 0$, with $w \in \mathbb{R}^m$ and $b \in \mathbb{R}$. The image classification problem is considered as an optimization problem and is represented as follows

$$M = \frac{1}{2} \|w\|_2^2 + C \sum_{i=1}^m \xi_i \qquad (3)$$

Subjected to $y_i <w, x> + b \geq 1 - \xi_i, \xi_i \geq 0, i = 1, 2, .., m$

Where C and ξ_i indicates regularization and penalizing relaxation variable. The Eq. (3) represent that

$$\begin{cases} w \times \phi(x_i) + b \geq +1 \, if \, y_i = +1 \\ w \times \phi(x_i) + b \geq -1 \, if \, y_i = -1 \end{cases} \qquad (4)$$

The non-linear classification technique in the input space can be defined in Eq. (5):

$$f(x) = sign\left(\sum_{i=1}^m \alpha_i^* \times y_i \times K(x_i, y_i) + b^*\right) \qquad (5)$$

To attain optimum performance, selections of some of the SVM parameters like regularization parameter C and the kernel parameter γ has to be proper. The efficiency of the SVM model is based on these parameters. Therefore, they have to be selected in a proper way for enhancing the classifier results of SVM. In this case, PSO algorithm employed for choosing the variables of SVM. The outcome is determined depending upon the classifier accuracy on the unseen testing data. At the procedure of learning, PSO based SVM model undergo training procedure for reducing the error level. Using the improvement in terms of error in the training procedure, the variables γ and C are managed by the use of PSO algorithm. These controlled variables with minimum error rate are treated as the proper ones. As a result, the optimal variables (C and γ) will be achieved. Upon the identification of optimum variables of SVM, it is applied to again train the SVM model. At the testing procedure, SVM is applied to identify fresh samples. The testing procedure will offer the proper outcome of the detection and classification of citrus plant diseases. When the optimal parameters of the SVM are found, it is employed to retrain the SVM model. During the training process, SVM is used to predict new samples in the testing stage. The testing stage gives the input to the trained SVM classification model to detect the presence of disease.

3   Performance Validation

The validation of the presented KPS model is evaluated using the Citrus Diseases Image Gallery dataset [11]. Some of the sample images from the dataset is displayed in Fig. 3 and the details are provided in Table 1. The dataset holds a maximum of 1K images including different kinds of citrus plant diseases. The dataset images include leaves as well as fruits of the plant with the dimensions of 100 × 150 pixels with 96 dbi. This paper utilizes a set of six citrus disease images namely "anthracnose, scab, black spot, melanose, greening, and canker". The selected diseases are initially segmented and then categorized into their corresponding classes.

Table 1. Dataset details

Citrus type	Total images	Training images	Testing images
Anthracnose	100	50	50
Black Spot	80	40	40
Canker	120	60	60
Scab	100	50	50
Greening	100	50	50
Melanose	70	35	35
Healthy	100	50	50

Fig. 3. Sample test images

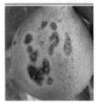  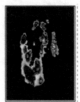 

a. Original Image   b. Enhanced Image   c. Mapped Image   d. Otsu Segmented Image

Fig. 4. Analysis of segmentation results

Table 2. Dataset details

Classifiers	Accuracy
Proposed	96.50
M-SVM	95.80
W-KNN	93.80
EBT	94.50
DT	94.50
LDA	93.50

Figure 4 shows the segmentation results attained by the Kapur's method. Figure 4a shows the provided input image, the enhanced version of the input image is shown in Fig. 4b, the mapped image is depicted in Fig. 4c and finally, the segmented image in binary color is shown in Fig. 4d.

The classification performance of the introduced KPS model is measured based on the essential classification measure called accuracy. Table 2 and Fig. 5 shows the detailed comparative analysis of various models with respect to accuracy. From the table, it is evident that optimal classification is achieved by the KPS method by obtaining a highest accuracy value of 96.50%. Next to KPS model, the existing M-SVM model showed near optimal classification by achieving a slightly lower accuracy value of 95.80%. At the same time, the EBT and DT models exhibit moderate classification. However, the W-KNN model tries to outperform the EBT and DT models. But, it leads to poor classification by achieving the accuracy of 93.80%. In addition, the existing LDA model showed worse classification by attaining the lowest accuracy of 93.50%. From the table, it is clear that the presented KPS model is an effective tool for the automatic identification of citrus plant diseases.

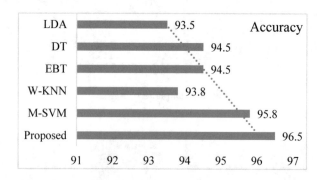

Fig. 5. Comparison of distinct classifiers on citrus disease

4 Conclusion

Presently, several studies begun to identify an automated computer vision based method to identify the existence of the citrus disease. This paper has introduced a new segmentation based classification model to identify the presence of citrus disease. The KPS model involves a two stage process namely Kapur's based segmentation and PSO-SVM based classification. PSO-SVM based classification process is executed which detects as well as classifies the type of citrus disease. Here, the PSO algorithm is applied for optimizing the variables of SVM to achieve proper classifier accuracy. From the simulation outcome, it is evident that optimal classification is achieved by the KPS method by obtaining a highest accuracy value of 96.50%.

References

1. Malik, Z., et al.: Detection and counting of on-tree citrus fruit for crop yield estimation. IJACSA. Int. J. Adv. Comput. Sci. Appl. **7**(5), 519–523 (2016)
2. Gómez-Sanchis, J., et al.: Automatic correction of the effects of the light source on spherical objects. An application to the analysis of hyperspectral images of citrus fruits. J. Food Eng. **85**(2), 191–200 (2008)
3. Omid, M., Khojastehnazhand, M., Tabatabaeefar, A.: Estimating volume and mass of citrus fruits by image processing technique. J. Food Eng. **100**(2), 315–321 (2010)
4. Kumar, C., Chauhan, S., Alla, R.N.: Classifications of citrus fruit using image processing-GLCM parameters. In: 2015 International Conference on IEEE Communications and Signal Processing (ICCSP) (2015)
5. Ali, H., et al.: Symptom based automated detection of citrus diseases using color histogram and textural descriptors. Comput. Electron. Agric. **138**, 92–104 (2017)
6. Pydipati, R., Burks, T., Lee, W.: Statistical and neural network classifiers for citrus disease detection using machine vision. Trans. ASAE **48**(5), 2007–2014 (2005)
7. Deng, X., et al.: Citrus greening detection using visible spectrum imaging and CSVC. Comput. Electron. Agric. **130**, 177–183 (2016)
8. Stegmayer, G., et al.: Automatic recognition of quarantine citrus diseases. Expert Syst. Appl. **40**(9), 3512–3517 (2013)
9. Zhang, M., Meng, Q.: Automatic citrus canker detection from leaf images captured in field. Pattern Recogn. Lett. **32**(15), 2036–2046 (2011)
10. Gavhale, K.R., Gawande, U., Hajari, K.O.: Unhealthy region of citrus leaf detection using image processing techniques. In: 2014 International Conference for IEEE Convergence of Technology (I2CT) (2014)
11. Citrus Diseases Image Gallery. http://idtools.org/id/citrus/diseases/gallery.php

Anomaly Detection Using Anomalous Behavior at Program Environment Through Relative Difference Between Return Addresses

Goverdhan Reddy Jidiga[1](✉) and P. Sammulal[2]

[1] Department of Technical Education, Government of Telangana,
Hyderabad, India
jgreddymtech@gmail.com
[2] JNTUH College of Engineering, Nachupally, JNTU University,
Hyderabad, Hyderabad, India
sammulalporika@gmail.com

Abstract. Information security is mandatory for the human population in all aspects of electronic gadgets usage. There are different kinds of attacks and anomalies found during the usage of latest applications, where it has the threat of losing the valuable credentials. The security experts have given different solutions to address various levels such as application and programming environment. The machine learning gives maximum solutions for finding anomalies at application level and tremendous outcomes will occur. But in case of programming, the coding exploits are still vulnerable and causes to create abnormal entries through security breaches which forces the program for malfunctioning. So in this paper we are presenting a new kind of anomaly detection to find different sequence of anomalies while running of infected program with help of different process tracing techniques. Here our proposed work uses Linux platform to grapple the anomalies by generating assembly code and tested various possibilities of attacks in program by modeling their original behavior. The virtual space contents such as address entries (return addresses) are helpful in our work to find any kind of anomaly. In this paper we are also improving the performance of anomaly detection by reducing the size of training and test datasets by computing the relative difference between return address entries. Here we have used standard tracing tricks and tools available in Linux platform and the experimental work done on 16 kinds of attacks, artificial datasets generated from normal runs of test programs and Linux commands, finally compared their performance on artificial datasets collected while program normal runs.

Keywords: Anomaly · Anomaly detection · Return address · Tracing tricks

1 Introduction

Anomaly detection is the method of detecting new objects (anomalies) and new behavior that are showing difference to previous context of original. At present days the information security is very important due to regular attacks and scope of creating vulnerable entries into the application programs creating huge loss to business organizations as well

© Springer Nature Switzerland AG 2020
A. P. Pandian et al. (Eds.): ICCBI 2019, LNDECT 49, pp. 240–251, 2020.
https://doi.org/10.1007/978-3-030-43192-1_27

as common public. So that security definitions are need to change every instant of time slots and every user need to know the advanced technical concepts such as anomalies, structure of anomalies to thwart the different levels of attacks.

The anomalies are commonly raised in execution of a bad program contains coding exploits. Now a day we are observing that different malicious definitions and codes are injecting into program definitions and applications. Those are executing in real mode with legitimate, but this kind of entries causes vulnerable actions and recognize them is difficult [13, 14]. In general there are two different ways to attacking on software's, one is at system coding and application levels. At present, expected work is not done to avoids on-the-fly coding exploits, but where as tremendous innovations are created in machine learning approaches to differentiate the cause of anomalies. Presently we are using some algorithms which are implementing at application level, but not able to recognizing the contracted anomalies such as point anomalies due to profile signatures are common. So here we have opted a technique system coding level to discover new anomalies. Here we uses the virtual memory process structure contents like the stack; it is providing a valuable source of data contents to model a new observations. The shell code injections, memory buffer overflows are leads to inject the error code into legitimate code through running of infected process by physical memory and this causes the control sequence to be out of main memory. At same time the call stack is an important source to detecting the anomalies scattered over the system's running programs injected with malicious programs, also exploited by unauthorized users. We observing that previous work presented based on system level showing large size of data extracted from running programs which may cause of long training period, so in this paper we are extracting the return addresses and their ordering of execution generated while calling the functions, system calls and libraries in the code. Our approach use many sets of input datasets such as return address (RA) set, relative difference (RD) set and call sequence paths (SP) set. In this process to collecting required data, we have selected some pre defined Linux trace tools and this is easy for collecting raw data to model the anomaly detection system and find the unknown behavior as anomalies. We tested 16 attacks on Linux platform by setting different threshold values while training.

In our work we are using different stack tracing tools, in that the stack trace is a kind of process tracing command used to consolidate the training data by invoking different return addresses by calling various system calls (SC), libraries and functions. From this the program counter values consider as RAs. The calling order is observed by creating stack frames for each level of calling. During this process the command and program based tracing techniques are useful to extract necessary data, In that the backtrace (bt) is a kind of tool for tracing of calling of routines. We take the support of Linux tracing methods such as LTRACE, DTRACE, PTRACE and STRACE for collecting trace data from all calls. So that, the detection of anomalies is conceded at system coding exploits is due to number of influential and quick tracing techniques presented well in experimental work. The practical work done on the different modules and heterogeneous collection of data extracted from running process. The cause of taking the different stack data combinations is to find out the exact contradictory behavior observed in process's code. The present work done here is giving the observation of the behavior dynamically with respect to fast work on profiles and signatures.

1.1 Anomaly Detection

The present anomaly detection (AD) is division of popular and traditional intrusion detection system (IDS) given by well recognized man Anderson [1, 3] for information security and has invented two different methods of fundamental IDS based on behavioral observations while running of programs. The anomaly is an abnormal observation and these anomalies are originated through either application level or alternative is on system data level. For this the anomaly detection criteria is designed for each one is different, but the extraction process of anomalies and definition is same for discovery of abnormal (strange) patterns are signifying pessimistic actions (behavior). So that the AD is a fine approach of disclosing anomalous and un-authorized prototypes (behavior) than signature based IDS.

The anomaly detection system (ADS) is well designed by past experts with multiple approaches at various levels to achieve the maximum security of information and they assure that no unauthorized, misuse, disclosure and modification of original contents [1]. The anomaly detections models are themselves are robust by training the well models with a innovative ideas of utilizing the behavior of anomalies contained in the data. Many of anomaly detection systems are designed with well training models by adorning machine learning approaches and which are investigate the latest anomalies consistently. But in case of system coding levels the machine learning approaches are playing a limited role to recognize coding exploits. The ADS at application level the enough work is presented and shown well exploration of both data mining and machine learning [2, 11]. We have given ADS with machine learning algorithms with good analysis due to its nature of adorned methods by working on different combinations in our previous work, which are well match for knowing the continuous and context anomalies [8, 11, 17, 18]. The machine learning(ML) algorithms in the ADS are presented more combinations in application point of view but not applicable more in system data levels and is not perfect to gathering global (point based) anomalies. The ADS is working on where the coding exploits raised in executable programs in sequence of return addresses caught in assembly code. This process may continue for data extraction from stack and the traced information is placed in the training sets. The similar observations are there in [4, 13, 14]. In this paper, maximum work is carried out on various levels of coding exploits due to the need of system data to observe the occurrences of anomalies of any kind which is not done at application level. So the system data is trained after placing into sets by normal runs of code and use the same for testing the next observations. This kind of development is well to use in our working model from learning of Linux debugging.

2 Related Work

The following section is giving the nature of fast work done at system coding level and merits or demerits of each one discussed. The early stage of work is on taking data maximum from UNIX operating system platform and system calls only taking into observations to find the anomalies in program level exploits deviate to normal behavior by latest one. In past work [7], presented their logic by taking the specific sequence order of occurrences of the system calls and its length. But the order of system calls is not providing solution to finding the point anomalies. In this case the anomalies are recognized in consecutive testing only due to normal profile data taking into observations. In similar lines of method is followed based on occurrences of system calls during exact time period of predefined window and modeling the system calls occur in that time period is little bit improved in detection of anomalies [8]. The function call based work is carried out by Feng et al. in [4] and Vt-Path (Virtual path) model developed for anomaly detection based on function call sequence. The functional combination of setjmp and longjmp used for holding the data during normal runs of programs and same to be used for testing to know the anomalous behavior in next runs. The problem of this procedure is resources may not released while use by those functions in case of opening files and allocation of memories. So this method is halting the system sometimes and creates deadlocks due to more use of above functions. It means to create flattening hazards [15] as usual described, but avoid it difficult due to running of recursive loops taking portions of memory.

The N-gram technique [7, 9] is one of oldest method also use of system calls modeling with machine learning traditional method causes lack in performance due to consuming more space and time. In this category of methods, most of the problems designed only system call data or function calls may not give solutions for present systems. Hence we plan for new approaches like taking sequence of function calls and system calls plotted into sets may give maximum unknown activities causes anomalous behavior. The maximum count of system calls raised in real time environment programs may cause of additional overhead to allocate memory during running of datasets [8, 9]. When the overhead time raised in training creates down in performance or plan for reducing the data sets size by using latest hash functions may give additional boost to performance. The STRACE and PIN tool is used in [12], to extract the system data from function calls achieved fine outcomes, but windows based tools mounted on the system causes delay in performance compare to tools used on Linux. We know that the stack is entry point in some cases for smashing the memory extensions due to vulnerable entries made by overflowing the memory buffers such as heap and stack by bad input given by programmers causes the system control is out of control and this state helps to divert the control to execute attacker's code. In stack based overflows are resolved in [6], but for entry in stack uses stack guard is overhead on memory, but any way experts in programming can handle this kind of issues.

The facts given in our work [10–16] encouraged how to use a simple and standard tools available in Linux platform are giving efficient results compared to traditional methods. We think that, this kind of adornments in work giving more elaborated way for any kind of anomalies. Here, this work contains better approach with help of different combinations of advanced tracing tricks.

3 Proposed Work: Programming Environment

Our proposed work is very interesting on selection of beautiful and powerful process tracing tools which are helpful to finding coding exploits and detect the cause of anomalies. In this paper, the Linux tracing approaches are used for better outcomes. These tracing tricks are supporting broad levels of options to extract necessary data from memory devices while running the program. So, most of required data collected from stack. The Linux platform supports different GUI and strong command based tracing methods of different types helps us to extract the system data from running process. We have used maximum command based techniques due fast nature of execution to speed up the progress of collecting data into training sets. The Fig. 1 is giving our new model to help us to bringing the fine work and evaluation of experimental work. The PTRACE (process trace), BACKTRACE (bt in GDB), LTRACE (library call trace), function trace (FTRACE), STRACE (system call trace) are very popular tools for tracing system data on Linux OS, so that we have opted the Linux. The other tools also available on UNIX based OS such KTRACE, DTRACE, TRUSS...etc., but not used in our work.

In programming environment, while running of program instructions which are related to sub-routine made an entry in the stack. The calling of functions, libraries, signals and system calls such are trap to interrupt routines store its PC values into the stack. The call-graph model shown in the Fig. 1 (left) developed for representing the order of calls (functions, libraries, system calls...etc.) initiates from start function. The call-graph model actually giving the hierarchical view of calls and which provides the basic knowledge from where we need to trace the PC points dropped in stack. The backtrace (bt) in Linux uses in our fast work explained more in [10], such cases are also found in GNU debugger earlier used in our work. Already said in above points that, only backtrace in not kind enough to trace required data, because it gives the thread of return addresses of all function calls in reverse collecting from last call to init() function. But we can use multiple combinations to define the new version of anomalies encountered during the normal course of runs. The backtrace process collecting various extra addresses similar to other calls such as start_up routine (or _init), which is parent of main function along with permutation or mixture of set of system calls (S_n), set of library call (L_n) and also various signals encountered. The return addresses are extracting from calling thread includes the return address of function backtrace() also, but we exempt it in testing case due to same address is replicate many times. Similarly do it for main () and init () also. Let assume a RA set is a collection of return addresses extracted from backtrace calling thread and store them in table through a buffer variable in program. The set of all the return addresses associate to sub routine call extracted by tracing tools stored into RA Set in training time. The RAP set contains return address paths (RAP) to represents the various virtual paths formed between RA's.

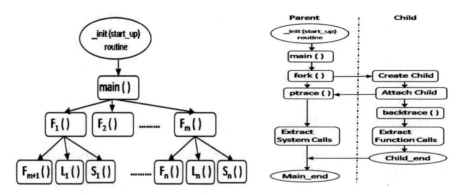

Fig. 1. The Call-graph (CG) view for working program (left), the PTRACE and BACKTRACE combination view of system model (right).

In our work, we use very powerful and popular method PTRACE for tracing of stack data. The PTRACE is itself is a system call and also perform best of tracing a process has additional advantages to collect maximum information compare to remaining traces. The PTRACE is highly independent tool helping to programmers to extract the lot of debugging information. It is tracing the stack includes functionality of BAKTRACE and STRACE. The PTRACE and BACKTRACE working combination is shown in Fig. 1(right). Generally the PTRACE is also extracting return addresses of function calls and system calls made an entry into memory stack. But in this we are working with combination is meant for exact verification of return addresses generated by both or not. The PTRACE is applied on an artificial program written exclusively which has a combination of functions. The PTRACE is giving lot of debugging information while program running, so here maximum necessary data gathered from this method. There is a possibility to collect various datasets from combination method and it is new approach in tracing techniques, so at a time we can extract all return addresses related to function calls, system calls, also a library calls and signals. All these are merging to training set and use them in testing for exploiting codes to discover anomalous behavior.

In our work, we also use STRACE and LTRACE tracing tools for better verification of datasets and tested on different Linux programs and other programs. This paper includes different datasets collected from training model during normal runs of programs and LTRACE, STRACE used here for getting better results. Because both are having different options of extracting valuable data at command prompt as well as in program by system() method. The simplicity of STRACE is like options setting while process tracing useful to collect system call timing, how many calls registered in stack thread can easy to measure. The other LTRACE is also verified whether the library calls are exactly identified or not, The LTRACE is good in identifying internet attacks on libraries such as forcing the mal functioning of library programs. All the tracing tools outcomes are stored into datasets for later usage in testing. In the testing (detection) process the stack data is also consider due to different test runs and compare with training sets. We know the stack is growing downward in memory corrupting total memory where as heap can do reverse manner also corrupt stack in the system memory shown in [16]. In testing phase we can find out anomalies through occurrences of different values registered in stack thread through buffer overflows, mistakes of programmer such as code copying from internet create instructions exploitations jump to attacker's code.

4   Datasets: Artificial Datasets

In the evaluation of our task, we have taken 16 different types of attacks and some of them are given in our previous work [10]. In this we are going to present the performance improvement by taking the relative difference between return addresses. We also created some fabricated programs to test four popular attacks by combination of PTRACE and BACKTRACE tracing tools. Also worked on popular Linux commands and extracted the program code for different commands such as ls, tar, ps, sty, ping, dash, and bash. We applied multiple tracing techniques to extract the data. All the datasets are collected from active system learning by dynamical running of programs. Training Set: (1) Return address set: this set includes return addresses related to all (2) Relative difference (RD) set (3) System call set (4) Library call set and (5) Signals set (6) Return address path (RAP) set also called sequence path sets. The suspected behavior of anomalous points noted through testing phase.

5   Implementation: Experiments and Results

The experiments (programs) are implemented on Linux platform. In this segment we have presented the working and implementation strategies corresponding to getting of results and analyzing test cases.. For this kind of progress, we have executed programs on latest processor (I3), installed Linux OS from Ubuntu source contain built in C. The Linux by default incorporates the tracing techniques (BACKTRACE, PTRACE, STRACE, LTRACE) helping in this work. The help part is considered from Linux materials [5]. The implementation outcomes are noted in tables in order of working shown in the Tables 1, 2 and 3. Here we have taken three different programs mentioned in abstract and all programs are assigned to each tracing process used in different manner to carry out experimental work. The first type of programs are showing the outcome of only BACKTRACE tracing method and are developed by writing an artificial programs written in C.

In the first category, we have done a lot of experimental work presented in [10], but in this paper we are going to give how the performance is improved in case of large size of datasets is generated shown in Table 1. If we observe in the Table 1, shows the change in size of datasets may directly affect the training and testing time. The main program is incorporating the mode of running program due to testing the affects of different types of attacks. Here the total no. of return addresses are counted including function calls, system calls and library calls. So that depending on the kind of attack, we need to consider individual datasets (RA set or RD set) or combined datasets (RA set and RD set). The relative difference (RD) is a kind of displacement between two consecutive return addresses. The programming running in attacked mode may change a size of dataset is more than in authentic (legitimate) mode. This is only a sample program tested on 16 kinds of attacks and maximum results presented in previous work [10]. The kind of anomalies such as RA anomaly, stack smashing, sequence path

Table 1. The comparison of performance during training and testing on the datasets generated through RA (return addresses), RD (relative differences) and overall changes in size.

S. No	Program name	Mode of running	No. of RAs noted	Size in bytes	No. of RDs noted	Size in bytes	% of dataset size reduced through RDs
1	passtest.c (Training time)	Authorized (legitimate)	3261	13312 bytes	2482	12288 bytes	7%
2	passtest.c (Testing time)	Un-authorized attacked mode	4872	19456 bytes	3652	14336 bytes	26%

Table 2. An observation of anomalies occurred in artificial programs shown by various tracing techniques. (* indicates that including of main (), init ())

Sl no	Program name	Nature of program used in testing	Training phase				Testing phase			
			No. of FC	No. of SC	No. of LC	Total no. of RAs	Total no. of RAs	PTRACE& BACKTRACE -o/p observation	STRACE- o/p observation	LTRACE- o/p observation
1	BO.c	Buffer overflow attack tested	152	26	32	*212	246	RA anomaly · Stack anomaly	SC Anomaly	SC, LC Anomaly
2	FS.c	Vulnerable function tested	148	38	27	*215	251	RA anomaly · Stack anomaly	SC Anomaly	SC, LC Anomaly
3	SCI.c	Shellcode injection tested	214	41	24	*281	328	RA anomaly · Stack anomaly	SC Anomaly	SC, LC Anomaly
4	HO.c	Heap overflow Attack tested	52	18	14	*86	109	RA anomaly · SO + HO	SC Anomaly	SC, LC Anomaly

anomaly, system call anomaly…etc. are generally occurred in programming exploits done through attacks. The sequence paths are well explained in [10], also helpful to in which call sequence has anomaly about the exact point whether before or after particular point of control line of program counter (PC).

The second category of programs is exclusively developed for four different types of attacks with combination of PTRACE and BACKTRACE. The Table 2 shows outcome of the experiment with various combinations of tracing tricks (PTRACE +BACKTRACE, STRACE, LTRACE) and four programs written exclusively such as BO.c, FS.c, SCI.c and HO.c with various kinds of attacks. Each program has incorporated by user defined functions, some standard C input, output functions for generating system calls and library calls. The BO.c is written for testing the buffer overflow attacks with different ways. Generally the intruders are attacking on the system by injecting malicious data into arrays of code that may cause of stack smashing and it leads to taking the system control into malware execution. These kinds of attacks are also done by programmers' mistakes while buffer allocation. So that this kind of security breaches may create to hold the system control by intruders. The program credentials are stored into datasets by storing the all the calls registered in stack threads. For all the programs the PTRACE and BACKTRACE are used to collect return addresses of each call is made by function or library or system call. Generally PTRACE trace tool or also called system call is powerful working at kernel based and easy to attaching process then collecting the process data through registers and stack. Similarly the different behavior such as new PC values observed in coding exploits treated a kind of anomaly. In all above cases the RAP (sequence path) anomaly also occurs due to RA anomaly.

All the programs are generating additional return addresses associated with main () and init (start-up routine) and explained good in [10]. The program FS.c is testing the formatting string vulnerability, in this the function such as scanf () is vulnerable to create exploiting the input buffer in memory stack by giving bad input. The LTRACE

Table 3. The Linux commands training and test case outcomes before and after modification code

		Training original (before modification)							Testing in attacked mode (% change after modification)			
S. no	File	1 PC(IP)	2 FC(R)	3 S.C	4 L.C	5 RAP	6 C1	7 C2	1 PC (IP)	6 C1	7 C2	Kind of anomaly
1	ls	5.4×10^5	1630	191	1557	210	1821	1840	5.20%	3.5%	40.28%	A_1, A_2, A_3, A_4
2	tar	1.3×10^6	10332	151	10332	450	10483	10782	10.89%	1.4%	85.26%	A_1, A_2, A_3, A_4
3	ps	4.4×10^5	359	1357	359	25	1716	384	4.56%	0.96%	48.23%	A_1, A_2, A_3, A_4
4	stty	2.2×10^5	118	122	118	8	240	126	3.28%	1.02%	63.21%	A_1, A_2, A_3, A_4
5	ping	1.4×10^5	6	41	6	2	47	8	3.56%	10.25%	50.00%	A_1, A_2, A_3, A_4
6	dash	4.5×10^3	125	82	100	15	207	140	2.12%	3.67%	32.57%	A_1, A_2, A_3, A_4
7	bash	3.9×10^6	451	128	450	30	579	481	12.03%	1.67%	59.38%	A_1, A_2, A_3, A_4

*PC(IP)-program counter (instruction pointer), FC-function call (R-Return Address), SC-system call, LC-library call, RAP - return address path, C1-combination of FC and SC, C2-combination of FC and RAP, A_1-Return address anomaly, A_2-System call anomaly, A_3-Library call anomaly. A_4 - Return address anomaly

and STRACE also giving similar kind of anomalies observed in Table 2. The SCI.c program is testing on popular attack called shell code injection. Generally the source codes are converting into object code where as the OBJDUMP option in Linux is useful for taking assembly code along with return addresses. The sequence of hexa-code values of assembly code can prepared as shell code and this can easy to embedding into executing process by many ways such as system() function or array creation in the memories. So this kind of process may chances to take the control into the new malicious target program by overflowing the buffer created in memory. So that in this program attacks the maximum new RAs are noted into the training datasets and more no. of changes in testing leads to various kinds of attacks shown in Table 2. The forth program HO.c is tested on allocation of heap for dynamic data. The heap is also leads to buffer overflow attack which smashes the linked list pointers and causes of both stack anomaly and heap overflow anomaly. In all above cases raising different types of attacks by stack smashing or heap smashing may create basic anomalies classified as SC anomaly, LC anomaly, FC anomaly, RA anomaly in testing.

The third categories of programs are shown in Table 3 taken from source code of most useful commands on Linux platform and same tracing techniques (PTRACE, STRACE and LTRACE) are applied on those commands converted into text. The trace data is simply storing into different datasets (PC set, RA set, SC set, LC set, and signal), it means that these sets act as training sets formed either individual or combined tracing for better accuracy of results. The command programs are executed by our own designed programs which has full control on it. Initially all commands are embedded execution in C program form the datasets before insertion of attacking code. Here combinational datasets C1 and C2 also computed for better results. The C1 (FC and SC) set, C2 (FC and RAP) sets are additional usage in this work to determine the what exact changes made before or after attack either individual or combined. The command program contains the no. of FC based on their present and calling order. Sometimes repeated calling creates an additional overhead on the memory and frames are formed for each call. Let take the snapshot of ls command program incorporate only 25 functions, which can continue the call sequence no. of times an average of 60 to 70 times. This scenario is continued then overall time complexity will be more, so that the repeated calls takes maximum time for training. The ls command reflect after attacked mode as follows 5.20%, 3.5%, 40.28% respectively for RAs, C1, C2. The C2 changes more due to RAP changes and all kinds of attacks observed in testing.

This kind of attacks are common to normal programs explained previously, but these commands are dropping millions of return addresses into stack thread takes lot of time to form the datasets in training. Let take tar program has maximum RAs, so that it can expose to maximum no. of anomalies due to same function call IP is used many times while formation of RAPs (virtual sequence paths) observed in Table 3. The commands ping has very less no. of calls identified in stack thread and in that maximum system calls are noted. The comparison of changes of code after modification of code for each command program can affects the datasets as usual, but observations of anomalies are showing similar for all. We have given maximum outcomes and shrinking into Table 3.

5.1 Discussions

The task identified in our work divided into three major parts such as prepare the model and way of designing the system to carry out the work, selection of tracing tools which works fast in response and formation of training sets and testing sets, and final one is classification of anomalies after testing outcomes. The Fig. 1 shows the view for working program that can help us to way of collecting raw data about stack entries. The method adapted in our work uses the mixture of PTRACE and BACKTRACE giving a view of our model to develop the code for each case. As mention in the abstract, the process of design of system and task execution time is minimized by reduce the sizes of datasets by taking the relative differences between consecutive return addresses. This kind of procedure is very useful in case of millions of RAs encountered in the program. Our program passtest.c is designed for thousands of return addresses noted datasets given in Table 1. Also we have taken small artificial programs for implementing the task specified and what kind of anomalies are encountered while execution of malicious programs shown in Table 2. There are multiple verifications of anomalies observed by use of expert tracing tools. The total no. of calls along with return addresses are shown in Table 2 based on only taking single instance and if we take multiple attacks in single instance may stop the execution of process some times. Finally the same experimental work is done on popular Linux commands, which are giving a huge no. of return addresses dropping into stack thread helpful to our work. Here the ratio of modifications after attacking the original code is mention in the Table 3 and kind of anomalies are shown in our observations.

6 Conclusion and Future Work

Our main objective of this work is improving the capability of disclosing anomalies in large programs contaminating anomalous behavior observed by malfunctioning. The program environment is itself creates security breaches to attackers, so that the incremental testing tools to be use to evaluate the strength of program and discloses the modification points in programs. In this paper we are giving an extension of work presented in [10, 17] and algorithms as well as sufficient results are given in that. We have used a separate technologies for each kind of work to knowing the anomaly detection and assuming that the scope of work is presented here is limited to program. But we have enough methods to identifying the system level anomalies given in the related work and if we extend the scope to applications then we expect beautiful outcomes will get. So that in future work, we are planning to extend the possibilities to combine the machine learning techniques with program level methods for better optimization of datasets, avoiding un-necessary training data which is not change for long time and improve the overall performance of anomaly detection.

References

1. Denning, D.E.: An intrusion detection model. IEEE Trans. Softw. Eng. **SE-13**, 222–232 (1987)
2. Axelsson, S.: IDS: A survey and taxonomy, Chalmers Univ'y, Technical report 99-15, March 2000
3. Anderson, J.P.: Computer security threat monitoring and surveillance, USA, Technical report 98–17, April 1980
4. Feng, H.H., et al.: Anomaly detection using call stack information. In: IEEE Symposium on Security and Privacy, 11–14 May 2003, pp. 62–75 (2003). ISSN 1081-6011, Print ISBN 0-7695-1940-7
5. http://www.linuxjournal.com, http://linux.die.net and http://www.gnu.org
6. Cowan, C., Grier, A.: Stack-Guard: automatic adaptive detection and prevention of buffer-overflow attacks. In: 7th-USENIX Security Symposium, San Antonio, TX (1998)
7. Forrest, S., Hofmeyr, S.A.: A sense of self for unix processes. In: IEEE Symposium on Research in Security and Privacy, Oakland, CA, USA, pp. 120–128 (1996)
8. Lee, W., Stolfo, S.J.: Data mining approaches for intrusion detection. In: 7th USENIX Security Symposium (SECURITY-98), Berkeley, CA, USA, pp. 79–94 (1998)
9. Hofmeyr, S.A., Somayaji, A., Forrest, S.: Intrusion detection system using sequences of system calls. J. Comput. Secur. **6**(3), 151–180 (1998)
10. Jidiga, G.R., Sammulal, P.: Anomaly detection using smarttracing tricks on system stack. In: IEEE International Conference on Convergence of Technology I2CT-2014, Pune, India, 6–8 April 2014, pp. 1–6 (2014). https://doi.org/10.1109/i2ct.2014.7092136, ISBN 978-1-4799-3759-2
11. Jidiga, G.R., Sammulal, P.: Foundations of intrusion detection systems: focus on role of anomaly detection using machine learning. In: ICACM - 2013 Elsevier 2nd International Conference, August 2013. ISBN 9789351071495
12. Peisert, S., Bishop, M.: Analysis of computer intrusions using sequences of function calls. IEEE Trans. Dependable Secure Comput. **4**(2), 137–150 (2007)
13. Sekar, R., Bendre, M., Bollineni, P., Dhurjati, D.: A fast automaton-based method for detecting anomalous program behaviors. In: IEEE Symposium on Security and Privacy, Oakland, CA (2001)
14. Wagner, D., Dean, D.: Intrusion detection via static analysis. In: IEEE Symposium on Security and Privacy, Oakland, CA (2001)

15. Praveena, A., Smys, S.: Anonymization in social networks: a survey on the issues of data privacy in social network sites. J. Int. J. Eng. Comput. Sci. 5(3), 15912–15918 (2016)
16. Yang, X., et al.: Eliminating the call stack to save RAM. In: Proceedings of the ACM Conference LCTES, Dublin, Ireland, 19–20 June, pp. 1–10. ACM. 978-1-60558-356-3/09/06
17. Jidiga, G.R., Sammulal, P.: Anomaly detection using new tracing tricks on program executions and analysis of system data. In: Proceedings of the First ICCII in AISC Springer Series, vol. 507, pp. 389–399, January 2017. https://doi.org/10.1007/978-981-10-2471-9_38
18. Jidiga, G.R., Sammulal, P.: Anomaly detection using light tracing tricks on system stack. In: National Conference NCPR-2014, 9–10 January 2014, pp. 72–77. JNTUHCEJ, India (2014). ISBN 9789383038176

Satellite Image Enhancement and Restoration

Vivek Giri and Sudhriti Sen Gupta[(⊠)]

Amity Institute of Information Technology, Amity University,
Noida, Uttar Pradesh, India
vkgiri7@gmail.com, ssgupta@amity.edu

Abstract. Image received from remote sensors like satellites are of low in value of contrast which hides main records & quality offers carried by way of the picture. Hence, photo restoration is very important and must in the photo processing area to surface up all of the details as it should be in the natural and original images. The low in quality contrast satellite image recovery is essentially based on adaptive histogram equalization.

1 Introduction

Image restoration attempts to reconstruct or get better a [1] photo that has been degenerate with the aid of a degradation phenomenon. Thus, restoration techniques are oriented in the direction of modeling the degradation and applying the inverse technique with a purpose to recover the unique picture taken from remote sensors like satellite. As in picture enhancement, the closing goal of healing strategies is to enhance the photograph in a few predefined experience. Satellite photograph processing is one of the thrust regions inside the field of computer science research. Images taken through satellites sensors likely degraded because of weather, climate and different factors. Satellite photo enhancement and restoration is scientifically possible by way of applying picture processing and different smooth computing strategies.

2 Literature Review

Pictures are the broadly powerful, common and effective mode of representing collective data and transmitting information. A digital image sometimes may worth more a thousand words. Humans also receives 80% of all given information in graphical form [2]. Restoration and Enhancement in satellite image processing is a widely effective method to enlarge the amount of pixels and improve the quality of the pixels as well in a digital image.

Nowadays variety of restoration techniques have been implemented in image processing to enhance the quality of image and it's resolution as well to upscale it with the wide range of screen displays. There are mainly three different types of image restoring strategies, which are nearest-object based, bicubic and bilinear. From these

© Springer Nature Switzerland AG 2020
A. P. Pandian et al. (Eds.): ICCBI 2019, LNDECT 49, pp. 252–258, 2020.
https://doi.org/10.1007/978-3-030-43192-1_28

three strategies, bicubic-restoration technique is more effective and most of the times gives results in smooth edges in outcoming image. On the hand noise removal and preservation of useful information carried through pixels are important aspects of image enhancement. Image enhancement is a strategy broadly focused on processing on an digital image in such a manner that the processed image is more observable than the taken image from optical sensors for the specific application. A number of enhancement strategies exist inside this spatial domain [9].

Applications
Satellite images and it's after imaging techniques have enlightened many fields and have applications in researching areas like oceanography, agriculture, forestry, meteorology, regional planning, biodiversity conservation etc.

Advantages
A single satellite image can store wide range of details and information through pixels. Photonic spectrum that these optical sensors use can be control to pick up very detail and moments on the earth's surface. For instance, an archaeologist can use these collective images to locate major variations in soils and can compare potential sites to target their projects. Railway and roadway based architectural engineers can find efficient and shortest route possible. An environmentalist can use these enhanced images to use them to detect variations in vegetation and moisture, with true detail updation over period of time. And [3] GIS based application can help rescue team to work on accidents.

3   Methodology

In this paper we will be using General Histogram Equalization and create Truecolor composite of input image in RGB format. This methodology proposes a effective image enhancement method that mostly deals with hue preservation to address details taken from satellite sensors. The results which we get from this method can produce equalized image of each RGB color channel (Fig. 2).

3.1   Contrast Enhancement

Contrast enhancement is most often referred as the crucial issues in image enhancement and processing [4]. The problem is to fine tuning the high dynamic contrast of an image in order to represent all the details in the images taken by satellites [5]. There are several other techniques to deal with this issue, some are listed below:

1. Local histogram equalization (LHE)
2. Decorrelation Stretching
3. General histogram equalization (GHE)
4. Linear Contrast Stretching

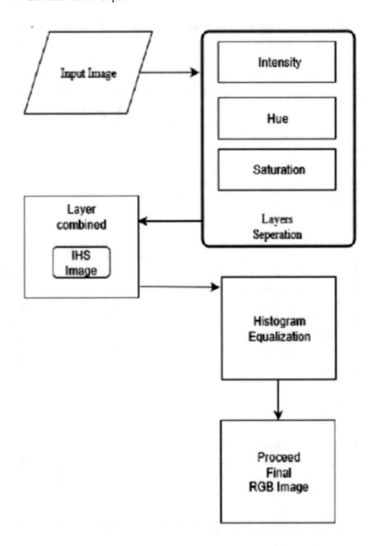

Fig. 1. Methodology for intensity equalization using IHS system

Methods & Techniques
Satellite image contrast enhancement of multispectral color composite and IHS (Intensity, Hue, Saturation) transformation and their algorithm and results (Fig. 1).

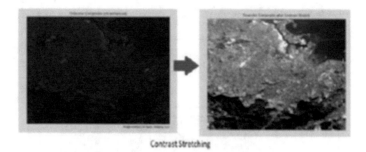

Contrast Stretching

Fig. 2. Contrast enhancing using truecolor composite

3.2   Intensity Equalization

The vital blocks of primary colors RGB (i.e. red, green and blue) is a well-known fundamental fact [6]. A different approach to color is the IHS system, where I stand for "intensity" H stands for "hue" and S stands for "saturation". It is very in use in image enhancement because it presents colors more near to the experience as the human observer look them out. The intensity i.e., I is responsible for the brightness variations and ranges from black to white i.e., 0 to 255 [7]. Hue i.e., H is responsible for the dominant wavelength of color. Saturation i.e., S is responsible for the purity of color and ranges from 0 to 255. A 0 value in saturation represents a completely impure color, whereas saturation with high values, represent as real as more intense and pure colors [8]. When any of the three illusory bands of a sensor data get combined to goes into the RGB system, the processed color images typically in some case lacks saturation, even though those bands have been gone through contrast-stretched.

Methods & Techniques

The IHS (intensity, hue and saturation) transformation can also be used for enhancing satellite images. In this case, we have to first separate the intensity component, hue component and saturation component from the RGB image and according to the requirement of the application we can enhance one or more components. After enhancing IHS image we have to convert it back to RGB for better visual perception:

1. Input RGB image
2. Separate the Intensity, Hue and saturation layer of the input images
3. Create IHS image
4. Histogram equalize the H, S and I images and the IHS image
5. Then convert the HSI image back to RGB.

4   Result

We have use Matlab coding language in GNU Octave v5.1.0 with Windows 10 machine having 8 GB RAM and intel i7 processor to execute the results. For input images we have taken them from QuickBird Open Source Satellite map.

As we visually can see that proceed output images have shown significantly higher and more rich in details than compare to input images, using our applied techniques in Table 1 given below

Table 1. Collective results of input images with their output images

Satellite Images	Input Images	Proceed Images
map 1		
map 2		
map 3		
map 4		
map 5		

We've also used the "entropy" for the quantitative analyzing the proceed and enhanced images came out from the proposed methods. Entropy is used to evaluate the overall quality of image. The higher the rate of an entropy of an image, the better it's quality and great in details. From the applied method, the entropies of the input images with respect to output image are (Graph 1):

Entropy:-
map1
input: 7.2350
output: 7.9741
map2
input: 7.3687
output: 7.8277
map3
input: 5.9524
output: 7.0835
map4
input: 7.2190
output: 7.4495
map5
input: 7.5045
output: 7.5765

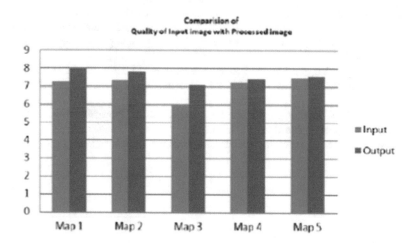

Graph 1. Comparison of input and output image entropy

5   Conclusion

This research paper have witness the true potential of after imaging techniques and its essential role in imaging field. There are different methods to implement each of the satellite image enhancement techniques. The selection of the method depends on the input image type, the sensor used to capture the image, the type of application for which it is to be used. These different methods for contrast enhancement, resolution enhancement, IHS transformation, density slicing, edge enhancement, digital mosaics, and synthetic stereo images are discussed and some are compared based on their performance. Among all above techniques, results of two techniques are shown viz. contrast enhancement of multispectral color composite and IHS (intensity, hue, saturation) transformation are shown and from these results, we can see that how satellite images get enhanced for better visual perception as well as for using them in various applications.

References

1. Rajeswari, K.R., Krishna, K.M., Naveen, V.J., Vamsidhar, A.: Performance comparison of wiener filter and CLS filter on 2D signals. In: 2009 60th International Conference on Information Technology: New Generations (2009)
2. Ashok, J., Rajan, E.G.: Multi-spectral image enhancement in satellite imagery. ACS Int. J. Comput. Intell. 1(1), 1–8 (2010)
3. Zhao, J., Qin, Q., Xie, C., Wang, J., Meng, Q.: An efficient method of predicting traffic noise using GIS. In: 2013 IEEE International Geoscience and Remote Sensing Symposium - IGARSS, Melbourne, VIC, pp. 3634–3637 (2013)
4. Ozcinar, C., Demirel, H., Anbarjafari, G.: Image equalization using singular value decomposition and discrete wavelet transform. InTech (2011). Chapter 5
5. Murao, T., Saitoh, K., Koshiba, M.: Realization of single-moded broadband air-guiding photonic bandgap fibers. IEEE Photonics Technol. Lett. 18(15), 1666–1668 (2006)
6. Koutsias, N., Karteris, M., Chuvieco, E.: The use of intensity-hue-saturation transformation of Landsat-5 Thematic Mapper data for burned land mapping. Photogramm. Eng. Remote. Sens. 66(7), 829–839 (2000)
7. Wang, T., Li, B., Yu, C., Cao, W.: Remote sensing image fusion based on Multiwavelet Transform and IHS Transform. In: 2006 8th International Conference on Signal Processing (2006)
8. Gonzalez, R.C., Woods, R.E., Eddins, S.L.: Digital Image Processing Using MATLAB. Prentice Hall, Upper Saddle River (2003). Chapter 11
9. Ahuja, S., Biday, S.: A survey of satellite image enhancement techniques (2018)

Role of AI in Gaming and Simulation

Shivam Tyagi[(✉)] and Sudhriti Sengupta

Amity Institute of Information Technology, Amity University, Noida, UP, India
Shivam.tyagi920@gmail.com, ssgupta@amity.edu

Abstract. The overall purpose pursued in this paper is to learn more about gaming and simulation in general and how artificial intelligence is being implemented into them and all the possibilities for future in this scenery. The mechanism of simulation and gaming go hand in hand where artificial intelligence has been coming in clutch to be related these two in all the aspects we can think of. All the possibilities for the AI in simulation have been discussed along with gaming while learning about the problems we face when trying to pester AI in applications as it obviously may have some drawbacks too. All the different roads that can be taken with these combinations are something the paper has focused on. The trend of virtual reality has taken off recently and thus it is only logical that AI may be implemented into that too therefore we discuss that here. AI helps in optimization of programs and also saves labor as the AI can be taught to learn and work alone. AI can imitate human ways while avoiding the drawbacks a human may have. Mapping environments, making levels and characters are just about some examples that are being helped by AI in gaming while in simulation the AI helps in human interaction and creating scenarios while trying to ignore the danger that an actual experiment might have against a simulation. Social realism in gaming and simulation is just another important thing that is discussed as it is something really important to have that if an AI can learn then we might go a long way.

Keywords: AI (artificial intelligence) · Simulation · Bio-sensing · VR (virtual reality)

1 Introduction

Simulation is an imitation of a process done in order to analyze the situation in a controlled environment. For example, automobile simulation where we can make a program that may act as an automobile that can be tested on without actually having possibilities of crashing that automobile in real life. It has proven to be a really useful mechanism and has been practiced for quite a few decades now. This can be done for almost any physical thing that needs to be tested as in Aircraft simulation, disaster simulation, military simulation etc.

More examples of simulation are Bio-mechanism, Futuristic classrooms, disaster preparedness, military, and project preparation. Computer simulation is the main focus of discussion here, now computer simulation has always been a part of the world ever since the computers started to exist. Simulation mainly was being worked on during World War 2 by nuclear scientist to simulate the bombing experiments. Simulation is

© Springer Nature Switzerland AG 2020
A. P. Pandian et al. (Eds.): ICCBI 2019, LNDECT 49, pp. 259–266, 2020.
https://doi.org/10.1007/978-3-030-43192-1_29

really useful for data preparedness where the quantity of data is entirely dependent. The variety is limitless and therefore there exist some simulation languages for programming too like MIDAS which later turned into some more equation oriented language called MIMIC.

Visualization of simulations is usually done by animations but it can be done through matrix representations too. To show data through regular charts and other ways is easier when compared to animations but in a simple simulation visualization, animations work better to teach someone with not enough understanding of the focused data [18].

There exist some risks with simulation too because the simulation cannot always be accurate and to be as accurate as people, one needs almost every possible data they can imagine in bigger simulations where risks can't be taken at any cost and all calibration has to be to the point too.

Gaming is just another simulation but is done for entertainment purposes and doesn't have to be an imitation of the real physical things present in our environment as it may be based on fantasies too. A gamer is anyone who plays a game be it role playing game, card game etc. and usually has been playing it for reasonable amount of time. There are a number of people indulged in gaming ever since the internet became so active. Age of a gamer usually varies from an average 12 year old to a 30 year old and they may play game Upton 12 h a day and even more for some people. Gamers may always be assumed to be young guys but nowadays a lot of variety can be seen among the number of gamers around the world.

This report focuses on how AI has been impacting for simulation and gaming and what is its future and we are also going to discuss the challenges and applications that we already have seen in this field.

AI or artificial intelligence is simply an intelligence that isn't natural and is made by humans for activities that wouldn't need a real human or maybe possibly cannot have a real human but does need to act like an actual human being. Doing things in a virtual environment is logically the safer way to check and test things out but sometimes we need to make situations where an unaffected or unbiased mind must be present in order to make the test real and successful which is why we need to implement an AI into that simulation.

2 AI in Simulation

AI can be used in multiple ways when implemented into simulation including testing, modeling, optimization etc. AI is currently being used in smart factories and a lot of different industries to maintain a swift and smart environment which is better when not done by humans who would be considered slow and more vulnerable to making mistakes in comparison to an AI. Machine learning and simulation are very different but can be combined to solve each other's problems [23]. In simulation, we try to create and study the physical attributes and working of the real world to create a prototype that may be used virtually and eventually be implemented for reality (Fig. 1).

If you are not familiar with what machine learning is then we can look at the words separately to make out what it might mean which is "Machine" as in a computer or any

Fig. 1. (Taken from open source) is an example of a diving simulation and is depicted from the user's POV [12]

non-human system and "Learning" for it to study the data and parse it to apply it into decision making [15]. Basically, we teach the machine by giving it loads of data and make out their own process of understanding decision making so they can make close predictions on their own.

One has to remember that deep learning is just a subset of machine learning where deep learning focuses on analysis structures that help out in giving out the bet conclusions for them [5].

AI in simulation would be mainly used for economic purposes to support the market strategies [25] and also to take the AI's help in finding out the best and most reasonable business decisions and the best use of simulation is to bring the safest and performing environment for humans. Machine learning will help us is unexpected system failures, maintenance and failure detection.

3   AI in Gaming

Social realism in games is a realism added through depicting real life experience into the game [17] for e.g. in call of duty games, world war has been the whole platform of levels and thus it makes the gaming more realistic because these events have already happened in the game. There are things that are socially accepted by the society and games are made on them to address this realism too for e.g. The butterfly sisters is a game which focuses on mental health and this mental health is thoroughly discussed in the game which may make the player feel more relatable to the game.

AI in gaming is built around making characters or situations, in board games like Hearthstone or Dota auto chess, prediction plays a big role and this is why neural networks are being implemented into these games. Gaming is a good platform for AI research as a lot of funding is being provided into this sector and these are a consistent player base that the AI can interact with [21].

Table 1. Applications of AI

Applications of AI in simulation
Prototyping
Decision making
Testing of elements
Optimization
Design and creation
Deep learning
Cognitive learning

The most basic type of AI in games is an NPC i.e. a non- playable character. These characters are based off of a script and therefore interact according to it too but a few things about them maybe tweaked in order to make them more realistic [4]. But, the best implementation of AI in games is in things like, Data Mining on user behavior; here researchers try to study the mind of a player by putting him/her into situations and then studying their reactions and problem handling techniques. Bio-sensing a technique used to make a gaming experience way more realistic where bio-sensing happens through gloves or other similar equipment [2]. Here movements are studied by the computer and then quickly interpreted into the game for a return of response. This is done by wearing the glove and then following instructions or actions of a game and then importing these responses accurately into the game. This experience is also called the immersive experience and immersion happens through gloves and bio-sensing.

4 Objective and Methodology

As we have thoroughly discussed earlier, AI can learn from human behavior and this behavior can be studied while the subject is playing games and put into different planned situations.

When an AI learns from this and gathers knowledge, this knowledge can be used to make game content through that AI. This may include map creation, character creation, level design etc. and this will also in turn save the labor cost. Making game characters includes a lot of things as in facial features, expressions, body language, body sync and reactions etc. By making an AI learn what a typical gamer likes about a character can result in making the almost perfect and most appropriate in game characters [22–26].

Map creation might prove to be a little complex as maps aren't supposed to just please a gamer but are simply required o be fair and suitable to the game's genre, cinematic and setting too [19].

When AI can make game levels better than the humans, then it is only good news for game developers although it might make human input lower and therefore lower the number of people required when making games or a simulation [6]. Some sectors of game-developer still might need some human interference but who knows, maybe that can be covered by AI in future too (Table 1 and Figs. 2 and 3).

Table 2. Application of AI

Application of AI in VR
Temperature change
Sound change and improvisation
Feeling of depth and distance
Motion control
Balance control
Climate change according to situation

New Atlas reviews the Oculus Rift with excellent Touch controllers (Credit: Will Shanklin/New Atlas)

Fig. 2. (Taken from open source) is to show an oculus rift being used by wearing the headset along with some hand motion sensors that come with it [13]

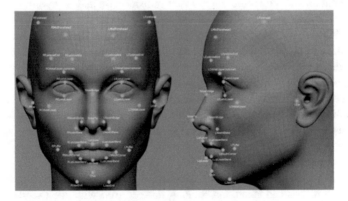

Fig. 3. (Taken from open source) is to show how multiple points of motion are added to a face for recognition by a regular image recognition software [14]

AI in VR

VR or virtual reality is a technique of making things more realistic by providing a visual and/or sensory experience that are physically handled by a person through equipment like glasses and movement controller as in Table 2. There is AR which is augmented reality and it is also a very creative way to game or to just mimic a virtual world into your own reality through the use of special equipment or phones even (Pokémon Go), AR helps in interaction with actual objects for interacting with computer based environment and may even help in rehabilitation [11] but we should not go too deep into that.

Mapping of environment through by making the AI study and environment and then implementing that environment on to a virtual world in hopes of being precise is a very accurate example of how AI will be used for VR, realistic motion tracked location navigation for people is something to be expected too and it would not require any heavy setup either as it'll be HMD based [9]. There can be many other examples along with this for example, character modeling is something that becomes very relevant to gaming and even simulation that is to use AI for perfection of a program, Also we can study Facebook when we study application of VR in AI as Facebook recently bought oculus rift which is a VR headset, Facebook did this to transit through pictures uploaded by people on their pages and depict those pictures for a more generative and believable model. When applying AI with VR, there comes a lot of challenges like some difficult relationships that might be spatial are not very easy to get the hang of by just 2D projections of provided data and the data can be too large which is a problem too [8].

To make a program optimum and relevant to the user by concise renders of models and other objects can be done to create realism [17]. A lot of models are under bad renders because of the reason that these models are world on by humans and if all of this was made by AI then the renders would be faster and more accurate with less expenses spent and also it would allow faster transmission and no loss of quality would be seen [20]. Now, with the help of AI we can teach the system to enhance the experience by adding realistic temperatures, sounds, depth feeling, echo, environment etc. These things can be done by adding equipment that may handle these things and all in all, the degree of this realism will be decided by the AI. It is already helping in making a good user experience through voice modulated assistant in phones and other smart devices that you may know of as cortana or siri. VR isn't just used for making a realistic gaming experience; it is also a tool to allow simulation to exist in VRs for e.g. VRs are used to simulate pilot training [10] and experience along with travel experiences.

Image Recognition Through AI

Neural networks programmed with deep learning are much harder to make use of as they are really difficult to teach and train. A dataset used with coco segmentation and coco detection (image recognition methods) can help in obtaining a very significant improvement and recognition of facial details into a 3d object and create characters with its help [1]. It helps in depth representation and to rotate a residual depth with the use of layers that may be going up to 152 layers even. Deep convolutional networks, end to end layer segmentation have been really beneficial for image recognition datasets. Sketch modeling tools, scaling models and their tools along with CAD applied modeling are some methods for environment representation which may further be used into image recognition [7].

Genetic Programming in AI

A game called MUGEN is a 2D fighting game that uses the automatic AI creating technique which doesn't require high qualification engineers in order to be functional [3]. This technique engages a tournament selecting algorithm to create a method of creating AIs that are harder and better than the previous ones. Even a survey was done to check the validity of the technique against the program and the results were very favorable to the automatic genetic programming AI generator [24]. This feature helped in saving the time to code every character individually and also avoided the requirement of checking internal code for appropriate formation of characters. The program didn't run long enough but can be revived to create even more complex characters for games that are known for their difficulty like Dark souls, Bloodborne, Sekiro: shadows die twice. The program can also create varying features for characters and then automatically increase difficulty as we keep on going further into the game. It can teach strategic skills to the AI which will eventually help in adaptive propagation in the characters [16].

5 Conclusion

So, as a final cord of the song we can say that AI is very beneficial towards most technological things and some non-tech related subjects too. The difficulties that we have with human labor can be avoided with the help of AI when it comes to time saving and when it comes to innovation, AI is still very supportive as machine learning is a big deal now and many people have been rushing towards this subject. AI is going to offer a lot of unconventional and unorthodox ideas while it will also prove to productive towards other technological fields but especially towards simulation and gaming. We will surely see a decrease in time span of doing jobs that humans take too long to do and we will definitely see an increase in better simulation productivity and functioning. Opinion on AI has been negative recently for the AI but handling of AI totally depends upon the developer and if done right then AI is nothing but useful to us.

References

1. He, K., Ren, S.: Deep residual learning for image recognition
2. Togelius, J., Yannakakis, G.: A panorama of artificial and computational intelligence in games
3. Woods, D., Cants, R.: Creating AI characters for fighting games using genetic programming
4. Richoux, F., Churchill, D.: A survey of real-time strategy game AI research and competition in StarCraft
5. Maire, F., Browne, C.: Evolutionary game design
6. Lucas, S.M.: Computational intelligence and AI in games: a new IEEE transactions
7. Pauwels, P., Samyn, K.: The role of game rules in architectural design environments
8. Hentschel, B., Wolter, M., Kuhlen, T.: Virtual reality-based multi-view visualization of time-dependent simulation data

9. Hodgson, E., Bachmann, E., Waller, D., Bair, A., Oberlin, A.: Virtual reality in the wild: a self-contained and wearable simulation system
10. Murphy, D.: Bodiless embodiment: a descriptive survey of avatar bodily coherence in first-wave consumer VR applications
11. Burke, J.W., McNeill, M.D.J., Charles, D.K., Morrow, P.J., Crosbie, J.H., McDonough, S.M.: Augmented reality games for upper-limb stroke rehabilitation
12. https://www.italiankubb.org/simulation-games-virtual-meets-reality/
13. https://newatlas.com/oculus-rift-review-touch-2017/46711/
14. https://savvycomsoftware.com/image-recognition-its-revolutionary-role-in-the-business-world/
15. Xue, M., Zhu, C.: A study and application on machine learning of artificial intelligence
16. Zufri, T., Hilman, D., Pratama, W.: Character design as bridging tools of ideological message in game
17. Fan, S.: Image visual realism: from human perception to machine computation
18. Mao, C., Yi, Z., JianGang, O., Guo-Tao, H.: Game design and development based on logical animation platform
19. Smys, S.: Virtual reality gaming technology for mental stimulation and therapy. J. Inf. Technol. 1(01), 19–26 (2019)
20. Bierre, K.: Implementing a game design course as a multiplayer game
21. Kim, M.-J., Kim, K.-J.: Opponent modeling based on action table for MCTS-based fighting game AI
22. Giraldo, F.A., Gómez, J.: The evolution of neural networks for decision making in non-cooperative repetitive games
23. Torrado, R.R., Bontrager, P., Togelius, J., Liu, J., Perez-Liebana, D.: Deep reinforcement learning for general video game AI
24. Camilleri, E., Yannakakis, G.N., Dingli, A.: Platformer level design for player believability
25. Li, Z., Yu, X.: Solution of pursuit/evasion Differential Games using genetic algorithms
26. Hedberg, S.R.: AI tools for business-process modeling

Analysis of MRI Images Using Image Processing Technique

Sudhriti Sengupta[(✉)], Ashutosh Dubey, and Neetu Mittal

Amity Institute of Information Technology, Amity University,
Noida, Uttar Pradesh, India
{ssgupta,nmittall}@amity.edu,
Ashutoshdubey2908@gmail.com

Abstract. Image processing is used in digital world to improve the quality of information present in the image. It is vital for creation of an automated diagnostic system for various type of disease. In a computerized system, edge deduction and analysis is of utmost importance. In this paper, we have proposal an edge deduction system for MRI images which will facilitates the development of an optimization system for MRI image analysis.

Keywords: MRI images · Histogram equalization · Filtering · Edge detection · Morphological structuring

1 Introduction

As understood from various resources, there is much advancement in the field of Imaging Technology. Many different technologies are being introduced from past decades that open up the way in analyzing and understanding the anatomy and the function of the body. The Imaging technology accesses the way to cure injuries, and anatomy which has been done with the aid of Magnetic Resonance Imaging Technique. The continuous advancements in the MRI lead to large amount of data with increasingly high quality levels. It has become tedious for Doctors and Clinicians to compered the problems who actually used to analyse information manually. With the advancement in the Technology, MRI has been used to overcome this problem. This manual analysis is much more time consuming and more prone to error due to lack of knowledge. Hence, Computerized Analysis Method should be introduced to improve diagnosis and testing. Nowadays, Computerized Method for Internal Images of Human Organs and various parts, i.e. MRI is being used to assist doctors to diagnose the patient.

Automated diagnostic system of MRI image will increase the efficiency of diagnostic. People in rural or fur-flung area may not get access to specialist. If an effective computerized is there to analysis the MRI image, then it will greatly help the patient to get customized and quick treatment with this motivated, we have proposed an Image processing and segmentation technique which helps to build the automatized diagnostic system for analysis of MRI images.

© Springer Nature Switzerland AG 2020
A. P. Pandian et al. (Eds.): ICCBI 2019, LNDECT 49, pp. 267–274, 2020.
https://doi.org/10.1007/978-3-030-43192-1_30

Medical image are the visual representation of inner body for medical analysis and medical interceding. Medical images reveal internal structure body hidden in skin and bones. Medical Images also proven a database of biology concerned with study of the structure of organisms and their park, and the scientific study of the function and mechanism, which work within a living system to make it possible to identify abnormalities.

2 Related Research Work

Chitradevi and Srimathi et al. address to enhance the raw images from the camera, the images are improved from the real time and extend to the field of science [1]. Kumar and Bhatia et al. address the feature like resizing, binarization, thresholding are applied on the image processing [5]. Philippe Burlina, Chellappa et al. address the paper attention on the mechanism of the interpretation for signature for better object configuration [7]. Gagnon and Jouan et al. address this study is comparative between WCS (wavelet coefficient shrinkage) filter and other several sparkle pictures [10].

3 MRI

MRI is Magnetic aided Resonance Imaging Scan that basically uses highly magnified Magnetic fields and radio waves to produce a detailed and elaborated Image. It is considered to be a medical Imaging Techniques that takes up the images from the anatomy of body and physiological processes to detect various Health Problems. A MRI is done on the patient's specific part to image the irregularity in it. The Magnetic field is applied to the body with a proper amount of Resonance Frequency. The Hydrogen atom is excited and then it emits a signal and received through a coil that may encode the Signal Information. Hence these signals produce the noise that leads to the formation of MRI Scan Image. The main components of MRI are: magnet, shim coil, gradient system. Magnet: The Magnets are used to polarise the part or the sample taken. Shim Coil: It is used maintaining the shifts in the magnetic field. Gradient System: used to localise the signals this excites the samples and detects the signal. MRI can be used for: Problems in Brain and Spinal Cord, Tumors and Cancers, Injuries in the Joint, Cardiac Problems, Pelvic Pain causing Fibrosis, etc.

4 Proposed Technique

The proposed Technique is a technique which is used to remove noises from MRI Images to make the image more clear and edgy. As an input we take a MRI image which needs to pre-processed to make the image clearer. In Pre-processing step, the contrast of image is improved followed by filtration which eliminates their noises. The next stage is Segmentation in which the image is edge detected. It's also helps in Boundary Extraction for clear image. The Proposed Workflow for MRI Image Analysis is given below:

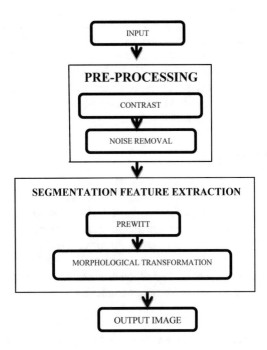

4.1   Pre-processing

Image pre-processing is the very important step in image processing technique. Many techniques like, the contrast enhancement, bright enhancement, sharpness enhancement; all this can be implemented at pre-processing level. In this paper we have used two stages of pre-processing; histogram equalization and median filtering. Histogram equalization is a method of improving the contrast of the image. Median filtering is used for filtering of noise. These operations are discussed in the subsequent section.

4.1.1   Histogram Equalization

Histogram Equalization is a technique for adjusting image intensity to enhance contrast. Histogram of an image can be drawn plotting pixel intensity and frequency of the pixel intensities or probabilities of the pixel intensities. One way of getting histogram is to plot pixel intensities v/s pixel frequencies. In matrix form, each element is a pixel of an image and value of the element represent intensities of the pixel. First step is to count the total number of pixel associated with each pixel intensity. Second step is to calculate probability of each pixel intensity in the image matrix. The next step is to calculate cumulative probability. Image before equalization and the same image after equalisation.

4.1.2   Median Filtering

It smoothens the image by utilizing neighboring of pixels. It is non-linear filter. It removes black and white noise from images In any case, with middle separating; the estimation of a yield pixel is dictated by the middle of the area pixels, as opposed to the mean. The middle is considerably less delicate than the intend to outrageous qualities. Median filtering is in this way better ready to expel these anomalies without diminishing the sharpness of the picture. Middle separating is a particular instance of request measurement sifting, otherwise called rank sifting.

4.2   Feature Segmentation

The feature of an image can be extracted by its content. Content like: Colour, Texture, Shape, Position, Dominant edges of images item and regions etc. As the image selected, next to Histogram Equalization level. It has a range of 256 colours or 64 colours. There are two problems in Histogram Equalization, first is Dimensionality curse and second is cross talk. In dimensionality curse, each image is having number or feature (feature of the 256 colours). So try to reduce the number of feature. Cross talk say the colour which are very close to each other needs to be considered by calculating the distance and it need to take more time complexity.

4.2.1   Prewitt

Prewitt is one of the stages of segmentation which is used to help in edge deduction. There are two ways to do prewitt operation, the first is horizontal edges and second is vertical edges. Edges are calculated by the different intensities of the pixels. There are masks for using the edge deduction which is also called derivative marks. Marks have some properties. There should be a presence of opposite signs in the masks. It should be equal to zero if there is the sum of masks to know the more edge deduction, there should be more weight. Both way of prewitt shows the different outcome. There is difference between horizontal edges and vertical edges. On the off chance that, square discovers edges by searching for the nearby maxima of the angle of the info picture. It computes the inclination utilizing the subsidiary of the Gaussian channel. The Canny technique utilizes two limits to distinguish solid and frail edges. It incorporates the feeble edges in the yield just on the off chance that they are associated with solid edges. Accordingly, the technique is increasingly powerful to commotion, and bound to distinguish genuine feeble edges.

4.2.2   Morphological Transformation

Morphology is essentially a subject which turned into used extensively in area of biology to talk about shape or form of human body. Iin image processing we used mathematical morphology as a tool for doing numerous structure and shapes related operations. To use this Morphological operation or, we ought to constitute an picture in form of a point set and all of the Morphological operations are not anything however one-of-a-kind units operation and the form of set operation that we ought to use rely on. What sort of Morphological transformation we want to do? And a Morphological transformation of given photo X describe a relation of the picture X with some other small point set, say b, but this small point set is what we name as structuring detail.

So, any mathematical Morphological operation is continually described with admire to a given structuring detail. there may be a Morphological Transformation set of rules which includes the boundary extraction, hit and misses transformation, place filling, skeletonization. also it has rescale morphology. In Boundary extraction, we have observed the boundary of an photograph. As there is an photo, the boundary covers the photo and proven in output is the boundary extraction.

5 Result and Discussion

The experiment were conducted using GNU Octave 4.2.1 and Matlab 2018 a Window 10, 64 bits operating system, x64-based processor with 8 GB RAM, Intel(R) Core(TM) i5-6200U CPU@ 2.30 GHz 2.40 GHz. Six MRI tested Images are taken from the open access website. There are different six images And each image have been processed by the GNU Octave software. In this paper, following body part are taken under consideration: Torso, Spinal cord, Brain Top View, Side View of Brain, Ankle and Neck. Entropy is selected for quantitative analysis of the resultant test images. Entropy is used to evaluate the quality of image. Higher value of Entropy means better image quality and better edge deduction.

Experiment 1
The experiment 1 is shown in Fig. 1. The quality of images are changed as per the technique and the value of all image are varies on their clarity. The greater value shows the better and clear image, which seems to the quality of being clear and easy to understand. As per the segmentation, image (d) and (e) are have better edge deduction and clear boundary. In Fig. 1(a) is original with 5.5595 values, (b) is Histogram Equalization with 4.3283 values, (c) is Median filtering with 5.4713 values.

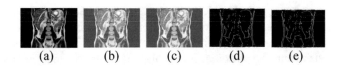

(a)   (b)   (c)   (d)   (e)

Fig. 1. (a) Original [4] (b) Histogram equalization (c) Median filtering (d) Prewitt (e) Morphological transformation

Experiment 2
The experiment 2 is shown in Fig. 2. The techniques have been applied and compared with each other As result, medfilt image is clear and have better value. As per the segmentation, image (d) and (e) are have better edge deduction and clear boundary. In Fig. 2(a) is the original image with 6.5466 value, (b) is Histogram Equalization with 4.6435 value, (c) is Median filtering with 6.9566 value.

272   S. Sengupta et al.

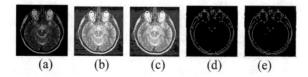

(a)   (b)   (c)   (d)   (e)

Fig. 2. (a) Original [5] (b) Histogram equalization (c) Median filtering (d) Prewitt (e) Morphological transformation

Experiment 3

The experiment 3 is shown in Fig. 3. The techniques have been applied and compared with each other. The image of spinal cord shown. Original image is better than the Histogram Equalization. And medfilt is better than original. As per the segmentation, image (d) and (e) are have better edge deduction and clear boundary. The result are in Fig. 3(a) is the original image with 6.7482 value, (b) is Histogram Equalization with 4.9360 value, (c) is Median filtering with 7.8327 value.

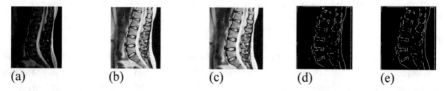

(a)   (b)   (c)   (d)   (e)

Fig. 3. (a) Original [6] (b) Histogram equalization (c) Median filtering (d) Prewitt (e) Morphological transformation

Experiment 4

The experiment 4 is shown in Fig. 4. The techniques have been applied and compared with each other. The image is shows the side view of the brain. As the result, Histogram Equalization is lower value from the other. The medfilt have better value from original and Histogram Equalization. As per the segmentation, image (d) and (e) are have better edge deduction and clear boundary. The result are in Fig. 4(a) is the original image with 5.9922 value, (b) is Histogram Equalization with 4.5091 value, (c) is Median filtering with 6.0771 value.

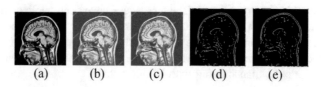

(a)   (b)   (c)   (d)   (e)

Fig. 4. (a) Original [2, 3] (b) Histogram equalization (c) Median filtering (d) Prewitt (e) Morphological transformation

Experiment 5
The experiment 5 is shown in Fig. 5. The techniques have been applied and compared with each other. The figure shows the image of ankle. The value of medfilt is high and its better after that original image value and at the end, Histogram Equalization have lower value, which means it have less clarity and quality is not been good. As per the segmentation, image (d) and (e) are have better edge deduction and clear boundary. As per result, Fig. 5(a) is the original image with 6.9490 value, (b) is Histogram Equalization with 5.043 value, (c) is Median filtering with 7.0387 value.

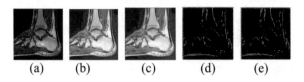

(a)   (b)   (c)   (d)   (e)

Fig. 5. (a) Original [4] (b) Histogram equalization (c) Median filtering (d) Prewitt (e) Morphological transformation

Experiment 6
The experiment 6 is shown in Fig. 6. The techniques have been applied and compared with each other. The figure shows the image of neck. The qualities of medfilt image have clarity and the value is being higher than other. After that original image come and then Histogram Equalization. As per the segmentation, image (d) and (e) are have better edge deduction and clear boundary. As per result, Fig. 6(a) is the original image with 6.7936 value, (b) is Histogram Equalization with 4.7953 value, (c) is Median filtering with 7.8485 value.

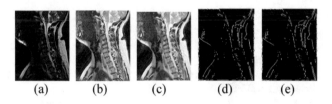

(a)   (b)   (c)   (d)   (e)

Fig. 6. (a) Original [5] (b) Histogram equalization (c) Median filtering (d) Prewitt (e) Morphological transformation

6 Conclusion

We have proposed a methodology to deduct the edges of an MRI images effeciently. There are two main steps in a proposed methodolgy. First step in Pre-processing which will filter and improve the quality of the input imput. In the second stage. We have used feature extraction which will deduct the edge of the MRI images. We have used entropy to major the performance of the preposing technique visual inspection satisfaction the border or edge analysis of the proposed methodology. Further, work will be done to implement classification in the methodology to categorized the different verity of MRI image.

References

1. Chitradevi, B., Srimathi, P.: An overview on image processing techniques
2. Rao, K.M.M.: Overview of image processing. Readings in Image Processing
3. Jain, A.K.: Fundamentals OF Digital Image Processing. Prentice-Hall, Upper Saddle River (1989)
4. Castleman, K.R.: Digital Image Processing. Prentice-Hall, Upper Saddle River (1996)
5. Kumar, G., Bhatia, P.K.: A detailed review of feature extraction in image processing systems. In: 2014 Fourth International Conference on Advanced Computing & Communication Technologies (2014)
6. Hal, E.L.: Computer Image Processing and Recognition. Academic Press, Cambridge (1979)
7. Chellappa, R.: Digital Image Processing, 2nd edn. IEEE Computer Society Press, Los Alamitos (1992)
8. Mansourpour, M., Rajabi, M.A., Blais, J.A.R.: Effects and performance of speckle noise reduction filters on active radar and SAR images. In: ISPRS. XXXVI/1-W41
9. Manikandan, S., Nigam, C., Vardhani, J.P., Vengadarajan, A.: Gradient based adaptive median filter for removal of speckle noise in airborne synthetic aperture radar images. In: ICEEA (2011)
10. Gagnon, L., Jouan, A.: Speckle filtering of SAR images: a comparative study between complex-wavelet-based and standard filters. In: SPIE (1997)
11. Karunakar, P., Praveen, V., Kumar, O.R.: Discrete wavelet transform-based satellite image resolution enhancement. Adv. Electron. Electr. Eng. 3(4), 405–412 (2013). ISSN 2231-1297
12. Byun, Y.G., Han, Y.K., Chae, T.B.: A multispectral image segmentation approach for object-based image classification of high resolution satellite imagery. KSCE J. Civil Eng. 17(2), 486–497 (2013)
13. Chakraborty, D., Sen, G.K., Hazra, S.: High-resolution satellite image segmentation using Hölder exponents. J. Earth Syst. Sci. 118(5), 609–617 (2009)
14. Lankoande, O., Hayat, M.M., Santhanam, B.: Segmentation of SAR images based on Markov random field model
15. Soh, L.-K., Tsatsoulis, C.: A feature extraction technique for synthetic aperture radar (SAR) sea ice imagery. IEEE (1993)
16. Vesecky, J.F., Smith, M.P., Samadani, R.: Extraction of ridge feature characteristics from SAR images of sea ice. IEEE (1989)
17. Karvonen, J., Kaarna, A.: Sea ice SAR feature extraction by non-negative matrix and tensor factorization
18. Zakhvatkina, N.Y., Alexandrov, V.Y., Johannessen, O.M., Sandven, S., Frolov, I.Y.: Classification of sea ice types in ENVISAT synthetic aperture radar images. IEEE (2012)
19. Wang, T., Yang, X.: A multi-level SAR sea ice image classification method by incorporating egg-code-based expert knowledge. IEEE (2012)
20. Kalra, K., Goswami, A.K., Gupta, R.: A comparative study of supervised image classification algorithms for satellite images. IJEEDC (2013)
21. Karvonen, J., Simila, M.: Independent component analysis for sea ice SAR image classification. IEEE (2001)

Recent Trends of IoT in Smart City Development

Disha Kohli[⊠] and Sudhriti Sen Gupta

Amity Institute of Technology, Amity University, Noida, Uttar Pradesh, India
Dishakohli1912@gmail.com, ssgupta@amity.edu

Abstract. The intelligent interconnection with the group of internet enabled devices is termed as the Internet of Things [IoT]. IoT has enhanced and deployed into diverse varieties of smart city applications such as smart homes, smart traffic, smart grids, smart healthcare etc. Smart city is also an application of Internet of Things [IoT] and is defined as a designation given to a city to implement advanced technology and upgraded lifestyle. This paper aims to discuss about the importance of Internet of Things [IoT] in smart city development.

Keywords: IoT · Smart city · Healthcare · Security

1 Introduction

In a historical perspective, a group of devices interrelated using the internet is coined as the internet of things. It was first introduced by Kevin Ashton, the co-founder of auto id center met in 1999. Also, Professor Neil Gershenfeld's book mentioned about this concept but with a different name. Internet of things has grown to have a lot of applications like smart healthcare, smart grid, smart homes etc. [1].

Smart city is a name given to a city which incorporates technology to daily life routine or example traffic control, garbage collection, etc. The aim of building a smart city is to provide a safe, upgraded, easy lifestyle to every individual. Market research states that SINGAPORE, BARCELONA, LONDON, SAN FRANCISCO, OSLO are the 5 smart cities across the globe. Many big companies like TELENSA are supporting the development of smart cities.

Telensa's planet, the market-leading Central Management System for streetlights, has over 50 city and regional networks in 8 countries and a project covering over 1 million streetlights, including the world's largest deployment in the US. The company has also planned for LED based street lights and it enables localized customization of lighting levels and also it provides a city wide platform for low cost sensor application [2]. IoT plays an important role in building smart cities as it provides all necessary sensors, networks and data to manage the city through technology. The sensors collect the data from the environment and then it is shared to the connected network of systems for performing the desired functions.

Smart cities enhance the lifestyle of the citizens by enabling traditional network and services in a more efficient way with the use of digital and telecommunication

A. P. Pandian et al. (Eds.): ICCBI 2019, LNDECT 49, pp. 275–280, 2020.
https://doi.org/10.1007/978-3-030-43192-1_31

technologies. The idea is to combine the variety of solutions to the various assets of the cities. This may include safety, parking systems, waste management, transportation, and lighting in the city infrastructure [3].

2 Internet of Things

Internet of things is network formed between a group of devices with the help of internet. IoT is used everywhere by gaining an increasing popularity. An IoT ecosystem consists of the web-enabled smart devices which gather information from their environment, process the data according to the need and then perform desired tasks. It provides a lot of benefits like security as the data should not be accessed by unauthorized users. It also provides comfort and efficiency in work as the user is just a click away from getting the work done. Also the data processed by sensors is processed thoroughly thus it provides a better decision making ability to the user [4] (Fig. 1).

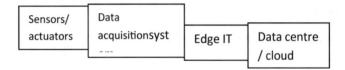

Fig. 1. Depicts the architecture of IoT

1. STAGE 1
   The stage 1 involves the sensor that gleans the information from the surroundings and conveys it to the gateways.
2. STAGE 2
   The data converted is conveyed to the edge and then forwarded to the cloud.
3. STAGE 3
   The data that needs more processing or needs feedback is transmitted to the cloud to have a deep analyzes and manage the data and to make it secure [5].

3 Smart City

The urban city that uses different technologies to make the lifestyle of the people easy and to improve the infrastructure and security of the city is known as the smart city. India has a goal of building 100 smart cities and till now Lucknow, Chandigarh, dharamshala, Faridabad and Raipur have been declared as smart cities.

For the method of smart city mission, the main objective is to promote cities which has: [6]

(1) Security
(2) Better infrastructure
(3) Advanced technology

(4) Smart usage of energy
(5) Smart transportation
(6) Smart healthcare

Smart city enhances the communication between the community and the infrastructure. There are some other concepts related to smart city and they are digital city, cyber Ville, teletopia etc. The city claims to reduce the traffic and improve safety. Also, Barcelona has utilized the analytics of the common traffic flows to design a new network for the bus. The vice president of UAE started the smart city project in 2013 which aimed to integrate private to access these sectors through smartphones [7].

4 IoT Supporting Security

The security enhancement in the smart cities is achieved by installing the CCTV's enabled with the internet all round and directing their footage to the PC's or the portable devices. The automatic incident detector are also engaged for the purpose security in the smart cities.

New security locks are introduced which eliminates the physical key and unlocks the doors with the code. This prevents chances of robbery as no one can enter without entering the right code [7].

The proposed work depicts the flow of IoT enabled automated detection and mitigation of accident. When an unfortunate incident like accident happens, the vehicle will send an SOS by enabling the IoT. This will led the response team to the site of accident to access and take care of the victims.

5 IoT in Healthcare

The involvement of the IOT in the medical industry ensure the better management of the sick in case of emergency and thus through the mobile app one can treat a patient and provide necessary first aid to save a person's life.

Also, the IoT gadgets gathers and exchanges wellbeing information like circulatory strain, oxygen and glucose levels, weight, and ECGs to the cloud and can be imparted to an approved individual, similar to a doctor, insurance agency, to enable them to take a gander at the gathered information paying little mind to their place, time, or gadget [8].

In Boston, Infants are given wristbands that enables remote system to find them whenever. If there is an instance that a child is taken excessively near the exit entryway without being marked out, lifts will naturally stop and leave entryways will be bolted. What's more, in the neonatal emergency unit, get basic cautions on clinic PDAs about their patients' ailments [9].

Figure 2 depicts some of the common area of where IoT is beneficial for providing healthcare. Some of them are better disease control, maintenance of medical device, better treatment result, lower expenses etc.

Fig. 2. Glimples of IoT in healthcare [2]

6 IoT in Energy Saving

Real time monitoring systems are developed which helps in monitoring the energy consumption. This makes the analyses quick and easy and also the results come out quick. Australian electrical wholesaler has planted a cloud based energy monitoring system which is expected to save energy up to 15% annually. An app is also made which analyzes the consumption of energy and helps to monitor the usage of energy [10, 11].

7 IoT in Providing Adequate Water Supply

IoT is also helpful in proper water supply for irrigation and also helps in preventing drought situations and these are done by real time systems, there are sensors places to remote places and the data is collected from those sensors and whenever there is a drought situation, the measures are taken to provide the necessary items.

Also, these sensors and real time tells the moisture level of the ground and thus tells when to stop the motor and saves electricity too. There are devices which help in improving the irrigation and they are CORPX and EEBE AG [12–14].

8 IoT in Transportation and Logistics

RFID, barcode scanners and mobile computers help in providing end to end visibility in the supply chains. The RFID is utilized in the logistics and the supply chains to have a faster and the speedy delivery at a minimized cost.

Warehouse and yard management are the core of the transport and logistics and IoT based mobile apps help in maintaining the stocks and incoming and outgoing items which in the end eases business [15].

9 IoT in Traffic Control

The traffic can also be controlled with IoT based systems with sensors in cars and on the traffic lights. Whenever the traffic is growing on one side of the road the lights changes accordingly and the jam can be controlled.

Also IoT is used for providing the information about the parking spaces in the mall parking and other public areas [17].

10 Challenges Faced in Building a Smart City

Many researchers have studied about smart city and have found out the challenges which are there for building a smart city and they are: [18–20].

(1) Lack of appropriate technology
(2) Digital security
(3) Legislation and policies
(4) Lack of confidence or reluctance shown by citizens
(5) Funding and business models
(6) Interoperability
(7) Existing infrastructure for energy, water and transportation systems

11 Conclusion and Future Scope

A group of devices interrelated using the internet is coined as the Internet of Things [IoT] and is a very important element for developing smart city as it provides all the necessary technology and hardware for the smart city development. There are some challenges in building the smart city with the appropriate technology, having enough funds, lack of confidence, existing infrastructure etc.

References

1. El Kouche, A.: Towards a wireless sensor network platform for the Internet of Things: sprouts WSN platform. In: 2012 IEEE International Conference on Communications (ICC), pp. 632–636. IEEE (2012)
2. Ferdoush, S., Li, X.: Wireless sensor network system design using Raspberry Pi and Arduino for environmental monitoring applications. In: The 9th International Conference on Future Networks and Communications (FNC-2014). Elsevier (2014)
3. Vujovic, V., Maksimovic, M.: Raspberry Pi as a wireless sensor node: performances and constraints. In: 2014 37th International Convention on Information and Communication Technology, Electronics and Microelectronics (MIPRO), 26–30 May 2014, pp. 1013–1018 (2014)
4. Pfister, C.: Getting Started with the Internet of Things. O'Reilly Media Inc., Sebastopol (2011)

5. Atzori, L., Iera, A., Morabito, G.: The Internet of Things: a survey. Comput. Netw. **54**(15), 2787–2805 (2010)
6. Karimi, K., Atkinson, G.: What the Internet of Things (IoT) needs to become a reality. White Paper, FreeScale and ARM (2013)
7. Taleb, T., Kunz, A.: Machine type communications in 3GPP networks: potential, challenges, and solutions. IEEE Commun. Mag. **50**(3), 178–184 (2012)
8. Bilodeau, V.P.: Intelligent parking technology adoption. Ph.D. thesis, University of Southern Queensland, Queensland, Australia (2010)
9. Li, T.S., Yeh, Y.-C., Wu, J.-D., Hsiao, M.Y., Chen, C.-Y.: Multifunctional intelligent autonomous parking controllers for carlike mobile robots. IEEE Trans. Ind. Electron. **57**(5), 1687–1700 (2009)
10. Faheem, Mahmud, S.A., Khan, G.M., Rahman, M., Zafar, H.: A survey of intelligent car parking system, October 2013
11. Alam, S., Chowdhury, M.M.R., Noll, J.: Senaas: an event-driven sensor virtualization approach for Internet of Things cloud. In: 2010 IEEE International Conference on Networked Embedded Systems for Enterprise Applications (NESEA), November 2010
12. Choeychuen, K.: Automatic parking IoT mapping for available parking space detection. In: Proceedings of the 5th International Conference on Knowledge and Smart Technology (KST), Chonburi, Thailand, 31 January–1 February 2013 (2013)
13. Schelby, Z., Hartke, K., Bormann, C.: Constrained application protocol (CoAP). CoRE Working Group Internet-Draft, 28 August 2013
14. Keat, C.T.M., Pradalier, C., Laugier, C.: Vehicle detection and car
15. Kumar, R.P., Smys, S.: A novel report on architecture, protocols and applications in Internet of Things (IoT). In: 2018 2nd International Conference on Inventive Systems and Control (ICISC), pp. 1156–1161. IEEE (2018)

Analysis of Biometric Modalities

Ayushi Wajhal[(✉)] and Sudhriti Sen Gupta

Amity Institute of Technology, Amity University, Noida, Uttar Pradesh, India
Ayushiashiwajhal29@gmail.com, ssgupya@amity.edu

Abstract. Biometrics refers to the system of metrics that is related to the human behavior and characteristics. Biometrics is an authentication process related to real life process used in form of access control and recognition. Biometric also identifies the individual behavior in the groups that are under surveillance and consideration. Biometric identifiers are utilized to quantify the particular conduct that is utilized to portray people and mark practices. Physiological characteristics is even identified with the state of human body. By using biometric system, people can be recognized in perspective on who they are rather than what they have (like token, data- card, scratch) or what they knows (card pin, mystery key). This paper, basically focuses on the principal of the distinctive biometrics and to analyze the various modalities of biometric.

Keywords: Biometrics · Biometric modalities · Facial acknowledgment · Fingerprints · Sound/voice acknowledgment · Retina scans · Handprints · Keystrokes · Gait · Signature

1 Introduction

Biometric is a digitalized technique in which we perceive an individual behavior on a physiological or the conduct trademark. Biometric field idea has been a turning point into the idea of establishment of profoundly secure ID and persons quality arrangements. As the cases raises and the system reaches the need of secure recognizable identity system and individual proof which leads to innovation of new ideas in biometric field. The biometric-based systems have accommodated the financial exchanges and persons information protection. The need of biometric has been vividly increased and could be found in every state, in military and in neighborhood governments, in business and even in colleges for student protection. The biometric system has now been into complete overflow in security frameworks, electronic banking system, government IDs, retail secure deals, wellbeing of people, other secure money related transactions and social administrations. The above systems have been in complete profit by these advancements. The vast system security framework based on authentication application leads with workstation system, single sign-in, application logins, insurance, access to assets and web security. This secure electronic exchange using biometric techniques leads to development of the worldwide economy. Used with different advance features, for examples, encryption and decryption keys, computerized marks, brilliant cards, biometric has leaded their concept in almost all part of providence and our day by day lives.

Using the advanced biometric system for individuals is getting to be more advanced and extensively more exact than present strategies, (for instant, the usage of security

A. P. Pandian et al. (Eds.): ICCBI 2019, LNDECT 49, pp. 281–289, 2020.
https://doi.org/10.1007/978-3-030-43192-1_32

PINS and passwords). The biometric system connects to a specific individual (a token or a password might be used by people other than the verified client), is superior (nothing to transport or recall), to the point (it accommodate constructive authentication), can give analysis trail and bend up socially, economically and moderate. Biometric is the structure which is especially used in china since fourteen century by involving finger-prints of the vendors to recognize the structure. At first of the 19th century, Bertillonage was created by collecting the human system estimates to memorize them by an anthropologist called Alphonse. Alphonse had recognized how human structure high-lights by involving them in classes named as weights, changing the length of hairs, and so on, and unaltered physical structures of human being like fingers length.

But later, this technique was disappeared due to acknowledgement of human as it is important that more than one person would have the same body estimations. After-wards, a method named fingerprinting was developed by Richard Edward Henry from Scotland. The proposition which stated the retinal proof was presented by Dr. Isadore and Dr. Carleton Simon in 1935. The main focus retina examining structure has been possible in 1981. In 199 at Cambridge university, John Daugman hand over iris description. BAT was presented in Kosovo that provides an unknown acknowledge-ment implies in 2001. Today, the biometric has completely discovered a self-deciding field which concentrates on advance level for building up individual characters.

2 Biometrics Framework

A biometric structure is a process which arranges the unique features of the picture (test picture) with the physiognomy of pre-secured pictures (presentation picture). For doing in that capacity, each biometric structure incorporates (a) Image verifying module:- it picks up the image of a biometric trademark and submits it to the system for further taking care of, (b) Features extraction module:- it shapes the secured picture accord-ingly removing the amazing or biased features, (c) Matcher module:- organizes expelled features of tested images with display picture to get the coordinate score. However, an introduced fundamental authority module rejects or checks the attested identity reliant on the coordinate score and (d) Database module:- this contains auto-mated depiction of as of late taken tests all the time named as designs and outlines the course of action of a typical biometric structure (Fig. 1).

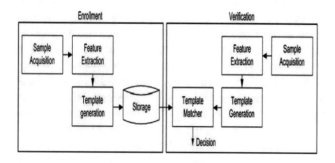

Fig. 1. Composition of biometric system

Biometric contraptions are of various sorts but altogether there are various different types of biometrics security which are generally used in biometric system. Biometrics is basically the affirmation of a person's personality that are different, unique and individualistic to every human being, which joins voice affirmation, facial affirmation, fingerprints, palm prints, retina checks and so on so forth. Biometric innovation is used to monitor and control the devices in the best way possible to deal with the assurance that people maintain a strategic distance from their productive information and assets, and will find that using any of these various biometrics security, contraptions is the marvelous technique to secure stuff and things [7] (Fig. 2).

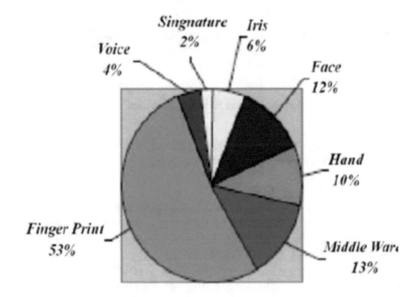

Fig. 2. Biometric market report

3   Biometric Modalities

Biometric modalities is only a class of a biometric framework relying on the sort of human feature it takes as input. The biometrics is to a great extent measurable. The more the information accessible from test, the more the framework is probably going to be one of a kind and dependable. It can word on different modalities relating to measurements of person's body and includes, and standards of conduct. The modalities are characterized dependent on the individual's trait.

To select biometric modalities for a particular system is challenge of the biometric system developer.

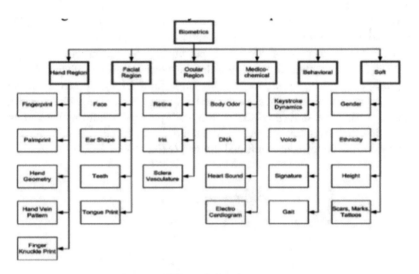

Fig. 3. Various biometric modalities

This segment gives an inside and out review of various biometric modalities according to the grouping gave in above Fig. 3.

1. Hand area modalities

Hand of human contains the rich surface data that gave the establishments to new acknowledgment frameworks dependent on fingerprints. Notwithstanding fingerprints, a few characteristics have been recognized and tried, for example, finger knuckle print, palmprint, geometry of hand, finger nail bed and vein design of hand. However, all the hand-based credits are an expansion to unique mark technology. A unique mark acknowledgment framework utilizes the surface of edges and prints present on the fingertips [2, 5] whereby the edge endings (details focuses) play out the acknowledgment task and the edge stream arrange the fingerprints into one of the five classifications, for example, curve, rose curve, left circle, right circle and whorl [6]. A computerized unique mark acknowledgment framework utilizes level 1, level 2 or dimension 3 highlights. However, a large portion of the industrially accessible frameworks utilize level 2 highlights or particulars focuses for acknowledgment though level 3 highlights offer powerful acknowledgment against low quality or inert unique mark pictures [7]. A mechanized palmprint acknowledgment framework thinks about the surface (Fig. 4).

For coordinating, the acknowledgment framework separates edges, particular focuses and details focuses from excellent picture, main lines, wrinkles and palm surface from low quality picture. As far as coordinating calculations, the framework applies line [10], sub-space [11], nearby and worldwide insights-based methodologies. A hand geometry framework considers the width and length, angle proportion of palm or fingers just as thickness, length, territory, skin overlap and wrinkle examples of hand [14, 15]. A framework of finger knuckle print uses the lines or patterns present on the external piece of fingers close phalangeal joint [16–18] though nail bed of finger alludes

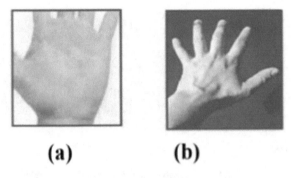

(a)   **(b)**

Fig. 4. (a) Palmprint [2] (b) hand geometry [4]

to the estimation of separation between dermal structures present present parallelly under the finger nail [19]. Hand vein acknowledgment framework uses the bifurcations of vein and endings underneath the skin of hand. The vast majority of the picture acquisition strategies in all the modalities based on hand are based on impression requiring optical, silicon, warm or ultrasonic imaging sensors. However, a few special cases are framework of hand geometry that utilize the camera to gain 3D picture of hand.

2. Facial region modalities

For researchers, facial area of humans has been a fascinating point due to the most customary biometric attribute of human face which is used to see individual creature since many years. Regardless of the way that it is the most widely recognized biometric trademark, the non- straight shape of human face makes it complex precedent affirmation issue similarly as a working domain of research in PC vision applications. Subsequently, the examination network put extensive endeavors in the advancement of strong mechanized face acknowledgment frameworks.

A mechanized face acknowledgment framework builds up the character of an individual dependent on the calculation of 3D or 2D highlights. We recognize that the facial-3D acknowledgment is the recent pattern proposed to redress the fundamental issues related with facial-2D acknowledgment, for example, picture debasement, poor arrangement and development of head and outward appearances. However, these frameworks can't ensure solid recognizable proof in nearness of antiquities, for example, use of beautifiers and plastic medical procedure. In addition, an individual's face may or may not be changed after some time which can significantly affect the exactness of such frameworks. Also, the costly imaging equipment is another factor which confines the utilization of such frameworks. So as to create strong face acknowledgment frameworks, the exploration network urged the possibility of the human acknowledgment dependent on the facial thermograph which demonstrates the warmth radiation examples of human face because of the nearness of vascular structure underneath the skin of human being. Hardly any examinations professed to accomplish better exactness as far as face discovery, restriction and division against the facial pictures procured in unmistakable light (Fig. 5).

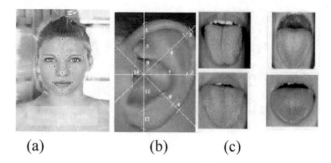

(a)   (b)   (c)

Fig. 5. (a) Face [2] (b) Ear [3] (c) Tongue print [2]

4   Voice, Gait and Signatures

Biometrics are social estimations of a person or some physiological. Such Biometrics can be either Physiological like Hand Geometry, Fingerprint, Iris, Ear, Face, Retina, DNA, and so forth or it tends to be Behavioral like Signature, Gait, Voice, Keystrokes and so forth. Utilization of Voice biometric is in high research now-a-days. Voice is the main biometric that enables clients to verify remotely. Preferences identified with voice biometric utilization resemble (i) non meddling, (ii) wide accessibility and simplicity of transmission, (iii) minimal effort, requiring little extra room and (iv) convenience, minimized for little electronic gadgets with mouthpiece and so on. In opposite, there are also a few burdens like (i) low perpetual quality, issues with maturing, hack cool, passionate changes, (ii) issues with high foundation and system clamor, (iii) Sensitivity to room acoustics and gadget confound and so on. Being a conduct biometric, human Voice isn't as remarkable as human DNA. Yet at the same time, with decisively structured degree and applications, it tends to be endeavored for explicit authentication prerequisites in our standard regular day to day existence. All these structure the premise or inspiration driving the testing errand of perceiving an individual's character utilizing just voice biometric, which is known as Automatic Speaker Recognition. Contingent on the issue determination, the errand can be either Automatic Speaker Identification (figuring out who is talking) or Automatic Speaker Verification (approving whether a similar individual is talking that has being asserted, or not) (Fig. 6).

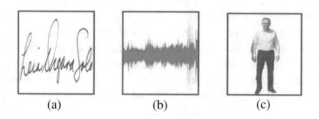

(a)   (b)   (c)

Fig. 6. (a) Signature [6] (b) Voice [2] (c) Gait [3]

5   Retina Scanner

A retinal sweep is one of the kind of biometric approach that makes utilization of somebody's retina to find them. Human retina is situated inside the back piece of the eye and is made up of a slender tissue made out of neural cells. All retina is extraordinary because of the vessels mind boggling state that fills the retina with blood. Inside the retina the system of veins is complicated to the point that is indistinguishable even in twins. Despite of the fact, that the styles of retinal can be modified in occasion of Glaucoma, or retinal degenerative disarranges, diabetes, retina normally stays unaffected from birth till kicking the bucket. Because of constant nature and one of a kind, retina is the most extreme exact and trustworthy biometric [8] explicit distinctive imprints. Using Retinal examination, focal points involve less inescapability of false positives, wonderfully nearly 0% fake horrendous charges, uncommonly reliable in light of the way that no people have a comparative retinal precedent, quick results: character of the issue is affirmed promptly [9, 10]. Dangers fuse estimation precision can be stricken by a contamination, for instance, cascades, estimation exactness likewise can be impacted by extraordinary astigmatism, canning strategy can be seen by few prominent, no longer very customer sincere, inconvenience being scanned must be close to tunnel cam optics, high gear cost (Fig. 7).

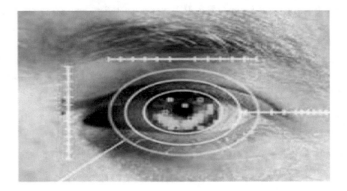

Fig. 7. Retina [7]

6   Unique Mark Scanner

On Human palm, fingerprints are the graphical skim like edges. Human finger edge designs are hard to change for the span of the life of an individual other than because of mishaps including wounds and cuts on the fingertips (Fig. 8).

Fig. 8. (a) Fingertip [6] (b) Fingerprint matching mechanism [8]

7   Conclusion

Biometrics or Biometric ID, alludes to the reliable recognizable proof of an individual dependent on their physiological (e.g., iris, face, DNA, hand, unique finger impression,) or social (e.g., voice, keystroke, signature) distinction. This procedure of distinguishing proof offers a few remunerations over conventional strategies including PIN numbers for different reasons or ID cards. Subsequently biometric frameworks are demonstrated exceptionally classified PC based security frameworks. Every single biometric framework is supportive and helpful and choice of biometric gadget depends on the application region, for example- biometric technology is going to be used in which region. For the most part it relies upon the amount of individuals, which will see the person in question and furthermore conditions. In case of confined individuals, we can use biometric technology as it takes less time and it is more verified than other biometric technology used where boundless individuals are perceived vastly however tad precision.

Compliance with Ethical Standards. All author states that there is no conflict of interest. We have used the dataset from this website: http://www.advancedsourcecode.com/

References

1. Alsaadi, I.M.: Physiological biometric authentication frameworks, focal points burdens and future improvement: a survey. Int. J. Sci. Technol. Res. **12** (2015)
2. Jain, A.K., Ross, A., Prabhakar, S.: An introduction to biometric recognition. IEEE Trans. Circ. Syst. Video Technol. **14**, 4–20 (2004)
3. Kaur, G., Singh, G., Kumar, V.: An audit on biometric acknowledgment. Univ. J. Bio-Sci. Bio-Technol. **4** (2014)
4. Kong, A., Zhang, D., Kamel, M.: A study of palmprint acknowledgment (2009)
5. Battaglia, F., Iannizzotto, G., Bello, L.: A biometric authentication framework dependent on face acknowledgment and RFID labels. Mondo Digit. **49**, 340–346 (2014)
6. Bowyer, W., Hollingsworth, K.P., Flynn, P.J.: A study of Iris biometrics look into: 2008–2010. In: Handbook of Iris Acknowledgment. Springer, London (2016)

7. Deriche, M.: Trends and challenges in mono and multi biometrics. In: 2008 First Workshops on Image Processing Theory, Tools and Applications, IPTA 2008 (2008)
8. Mach. Intell. **33**, 88–100 (2011)
9. Kong, A.W.K., Zhang, D., Lu, G.: A study of identical twins' palmprints for personal verification. Pattern Recogn. **39**, 2149–2156 (2006)
10. Pattern Recogn. **41** (2008)
11. Roberts, Chris. Biometrics. Accessed 11 June 2009
12. Shang, L., Huang, D.S., Du, J.X., Zheng, C.H.: Palmprint recognition using FastICA algorithm and radial basis probabilistic neural network. Neurocomputing **69**, 1782–1786 (2006)
13. You, J., Kong, W.K., Zhang, D., Cheung, K.H.: On hierarchical palmprint coding with multiple features for personal identification in large databases. IEEE Trans. Circ. Syst. Video Technol. **14**, 234–243 (2004)
14. Li, J., Yau, W.-Y., Wang, H.: Combining singular points and orientation image information for fingerprint classification. Pattern Recogn. **41**, 353–366 (2008)
15. Yampolskiy, R.V.: Biometrics: a survey and classification. Biometrics **11**(1) (2008)
16. Kumar, B.S., Debnath, B., Swarnendu, M., Debashis, G., Das, P.: Statistical approach for offline handwritten signature verification. J. Comput. Sci. **4**(3) (2008)

Comparative Analysis of Contrast Enhancement Techniques for MRI Images

Sudhriti Sengupta$^{(\boxtimes)}$ and Astha Negi

Amity Institute of Information Technology, Amity University,
Noida, Uttar Pradesh, India
ssgupta@amity.edu

Abstract. Contrast Enhancement is one of the most essential stages of image processing. This method changes the distribution of image value over a wider range so that images can be perceived in a better way. In this paper we have studied and analyzed the best contrast enhancement technique which can be applied to MRI images. MRI images are very important for diagnosis and understanding the pathology of different organs. Proper contrast enhancement will lead to better understanding of these images. We have comparatively analyzed three different contrast enhancement techniques and interpreted the best technique for MRI images.

1 Introduction

An image provides the ability to interpret the nature of the subject. Images are of various types accumulated in different domain like remote sensing, computer aided visualization, satellite images, medical image etc [1]. Medical image is very significant as it helps the medical practitioners to provide better healthcare to patients. Medical images depict the structure of the human body. The different varieties of medical images are CT, MRI, skin lesions, nuclear medicine etc. MRI or Magnetic resonance imaging is a process of utilizing magnetic fields and radio waves to deliver definite pictures of within the body. Medicinal services experts use MRI outputs to analyze an assortment of conditions, from torn tendons to tumors. One of the issues in analysis of the MRI images is the improper illumination of the images [2]. This can be improved by contrast enhancement process. In 2016, Kaur et al. has used different histogram equalization technique for enhancement if MRI brain images using [4]. Global histogram equalization, Local histogram equalization, and Adaptive histogram equalization for different objective quality is done by Senthilkumaran et al. [5].

Improvement of border information of edge by applying the gradient information of a highly contrast image is done by Zheng et al. [11].

Unsampled technique which applies high resolution weighted image for vowel-wise analysis of dynamic contrast enhanced MRI was proposed by Khouzani et al. [12]. In 2017, Ti et al. proposed a contrast measurement technique for MRI images. This method used the variation in histogram of second order derivation [13]. From the discussion in the above papers, we can conclude that histogram equalization and its different variations are popular for contrast enhancement of MRI image. The efficiency

© Springer Nature Switzerland AG 2020
A. P. Pandian et al. (Eds.): ICCBI 2019, LNDECT 49, pp. 290–296, 2020.
https://doi.org/10.1007/978-3-030-43192-1_33

of feature extraction of area of interest is depending on good pre-processing step. Improvement in the quality of image is done by contrast enhancement techniques such as: Intensity value, Histogram equalization and Adaptive histogram equalization. In this paper, we have organized our paper as follows; Section two discusses the Methodology, Section three and Section four discusses the result conclusion of the paper respectively.

2 Methodology

There are many contrast enhancement techniques applied in image processing, but some methods are not suitable for MRI images because of some characteristics of the images. One of conventional technique in the histogram equalization. Distributing pixel values throughout the region of interest is done in the Histogram equalization. This enhance the images so that better quality image is obtained. Another popular method of contrast enhancement is the adaptive histogram equalization [6, 7]. In this method image intensity is spread throughout the sub images and this sub images are combined to get the whole image. The contrast of an image can also adjust by mapping values of intensity of an image to a new value. These three methods are applied in 12 MRI images. The output image is quantitatively compared to find the best method of contrast enhancement out of this three. Details about the method is mentioned below:

A. Intensity Adjustment
In this process the range of the intensity value is specified in the output image. Remapping of the data value is used to fill the intensity range from 0–255. This leads to increase in contrast in low-scale image. Here, new value in the output image is calculated by saturating the 1% of pixel value from top and bottom pixel. The two steps in intensity value adjustment are:

1. Finding the intensity value limit by histogram of the image.
2. Changing the limit in a range between 0.0 to 9.0, so that it can be passed in high and low vector.

The application of this method on test images is shown in Fig. 1b. Figure 1a represents the original image.

B. Histogram Equadulization
By using the histogram equalization, we can automatically adjust intensity value of an image. This process transforms the intensity values so that the output image's histogram is comparable with a histogram having equally spaced bins where the range of intensity value range from 0–1. In histogram equalization, we enhance the contrast by adjusting the intensity of image. The main application of histogram equalization is to increase the global contrast of any image. To perform the histogram equalization of an image we need to calculate the probability mass function and cumulative distribution function. Histogram equalization is most applicable in cases where both background and foreground of the images are either dark or both bright. For example, X Ray images and under-exposed photographs. The main aim of histogram equalization

method is to distributes gray levels in dynamic range. One risk involving histogram equalization is that it cannot differentiate noise, and there is a risk of increasing noise in the image [8]. The result of application of this method is shown in Fig. 1c.

C. Adaptive Histogram Equalization
One variation of the standard histogram equalization is the adaptive Histogram Equalization, where small subarea in the image is considered. In this method, the image is divided into smaller regions called tiles. Then, enhancement of each tiles is done and finally, the neighboring tiles are merged to get the enhanced output image. The merging of tiles is done by bilinear interpolation. AHE is a modification of HE which is applied on image contains area of low brightness or darker. This method divides the whole image into tiles having brightness and darkness then histograms are computed based on this. These individual histograms are used to redistribute the value of light in image. After this, the combination of the neighboring tiles is done using the process of bilinear interpolation which eliminate all the artificially created boundaries. This version of AHE is called contrast limited adaptive histogram equalization (CLAHE) [9].

Figure 1d depicts the result of this method when applied in images depicted in Fig. 1a.

3   Result and Discussion

The experiments were conducted using GNU Octave 4.2.1 and Mathlab 2018a Windows 7, 64 bits OS with Intel Core TM. 1.90 GHz CPU and 4.00 GB RAM. Twelve test images of different types are taken from open access websites. We have taken images of different body parts to show the variety of MRI modalities. In this paper, following body parts are taken under consideration: Knees, neck, abdomen, brain, hips, ankle etc. Entropy is selected for quantitative analysis of the resultant test images. Entropy is used to evaluate the quality of an image. Higher value of entropy means better image quality. Entropy can be calculated from histogram of the image. It is the measure of unpredictability of an image. It can be used characterize the image content. Entropy is given by:

$$E = -sum(p. * \log_2 p) \qquad (1)$$

where p contains the normalized histogram count.

The results of application of the three techniques of contrast enhancement of twelve MRI images are shown in Fig. 1.

Figure 1(a) Original MRI image [3]
Figure 1(b) Image obtained by intensity value adjustment
Figure 1(c) Image obtained by histogram equalization
Figure 1(d) Image obtained by adaptive histogram equalization.

We have calculated the Entropy of the original image and subsequent output is shown in Table 1.

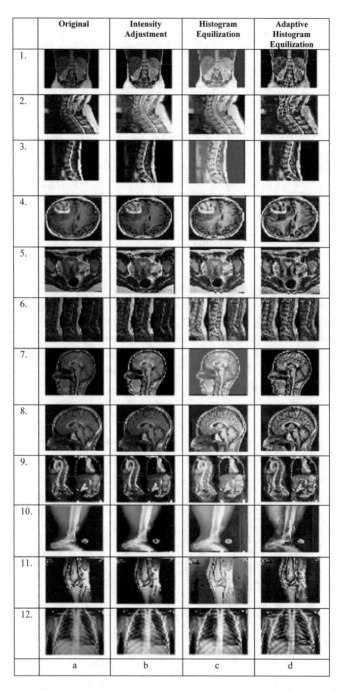

Fig. 1. Original MRI image and corresponding enhanced images obtained by the discussed techniques

Table 1. Entropy of the original image and the resultant images

Test image	Intensity adjustment	Histogram equilization	Adaptive histogram equilization
1	6.1951	5.8433	4.3105
2	7.8886	7.8886	3.4771
3	5.2452	5.3147	3.4793
4	7.3342	7.7377	3.5409
5	7.8072	7.9605	3.6089
6	7.2466	7.9373	3.6213
7	5.4806	5.4878	3.6265
8	7.2106	7.675	3.6445
9	7.3642	7.8294	3.6612
10	6.78	7.2521	3.6633
11	6.5482	7.0522	3.6817
12	7.9314	7.9497	3.6858

From Fig. 2 we can see that there is the uniformity in the performance of adaptive histogram equalization for all images. Though the entropy of other techniques is higher, there is no uniformity. Thus, we conclude that Adaptive Histogram Equalization is effective, however we have also seen that Histogram Equalization produces the best entropy in all the 10 test images. So, we can conclude that histogram equalization is producing the best result. Further, work can be done on creating some method of enhancing the histogram equalization method to make its application uniform for all images.

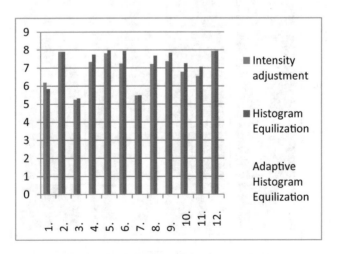

Fig. 2. Result analysis of the three methods.

4 Future Work

Future Work in image processing includes developing of techniques to preserve edges after contrast enhancement. Decrease the sensitivity towards noise and preserving sharpness of image is another goal of MRI image processing. Another important agenda is addressing of single uniform technique which can be applied effectively and efficiently in all different types of MRI images [10].

5 Conclusion

The main goal of this paper is to examine the intensity value based on contrast enhancement methods. The three Contrast enhancement techniques re applied on twelve MRI images and the entropy of all the resultant images along with the original image is also determined. From the result, we can state that adaptive histogram equalization provides the most uniform result in comparison to the other two testing techniques. However, the histogram equalization produces the best entropy in maximum case. Further, we can lead to future study and research which will aim to develop a system which is efficient as histogram equalization and uniform as Adaptive Histogram Equalization.

References

1. Gonzalez, R.C., Woods, R.E.: Digital Image Processing, 4th edn. Pearson (2018)
2. Ganguly, D., Chakraborty, S., Balitanas, M., Kim, T.: Medical Imaging: A Review (2010). https://doi.org/10.1007/978-3-642-16444-6_63
3. Patil, C., et al.: Using image processing on MRI scans. In: 2015 IEEE International Conference on Signal Processing, Informatics, Communication and Energy Systems (SPICES), Kozhikode, pp. 1–5 (2015). https://doi.org/10.1109/SPICES.2015.7091517
4. Kaur, H., Rani, J.: MRI brain image enhancement using Histogram Equalization techniques. In: 2016 International Conference on Wireless Communications, Signal Processing and Networking (WiSPNET), Chennai, pp. 770–773 (2016)
5. Senthilkumaran, N., Thimmiaraja, J.: Histogram equalization for image enhancement using MRI brain images. In: 2014 World Congress on Computing and Communication Technologies, Tiruchirappalli, pp. 80–83 (2014)
6. Yelmanova, E.S., Romanyshyn, Y.M.: Adaptive enhancement of monochrome images with low-contrast objects. In: 2017 12th International Scientific and Technical Conference on Computer Sciences and Information Technologies (CSIT), Lviv, pp. 421–424 (2017). https://doi.org/10.1109/stc-csit.2017.8098820
7. Muniyappan, S., Allirani, A., Saraswathi, S.: A novel approach for image enhancement by using contrast limited adaptive histogram equalization method. In: 2013 Fourth International Conference on Computing, Communications and Networking Technologies (ICCCNT), Tiruchengode, pp. 1–6. (2013). https://doi.org/10.1109/icccnt.2013.6726470
8. Hummel, R.A.: Image enhancement by histogram transformation. Comput. Graph. Image Process. **6**, 184–195 (1977)

9. Acharya, T., Ray, A.K.: Image Processing: Principles and Applications. Wiley-Interscience (2005). ISBN 0-471-71998-6
10. Kumari, L.S., Kanhirodan, R.: Contrast enhancement of medical radiography images using edge preserving filters. In: 2018 Fourth International Conference on Biosignals, Images and Instrumentation (ICBSII), Chennai, pp. 206–212 (2018)
11. Zhang, Y.H., Li, X., Xiao, J.Y.: A digital fuzzy edge detector for color images. Comput. Vis. Pattern Recognit. **1**, 1–7 (2017)
12. Jafari-Khouzani, K., Gerstner, E., Rosen, B., Kalpathy-Cramer, J.: Upsampling dynamic contrast enhanced MRI. In: 2015 IEEE 12th International Symposium on Biomedical Imaging (ISBI), New York, NY, pp. 1032–1035 (2015). https://doi.org/10.1109/ISBI.2015. 716404
13. Ti, C.W., Swee, S.K., Abas, F.S.: Contrast measurement for MRI images using histogram of second-order derivatives. In: 2017 International Conference on Robotics, Automation and Sciences (ICORAS), Melaka, pp. 1–5 (2017). https://doi.org/10.1109/ICORAS.2017. 8308080

Global User Social Networking Ranking Statistics

P. Ajitha[1]($\boxtimes$), Bana Suresh Kumar Reddy[2], and Balina Prudhvi[2]

[1] Department of Information and Technology,
Sathyabama Institute of Science and Technology, Chennai, India
ajitha.it@sathyabama.ac.in
[2] Information Technology, Sathyabama Institute of Science and Technology,
Chennai, India
sureshbanal997@gmail.com, prudhvibalina@gmail.com

Abstract. Long range social correspondence affiliations are open at a couple on-line systems like Twitter.com and Weibo.com, wherever a couple of customers keep coordinating with one another dependably. One captivating and principal trouble inside the long range social correspondence affiliations is to rank customers kept up their centrality promisingly. Frill in nursing right masterminding once-over of customer centrality may benefit a couple of get-togethers in social connection affiliations like the types of progress suppliers and site heads. Regardless of the way in which that it's incredibly promising to get a significance based organizing once-over of customers, there square measure a couple of express troubles because of the enormous scale and fragments of individual to particular correspondence information. In the midst of this paper, we will by and large propose a lone perspective to achieve this goal is studying customer hugeness by separating the dynamic encouraged undertakings among customers on accommodating affiliations.

Keywords: Passed on structures · Watching data · Social affiliations · Customer development · Significance

1 Introduction

Person to individual communication association was daunting at various online stages with the enhancement of internet development. The long range agreeable contact association strengthens the functioning of social affiliations or social relationships between customers who share interest, interests, and physical affiliations, for example. Through such connection, customers can stay connected to each other and be told about the actions of accomplices, such as posting at a point, and being influenced by each other along these lines. For example, a customer can get the minute revives about the posts of his related accomplices in the current Twitter and Weibo, and can help retweet or comment on the posts. Within a span of time, countless can take explicit exercises, such as posting and retweeting to specific correspondence sites at these individuals. One spell-binding and fundamental problem is how consumers can be ranked with clear data based on their centrality. An exact centrality orchestrating of customers will give incomprehensible making sense of how to various applications in most online individual to solitary correspondence goals. For instance, online promotions providers may improve method for passing on their degrees of progress by systems for considering the organized centrality of customers; website page executives may setup better practices for online campaigns (e.g., online audit) by strategies for using the masterminding plot. While it is promising for several social gatherings to give a criticalness organizing of customers, there are diverse explicit bothers to deal with this issue.

In this paper, we isolate Network ranking information by using nearby hadoop contraction of certain hadoop conditions such as hdfs, mapreduce, sqoop, hive and pig. By using these tools, we can process no data barrier, no data lost problem, we can achieve high efficiency, bolstering costs in the same way.

2 Literature Survey

In any case, an undirected structure is the easy-going relationship found in our pressure and the collaboration between two clients is also symmetrical. Second, given the measurement of cooperation between all customers' courses of action, we can test for each customer the measurement of all joint efforts and rate them based on the check. In any case, given the assessment of joint efforts between two focuses (clients), it is attempting to reason which one is contributing how much to all coordinated efforts. All things considered, it may not be clear to rank all clients dependent on the aggregated check everything being proportional. Third, this issue isn't actually proportionate to many existing focus point arranging issues, for example, site page arranging. Most focus point arranging calculations couldn't be unmistakably utilized for this issue in light of the fact that the objective is to rank focus focuses dependent on the dynamic joint endeavors that really advance over times. A regression based framework for the craving errand. Raised tests on two true illuminating records that are collected from various spaces plainly demonstrate the reasonableness of our arranging what's more, want systems.

In the middle of February 9–12, 2011, the Fourth International Conference on Web Search and Data Mining was held in Hong Kong. WSDM was first conducted at Stanford University in 2008 and at that limit in Barcelona, Spain and New York, USA. This is the first time the system of social affairs has been dealt with in the Asia Pacific, leading to the development of web search and data burrowing study imperative for this district. Regardless of how it started late for two or quite a while since its inception, WSDM has quickly made the head wrap up when everything is said in Web search and Web data mining collection. WSDM focuses on mystery and data mining are not proportionate to WWW. WSDM promotes funding and collaboration for the academic environment as well as business. It encourages main and theoretical work on Web examining for and Web data mining, as well as the region's fair use. It is perceptible that Web look and Web information mining have made a director twist among the most striking areas of computer science and construction examination. It is continuously and increasingly changing the work and life of individuals and has a strong and gigantic connection with all that is considered when masses are in question.

We present Google in paper [2], a prototype of a far-reaching web search engine that makes imperative use of the hypertext framework. It is suggested that Google crawl and document the Internet sensibly and create and generate archived documents that are more convenient than existing systems. The model with a full substance and hyperlink database of something almost 24 million pages is open at http://google. stanford.edu/. To produce a web list is an endeavoring errand. Web records list tens to an epic number of webpage pages incorporating an in each pragmatic sense indistinct number of express terms. They answer countless reliably. Despite the centrality of far reaching scale web search for instruments on the web, by no informative research has been done on them. Likewise, in context on quick improvement in movement and web extension, making a web report today is total not actually equal to three years sooner. Author [3] in his paper offers an all-around structure of our tremendous web-looking device–the fundamental point-by-point description we believe to be going so far. In addition to the issues of scaling generic interest methodologies to details of this enormity, the use of the additional information present in hypertext to communicate better summary things requires new specific troubles.

As PC clashed with communication advantages in the area of Web 2.0, author [4] suggested digital platforms were the essential spots for sharing information and interactions with Web clients. The strategies for discovering pro clients in the framework are one of the fundamental research issues in online frameworks. In this paper, we explore the wellness that clients showed up in online frameworks, particularly in talk parties and propose a pivotal master arranging check, which joins both talk string substance and easygoing affiliation ousted from enormous social affiliations. We present a model of vector space for the substance criticality portion and a calculation of PageRank style for the pro make part. This can ensure that the arranged stars are both perfectly suited to the specific request and exceptionally accurate in related areas.

PageRank is routinely figured from the power of progress structure in a MarkovChainmodel. It is as such computationally expensive, and persuading assessment methods to vivify the count are essential, especially with respect to broad graphs. In this paper, we propose two testing implies PageRank valuable estimation: Direct assessing and Adaptive looking at. The two methods test the progress structure and use the perspective in PageRank count. Direct sampling method tests the change framework once and uses the perspective indisputably in PageRank estimation, while versatile seeing method tests the movement grid on various events with an adaptable model rate which is adjusted iteratively as the enrolling structure proceeds. This versatile perspective rate is typical for an unrivaled than ordinary trade off among accuracy and sufficiency for PageRank check. We give point by point theoretical examination on the ruin furthest reaches of both methods Experimental results of author [5] on a few legitimate world datasets show that our systems can achieve on a very basic level higher capacity while accomplishing basically indistinguishable precision than cutting edge strategies.

3 Proposed System

Hadoop Framework and Spark are used by the Proposed System. Hadoop is an open source framework that has been developed by the apache software foundation and is used to focus on and handle titanic datasets with a social affair of gear. Hadoop mechanical get-together involves two items that are hdfs and mapreduce. We also use proprietary frameworks such as sqoop, hive and pig from Hadoop.

(1) Data Importing Ranking Database
The panel structure is a shared server in MySQL. RDBMS uses relationships or tables to store User criticality coordinated by parts with specific keys and internal keys as a row structure. Using MySQL language, it is usually possible to collect, verify, monitor, recoup, delete and manage user vitality orchestrating in tables for business reasons. Existing thought oversees giving backend by using MySQL that involves some hindrances i.e. critical data is that preparation time is high when the data is enormous and once data is lost we can't recover so by using Hadoop contraction we suggest thinking (Fig. 1).

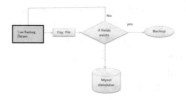

Fig. 1. Flow of data importing ranking system

(2) Storage
Sqoop is an interest line interface application for transferring User criticalness masterminding between social databases (MySQL) and Hadoop (Fig. 2).

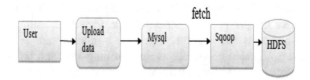

Fig. 2. Flow of storage system

(3) Analyzation of Query
Hive is a data thing house structure for Hadoop that runs SQL like interest called HQL (Hive request language) which gets inside changed over to depict jobs. In Hive, User importance masterminding tables and databases are made first and a brief time allotment later data is stacked into these tables. Hive as User centrality orchestrating stockroom expected for planning and keeping an eye on just made data that is confirmed in tables. Hive organizes User vitality orchestrating tables into packs. It is a system for isolating a table into related parts subject to the estimations of distributed

sections. Using territory, it is surely not hard to request a bit of the User significance ranking. Tables or packs are sub-designated into holders, to give extra structure to the User centrality masterminding that may be used for tenaciously beneficial tending to. Bucketing works reliant on the estimation of hash limit of some portion of a table (Fig. 3).

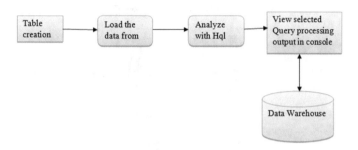

Fig. 3. Flow of query system

(4) Multi-query Approach

To evaluate user significance orchestrating using Pig, programming engineers need to outline substance using the language of Pig Latin and use the Grunt shell to execute it in normal mode. Within, these materials are converted into tasks of map and reduction. You can run your Pig material into the shell after conjuring the Grunt shell. Except for LOAD and STORE, Pig Latin assertions interpret a relationship as information and generate another relationship as yield when carrying out each other's development. Once you reach the Grunt shell's Load illumination, the linguistic clear look will be passed on. You need to use the official Dump to see the content of the association. The MapReduce work to stack the data into the record system will be done, basically in the wake of playing the landfill upgrade. Pig offers numerous trademark managers to assist with software rehearsals such as get-together, platforms, applications, etc. (Fig. 4).

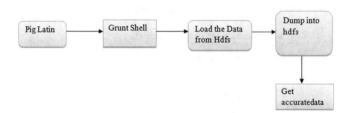

Fig. 4. Flow of multi query system

(5) Processing

MapReduce is a framework that helps us to form applications for processing epic degrees of user centrality, organizing unambiguously on far-reaching get-togethers of equipment. MapReduce is a supervisory structure and a program appears for appropriate selection subject to java. The MapReduce figure contains two key endeavors,

Map and reduce unequivocally. Map Reduce system runs in three steps, specifically direct sorting, mixing main thrust, and reduction phase. The basic task of the guide or mapper is to process the data. It is reviewed in the Hadoop record structure (HDFS) as account or inventory all around the software. The data record is moved line by line to the mapper job line. The mapper is recording the data and making a few bits of data. The commitment of the Reducer is to process the data starting from the mapper. When it happens to handle, it makes another yield project that will be tested in the HDFS (Figs. 5 and 6).

Fig. 5. Flow of processing system

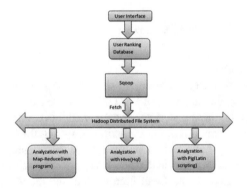

Fig. 6. Architecture of ranking system

Algorithm 1. MapReduce Execution

---

   1.   Class MAPPER

   2.   method Map(prid a, prname d)

   3.   For all term t $\in$ doc d do

   4.   Emit(term t, count 1)
       class Reducer

       i.   method Reduce(term t, counts [c1, c2, . . .])

       ii.   method Reduce(term t, counts [c1, c2, . . .])

       iii. sum $\leftarrow$ 0

       iv.   for all count c $\in$ counts [c1, c2, . . .] do

       v.   sum $\leftarrow$ sum + c

       vi.   Emit(term t, count sum)

For each word in a file, the mapper transmits a brilliant overall key-respect enter.

4 Result and Discussions

In this paper, we suggest two kinds of focus point centrality organizing tallies which concentrate everything considered on a gander at the vitality of all. Second, for a middle A that has different joint endeavors with its accomplices over a period of time, if the vast majority of its partners do not have different joint endeavors with their partners, it is unfathomably likely that the middle point A will be highly important. In this context, we define two calculations to determine the centrality estimate of each middle and suggest the basic figure. Second, by managing the mutual centrality dependence among all customers within an easy association, we suggest the second check which iteratively translates the customer's essentialness estimate. Through the concentration, the estimates of all focuses affect each other through the structure and impact. Accordingly, the second count is designed to dismantle all deemed to dismantle the tremendousness rating by thinking about the whole system. In addition, we suggest an updated model to predict the monstrosity of consumers on our full understanding of consumer centrality. Furthermore, the successful desire findings will support various applications on long-range social communication objectives. Ultimately, they perform clear tests on both consumer centrality systems and want veritable instructive meetings of two vital scale. The test results show our frameworks' common sense and breadth.

An essential bit of the quantifiable examination depends on numerical strategy, for example, sureness between times, theory testing, fall a long way from the certainty examination, etc. If all else fails, these systems depend on suppositions about the information being utilized. One approach to manage supervise pick whether information fit in with these request is the Graphical Data Analysis with R, as a configuration can give different bits of learning into the properties of the plotted dataset. Charts are tremendous for non-numerical information, for example, tints, flavors, brand names, no two ways about it. Totally when numerical measures are troublesome or difficult to choose, plots imagine a fundamental occupation. Sound managing is finished with the course of action to make brilliant amusement arrangements (Fig. 7).

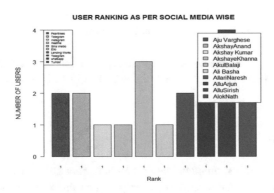

Fig. 7. Performance comparison

Here pie blueprints to organize the general beautifying administrators of a model in presentations had been used. Here the purpose of imprisonment pie takes a vector of numbers and changes them into degrees. It by then isolates the float dependent on those degrees. It demonstrates relationship inside a gathering. One cannot use a pie outline to demonstrate changes over the time span (Fig. 8).

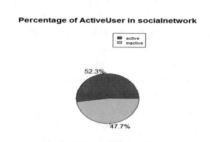

Percentage of ActiveUser in socialnetwork

Fig. 8. Pie chart comparison

The rating data is compiled and a database is generated and additional HDFS name center point is obtained as shown below (Fig. 9).

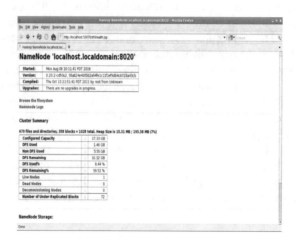

Fig. 9. Cluster summary

Throughput is the degree of data experienced sensibly in a given time. It can be implied as bits dependably [bps], uber bits dependably [mbps] giga bits dependably [gbps].

5   Conclusion

In this paper, we exhibited an examination on client centrality sorting out and need in decent correspondence relationship, for example, microblog application. To examination the user significance coordinating information in hadoop common structure.

Hadoop trademark system is hive, pig, mapreduce, in the event that you need examination to locate the some fundamental examination the dynamic shaped endeavors among clients on pleasant structures. In future the brilliance on different events speedier than hadoop, it is effectively examination snappier.

Apache Spark is an open source overseeing engine worked around speed, example of utilization, and examination. If you have a ton of data that requires low gradualness managing that a standard Map Reduce program can't give, Spark is the alternative. Shimmer gives in-memory cluster choosing to astoundingly brilliant speed and reinforce Java, Scala, and Python APIs for straightforwardness of development.

References

1. Bakshy, E., Hofman, J.M., Mason, W.A., Watts, D.J.: Everyone's an influencer: quantifying influence on twitter. In: Proceedings of the Fourth ACM International Conference on Web Search and Data Mining, pp. 65–74. ACM (2011)
2. Brin, S., Page, L.: Reprint of: the anatomy of a large scale hypertextual web search engine. Comput. Netw. **56**(18), 3825–3833 (2012)
3. Brown, R.G.: Smoothing, Forecasting and Prediction of Discrete Time Series. Courier Corporation (2004)
4. Campbell, C.S., Maglio, P.P., Cozzi, A., Dom, B.: Expertise identification using email communications. In: Proceedings of the Twelfth International Conference on Information and Knowledge Management, pp. 528–531. ACM (2003)
5. Cha, M., Haddadi, H., Benevenuto, F., Gummadi, K.P.: Measuring user influence in twitter: the million follower fallacy. In: ICWSM 2010, pp. 10–17 (2010)
6. Domingos, P., Richardson, M.: Mining the network value of customers. In: Proceedings of the Seventh ACM SIGKDD International Conference on Knowledge Discovery and Data Mining, pp. 57–66. ACM (2001)
7. Brown, P.E., Feng, J.: Measuring user influence on twitter using modified k-shell decomposition (2011)
8. Jiao, J., Yan, J., Zhao, H., Fan, W.: Expertrank: an expert user ranking algorithm in online communities. In: International Conference on New Trends in Information and Service Science, NISS 2009, pp. 674–679. IEEE (2009)
9. Kleinberg, J.M.: Authoritative sources in a hyperlinked environment. J. ACM (JACM) **46**(5), 604–632 (1999)
10. Kumar, S., Morstatter, F., Liu, H.: Twitter Data Analytics. Springer (2014)

Detecting Spam Emails/SMS Using Naive Bayes, Support Vector Machine and Random Forest

Vasudha Goswami[(✉)], Vijay Malviya, and Pratyush Sharma

RGPV, Bhopal, India
vasudha.goswami30@gmail.com,
{vijaymalviya,hod_cs}@mitindore.co.in

Abstract. SMS spams are dramatically increasing year by year because of the expansion of movable users around the world. Recent reports have clearly indicated an equivalent. Mobile or SMS spam may be a physical and thriving drawback because of the actual fact that bulk pre-pay SMS packages are handily obtainable and SMS is taken into account as a private and trustable service. SMS spam filtering may be a relatively recent trip to deal such a haul. The amount of information traffic moving over the network is increasing exponentially and therefore the connected devices are considerably vulnerable. Here network security plays a vital role in this context. In this paper, a SMS spams dataset is taken from UCI Machine Learning repository, and after pre-processing, different machine learning techniques such as random forest (RF), Naive Bayes (NB), Support Vector Machine (SVM) are applied to the dataset inorder to compute the performance of these algorithms.

Keywords: Data mining · Review spam · Classification · Comparative analysis · Spam detection · Sentiment analysis

1 Introduction

Short Message Service (SMS) is that the most often and widely used communication medium. The term "SMS" is employed for each the user activity and every one sorts of the short text electronic messaging in several components of the planet. It has become a medium of promotion and promotion of product, banking updates, agricultural data, flight updates and net offers. Typically SMS promotion could be a matter of disturbance to users. These types of SMSs area unit referred to as spam SMS. Spam is one or a lot of uninvited messages, that is unwanted to the users, sent or announced as a part of a bigger assortment of messages, all having considerably identical content. SMS spamming gained quality over alternative spamming approaches like email and twitter, thanks to the increasing quality of SMS communication. However, gap rates of SMS area unit beyond ninetieth and opened inside quarter-hour of receipt whereas gap rate in email is just 20–25% inside twenty four hours of receipt. Thus, a correct SMS spam detection technique has been an important necessity for all the mobile gadget users.

© Springer Nature Switzerland AG 2020
A. P. Pandian et al. (Eds.): ICCBI 2019, LNDECT 49, pp. 306–313, 2020.
https://doi.org/10.1007/978-3-030-43192-1_35

There are many researches on email, twitter, net and social tagging spam detection techniques. However, only few researches are conducted on SMS spam detection.

2 Literature Review

According to [1], conducted a survey on ways and applications for detection and filtering uninvited advertising messages or spam in a telecommunication network [2]. Building up a classification algorithmic [3] program that channels SMS spam would provide useful equipment for portable suppliers. Since naïve mathematician has been used effectively for email spam detection, it seems to be expected that it may likewise be accustomed and build SMS spam classifier. With reference to email spam, SMS spam represents further difficulties for automatic channels [4, 6].

3 Problem Definition

The fact that an email box can be flooded with unsolicited emails makes it possible for the account holder to miss an important message; thereby defeating the purpose of having an email address for effective communication. These junk emails from online marketing campaigns, online fraudsters among others is one of the reasons for this paper. We try to obtain the feature sets that can best represent and distinguish the spams from ham(non-spam). We then follow both supervised and unsupervised methodology to obtain spams from the dataset. We also include sentiment analysis methodology into our spam detection. Lastly, we compare our analysis obtained from taking various types of feature sets based on text, sentiment scores, reviewer features, as well as the combined method.

4 Proposed Work

A machine learning techniques have been proposed to detect and classify the reviews through various processing steps which is shown in Fig. 1.

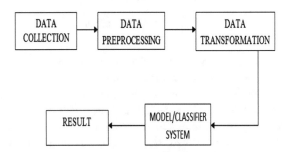

Fig. 1. Proposed flow diagram

Our Steps or Algorithm Steps will follow:

1. Data Collection: We will use a dataset from the dataset repository of Centre for Machine Learning and Intelligent Systems at the University of California, Irvine.
2. Data Preprocessing: Data preprocessing is the most important phase in detection models as the data consists of ambiguities, errors, redundancy which needs to be cleaned beforehand.
3. Data Transformation: Data is transformed into lowercase and change the data types according to algorithm needs.
4. Classification System: The attributes are identified for classifying process and system perform feature extraction and then these classification system classify the content into spam or ham.

5 Experimental Analysis

The experimental and result analysis is done by using intel i5-2410M CPU with 2.30 GHz processor along with 4 GB of RAM and the windows operating system is running. For result analysis we use R and R studio for processing the data and then we load the sms dataset which consist a 5574 observation with no missing are present in the dataset. Figure 2 shows the dataset has been loaded.

```
> data_text <- read.delim("SMSSpamCollection", sep="\t", header=F, colClasses="chara
cter", quote="")
> str(data_text)
'data.frame':   5574 obs. of  2 variables:
 $ V1: chr  "ham" "ham" "spam" "ham" ...
 $ V2: chr  "Go until jurong point, crazy.. Available only in bugis n great world la
e buffet... Cine there got amore wat..." "ok lar... Joking wif u oni..." "Free entr
y in 2 a wkly comp to win FA Cup final tkts 21st May 2005. Text FA to 87121 to recei
ve entry question("| _truncated_ "U dun say so early hor... U c already then sa
y..." ...
>
> head(data_text)
    V1
1  ham
2  ham
3 spam
4  ham
5  ham
6 spam
                                                                              V2
1                         Go until jurong point, crazy.. Availab
le only in bugis n great world la e buffet... Cine there got amore wat...
2                                                       ok lar... Joking wif u oni...
3 Free entry in 2 a wkly comp to win FA Cup final tkts 21st May 2005. Text FA to 871
21 to receive entry question(std txt rate)T&C's apply 08452810075over18's
4
5                        U dun say so early hor... U c already then say...
                         Nah I don't think he goes to usf, he lives around here though
6           FreeMsg Hey there darling it's been 3 week's now and no word back! I'd like
some fun you up for it still? Tb ok! xxx std chgs to send, £1.50 to rcv
```

Fig. 2. Loading a dataset

After loading, For easy identification of the columns, we rename V1 as Class and V2 as Text. And we have to also convert the Class column from Character strings to factor. Data often come from different sources and most of the time don't come in the right format for the machine to process them. Hence, data cleaning is an important aspect of a data science project. In text mining, we need to put the words in lowercase, remove stops words that do not add any meaning to the model etc. Various data cleaning steps are shown in Fig. 3.

```
            ham      spam
0.8659849 0.1340151
> library(tm)
Loading required package: NLP
>
> library(snowballc)
> corpus = VCorpus(VectorSource(data_text$Text))
> as.character(corpus[[1]])
[1] "Go until jurong point, crazy.. Available only in bugis n great world la e buffet... Cine th
ere got amore wat..."
>
> corpus = tm_map(corpus, content_transformer(tolower))
> corpus = tm_map(corpus, removeNumbers)
> corpus = tm_map(corpus, removePunctuation)
> corpus = tm_map(corpus, removewords, stopwords("english"))
> corpus = tm_map(corpus, stemDocument)
> corpus = tm_map(corpus, stripwhitespace)
> as.character(corpus[[1]])
[1] "go jurong point crazi avail bugi n great world la e buffet cine got amor wat"
>
>
> #Creating the Bag of words for the model
>
> dtm = DocumentTermMatrix(corpus)
> dtm
<<DocumentTermMatrix (documents: 5574, terms: 6981)>>
Non-/sparse entries: 43801/38868293
Sparsity            : 100%
Maximal term length: 40
weighting           : term frequency (tf)
>
> dtm = removeSparseTerms(dtm, 0.999)
> |
```

Fig. 3. Data pre-processing

In text mining, it is important to get a feel of words that describes if a text message will be regarded as spam or ham. What is the frequency of each of these words? Which word appears the most? In other to answer this question; we are creating a DocumentTermMatrix to keep all these words. We want to words that frequently appeared in the dataset. Due to the number of words in the dataset, we are keeping words that appeared more than 60 times. We will like to plot those words that appeared more than 60 times in our dataset. Figure 4 shows the wordcloud of the dataset.

Fig. 4. Presenting the word frequency as a word cloud

Usually in Machine Learning is to split the dataset into both training and test set. While the model is built on the training set; the model is evaluated on the test set which the model has not been exposed to before. We will be building our model on 3 different Machine Learning algorithms which are Random Forest, Naive Bayes and Support Vector Machine for the purpose of deciding which perform the best.

Random Forest

The Random Forest Model is an ensemble method of Machine Learning with which 300 decision trees were used to build this model with the mode of the outcomes of each individual trees taken as the final output. So we can train the random forest model on training dataset and then test the performance of the model on testing dataset which is shown in Fig. 5.

```
>
> # Predicting the Test set results
> rf_pred = predict(rf_classifier, newdata = test_set[-1210])
>
> # Making the Confusion Matrix
> library(caret)
Loading required package: lattice
>
> confusionMatrix(table(rf_pred,test_set$Class))
Confusion Matrix and Statistics

rf_pred  ham spam
   ham  1191   36
   spam    4  154

              Accuracy : 0.9711
                95% CI : (0.9609, 0.9793)
   No Information Rate : 0.8628
   P-Value [Acc > NIR] : < 2.2e-16

                 Kappa : 0.8687

Mcnemar's Test P-Value : 9.509e-07

           Sensitivity : 0.9967
           Specificity : 0.8105
        Pos Pred Value : 0.9707
        Neg Pred Value : 0.9747
            Prevalence : 0.8628
        Detection Rate : 0.8599
  Detection Prevalence : 0.8859
     Balanced Accuracy : 0.9036

      'Positive' Class : ham
```

Fig. 5. Performance measure of Random Forest

Naive Bayes

Naive Bayes Classifier is a Machine Learning model that is based upon the assumptions of conditional probability as proposed by Bayes' Theorem. It is fast and easy and the performance outcomes of the model on testing dataset are shown in Fig. 6.

```
> nb_pred = predict(classifier_nb, type = 'class', newdata = test_set)
>
> confusionMatrix(nb_pred,test_set$Class)
Confusion Matrix and Statistics

          Reference
Prediction  ham spam
      ham  1195    7
      spam    0  183

               Accuracy : 0.9949
                 95% CI : (0.9896, 0.998)
    No Information Rate : 0.8628
    P-Value [Acc > NIR] : < 2e-16

                  Kappa : 0.9783

 Mcnemar's Test P-Value : 0.02334

            Sensitivity : 1.0000
            Specificity : 0.9632
         Pos Pred Value : 0.9942
         Neg Pred Value : 1.0000
             Prevalence : 0.8628
         Detection Rate : 0.8628
   Detection Prevalence : 0.8679
      Balanced Accuracy : 0.9816

       'Positive' Class : ham
```

Fig. 6. Performance measure of Naive bayes

Support Vector Machine

The Support Vector Machine is another algorithm that finds the hyperplane that differentiates the two classes to be predicted, ham and spam in this case; very well. SVM can perform both linear and non-linear classification problems. The figure shows the performance outcomes of the model on the testing dataset (Fig. 7).

```
> svm_pred = predict(svm_classifier,test_set)
>
> confusionMatrix(svm_pred,test_set$Class)
Confusion Matrix and Statistics

          Reference
Prediction  ham spam
      ham  1195  189
      spam    0    1

               Accuracy : 0.8635
                 95% CI : (0.8443, 0.8812)
    No Information Rate : 0.8628
    P-Value [Acc > NIR] : 0.4882

                  Kappa : 0.009

 Mcnemar's Test P-Value : <2e-16

            Sensitivity : 1.000000
            Specificity : 0.005263
         Pos Pred Value : 0.863439
         Neg Pred Value : 1.000000
             Prevalence : 0.862816
         Detection Rate : 0.862816
   Detection Prevalence : 0.999278
      Balanced Accuracy : 0.502632

       'Positive' Class : ham
```

Fig. 7. Performance measure of SVM

Performance Measure

We used accuracy, which are derived using confusion matrix (Tables 1, 2 and Fig. 8).

Table 1. Confusion matrix

	Classified as normal	Classified as attack
Normal	TP	FP
Attack	FN	TN

Table 2. Accuracy of the models

Model	Accuracy
Random Forest	97.11%
Naive Bayes	99.49%
Support Vector Machine	86.35%

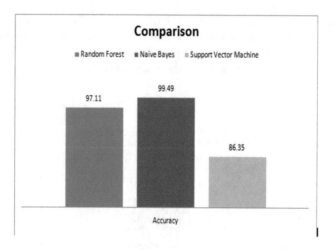

Fig. 8. Accuracy comparison

6 Conclusion

In this paper, we propose a machine learning technique for SMS Spam filtering based on algorithms namely Naïve Bayes, Support vector machine and Random Forest. The dataset that we have used in our work consists of 5574 observations of 2 variables. The first variable is the content of the emails and the second variable the target variable, which is the class to be predicted. In this paper, The Random Forest and Naive Bayes performed exceptionally well as compared to SVM.

References

1. Navaney, P., Dubey, G., Rana, A.: SMS spam filtering using supervised machine learning algorithms. IEEE (2018)
2. Shrivastava, S., Anju, R.: Spam mail detection through data mining techniques. IEEE (2017)
3. Abdulhamid, S.M., Latiff, M.S.A., Chiroma, H., Osho, O., Abdul-Salaam, G., Abubakar, A.I., Herawan, T.: A review on mobile SMS spam filtering techniques. IEEE (2017)
4. Jindal, N., Liu, B.: Mining comparative sentences and relations. In: AAAI, vol. 22 (2006)
5. Jindal, N., Liu, B.: Review spam detection. In: Proceedings of the 16th International Conference on World Wide Web, Canada, pp. 1189–1190. ACM Press, New York (2007)
6. Jindal, N., Liu, B.: Opinion spam and analysis. In: Proceedings of the 2008 International Conference on Web Search and Data Mining, pp. 219–230. ACM Press, New York (2008)
7. Xie, S., Wang, G., Lin, S., et al.: Review spam detection via temporal pattern discovery. In: Proceedings of the 18th ACM SIGKDD International Conference on Knowledge Discovery and Data Mining, pp. 823–831. ACM Press, New York (2012)
8. Lim, E.-P., Nguyen, V.-A., Jindal, N., et al.: Detecting product review spammers using rating behaviors. In: Proceedings of the 19th ACM International Conference on Information and Knowledge Management, pp. 939–948. ACM Press, New York (2010)
9. Jindal, N., Liu, B., Lim, E.-P., et al.: Finding unusual review patterns using unexpected rules. In: Proceedings of the 19th ACM International Conference on Information and Knowledge Management, pp. 1549–1552. ACM Press, New York (2010)
10. Wang, G., Xie, S., Liu, B., et al.: Identify online store review spammers via social review graph. ACM Trans. Intell. Syst. Technol. 3(4), 61.1–61.21 (2011)
11. Indurkhya, N., Damerau, F.J.: Handbook of Natural Language Processing, 2nd edn. Chapman and Hall/CRC, London (2010)
12. Ohana, B., Tierney, B.: Sentiment classification of reviews using SentiWordNet. In: 9th IT & T Conference, vol. 13 (2009)

Detection of Flooding Attacks Using Multivariate Analysis

Priyanka Meel$^{(\boxtimes)}$ and Tanmay Singh

Department of Information Technology, Delhi Technological University, New Delhi, India
{priyankameel, tanmay_bt2k16}@dtu.ac.in

Abstract. In this paper, we propose a multivariate statistical analysis method namely the Hotelling's T^2 Method for the analysis of common network flooding attacks. The method analyses the behavior of system resources and network protocols and builds a baseline profile for its normal operation. We validated the proposed mechanism by carrying out flooding attacks on a wired network with Windows. We generated and sent attack packets through codes to a host machine, analyzed them (using Wireshark) and used a multivariate statistical method for testing the attack. This method effectively differentiates between normal and attack traffic and sets an alert in case of any abnormality in behavior.

Keywords: Abnormality distance metric · Anomaly detection · Multivariate statistical analysis

1 Introduction

In the modern world all the services are dependent on the Internet and any disruption in it can cause havoc in the operation of these services. The security of the Internet is a major issue considering the pivotal role it plays in this technological era [8]. The denial of service attacks and spam emails are the most ubiquitous threat attacks for users. The detection and protection against such attacks and the mitigation of its effects are important for the use of the internet. As these attacks are a form of request unlike malware attacks, so the behavioral pattern in which the attack takes place helps in fighting against such attacks. Two types of attack detect strategies are used nowadays [1]. Rule-based attack detection strategy in which the observed behavior is compared with a signature of known attack to detect an anomaly. But the limitation of this approach is that it will not be able to detect new attacks. In the anomaly-based approach, a baseline profile for normal network behavior is built and any deviation from this is detected as an attack. This approach can detect both known and unprecedented attacks. A single measuring attribute is not sufficient to describe the operational state of a system accurately so in this paper additional metrics such as memory usage and CPU usage are used to measure a reduction in performance of the system which is proportional to the intensity of the attack. In this paper, multivariate analysis methods are used to detect abnormal behavior of the system. Various network elements are quantified using tools like Wireshark and if the distance of the current state is more than a threshold value then abnormal behavior is predicted. The paper has been organized as follows.

© Springer Nature Switzerland AG 2020
A. P. Pandian et al. (Eds.): ICCBI 2019, LNDECT 49, pp. 314–324, 2020.
https://doi.org/10.1007/978-3-030-43192-1_36

Section 2 portrays the attempts done on similar concepts. Section 3 presents an algorithm with details about the computation of the baseline profile metric and the calculation of abnormality distance using multivariate statistical analysis for the detection of network attacks. Section 4 provides the implementation of the approach proposed in the paper along with the configuration details. Section 5 presents proof of the concept of the suggested method based on experimental results and demonstrates the effectiveness of the approach for detecting a wide range of network attacks. Section 6 presents the conclusion and discussion of the possible extension of this research.

2   Related Work

In previous papers, there were many flooding attack detection algorithms that had been researched and worked upon. In a recent paper [12] Chinese Remainder Theorem based Reversible Sketch Algorithm is proposed which overcomes the drawback of using sketch data structures. Sketch data structures are used for compressing and fusing network traffic. As hash functions are an irreversible reconstruction of keys that exhibit abnormal behavior is difficult. Using CRT-RS to generate traffic records the paper proposed a Modified Multi-chart Cumulative Sum (MM-CUSUM) algorithm to detect Distributed DoS flooding attacks which are protocol independent. In another approach [13] unmodified commercial SDN switches were used to detect DDOS attacks via adaptive correlation analysis. It detects attacks by identifying attack features in suspicious flows and locates the attackers or victims. In this paper [14] intrusion factors for a DDOS attack are analyzed and data is collected using Honeynet System, this data is used to forecast the DDOS attack using the statistical approaches of correlation and regression analysis [16]. Another paper [15] proposed a statistical measure called Feature Feature Score (FFS) for multivariate data analysis to detect anomaly in network behavior. Three network features namely packet rate, a variation of source IP's and entropy of source IP's are used to detect DDOS attacks [3]. The authors of [17–19] proposed two anomaly detection algorithms to detect SYN flood attacks. The first, Cumulative Sum Algorithm which uses Hypothesis Testing to detect the anomalous period. When the mean value of samples supersedes the threshold, change points are registered. Second, the approach in which violations of a threshold are detected by using the Adaptive Threshold Algorithm which is based on recent traffic measurements.

3   Abnormality Analysis Methodology

In a network, many protocols and services work simultaneously to provide functionality and services. For every node, each protocol is analyzed and the measuring attributes are assumed to follow the multivariate normal distribution. We use Wireshark for capturing network packets and Resource Monitor for CPU and Memory usage of the host machine. Using this data we computed the Sample mean matrix and inverse covariance matrix for every attack. This process is involved in the creation of the baseline profile metric. Now for every packet received we will find its variation from the baseline profile using abnormality distance as the metric. This value should lie within the normal region center of the baseline profile for the normal behavior of the network. For a better understanding of the algorithm, Fig. 1 can be referred to.

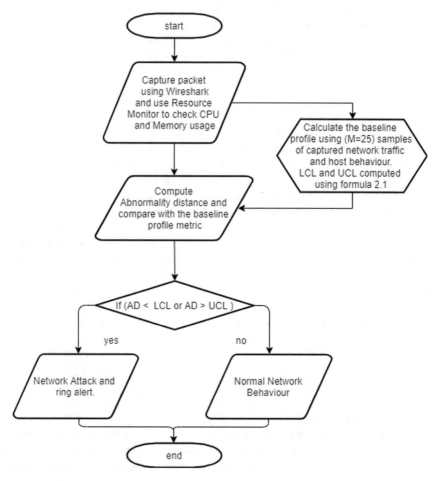

Fig. 1. Flowchart for the algorithm used for detecting flooding attacks.

3.1   Multivariate Statistical Analysis

Multivariate statistics is a very desirable approach as researchers hypothesize that a given outcome can be influenced by more than one factor. We have used multivariate analysis as we need multiple measuring attributes to accurately define the operation state of a network (Table 1).

Table 1. Measurement attributes

Impacted protocols	Measurement attributes	Observed behaviours
Application Layer HTTP, DNS, SMTP, POP	Number of incoming Packets (NIP) Number of outgoing packets (NOP) Invocation rate (IR)	IR increases 2 or 3 in order of magnitude
Transport Layer TCP, UDP	NIP NOP	NIP/NOP increases 2 or 3 in order of magnitude
Network Layer IP, ICMP, ARP	NIP NOP ARP request rate (AR)	AR increases 2 or 3 in order of magnitude

3.2 Hotelling's T^2 Method

The Hotelling's T^2 control chart is used for anomaly detection and performs root cause analysis based on multiple measurement attributes to generate a normal region profile. Firstly the baseline profile needs to be determined. According to the aforementioned method, we need M (M > 20) observations. If we have P measurement attributes then the normal behavior can be expressed as:

$$MA = \left\{ \begin{pmatrix} MA_1(t) \\ MA_2(t) \\ \vdots \\ MA_P(t) \end{pmatrix} \begin{pmatrix} MA_1(t+1) \\ MA_2(t+1) \\ \vdots \\ MA_P(t+1) \end{pmatrix} \cdots \begin{pmatrix} MA_1(t+M) \\ MA_2(t+M) \\ \vdots \\ MA_P(t+M) \end{pmatrix} \right\}$$

On the basis of these normal measurement attributes two control limits upper and lower can be determined using these M preliminary blocks for obtaining in-control data to estimate the sample mean MA and covariance matrix S. A normal region is determined with respect to the P measurement attributes. The normal region center (NRC) is determined from the sample mean MA whereas the shape of the normal region is determined by the sample covariance matrix S. For a pre-defined Type-I error α, the lower and upper control limits are computed:

$$LCL = 0$$
$$UCL = [(n-1)p \div (n-p)]F_{p,n-p}(\alpha) \tag{2.1}$$

3.3 Mean Vector and Covariance Matrix

The first step in the analysis of multivariate data is to compute the mean vector and the variance-covariance matrix. The formula for the computation of the covariance of the variables X and Y is as follows:

$$COV = \frac{\sum_{i=1}^{n}(X_i - \bar{x})(Y_i - \bar{y})}{n-1} \tag{2.2}$$

where $\bar{x}$ and $\bar{y}$ denote the means of X and Y respectively. The mean vector is called the centroid and the variance-covariance matrix is called the dispersion or dispersion matrix.

3.4 Attack Detection Approach Using Abnormality Distance Metric

Abnormality distance metric (AD) has been used on multiple measurement attributes to accurately quantify the network system state or network attack impacts on a network system and its various components. It will improve our accuracy and precisely define the operation state of the network. AD metric is computed as the distance between the operation state of the network node from the NRC at time t. The AD metric with respect to multiple measurement attributes is defined as a function of time t as shown below. The Hotelling's T^2 control chart is used for anomaly detection and performs root cause analysis based on multiple measurement attributes to generate a normal region profile. Firstly the baseline profile needs to be determined. According to the aforementioned method, we need M (M > 20) observations. If we have P measurement attributes then the normal behavior can be expressed as:

$$AD_{MA}(t) = \left(MA - \overline{MA}\right)^t S^{-1}\left(MA - \overline{MA}\right) \tag{2.3}$$

Algorithm Used

```
Step1: For (i=1 to M) do //i=No. of Normal Sample
Step 2: Input (MA₁ (i), MA₂ (i),....MAₚ(i))
            //P=No. of Measuring Attributes
      End for
Step 3: Calculate the Baseline Profile
Step 4: Compute LCL and UCL.
Step 5: Input (MA₁, MA₂, ....MAₚ) //for attacked profile
Step 6: Calculate AD.
Step 7: If (AD < LCL or AD > UCL), Then
          Identify as Network Attack and ring Alert
      Else
            Detect as Normal Behavior
Step 8 End if
```

4 Implementation

4.1 Configuration

We use two host machines for our experiment with Windows 7 installed in them as it is more vulnerable to flooding attacks (Tables 2 and 3).

We use Wireshark along with WinPcap for packet sniffing on the target machine [11]. We can use the resource monitor on the target machine to monitor the required parameters such as CPU utilization, memory usage, etc. during normal and abnormal traffic. The local wired network configuration used to connect the two hosts is given in the figure below (Fig. 2).

Table 2. The attacking machine host A

OS name	Ubuntu
Version	10.04 (lucid)
Processor	Intel® Core™ 2 DUO
Processor speed	1.80 GHz
Physical memory	2.9 GB

Table 3. The target machine host B

OS name	Microsoft Windows 7 Ultimate
Version	Service Pack 1
Processor	Intel core 2 duo
Processor speed	2.20 GHz
Physical memory	4 GB

Network configuration:

Interface (Host A): eth0

Interface (Host B): Realtek RTL8168D/8111D PCI-E Gigabit Ethernet

IP Configuration (static):

Host A: 192.168.66.73

Host B: 192.168.66.108

DNS Server: 192.168.1.1

Subnet mask: 255.255.255.0

Fig. 2. A local wired network configuration.

4.2 Generating and Sending Flooding Attack Packets

We use C source codes of the attacks running on gcc compiler for generating UDP, TCP-SYN and ICMP Smurf attacks packets [6]. The codes used can be run on any Linux machine. As our attacking machine has Ubuntu OS, we used it for generating the flood packets for carrying out the desired attack on the destination IP.

4.3 Capturing the Attack Packet

- *Step 1:* To ensure that capturing packets is allowed in the network.
- *Step 2:* Setting up the target machine's configuration for capturing.
- *Step 3:* To ensure using the right interface that local traffic can be captured.
- *Step 4:* Switch promiscuous mode on in target machine to capture traffic of other machines.

4.4 Testing the Attack

We developed a small GUI based Network Attack Analyzer application using Java [4] programming language and Netbeans IDE for implementing and testing the algorithm. An overview of the GUI is given in Fig. 3.

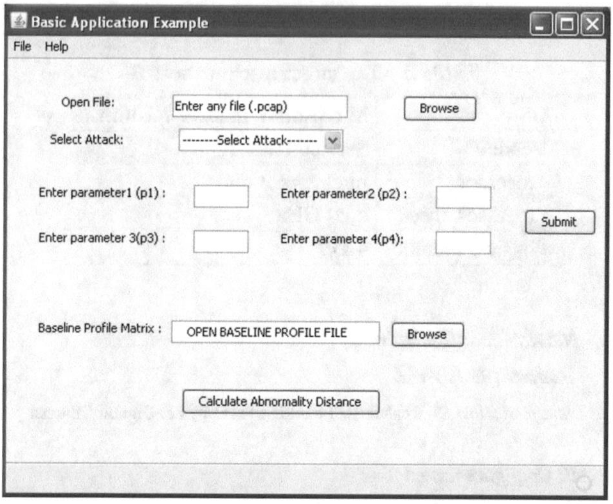

Fig. 3. An overview of GUI application for detecting network attacks.

We select the captured packet to be checked and use the predetermined LCL and UCL values using the baseline profile metric entered by the user. Using the baseline profile metric and the metric made from the newly captured Wireshark packet, we determine the abnormality distance (AD) of the new metric for the desired attack detection. If the value comes out to be beyond the control limits (LCL and UCL) then an alert message prompts indicating an attack. The code is tested for the UDP, TCP-SYN and the ICMP Smurf flooding attacks. Even though the three protocols have a different approach in its implementation, they utilize the same resources and disrupt the services in a similar manner [2].

5   Results and Discussion

We set up an experimental model by using two systems one which is the attacker (having Linux OS) and another is the host (having Windows OS), connected to the LAN, resource monitor (for monitoring host behavior) and network packet capturing tool Wireshark (for monitoring incoming packet behavior). Firstly under the normal mode of operation host behavior and network behavior are captured for making the baseline profile, then under the attacked condition, the same attributes are measured. For calculating the baseline profile for the attack we first take 25 (M = 25) samples of captured network traffic, each data is of 20 s and also monitored the host behavior

(CPU utilization, Memory uses) for these 20 s. The upper control limit is computed using predefined type-I error with the value of α taken to be 5%. We used four parameters (P = 4) for multivariate analysis:-

$$p_1 = \text{CPU usages} \qquad p_2 = \text{Memory usages}$$
$$p_3 = \text{Total Packets} \qquad p_4 = \text{Total TCP packets}$$

Various protocols are used on the internet. In this paper, we have used the aforementioned method to detect flooding attacks on TCP UDP and ICMP protocols. The simulation results of the above method based on the experimental data are analyzed and the results are drawn.

5.1   TCP-SYN Attack Results

From a given sample set, we calculate the mean matrix and the inverse of the covariance matrix to calculate the baseline profile in case of a TCP-SYN attack [10].

$$(\overline{MA}) = \begin{bmatrix} 83.52 & 54.68 & 997.68 & 834.6 \end{bmatrix}$$

$$S^{-1} = \begin{bmatrix} 0.006418278 & -0.001063803 & -0.000279508 & 0.000233366 \\ -0.001063803 & 0.006561646 & 0.000282863 & -0.000294525 \\ -0.000279508 & 0.000282863 & 0.000146855 & -0.000141159 \\ 0.000233366 & -0.000294525 & -0.000141159 & 0.00013765 \end{bmatrix}$$

Using the data computed above we calculate the LCL = 0, UCL = 12.9828, and AD = 537776.049. To verify the correctness of the detection method and the limits calculated we take normal samples of 20 s, extract the four measuring attributes and calculate the abnormality distance.

Normal Packet:

$$(MA) = \begin{bmatrix} 80 \\ 44 \\ 132 \\ 8 \end{bmatrix} \quad AD = 2.41888303$$

The AD is in between LCL and UCL, indicates the normal behavior of system i.e. it is not an attack [7].

5.2   UDP Attack Results

From a given sample set, we calculate the mean matrix and the inverse of the covariance matrix to calculate the baseline profile in case of a UDP attack [5].

$$(\overline{MA}) = [\,83.52 \quad 54.68 \quad 997.68 \quad 130.8\,]$$

$$S^{-1} = \begin{bmatrix} 0.006420483 & -0.001065809 & -4.73958E{-}05 & -0.000312488 \\ -0.001065809 & 0.006563198 & -1.00941E{-}05 & 0.000393773 \\ -4.73958E{-}05 & -1.00941E{-}05 & 2.22839E{-}06 & 5.65676E{-}06 \\ -0.000312488 & 0.000393773 & 5.65676E{-}06 & 0.000245445 \end{bmatrix}$$

Using the data computed above we calculate the LCL = 0, UCL = 12.9828, and AD = 29.2614653. To verify the correctness of the detection method and the limits calculated we take normal samples of 20 s, extract the four measuring attributes and calculate the abnormality distance.

Normal Packet:

$$(MA) = \begin{bmatrix} 78 \\ 52 \\ 184 \\ 129 \end{bmatrix}$$

$$AD = 1.18250118$$

The AD is in between LCL and UCL, indicates the normal behavior of system i.e. it is not an attack.

The Hotelling's T2 control chart is used for anomaly detection and performs root cause analysis based on multiple measurement attributes to generate a normal region profile. Firstly the baseline profile needs to be determined

5.3   ICMP Attack Results

From a given sample set, we calculate the mean matrix and the inverse of the covariance matrix to calculate the baseline profile in case of an ICMP attack [9].

$$(\overline{MA}) = [\,83.52 \quad 54.68 \quad 997.68 \quad 13.88\,]$$

$$S^{-1} = \begin{bmatrix} 0.006033837 & -0.00060735 & -4.00721E{-}05 & -0.000294814 \\ -0.00060735 & 0.006095603 & -1.96356E{-}05 & 0.001128738 \\ -4.00721E{-}05 & -1.96356E{-}05 & 2.09934E{-}06 & -3.20671E{-}06 \\ -0.000294814 & 0.001128738 & -3.20671E{-}06 & 0.007761848 \end{bmatrix}$$

Using the data computed above we calculate the LCL = 0, UCL = 12.9828, and AD = 24793.9572. To verify the correctness of the detection method and the limits calculated we take normal samples of 20 s, extract the four measuring attributes and calculate the abnormality distance.

Normal Packet:

$$(MA) = \begin{bmatrix} 83 \\ 72 \\ 172 \\ 22 \end{bmatrix}$$

AD = 4.47936729

The AD is in between LCL and UCL, indicates the normal behavior of system i.e. it is not an attack.

6 Conclusion and Future Work

Denials of service attacks are a primary concern in internet security, many approaches are proposed to counter them. In this paper, we used the Hotelling T^2 method, a multivariate statistical tool to detect DOS attacks. The study introduced the advantage of multivariate analysis, the Hotelling T^2 method, the use of network packet capturing tool Wireshark and resource monitor. The results obtained on using a threshold value for the detection of abnormal behavior are also discussed in the study. If the calculated abnormality distance value is in between LCL and UCL it is considered as a normal network behavior and if the AD value exceeds UCL it is considered as an attacked network condition. Future work includes the implementation and testing of this model in real-time where network packets are collected and analyzed for attacks. It might include using some other network attributes and the method can also be tested to detect an increased number of network attacks other than Denial of Service Attacks.

References

1. Forouzan, B.A.: Data Communications and Networking (McGraw-Hill Forouzan Networking). McGraw-Hill Higher Education (2007)
2. Biswas, A.: Impact Analysis of System and Network Attacks (2008). All Graduate Theses and Dissertations. Paper 199. http://digitalcommons.usu.edu/etd/199
3. Li, M., Chi, C.H., Jia, W., Zhao, W., Zhou, W., Cao, J., Long, D., Meng, Q.: Decision analysis of statistically detecting distributed denial-of-service flooding attacks. Int. J. Inf. Technol. Decis. Making **2**(3), 397–405 (2003)
4. Schildt, H.: The Complete Reference Java 2, 5th edn (2002)
5. Xiaoming, L., Sejdini, V., Chowdhury, H.: Denial of Service (DoS) attack with UDP Flood (2010)
6. Ligh, M., Adair, S., Hartstein, B., Richard, M.: Malware analyst's cookbook and DVD: tools and techniques for fighting malicious code (2011)
7. Alam, M.O., Adnan, A., Aktaruzzaman, A.K.M.: TCP SYN Flood DoS Attack Experiments in Wireless Networks (2007)
8. Noureldien, N.A.: Protecting web servers from DoS/DDoS flooding attacks. A technical overview (2002)
9. Antoniou, S.: The ping of death and other dos network attacks, 14 May 2009

10. Eddy, W.M., Verizon Federal Network Systems: Defenses against TCP SYN flooding attacks. Internet Protocol J. **9**(4), 2–16 (2006)
11. Wireshark User's Guide (2008)
12. Jing, X., Yan, Z., Jiang, X., Pedrycz, W.: Network traffic fusion and analysis against DDoS flooding attacks with a novel reversible sketch. In: Information Fusion 2019 (2019)
13. Zheng, J., Li, Q., Gu, G., Cao, J., Yau, D.K., Wu, J.: Realtime DDoS defense using COTS SDN switches via adaptive correlation analysis. IEEE Trans. Inf. Forensics Secur. **13** (7), 1838–1853 (2018)
14. Kwon, D., Kim, H., An, D., Ju, H.: DDoS attack volume forecasting using a statistical approach. In: 2017 IFIP/IEEE Symposium on Integrated Network and Service Management (IM) (2017)
15. Hoque, N., Bhattacharyya, D.K., Kalita, J.K.: A novel measure for low-rate and high-rate DDoS attack detection using multivariate data analysis. In: 2016 8th International Conference on Communication Systems and Networks (COMSNETS) (2016)
16. Tan, Z., Jamdagni, A., He, X., Nanda, P., Liu, R.P.: A system for denial-of-service attack detection based on multivariate correlation analysis. IEEE Trans. Parallel Distrib. Syst. **25**, 447–456 (2014)
17. Bogdanoski, M., Shuminoski, T., Risteski, A.: Analysis of the SYN Flood DoS Attack (2013)
18. Tan, Z., Jamdagni, A., He, X., Nanda, P., Liu, R.P.: Denial-of-service attack detection based on multivariate correlation analysis. In: Lu, B.L., Zhang, L., Kwok, J. (eds.) Neural Information Processing, ICONIP 2011. LNCS, vol. 7064. Springer, Heidelberg (2011)
19. Praveena, A., Smys, S.: Anonymization in social networks: a survey on the issues of data privacy in social network sites. J. Int. J. Eng. Comput. Sci. **5**(3), 15912–15918 (2016)

Human Gait Recognition

(Analysis and Classification)

Vaishnavi Ahuja$^{(\boxtimes)}$ and Rejo Mathew

Department of IT, Mukesh Patel School of Technology Management
and Engineering, NMIMS (deemed-to-be) University, Mumbai, India
Vaishnavi.ahuja05@nmims.edu.in, Rejo.Mathew@nmims.edu

Abstract. Gait State Analysis refers to the way in which an individual walks. With every individual having a unique walking pattern, it can be used as a biometric in many domains like sports, medicine, and many more. The gait state analysis can be used as a biometric without any contact with the individual, making it a very versatile methods for many purposes. Gait recognition has two major steps, first the data is to be collected and stored, then from the parameters calculated using that data, mechanisms to identify an individual are applied. The first step can be done by using two methods: Wearable Sensors and Non-Wearable Sensors. Types of Wearable sensors explained are: Floor Sensors and Image Processing Sensors; Types of Non-Wearable sensors explained are: Force and pressure sensors, Sensing Fabric and Electromagnetic Tracking System. Once the data is collected and stored, it can be used for the recognition of an individual by means of ANN (Artificial Neural Network Deep Learning), Inductive Machine Learning, and Fuzzy Logic mechanisms.

Keywords: Gait state analysis · Wearable Sensors · Non-Wearable Sensors · ANN · Inductive Machine Learning · Fuzzy Logic

1 Introduction

Human Gait refers to an individual's walking pattern, which stays unique for every individual. Hence, it can be used as a biometric. With the help of Gait State Analysis, one can identify an individual from a distance and without any contact with the individual. It has been significantly used in: Sports, security purposes and medicine. In sports, it can be used for the identification of the forces which were exerted on the muscle with the help of EMG. In case of security, with the help of this technique, it is now possible to identify the culprit with the help of the person's unique gait features (with the help of non-wearable sensors). It can be used for the diagnosis and treatment of Parkinson's disease, multiple sclerosis, and other laboratory techniques. There are two methods used for analyzing the gait stages: Wearable sensors and Non Wearable Sensors.

In case of Non-Wearable Sensors (NWS), the gait analysis is done under controlled research facilities [1] and the sensors used are located in the surroundings such as on the pre-defined walkway(i.e. floor sensors), or mounted on a wall. In case of Wearable

© Springer Nature Switzerland AG 2020
A. P. Pandian et al. (Eds.): ICCBI 2019, LNDECT 49, pp. 325–332, 2020.
https://doi.org/10.1007/978-3-030-43192-1_37

Sensors (WS), the gait analysis is done with the help of sensors which are mounted on the individual whose gait state analysis is being performed.

Once the data is obtained with the help of sensors, the process of converting it into a format that is understandable is necessary. Hence, with the help of techniques like ANN, Inductive Learning Technique, Fuzzy Logic, it is possible to compute and interpret the collected data accurately. Section 2 focuses on wearable and non-wearable sensors. Section 3 focuses on the interpretation done with the help of the gait parameters, followed by Analysis, Conclusion and Future Work.

2   Techniques Used for Obtaining Gait Parameters

Different methods are used for the analysis of different gait parameters. With the help of these methods, one can find out different parameters which affect the human gait analysis like velocity, step length, joint angle, as explained [1]. The gait stages can be analyzed by two methods: With the help of wearable sensors; and with the help of Non-Wearable Sensors (Fig. 1).

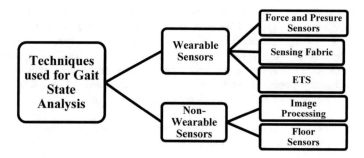

Fig. 1.   Gait state analysis methods

2.1   Non-Wearable Sensors

2.1.1   Image Processing
In the case of NWS, with methods like image processing and floor sensors, an individual's gait state analysis can be done. This is the case of Image Processing.

2.1.1.1 Image Pre-processing
With the help of sensors, the picture is captured (i.e., the data is captured), in the next stage, the image pre-processing and Human Modeling is done to enhance the feature of the human 2D model, here, the image is first passed through a body silhouette where the background segmentation is done, and for further processing erosion and dilation are performed.

2.1.1.2 Human Modeling
Human Modeling is used to get a sharper 2D image of the individual. It can be done by two methods: with the help of Model Based approach; and Non-Model based approach as explained in [4]. In case of model based approach, the basic features like color of the

individual, height, are taken into consideration. Whereas in case of Non-Model based approach, the calculations depend on an already existing model for its comparison. For the purpose of gait analysis, the non-model based approached is used.

2.1.1.3 Feature Extraction

In the next stage, Feature Extraction is done. A skeleton figure is used in compassion with the body silhouette prepared in the earlier stages. With the head of the individual as origin, parameters that affect the gait state are calculated.

2.1.2   Floor Sensors

In this type of system, a walkway is developed for the gait analysis of an individual. The gait features are taken in account with the help of Pressure and Force Sensors along with moment transducers which are placed in the walkway. They are of 2 types: pressure measurement and force measurement systems. Pressure measurement systems quantify the value of pressure exerted under the foot, but isn't capable of quantifying different components of the force exerted by the foot [13].

The following shows the two kinds of NWS:

Image processing	Floor sensors
It is done by performing Image Pre-processing, Human Modelling and Feature Extraction	Floor Sensors are the Force and Pressure sensors encapsulated walkways, which calculate force/pressure of an individual walks on them, thereby calculating the individual's gait features

2.2   Wearable Sensors

With the help of Wearable Sensors, it is possible to obtain the gait parameters of an individual in the natural surroundings. They are mounted on an individual and do not require a proper set-up (like a laboratory).

2.2.1   Pressure and Force Sensors

The Pressure and Force Sensors make it possible to capture the Gait features of the individual with the help of sensors placed in the shoes of the individual. These force sensors calculate the GRF (Ground Reaction Force) under the foot, and then they send proportional currents and voltages. One of the most important things while using this technology to calculate the gait features is to take into account that this method doesn't take the components of the force applied on its entire axis.

There are three types of commonly used Pressure and Force Sensors: Capacitive Sensors, Resistive Sensors, and piezoelectric sensors. In case of Resistive Sensors, the electric resistance is inversely proportional to the weight applied on the sensors, i.e., as the weight increases, the electric resistivity decreases and vice-a-versa. In case of Piezoelectric Sensors, three deformation metals are placed in orthogonal position over silicone gel. Whenever pressure is applied on it, the silicone gel gets deformed, and the metals calculate the deformation caused; and hence with the help of factors of gel characteristics and metal deformation one can calculate the total pressure applied. The Capacitive sensors are those sensors which work on the principle that, with change in parameters like distance between two electrodes, the capacity of the condenser varies.

With the help of these types of sensors, many studies on 'wearable gait analysis systems' are performed. They are integrated into human shoes, or into certain insoles [6]. In [7], the GRF is calculated with the help of twelve wearable sensors, integrated into the insole if an individual.

Thus, a data set containing the position of gait stages of an individual can be obtained with by calculating of the induced voltages that are generated as the individual moves [3].

The following shows the three kinds of WS discussed:

Force and pressure sensors	Sensing fabric	ETS
Sensors are placed inside shoes, which helps in fetching data by calculating GRF and sends proportional vales of voltage and current	These are the kind of composite fabrics which have sensing technologies for the calculation of gait data	They work on the Faraday's Law of Induction. A transmitter and receiver work together to give values for the gait states of an individual

3   Detection of Gait Events

3.1   ANN Classifier

The collected data from different types of sensors is passed to the ANN classifier (Artificial Neural Network). In the ANN classifier, there are two possibilities. Firstly, if the individual's Gait features were to be added to the database for future recognition; and secondly, if the person's identity have to be revealed with the help of the data stored in the database [2]. It has an input layer, an output layer and hidden layers. The hidden layer iteratively weights to map the inputs to the output. The ANN has majorly two parts, the first one is Training Data Set Collector, and the second is Testing Data Analyzer.

3.1.1   Training Data Collector

In the first case, once the input data is collected and stored, then fed to the training data collector of the ANN Classifier. The data is collected and the final resultant is stored in the database for future usage.

3.1.2   Training Data Analyzer

3.1.2.1 Data Recognition

In the second case, the process of recognition is to be performed. The picture is captured, body silhouette is made. Then the process of clustering is done. Clustering of data refers to grouping together similar data in the same class, which is done with the help of K-means clustering. (In K-means, a cluster is formed according to the related data surrounding it. It is a method of vector quantization, which is popular for cluster analysis due to its versatile nature).

3.1.2.2 K-means Clustering

With the help of K-means clustering, the body silhouette created is matched to the existing data base (in the hidden layer), clustering is used to check if the body silhouette matches any existing model in the database; and the process of recognition is done (Fig. 2).

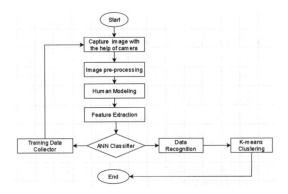

Fig. 2. ANN Classifier and Image Processing using wearable sensors.

Figure 3 represents the flow of processes in the case of Gait State analysis with the help of Non-Wearable Sensors (Image Processing).

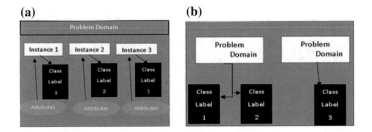

Fig. 3. Inductive learning techniques

3.2   Inductive Learning Technique (Decision Tree)

In the case of inductive learning technique, a set of rules (also known as the decision tree) is defined to determine the accurate position of the gait features, with the minimal use of sensors. Inductive inference defines the process of deriving general laws from particular instances; whereas Inductive learning is the process of searching for the general laws that are derived from the particular instances [12]. For inductive learning, every instance is defined with attributes (i.e. dimension variation of an instance) within a Problem Domain, and every attribute is associated with a class label. Class labels are subsets of the problem domains, and the domain is further divided into a number of such classes. A decision tree is then made in accordance with the rule based upon the attributes of every instance.

The decision rule puts each description of an instance to where it belongs. A Training Set is developed to gather informative data, and a Data analyzer to analyze the data further. The decision tree continues to grow until the last node of the tree has members under only a single class label.

3.3   Fuzzy Logic

The Fuzzy logic works on the degree of correctness, and not on the Boolean Logic, i.e. 0 s and 1 s. It takes input as variables which lie between 0 and 1. On the basis of extracted gait features we can identify the person using this method [14]. The purpose of using this method is to map the inputs (lying between 0 and 1) using a set of input membership functions (a membership function represents the input data elements 'x' in a set A) They are then used to define the fuzzy concepts into categories like 'very small', 'small', 'average', 'large' and 'very large'. Maximum aggregation technique is chosen to combine various fuzzy sets into a single set. Then by the Mean of maximum method, the fuzzy controller chooses the typical value of the consequent term of the most valid rule as the crisp output value.

4   Analysis and Comparison

Comparison Tables 1 and 2 are the two comparison tables. Comparison Table 1 depicts the comparison between Wearable and Non-Wearable Sensors.

Table 1. Wearable and Non-Wearable Sensors

Features	Wearable Sensors	Non-Wearable Sensors
Placement of sensors	Mounted on the individual	Placed in cameras or floor, etc.
Need for controlled environment	No need, as can be done in natural surroundings	Required, as the scope is limited to the laboratory
Power limitations	Yes, Due to limited battery life of the sensors	No, As done inside the laboratory
Cost	Cost effective, Due to bulk manufacturing	Expensive, As the setup is to be organized in a laboratory
Usage	Extensive, Helps in taking down everyday activities of the individual	Limited, As it has to be done inside a Laboratory, on fixed walkways where the sensors could capture data
Complexity	Complex algorithms	Complex analysis
External interference	Susceptible to external factors such as noise	Less susceptible, as the process is done in a controlled environment

Table 2. ANN, Inductive Learning Technique, Fuzzy Logic

Features	Artificial Neural Network [2]	Inductive Learning Technique [12]	Fuzzy Logic [14]
Methodology used	K-means clustering	Decision tree	Membership functions, maximum aggregation & mean of maximum methods
Accuracy	Very High	High	Improved
Training Set Generation	Yes	Yes	No
Time efficient	Less	Medium	High
Usability	Complex	Less Complex	Easiest to implement
Number of layers	Multiple-Layer (Input, Hidden, Output Layers)	Single layer (Possible with multiple layers)	Three layers (Fuzzification (input), Rule Matrix, Defuzzification (output))

Once the gait parameters are collected, they are to be quantized and compared for gait recognition, which is done by ANN, Inductive Learning Technique and Fuzzy logic Technique. The comparison Table 2 shows how these methods differ from one another.

5   Conclusion and Future Work

Initially when Gait Recognition was initiated, Non-Wearable Sensors were widely used for gait recognition. With the advancement of technology, over the years wearable sensors have proved out to be more useful and less expensive as compared to Non-Wearable Sensors. In case of NWS, as the studies are performed in the premises of the laboratory, no external factors affect the outcomes of the gait analysis, but the major drawback is that as they are performed within the laboratory, it only gives its conclusions for a short interval which might differ in everyday life. With the help of WS, we have an added advantage of using the sensors during the user's everyday activities. But the disadvantages are that as the sensors are to be placed on the body of the individual, they can be uncomfortable or even intrusive [2, 8]. For gait recognition, methods like ANN, Inductive Machine Learning and Fuzzy logic can be used. With ANN giving the highest accuracy, it adds to the complexity of its algorithm and implementation.

Future work includes development of new experimental techniques that can perform gait recognition in real-time motion measurements. Techniques to visualize gait data intuitively are yet to be introduced. With the advancement of technology, it will be required to enhance the ability of the systems for faster identification of critical data set, proceeded by faster interpretation, prediction and quantization of gait data. Another advancement can be allowing one system to do gait recognition of multiple users at one instance. This will enhance the efficiency and productivity of the system.

References

1. Muro-de-la-Herran, A., et al.: Gait analysis methods: an overview of wearable and non-wearable systems, highlighting clinical applications. Sensors **14**(2), 3362–3394 (2014)
2. Sharma, T., Dub, S., Gupta, B.: Performance analysis of ANN based gait recognition. Int. J. Sci. Res. Sci. Technol. 112–125 (2018)
3. Tao, W., et al.: Gait analysis using wearable sensors. Sensors **12**(2), 2255–2283 (2012)
4. Kong, W., Saad, M.H., Hannan, M.A., Hussain, A.: Human gait state classification using artificial neural network. In: 2014 IEEE Symposium on Computational Intelligence for Multimedia, Signal and Vision Processing (CIMSIVP), Orlando, FL, pp. 1–5 (2014)
5. Castermans, T., et al.: Towards effective non-invasive brain-computer interfaces dedicated to gait rehabilitation systems. Brain Sci. **4**(1), 1–48 (2013)
6. Forner-Cordero, A., Koopman, H., van der Helm, F.: Use of pressure insoles to calculate the complete ground reaction forces. J. Biomech. **37**, 1427–1432 (2003). https://doi.org/10. 1016/j.jbiomech
7. Howell, A.M., Kobayashi, T., Hayes, H.A., Foreman, K.B., Bamberg, S.J.M.: Kinetic gait analysis using a low-cost insole. IEEE Trans. Biomed. Eng. **60**(12), 3284–3290 (2013)
8. Tarnita, D.: Wearable sensors used for human gait analysis. Rom. J. Morphol. Embryol. **57**, 373–382 (2016)
9. Sawhney, A., Agrawal, A., Patra, P., Calvert, P.: Piezoresistive sensors on textiles byinkjet printing and electroless plating. In: Proceedings of Material Research Society Symposium, Boston, MA, USA, vol. 920, pp. 103–111 (2006)
10. Kim, K.J., Chang, Y.M., Yoon, S.K., Hyun, J.: A novel piezoelectric PVDF film-based Physiological sensing belt for a complementary respiration and heartbeat monitoring system. Integr. Ferroelectr. **107**, 53–68 (2009)
11. Taya, M., Kim, W.J., Ono, K.: Piezoresistivity of a short fiber/elastomer matrix composite. Mech. Mater. **28**, 53–59 (1998)
12. Kirkwood, C.A., Andrews, B.J., Mowforth, P.: Automatic detection of gait events: a case study using inductive learning techniques. J. Biomed. Eng. **11**(6), 511–516 (1989)
13. Rowe, P.J., Nicol, A.C., Kelly, I.G.: Flexible goniometer computer system for the assessment of hip function. Clin. Biomech. **4**, 68–72 (1989)
14. Narula, P., Srivastava, S., Chawla, A., Singh, S.: Human gait recognition using fuzzy logic. **412**, 277–287 (2015)

Reduction of Traffic on Roads Using Big Data Applications

P. Ajitha[1(⊠)], A. Sivasangari[1], G. Lakshmi Mounica[2],
and Lakshmi Prathyusha[2]

[1] Department of Information and Technology,
Sathyabama Institute of Science and Technology, Chennai, India
ajitha.it@sathyabama.ac.in
[2] Computer Science and Engineering,
Sathyabama Institute of Science and Technology, Chennai, India
mounicagovardhanagiri@gmail.com,
prathyusha.chelemella@gmail.com

Abstract. Mapreduce implementation simplifies big amount of data implementation on complex and large datasets by using parallelization of map tasks and reduce tasks. Several contributions and methodologies are made to eradicate the functionality and usage of Mapreduce tasks, the data generated in the shuffle phase should be ignored, which contributes as an important feature in functionality development. Shuffle phase uses hash aggregator to merge information into clusters which is useless in managing traffic data creating a bottleneck. Improvisation of the execution of system traffic in the shuffle stage is fundamental to enhance the execution of the tasks. Main objective of slacking off the structured dataset is developed using utilisation tasks and total. The implemented course of action is known to limit and sort out traffic cost in Mapreduce. The hash aggregators are implemented in various tasks, where every aggregator can diminish overall data traffic from different tasks. This dispersed estimation is supposed to maintain the extensive data organizing problem for huge data implemented applications. Besides, an integrated computation is proposed to change the data and assemble successfully in a proper orientation. In this paper, a study of traffic cost for a Map reduce tasks by determining new intermediate data division schema is carried out. We consider data lifting and other applications that have all around attributes from past data that is collected bit-by-bit using Map reduce. For example, a count that has all the means of taking after a word checking at the key signal may wrap up having all around various attributes, for instance, the numbers and transitional results, in that limit, we recognize the startling execution. Grabbing the data from various applications, we review the detailed report for bolstering data focused and implement the new data applications on server ranch scale systems. The promising results may include improving a business control structure with a medley of connections and function that support in a Map reduce structure. Finally, the outcome exhibits that our manifesto can tremendously diminish traffic congestion beneath any conditions.

Keywords: Map reduce · Cluster · Aggregator

© Springer Nature Switzerland AG 2020
A. P. Pandian et al. (Eds.): ICCBI 2019, LNDECT 49, pp. 333–339, 2020.
https://doi.org/10.1007/978-3-030-43192-1_38

1 Introduction

Big Data is usually used with multiple data sources to analyze and process massive, integrated, and complex, through data [1–5]. The main challenge for the information resources is to find a technique and proper methodology that will be used to interpret enormous data extensions and extricate only the knowledge required for the time to come. Mapreduce is a broad technique execution system that refers to its simple model of programming and semantic execution. Map reduction and a lot of huge corporations like Yahoo hire Hadoop! Of example, Google and Twitter. Computation is implemented in Map Reduction by using two phases, called 'map' and 'reduce' within the data analysis section, where the data is organized and synchronized so that the necessary execution can be accomplished by implementing one method on few or no data components. Map reduction is known to cause data back to scale. Since each stage is prepared for large-scale storage and evaluation, it is understood that Map reduction results in data backup scale [6–10]. Since each stage is prepared for comprehensive processing and evaluation.

Map reduction tasks are accepted, it is easy to look at a Map reduction task by broad cluster scales could make up most of the enormous amount of computation costs. Once the Map reduction tasks feature is accepted, a Map reduction process that has three phases instead of combining other phases is easy to look at. In addition to the transfer of information, 'shuffle' is used to reintegrate the end result of the map step and then move it to the subtle cipher nodes to provide and execute the corresponding reduction operations. Although many efforts are made to help Map's reduced tasks in usability and reliability, they neglect the traffic decided in the shuffle process, which plays an important role in improving the functionality. In ancient times, a hash function is used to separate primary data between reduced tasks, which, however, is not efficient in traffic as a result of a nursing tendency associate that does not ruminate the configuration and the data size associated with each value. In this paper, we have inculcated a reduction in traffic costs by introducing an extremely distinctive primary data.

2 Ease of Use

Section 3 gives the literature review of past studies. Section 4 explains the system that is used to analyze data and store it in the filesystem. Table 1 gives the data attributes used in the query processing of Sqoop and Hive. Section 4 explains the system architecture and Fig. 1 shows the architecture structure. Figure 2 indicates the graph diagram where a list of CCTV footage status in various locations. We conclude the paper in Sect. 6.

3 Literature Survey

Data is processed as huge chunks used as crawled files, data request logs, etc., to execute various types of processed data, such as indices, a list of immediate queries used in a given day, etc. The PC file is usually large, however, and the incorporation of the execution processes tends to be split into hundreds or thousands of machines and thus finishes in less time. As a consequence, the current reliability of parallelization, fault tolerance, information distribution, etc. irregular and unstructured data specifics can be achieved by applying a map operation to each logical record output, giving key-value pairs [1]. Using trace analysis on some implementation of Mapreduce clusters, which display associates in nursing large portions of jobs in map tasks, simple queries from a random network can be answered. The computing clusters were sculpted as time-plotted computing servers, throughout which works consisting of map tasks with some stochastic methodology of each time plot [2].

An related equation is built for combined process programming and shuffling phases which provide an estimated result based on past studies in which it is presumed that the servers are allocated, i.e. the jobs are first assigned to the servers and then their programming is applied to handle the map sequences and reduce tasks in each job [3]. The existing system uses primarily hash-based functions which are typically used to divide all intermediate tasks, the projected systems deal with the distributed rule for handling massive scale problems and therefore the security of sensitive information is provided, sensitive data are those where data sets from a user's GPS location and therefore data sets from various CCTVs are required.

Mapreduce provides a recent system for dealing with a significant amount of knowledge as a result of which they need a clear programming model and together they consist of automated management that is used for parallel executions [4]. Less practical usage enables the aggregation of intermediate data generated on a database using a combination process []. It has been shown that performance can be further improved by operation in the second stage of partial aggregation at a rack-level in sold-out clusters. The main reason is that the separating server's data connection is distributed equally between all alternative servers and the rack which contributes to a bottleneck [5]. Some of the data and compute intensive applications have a wide variety of features taken from the Mapreduce applications that have been previously recorded [9].

It is quite difficult to find a system that is semantically correct, to achieve good results [7]. For example, a computation that at first stage looks very closely related to the word count function may turn out to have very different features and results, with few of them being the number of variance of intermediate outcomes, giving an unexpected performance. Exploration of the design space to support computer-scale data center-intensive and compute-intensive applications [6].

4 Proposed System

We use a database of traffic record information and analyze road traffic using Hadoop software along with some Hadoop ecosystems such as HDFS, Sqoop, Hive, and Pig. We can process large chunks of data by using these devices, we can get high throughput, and it's an open access technology. A very large task is typically divided into smaller tasks and performed by Mapreduce tasks. To track daily and meanwhile road traffic crossing each stage of traffic issues that can be evaluated over several years and implemented in Transport Research Wing (TRW).

The data were carefully processed and transferred to MySQL backup. To get the data protected analyzed document, Mapreduce is implemented and processed. Using this method, it is estimated that traffic will indicate when a particular area will be high and low in a day. Prediction is made so that a pattern that can be used for further purposes is registered and reported. We use the Prediction algorithm to determine if the road is busy or idle and then implement the execution algorithm for Mapreduce. The acronyms and abbreviations are:

4.1 HDFS

Hadoop Distributed File System is a disseminated file system that has been developed to work on different applications. It uses architecture of the Name Node and Data Node. It's vaguely forgiving of blame. It is planned for lower price hardware implementation. It duplicates or copies multiple times that piece of data, putting at least one copy in the filesystem on a different database rack. This gives the system information a higher throughput.

4.2 Sqoop

Sqoop is a linking framework between Relational Databases and HDFS for data transmission. It takes up high quantities of a unit table or a free form request along with performed jobs that are executed multiple times to import changes to the information that have been made to a server since the last import.

4.3 Hive

Apache Hive is characterized as a software system providing data query for data warehouse. It offers a SQL query language that is used with visibility read schema and converts queries into tasks of map and reduction. It has a filesystem that is compatible.

4.4 Pig

It is a base of a high level for the development of programs implemented on Hadoop. It abstracts Java Map Reduce idioms programming. It has a special relational database and uses lazy validation process, extracting, transforming, and loading (ETL).

4.5 MySQL

It is a free open-source program that many database-driven web applications such as WordPress, phpBB, Drupal, etc. use.

The first step in the architecture is the Pre-processing of Road traffic data into the database. The data is analyzed with various kinds of fields and is later converted into comma delimited format known as CSV (comma separator value). The data stored in the files are moved and backed up through MySQL. The next step is to load the data in HDFS. The backup data that has been stored in MySQL is imported to HDFS by implementing sqoop commands. The data that is stored in the HDFS is now ready to

Table 1. Attributes used in query

Reference	Timing	Route 1	Route 2	Route 3
12I0740	1:25 Pm	Heavy traffic	Normal traffic	Light traffic
12I0834	6:02 Pm	Light traffic	Heavy traffic	Normal traffic
12J1142	11:18 Am	Normal traffic	Less traffic	Heavy traffic
12K1171	10:20 Am	Light traffic	Normal traffic	Heavy traffic
12M0800	9:30 Pm	Heavy traffic	Less traffic	Normal traffic

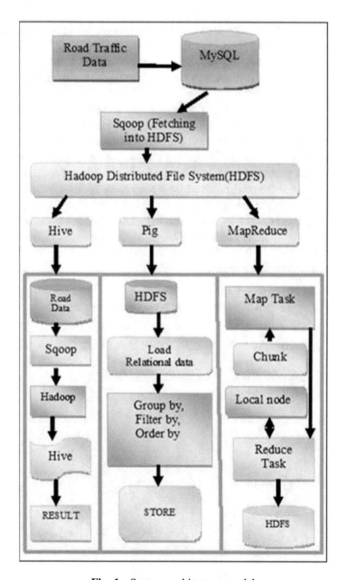

Fig. 1. System architecture model

get processed by Hive. The third step in the architecture model is analyzing query using Hive. Here, only the structured data obtained by cleansing is analyzed and the unstructured data is neglected. The fourth step is to script the data using Pig.

The Table 1 above provides data on the attributes used in the handling of queries. These attributes provide information on various data such as timing, route 1, etc. These attributes are used by implementing commands in query format in the sqoop to determine the area's condition.

We load the data into Hive afterwards. In Hive, we build tables and databases for easy file system processing and access to data. System architecture includes various system components and operates internally in the form of Java code, which works together to implement the large data handling system as a whole. The various components and their functions include storing server database into the file system and extracting information from the archive and matching data using Hive, Pig, and Mapreduce. The information is used to map and through stages, and the result is sent back to HDFS after encryption.

Pig is used for the execution of data plans and workflow. Using the Grunt Shell mode, the data is usually translated to map and tasks are reduced and the scripts are executed in the shell. Pig reads the data as a relationship except for loading and storing the data and generates another relationship as its output. A semantic test is performed as soon as a loaded statement is inserted in the grunt container. Dump operator is used to view the contents of the schema and Mapreduce tasks are later used to load the data into the filesystem. Embedded operators are used in this phase to support data operations such as grouping, sorting, and ordering. The resulting data is taken out for storage and is highly tolerant to fault. The fifth and final step is the programming process to be implemented. Mapreduce is a framework for analyzing and reducing the phase of large and complex shuffles.

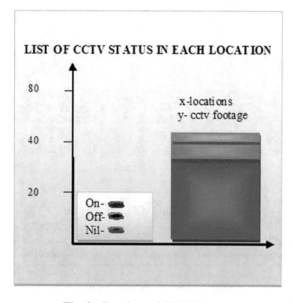

Fig. 2. Bar chart of CCTV footage

Map tasks implement user-specified information. User data is generally present in a file format or record in the filesystem. The user data file is transferred to a map function that processes the data line by line and produces a larger quantity of small chunks of data. This step is a shuffle collaboration and reduction phases. The Tasks Reduce analyzes the data received from the map function, creating a new record of the resulting data and being stored in the filesystem.

5 Conclusion and Future Enhancements

Using Hadoop, introduced the transport analysis wing that can identify issues behind the road traffic and it will also be helpful to find the reasons in advance by forecasting using the Big Data analytical report. We generate a detailed report that provides us with road traffic data and implementation of Sqoop and Hive applications, we will know which road is free at a particular time and for which area at that particular time was listed. We can also use Spark, an Apache software platform, a thousand times faster and more to implement and process the data. In this paper, we presented with a road traffic report by generating real-time data using Zenodo (a general-purpose open access repository which allows depositing data sets, reports, and any other research-related artifacts) that backs up two years of data, we can predict the traffic cost for particular areas and at a certain time and the road with high traffic can be avoided likewise to save time. This paper study is useful for prediction purposes which can be helpful to indicate and perform proper maintenance.

References

1. Deanand, J., Ghemawat, S.: MapReduce: simplified data processing on large clusters. Commun. ACM **51**(1), 107–113 (2008)
2. Wang, W., Zhu, K., Ying, L., Tan, J., Zhang, L.: Maptaskscheduling InMapReduce with data locality: throughput and heavy-traffic optimality. In: 2013 Proceedings of the IEEE INFOCOM, pp. 1609–1617. IEEE (2013)
3. Chen, F., Kodialam, M., Lakshman, T.: Joint scheduling of processing and shuffle phases in MapReduce systems. In: Proceedings of the IEEE INFOCOM, pp. 1143–1151. IEEE (2012)
4. Wang, Y., Wang, W., Ma, C., Meng, D.: Zput: a speedy data uploading approach for the distributed file system. In: 2013 IEEE International Conference on Cluster Computing (CLUSTER), pp. 1–5. IEEE (2013)
5. White, T.: Hadoop: The Definitive Guide: The Definitive Guide. O'Reilly Media, Inc., Sebastopol (2009)
6. Chen, S., Schlosser, S.W.: Map-reduce meets wider varieties of applications. Intel Research Pittsburgh, Technical report IRP-TR-08-05 (2008)
7. Saravana Kumar, P., Athigopal, M., Vetrivel, S.: Extract transform and load strategy for unstructured data into data warehouse using map reduce algorithm and big data analytics. IJIRCCE **02**, 7456–7462 (2014)
8. Sivasangari, A., Ajitha, P., Indira, K.: Air pollution monitoring and prediction using multi view hybrid model. Int. J. Eng. Adv. Technol. (IJEAT) **8**(2S), 1370–1372. ISSN: 2249–8958
9. Smys, S., Wee, H.-M., Joo, M.: Introduction to the Special Section on Inventive Systems and Smart Cities, pp. 32–33 (2018)
10. Brumancia, E., Samuel, S.J., Gomathi, R.M., Dhas, Y.M.: An effective study on data fusion models in wireless sensor networks. **13**(2) (2018)

Survey on Utilization of Internet of Things in Health Monitoring Systems

R. Jane Preetha Princy[1(✉)], B. Brenda Jennifer[1], M. J. Abiya[1],
B. Thilagavathi[1], Saravanan Parthasarathy[2],
and Arun Raj Lakshminarayanan[2]

[1] Karunya Institute of Technology and Sciences, Coimbatore, India
{janer,brendajennifer,abiyaj}@karunya.edu.in,
thilagavathib@karunya.edu
[2] B.S. Abdur Rahman Crescent Institute of Science and Technology,
Chennai, India
{saravanan_cse_2019,arunraj}@crescent.education

Abstract. The development of IoT has revolutionized the healthcare in a constructive way. It allows the Health providers to collect the data effectively,
automated the workflow, improved the accuracy of diagnosis and reduces the risk
of errors. The implementation of IoT in Health Monitoring Systems saves many
lives every day. In India, the healthcare industry is started using the IoT enabled
devices to track the health condition of elderly and diseased persons. This survey
explores the existing IoT enabled Health Monitoring Systems and their benefits.
Since the hypertension is identified as a threat to notable population in India, the
relationship between the blood pressure and lifestyle has to be navigated.
Specifically, the millennial human beings are sleeping very minimal duration of
time and their quality of sleep is also vulnerably low. These issues hurt the human
body's ability to regulate stress hormones. That leads to high blood pressure,
which paves the way to heart problems and other health risks. Self-monitoring the
sleeping pattern might help to resolve this issue. Nevertheless, the wearable
health monitoring devices could only be affordable by part of population. Hence,
there is a need to develop an affordable electronic wearable device to measure the
blood pressure and sleep pattern of the individual.

Keywords: Internet of Things (IoT) · Healthcare · Health monitoring systems ·
Data mining · Cloud storage

1 Introduction

In yesteryears, monitoring a person's health throughout the day is considered as an
incredible task. The association of IoT with healthcare systems makes ease the lives of
patients, healthcare providers, doctors and nurses across the globe. The scope of IoT in
healthcare is extremely eclectic and is rapidly becoming a part of life. IoT enables
health providers to collect, store and share the patient's real-time data. It reduces the
turnaround time, simplifies the process to identify chronic diseases and helps to mitigate the risk. The built-in sensors in these devices are connected to IoT platform. When
these wearable devices are used in medical applications, they are classified as Internet
of Medical Things (IoMT).

© Springer Nature Switzerland AG 2020
A. P. Pandian et al. (Eds.): ICCBI 2019, LNDECT 49, pp. 340–347, 2020.
https://doi.org/10.1007/978-3-030-43192-1_39

Sleeping habit plays a significant role in the existence of every creature; especially in human life. Attaining sufficient, quality sleep at the correct time could help to maintain the mental and physical wellness, quality and goodness of life. People are now sleeping less than they did in the past and quality of sleep has also drastically decreased. Certain medical conditions including obesity, diabetes, hypertension, stroke or transient ischemic attack, melancholia, Coronary artery disease, cardiac failure and hyperkinetic disorder are associated with sleeping disorders. Among these diseases, blood pressure is considered as a most vulnerable one in India. Blood pressure or hypertension eventuates when the body has built up too much resistance in your arteries, so spouting blood throughout the body becomes acutely tough. Hence, the heart tries to work hard to pump the blood, it results in hypertension. Over the time, this condition could seriously affect the health and causing heart disease, stroke, heart failure and many other problems.

In a recent study [1], researchers ascertained that those who slumbered less than six hours a night were 20% more probably to have hypertension. Another study was conducted on 240 adolescents. In that, 210 (87.5%) persons were normotensive and 30 (12.5%) persons were hypertensive [2]. This indicates that quality of sleep is correlated with hypertension in youngsters.

2 Survey Classification

An IoT system comprises of sensors and devices which were connected to the cloud. When the data are acquired by the cloud, it would be processed by applications and then predetermined action would be initiated. That might be an alert sent to users or reply to the respective sensors or devices without any human intervention. A broad view about piebald researches in IoT is dispensed here. This assessment involves the identification of methods used, approaches implemented, suppositions obtained and imminent routings directed in the earlier works. The listed articles are categorized based on the following aspects:

2.1 Collecting the health data from individuals (i.e. Blood pressure, Sleep quality, Temperature)
2.2 Identifying the association between various features using different methodologies.
2.3 Ensuring the privacy and security of the patient by providing fortified transactions.
2.4 Addressing the technological challenges and predicting anomalies by analyzing the patient data.

3 Literature Survey

Shreshth et al. [3] suggests how heart diseases can be analyzed in real time by reducing latency, delay or response time by integrating edge computing and ensemble deep learning techniques. Since cloud computing has high time delay, technologies like fog computing, edge computing and Big data have been integrated together to provide computation, storage, enhance privacy, security, low latency and network bandwidth.

Heart problems are difficult to detect and can only be identified by experienced doctors who are very limited in numbers. Hence there is a need for data to be sent at high speed and accuracy as possible. This is achieved by integrated Edge-Fog-Cloud based computational model, which provides healthcare with high accuracy, minimum energy consumption and low response time. The health fog architecture is designed to manage data which is productive and has low overheads.

The customer's impression on technology would play the major role on success of products. Mansour et al. [4] piloted a survey on user's attitude in receiving healthcare through IoT technology. It was conducted among the users of Oman about their trust and attitude towards using IoT keeping risk perception as a mediator. The data was collected by using SPSS 25 and AMOS 25 statistics software and 387 responds were analyzed. The results showed that Privacy, Security and Familiarity affected the trust of users and how trust in turn affects the risk perception and attitude towards the usage of IoT. The result recommends improving Privacy and security of IoT devices by conducting awareness, where users must be taught how to use the IoT devices safely and thereby reducing the perception of risk.

Sodhro et al. [5] addressed the unembellished problem faced by Medical Internet of Things (MIoT). They are energy famished media transmission and strolling evolvement of battery technologies which has made media transmission in emergency situations a challenging and crucial task. This is being resolved by proposed Green Media Transmission Algorithm (GMTA) and a 8-min medical media stream called, 'Navigation to the Uterine Horn, transection of the horn and re-anastomosis' to efficiently manage energy related issues in MIoT. This aids in viable dissemination of crucial health information from patient to doctor and vice versa. Through this approach, 41% of energy could be saved.

Contriving an effective connectivity mechanism and constructing an efficient energy management system are considered as herculean tasks in IoMT. Arun et al. [6] scrutinized these challenges and come up with the solution. The former could be resolved by integrating IoMT with Product Lifecycle Management (PLM) and the latter by Battery Recovery based Algorithm (BRA), Joint Energy Harvesting and Duty-cycle Optimization based (JEHDO) algorithm. PLM regulates the transfer of information from one entity to another in efficient and accurate way. BRA and JEHDO ensure the battery lifespan and efficacious energy management of tiny wearable devices respectively.

Fog Assisted – IoT Enabled Patient Health Monitoring system was proposed by Prabal et al. [7]. They addressed the hindrance instigated while transferring information to and back to the cloud. This could be overcome by introducing a fog layer between the cloud and device layer. They have also used few techniques like data mining, distributed storage and notification services, Event Triggering based temporal mining to address this issue. This Fog assisted system is used to deal with the real time patient data and gives high precision and response time while ascertaining the state of an incident. Fog layer has many Fog nodes which communicate with each other to initiate an action. The architecture of the intended system consists of Data acquisition layer, Event classification layer, Information mining layer, Decision making layer and Cloud storage layer. These layers would dispense consistent outcomes by performing the desired functions.

Himadri et al. [8] recommended a system where patient's health is traced with the help of sensors and internet. The patient's pulse and heart rate, pressure level, internal heat and few other factors are tracked by the corresponding sensors. The collected data are transferred to cloud environment for further processing. The results could be viewable in the device. If there is any drag or abrupt changes occurred in heart beat or temperature, the system alerts the concerned patient or doctor. The authenticated users could access the patient's health records via internet.

A System for prompting medication schedule and monitoring the patient was recommended by Samir et al. [9] They have used IoT to monitor and transmit the data. This was developed by keeping in mind the aged people and sufferers of chronic illnesses, who miss out their regular intake of medicine due to dementia. The faraway specialist care is made possible by biomedical widgets which measures and transmits information via Bluetooth or ZigBee. In this system, patients are identified with the unique RFID. The obtained data are stored and shared to both the patient and the doctor.

The importance of compactness, IP connectivity, low power consumption and security of the mobile-health devices were studied by Sultan et al. [10]. That system consists of three layers, data collection, data storage and data processing. The m-health has wireless connection and transmission of data, and each device has unique IP address. The following parameters are analyzed by m-health, Blood sugar, ECG, Blood Pressure and Asthma. The patterns are examined and according to the condition of the patient's different level of alerts, normal, cautious and emergency are generated. Quick diagnosis, remote monitoring and home rehabilitation have been made possible by m-health.

Dimiter et al. [11] postulates how the combination of MIoT and Big Data could transform the healthcare industry by making it participatory, customized, extrapolative and pre-emptive field. This methodology enables the patients to share their health data across various service providers which lead to a new kind of service platform and business models. The Electronic Health Recorder stores the laboratory results, clinical and treatment histories of the patients. Hence the physicians and other healthcare providers could obtain these data from anywhere provided if the user gives the permission to access. It reduces cost and eliminates the need of paper records.

Dimitra et al. [12] came up with an approach, which Analyze ECG Signals and Detect Arrhythmia by using wearable IoT medical devices. A specialized algorithm was developed for analyzing and classifying the ECG signals. The analysis and classification was done by using Discrete Wavelet Transform (DWT) and Support Vector Machine (SVM) classifier respectively. The enactment of algorithm was done on Galileo board where filtering, heartbeat detection, heartbeat segmentation, feature extraction and classification are performed. This approach delivers the results with the accuracy of 98.8%. IoT is employed for promoting 24 h continuous remote health monitoring (Table 1).

Table 1. Key findings and interpretation

S. No.	Author(s)	Methodology used	Advantage	Disadvantage
1	Tuli et al. [3]	Integrated Edge-Fog-Cloud based computational model	High accuracy, minimum energy consumption and low response time	High cost, not user friendly, complex and confusing, increased risk
2	Alraja et al. [4]	SPSS25 and AMOS 25 statistics software	Understanding Trust and users attitudes towards using the IoT	Less sophistication, security and privacy in IoT devices, low quality, less awareness programmes, increased perception of the risks
3	Sodhro et al. [5]	Green Media Transmission Algorithm (GMTA), Navigation to Uterine Horn, transection of the horn and reanastomosis	Fast media transmission, reduce the power consumption	Slow execution process, very expensive
4	Sodhro et al. [6]	Product Lifecycle Management, Battery Recovery based Algorithm, Joint Energy Harvesting and Duty-cycle Optimization based (JEHDO) algorithm	Efficient and accurate data transfer, efficient battery lifespan and power utilization	Inter-Operability issues, undefined algorithms, complicated system
5	Verma and Sood [7]	Data mining, distributed storage, and notification services, Event Triggering based temporal mining	Fog layer, High accuracy and response, power efficiency, low latency, efficient service	Complicated system, Additional expenses, limited scalability
6	Saha et al. [8]	IoT based Non-invasive health monitoring system	Helps to keep the track of patient's health	Network connection dependency, Security issues
7	Zanjal and Talmale [9]	Remote monitoring system	Improved medicine intake, remote monitoring, RFID identity detection	Security, network, technical issues
8	AlMotiri et al. [10]	Multilayer IoT system, RFID Technology	Moderates hospitalization cost, optimized energy consumption, confidentiality, privacy and security, quick remote monitoring	Low security and data privacy, less accuracy
9	Azariadi et al. [12]	Discrete Wavelet Transform, Support Vector Machine classifier, Galileo board	24 h continuous remote health monitoring, accuracy of 98.8%	Higher cost and Higher power requirement
10	Dimitrov [11]	MIoT, Artificial Intelligence (AI) Electronic Health Recorder (EHR)	Remote real time monitoring, Prevention services, error minimization, solves inter- operability issue	Security and privacy issues, fiscal and policy issues, hard to collect data

4   Research Gap

A smart system was proffered by Diana et al. [13–17] has two kinds of processing. The fog computing based data pre-processing sends real time notifications and batch processing performs analysis and prediction. This method is proved to be 93.3% effective. The motion sensors, accelerometer and gyroscope are availed to observe the sleep

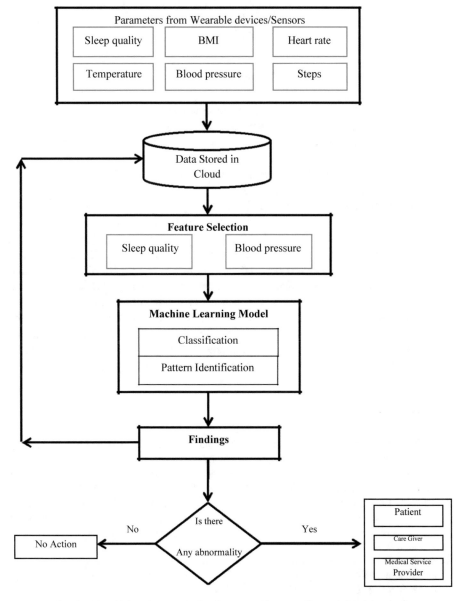

Fig. 1. Identifying the correlation between Sleep quality and Blood pressure

quality and other related parameters like heart rate, temperature, steps, BMI etc. The fog layer enables interoperability, storage and processing. Big Data Analyzer evaluates the information from the fog layer.

In this research work, the importance has been given to sleep and heart rate parameters. Nevertheless, blood pressure is considered as a primary factor to measure the risk of heart attack and other cardio diseases. Therefore, our aim is to monitor blood pressure along with sleeping parameters and find a correlation between them. If there is any variation in these factors, a warning will be sent to the respected individual or medical service provider. This will help to keep the patient health in track and prevent the complications and emergency situations (Fig. 1).

5   Conclusion and Future Work

The wearable health monitoring systems are employed to improve the individual's quality of life [14, 15]. They ease the process of collecting and storing patient's data. These advancements are improving the quality and efficiency of healthcare. The gap between physicians and patients got reduced while the health devices are connected through the internet. It permits to remotely receive and examine the biological signals without impinging the patient's regular life.

The IoT technology is refining rapidly day by day. Since it is an interoperable field, the advancements in related fields like Data mining, Cloud computing, Electronic systems and Sensors are directly making impact on the same. In the proposed approach, the blood pressure and sleep quality would be measured, analyzed and correlated. The machine learning models would be used in the process of classification of data and predicting the cardiovascular diseases. Whenever the anomaly got identified, that would be sent it as an alert to the patient and the respective healthcare providers. This instant alert approach would give sufficient time to take necessary steps to save the patient and also reduces the risk of mortality. In the meantime, the results would be stored in the cloud environment for future reference. This IoT based approach would be very supportive to progress the e-Health service oriented ecosystem.

References

1. Battling Hypertension? Getting More Sleep Can Ease High Blood Pressure by Ryan Morton. https://amerisleep.com/blog/sleep-affects-hypertension/
2. Nasution, A.T.P., Ramayati, R., Sofyani, S., rRna Ramayani, O., Siregar, R.: Quality of sleep and hypertension in adolescents. Paediatr. Indones. **56**(5), 272–276 (2016)
3. Tuli, S., Basumatary, N., Gill, S.S., Kahani, M., Arya, R.C., Wander, G.S., Buyya, R.: HealthFog: an ensemble deep learning based smart healthcare system for automatic diagnosis of heart diseases in integrated IoT and Fog computing environments. Future Gen. Comput. Syst. **104**, 187–200 (2019)
4. Alraja, M.N., Farooque, M.M.J., Khashab, B.: The effect of security, privacy, familiarity, and trust on users' attitudes toward the use of the IoT-based healthcare: the mediation role of risk perception. IEEE Access **7**, 111341–111354 (2019)

5. Sodhro, A.H., Sangaiah, A.K., Pirphulal, S., Sekhari, A., Ouzrout, Y.: Green media-aware medical IoT system. Multimedia Tools Appl. **78**(3), 3045–3064 (2019)
6. Sodhro, A.H., Pirbhulal, S., Sangaiah, A.K.: Convergence of IoT and product lifecycle management in medical health care. Future Gen. Comput. Syst. **86**, 380–391 (2018)
7. Verma, P., Sood, S.K.: Fog assisted-IoT enabled patient health monitoring in smart homes. IEEE Internet Things J. **5**(3), 1789–1796 (2018)
8. Saha, H.N., Auddy, S., Pal, S., Kumar, S., Pandey, S., Singh, R., Singh, A.K., Sharan, P., Ghosh, D., Saha, S.: Health Monitoring using Internet of Things (IoT). In: 2017 8th Annual Industrial Automation and Electromechanical Engineering Conference (IEMECON), pp. 69–73. IEEE, August 2017
9. Zanjal, S.V., Talmale, G.R.: Medicine reminder and monitoring system for secure health using IoT. Procedia Comput. Sci. **78**, 471–476 (2016)
10. Almotiri, S.H., Khan, M.A., Alghamdi, M.A.: Mobile health (m-health) system in the context of IoT. In: 2016 IEEE 4th International Conference on Future Internet of Things and Cloud Workshops (FiCloudW), pp. 39–42, August 2016
11. Dimitrov, D.V.: Medical internet of things and big data in healthcare. Healthcare Inform. Res. **22**(3), 156–163 (2016)
12. Azariadi, D., Tsoutsouras, V., Xydis, S., Soudris, D.: ECG signal analysis and arrhythmia detection on IoT wearable medical devices. In: 2016 5th International Conference on Modern Circuits and Systems Technologies (MOCAST), pp. 1–4. IEEE, May 2016
13. Yacchirema, D.C., Sarabia-Jácome, D., Palau, C.E., Esteve, M.: A smart system for sleep monitoring by integrating IoT with big data analytics. IEEE Access **6**, 35988–36010 (2018)
14. Valanarasu, Mr R.: Smart and secure IoT and AI integration framework for hospital environment. J. ISMAC **1**(03), 172–179 (2019)
15. Lee, J., Kim, D., Ryoo, H.Y., Shin, B.S.: Sustainable wearables: wearable technology for enhancing the quality of human life. Sustainability **8**(5), 466 (2016)
16. 5 Ways Technology is Improving Health. https://healthinformatics.uic.edu/blog/5-ways-technology-is-improving-health/
17. Smys, S., Raj, J.S.: Internet of things and big data analytics for health care with cloud computing. J. Inf. Technol. **1**(01), 9–18 (2019)

AgroFarming - An IoT Based Approach for Smart Hydroponic Farming

Aashray Mody$^{(\boxtimes)}$ and Rejo Mathew

Mukesh Patel School of Technology Management and Engineering,
NMIMS (Deemed to be University), Mumbai, India
aashraymody@gmail.com, rejo.mathew@nmims.edu

Abstract. With world temperatures on the rise and climate change appears as a major factor affecting food production the traditional agriculture techniques may not be the answer to the ever increasing food demand across the globe. One of the key solutions to combat these problems is hydroponics. A lot of research has been done on varied solutions offered by smart hydroponics. A comparative study of the different solutions has not yet been documented. This paper aims to provide a critical review of the current hydroponic systems considering various parameters such as energy efficiency, cost, and type of design. Some of the solutions are PlantTalk Robot Farm and MircoCEA are discussed here.

Keywords: Hydroponics · Smart farming · Sensors · Mobile · IoT

1 Introduction

The world currently produces more than enough food to feed everyone, however, due to food wastage roughly 11% of the global population which about 820 million of the population went hungry in 2016, according to the U.N [1] from the present population of 7.7 billion [1]. Each of the plant species in the traditional earth farming methods is under severe risk due to constant climate change and overexploitation of land. Currently, 12 crops constitute roughly about 75% of human calories if a pest or pathogen arises it can lead to severe food shortages [2]. The aim of the current researchers in this field is to enable households to grow plants autonomously and support food practices at home with little effort. In addition, to contribute to food security. The solution to these problems is hydroponics, which allows automation and uses fewer resources than traditional farming. Hydroponic crops are not vulnerable to many diseases, insect attacks which makes it superior to traditional farming. The basic idea of using IoT with smart farming is that IoT allows multiple uses of sensors with and the ability to transfer data over a network. Hydroponic crops are not vulnerable to many diseases, insect attacks which makes it superior to traditional farming The basic idea of using IoT with smart farming is that IoT allows multiple uses of sensors with and the ability to transfer data over a network. The paper is divided into six sections. Sections 2 and 3, which lays out the problems Sect. 4 shows the solutions to the problems mentioned in Sect. 3. Section 5 gives a comparison of the various solutions discussed in Sect. 6 is the conclusion.

© Springer Nature Switzerland AG 2020
A. P. Pandian et al. (Eds.): ICCBI 2019, LNDECT 49, pp. 348–355, 2020.
https://doi.org/10.1007/978-3-030-43192-1_40

2 Related Work

Though many solutions are existing however many of them need to be modified. As Lee [4] has suggested that systems should change from using LED's to sunlight regularly to save energy. Palande [5] suggested a hydroponic system for automatic growth and monitoring of plants. Kobayashi [6] studied the impact of various LEDs on lettuce. Siregar [7, 8] allowed the user to monitor data with a smartphone but each sensor was still controlled manually. AeroGarden [3] Harvest, has devised a hydroponic solution, which has a self-watering tank to grow fruits and vegetables along with an LCD screen to remind users to add water, and the nutrient solution. Aero-garden has multiple modules such as harvest, Herbie sprout that has multiple pods depending on the size.

3 Shortcomings in the Previous Solutions, Which Were Not Addressed

3.1 No Real Time Notifications to the User [7, 8]

Real time updates to the user are given; if the values breach the set point then it can be disastrous to the plants and can destroy the harvest. None of the current solutions provides real time updates and notifications to the user. Real time notifications are used to alert users so that the users can take corrective steps to rectify the problem.

3.2 No Customization of Sensors Available [12]

None of the current solutions in the market allows customization of the sensors, which is an important factor as suggested in plant-talk [12]. None of the previous work allowed addition and configuration of sensors or actuators in the system, which are needed if the user are specialized users. For example systems such as Aero-garden [3] does not have CO_2, O_2 and PH sensors, which would be necessary if the system needs to be upgraded to industrial standard and for specialized users.

3.3 Addition of the Off-Shelf Camera Is Not Possible in Any of the Current Solutions [11]

Addition of off the shelf camera is not possible in any of the solutions in the market. Off the shelf camera adds value to the product by allowing the user to monitor the health of the plants and off the shelf camera to be added for remote sensing and remote monitoring of the plant based systems. Off the shelf camera allows the user to monitor the yield.

3.4 Different Modes of Operation Is Not Present [11]

As suggested by Gonçalo Marques [11] different modes of operation allow greater energy efficiency. In many of the present solutions, there are no customizable modes available for different users meaning a farm owner will want a customable mode with pre-set conditions based on his own e recipe for greater efficiency. While a homeowner would want the mode to be on the e recipe provided by the module.

3.5 E-Recipes for Different Plants Is Not Present [13]

As suggested by robot farm [13] for naïve users an automated recipe is already inbuilt. All the sensors should be controlled autonomously without the need of user intervention. For different plants there should be a pre-set recipe which should be used for inexperienced users especially homeowners. This feature allows a naïve user to spend less time focusing on the sensors and more time on growing food.

3.6 Open Source Is Not Present

[14, 15] Believe that open source is a key requirement to make a good end-user platform for agriculture. Open source is not present in any of the systems which can be needed if the system is designed for different types of users.

3.7 No Suitable Size

None of the sizes are designed in a manner that they would fit or work outside a lab.

3.8 No User Friendliness

The communication with the system should be intuitive, simple, and efficient. Plug and play connections are needed to allow ease of use in the system [18].

4 Current Solutions in the Market

4.1 iHydroMobile

Enhanced Hydroponic Agriculture an Environmental Monitoring by Gonçalo Marques [9] proposes iHydroMobile.

[9] is a hydroponic approach to allow the user to see and control sensor data. This method allows and offers real time notifications to the farm manager when set points are not met and pre-existing conditions are exceeded. The data is stored using Plotly [14], which is a data analysis tool. Data analytics and visualization tools are used to plot graphs and provide meaningful data to the user. The product is right now is made up of a prototype designed for data collection with the IOS system providing real time analysis. Push notifications are sent to the user to notify that certain parameters have exceeded the set points. The application should allow the user to access all the sensor data including luminosity and electrical conductivity. iHydroMobile [9] not only has O_2 and CO_2 and other sensors, which a normal market module has but also has water level sensors for, enhanced hydroponics. Some of the sensors used in iHydroMobile [9] are Arduino Uno as a microcontroller and BLE model for communication. However, the sensors are chosen on merely cost and not on the energy efficiency of the sensors. There are two working modes in iHydroMobile which are manual mode and automatic

mode. The automatic mode is mainly used as a power saving and is used to maximise agriculture productivity. The features of this makes it perfect suited for managing data be it historical or current. The application and sensors joints make it possible to supervise and identify poor water quality. Due to this feature, it would becomes easy to predict any unfavourable condition and technical intervention will be used to solve this. iHydro-Mobile allows easy scalability meaning that modules can be added without affecting the inherent working of the system. The main users of iHydroMobile are large greenhouse owners and not for households. Future scope of iHydroMobile is that the sensors used will be energy efficient. This solution allows flexibility meaning that the user can start with one iHydroMobile unit and can add more units irrespective of the size of the greenhouse. iHydroMobile solves the problems mentioned in Sects. 3.1 and 3.4. iHydroMobile solves the problem of No real time notifications by providing push notifications, maximises energy efficiency using different modes, and provides easy scalability.

4.2 Plant Talk

Plant Talk [10] is an IoT-based solution for smart hydroponic farming where the device is controlled via a smartphone using the current 5 g technology. Historical data can be seen and can be used by the user to provide meaningful insights to the system. An off the shelf camera can be added so that the users can monitor the heath of the plants remotely and this feature is supported in the application bar by a video tab which allows the user to zoom rotate the camera to get the best possible view of the plants. Plant Talk allows board and vendor diversity by can supporting multiple control boards such as Arduino, ESP8266 ESP-12F, and ROHMT. This is because plant talk is open platform and provides API to specialized users who wish to add control boards and sensors. Plant talk allows sharing of RO applications if the user has a smart aquarium and Plant Talk and the mobile application allows the user to switch between them this allows greater energy efficiency and water sharing capacity.

Each of the sensor data can be sent via 5 g and LTE technology to the smartphone the data is real time in nature. For the measured data of each sensor, Plant Talk logs the history, and the user can see the time series chart of a sensor by clicking this sensor's icon in the dashboard. AgriTalk [11], which is, developed which is an extended version of Plan talk for factories. Plant-talk is easily scalable meaning that multiple models can be added. Agri-talk [11] is one of the examples where plant talk has been scaled to industrial levels to make food at a larger level.

The Main users of this system are large-scale factory owners who wish customizability and want API, which allows them to use plant-talk since it allows vendor diversity. Plant-talk allows each scalability to industrial levels so that production can be maximised. Plant Talk intelligence can be easily created and maintained, which allows the users to know more about their plants than what the sensors can provide. Plant Talk is the one of solution that provides smartphone-based intelligence to the problems mentioned in Sect. 3 and provides a way of remote monitoring of the systems. Plant Talk provides a solution to problems mentioned in Sects. 3.1, 3.2 and 3.6 by using open source and scalability. This solution solves these problems due to it supporting vendor diversity.

4.3   Robot Farm

Robot Farm [12] is a hydroponic module specifically designed for homeowners who wish to grow their plants at home and designed for people who have no experiences of agriculture. This particular solution helps homeowners grow their own food at a micro level throughout the year. Robot farm is in line to the behavioral profile of the so-called prosumers which a neologism coined to indicate the growing trend of users wishing to participate directly in the production of what they consume [16]. The light source in the robot farm mimics the midsummer sun. An electronic cultivation recipe has been designed so that the products can be grown in an environment such as a kitchen without the need of human intervention [12]. Throughout the entire life of the plant, the climate and irrigation features are automatically controlled meaning robot farm is a closed system. Due to Robot farm being a closed system meaning that individual sensors cannot be configured. This allows total autonomy of the system from the seedling to the harvesting of the crop. The robot farm is sterile as it consists of UV lamps. Time taken from sowing seed to yield is usually in 25 days. Robot Farm has a display and buttons to control all the features like a normal kitchen appliance. The future scope of Robot Farm is to have an app to minimize the buttons. Robot Farm has e-recipes these allow automated growth of the plants without any human intervention. Robot farm design is like a standalone kitchen appliance such as a refrigerator, it has a similar set of display consisting of these push buttons to mimic a normal kitchen appliance. The appliance is made of two cultivation 25 plants [12]. These 25 trays can be used to accommodate lettuce microgreens and other fruits. The e recipes are right now limited to micro greens and lettuces. Robot Farm is not easily scalable as it is like a standalone kitchen appliance. The 25-day life cycle allows the plant from a seedling to a plant and allows easy harvesting with th user only has to fill the water level if it goes below. The future scope of robot farm is to allow or integrate smartphone intelligence in the system to allow push notifications to the user and allow the removal of the buttons per say. Robot Farm solves the problems mentioned in Sect. 3.4 as it has been specially designed for homeowners.

4.4   MicroCEA: Developing a Personal Urban Smart Farming Device

Micro Controlled Environment Agriculture (MicroCEA) [17] is a sealed micro plant factories which means it is a self- growing sealed box once plants are planted that doesn't require any external intervention. MicroCEA allows the ease of IoT through the use of cloud for monitoring, controlling and data analysis using a GUI. MicroCEA has been built upon the existing Food Computer (FC) [18, 19] by MIT.Microbe has LED lights that can be used to control the light intensity in the system. MicroCEA has humidity and water sensors these are used to control the humidity and air. CO_2 and O_2 sensors are present which allows for enhanced monitoring of the system. MicroCEA uses the MQTT server for communication between the sensor and the web browser. The server of MQTT is used to send and receive data across the internet. All the data is transmitted and stored in JSON format. Making it easy to extract and analyse [17]. MicroCEA uses RPiv2 for computation. MicroCEA uses a web browser as its GUI, which shows real time and live updates of the sensor data. MicroCEA is intended for homeowners especially in UAE with conditions designed to match the climate in UAE.

MicroCEA allows the users to easily integrate and install software's and allows cloud accessibility with data analysis features.

Future scope of the system is to focus more on the functionality of each of the sensors and not focus on vendor diversity and focus on finding out requirements of each of the plants.

4.5 Personal Food Computer (PFC):- A New Device for Controlled-Environment Agriculture [19]

OpenAg™ Personal Food Computer enables its client to make, store, and offer the information created during the development cycle. In this manner, giving the probability of making atmosphere plans and enabling other reasonable gadgets to reproduce the equivalent natural conditions, improving the reproducibility of the examinations. The PFC is an open-source openhardware platform; its design prioritizes low cost and user friendliness by making it an easy to understand gadget. The user can see the real time health of the plants as PFC has an off shelf camera option which allows users to see time-lapse videos and visualization of camera feeds. The use of SBC provides flexibility due to it using Universal Serial Bus (USB). The hardware components are used using an on-board single board computers (SBC) switch Raspberry Pi. The sensors used in PFC include PH, water level CO_2 O_2 and electrical conductivity and temperature sensors.

5 Comparison of the Hydroponic Solutions

Features	iHydroMobile	Plant Talk	Robot Farm	MicroCEA	Personal Food Computer
Users	• Farm managers	• Households • Farm managers	• Households	• Household	• Households
E recipes for automated growth	• Not present	• Not present	• Present as its for naïve users	• Not present	• Present can be made by user
Addition of camera	• Doesn't allow addition of components	• Yes off the shelf camera can be added	• Not present	• Can't be added	• Yes present
Display of data shown	• Real time • Historical.	• Real time • Historical	• Data not shown to the user	• Real time	• Real time • Historical
Vendor diversity support	• Only Arduino supported	• Multiple control boards are supported	• Not supported.	• Arduino • RPiv2	• Arduino
App based control	•Sensor data is displayed there is no user control	• Sensor data is displayed and can be remotely controlled	• Has push buttons so doesn't support app based	• Sensor data is displayed and can be remotely controlled	• Sensor data is displayed and can be remotely controlled

(continued)

(continued)

Features	iHydroMobile	Plant Talk	Robot Farm	MicroCEA	Personal Food Computer
Scalability	• Easy to scale due to multiple	• Easy to scale AgriTalk is an example of scaled version	• Tough to scale due to cost	• Easy to scale	• Easy to scale
Language of programming used	• Python has been used	• Python has been used	• Not needed as it's a standalone kitchen appliance	• Elixir	• Recipes are in JSON format

6 Conclusion

With veganism on the rise and more than one in three individuals in Great Britain being vegan [20] smart hydroponics will be beneficial as the vegan population will be able to grow their own food with minimal effort. Humanity is projected to need double the amount of food, fiber, and fuel generated to meet global demands in the coming decades. However, it is predicted that growing seasons will become more volatile, and due to global warming, 80% of arable land is already being used [21]. Concurrently, public and private institutions are starting to take an increasing interest in producing specific compounds and chemical elements using innovative agricultural platforms for several applications such as medical and environmental [23, 24]. This paper gives a comparison of the various solutions and its sensors and various features. Hydroponics is the way forward as it has many advantages over traditional farming. Choosing of the solution will depend upon on the type of user if the homeowner or a specialized. Only one factor cannot be considered if one solution has greater advantages over the other the solutions. Energy efficiency and productivity will be some of the factors will be key driving factors in choosing the solution with ease of use of the software application. Plant Talk and PFC are comprehensive solutions as it supports vendor diversity and has an option of on the shelf camera. Further work, more solutions need to be compared with various other parameters.

References

1. World Population Prospects - Population Division - United Nations, Population.un.org (2019). https://population.un.org/wpp/. Accessed 27 Aug 2019
2. https://time.com, Time (2019). https://time.com/5216532/global-food-security-richard-deverell/. Accessed 27 Aug 2019
3. AeroGarden Official Store - Shop and Save on Aero Gardens, Seed Kits, Grow Bulbs & More, AeroGarden Official Store (2019). https://www.aerogarden.com/. Accessed 02 Sept 2019

4. Lee, S., Park, S.: Energy savings of home growing plants by using daylight and LED. In: Proceedings of the IEEE Sensors Applications Symposium, Galveston, TX, USA, 19–21 February 2013
5. Palande, V., Zaheer, A., George, K.: Fully automated hydroponic system for indoor plant growth. Procedia Comput. Sci. **129**, 482–488 (2018)
6. Kobayashi, K., Amore, T., Lazaro, M.: Light-emitting diodes (LEDs) for miniature hydroponic lettuce. Opt. Photonics J. **03**(01), 74–77 (2013)
7. Siregar, B., Efendi, S., Pranoto, H., Ginting, R., Andayani, U., Fahmi, F.: Remote monitoring system for hydroponic planting media. In: 2017 International Conference on ICT for Smart Society (ICISS), Tangerang, pp. 1–6 (2017)
8. Marques, G., Alexia, D., Pitarma, R.: Enhanced Hydroponic Agriculture Environmental Monitoring: An Internet of Things Approach (2019)
9. Van, L., et al.: PlantTalk: a smartphone-based intelligent hydroponic plant box. Sensors **19** (8), 1763 (2019)
10. Chen, W.-L., Lin, Y.-B., Lin, Y.-W., Chen, R., Liao, J.-K., Ng, F.-L., Chiu, C.-H., et al.: AgriTalk: IoT for precision soil farming of turmeric cultivation. IEEE Internet Things J. **6**, 5209–5223 (2019)
11. Angeloni, S., Pontetti, G.: RobotFarm: a smart and sustainable hydroponic appliance for meeting individual and collective needs. In: Barolli, L., Xhafa, F., Hussain, O. (eds.) Innovative Mobile and Internet Services in Ubiquitous Computing, IMIS 2019. Advances in Intelligent Systems and Computing, vol. 994. Springer, Cham (2020)
12. Modern Analytic Apps for the Enterprise, Plotly (2019). https://plot.ly/. Accessed 02 Sept 2019
13. Harper, C., Siller, M.: OpenAG: a globally distributed network of food computing. IEEE Pervasive Comput. **14**(4), 24–27 (2015)
14. Stočes, M., Vaněk, J., Masner, J., Pavlik, J.: Internet of things (IoT) in agriculture - selected aspects. AGRIS on-line Papers Econ. Inform. **8**(1), 83–88 (2016)
15. Toffler, A.: The Third Wave. William Morrow, New York (1980)
16. Stevens, J.D., Shaikh, T.: MicroCEA: developing a personal urban smart farming device. In: 2018 Second International Conference on Smart Grid and Smart Cities (ICSGSC), Kuala Lumpur, pp. 49–56 (2018)
17. Raj, J.S., Ananthi, J.V.: Automation using IoT in greenhouse environment. J. Inf. Technol. **1** (01), 38–47 (2019)
18. Group Overview "Open Agriculture (OpenAg) – MIT Media Lab", MIT Media Lab (2019). https://www.media.mit.edu/groups/open-agriculture-openag/overview/. Accessed 26 Sept 2019
19. Ferrer, E.C., Rye, J., Brander, G., Savas, T., Chambers, D., England, H., Harper, C.: Personal Food Computer: A new device for controlled- environment agriculture. (Submitted on 15 Jun 2017 (v1). Accessed 24 June 2017
20. Ipsos Mori survey, commissioned by The Vegan Society, and The Food & You surveys, organised by the Food Standards Agency (FSA) and the Centre for Social Science Research (Natcen) (2018)
21. The state of food and agriculture: Climate change agriculture and food security, Food and Agriculture Organization of the United Nations (FAO), Rome, Technical report (2016)
22. Lee, K.: Turning plants into drug factories. Sci. Am. (2016)
23. Fox, J.L.: Turning plants into protein factories. Nat. Biotechnol. **24**(10), 1191–1193 (2006)
24. Olinger, G.G., Pettitt, J., Kim, D., Working, C., Bohorov, O., Bratcher, B., Hiatt, E., Hume, S. D., Johnson, A.K., Morton, J., Pauly, M., Whaley, K.J., Lear, C.M., Biggins, J.E., Scully, C., Hensley, L., Zeitlin, L.: Delayed treatment of Ebola virus infection with plant-derived monoclonal antibodies provides protection in rhesus macaques. Proc. Natl. Acad. Sci. U.S.A. **109**(44), 18030–18035 (2012)

Review on Dimensionality Reduction Techniques

Dhruv Chauhan$^{(\boxtimes)}$ and Rejo Mathews

Mukesh Patel School of Technology Management and Education,
NMIMS, Vile Parle West, Mumbai 400056, India
chauhandhruv78@gmail.com, rejo.mathew@nmims.edu

Abstract. With the increasing use of Machine day by day, data analysts' job has increased drastically. With the data gathered from millions of machines and sensors, modern day datasets becomes wealthier in information. This makes the data to be high dimensional and it is quite common to see datasets with hundreds of features. One of the biggest problems that data analysts face is dealt with high dimensional data. Without a major loss of information, data can be effectively reduced to a much smaller number of variables. This method of reducing variables is known as Dimensionality Reduction. The objective of this paper is to review methods used for reducing Dimensionality.

Keywords: Data Mining · Dimensionality reduction · Machine learning · Principle component analysis

1 Introduction

Data plays an indispensable role in all business processes and to analyse large sets of data, Machine Learning is used. In machine learning, to catch useful indicators and obtain a more accurate result, users tend to add as many features as possible. With the increase in the number of features, the existing samples also increase in number which further increase proportionally and makes the model more complex [3]. After a certain point, the performance of the model will decrease with the increasing number of features. This phenomenon is often referred to as "The Curse of Dimensionality." The curse of dimensionality can be said as all the issues that occur when dealing high dimensions of data that were not present in the lower dimensions. Dimensionality reduction is the process of reducing the dimensionality of the features, without the loss of data [5]. While some specific type of information is lost, critical information is still present in the low dimensional reduced data.

2 Problems Faced While Handling High Dimensional Data

Adding of new features for the analysis of data isn't always a good idea. The constant adding of new features only results in the increase in dimensions without any significant benefit. An increase in the dimensionality of data results in the scattering of data. Simply put, a small increase in the dimensionality would require a large increase in the

© Springer Nature Switzerland AG 2020
A. P. Pandian et al. (Eds.): ICCBI 2019, LNDECT 49, pp. 356–362, 2020.
https://doi.org/10.1007/978-3-030-43192-1_41

volume of the data to maintain similar level of performance in tasks such as clustering, regression etc. For example, supposing there are accurate results for a case study with 5 variables and with 100 inputs, now to do the same case study with 8 variables then at least 200 inputs are needed for the same level of accuracy. Sometimes even after taking more than 200 inputs there might not be the same level of accuracy as some of the algorithms only perform well in lower dimensions. Hence, dimensionality reduction forms the basis of machine learning.

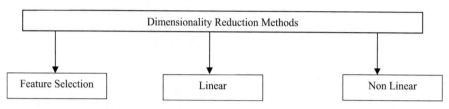

3 Various Methods for Dimension Reduction

3.1 Feature Selection

A feature is good if it is unique and contains unique information about the different classes in the data. In feature selection method, the "good" features are identified, which might prove to be helpful. The aim of feature selection is to look for a number of important features to shape a subset of features that can define the main property of all the original features [4]. Feature selection is the simplest of dimensionality reduction methods.

3.1.1 Variance Threshold

Variance Threshold is a basic approach in Feature Selection. In this method, features are dropped wherein the variance through a particular column does not exceed a certain value of threshold. The basic concept is that if a feature is carrying very similar information, it has very little predictive power [5]. For example consider the following table.

Subject/name	A	B	C	D
Maths	75	52	81	55
English	71	73	80	68

In the above table, the variance of the marks of Maths is higher than the variance of the marks of English. The subject English would not be of much help as its variance is very low and it does not help in predicting future data or analysing the pattern.

3.1.2   Feature Ranking

Feature ranking assigns a rank to each variable and uses a scoring function F(i). Features are ranked according to various methods. A low scores means it is less valuable to the outcome, hence the features are sorted in decreasing order of F(i).

The least ranked features, the ones that are last in the table are removed from the feature set and the accuracy of the model is again calculated. This process of removing the least important feature is continuously done until the accuracy of the model is significantly lost. This approach is only effective in supervised learning and is effective only with low number of features. Also to find a ranking algorithm which works on all the datasets is another problem faced in this technique [4].

3.1.3   High Correlation Factor

High similarities of trends between two features specifies that they are same in pattern and mostly carry similar information. For examples, time walked on treadmill and calories burnt are heavily correlated to each other, hence dropping one of these would not affect the final result. The correlation can be calculated between independent features that are numerical in nature. If the correlation coefficient crosses a certain threshold value, then one of the features can be dropped. Consider the previous example [3].

Subject/name	A	B	C	D
Maths	75	52	81	55
English	71	73	80	68
History	90	93	99	92

In the above table, History and English carry similar type of data hence dropping any one out of this would not affect the result.

Drawbacks and Assumptions for Feature Selection Method: Feature selection only works on numerical data. It cannot be used for Face Recognition and non-numerical set of data. Feature selection is not suitable for unsupervised learning. And it works well only on simple data, presence of any complexity will result into inaccurate results.

3.2   Linear Dimensionality Reduction Algorithms

These techniques generate a linear, low dimensional mapping of the initial high dimensional data that retains the valuable information. Usually bringing it down to a 2D or a 3D model. One of the misconception faced is that almost all the problems related to the reduction of linear dimensionality can be reduced to problems of Eigenvalue [7].

3.2.1   Principle Component Analysis

PCA is a feature extraction unsupervised method for dimensionality reduction. It combines input variables in a specific way, that the "least important" variables can be dropped while still being able to retain the most valuable parts of all of the variables. Generally, high variance can be identified using only a small number of main components. PCA provides a low-dimensional data representation that defines as much of the data variation as possible. PCA reduces dimensions by focusing on the features with most variation.

PCA tends to establish a new set of features from existing large set of variables. These newly derived features are called Principal Components. Thus, Principal component is derived from the original variables. From the co variance matrix of original data, all the eigenvalues are calculated. From this Principal Components are derived. Principal Components are those eigenvectors that correspond to the largest eigenvalues. The largest eigenvector with the largest eigenvalues represents maximum variance and this is principal component one. The eigenvector which represents the second most variance is called principal component two [6].

The third main component attempts to explain the variability that the first two main components do not explain, and so on [7]. X1 first principal component and it goes in the direction of most variance.

X2 second principal component, orthogonal to the first and with the greatest variance in the second direction. The information are based around the origin without the lack of generality [2]. PCA is still by far the most popular dimensionality reduction technique and it finds applications in almost all fields. PCA is also an effective method to analyse high-throughput biological data at which other methods fail [9].

Drawbacks and Assumptions for PCA:

- Input variables are assumed to be independent of one another.
- PCA assumes that the principle components are a linear combination of the original features. If this is not true, PCA will not give you sensible results.
- The eigenvectors of the PCA have contributions from all input variables. This makes their interpretation difficult since a large feature value cannot be attributed to few (ideally a single) input values.

3.2.2 Linear Discriminant Analysis (LDA)

The goal of Linear Discriminant Analysis is to answer the question as to whether the dimensions that provide maximum variance are actually the useful ones. It tries to project data in such a way that it maximizes the class separation. It considers the scatter between the classes as well scatter within the classes itself and then seeks to find the axis along which the classes are separated the most. Transformation is done in a way that the scatter between the classes is maximum while the scatter inside the class is minimum [5].

Some primary constraints of classical LDA are as follows, each data point has to be labelled with a class, each data point has to be labelled with only one class, and the class boundaries are represented as a row. That is there should be no missing observations and no mixed memberships. To reduce dimensionality of a dataset, practically first PCA is applied to find the principal components followed by LDA to find the axis which maximises the scatter between the classes [4].

Drawbacks and Assumptions for LDA

- Input variables are assumed to be independent of one another.
- Assumes equal class co variance
- Data is assumed to be normally distributed.

3.3   Non Linear Dimensionality Reduction

If the associated data map is non-linear, then the dimension reduction technique is known as a non-linear dimension reduction technique. This technique converts data into low output zone and uses various techniques for its implementation [7].

3.3.1   Kernel PCA

Another approach of capturing non-linear interactions is kernel PCA. It uses a function ø transforming the input vector X into a new vector ø(x) whose dimension is larger than dataset [7]. This may seem like dimensionality expansion, however if ø is chosen well enough, the data in the new larger dimension becomes more linear. Now in this new space standard PCA can be performed. The Kernel PCA is not interested in the main components in the input space, but in the main components of variables that are not linearly related to the input variables. With the use of relationship between MDS and PCA, there is no need to find co-variance, simply their inner products could be found out. The dataset can be described accurately using only one principle component as the modified version follows a linear relationship [8, 9].

3.3.2   Isomap

Isomap is the short form for isometric mapping. Isomap is a non-linear method of reducing dimensionality based on the spectral principle which tries to retain the lower dimensional geodesic distances. At the first step, a neighbourhood network is created. After that, it tries to find approximate geodesic distance between all pairs of points with the help of graph distance. During the third step, the eigenvalues are decomposed of the geodesic distance matrix to find the low dimensional embedding of the dataset. Unlike t-SNE, it does not over emphasise the distance between clusters and hence calculates more appropriate distance measure between different classes to analyse different trajectories, which do not normally fall under a typical cluster. Geodesic distances are better and more accurate while reducing dimensionality in non linear data than the Euclidean distance [10–16].

Isomap doesn't lose nonlinear relationships between cells. It has the special property of extracting trajectories between cells [9].

4 Analysis/Review

See Tables 1 and 2.

Table 1. Comparison between PCA and LDA

	PCA	LDA
Approach used to reduce dimensionality	Focuses on the data with the most variance	Tries to maximize the class separation
Based on supervision	Unsupervised	Supervised
Performance	Works better with the datasets having less number of samples per class	Is more effective with large dataset having many classes

Table 2. Comparison between different techniques

	Feature selection	Linear algorithms	Non linear algorithms
Based on the data	Linear and non linear	Linear	Non linear
Complexity	It simply works on the variance of the data. Hence it is the simplest of all	Involves high mathematics makes it complex but easier than non linear methods	The most complex of all
Efficiency	Efficient enough for simple data	Provides best analysis for the data which are sparse and when the intrinsic dimensionality is high	Provides better analysis than Linear Methods if there is a lot of data and the data doesn't have any gaps in it, i.e. if the data is evenly distributed
Time required	Quickest and easiest of all as it selects some features	Relatively high	Relatively high
Predictive power	Very low predictive power. Since it doesn't consider the relationship of features with the target variable	It uses linear subspace to analyze data, but doesn't adequately cover everything	It represents the data underlining manifold which can't be represented by linear subspace, hence very useful

5 Conclusion

This paper tried to review the already well-established methods for dimensionality and tried to explore the differences on when and where these different methods are used. Linear Techniques are still the center of all the big data analysis. The most popular and widely used techniques such as PCA etc. have been fundamental in Dimensionality Reduction, however due to the nonlinear nature of the data in recent times, analysts have started shifting towards analyses made through projections. Non Linear methods are only used when absolutely necessary due to their complex nature, however the new requirement of image data analysis has increased the use of nonlinear methods, as they adequately represent these type of data hence for some cottage industries the nonlinear methods are predominately used.

References

1. Maaten, L.V.D., Postma, E., Herik, J.V.D.: Dimensionality reduction: a comparative review. Tilburg centre for Creative Computing, 26 October 2009
2. Saini, O., Sharma, S.: A review on dimension reduction techniques in data mining. Comput. Eng. Intell. Syst. **9**(1), 7–14 (2018)
3. Singh, A.G., Asir, D., Leavline, E.J., Appavu, B.S.: An empirical study on dimensionality reduction and improvement of classification accuracy using feature subset selection and ranking. J. Theoret. Appl. Inf. Technol. (2012)
4. Shi, C., Chen, L.: Feature dimension reduction for microarray data analysis using locally linear embedding. In: International Symposium on Bioinformatics Research and Applications APBC, pp. 211–217 (2005)
5. Venkat, N.: The curse of dimensionality: inside out (2018)
6. Cunningham, J.P., Ghahramani, Z.: Linear dimensionality reduction: survey, insights, and generalizations. J. Mach. Learn. Res. **16**, 2859–2900 (2015)
7. Ebied, H.: Feature extraction using PCA and kernel-PCA for face recognition. In: International Conference on Informatics and Systems (2012)
8. Sorzano, C.O.S., Vargas, J., Motano, A.P.: A survey of dimensionality reduction techniques. ArXiv (2014)
9. Konsorum, A., Jackel, N., Vidal, E., Laubenbacher, R.: Comparative analysis of Linear and Non Linear Dimension Reduction Techniques on Mass. Cold Spring Harbor Laboratory (2018)
10. Krivov, E., Belyeav, M.: Dimensionality reduction with isomap algorithm for EEG covariance matrices. In: International Winter Conference on Brain Computer Interface (2016)
11. Griparis, A., Faur, D., Datchu, M.: Feature space dimensionality reduction for the optimization of visualization methods. In: 2015 IEEE International Geoscience and Remote Sensing Symposium (IGARSS) (2015)
12. Thomas, D., Oke, O., Smartt, C.: Statistical analysis in EMC using dimension reduction methods. In: 2014 IEEE International Symposium on Electromagnetic Compatibility (EMC) (2014)
13. Sancheti, P., Shedge, R., Pulgam, N.: Word-IPCA: an improvement in dimension reduction techniques. In: 2018 International Conference on Control, Power, Communication and Computing Technologies (ICCPCCT) (2018)
14. Jatram, A., Biswas, B.: Dimension reduction using spectral methods in FANNY for fuzzy clustering of graphs. In: 2015 Eighth International Conference on Contemporary Computing (IC3) (2015)
15. Pei, Z.H., Shen, Q.: Local linear dimensionality reduction algorithm based on nonlinear manifolds decomposition. In: 2017 International Conference on Network and Information Systems for Computers (ICNISC) (2017)
16. Raj, J.S.: A comprehensive survey on the computational intelligence techniques and its applications. J. ISMAC **1**(03), 147–159 (2019)

Security Issues in Cloud Computing

Shivani Goyal$^{(\boxtimes)}$ and Rejo Mathew

Mukesh Patel School of Technology Management and Education,
NMIMS, Vile Parle West, Mumbai 400056, India
shinugoyal12@gmail.com, Rejo.Mathew@nmims.edu

Abstract. Cloud Computing is the resource provided over the Internet for the use of data storage by the user. It is an environment used to enable a cost-effective storage of a large amount of shared pool of resources over the Internet. It has a rapid expansion over the few years for large companies and private sectors. Cloud computing with its numerous benefits also faces data privacy and security issues that causes threats in the clouds. This paper, hence, lists the various security issues in the clouds based on their confidentiality, integrity and availability and the corresponding solutions to those threats and issues in the cloud environment. It also lists a comparative study of all the solutions to the security issues so that the best-fit solutions can be identified based on the CIA triad. This paper analyses the solutions of security issues faced by the cloud users and providers.

Keywords: Access · Administration · Assaults · Attack · Authentication · Availability · Client · Cloud confidentiality · Data security · Hackers · Information · Integrity · User

1 Introduction

Cloud Computing uses different services such as software or servers over the Internet for data storage and management over the Internet. It allows a user to store data in a private location or with a third party user. It is a convenient, faster and on access demand of IT resources. Cloud computing services can be classified as Infrastructure as a service (IaaS), Software as a service (SaaS) and Platform as a service (PaaS) [1]. Some of the well-known cloud services of today's time are Dropbox, Gmail and Amazon, which can be accessed through the web. It is known for its scalability and reliability, but still has various security issues with respect to the sensitivity of the individual's data. Although, various issues concerning to the security are the information credibility, secrecy of the client and administrative clients, while the threats includes sensitive data loss, cloning and data leakage. The next section of the paper enlists issues and the various solutions which can prevent or reduce the threats to the security of cloud computing. Cloud environment is exploited which has various issues in different aspects, one of the major is based on the CIA triad where cloud faces major physical and data related attacks which can be either mitigated or detected at earlier stage to protect the cloud environment. Confidentiality of cloud is to provide with the authenticated user access, integrity to not have any modifications to the data that the

© Springer Nature Switzerland AG 2020
A. P. Pandian et al. (Eds.): ICCBI 2019, LNDECT 49, pp. 363–373, 2020.
https://doi.org/10.1007/978-3-030-43192-1_42

user has and availability is the presence of information without any leaks and timely updates. Based on these the cloud issues are listed and appropriate solutions have been determined [2, 4].

2   Issues in Security of Cloud Computing

One of the major drawback that cloud computing faces is to build a secure environment for the business software. There should be a trusted network for the cloud security otherwise the hardware and the software face a lot of issues [1] (Fig. 1):

Fig. 1.   Cloud attacks based on CIA

a. Confidentiality

Confidentiality is the protection of one's data from unauthorized access, such users can refer to the clients in cloud computing. A cloud's confidentiality can be affected by viewing or tampering the client's information by another unauthorized client or the administrator. Thus, the dishonest access on the cloud can be done through various methods some the categories are listed below: [11].

2.1   Data Breaches

Hackers breach the data on cloud providers as large amount of data is stored on the cloud which takes away the confidentiality of the cloud network. This data breach can lead to data loss of important health, financial or personal data. Thus the cloud suppliers have to convey the security control to ensure their own information in the cloud. The information could be of financial use or any other important confidential private information [1, 2].

2.2   Reduced Network Security

The transfer of data takes place through the internet in cloud computing hence must be highly secured so that there is no leakage of the data. It is because of an enterprise's lack of visibility to the network which leads to suspicious activity in the cloud [2].

2.3   Advanced Persistent Threat Parasite (APT Parasite)

It is the cyber-attack caused through a USB gadget or other PCs, though once they are in the system they behave normally like the system moves but the hackers can use and destroy the system without the knowledge of the user [7, 13].

2.4   Side Channel Attack

An attacker places a virtual machine in the vicinity of the target cloud computing system to perform a side channel attack. This attack is due to exploitation of the computer system when it performs its cryptographic operations, thus it is not due to the weakness inside the system but due to monitoring of computer's electromagnetic radiation and power emissions. Side channel attacks take two major steps to attack the system, Virtual Machine's Co-Residence and VM's Extraction to firstly place the attacker's instance on same machine as the target machine and later to let the malicious instance to utilize the information of co-resident channels through the side channel [10].

b. Integrity

Integrity is to protect the user's data by not allowing any modification by any third party user or the administrator itself. It is used to ensure that the data present is accurate and updated with respect to the data entered by the user or the client. Integrity is thus an important aspect which if not present might delete, append or edit the information due to the presence of various attacks: [11].

2.5   No Recovery

When the cloud server is down, if the users do not have complete surety of their data it can be used by a third party if the data is not restored properly. This third party can modify the data, make it less reliable and introduce different malicious links. Thus recovery of the old state of the system is not possible [8].

2.6   Wrapping Attack

It is also called as XML signature wrapping attack. During the translation of SOAP messages at transport layer, the signatures as well the content of the message is sent to the server as a valid user. Thus while executing the cloud storage is interrupted by the attacker and the malicious activities are performed on it [10].

2.7   Malware Injection Attack (Meta Data Spoofing Attack)

In this attack, the attacker inserts a malicious code which acts like the existing service to the cloud. Attackers steal information and force the users to download malicious link without any knowledge of the user. This could be done for data modifications, blockings or full reverse of the user's data, the attacker thus has full control over user's data [10].

2.8   Cross Scripting Acting

The hackers insert malware scripts such JavaScript, HTML and VBScript into the vulnerable web pages of the cloud environment which gets executed on the user's browser and thus the access is received to the third party which leads to the changes in the user's data [11, 12].

c. Availability

Availability is the data that should be available in the cloud storage in its original or intact form. A slightest downtime will result in the irrecoverable loss such as monetary losses. Availability can suffer due to various downtimes which mainly consists of: [11, 14].

2.9   Breakage of Information

Data clouds are accessible from anywhere and at anytime, cloud storages can face issues with the access as information can break and different types of assaults can occur, thus we have to assure solid passwords and multi figure validation. Clouds are a target for such assaults as they store a huge amount of information which can easily be broken down [3, 4].

2.10   Permanent Data Loss

Permanent data loss is caused due to unintentional cancellation of information by the user or admin or due to natural loss such as breakdown of operating system or damage to the hardware [4].

2.11   DOS Attacks

DOS assaults is consuming of all the resources of the cloud service so the users do not get any information and resources. It is done on public clouds.DOS attacks prevents authenticated users from accessing the data by flooding the network or disrupting the service which has been provided to the user. The attacker sends a large amount of network bandwidth which is then consumed by the user. The major types of impact on DOS attacks are direct and indirect DOS attacks [7].

2.11.1   Direct Denial of Service
When there is a high workload on the cloud computing storages, it attempts to give computational power to other systems like the Virtual Machine or Service Instances. This is to cope up with the extra workload, providing cloud protection.

2.11.2   Indirect Denial of Service
One of the side effect of flooding attack is that services that are provided on the same hardware servers may have a heavier workload due to flooding. Thus if a service runs on the same server with another service instance which has been flooded it will affect the availability of the cloud storage system [7].

2.12   Flooding Attack

In a flooding attack a third party creates fake data and thus whenever the server is overloaded it assigns the job to the nearest server which is not loaded or is offload. When the requests are sent the server checks the legitimacy of the system, authentication and the CPU consumption and use of memory space causing a flooding. The server allocated offers a quicker and more capable request answer thus the server is assigned to the user [10, 15].

3   Solutions to Security Issues

Fig. 2.   Cloud attack solutions based on CIA

To make clouds more secure there are various solutions to issues of security issues faced by the clouds. They include (Fig. 2),

a. Confidentiality

Confidentiality of the data is preserved by the help of authentication and prevention of the various attacks by the third party that the system is vulnerable to, it helps to allow only the legal users to access their data with the various enlisted solutions mentioned below:

3.1   Authenticated Access to Data

Data should be accessible to the administrator only and not the users. To avoid security breaches, all the different users should not be given access to the user's account. Authentication is done by computer-human interaction through a valid username and password created. The credentials used are compared to the one stored in the database or on any OS, then compared, if they match the user is given access to the cloud [1]. Token Based Authentication uses tokens instead of session IDs, stores our identity in the server or the memory disk instead of the database. These tokens are generated after user logs in and are passed to another state without user requiring to authenticate themselves again. It thus lowers the load on the server and is stateless [1].

3.2   End to End Encryption

It prevents data from being read by any other person except for true users and receivers. Data is thus, decrypted on the user's personal computer or the phone and not on the cloud server. The messages stored on the server are encrypted so that no Man in the Middle Attacks can be performed on it. Network security can be obtained by end to end encryption of the transport layer so that when an enterprise sends and stores a message on cloud providers, it is encrypted and there is no possibility of data leakage. Cloud providers like Apple where Icloud is a provider uses end to end encryption [9].

3.3   Image Steganography

Steganography refers to hiding information in some other information so that the secret data that exists is not known to the third party users. In image steganography an image is used to cover the information which is to be kept. The image is like a cover in which all the information is embedded which is termed as Stego-image. The information or data is embedded into the image by detecting the edges of the image through the pixel key pattern. Image pixels are selected at random and then are replaced by the text ASCII values. Hence the pixel values which are made up of three bits are replaced and merged to form a new matrix which is secret and can be extracted later by the administrator [2].

3.4   Prevention of APTs (Advanced Persistent Threat)

APT attack can be prevented by the use of antivirus software and scan them frequently. APTs can also be prevented by use of less gadgets in cloud computing [13].

Install a Firewall-Installation of firewalls such as hardware, software or cloud firewalls helps to prevent the APT attack. Install a VPN (Virtual Private Network)-It is an encrypted tunnel which is between the employee and the admin, preventing the hackers to gain access and snoop through the network [13].

3.5   Solution to Side Channel Attack

A combination of virtual firewall appliance and at random encryption decryption is done to achieve security at the front as well as the back end of the cloud computing system. A Virtual Firewall Appliance prevents the placement of an instantiate new Virtual Machine to the targeted Virtual Machine to extract confidential information from the cloud. Through random encryption and decryption the extraction of the side channel attack is prevented, confusion-diffusion occurs to make the relationship between plaintext and cipher text difficult to know and later diffusion occurs to dissipate the redundancy of the text [10, 15].

b. Integrity

Integrity of the data is achieved by no modifications to the data present and maintaining the accuracy of the data entered which is done by preventing the attacks, securing the leakage or destruction of data and various other methods listed below:

3.6   Secure the Data Destruction

Secure data destruction means to securely remove the unnecessary data from the cloud as and when needed so that the hacker does not retrieve the data, thus the information needs to be pulverised appropriately [1]. To maintain data transparency- Data transparency is to easily access the data and make changes in the data irrespective of the data's location. Data transparency can be maintained by the use of various orchestration and management tools [1, 3]. To establish documentation methods-Chain of custody is required for data to be accessed and used by authorized users only and they cannot tamper the hard drive. It is on the cloud providers to decide whether to give permissions for its users or businesses to allow making changes in the data or allowing data sanitization [3].

3.7   Wrapping Attack Solution

A SOAP (Simple Object Access Protocol) message has the structural information which stores the data that has to be exchanged between the server and the browser. The attack is thus prevented by increasing security through SOAP messages. STAMP, called as an extra bit is appended to the signature as a SOAP header through the Transport Service Layer so that the hacker cannot attack as the value of the signature is altered. It uses XML signature wrapping [10].

3.8   Malware Injection Attack Solution

The cloud user creates an account and the provider thus, creates a virtual machine (VM) image in the cloud image storage system. File Allocation Table (FAT) is used to check the current code that is being run in the user's VM. From the FAT table the information that the customer is going to run or runs is fetched. The previous instances can be checked that has already been executed in the customer's machine to know validity and integrity of the instance [9, 10].

3.9   Cross Scripting Attack Solution

Cross Scripting Attack can be mitigated through various techniques such as input validation, HTTP cookies and encoding, by following standards and guidelines. Input Validation checks on the input data by converting the certain characters into the format of the language such as HTML or JavaScript. In HTTP cookies, a cookie is sent as a response header with the HTTP which reduces the risk of cross scripting by sending the cookie which does not interact with the malicious script. Encoding is the other way which converts the data into HTML quote or encodes the alpha numeric code to prevent them to be interpreted as HTML or JavaScript language [12].

c. Availability

Availability of data is necessary to keep up with the updates of the system and the data, which when affected lags the cloud environment. To ensure the consistent availability of data various steps or methods can be followed:

3.10   Multifactor Authentication

Authentication is to ensure that authorized users only access the application and their identity is verified before the access is granted for the use of any services. Multifactor authentication such as one time password and mobile verification codes [5] can be provided to increase the security on the network. If someone tries to breach into someone's account then the code and information will be sent to user's account so that he authenticate the data. Two factor authentication makes it more difficult for the unauthorized users to access the application since Public Key Infrastructure (PKI) is combined with mobile verification [6, 14].

3.11   Encryption of Backups

The backup data should be encrypted otherwise encryption of data would not have any meaning as the data can be hacked and the backups can be accessed. Thus with the encryption all the lost data by fault or otherwise can be retained [1, 4].

3.12   Prevention of Dos Attacks

This can be prevented by already knowing the vulnerabilities of the system and working on it, also by maintaining a separate backup of Internet connection with setoff IP addresses so that an alternative path is given to the users [7]. Upstream Filtering routes the data through the means of network cleaning in which the malicious data is separated from the important. It blocks the entire traffic and then separates the malicious traffic. Set router level security and install firewall by which the threats are prevented, which encapsulates firewalls, VPNs, anti-spam and content filtering. It also includes the traffic management to check for the attacks and system should be updated in time [7].

3.13   Solution to Flooding Attack

Flooding attack can be solved by allocating each server to do a specific task while all the servers can internally communicate with each other with the help of message passing. Thus, when a server is overloaded a new server acts as the final platform or destination of all the requests from the overloaded server. To validate that the request is coming from a specific known user PID (Patient Identification Segment) is added to the message to check for the authentication, this PID is also encrypted through public key cryptography such as RSA [10, 14].

4 Analysis/Review

(See Table 1).

Table 1. Comparison table for confidentiality solutions

Based On Confidentiality

	Authenticated Access To The Data	End to end encryption	Image Steganography	Prevention of APTs	Solution To Side Channel Attack
Load On Server	• Identity is already stored in database. • Hence less load on server.	• Encrypted data is stored on the database. • Heavy load on the server.	• Data is stored inside the other encrypted data. • Thus heaviest load on server.	• Least load on server until too many antiviruses are installed.	• Encryption decryption and firewall machines present. • Thus heavy amount of load.
Ease of Login	• Single log-in as tokens are generated after first login.	• Double login authentication required.	• Single log-in, as the right image is selected then is authenticated.	• Double login authentication required.	• Double login authentication required.
Access to Data	• The administrator and server have access to data.	• Only the end users have access to the data.	• Only the users have access to the data if there is no breaching of image encryption.	• The end users and servers have the access to the data.	• Only users who are authenticated have data access.
Software/ Algorithm Used	• Requires biometric or SMS verification.	• Does not depend on software as encryption is provided.	• Edge detection algorithm is used through the Pixel Key Pattern.	• Uses strong antivirus software, along with Firewall or VPN (Virtual Private Network).	• Public key cryptography is used.
Data leakage	-	• No possibility of data leakage.	-	-	-

Based On Integrity

	Secure Data Destruction	Solution to Wrapping Attack	Solution to Malware Injection Attack	Solution to Cross Scripting Attack
Supported with	• No such specific support is required.	• Uses XML wrapping • Based on SOAP for the storage of information.	• Supported through the Virtual Machine. • File Allocation Table (FAT) is used for storage.	• Supported with HTTP cookies and HTML or JavaScript.
Recovery of Data	• Data can be recovered through the documentation methods.	• Recovery of Data is not possible. •	• Recovery of Data is not possible. • Only prevention can safeguard the data.	• Recovery of Data is not possible.
Amount ofStorage	• It keeps on removing the unnecessary data from the cloud. • Thus maintains the storage space in cloud.	• SOAP messages when appended with extra bits add more to the storage space in the cloud.	• The File Allocation Table (FAT) keeps a record of all the data, thus requires a huge amount of storage.	• Storage required for HTTP cookies, encoded data, input validation data.

5   Conclusion

The paper overviews the cloud environment security and what are the major attacks it could face in today's time, along with the solutions to such attacks and the best possible solutions to such attacks for the user to determine which solution could be followed. It stated about the protection of one's computer or cloud environment from the malicious attacks that can be faced by the client and the reasons behind it. In the near future such attacks can be lowered by the nown knowledge of both the users and the service providers providing a safer environment in the cloud for the storage and transmission of data. Thus, data security can be provided on the basis of data confidentiality, integrity and availability. For data confidentiality, one of the best solutions to the threats is authentication of the data present and the end to end encryption level, under the condition that they should be given adequate data protection. For data integrity, there should be secure data destruction so that blackouts of the data are avoided and there is no unwanted modification of data. Data availability can be maintained by encryption of the backups if the cost factor is considered, also high multifactor authentication is one of the best solutions' since it is very cost effective and affects the system positively the fastest. Thus, cloud computing solutions prove to be a great help for the cloud providers and users to maintain the usage of the cloud.

References

1. Tabassam, S.: Security and privacy issues in cloud computing environment. J. Inf. Technol. Softw. Eng. (2017)
2. Kaur, R., Kaur, J.: Cloud computing security issues and its solution. In: 2nd International Conference on Computing for Sustainable Global Development INDIACom. Punjabi University, Patiala (2015)
3. Sinjilawi, Y.K., AL-Nabhan, M.Q., Shanab, E.A.A.: Addressing security and privacy issues in cloud computing. J. Emerg. Technol. Web Intell. 6(2) (2014)
4. Bokhari, M.U., Shallal, Q.M., Tamandani, Y.K.: Security and privacy issues in cloud computing. In: International Conference on Computing for Sustainable Global Development INDIACom. Aligarh Muslim University, India (2016)
5. Gayathri, M., Srinivasan, M.K.: A survey on mobile cloud computing architecture, applications and challenges. Int. J. Sci. Res. Eng. Technol. IJSRET 3(6) (2014)
6. Fernando, N., Loke, S.W., Rahayu, W.: Mobile cloud computing: a survey. Future Gener. Comput. Syst. 29, 84–106 (2013)
7. Mehta, R.: Distributed denial of service attacks on cloud environment. Int. J. Adv. Res. Comput. Sci. 8(5), 2204–2206 (2017)
8. Chandrahasan, R.K., Priya, S.S., Arockiam, L.: Research challenges and security issues in cloud computing. Int. J. Comput. Intell. Inf. Secur. 3(3), 42–48 (2012)
9. Gentry, C.: Fully homomorphic encryption using ideal lattices. In: ACM Symposium on Theory of Computing, STOC, pp. 169–178. Stanford University and IBM Watson (2009)
10. Kumar, P.: Cloud computing: threats, attacks and solutions. Int. J. Emerg. Technol. Eng. Res. IJETER 4(8) (2016)
11. Basu, S., Bardhan, A., Gupta, K., Saha, P., Pal, M., Bose, M., Basu, K.: Cloud computing security challenges & solutions-a survey. In: PIEEE 8th Annual Computing and Communication Workshop and Conference. Institute of Engineering & Management, Kolkata (2018
12. Sheeta, N., Alam, M., Singh, M.: Web based XSS and SQL attacks on cloud and mitigation. J. Comput. Sci. Eng. Softw. Test. (2015)
13. Vance, A.: Flow based analysis of advanced persistent threats: detecting targeted attacks in cloud computing. In: IEEE International Scientific Practical Conference, Problems of Infocommunications, Science and Technology, Kharkiv, Ukraine, pp. 173–176 (2014)
14. Kumar, D.: Review on task scheduling in ubiquitous clouds. J. ISMAC 1(01), 72–80 (2019)
15. Praveena, A., Smys, S.: Ensuring data security in cloud based social networks. In: 2017 International Conference of Electronics, Communication and Aerospace Technology (ICECA), vol. 2, pp. 289–295. IEEE (2017)

A Review on Predicting Cardiovascular Diseases Using Data Mining Techniques

V. Pavithra[✉] and V. Jayalakshmi

School of Computing Sciences, Vels Institute of Science,
Technology and Advanced Studies (VISTAS), Chennai, India
vpavithra.1989@gmail.com, jayasekar1996@yahoo.co.in

Abstract. The main objective of the work is to analyze various data mining techniques in the health care field that can be employed in "predicting cardiovascular diseases and their efficient diagnosis. Cardiovascular system is the first organ system to become fully functional in uterus. Cardiovascular diseases is one of the major common diseases that cause death all around worldwide. People of all ages are affected by this disease particularly the elderly people. It is essential to predict the group of people commonly affected and identifying risk factors like age, sex, lifestyle that will be helpful in early diagnosing and prevention of heart diseases. At present huge number of people are pretentious to heart diseases and hence it is quite difficult to predict accurately. Proper mining methods can save enormous number of people from mortality due to heart diseases. This paper analyses various types of heart disease and prediction techniques used in heart disease prediction.

Keywords: Data mining · Cardiovascular diseases · Data mining techniques · Data mining tool

1 Introduction

Data mining is the process of exploring and analyzing large amounts of data to identify valuable patterns and trends. There are plenty of hidden information that are largely untapped, with the help of data mining the data can be transformed into useful information. Over the past 10 years data mining techniques play a vital role in the health care industry to diagnosis the heart diseases. Large amount of data is generated by health care organization and to extract the data from the same "intelligent healthcare information system" was used. With the help of data mining techniques, we can extract the hidden data from large number of datasets available [1]. The health care industry mainly focuses on quality, services, diagnosis or predict the diseases in efficient manner.

Medical diagnosis is often a challenging and complex task because many of the science are non-specific. one of the most interesting aspects of the evolution of medical technologies is the constantly increasing involvement of mechanization in various medical operations [2]. All doctors do not possess expertise in every subspecialty and automation helps to get rid of these limitations to achieve accurate results. Data mining in the health care field helps to extract the knowledge and helps health care services professionals to take intelligent clinical decisions.

© Springer Nature Switzerland AG 2020
A. P. Pandian et al. (Eds.): ICCBI 2019, LNDECT 49, pp. 374–380, 2020.
https://doi.org/10.1007/978-3-030-43192-1_43

ML (machine learning) is the process of automatically identifying trends and making intelligent decisions based on qualified data samples. Machine learning methods play a vital role in diseases diagnosis, prediction and other medical related aspects with help of data mining techniques. IN health care industry diseases prediction and decision making plays an important role and falls under the umbrella of computerized reasoning. It is the capacity of a machine to gain from a huge arrangement of information and predict, group or order similar data. Some well-known ML strategies incorporate Artificial Neural Network (ANN), Support vector machine (SVM), Decision Tree, self- association guide, and means K-means clustering, SKNN and so on [3, 4].

This paper reviews with various automation technology (data mining) for the prediction of heart diseases with the help of machine learning techniques. The rest of the paper includes three sections. Section 2 describes various types of cardiovascular diseases and its risk factor. Section 3 presents a brief review on predicting cardiovascular diseases using data mining techniques. Finally, Sect. 4 discuss the conclusion and future work.

2 Cardiovascular Diseases and Its Risk Factors

One of the main sources of death all around the globe is heart disease, in recent research also it was reported all around the world 75% of the people die yearly from cardiovascular diseases and specifically 25% of people are affected by cardiovascular diseases in United States alone. Cardiovascular diseases are the most common problem faced by the community at present and it was viewed as a notable disease in old and medieval times. Almost more than half of the men alone at the age of 60 suffer from some form of heart Disease and once affected there are explored to lifelong treatment. There are various kinds of heart diseases and each type has its own symptoms and treatment differ based on the symptoms observed. For instance, just lifestyle changes and medicine alone can improve the health of the person and for some surgery is mandated to make the ticker work well.

Coronary Artery Disease (CAD): Is the most common type of heart problem occurs due to blockage in Coronary Arteries. This artery supplies nutrients and oxygen in the form of blood to heart muscles, if the arteries get affected due to plaque deposits due to cholesterol, inactive lifestyle, high blood pressure, smoking and stress. There will be increased risk for a heart attack once the artery is completely blocked. In United States 16.5 million people over age 20 are affected by CAD. [6]

Heart Arrhythmias: It means heartbeat is irregular i.e. the heartbeat count is either too fast or slow. Irregular heartbeat is caused due to dizziness, breathlessness, and palpitations. Some of the arrhythmias are harmless and some of them are abnormal. The damaged arrhythmias cause serious problem and even healthy people with no medical history also get affected by elongated term arrhythmias due to outside cause such as abusing a drug or unwanted medicine or electric shock.

Heart Failure: It occurs when heart not able to pump required amount of blood to all organs of the body, it can affect both side of the heart. It can be either be long-term or short-term conditions. Short- term (Acute) condition occurs only when there is a problem in heart valves and this condition occurs after a heart attack. The symptoms appear suddenly and can improve over time however in chronic or long-time conditions the symptoms will be observed over long period of time. Majority of heart failure cases are of type chronic [7]. In recent research reported by the "center for diseases control and prevention" over 5.7 million American people get affected by heart failure in that most of them are men. The main reason for this could be CAD, Thyroid issuer certain other conditions.

Congenital Heart Disease: This is type of heart disease where heart formation is not proper during the development of baby in the womb. Holes will be observed in between two sides of the heart or blood flow will be minimal to the lungs. It can be detected through ultrasound and sometimes it cannot be properly diagnosed till birth. In recent article it was reported 10% of babies are born with CHD defects globally. The common cause of death is due to improper diagnosis and wrong treatment. The most important fact that is essential to control the death rate is proper intervention and it can be achieved by collecting adequate information about heart disease.

2.1 Risk Factors and Symptoms in Cardiovascular Diseases

There is a number of risk factor in cardiovascular diseases and it can either be controllable or uncontrollable and below are the leading risk factors for cardiovascular heart diseases [5]. The definitions and attributes are summarized in Table 1.

Table 1. Risk factor in cardiovascular diseases

No.	Attributes name	Description
1.	Age	Age of the patient in a year
2.	Sex	Gender of the patient
3.	Cp	Chest pain type
4.	Trestbps	Blood pressure
5.	Chol	Cholesterol
6.	Restecg	Resting results of electrocardiography
7.	Ca	Number of major vessels that colored by flourosopy
8.	Fbs	Fast blood sugar
9.	Oldpeak	ST segments that induced by exercise relative to rest
10.	Slope	Peak exercise ST segments
11.	Thal	Defect values
12.	Exang	Exercise induced angina

3   Review on Cardiovascular Diseases Prediction Using Data Mining Techniques

Tarawneh and Embark [8] proposed an algorithm that combines all the techniques in a single algorithm (hybrid algorithm) of the systems into one. This article is to analyze different classification and prediction techniques to predict cardiovascular diseases and consider them, by then, merge the result from all of them to get exact results. The diagnoses of heart disease with various methodologies and different classification accuracy. Here we have examined the majority of the classification techniques in data mining and apply them to the Cleveland informational collection. A portion of the procedures performs better in every case, for example, Naïve Bayes and SVM, while others rely upon the chose features. Here the efficiency of the algorithm can be improved by hybridizing the systems (KNN, ANN, NAVIE'S BAYES, DECISION TREE) into a single algorithm. This hybridization calculation can be utilized as a specialist framework in medical clinics to help specialists to analyze coronary illness rapidly and spare an actual existence.

Bashir et al., concentrates on feature selection techniques and algorithms where multiple heart disease datasets are used for experimentation analysis and also to show the accuracy improvement. The work is implemented with the help of Rapid miner as a tool. DT, LR, Logistic Regression SVM, NB, and RF algorithms are used for feature selection techniques. This research achieves the goal, which was as per expectation and accuracy have improved from previously mentioned values in a literature review. The outcome of the paper shows clearly that the increase in efficiency has achieved in 2 techniques (SVM) and (NB) are applied through this paper. This paper proposed with Logistic regression as higher accuracy, and it has as the best feature selection technique for predicting heart disease [9].

Hybrid Recommendation System for the diagnosis of heart disease [10] in which the author addressed the diagnosis of heart disease on the basis of two deep learning techniques Multiple Kernel Learning with Adaptive Neuro Fuzzy Inference System (MKL with ANFIS). The MKL technique is contrasted with distinct methods in the deep learning algorithm in order to provide a precise amount of heart prediction. The suggested MKL method with ANFIS to classify patients with heart disease. In this work, the MKL method utilizes two techniques of multiple cross validations, Resulting in elevated precision relative to the existing research.

Ambedkar et al., the analyst investigated the information about whether the patient is having heart diseases or not and also to predict the risk level of the patients in CV. Examining information assumes a significant job in dealing with a lot of information in the medicinal services industry. Be that as it may, the exactness diminishes when the medicinal information is in part absent. To conquer the issue of missing medical information, we perform information preprocessing procedure to change the missing information into finish information. This work manages the CNN-UDRP calculation to predict the risk level of the patients is 65% and for anticipating heart illnesses they utilized two strategies NB, KNN, they look at the outcomes among NB and KNN. The exactness level of NB is high, which is more than KNN [11].

Chavda et al., focuses on machine learning algorithm to predict the occurrences of cardiovascular diseases another thought process to build up the proposed framework was that, in the present framework an individual who expects to check his cardiovascular wellbeing needs to visit the specialist routinely and complete all the therapeutic tests according to instruct regarding the specialist and demonstrate the separate reports to him. Contingent upon the status of the reports, the specialist analyzes the patient and whenever found tricky, the patient is dealt with appropriately. The proposed technique based on mobile application that uses sensor to retrieve human body parameters and applies the attributes in decision tree classification algorithm in machine learning to predict the risk of CV [12].

A neural network technique [13] has been presented to monitor the health care system. The proposed work of these paper is to identify the diseases in patients with more accuracy, and also compare the results with existing work. Here ANN plays a vital role in managing and working in the medical diseases' dataset information. They used MATLAB software to find accuracy in the techniques.

Bojin et al., proposes EHR sequential modeling to predict the risk of heart failure. This method of heart failure prediction model is very effective. The work is implemented using neural network that can be engaged with one-hot encoding and word vector techniques to represent the patient diagnostic events. To build the network model they propose LSTM method and compared with all existing techniques and shows the accuracy level high in proposed system.

Wiharto et al., proposes Improved system diagnosis of coronary heart diseases. The improvement can be done in the feature selection. In most of the work the feature selection is not done properly. This paper focuses on selecting the features in very efficient manner with the help of tiered multivariate technique. Analytical method is done in logistic regression and for the classification they uses multi-layer perceptor neural network. The proposed system shows the way of selecting the good attributes and improved the accuracy [15, 16]. Different researchers used different data mining techniques and factors to find the accuracy of CAD prediction. The below tabular column represents the same (Table 2).

Table 2. Different data mining methods to predict cardiovascular diseases.

Year	Author	Purpose	Techniques	Accuracy
2016	Marjia et al.	Heart diseases prediction using WEKA tool and 10 fold cross validation	KSTAR	75%
			J48	86%
			SMO	89%
			MULTILAYER PRECEPTRON	86%
2017	Emrana Kabir Hashi	An expert clinical decision support system to predict disease using classification techniques	C4.5	90%
			KNN	76%

(*continued*)

Table 2. (*continued*)

Year	Author	Purpose	Techniques	Accuracy
2019	Yin Kia Chiam 1, Kasturi Dewi Varathan2	Identification of significant features and data mining techniques in predicting heart disease	Vote (a hybrid technique with Naive Bayes and logistic regression)	87.4%
2017	Syed Muhammad Saqlain Shah et al.	Analysis of heart disease Diagnosis based on feature extraction using K-Fold cross-validation	SVM	91%
2016	J.S. Phul, S. Vatta	Diagnosis and predict diabetic heart diseases	J48, Navies Bayes	95.53%
2016	Dhivya s E. M. Mercy	Heart diseases prediction strategy using combinational and genetic algorithm	SAM algorithm	96.8%
2016	Oluwarotimi Williams Samuel	An integrated decision support system based on ANN and fuzzy AHP for heart failure risk prediction	AHP-FUZZY_AHP	91.10%

4 Conclusion

Cardiovascular diseases are one of the leading diseases at present worldwide, in this paper various data mining methods were discussed for predicting Cardiovascular disease. Various Machine learning algorithms were used to increase the accuracy of the heart disease prediction by using different attributes for the same. Various data mining techniques like Naïve Bayes, Decision tree, SVM, KNN etc. were discussed in the review for increasing accuracy. Our future work will be based on different data set and applying different data mining technique to improve the accuracy by selecting different attribute.

References

1. Khan, Y., Qamar, U., Yousaf, N., Khan, A.: Machine learning techniques for heart disease datasets: a survey. In: Proceedings of the 2019 11th International Conference on Machine Learning and Computing, pp. 27–35. ACM (2019)
2. Verma, C.V., Ghosh, S.M.: Review of cardiovascular disease in diabetic patients using data mining techniques (2017)
3. Ansari, H.F., Namdeo, V.: An efficient SKNN based approach for heart disease classification. Int. J. Adv. Technol. Eng. Explor. 6(53), 101–106 (2019)
4. Raman, M., Sharma, V.K.: Classification utility & procedures for recognition of heart disease: a review. Int. J. Sci. Res. Sci. Technol. (IJSRST) 3(8), 383–387 (2017)
5. Musunuru, K., Kathiresan, S.: Genetics of common, complex coronary artery disease. Cell 177(1), 132–145 (2019)

6. Guo, J., Erqou, S.A., Miller, R.G., Edmundowicz, D., Orchard, T.J., Costacou, T.: The role of coronary artery calcification testing in incident coronary artery disease risk prediction in type 1 diabetes. Diabetologia **62**(2), 259–268 (2019)
7. Michael, F.G., Mann, D.L.: Heart Failure: A Companion to Braunwald's Heart Disease E-Book. Elsevier Health Sciences, Berlin (2019)
8. Tarawneh, M., Embarak, O.: Hybrid approach for heart disease prediction using data mining techniques. In: International Conference on Emerging Internetworking, Data & Web Technologies, pp. 447–454. Springer, Cham (2019)
9. Bashir, S., Khan, Z.S., Khan, F.H., Anjum, A., Bashir, K.: Improving heart disease prediction using feature selection approaches. In: 2019 16th International Bhurban Conference on Applied Sciences and Technology (IBCAST), pp. 619–623. IEEE (2019)
10. Manogaran, G., Varatharajan, R., Priyan, M.K.: Hybrid recommendation system for heart disease diagnosis based on multiple kernel learning with adaptive neuro-fuzzy inference system. Multimedia Tools Appl. **77**(4), 4379–4399 (2018)
11. Ambekar, S., Phalnikar, R.: Disease risk prediction by using convolutional neural network. In: 2018 Fourth International Conference on Computing Communication Control and Automation (ICCUBEA), pp. 1–5. IEEE (2018)
12. Chavda, P., Bhavsar, H., Pithadia, Y., Kotecha, R.: Early detection of cardiac disease using machine learning. Available at SSRN 3370813 (2019)
13. Maji, S., Arora, S.: Decision tree algorithms for prediction of heart disease. In: Information and Communication Technology for Competitive Strategies, pp. 447–454. Springer, Singapore (2019)
14. Joseph, S.I.T.: Survey of data mining algorithm's for intelligent computing system. J. Trends Comput. Sci. Smart Technol. (TCSST) **1**(01), 14–24 (2019)
15. Jin, B., Che, C., Liu, Z., Zhang, S., Yin, X., Wei, X.: Predicting the risk of heart failure with EHR sequential data modeling. IEEE Access **6**, 9256–9261 (2018)
16. Wiharto, W., Kusnanto, H., Herianto, H.: Hybrid system of tiered multivariate analysis and artificial neural network for coronary heart disease diagnosis. Int. J. Electr. Comput. Eng. **7**(2), 1023 (2017)

Adaptive Object Tracking Using Algorithms Employing Machine Learning

Shubham Rai$^{(\boxtimes)}$ and Rejo Mathew

I.T. Department, Mukesh Patel School of Technology Management
and Engineering, Mumbai, India
shubham.rai46@nmims.edu.in, rejo.mathew@nmims.edu

Abstract. Object tracking is considered as one of the most crucial areas of research. Real time applications of object tracking include video surveillance, vehicle tracking and so on. In this paper, solutions for adaptive tracking of moving objects using algorithms that employ machine learning are discussed. This paper discusses the following machine learning tracking algorithms; Multiple Instance Learning Track (MIL Track), Tracking-Learning-Detecting (TLD), Multi Domain Network (MDNet) and Circulant Structure Kernel tracker (CSK). MIL Track algorithm uses a special classifier to generate an adaptive appearance model. TLD is a tracking framework that is based on the processes of tracking learning and detection. MDNet uses Convolutional Neural Network for training the tracker. The depth analysis method employs CSK to handle occlusions. These tracking algorithms use techniques that have proven to be successful in overcoming the adaptive object tracking problems. These problems and their corresponding solutions are discussed further in this paper.

Keywords: MIL Track · TLD · MDNet · Long–term tracking · Semi-supervised

1 Introduction

Traditional tracking algorithms are not full proof and suffer from the following issues,

- The tracking results are easily disrupted by similar objects.
- Trackers updated with sub-optimal examples leads to drifting.
- The tracking window shrinks to a small size when the target object gets lost.
- Reappearance of objects with changed appearance makes the initial frame reference irrelevant.
- If the target object disappears from the view frame for an extended period of time, then the tracker confuses the target with its surroundings after the target's reappearance.

Some of the issues faced by the traditional object tracking algorithms occur because they employ human labelers, and it is tough for the labeler to be consistent with cropping of the examples, since the exact locations of the target object are unknown. Tracking algorithms that employ machine learning help overcome these issues by eliminating the need for a human labeler and by developing an adaptive appearance

© Springer Nature Switzerland AG 2020
A. P. Pandian et al. (Eds.): ICCBI 2019, LNDECT 49, pp. 381–388, 2020.
https://doi.org/10.1007/978-3-030-43192-1_44

model. The adaptive appearance model is capable of creating real time examples depending on the motion of the target object; no other technique apart from machine learning has proven to be as successful as machine learning for this purpose. In Multiple Instance Learning Track (MIL Track) [1], the adaptive appearance model is generated by using the MIL classifier. The MIL Track algorithm is based on MIL Boost algorithm [2] and Online-AdaBoost algorithm [3]. MIL Track is analyzed further in Sect. 3.1. The Tracking-Learning-Detecting (TLD) framework [4], achieves a long-term tracking task by overcoming detector errors (explained in Sect. 2.2) and employing a learning technique known as P-N Learning. Section 3.2 discusses the P-N Learning technique in detail. MDNet (Multi Domain Network) [5], uses symbols from an independently trained Convolutional Neural Network (CNN). MDNet employs the Stochastic Gradient Descent technique for creating a more robust data set for CNN (discussed further in Sect. 3.3). Circulant Structure Kernel Tracker [14] (discussed in Sect. 3.5) uses RGB and depth space synchronization to overcome the occlusion problems (discussed in Sect. 2.5).

2   Problems in Adaptive Object Tracking

2.1   Selecting Sub-optimal Examples for Training

Training examples/samples are used to train the tracker in recognizing the target object. This can be done in a supervised fashion (examples are selected explicitly) or in a semi-supervised fashion. Supervised fashion shows poor results as compared semi-supervised fashion. Semi-supervised training uses real-time examples for training but suffers from drifting when the examples used for training are sub-optimal [1]. Rectangular bounding boxes are generally used for extracting examples for training, when these bounding boxes only include a part of the target object or if an occlusion is present at the time of sampling the extracted examples are said to be sub-optimal. Also, if too many bounding boxes containing positive examples are present in close proximity then this confuses the tracker.

2.2   Detector Errors

The detector in a tracking algorithm treats each frame independently and confines all the observed appearances to update the tracker location accordingly. Detectors in any tracking algorithm suffer from two problems: false positives and false negatives [4]. False negatives mean the examples that miss the target object even when the target object is present in the video frame. Whereas, false positives mean the examples that detect the target object in a frame even when the target object is absent. False positives also include confusion with similar objects and confusion with background. If the bounding box is overlapped with an object of similar category then the example is taken as positive, whereas highly textured backgrounds also lead to false positives. When the target object reappears in the frame after being out of it for an extended period of time then the tracker often misses it leading to false negatives.

2.3   Training Data Set for Convolutional Neural Network

In real time object tracking, training a data set for CNNs becomes tough because the target object suffers from variations from one frame to another [5]. For e.g. the same kind of object can be considered as a target in one frame and as a background in another frame. As a result of such inconsistencies, traditional learning techniques that are based on standard classification task are not suited for training CNNs. Thus, a different approach is required to capture the domain independent information that will lead to a better representation of the target object. CNNs that are pre-trained offline and then transferred onto a large-scale data set created for image classification [9, 10] provide poor accuracy in object representation because of the basic difference between the classification and tracking tasks.

2.4   Drifting in Stochastic Gradient Descent

Out of all the examples considered for training only a few of the negative examples are used (these are called as distractors, they are usually false positives), since majority of the negative examples are redundant and do not carry any useful feature/information. Traditional SGD techniques for training a CNN involve an even contribution of all the examples extracted. Due to this, the contribution of the distractors is not used efficiently. As a result of this, the CNN is trained inadequately, and this leads to drifting. Also, the lack of effective drifter contribution can lead to the problem of confusion between the target object and the background/similar objects.

2.5   Target Object Occlusions

The problem of occlusions is the most common hurdle faced by any tracking algorithm. Occlusions occur when the target object or the primary features of the target object (that are used to recognize the object) are not available for a camera sensor to track the object, even when the target object is present in the frame [12, 13]. Occlusions can occur in the following situations:

- Self-Occlusion: One section of the object is occluded by another section of the same object; this is common when jointed objects are being tracked.
- Inter-Object Occlusion: Occurs when multiple objects are being tracked, in this case target object covers another target object.
- Background Occlusion: A texture/feature of the background occludes the target object.

Severity of the occlusion depends on the extent of occlusion. When only a part of the target object is occluded either due to self-occlusion, inter-object occlusion or background occlusion, it is known as partial occlusion. In partial occlusion, not enough features are visible; hence object tracking becomes difficult in subsequent frames. Whereas, when the object remains in the frame but is completely covered it is known as full occlusion. Object tracking is toughest in the case of full occlusion because the tracker now has no relevant information about the object location from the current frame.

3 Solutions/Countermeasures

3.1 Multiple Instance Learning Track (MIL Track)

The MIL Track algorithm uses a classifier that creates a data set for training the tracker by dividing the examples extracted from each frame into positive and negative sets/bags. This approach of learning is more complex because very little information is provided to the learner, but this approach still provides better results than supervised learning algorithms that use explicit bounding boxes. A bag is said to be positive if at least one positive instance is present, else the bag termed as negative. While creating the required data set multiple examples can be termed positive, this happens because the algorithm crops out several bounding boxes around the point of the target object that has been selected by the human labeler (multiple positive instances can help in extracting more useful feature vectors of the target object). This information is then passed onto the learning algorithm that determines the most appropriate example. The MIL Track algorithm makes use of Haar-like features [1, 2] for depicting the object which are computed for each frame. Haar-like features provide a much better computation speed as compared to other features because they use integral images. The appearance model created by the MIL Track algorithm overcomes the problem of selecting sub-optimal problems for training. The MIL Track algorithm works as follows:

- Crop out examples from each frame that are within the search radius of the present tracker position as p(y|x), where x represents the example and y is the binary number representing whether or not the target object is present in the corresponding example (y = 1 implies object present).
- The MIL classifier is used to compute p (y|x) for every example.
- Now a greedy strategy is used to update the tracker location, i.e. instead of keeping a record of the target object's location in each frame a motion model is used where the tracker at any time 't' is likely to be present within the radius of the tracker position at time 't−1.'
- Now the cropped-out examples are classified into Xp positive bags and Xn negative bags. Since the bags contain a large number of examples, a random subset of the examples is selected for training.
- The appearance model for the MIL Track algorithm is updated with one positive bag and Xn negative bags.

Thus, by classifying the examples into positive and negative bags and using Haar-like features only the optimal examples are selected for training the tracker. Also, as a subset of the examples is chosen for training, the problem of too many positive examples in close proximity is also solved.

3.2 P-N Learning in TLD Framework

The TLD tracking framework makes use of a special technique, called the P-N Learning technique to overcome the errors faced by the detector. The basic idea behind P-N learning is that the detector errors can be recognized by two types of 'experts', the

P-Expert and the N-expert. The P-Expert only deals with the false negatives, whereas the N-Expert deals with only the false positives. Both of the experts can also have errors, but as their operation is mutually independent, the errors faced by them are compensated. The P-N Learning technique has for elements:

- A classifier that is to be trained.
- A data set containing the training examples.
- Training the classifier in a supervised fashion using the above data set.
- Functions to create positive and negative examples during learning, i.e. P and N Experts.

The first step in the learning process involves inserting a labeled set 'L' to the training set. This training set is then passed on, to train the classifier in a supervised fashion. Here, the classifier estimates the initial parameters, such as contour, illumination, background texture, etc. The further steps of the learning process occur through iterative bootstrapping. Now the classifier trained in iteration j−1, classifies the unlabeled data set in iteration j. This classification is then analyzed by the P and N Experts, which identify the examples that have been incorrectly labeled. The labels of these incorrectly labeled examples is changed and then added to the training data set. This iterative process continues till convergence. Recognizing the errors of the classifier is an important part of the P-N Learning technique. This is achieved by separating the estimations of false positives from the estimations of false negatives. Thus, the unlabeled data set is split into two parts based on the current classification and each of these parts is analyzed by the experts independently. P-Expert only analyzes the examples labeled as negative to identify the false negatives within this set. The P-Expert then adds these examples to a positive labeled training set. Whereas, the N-Expert analyzes only the examples labeled as positive to identify the false positives in this set. These examples are then added to a negative labeled training set.

As the classifications are being analyzed by two independent 'experts', the classifier is updated accordingly which results in fewer errors in detection.

3.3 Stochastic Gradient Descent in MDNet

Multi Domain Network (MDNet) involves a learning technique that aims to train the CNN employed by MDNet in a manner that makes it possible for the tracker to distinguish between the target object and the background in any given frame, this distinction should be unambiguous. What makes this learning technique tough is the fact that, data from different domains has different perceptions of the target object and the background. However, properties like motion blur, scale variations, etc. still have sufficient amount of commonality to be used to represent the target. The useful features from the properties are obtained by separating the domain specific from the domain independent information. MDNet uses Stochastic Gradient Descent (SGD) to train the CNN and achieve the above result. SGD is an optimization technique used in machine learning that uses a few randomly selected samples instead of the entire data set. This helps in achieving a faster computation time at the cost of more noise. In every iteration, SGD deals with a separate domain independently. In an iteration i, CNN is trained using a mini-batch that contains samples from the (i mod I) th sequence (i.e. a set of domains). This process continues until the network congregates or the predetermined number of iterations (I) is reached. As a result of this learning-technique, the domain independent knowledge is represented in the shared layers from which the useful features for target representation are obtained.

3.4   Hard Mini-Batch Mining

SGD technique employed by MDNet for training the CNN uses extensive distractor contribution to overcome the problem of drifting. This is achieved by using a technique called hard negative mining [7], which is employed for each mini-batch that holds the examples for training. Here, to identify the hard-negative examples, there is a constant switching occurring between training and testing processes. Each mini-batch in the iterative SGD technique contains M+ positive examples and Mh− hard negative examples. To identify the hard-negative examples, the negative examples are tested and those examples are selected that have the highest scores. Hard mini-batch mining tests a predetermined number of examples and identifies the distractors from these examples without having to use a detector to identify the false positives.

3.5   Handling Target Object Occlusions

Occlusion handling is an inevitable aspect of object tracking. The occlusion problem can either be tackled during or after the occlusion has occurred. Methods used to handle the occlusion problem are generally called as occlusion recovery mechanism [13]; two of them are discussed here:

3.5.1   Depth Analysis Method

Depth feature calculations have proven to be extremely useful in safe guarding machine learning tracking algorithms against noise due to occlusions. In [13], the depth attributes are effectively included in the Circulant Structure Kernel Tracker (CSK) [14]. This algorithm uses a 3D approach to track objects that combines the depth attributes with the RGB features to generate a single tracking framework. The CSK tracker is selected as the core tracker in this framework. The above algorithm progresses through the following steps:

- The tracker is initialized with the coordinates of the target object in the first frame and then a handshake process is used to synchronize the RGB and depth spaces.
- Now, Circulant tracking starts and training examples are generated for the classifier. At the same time, a patch-based appearance modeling of the target object is done in the depth space.
- Here, Gaussian Mixture Modeling [14–16] is used for depth space optimization. The patch-based modeling is now applied to the optimized depth space.
- The patch-based model obtained from the depth space is then used by a robust occlusion estimator to detect the occlusions and update an adaptive model.
- Occlusion recovery becomes feasible in the next frames due to the adaptive model generated above.

The adaptive model mentioned above is meant to effectively capture the true feature representations of the object in both occluded and non-occluded states.

4   Analysis/Review

4.1   Comparison Table

See Table 1.

Table 1. Comparison between MILTrack, P-N Learning, SGD and CSK-3D

	MIL Track	P-N learning	SGD	CSK-3D
Training data set	Created by classifying the examples into positive and negative bags/sets	Robust training data set created by identifying false positives and negatives	Hard negatives used to create a more informed training data set	Co-ordinates of the object are provided for initialization. Further steps progress based on the updated adaptive model
Solution progression	Iterative progress without bootstrapping	Progress iteratively with bootstrapping	Progresses iteratively with bootstrapping	Progresses iteratively with bootstrapping
Target object representation	Uses Haar-like features for target representation	No special ways for representing target object	Representation of target object is based on invariant features	RGB and depth attributes are used for representing target objects
Optimization	MIL classifier helps in optimizing each iterative step through robust classification	Tracker and detector work simultaneously to optimize each output by benefiting from each other	Uses hard mini-batch mining to optimize the SGD outputs in MDNet algorithm	Gaussian Mixture Modeling is used for depth space optimization

5   Conclusion and Future Work

This paper analyzed various issues related to real-time object tracking and the corresponding solutions for these problems. A broad field for research is still untouched in incorporating machine learning with object tracking. However, no other approach proves to be as effective and efficient in developing a real time long-term object tracker as machine learning. Out of the solutions discussed above the TLD framework proves to be comparatively more effective in developing a real time tracker, due to the P-N

Learning technique it employs. However, if target object does not suffer from a large number of variations then MDNet algorithm along with its SGD learning technique can outperform the TLD framework. MIL Track on the other hand makes the creation of training data sets a lot easier when compared to the TLD and MDNet algorithms. MIL Tracker can be extended if used with techniques like particle filter [11]. Whereas in the depth analysis approach the major drawback is of depth holes/depth shadows. Thus, by understanding the advantages and limitations of each tracking approach we can select the most appropriate approach for any given tracking scene.

References

1. Babenko, B., Yang, M., Belongie, S.: Visual tracking with online Multiple Instance Learning. In: IEEE Conference on Computer Vision and Pattern Recognition, pp. 983–990 (2009)
2. Viola, P., Platt, J.C., Zhang, C.: Multiple instance boosting for object detection. In: Conference on Neural Information Processing Systems, pp. 1417–1426 (2005)
3. Oza, N.C.: Online ensemble learning. Ph. D. thesis, University of California, Berkeley (2001)
4. Kalal, Z., Mikolajczyk, K., Matas, J.: Tracking-learning-detection. IEEE Trans. Pattern Anal. Mach. Intell. 34(7), 1409–1422 (2012)
5. Nam, H., Han, B.: Learning multi-domain convolutional neural networks for visual tracking. In: IEEE Conference on Computer Vision and Pattern Recognition (2016)
6. Wang, N., Li, S., Gupta, A., Yeung, D.-Y.: Transferring rich feature hierarchies for robust visual tracking. ArXiv preprint arXiv:1501.04587 (2015)
7. Hong, S., You, T., Kwak, S., Han, B.: Online tracking by learning discriminative saliency map with convolutional neural network. In: International Conference on Machine Learning (2015)
8. Doll'ar, P., Tu, Z., Tao, H., Belongie, S.: Feature mining for image classification. In: Conference on Computer Vision and Pattern Recognition (2007)
9. Viola, P., Jones, M.: Rapid object detection using a boosted cascade of simple features. In: Conference on Computer Vision and Pattern Recognition (2001)
10. Sung, K., Poggio, T.: Example-based learning for view-based human face detection. IEEE Trans. Pattern Anal. Mach. Intell. 20(1), 39–51 (1998)
11. Wang, J., Chen, X., Gao, W.: Online selecting discriminative tracking features using particle filter. In: Conference on Computer Vision and Pattern Recognition, vol. 2, pp. 1037–1042 (2005)
12. Lee, B., Liew, L., Cheah, W., Wang, Y.: Occlusion handling in videos object tracking: a survey. In: IOP Conference Series (2018)
13. Bashar, A.: Survey on evolving deep learning neural network architectures. J. Artif. Intell. 1(02), 73–82 (2019)
14. Liu, C., Huynh, D.Q., Reynolds, M.: Toward occlusion handling in visual tracking via probabilistic finite state machines. In: IEEE Trans. Cybern. 1–13 (2018)
15. Henriques, F., Caseiro, R., Martins, P., Batista, J.: Exploiting the circulant structure of tracking-by-detection with kernels. In: Proceedings of 12th European Conference on Computer Vision (2012)
16. Stauffer, C., Grimson, W.: Learning patterns of activity using real-time tracking. IEEE Trans. Pattern Anal. Mach. Intell. 22(8), 747–757 (2000)

Emotion Recognition
Using Physiological Signals

Mrigank Sharma$^{(\boxtimes)}$ and Rejo Mathew

I.T. Department, Mukesh Patel School of Technology
Management and Engineering, Mumbai, India
mriganksharma03.22@gmail.com, rejo.mathew@nmims.edu

Abstract. Emotion Recognition is the process of identifying human emotions. Facial expression recognition and human speech are the most common methods used for this process. However, one of the most recent development in this field is the use of physiological signals to recognize human emotions. In this paper an overview of emotion recognition using physiological signals is presented. These signals include Electrocardiogram, Electromyogram, Galvanic skin response etc. The problems encountered during various stages of emotion recognition are discussed along with their solutions. These solutions are then compared with each other.

Keywords: Artificial intelligence · Emotion recognition · Deep learning · Physiological signals · Neural networks · DBN · LSTM · CNN

1 Introduction

Emotions are an integral part of human life. Our emotions help us to convey our thoughts and feelings in the best possible manner. They can both positive and negative. Positive emotions such as joy, excitement increase our productivity while negative emotions like sadness, anger may decrease our work efficiency and lead to bad health. Emotions are very complex, dynamic and short lived in nature. They are produced by limbic system activity in response to a spur which leads to the activation of somatosensory system [1]. Due to this complexity emotions are represented in two ways:

The Discrete Model [2] – This was proposed by Ekman. He says that there are six basic emotions (Anger, Fear, Disgust, Happiness, Sadness and Surprise). Each of these emotions have a particular characteristic which allows them to be expressed in various degrees.

Dimensional Model [4] – This method differentiates between emotions on the basis of 2 dimensions - Arousal and Valence. Valence describes the positivity or negativity of the emotions while arousal tells us the intensity (high or low) of the emotion [5]. For example – Excitement is characterized by positive valence and high arousal.

As technology becomes more advanced it becomes important that the machines are able to recognize human emotions. Thus, Emotion Recognition is a huge step in reducing the gap between "Human-Computer" interaction. Emotion recognition methods are classified in 2 main categories:

A. P. Pandian et al. (Eds.): ICCBI 2019, LNDECT 49, pp. 389–396, 2020.
https://doi.org/10.1007/978-3-030-43192-1_45

The first category involves recognition through facial expression [6], speech [7], gestures and other physical signals. These methods are not very successful when used outside of controlled environments i.e. real world. This is because it is easy for humans to mask these physical signals to hide their real emotions.

The second category uses physiological signals to recognize emotions. This helps to overcome the problem of social masking as these internal signals are controlled by human nervous system. The signals include Electroencephalogram (EEG), Electromyography (EMG), Blood Volume Pressure (BVP), Galvanic Skin Response (GSR), Respiration (RSP) etc.

Data Acquisition, Pre-processing, Feature extraction and selection and then classification are the major steps involved in emotion recognition. This paper focuses on the different problems encountered during each step. Different solutions are discussed while comparing their efficiency based on different features. This paper is organized as follows: Sect. 2 describes and lists down the problems involved in the field of emotion recognition using physiological signals. Section 3 presents the algorithms used to solve these problems. A brief description of each algorithm is also given. Section 4 consists of an analysis of the problems and solutions and compares them in a tabular format. Finally, Sect. 5 presents a conclusion of the paper.

The general process of Emotion Recognition is divided into multiple steps as shown in Fig. 1 [8].

Fig. 1. Steps in emotion recognition

The first step i.e. Emotion Elicitation is one of the most important steps in this process. It involves elicitation/provoking of emotions in the subject through different media. These may include images, videos, songs etc. The second step involves recording these emotions and then processing them so as to remove any external noise that has been captured with them. This ensures that our model is trained with the appropriate data. The next three steps are a part of feature engineering. A feature can be defined as a property that is present in all the independent units on which analysis and classification has to be done. Feature engineering involves extraction of useful features from the input signal data, processing the and then classifying them into different classes.

2 Problems in Emotion Recognition

2.1 Presence of Outliers

Outliers refer to data points that are far away from other data points in the input. These are extreme values that deviate from other observations. Outliers can be both multivariate and univariate. Outliers present in single feature space are known as univariate while those present in n dimensional feature space are known as multivariate. Presence of outliers in the input can mislead the model and hamper its training process causing longer training times. This may lead to less accurate models and poor results.

2.2 Vanishing Gradient

The vanishing gradient problem is generally found in deep learning algorithms which use gradient-based methods and backpropagation techniques for training the network. This problem increases the training time of the networks and decreases their performance as the number of layers increase in the network. These networks use certain activation functions which squish a large input space into a small input space. So, a large change in input parameters in early layers will only cause a small change in the final output of the network. Gradient-based methods are trained by noticing the change in the output with respect to the change input. So, it may happen that the network will not learn some parameters as they do not cause a change in the output.

2.3 Local Minima

Most of the Feed Forward Neural networks use Backpropagation algorithm (BP) for the supervised learning. This algorithm calculates the rate of change of networks error with respect to the modifiable weights in the networks. Using a gradient descent, the network weights are adjusted by following the local slope of error surface. Problem occurs when our error surface contains multiple grooves and while performing the gradient descent, we fall into a groove but it is not the lowest possible groove. This leads to wrong estimate of the lowest possible weight in the network and leads to increased error rate in final output.

2.4 Manual Feature Extraction

During the process of emotion recognition, the input data is very large and may have some redundancies. This makes it important to select only the desired features from the input. Manual feature extraction refers to the problem of using separate algorithms specifically for the process of Feature extraction. This increases the complexity and the memory requirements of the emotion recognition models.

3 Solutions

3.1 PNN

Probabilistic Neural Networks (PNN) is a type of feed forward neural network [3]. It is based on the statistical network and Kernel Fisher discriminant analysis. PNN has the advantage of simple structure and fast learning ability which increases the classification accuracy and make sit more tolerant to noises and errors. PNN are much faster as compared to multilayer perceptron networks and are also more accurate than them.

PNN consists of 4 feedforward layers:

Input Layer - After the input in the form of physiological signals is taken, this layer finds the distance between the input vector and the training vector. Each neuron in this layer is a predictor variable. Its main aim is to standardize the range of values.

Pattern Layer – This layer has one neuron for each case in the training data. It has gaussian functions formed by using the input dataset as center points. This neuro stores the values of the predictor variables and the target value. A hidden neuron is used to find the test case form the center of neuron. Also, RBF kernel is applied by this neuron using the sigma values.

Summation Layer - Its main function is to perform summation of all the outputs from the second layer for all the available classes in input data. The actual target category of each training case is present in each hidden neuron.

Output Layer - This is the last layer in an PNN network. It is responsible for comparing the weights of each target category collected using the pattern layer and uses the biggest vote predict the category.

A radial basis function is used by the PNN to find the weight/influence of each point on the target. Since all neighbors are considered by PNN, outliers have little to no influence over the final outcome.

3.2   LSTM (Long Short-Term Memory)

Long short-term memory networks are a special kind of RNN that can learn long-term dependencies present in the data. Unlike other networks which struggle to learn information for long term, LSTM can easily remember the important information for sufficient period. A LSTM network is made up of multiple cells which are responsible for passing and processing the information. The main concept of LSTM is its cell states and various gates. The cell state is responsible for carrying information from initial layers to the final layers of the network. Information is added or remove to cell state through the use of gates (different neural networks). These also decide which information is to be remembered by the network [10].

Gates have sigmoid which are responsible for squishing inputs/parameters between 0 and 1. Any value multiplied by 0 is 0. So, these values disappear or are forgotten while any value multiplied by 1 stays the same and is thus remembered by the network. LSTM has three kinds of gate the regulate flow of information:

Forget Gate – It is responsible for deciding which information is irrelevant and has to be discarded. The current input and the previous hidden state (It refer to the output of the previous cell) are passed to the sigmoid function. If the output is 1 then the info is to be remembered by the network and if the output is zero then the information has to be discarded.

Input Gate – This is responsible for deciding which information has to be added to the cell state. The previous hidden state and the current input are passed to the sigmoid function and information relevant to network is found out. Also, these values are passed to the tanh function which squishes the values between −1 and 1. Then the tanh output is multiplied by the sigmoid output. This final output decides the information that has to be updated.

Output Gate – It is responsible for deciding the next hidden state. The current input and the previous hidden state are passed to sigmoid function. Then the modified cell state is passed to the tanh function and output of tanh and sigmoid function are multiplied to decide the information to be carried by the hidden state. The presence of the tanh distributes the gradients and hence prevents vanishing.

3.3 Deep Belief Networks

Deep Belief Network is made up of simpler Restricted Boltzmann Machine (RBM) models. Each RBM layer communicates with both the previous and the next layers in the network. Also, there is no communication between the nodes present in one layer. A DBN consists of 1 input, 1 output and multiple hidden layers. The input layer represents the raw sensory inputs and passes it to the next hidden layer. The hidden layers learn representations of the input and finally the output layer is used of network classification. Before we understand the working of DBN, we need to discuss RBM which act as the building block for Deep belief Networks.

Restricted Boltzmann Machine – RBM has simple network composition i.e. they are made up of only 2 layers. The first layer is the visible/input layer while the second layer is the hidden layer. There is no intralayer communication in RBM. RBM uses 2 bias, one for hidden layer and one for the input layer. The hidden layer bias helps in the activation when passing the data to next layer while the input layer bias helps in the reconstruction during backward passing of data [14].

Deep Belief Networks work in the following manner:

- The first layer of RBM learns the input structure and generates the activation to be passed to the hidden layer. Each RBM layer has its own set of bias and weights and these help in learning complex characteristics of data.
- Training in DBN occurs in two steps: Unsupervised pretraining and Supervised fine tuning.
- In unsupervised pretraining, each RBM layer is trained to recreate it's input. Greedy Algorithm is used for the purpose of pre training.
- The first hidden layer is trained from the input data greedily while all other layers are feezed. Then the output from this hidden layer acts as an input for the next hidden layer. This process is continued until all the layers have been pre-trained.
- This pre-training step overcomes the problem of Local minima and reduces the error rate in the output.
- Pre-training is followed by Supervised fine tuning. Backpropagation Algorithm is used for fine tuning purpose.
- In this phase the output nodes are assigned labels. This helps to understand what these nodes mean with respect to the network.
- Node weights are also adjusted during fine tuning to improve accuracy of the model.

3.4 Convolutional Neural Networks

CNN (Convolutional Neural Networks) are a class of artificial neural network architecture in machine learning. CNN consists of multiple layers which are interconnected with each other. Every layer is made up of multiple neurons. It consists of some hidden convolutional layers inside it. Each layer inside the CNN receives an input, transforms it and then passes it to the next layer. CNN are highly efficient in extracting patterns from the input data. Unlike regular neural networks, CNN layers are organized in 3 dimensions i.e. Height, Width and Depth. Also, neurons in one layer are not connected to all the neurons in the other layer [9].

CNN architecture is composed of 2 major parts [15]:
Feature Extraction – In this part the neural network performs convolutions and pooling operations on the input data. It has the Convolution, pooling and ReLU layer.

- **Convolutional Layer** – It contains a set of filters. When an input is received each filter is convolved across the width and height of input volume. A 2D activation map is generated and stacking of these activation maps forms the full output of this layer. Local connectivity, Spatial arrangement and Parameter sharing are important features in this layer.
- **ReLU** – It stands for Rectified Linear Unit. Its main function is to remove negative values from the activation map and increase non-linear properties of the overall network.
- **Pooling Layer** – This layer is useful for down-sampling of the input data. Its main function is to reduce the spatial size of the input, reduce the number of parameters and the amount of computation. This also reduces overfitting in the model.

Classification - In this part fully connected layers are used as a classifier for the extracted features and help in detecting emotions. It has the FC and Loss layer.

- **FC Layer** – The Fully connected layer is responsible for connecting all the neurons in one layer to the neurons in another layer. The high-level reasoning of the neural network and classification is done in this layer.
- **Loss Layer** – This Layer helps in penalizing the deviation of the current output from the expected output. Loss functions are used for this purpose.

CNN uses the following steps to perform Emotion Recognition:

- Pre-processed data in the form of signals/images or text is taken as an input by the CNN model.
- Convolution is performed on this input data using kernel or filters. This happens at the convolutional layer. Filters present in the convolution layer pass over the input data and each filter is convolved across the width and height of the input.
- Padding is done on the convolved features to reduce their dimensionality.
- The padded output is passed over an activation function called ReLU. It converts all the negative values to zero and retains the positive values.
- The features are then passed to the pooling layer which is responsible for extracting dominant features. Pooling also shortens the training time of the model.
- Finally, the FC layer (Fully connected layer) is utilized for classification of these dominant features. Loss layers can also be added to penalize the model for deviation from the expected output.

4   Analysis of Solutions

4.1   Comparison Table

See Table 1.

Table 1. Comparison between the provided solutions

Parameters	PNN	DBN	LSTM	CNN
Recognition accuracy	81.28% for valence [14]	78.28% for valence [11]	72.06% for valence [12]	85.83% [13]
Memory requirements	High	Low	Very high	High
Real world application	Can be easily implemented in real world scenario	Can be easily implemented in real world scenario	Unsuitable as it requires huge amount of resources to train	Suitable as it can adapt to unstructured data
Training time	Fastest to train	Moderate	Slowest	Moderate
Training approach	Unsupervised training is used	Unsupervised training is used	Sequential training is used	Unsupervised training is used
Activation function used	Gaussian function	Softmax function	Sigmoid and tanh	Rectified Linear Unit

5   Conclusion

Among all the methods discussed above Convolutional Neural Networks is the most efficient method for emotion recognition in a real-world scenario. They are self-sufficient and require very less human intervention. One of the major strengths of CNN is that they are self-sufficient and very efficient for feature engineering. The ReLU function used in CNN makes it easy to converge the network and allows for back-propagation. Although the pooling layer present in CNN helps to counter overfitting, they can be supported with another algorithm like dropout to further reduce the problem of overfitting. They use special convolution and pooling operations and perform parameter sharing. This makes CNN models universally attractive and allows them to run on any device.

References

1. Lalanne, C.:La cognition: l'approche des neurosciences cognitives.Rapport, Département d'informatique, pp. 26–28. René Descartes University, Paris (2005)
2. Ekman, P., Friesen, W.V., O'Sullivan, M., Chan, A., Diacoyanni-Tarlatzis, I., Heider, K., Krause, R., LeCompte, W.A., Pitcairn, T., Ricci-Bitti, P.E.: Universals and cultural differences in the judgments of facial expressions of emotion. J. Pers. Soc. Psychol. **LIII**(4), 712–717 (1987)

3. Specht, D.F.: Probabilistic neural networks and the polynomial Adaline as complementary techniques for classification. IEEE Trans. Neural Netw. I(1), 111–121 (1990)
4. Lang, P.J.: The emotion probe: studies of motivation and attention. Am. Psychol. 50, 372–385 (1995)
5. Kim, J., André, E.: Emotion recognition based on physiological changes in music listening. IEEE Trans. Pattern Anal. Mach. Intell. 30(12), 2067–2083 (2008)
6. Zhang, Y.D., Yang, Z.J., Lu, H.M., Zhou, X.X., Phillips, P., Liu, Q.M., Wang, S.H.: Facial emotion recognition based on biorthogonal wavelet entropy, fuzzy support vector machine, and stratified cross validation. IEEE Access 4, 8375–8385 (2016)
7. Mao, Q., Dong, M., Huang, Z., Zhan, Y.: learning salient features for speech emotion recognition using Convolutional Neural Networks. IEEE Trans. Multimedia 16, 2203–2213 (2014)
8. Shu, L., Xie, J., Yang, M., Li, Z., Li, Z., Liao, D., Xu, X., Yang, X.: A review of emotion recognition using physiological signals. Sensors 18, 2074 (2018)
9. Albawi, S., Mohammed, T.A., Al-Zawi, S.:Understanding of a Convolutional Neural Network. In: 2017 International Conference on Engineering and Technology (ICET), Antalya, pp. 1–6 (2017)
10. Chao, L., Tao, J., Yang, M., Li, Y., Wen, Z.: Long short term memory recurrent neural network based encoding method for emotion recognition in video. In: 2016 IEEE International Conference on Acoustics, Speech and Signal Processing (ICASSP), Shanghai, pp. 2752–2756 (2016)
11. Kawde, P., Verma, G.K.: Deep belief network based affect recognition from physiological signals. In: Proceedings of the 2017 4th IEEE Uttar Pradesh Section International Conference on Electrical, Computer and Electronics (UPCON), Mathura, India, pp. 587–592 (2017)
12. Li, X., Song, D., Zhang, P., Yu, G., Hou, Y., Hu, B.: Emotion recognition from multi-channel EEG data through convolutional recurrent neural network. In: Proceedings of the IEEE International Conference on Bioinformatics and Biomedicine, Shenzhen, China, pp. 352–359 (2016)
13. Salari, S., Ansarian, A., Atrianfar, H.: Robust emotion classification using neural network models. In: Proceedings of the 2018 6th Iranian Joint Congress on Fuzzy and Intelligent Systems (CFIS), Kerman, Iran, pp. 190–194 (2018)
14. Zhang, J., Chen, M., Hu, S., Cao, Y., Kozma, R.: PNN for EEG-based Emotion Recognition. In: Proceedings of the 2016 IEEE International Conference on Systems, Man, and Cybernetics (SMC), Budapest, Hungary, 9–12 October, pp. 2319–2323 (2016)
15. Martinez, H.P., Bengio, Y., Yannakakis, G.N.: Learning deep physiological models of affect. IEEE Comput. Intell. Mag. VIII, 20–33 (2013)

Prediction of Sudden Cardiac Arrest Due to Diabetes Mellitus Using Fuzzy Based Classification Approach

K. G. Rani Roopha Devi[1(✉)] and R. Mahendra Chozhan[2]

[1] Madurai Kamaraj University, Madurai, India
raniroopha@gmail.com
[2] Chozhan Dental Clinic, Kodaikanal, Periyakulam, Lakshmipuram, India
drchozhanbds@gmail.com

Abstract. In the past few decades, Diabetes mellitus has been considered as a chronic disease and one of the foremost serious health confronts all over the world. Based on International Diabetic Federation, approximately over 250 million diabetic patients all over the world, and expected to rise to 350 million in 2022. Moreover, 3.8 million deaths are due to diabetic complications. Around, 80% of death is due to diabetic mellitus (Type II), which can be prevented by prior detection of people with this risk. However, now machine learning approaches are utilized for diagnosis of diabetics more accurately. In this investigation, an effectual Fuzzy based classification using variable updation and normalization is anticipated as an intelligent representation of Fuzzy diagnosis (classification) decision. An efficient effort is made to identify cardiac death at initial stage arising from the severity of diabetic mellitus; where feature prediction prior to heart rate variability analysis is performed. The features of diabetic mellitus were examined from diabetic's database to identify the cause of sudden cardiac arrest. Various derived/performance measures of accuracy with Fuzzy classifier confirm strongly the cause of sudden cardiac arrest in prior stage. For the purpose of clinical applications, incorrect detection of heart rate (BPM) is considered as significant and assessed here. Simulation was carried out in MATLAB environment, real time dataset demonstrates that Fuzzy classifier offer a promising solution for cardiac arrest prediction due to diabetics mellitus. A rule set has been generated using Mamdani membership function with prediction accuracy of 95%, sensitivity of 94% and specificity of 95%. Here, extracted rules are effectual and outcomes are relevant to diabetic and cardiac medical studies.

Index Terms: Diabetes mellitus · Machine learning · Medical diagnosis · Sudden cardiac arrest · Fuzzy · Mamdani membership

1 Introduction

In medical field, diabetes is disease primarily related with a raise in blood glucose level (hyperglycaemia) [1]. The initial cause for hyperglycaemia is deficiency of insulin, wherein beta cells in pancreas is unsuccessful in secreting insulin. It is generally termed as Type I diabetes. Common diabetes is known as Type II, where body lacks effective utilization of insulin [2].

© Springer Nature Switzerland AG 2020
A. P. Pandian et al. (Eds.): ICCBI 2019, LNDECT 49, pp. 397–411, 2020.
https://doi.org/10.1007/978-3-030-43192-1_46

In general, individuals with chronic hyperglycaemia have higher risk of microvascular damage, which causes nephropathy, retinopathy and neuropathy. Hence, diabetes leads to the major cause of visual impairments and blindness in adults of developing countries [2] and accountable for around 1 million amputations. People with diabetes are more prone towards the macro-vascular complications risk, in which the individual is likely to have two to four times the chance of cardio-vascular disease (CVD) than people without diabetes. Owing to these difficulties, it is seen as leading death cause globally.

The occurrence of Type II diabetes is heightened owing to obesity, and in specific, due to physical inactivity, unhealthy dietary habits [3]. Prior diabetes detection would have a superior value proved by more than 50% to 80% in certain countries, people with diabetes are unaware of health condition and seems as an unaware manner till complication arises. Recent investigations have demonstrates that 82% of type II diabetes complications is delayed or prevented by prior intervention and identification in people who at risk [4]. For instance, altering individuals [3] or with therapeutic techniques. Data analysis like machine learning approaches is valuable for prediction of those people. Sometimes, diabetes leads to sudden cardiac arrest if not treated properly. For this purpose, numerous ML approaches is anticipated for management and diagnosis of diabetes [5]. It has been also depicted that utilizing CAD as 'second option' leads to enhanced diagnosis decision and Fuzzy classifier have shown remarkable success in classifying individuals with the higher risk of cardiac arrest.

In this investigation, a novel Fuzzy based classification for medical diagnosis, is utilised, with identifying variation, scoring and with rule generation with Machine learning approach. To be more specific, an unsupervised approach is utilized for normalization and constructing the model correspondingly, which is then followed by rule based explanation component. Moreover, this model is validated with real time dataset for type II diabetes prediction. Also, it shows that this approach is competent with predicting diabetes with higher sensitivity, specificity and accuracy which outperforms the prevailing techniques. It also depicts that diagnostic criteria based rules generated by this model is valid from medical point and other outcomes are provided by medical investigations (Fig. 1).

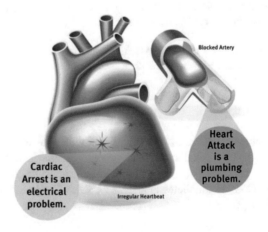

Fig. 1. Sudden cardiac arrest representation [3]

Investigational idea is organized as: A brief literature on existing machine learning classifier for predicting diabetes type II and sudden cardiac arrest in Sect. 2. Experimentation is explained in Sect. 3 which is depicted by numerical outcomes and discussions in Sect. 4. Rules with conclusion and future direction are drawn in Sect. 5.

2 Related Works

In [6], Tarawneh et al., investigated the popular classification approach in DM and utilized them over Cleveland dataset. Classification approaches performs well in this case. They are SVM and Naive Bayes classifier, whilst others rely over selected features. The significant objective is to investigate obtainable DM approaches to identify heart disease, to attain most appropriate outcome. The foremost focus is on prediction and classification techniques. Accuracy of algorithms can be enhanced by combining the algorithm into single powerful algorithm. In [7], Devi et al., carried out cardiac arrest death prediction at extremely earlier stage, that is, one hour before the occurrence which is helpful in saving the patients' life. Sudden cardiac death prediction at earlier stage is not done by any other investigators. For clinical study, ECG signals is chosen, methodology and simulation, feature extraction with diverse variability heart rate, which is confirmation with Task Force of European Societies guidelines for heart rate variability.

In [8], Liu et al., anticipated risk score prediction scheme with HRV factors and other essential signs, where geometric distance is used as a key factor. Intelligent scoring scheme has illustrated its capability to construct understandable human risk scores, and illustrates efficacy to an effectual predictor of cardiac arrest in 72 h. This work predicts potentially enlarged with scoring system to identify clinical outcomes. In [9], Barakat et al., constructed hybrid scheme for medical diagnosis. In specific, this work cast off SVM for prediction and diabetic's diagnosis, where rule based component explanation is cast off to offer comprehensibility. Rules and SVM extracted are intended to function as opinion for diagnosis and tool to identify diabetes with high risk.

In [10], Sharma et al., depicted fuzzy logic approach, with featured inputs which are provided as basis to the generated output. In this investigation, expert system is modelled to identify heart disease. Here, 19 input factors are utilized to determine output where heart disease prediction, may be low, moderate and high. To raise expert system efficacy, ECG will be provided as input for other parameters identify heart disease calculation. In [11], Susan et al., depicted statistical modelling are not necessary for disease prediction as incapable of preserving categorical values and huge amount of missing values. Here, ML algorithm is utilized. It is cast off to merge data mining approaches to categorize diverse types of diseases sourced on patients' record. Accuracy is essential for DM in medical field.

In [12], Chauhan et al., anticipated an investigation where DM approaches are utilized for patients with heart disease for disease prognosis. Anticipated work discusses diverse classification approaches for heart attack prediction and depicts finest classifier among them. Outcomes validate effectual and well-organized for knowledge discovery among classified. In [13–17], Isasi, initiates ML for decision making with piston driven chest compressions. This algorithm improves accuracy of finest solutions with BAC points with added 5-fold reduction in computational cost. It leads to more effectual and extremely accurate. There are two main causes for this enhancement [18–21]. Initially, feature extraction sourced on stationary wavelet analysis outcome in novel and enhanced discriminating features. Subsequently, feature extraction after hauling out CPR artefact and then feed those features to SVM, enhances accuracy significantly, as machine learning algorithm, is competent to learn filtering residuals features. The outcome depicts that this model facilitates feature relaxing with artefact compression filters.

3 Proposed Work

3.1 Dataset

In this study, the investigation is carried out in MATLAB. It is equipped with machine learning based rule generation. Diabetes information in UCI is initiated [14]. There are totally about 270 instances and 13 attributes. So as to carry out performance in MATLAB, dataset should be in Attribute – Relation File Format (ARFF). It assists in pre-processing and filtering based on diverse useful attributes. Finest classifying attribute is chosen. In the end, results are compared amongst diverse algorithms based on accuracy, sensitivity, specificity, Mean square error and so on [15]. Conclusion is generated from this. Trained model is utilized for prediction of sudden cardiac arrest and resourcefully help to diagnose individuals with risk of sudden cardiac arrest and death due to type II diabetes. This diminishes number of tests. Therefore, disease is treated in appropriate time and lives of more people are saved. Figure 2 depicts work functionality of anticipated approach. Initially, inputs have to be collected from available dataset. These attained inputs comprises of noisy data that degrades system performance, therefore it has to be removed with filtering. Followed by this, essential features have to be extracted and chosen for classification purpose. Classification is done with Fuzzy based variable selection an rank scoring model. Effectual outcomes are attained and compared with existing model. The outcomes were analyzed for demonstrating probability of cardiac arrest suddenly. Detailed explanation is provided in Methodology section.

3.2 Methodology

This investigation anticipates a Fuzzy model to design an efficient and an effectual modelling approach that can classify class data for future decision making purpose. It is to construct predictive replica for appropriate class and describe classifier assists in diagnosis. Classification model utilized in this investigation is Fuzzy based classifier; Diabetes and heart dataset is analysed from UCI repository for demonstrating better suited for retrieval of efficient and effectual patterns. Anticipated model concentrates to retrieve effectual patterns that generalises cardiac arrest to assists patients for disease diagnosis. Pattern discovery leads to automated procedure where classification is perspired to categorize data. Moreover, classification are examined amongst others and described to provide maximum diagnosis accuracy of disease. Table 1 shows the attributes related to diabetes and sudden cardiac arrest.

Table 1. Attributes related to Heart

Attributes	Explanation	Values	Type
Age	Individuals' age in years	–	Numeric_type
Chest_pain	Pain in individuals chest	Typical angina Atypical angina Non-angina pain Asymptomatic	Numeric_type
Rest_blood pressure	Blood pressure during resting (mm Hg during hospital admission)		Numeric_type
Blood_sugar	Blood sugar during fasting > 120 mg/dl	1 = true; 0 = false;	Numeric_type
Rest_electrocardiography	Electrocardiographic outcomes in rest state	Normal Hyper_left_vein Wave_abnormality state	Numeric_type

Intelligent Fuzzy prediction is anticipated to evaluate risk of individuals' clinical outcomes, using HRV factors and essential signs. Scoring is constructed on geometric distance calculation between feature set vectors attained from records of various patients. Anticipated score prediction model is explained below and details of HRV are as follows:

Attribute/Variable selection: This work considers 24 variables (16 HRV parameters, 8 vital signs) to investigate probable relation with cardiac arrest risk in 72 h. Significant signs are temperature, heart rate, diastolic blood pressure, respiratory rate, systolic blood pressure, GCS and oxygen saturation.

Before the computation of risk score computation, feature set is provided to interval $[-1, 1]$ executing min-max normalization over original data. Consider dataset $X = [x_1, x_2, \ldots, x_N]$ where ever 'x' specifies patients, let $\min_A$ and $\max_A$ specifies minimum and maximum attribute vector values where $A = [x_1(m), \ldots, x_N(m)]$ $m = 1, 2, \ldots, 24$ and 'N' is amount of samples. Min-Max normalization plots value 'v' with range $[\min_A \text{ and } \max_A]$ by evaluating the formula given below Eq. 1:

$$v' = \frac{v - \min_A}{\max_A - \min_A} \left(\max'_A - \min'_A\right) + \min'_A \tag{1}$$

Normalization process is competent to preserve relationship between original data values, henceforth it provides geometric distance computation along with risk prediction.

Preliminary risk score computation: Geometric feature computation based risk score is determined. Initially, cluster values of $+ve$ and $-ve$ samples are evaluated in Euclidean space, in which $+ve$ samples are individuals with sudden arrest in 48 h and $-ve$ samples are patients devoid of sudden cardiac arrest.

3.3   Classification on Risk Score

In numerous patients' condition, samples from diverse classes are generally grouped as depicted in Fig. 2. It is thereby complex to differentiate one from other based on geometric features based risk predictor. So as to overcome complexity, score updating approach is anticipated as in Figs. 3 and 4. It comprises of two components: classification and normalization. Outcomes of conventional classifier are binary prediction on arrest in 48 h. Even though, prediction outcomes are helpful in improving risk score computation (Table 2).

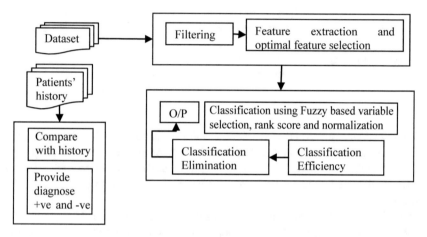

Fig. 2. Flow diagram of Fuzzy classification

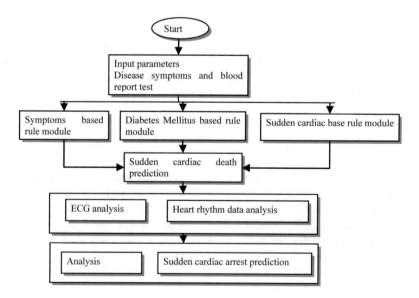

Fig. 3. Prediction of sudden cardiac arrest

Table 2. Diabetes parameters

S. No.	Parameter
1	Fasting blood sugar
2	Random blood sugar
3	AC1 test
4	Oral glucose test

3.4 Fuzzification

In general, fuzzifier transforms value to membership degree by associated membership functions. Usually, it describes certainty with crisp value which is related to specific linguistic functions. Membership function possesses diverse shapes. Mamdani membership function is utilized. Most frequently utilized membership functions are triangular, Gaussian and trapezoidal shape as in Fig. 4. Specifically, membership function is determined by relying over domain knowledge or via diverse learning applications.

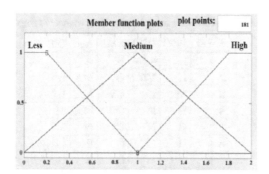

Fig. 4. Mamdani membership function

3.5 Rule Based Classification

Rule base is linguistic statement set termed rules. They are IF and THEN format, consequently premises is built with fuzzy input variable associated with logical functionality (for instance, AND, NOT, OR) and fuzzy output variable is attained consequently. It is generally constructed as exhaustive set with probable value-which is a combination of input linguistic variable that constitute the variables. For determining the membership function, rule base is attained from either dominance via ML approaches. It is obtained by performing rule based system. If-then rules with 'n' attributes are provided as Rule 'R_j': If x_1 is A_{j1} and x_n is A_{jn}, then class C_j, For $j = 1,\ldots, N$, where $X = [x_1, x_2,\ldots, x_n]$ is n-dimensional vector, A_{ji}, where $(i = 1,\ldots, n)$ is antecedent linguistic value associated with R_j, similarly C_j is class and 'N' is generated fuzzy rules count.

3.6 Defuzzification

After extracting fuzzy rules which offers numerous shapes (features of cardiac arrest) specifying modified membership functions, de-fuzzification is carried out. De-fuzzification is percentage set transformation to crisp value. In general, centre of gravity is utilized for de-fuzzification. Owing to this performance, it is partitioned into three type's low, moderate and higher risk of sudden cardiac arrest.

4   Numerical Results

Numerical results and discussion of anticipated Fuzzy based classification method is explained here. The simulation was carried out in MATLAB 2018a environment. The high risk prediction of sudden cardiac arrest for diabetes mellitus is determined and evaluated as in Fig. 5. The prediction with this anticipated model provides better efficiency in contrast to prevailing approaches like SVM. Table 3 is attributes of diabetes samples.

Table 3. Attributes related to various diabetes samples

S. No.	PGC	DBP	BMI	Age	O/P
1	L	——	N	A	L
2	L	VL	OW	MA	L
3	M	L	OW	MA	L
4	H	N	OW	MA	M
5	H	N	O	E	M
6	H	H	O	E	M
7	VH	H	O	E	H
8	VH	H	EO	E	H
9	VH	VH	EO	E	H
10	VH	N	N	OA	H
11	H	H	O	E	H
12	H	H	OW	OA	H
13	M	H	O	E	I
14	M	N	O	E	H
15	M	N	OW	E	I
16	H	——	——	E	L
17	L	VL	OW	MA	H
18	VH	N	——	MA	H

Note: L-Low, M-Moderate, H-High, VH-Very High, VL-Very Low, N-Normal, OW-Over Weight, O-Obesity, EO-Extreme Obesity, A-Adult, MA-Middle aged, E-Elder, OA-Over aged, In-Intermediate.

TP: Diabetic patients with sudden cardiac arrest risk
TN: Non-diabetic patients not with sudden cardiac arrest risk
FP: Non-diabetic patients with sudden cardiac arrest risk
FN: Diabetic patients not with sudden cardiac arrest risk

$$Sensitivity = \frac{TP}{(TP + FN)} \qquad (2)$$

$$Specificity = \frac{TN}{(TN + FP)} \qquad (3)$$

$$PPV = \frac{TP}{(TP + FP)} \qquad (4)$$

$$NPV = \frac{TN}{(TN + FN)} \qquad (5)$$

Table 4. Samples of death rate reduction

Samples	10	20	30	40	50
Efficiency %	1%	2%	3%	4%	5%

Table 5. Significant attributes of SCA

S. No.	Attributes
1	Age
2	Gender
3	Chest pain
4	Cholesterol
5	Rest SBP
6	Rest ECG
7	Fasting blood

Percentage of efficiency computation for every 10 iteration is provided in Tables 4 and 5 shows the death rate reduction and attributes of SCA respectively. The anticipated model shows 99% accuracy, 99% sensitivity and 72% specificity.

Table 6. Accuracy computation

Method	TP	TN	FP	FN	Sensitivity	Specificity	Accuracy
Fuzzy classifier	265	6	3	2	99	72	99

Table 7. Comparison of performance metrics

Measure	Proposed Fuzzy	SVM-LIN	SVM-RBF	GLM	PSO
Sensitivity	99%	73.1%	61%	63%	78%
Specificity	72%	80.8%	80%	80.8%	80%
PPV	85%	79.2%	76%	76%	80%
NPV	84%	75%	67%	68%	79%
MSE	0.198	0.439	0.378	0.378	0.206

Tables 6 and 7 depicts performance metrics such as sensitivity, specificity, PPV, NPV and MSE respectively. Sensitivity is 99% for Fuzzy which is higher than SVM-LIN, SVM-RBF, GLCM and PSO. MSE is lesser Fuzzy while compared to prevailing approaches.

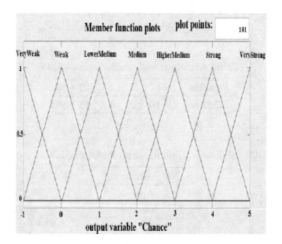

Fig. 5. Output chance of SCA

Cardiac arrest is considered as complex prediction strategy. There exist proofs that validate heart rate variability as non-invasive computation of cardiac arrest that may identify arrest with indexes from spectral, temporal and non-linear series analysis of inter beat. If this is reliable, heart rate variability indices (VI) are merged with life saving models by bed-side monitoring. Moreover, cardiac arrest computation based on VI is sub-optimal and performed with clinical events, as it is complex to predict when applied in huge population. Therefore, finest science for treatment and prevention of cardiac arrest has to be modelled effectually.

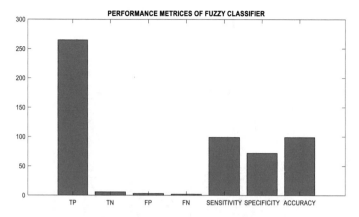

Fig. 6. Performance metrics of proposed fuzzy classifier

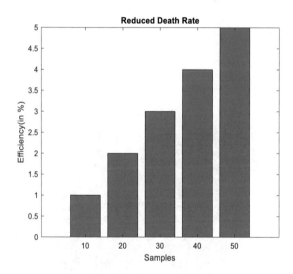

Fig. 7. Efficiency computation

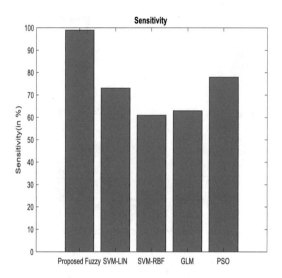

Fig. 8. Sensitivity computation

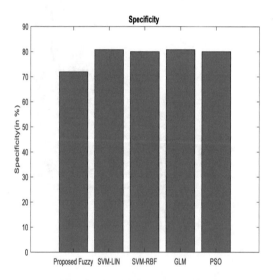

Fig. 9. Specificity computation

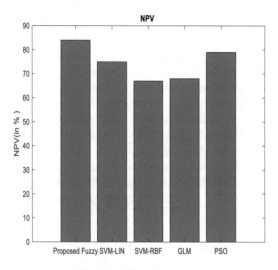

Fig. 10. NPV computation

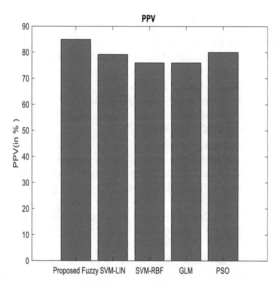

Fig. 11. PPV computation

In general, PPV and NPV in Figs. 10 and 11 are termed as $+ve$ and $-ve$ predictive value correspondingly. These parameters are also specified for evaluating the performance of risk prediction. With these parameters, TP specifies patients with sudden cardiac arrest within 48 h; FP specifies healthy person who are inappropriately predicted with sudden cardiac arrest within 48 h; TN specifies, healthy persons, appropriately predicted as healthy; and FN specifies person with sudden cardiac arrest in 48 h inappropriately identified as healthy as in Figs. 6, 7, 8 and 9. Usually, higher specificity, sensitivity, NPV and PPV are desirable for scoring system.

5 Conclusion

In this investigation, expert system is modelled to generate output in accordance to provided input. FL is utilized to model effectual expert system. Certain specific number of inputs is provided on generated output basis. Expert system is constructed to predict sudden cardiac arrest for diabetes mellitus patients. 27 input factors are utilized to acquire output system in which chance of higher risk of sudden cardiac arrest is provided as low, moderate and higher. So as to increase the efficacy of expert system, ECG parameters is provided as an input with other related parameters which predicts the higher chance of sudden cardiac arrest. Here, accuracy of prediction is 99%. The proposed model shows better trade off in contrast to prevailing techniques. In future, machine learning approach can be hybridized for reducing the complexity of computation.

References

1. Kalaiselvi, C., Nasira, G.: Prediction of heart diseases and cancer in diabetic patients using data mining techniques. Indian J. Sci. Technol. **8**(14), 1 (2015)
2. Santhanam, T., Ephzibah, E.: Heart disease prediction using hybrid genetic fuzzy model. Indian J. Sci. Technol. **8**(9), 797–803 (2015)
3. Yeh, Y.-C., et al.: A reliable feature selection algorithm for determining heartbeat case using weighted principal component analysis. In: 2016 International Conference on System Science and Engineering (ICSSE). IEEE (2016)
4. Dubey, V.K., Saxena, A.K.: Hybrid classification model of correlation-based feature selection and support vector machine. In: IEEE International Conference on Current Trends in Advanced Computing (ICCTAC). IEEE (2016)
5. Krishnaiah, V., Narsimha, G., Chandra, N.S.: Heart disease prediction system using datamining technique by fuzzy K-NN approach. In: Satapathy, S., Govardhan, A., Raju, K., Mandal, J. (eds.) Emerging ICT for Bridging the Future - Proceedings of the 49th Annual Convention of the Computer Society of India (CSI), vol. 1, pp. 371–384. Springer, Cham (2015)
6. Tarawneh, M.: Hybrid approach for heart disease prediction using data mining techniques. In: Barolli, L., et al. (eds.) AG 2019, EIDWT 2019, LNDECT 29, pp. 447–454. Springer (2019)
7. Devi, R., Tyagi, H.K.: Early stage prediction of sudden cardiac death. IEEE (2017)
8. Liu, N., Lin, Z.: An intelligent scoring system and its application to cardiac arrest prediction. IEEE Trans. Inf Technol. Biomed. **16**(6), 1324–1331 (2012)
9. Barakat, N.H., Bradley, A.P.: Intelligible support vector machines for diagnosis of diabetes mellitus. IEEE Trans. Inf Technol. Biomed. **14**(4), 1114–1120 (2010)
10. Sherubha, P.: A detailed survey on security attacks in wireless sensor networks. Int. J. Soft Comput. **11**(3), 221–226 (2016)
11. Sherubha, P., Chitra, M.B.: Multi class feature selection for breast cancer detection. Int. J. Pure Appl. Math. **118**, 301–306 (2018)
12. Sherubha, P., Amudhavalli, P., Sasirekha, S.P.: Clone attack detection using random forest and multi objective cuckoo search classification. In: International Conference on Communication and Signal Processing (2019)

13. Sherubha, P., Mohanasundaram, N.: An efficient intrusion detection and authentication mechanism for detecting clone attack in wireless sensor networks. J. Adv. Res. Dyn. Control Syst. **11**(5), 55–68 (2019)
14. Sherubha, P.: An efficient network threat detection and classification method using ANP-MVPS algorithm in wireless sensor networks. Int. J. Innov. Technol. Explor. Eng. **8** (2019)
15. Sharma, S.: Heart disease prediction using fuzzy system. In: Luhach, A.K., et al. (eds.) ICAICR 2018, CCIS 955, pp. 424–434. Springer (2019)
16. Rezaeieh, S.A.: Microwave system for the early stage detection of congestive heart failure. IEEE Access **2**, 921–929 (2014)
17. Chauhan, R., Jangade, R.: Classification model for prediction of heart disease. In: Pant, M., et al. (eds.) Soft Computing: Theories and Applications. Springer (2018)
18. Isasi, I., Irusta, U.: A machine learning shock decision algorithm for use during piston-driven chest compressions. IEEE Trans. Biomed. Eng. **66**(6), 1752–1760 (2018)
19. Vaillancourt, C., et al.: The impact of increased chest compression fraction on return of spontaneous circulation for out-of-hospital cardiac arrest patients not in ventricular fibrillation. Resuscitation **82**(12), 1501–1507 (2011)
20. Kwok, H., et al.: Adaptive rhythm sequencing: a method for dynamic rhythm classification during CPR. Resuscitation **91**, 26–31 (2015)
21. Raj, J.S.: A comprehensive survey on the computational intelligence techniques and its applications. J. ISMAC **1**(03), 147–159 (2019)

Suggesting a System to Enhance Decision Making in Location Based Social Networks

R. Sridevi$^{(\boxtimes)}$, G. Bhavani, and R. Meena

Department of Computer Science and Engineering,
K Ramakrishnan College of Engineering, Tiruchirappalli, India
sridevivelon@gmail.com

Abstract. The vast increase in mobile technology and more number of location-aware mobile devices used increases the percentage of data, which is easily available from smart phones and mobile platforms. The combination of both location and social data used to frame the constructive results either in business or social which can be constructed from the *geo-social data*. The *geo-social queries*, the use of collaborative spatial computing in database has attracted both developed and intellectual communities. This paper suggested the method to improve group decision making, travel recommendation and spatial task outsourcing based on geographical locations. When the user gave the query, then the details of that query directly fetched from the server. For example, if the user try to analyze the query like face book, twitter and Google plus. It retrieves the data like social data and location data.

The location data fetched based on user latitude and longitude values.

Keywords: Collaborative spatial computing · Latitude and longitude values · Ubiquitous · Internet access · Social computing · Social network

1 Introduction

1.1 Data Mining

Data mining deals with extracting essential data from huge datasets, in other words, this means effective practice of mining needed knowledge from the collected or available data source. The process of data mining comprise of numerous successive steps including the option from the preparedness of data obtained from different sources and finish with the obtained results [1].

The data learned can be practiced in various applications including marketing strategy Analysis, Identifying Fraudulent activities, retaining the existing customer, identifying the risk in management and extensively used in the areas of production control, sports, astrology and Internet Surfing. Data mining suggest the group of patterns incurred in data, and the data mining challenges which can be fall into two classes: First one Descriptive - It depict the frequent features in the database and the second one Predictive - It achieve suggestion in the active data set to make predictions [2].

© Springer Nature Switzerland AG 2020
A. P. Pandian et al. (Eds.): ICCBI 2019, LNDECT 49, pp. 412–419, 2020.
https://doi.org/10.1007/978-3-030-43192-1_47

1.2 Market Analysis and Management

In today's technological era the customers with lots of choices and opinions existing so it impossible to categorise them in a massive communities to promote promotion policies and the marketing business plans like advertisement based on result in meagre reaction ratio and increases the cost involved in investment. In Database marketing, usually done by the information obtained from the transaction databases and customer databases built databases based on their past transactions. The suitable tools and techniques to handle huge database for customer information and business schemas found to be less so the knowledge-based market appropriate for data mining to addresses the requirement to get knowledge hidden in databases by means of three ways like Customer profiling, Deviation analysis and Trend analysis which also includes the sub areas like Customer Requirement Identification, Cross Market Analysis, Target Marketing, Providing Summary Information used in the field of market [3].

Customer relationship management consists of four dimensions: (1) Customer Identification; (2) Customer Attraction; (3) Customer Retention; (4) Customer Development. These four dimensions can be seen as a closed cycle of a customer management system [4, 5].

Data Mining [14] – Issues
Data mining not found to be a simple process; the algorithms used get very difficult and data usually unavailable at single source. It apparent to integrate data from diverse data sources which generate several different issues [6] which can be resolved by choosing appropriate tools for extracting and identifying useful information and knowledge from huge databases were done with the help of Customer relationship management decisions [7].

1.3 Spatial Database

A spatial database efficient to retain and query data's which available in a geometric space; the spatial databases let simple geometric objects of points, lines and polygons and only a meagre databases can handle complex like 3D objects, coverage area, linear networks, whereas databases built to handle a variety of purposes like numeric as well as character types data to process spatial data types efficiently.

Features of Spatial Databases
Spatial index were used in Database systems to quickly look up values in it, which can able to perform operations like typical SQL query statements. A spatial database uses spatial measurements to compute length of the line, to find area of the polygon and to measure the distance between geometries, etc. either from existing or generate new adopted by specifying the points or nodes for defining the shape [8].

1.4 Techniques

Several different techniques were invoked to rank the systems and usually chosen query answers from different types of existing information; the end-users expected to get the exact answers for the queries which available in available data [9]. The applications which developed can suits the efficient and effective support of query accessing and the meta-search engines in web used to get by merging rankings from dissimilar search engines related by means of efficient rank aggregation methods. [10]. There were several applications available in the information retrieval and data mining context and generally the applications work out queries with aggregating multiple inputs to offer users the optimum results with the help of scoring function which categorize the top-k objects in scoring. An object grade takes into account in assessing object based on its features and generally assessed with multiple scoring which gives the total score of the object and usually the scoring normally aggregate all the partial scores together [11].

1.5   R-Tree Index

R-trees predominantly served for the purpose of indexing the data with multiple dimensions of geographical information by using different positional objects like rectangles and polygons and generally the data which present in R-trees is organized as number of pages along with different entries. The principal idea behind is to unite together all the objects which presents nearby and form a least bounded rectangle in higher level of tree and at lower level depicts a single object. R-tree which resembles the B-tree in balanced search and ensures minimum fill away from each other from the root with 30%–40% of entries. In most of trees, the searching algorithms use the bounding boxes to make a decision either or not to search inside a sub tree, this is the reason that R-trees suitable for large data sets and databases. For indexing R-tree is chosen for the reasons of saving both disk space as well as the time required to build and maintain the index, it improves the build performance, reduces the space needed to update the index [12].

2   Need for the Study

When the user gave the query, then the details of that query directly fetched from the server. For example, if the user try to analyze the query like face book, twitter and Google plus. It retrieves the data like social data and location data and the location data accessed with the help of user latitude and longitude values.

2.1   Issues

- A query raised by the user can generate request and returns a set of neighbour users with rigid social associations.
- The size of the workgroup found to be very high.
- To evaluate Social entropy and community score.
- To provide approximate solutions (Values) Geo-social data.
- Unsupervised grouping
- Pay no attention about the people who are related [13].

2.2 Objective of the Study

To improve collaborative spatial computing in location and social factors to meet out the requirement to make decision in providing travel recommendation and outsourcing spatial task and also to protect the user from unwanted messages send by another user and implemented social activities in network based on geographical locations which clearly stated in system architecture Fig. 1.

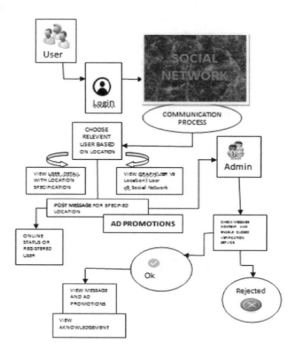

Fig. 1. System architecture

3 Proposed Flow Diagram

Figure 2 states the flow diagram of the proposed method. The first step deals with the identification of the creator in given social network; creators known through exploiting data from their social graph, which indicates the status conditionality on trust values, type and depth of the group creators concerned and implement the explicit rules. All the possibilities learned from the view of creator specification. The next stage specifies implementation of filtering rules. When determining the language for FRS (Filter Rule Specification), three main items need to be anticipated which must manipulate a message filtering decision. In day by day existence, the identical message got various meanings and significance relies on who write it, as a consequence, FRS should allow users to state constraints on message creators. Creators on which a FR concurs can be attained based on several different criteria; one of the most relevant conditions relies on their profile attributes and possible to depict rules enforcing hardly to young creators or to creators.

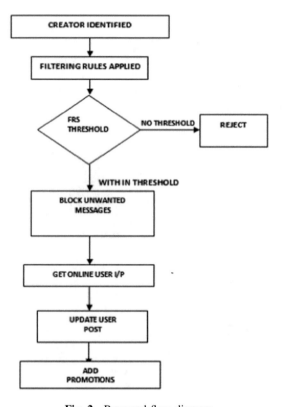

Fig. 2. Proposed flow diagram

The next stage OSA (Online Setup Assistant) used to get the set of messages selected from the dataset. Upon receiving each message, the user directs either accepts or rejects the message and resolving the addressing issue by assigning thresholds to filter rules with the help of an Online Setup Assistant (OSA) procedure. The collection and processing of user decisions on an ample set of messages disseminated over the entire module allows computing customized thresholds represent in accepting or rejecting certain contents. The messages are chosen based on condition of fair messages from subset of the dataset which not belongs to both training, test sets. The next stage deals with blocking of unwanted messages. Users banned based on their previous history profile data along with the relationships exists in OSN directly or indirectly by the opinions of a person and the process of banning can be applied either for a specific predefined time or undefined period of time. Users' bad behaviour decided based on two factors, the first is associated to the standard that if within a given time period a user has been incorporated into a number of times greater than a specified threshold, he/she might entitled to stay for another. This principle works for those users that have been already inserted and considered at least one time. The next stage followed gets the online user details, in which the user can have a chance to know the user details those who are all online. By way of this sort of process the user can straightforwardly communicate among their friends through chatting. These types of process make the application more users friendly. The updating work carried out in the next stage here, the user have right to post any data based on their requirement whatever they are need to process. These posts are accessed by other users based on their posted person location. For example, if user posts any data and his/her location is Trichy means, those who are all in Trichy only can view the post. The final stage deals with adding Ad promotions. If any user wants to promote any type of ad on social media they can put any data on social network through what they register in their network. These types of promotions can view to the admin. Then the admin can classify the promotion with group cost and details what they are added.

4   Results and Discussion

In this work, three different synthetic datasets of 150 users in simulating scenarios in a big city created, and designate locales to persons who are all active based on two principles: (i) two users probable to check-in adjacent spaces, (ii) locate people who are socially related to particular location. To a greater extent specifically, the task starts with traversing the entire graph in a BFS fashion by taking a random user from random location and assigned locations based on their Euclidean distances to their friends. The resultant sign spread in an overall area of 200 km. In experimental setup Intel core i7 2.8 GHz CPU with 4 GB memory on Windows 7 64-bit operating system used and performed the experiments on PostgreSQL with an extension of PostGIS with R-Tree. By dividing the queries according to their categories and here identify by their variable. Execution time in milliseconds for each category without index and indexed in all three indexing structures, the execution time of each category represented with the help of histograms. Each bar of the histogram will represent the time taken (in ms) by each query. The following table indicates Comparison of response time with and without R-Tree Indexing, starting from the first category i.e. Simple SQL (Table 1 and Fig. 3).

Table 1. Comparison of response time with and without R-Tree Indexing

Query	Without R-Tree Index	With R-Tree Index
1	61	11
2	31	11
3	11	12
4	99	22
5	20	11
6	32	31
Average	42	16

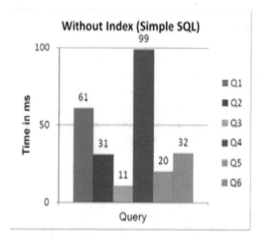

Fig. 3. Without indexing method

5 Conclusion and Future Enhancement

This paper suggested the method to improve group decision making, travel recommendation and spatial task outsourcing based on geographical locations. When the user gave the query, then the details of that query directly fetched from the server. It retrieves the data like social data and location data. The location data fetched based on user latitude and longitude values. Comprehensive experimentation on synthetic datasets exhibits the efficiency and effectiveness of the solution and achieves desirable query performance. In future work the algorithm can extended to process the query with dynamic R+ index structure to reduce the insertion and deletion operation overhead in R-tree Index structure.

References

1. Sridevi, R., Lakshmi, K.: Signature analysis of UDP streams for intrusion detection using data mining algorithms. Int. J. Comput. Sci. Eng. **02**(07), 2461–2465 (2010)
2. Ziafat, H., Shakeri, M.: Using data mining techniques in customer segmentation. Int. J. Eng. Res. Appl. **4**(9), 70–79 (2014). ISSN 2248-9622 (Version 3)
3. Shaw, M.J., Subramaniam, C., Tan, G.W., Welge, M.E.: Knowledge management and data mining for marketing. Decis. Support Syst. **31**(1), 127–137 (2001)
4. Au, W.H., Chan, K.C.C., Yao, X.: A novel evolutionary data mining algorithm with applications to churn prediction. IEEE Trans. Evol. Comput. **7**, 532–545 (2003)
5. Ngai, E.W., Xiu, L., Chau, D.C.K.: Application of data mining techniques in customer relationship management: a literature review and classification. Expert Syst. Appl. **36**, 2592–2602 (2009)
6. www.tutorialspoint.com/dm/dm_overview.htm
7. Ghaderi, H., Fei, J., Shakeizadeh, M.H.: Data mining practice in SMEs: a customer relationship management perspective. In: ANZAM 2013 (2013)

8. https://ipfs.io/ipfs/Spatial_database.html
9. Ilyas, I.F., Beskales, G., Soliman, M.A.: A survey of top-k query processing techniques in relational database systems. ACM Comput. Surv. **40**(4), 1–58 (2008). Article no. 1
10. Salunke, A.B., Kasar, S.L.: Personalized recommendation system for medical assistance using hybrid filtering. Int. J. Comput. Appl. **128**(9), 0975–8887 (2015)
11. Kashem, M.A., Chowdhury, A.S., Deb, R., Jahan, M.: Query optimization on relational databases for supporting top-k query processing techniques. JCIT **1**, 53–58 (2010). ISSN 2078-5828 (Print), ISSN 2218-5224 (Online), Manuscript code 1007
12. Sao, S.S., Shiekh, R.: A review: fast nearest neighbour search with keywords. Int. J. Sci. Eng. Res. (IJSER) 52–54. ISSn 2347-3878
13. Li, Y., Chen, R., Xu, J., Huang, Q., Hu, H., Choi, B.: Geo-social k-cover group queries for collaborative spatial computing. IEEE Trans. Knowl. Data Eng. **27**(10), 2729–2742 (2015)
14. Joseph, S., Thanakumar, I.: Survey of data mining algorithm's for intelligent computing system. J. Trends Comput. Sci. Smart Technol. (TCSST) **1**(01), 14–24 (2019)

Security Challenges in NoSQL
and Their Control Methods

Mahiraj Parmar$^{(\boxtimes)}$ and Rejo Mathew

I.T. Department, Mukesh Patel School of Technology
Management and Engineering, Mumbai, India
mahirajparmar9@gmail.com, rejo.mathew@nmims.edu

Abstract. NoSQL databases have become popular due to their scalability, ease of implementation of JSON (JavaScript Object Notation). This document analyzes the maturity of NoSQL safety policies compared with the SQL database, addressing the new access and query mechanisms. Analysis of the attacks and vulnerabilities is done along with mentioning the methodologies to mitigate them. It states how this newly developed technology lacks in security and awareness, which was an issue over the years in SQL systems.

Keywords: NoSQL · NoSQL injection · MongoDB · Cassandra · Redis · NoSQL attacks · NoSQL solutions

1 Introduction

Database Security is one of the most critical issues in Information Security. Databases like SQL and NoSQL are used for storage and retrieval of data. NoSQL stores data in the form of chunks which support JSON rather than rows and columns [4]. The fact that it is horizontally scalable helps in increasing the performance by the addition of more servers. NoSQL provides Consistency, Availability, Partition Tolerance [5] properties. It provides a solution to the common injection attacks that occurred in SQL. Some of the popular databases are MongoDB [1], Redis [2] and Cassandra [3] (ranked among the top 10 most popular databases). This does not imply that NoSQL is immune to injection attacks. Recent study shows that even after improvement in the query language there are still techniques that result in the injection of malicious queries [13].

2 Vulnerabilities in NoSQL

The following security issues and attacks have been observed taking into consideration the two most popular NoSQL databases Mongo DB and Cassandra.

2.1 Lack of Encryption

In MongoDB and Cassandra, a file is never encrypted by default. This void can be exploited by the attackers by adding a malicious code when information is to be transferred to database layer. When the older databases are replaced by the newer ones

© Springer Nature Switzerland AG 2020
A. P. Pandian et al. (Eds.): ICCBI 2019, LNDECT 49, pp. 420–427, 2020.
https://doi.org/10.1007/978-3-030-43192-1_48

which lack in security, sensitive information can be exposed if permissions are not implemented [6, 13].

2.2 Malware

Hackers and cyber criminals use different tactics like malware and phishing emails in order to breach the security algorithms and capture the important data of an organization. The users are not aware of their system being affected by malware and continue accessing the sensitive data [13].

2.3 No Authentication

Authentication is disabled by default in NoSQL databases. No username and password will be checked during the login procedure which can result in access to any user [6].

2.4 No Authorization

Authorization is disabled by default in NoSQL databases. This will allow access to all irrespective of their roles. For example: In the absence of authorization, an employee would have access to the salary management as well as the important documents of the company which he or she is not allowed in an ideal situation [6].

2.5 PHP Array Injection Attack

Consider an application implemented with a PHP backend, uses JSON format to query the data. Now, let us consider that the PHP backend where users log-in entering the required email and password.

PHP allows the following code:

```
email[$ne]=1&pass[$ne]=1

db.login.find({ email: { $ne: 1 }, pass: { $ne: 1 } }
```

'$ne' means not equal to in MongoDB. The output of the following code would be a list of all the users linked with the organization. Thus, the attacker would successfully login without email and password [7].

2.6 SQL 'OR' Injection Attack

One of the main causes of Injection is building the query with user input and not using proper encoding. This type of attack is difficult to execute in modern databases but there is a possibility of it to occur. Assume a login form which sends username and password through a form (POST). The query of this scenario can be written in a way that the username is checked by the compiler while the password will be ignored and be mentioned in the redundant part of the code [7].

Example:

> {username: 'admin', $or: [{}, {'okay': 'okay', password: ''}], $comment: 'Vulnerability is compromised'}

2.7 NoSQL JavaScript Injection Attack

NoSQL database allows JavaScript to run to perform complicated transactions like MapReduce. JavaScript execution exposes a dangerous attack because of user input. Let's take an example, a store has a collection of items and each item has a price. To get the sum or average of the fields, the developer writes a MapReduce function which takes name as a parameter from the user [7].

> $map = "function() {   for (var i = 0; i < this.items.length; i++) {   emit(this.name, this.items[i].$param); } }"; $reduce = "function(name, sum) { return Array.sum(sum); }"; $opt = "{ out: 'totals' }"; $db->execute("db.stores.mapReduce($map, $reduce, $opt);");

This program sums up the field provided by $param in the input. The quantity or cost of the above code is anticipated to be received, but a malicious entry can create a distinction. The payload closes the initial feature of MapReduce; attackers can run any required JavaScript on the database. A fresh MapReduce feature is called to balance the initial code and we will receive the following after adding it to the initial code. The following attack is similar to SQL injection attack and is quite hard to achieve in modern databases like MongoDB and Cassandra.

2.8 Denial of Service Attack

Due to no authentication and authorization by default in NoSQL databases, an attacker can send multiple requests. As a result of these attacks the system can get crashed due to Denial of Service (DoS) attack. Also, an intruder may use valid credentials and he/she does not need to be the administrator to perform such an assault [6].

3 Solutions to the Vulnerabilities in NoSQL

3.1 Mitigate the Injection Attacks

In order to mitigate the occurring injection attacks one must have a good native coding and use out of the box tools while writing a query. The security vulnerabilities and weaknesses of a particular web application must be identified in order to keep it free from attacks. It can be done in many ways. Two methods are discussed below: [7, 9].

3.1.1   Static Application Security Testing

The analysis of a certain program is performed without the execution of its code. The analysis is performed after checking the version of the source code and in some cases the object code is checked for the same [7].

3.1.2   Dynamic Application Security Testing

Dynamic Application Security Testing unlike Static Application Security Testing tools do not have access to the source code and therefore detect vulnerabilities by actually performing attacks [7, 9].

3.2   Access Control

In order to implement security in a database, controlling access to resources is a crucial step to be taken. Access control will protect the resources of a particular organization from unauthorized access and will provide the same usage to the legitimate users. They can be implemented in the following ways:

3.2.1   Mandatory Access Control

In Mandatory Access Control (MAC) the access is granted on labels. The objects are labeled on the basis of classification and subjects on the other hand are classified on the basis of clearance. This approach is centralized and preferred for access control mechanism [13].

3.2.2   Discretionary Access Control

In Discretionary Access Control (DAC) the access is granted through identity. It allows the owner of a particular object to decide who gets the access to it. DAC is implemented using Access Control Lists (ACLs) on the objects [13].

3.2.3   Role-Based Access Control

In Role-based Access Control (RBAC) the access is granted on the basis of roles or jobs. An employee may have different roles and one role may be assigned to a number of employees. The roles here can be mutually exclusive. RBAC supports MAC and DAC [13].

3.3   Protection Against Malware

3.3.1   Firewall

A firewall is intended to filter out the network-to-computer IP packets. It guarantees that all traffic from outside to inside is transferred or vice versa through it. It has the following three types:

3.3.1.1.   Packet Filters

When a packet arrives, packet filters apply a set of rules on it and then based on the outcome the packet is either forwarded or is discarded. Its security can be violated using IP spoofing [11].

3.3.1.2. Application Gateway
Application Gateway is also referred as Proxy Server. It hides the source IP and also decides the flow of the application level traffic [11].

3.3.1.3. Circuit Gateway
Circuit Gateway is similar to Application Gateway with some advanced features like establishing a new connection with the host and can change the source IP address of a particular packet [11].

3.3.2   Antivirus
It is an application software providing security against internet-based malicious programs. However, preventing them from being completely connected to the internet from the world is extremely difficult or nearly impossible [12].

3.4   Mitigate Denial of Service Attack

This attack occurs due to lack of proper authentication and role management. Today, it is possible to implement RBAC authorization and proper authentication in NoSQL. They help in the implementation of least privilege and thus prevent the escalation of attacks as only legitimate users are allowed. Example: data accessed via a web application, normal entries will be visible to 'the user' while sensitive entries require 'admin role' [7].

3.5   Implement Encryption in NoSQL

In order to encrypt data at rest, MongoDB Enterprise and Cassandra offer native storage-based file symmetric key encryption, which means that only one key will be used in the process of encryption and decryption. The user at the storage level can use transparent data encryption (TDE) to encrypt whole database files [6].

4   Analysis/Review

4.1   Comparison Table

Table 1. Comparison between static application security testing and dynamic application security testing

Parameters	Vulnerability-injection attacks in NoSQL (2.5., 2.6., 2.7.)	
	Static application security testing (3.1.1.)	Dynamic application security testing (3.1.2.)
How does way of implementation affect the output?	White Box Testing, access to the design, code and framework	Black Box Testing, no knowledge of the framework and design
Input	A developers approach and requires the source code for analysis	A hacker approach, requires a running application does not require the source code

(continued)

Table 1. (*continued*)

Parameters	Vulnerability-injection attacks in NoSQL (2.5., 2.6., 2.7.)	
	Static application security testing (3.1.1.)	Dynamic application security testing (3.1.2.)
Cost	Vulnerabilities are found in the early stages of Software Development Life Cycle so it is less expensive to fix them	Vulnerabilities are found at the end of the Software Development Life Cycle so it is expensive to solve them
Does it support all types of software?	Yes	No, can scan only through web applications and services
Issue detection	Cannot discover the environment and run-time related issues	Can discover the environment and run-time related issues

Table 2. Comparison between packet filters, application gateway and circuit level gateway

Parameters	Vulnerability-Malware (2.2.)		
	Packet filters (3.3.1.1.)	Application gateway (3.3.1.2.)	Circuit level gateway (3.3.1.3.)
Security	Least secure	Most secure	More secure than Packet Filters but less secure than Application Gateway
Does every router provide this functionality?	Yes	No, requires a unique program for execution	No, requires relay TCP connections for execution
Merits	Defined rules passes and rejects packets based on the rules	Mechanism is good for authentication and logging	Permission is granted through port addresses
Demerits	Difficult to manage	Not always transparent to all the users and can cause problems	No application level checking

Table 3. Comparison mandatory access control, discretionary access control and role-based access control

Parameters	Vulnerability-no authorization in NoSQL (2.4.) and Denial of Service Attack (2.8.)		
	Mandatory Access Control (3.2.1.)	Discretionary Access Control (3.2.2.)	Role-based Access Control (3.2.3.)
Authority	Access according to the decided parameters	Access decided by the owner of object	Access decided a central authority
Access criteria	based on security level Object-classification Subject-clearance	Based on the identity of the subject	Based on roles and job
Mechanism	Static	Flexible	Flexible
Merits	Secure as a subject with highest clearance is also asked about relevance of the object during the procedure	Easy implementation and flexible response	Management is easy. It is mainly used in multiple databases having multiple objects
Demerits	Cost is high and difficult to implement	Not as secure as the other two as violation can occur	Conflict roles may exist which disrupts the working flow

5 Future Scope

NoSQL consolidates enterprise versatility and efficiency in large data collection, retrieval and processing. It manages large data files and sets that provide enhanced performance, scalability and flexibility in real time. Modern companies are overwhelmed every second with loads of information from a variety of sources that access the internet. Such data can be further analyzed and documented for market analysis and future predictions. NoSQL's benefits include flexible design, even scaling, and better accessibility control. This database really helps web organizations in the fast-paced world to achieve their analytical goals [14]. Also, NoSQL offers a cheaper alternative for data storage and retrieval. Compared to the other NOSQL server, Mongo DB offers better results. Combining advantages from both NoSQL and SQL databases to further expand research horizons and optimize profitability is the best approach to balance the pros and cons (Tables 1, 2 and 3).

6 Conclusion

In terms of scalability and safety measures, NOSQL is more efficient than SQL, but it is not 100% secure. It faces the same problems as SQL. Some of the low-level vulnerabilities and rules have changed, but NoSQL systems just like the SQL systems still have high risks of injection, network exposure and improper access control management [15]. So, one must use mature databases having well defined security measures. But even the most secure database has a chance of being attacked.

References

1. Aboutorabi, S.H., Rezapour, M., Moradi, M., Ghadiri, N.: Performance evaluation of SQL and MongoDB databases for big e-commerce data. In: 2015 International Symposium on Computer Science and Software Engineering, CSSE (2016)
2. Ji, Z., Ganchev, I., O'Droma, M., Ding, T.: A distributed Redis framework for use in the UCWW. In: 2014 International Conference on Cyber-Enabled Distributed Computing and Knowledge Discovery (2014)
3. Aniceto, R., Xavier, R., Guimarães, V., Hondo, F., Holanda, M., Walter, M.E., Lifschitz, S.: Evaluating the Cassandra NoSQL database approach for genomic data persistency. Int. J. Genomics 2015 (2015). Article ID 502795
4. Gilbert, S., Lynch, N.: Brewer's conjecture and the feasibility of consistent, available, partition-tolerant web services. SIGACT News 33, 51–59 (2002)
5. Brewer, E.: Pushing the cap: strategies for consistency and availability. Computer 45, 2329 (2012)
6. Shahriar, H., Haddad, H.M.: Security vulnerabilities of NoSQL and SQL databases for MOOC applications. Int. J. Digit. Soc. (IJDS) 8(1) (2017)
7. Ron, A., Shulman-Peleg, A., Bronshtein, E.: No SQL, No Injection? Examining NoSQL Security (2015)

8. MacDonald, N.: Static or dynamic application security testing? (2011). http://blogs.gartner. com/neil_macdonald/2011/01/19/static-or-dynamicapplication-security-testing-both
9. Yadav, P., Parekh, C.D.: A report on CSRF security challenges & prevention techniques. In: 2017 International Conference on Innovations in Information, Embedded and Communication Systems (ICIIECS) (2018)
10. Yubin, G., Liankuan, Z., Fengren, L., Ximing, L.: A solution for privacy-preserving data manipulation and query on NoSQL database. J. Comput. **8**, 1427–1432 (2013)
11. Osawaru, E., Ahamed, A.H.R.: A highlight of security challenges in big data. Int. J. Inf. Syst. Eng. **2**(1), 2265–2289 (2014)
12. Malik, M., Patel, T.: Database security-attacks and control methods. Int. J. Inf. Sci. Tech. (IJIST) **6**(1/2), 175–183 (2016)
13. Kulkarni, S., Urolagin, S.: Review of attacks on databases and database security techniques. Facil. Int. J. Eng. Technol. Database Secur. Tech. Res. **2**(11), 253–263 (2012)
14. Singh, S., Rai, R.K.: A review report on security threats on database. Int. J. Comput. Sci. Inf. Technol. **5**(3), 3215–3219 (2014)
15. Das, D., Sharma, U., Bhattacharyya, D.K.: An approach to detection of SQL injection attack based on dynamic query matching. Int. J. Comput. Appl. **1**(25), 28–34 (2010)

Linux Server Based Automatic Online Ticketing Kiosk

Arvind Vishnubhatla[(⊠)]

Electronics and Communication Department,
Gokaraju Rangaraju Institute of Engineering and Technology,
Hyderabad, India
vainfo66@gmail.com

Abstract. Online ticketing kiosk is an embedded system used to book the tickets through an internet portal apart from delivering the tickets from its regular window. The objective of this project is to avoid long and rush waiting queues at the booking counters thereby making ease of access, and save the valuable time and energy of the users. The project is implemented by using a processor, the design of Graphical User Interface is made using Qtopia and MYSQL is used as the database server.

Keywords: Qtopia · MYSQL · LINUX

1 Introduction

The usage of kiosks is increasing in modern days because kiosks provide a quicker access to the users when compared to the traditional booking systems. Kiosks save user's valuable time and energy, apart from these users have a wide choice in their decisions. Kiosks are unmanned machines used to deliver tickets to the users, kiosks can be utilized for travel industry where vending of a ticket is a prime criteria. Kiosk provides a way to automate the traditional booking system.

The message attributes for each of the three core network types (Point of Wait, Point of Sale, and Point of Transit) are referred in [1]. A kiosk is computer terminal where access to communication, commerce, entertainment and education is described in [2]. The use of a kiosk to pay the electricity bill is highlighted in [3]. The use of a wayfinding kiosk for Orientation, Route decision, Route monitoring, Destination recognition is described in [4]. Video kiosk integrates **video conferencing** and collaboration capabilities [5]. Telemedicine Kiosks Bring Doctors And Patients Together For Healthy Outcomes [6]. The ICICI insta banking kiosk is described in [7]. The use of security to make kiosks hack free is illustrated in [8]. Locking down your computer to prevent hacking is shown in [9]. Protecting public facing devices from hackers, inadvertent misuse, and unauthorized access is discussed in [10]. Self checkout through semi-attended customer-activated terminal, SACAT are discussed in [11]. Machines where the customer performs the job of the cashier themselves, by scanning and applying payment for the items is discussed in [12].

© Springer Nature Switzerland AG 2020
A. P. Pandian et al. (Eds.): ICCBI 2019, LNDECT 49, pp. 428–439, 2020.
https://doi.org/10.1007/978-3-030-43192-1_49

1.1   Methodology

The methodology used in this project is spiral methodology In the spiral methodology initially the product is evaluated with some minimal features and slowly the development takes on incrementally which means after initial development the existing stage would be tested and depending upon the results obtained the product would be modified and the next phase is entered, the advantage of this method is the analysis of the product would be made in each and every stage which makes it easy for the designers the spiral methodology is shown pictorially further and explains the same concept in the Figure 1 shown further.

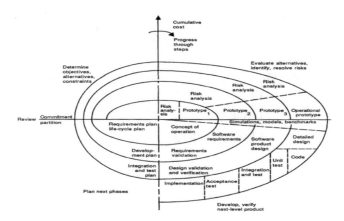

Fig. 1. Spiral model

1.2   Contemplated Requirements

- Manage Touchscreen
- Manage keypad (buttons)
- Dispense Tickets
- Manage Database transactions
- Manage Hardware Diagnostics

2   Project Methodology

The connection between the ticketing kiosk (client) and the database (server) would be a wired medium, of ethernet cable. The Internet Protocol address (IP address) of the main server would be installed in the ticketing kiosks, and the IP address of the ticketing kiosks would also be stored in the main server through which the transactions get carried, depending upon the IP address of the remote kiosk(s) the server would identify the remote kiosk(s) and establish a connection with it through the socket programming. The ticketing kiosk is used to interact with the user, who may perform several actions such as ticket booking, cancellation and know his current status. The database contains the information related to these queries, the interaction between the server and client is implemented through Transmission Control Protocol/Internet Protocol (TCP/IP) packets (Figs. 2 and 3).

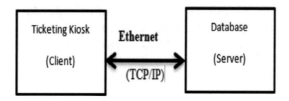

Fig. 2. Block diagram

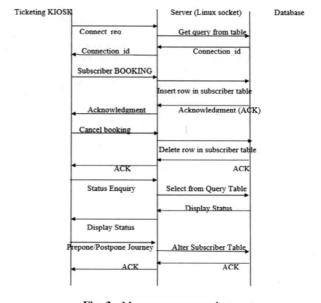

Fig. 3. Message sequence chart

The above figure represents message sequence chart between ticketing kiosk and database through sockets. The user interacts through kiosk where he has different options available such as booking, cancellation and status. Initially a connection request is sent from kiosk to server using the sockets in response to this request the server sends a connection id to the kiosk. When subscriber wants to book a ticket through book option available he books it this message reaches the server and it will add a table to the database which has been created earlier, similarly as shown above corresponding actions takes place depending upon the requests send by the user (Figs. 4, 5, 6, 7, 8 and 9).

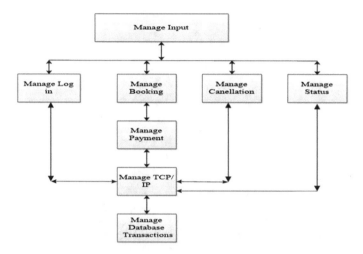

Fig. 4. State diagram of the project

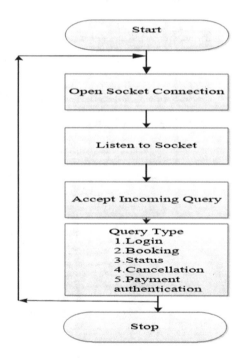

Fig. 5. Server flow chart

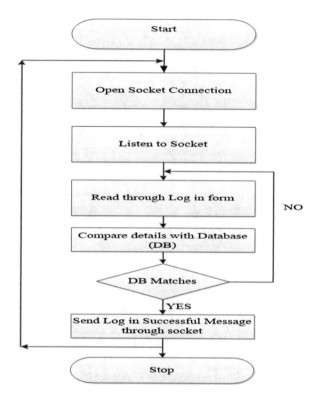

Fig. 6. Log in flow chart at server side

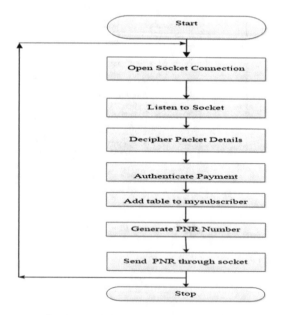

Fig. 7. Booking flow chart at server side

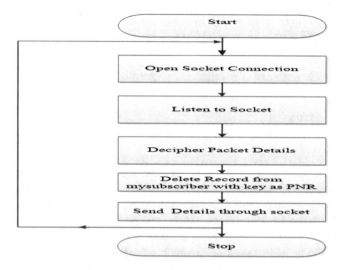

Fig. 8. Cancellation flow chart at server side

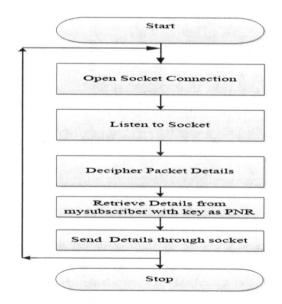

Fig. 9. Reservation status flow chart at server side

2.1 GUI Design

This project is implemented by the concept of server-client model, the user becomes the client while the database forms the server. The implementation on the client side is a Graphical User Interface (GUI) which is programmed in qtopia which consists of forms intended for the user to carry out his transactions. The server in this project is a database which is implemented in MYSQL.

Client Design

The client side is a user who has different options such as

1. Log in form
2. Booking form
3. Cancellation form
4. Status form

Log in form: This form is designed to provide log in for the user, in this project the user is required to initially register himself by which he will be provided a user name and a password. During the process of registration the credit card details are collected from the user which is required for cash payment purposes. In this form the user is required to enter his user name and password. He is provided with 2 options login and clear, clear is used to clear the fields in case if he forgets his details. When the user enters login his details get compare with the records and if they match successfully he will proceed further. The screen shot illustrates the same which is shown further (Fig. 10).

Fig. 10. Log in form screen shot

Booking Form: This form is designed to provide booking facility for the user, initially the user has to select his city and location and proceed further the next form again consists of several fields required to book a ticket once the user fills all these, details he is permitted further The screen shot illustrates the same which is shown further (Figs. 11, 12 and 13).

Fig. 11. Location record screen shot

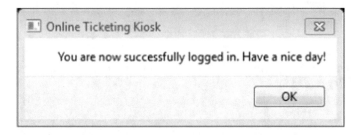

Fig. 12. Successful log in message screen shot

Fig. 13. Booking form screenshot

After completing all the fields show above fields in booking form the user has to checkout for payment. The screenshot of payment is shown further (Fig. 14).

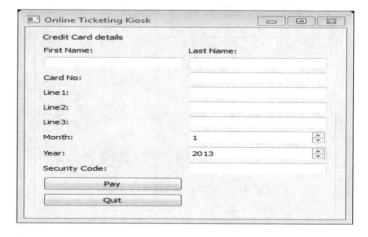

Fig. 14. Payment screenshot

In this phase if the payment is successful then the user gets his Passenger Name Record (PNR) (Fig. 15).

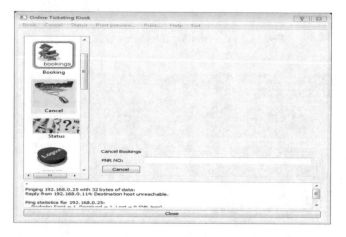

Fig. 15. Cancellation screenshot

If the user wishes to cancel his ticket he enters his PNR and then enters cancel option (Fig. 16).

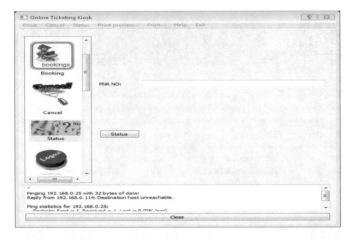

Fig. 16. Status screenshot

The above field is used to know the status of the user, the user is required to enter his PNR number and enter status button which displays his current status of his booking.

2.2 Server Design

The server side is a database which is designed in MYSQL The database consists of tables related to user information and credit card details all these details are shown further in screen shots further (Figs. 17, 18 and 19).

Fig. 17. Database screen shot

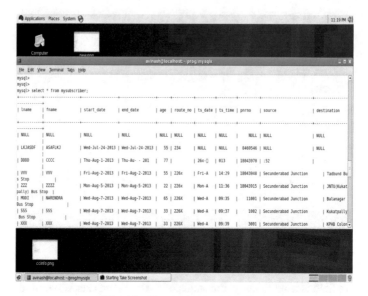

Fig. 18. Users records screenshot

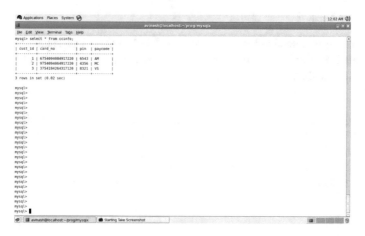

Fig. 19. Credit card details screenshot

3 Discussion

As planned earlier client and server were successfully implemented, the transactions as outlined in message sequence diagram were rigorously put to test. The output was tested and corrected to give a robust performance. Hence I strongly feel the objective set forth have been successfully achieved.

References

1. Kielsen, K.: Unleashing the Power of Digital Signage: Content Strategies for the 5th Screen. Focal Press. Information Stations in VA State Parks (2010). ISBN 0-240-81302-2
2. "Outdoor Digital LCD Displays", Networld Media Group: "Department for Work and Pensions: Communicating with customers" (PDF). National Audit Office, 5 May 2009. p. 6. Accessed 20 Feb 2014
3. A convenient way to pay electricity bill? The Hindu, Bangalore, 22 December 2013. Accessed 6 July 2015
4. Raven, A., Laberge, J., Ganton, J., Johnson, M.: Wayfinding in a hospital: electronic kiosks point the way, UX Magazine 14.3, September 2014
5. Employee-Free Bank Branch Powered by TrueConf - Video Conferencing Blog. Video Conferencing Blog, 18 December 2017. Accessed 13 Nov 2018
6. LLC, TelaCare Health Solutions: Telemedicine Kiosks Bring Doctors And Patients Together For Healthy Outcomes. www.telacare.com. Accessed 13 Nov 2018
7. ICICI insta banking kiosk. https://www.icicibank.com/Personal-Banking/insta-banking/insta-banking.page
8. The Next Web: "Brinks has a safe that runs Windows XP and hackers say they can crack it in 60 seconds"
9. Kiosk Software vs. Kiosk Mode: "Windows Kiosk Mode Uses & Limitations"
10. GPO vs Kiosk Software: "Using Group Policy Object (GPO) for Device Security"
11. "The Pros and Cons of Using Self-Checkouts - BusinessBee". BusinessBee, 7 August 2013. Accessed 5 June 2016
12. Zimmerman, A.: Check out the future of shopping. Wall Street J. (2011). Accessed 7 Dec 2016

A Reliable Automation of Motorized Berth Climb

Arvind Vishnubhatla[(✉)]

Electronics and Communication Department,
Gokaraju Rangaraju Institute of Engineering and Technology, Hyderabad, India
vainfo66@gmail.com

Abstract. The traditional seat allocation problems in Indian railways has led to one of the main bottlenecks called berth scheduling. The railway berths play a crucial role in the night time journey. The automated filling of berths will proportionally dependent on the equal weight allocation in the railway terminal. An alternative way for increasing the ease of transport for the aged population travelling in the train is by automatically shifting them instead of climbing to their respective upper berths. The proposed method using precision engineered column TV lift linear actuators will enhance the berth usability and allows the passengers to utilize the upper berths in an efficient manner.

Keywords: Railway · Upper berth · Aging · Actuators

1 Introduction

For a planned railway journey, we already have a convenient computerized online booking however it is not an easy task to get the lower berths at all the time. Even if you book in advance there is no guarantee that you would get the seats of your choice on the Indian train. In particular, senior citizens will always have a preference of lower berth. Eventhough there is an upper birth availability, the preparation to get up there is an ordeal. This task is more complicated if you do not have a good cardiovascular condition or if you are a little overweight.

The climb to the upper berth is far from a comfortable travelling zone where you are required to use a set of metal rods to hurl yourself up. As people get old they lose their ability to do simple things. It is not just because of aging but they're no longer active. There is a constant fear of getting injured like muscle pulls and strains. Some people may feel that it is a needless attempt to push their limits (Fig. 1).

This makes people to feel intimidated and discouraged. Eventhough the exercise looks simple, it requires endurance. One's inherent ability and flexibility is put to test. For aged population, it will affect their functional mobility and the confidence (Fig. 2).

Even if you get up the seat, conditions like diabetes may cause the bladder to be disturbed and you have an urgent need to climb down from the berth. This may be particularly uncomfortable because people in the lower berth will not feel comfortable all the times. The task becomes particularly difficult if you are suffering from arthritis.

© Springer Nature Switzerland AG 2020
A. P. Pandian et al. (Eds.): ICCBI 2019, LNDECT 49, pp. 440–448, 2020.
https://doi.org/10.1007/978-3-030-43192-1_50

Fig. 1. Lower and upper berths in railways [3]

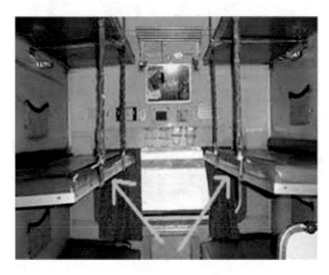

Fig. 2. The inconvenient climb [5]

By keeping all these drawbacks into consideration, An "Eight-Maglev" topological structure using electromagnetic guidance system is referred in [1]. A lifting payload method is shown in [2]. A two-phase permanent-magnet tubular linear motor is discussed in [3]. A novel load-sensitive linear actuator with a wide transmission range is discussed in [4]. Multi-car elevators realized efficiently by linear motors are described in [5]. Model-Based Networked Control Systems (MB-NCS) are discussed in [6]. A comprehensive review of the current advances in biomedical untethered mobile milli/microrobots is shown in [7]. Design and implementation of a forklift with dynamic stability is studied in [8]. Various mechanisms and devices are studied in [9]. Examples of some recent and useful actuators are available in [10, 11] and [12]. These research works makes it evident that the proposed work will overcome all the traditional bottlenecks in Indian railway journeys.

2 Methodology

It is believed that the railways is revolutionizing the existing ladder with a new design. In this connection it would be desirable to suggest few alternatives.

One of the proposed alternatives is the precision engineered column TV lift linear actuators from Firgelli automations.

The proposed lift based architecture is very simple, efficient, durable and reliable in nature. There are no exposed tracks. There are no gears or cables. The lift column technology is much better than rack and pinion lifts. They are superior to actuator lifts that tend to be very complicated and have poor reliability. The actuator lifts are also very noisy. Shown below in Fig. 8 is the operation of the proposed lifts with a remote control (Fig. 3).

Fig. 3. Programmable controller [6]

A programmable controller will help the user to negotiate lifting column through a button between four preset positions and ease of programming.

The down button can be used to fully retract the column. The up button can be used to negotiate between four heights.

3 Discussion

Rombout Frieling is a designer based in Eindhoven has designed a vertical walking system called Vertiwalk. This is a human powered alternative that allows people with disabilities to manually lift themselves to different floors. The user balances himself between a footplate and a seat and then perform a rowing motion to move himself to a different floor. The system does not rely on external power sources and it can be deployed in places without any access to electricity (Figs. 4, 5 and 6).

Fig. 4. a, b, c, d, e, f Motorized column lift actuator (ascent)

A hybrid pulley operated elevator where the arms and legs are used to propel oneself to the upper berth without having to negotiate the difficult staircase.

The inventor has successfully tested the prototype in various applications.

Rombout connects to a large international network and is keen to collaborate with others to propagate the innovation (Fig. 7).

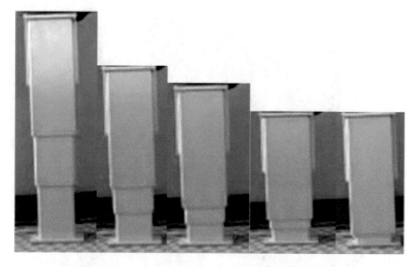

Fig. 5. (g, h, i, j, k) Motorized column lift actuator (descent)

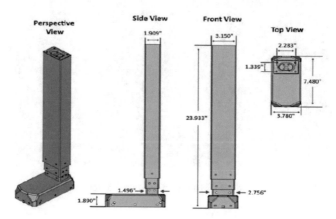

FA-SCL-650

Fig. 6. Technical drawings [7]

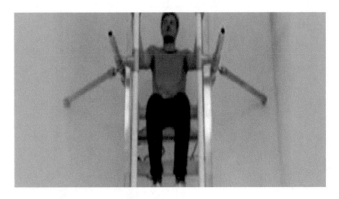

Fig. 7. Vertiwalk [5]

Fig. 8. Pulley driven mechanism [6]

Another alternative is to use a chair like device where straight line motion is created by a conventional electric motor. Different manufacturers tend to have proprietary methods of implementing an electromechanical linear actuator. This is illustrated in Figs. 9a, b, 10 and 11.

(a)

(b)

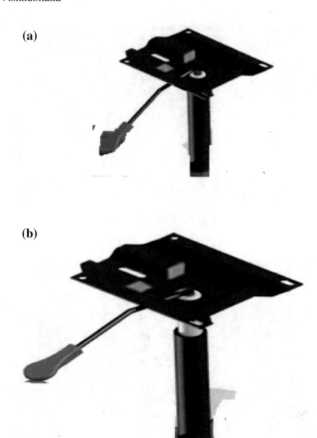

Fig. 9. a, b Adjustable motorized seat mechanism with a linear actuator [4]

Fig. 10. Adjustable motorized seat mechanism with a linear actuator [4]

Fig. 11. Adjustable positions of a linear motorized actuator [4]

4 Conclusion

The new method of having a motorized column actuator or the vertical lift eases the climbing operation to the upper berth and provides relief to the senior citizens who are analyzed with restricted medical conditions. It is easy to implement and helps to alleviate the woes of the common man in the traditional railway journey.

References

1. Hu, Q., Wang, H., Yu, D.: H∞robust control for electromagnetic guiding system suspension altitude of linear elevator. In: 2012 24th Chinese Control and Decision Conference (CCDC) (2012)
2. Jeong, D.-H., Choo, J., Jeong, S., Chu, G.: Attaching sub-links on linear actuators of wearable robots for payload increase, pp. 1296–1301 (2014)
3. Urban, C., Gunther, R., Nagel, T., Richter, R., Witt, R.: Development of a bendable permanent-magnet tubular linear motor. IEEE Trans. Magn. **48**(8), 2367–2373 (2012)
4. Hagiwara, T., Hirose, S.: Development of dual mode X-screw: a novel load-sensitive linear actuator with a wide transmission range. In: Proceedings 1999 IEEE International Conference on Robotics and Automation (Cat. No.99CH36288C), vol. 1, pp. 537–542 (1999)
5. Markon, S., Komatsu, Y., Yamanaka, A., Onat, A., Kazan, E.: Linear motor coils as brake actuators for multi-car elevators. In: 2007 International Conference on Electrical Machines and Systems (ICEMS), pp. 1492–1495 (2007)
6. Garcia, E., Antsaklis, P.J.: Model-based control using a lifting approach. In: 18th Mediterranean Conference on Control and Automation, MED 2010, pp. 105–110 (2010)

7. Qiu, T., Palagi, S., Sachs, J., Fischer, P.: Soft miniaturized linear actuators wirelessly powered by rotating permanent magnets. In: 2018 IEEE International Conference on Robotics and Automation (ICRA), pp. 1–6 (2016)
8. Sarker, A., Al Amin, S., Tamzid, S.M.T.H., Chisty, N.A.: Design and implementation of a forklift with dynamic stability. In: 2017 IEEE Region 10 Humanitarian Technology Conference (R10-HTC), pp. 658–663 (2017)
9. Sclater, N., Mechanisms and Mechanical Devices Source Book, 4th edn., vol. 25. McGraw-Hill (2007)
10. Electric Micro Linear Actuators: Actuonix Motion Devices. www.actuonix.com. Accessed 10 Sept 2017
11. Firgelli voted best Linear Actuators: TV Lifts & desk Lifts
12. Linear Actuator Guide: Anaheim Automation. Accessed 12 May 2016

Requirement Gathering for Multi-tasking Autonomous Bus for Smart City Applications

Arvind Vishnubhatla$^{(\boxtimes)}$

Electronics and Communication Department,
Gokaraju Rangaraju Institute of Engineering and Technology, Hyderabad, India
vainfo66@gmail.com

Abstract. The innovation of autonomous buses develops more advantages into the smart city infrastructure by reducing the cost and increasing its reliability. This can also enhance the transit capability and accessibility in a cost-effective manner. The main objective of this paper is to gather requirements for developing an autonomous bus which can propel itself from place to place without human intervention. This research study sets a new platform for decreasing labor costs and increasing safety of passengers in transit. The project is supposed to be implemented as a miniature prototype where the major functionality is implemented and demonstrated. The proposed methodology will majorly benefit the public transport with more flexible modes that can generate more advantages to metropolitan residents.

Keywords: Autonomous bus · Miniature · Prototype · Automation

1 Introduction

The increasing benefits of autonomous vehicles are strongly encouraging a rapid technological innovations in autonomous buses. A set of characteristics is used to define the problem in order to make the customer happy are to be determined. Without proper planning and a strategy the most sophisticated projects can end up in jeopardy.

A project is a goal oriented activity using a sequence of steps called algorithms [1]. To execute the required project we understand nature of software engineering as a profession, as a discipline and as a culture [2]. A common vocabulary is applicable to all systems and software is available [3]. The terms currently in use in software engineering domain are available in [4]. The roots of software engineering are available in [5]. The application of a systematic, disciplined, quantifiable approach to the development is illustrated in [6]. A report on undergraduate curriculum on software engineering for major embedded and computer systems is mentioned in [7]. The disciplined use of scientific and technological methods, knowledge and experience can be imbibed through [8]. The building of intelligent systems and technologies is available in [9–12].

Requirements engineering requires defining, managing and documenting the user needs. This is a way of using domain knowledge of the problem at hand and brainstorming the knowledge to consolidate user needs. All functional and non-functional needs are specified in detail. These are then captured using a suitable model e.g. the

A. P. Pandian et al. (Eds.): ICCBI 2019, LNDECT 49, pp. 449–456, 2020.
https://doi.org/10.1007/978-3-030-43192-1_51

waterfall model or the spiral model. The requirements so captured are verified and validated. If requirements are not properly consolidated, errors propagate into other stages of the project resulting in costly modifications and rework.

The proper documenting and tracking of requirements helps in establishing a clear communication path with all the stakeholders. A well maintained software repository helps to modify the requirements at later stages and incorporate changes if necessary.

A proper life cycle helps the product to exceed customer expectations. A disciplined process helps in defining in detail the deliverables expected from the next phase of the life cycle. This helps to ensure that code is commensurate with the design.

This also ensures a fully tested product so that the product can be deployed with minimal amount of error.

The key stakeholders are Project Manager, Business analyst and other senior team members.

The requirements phase is followed by a highlevel design and a lowlevel design. Here care must be taken to see that every feature of the product is workable. At the end of this phase coding is done. This is followed by a test phase where either manual or automated testing is performed. The development does not end even after deployment. The released product can be properly maintained if a proper development life cycle is followed (Figs. 1, 2 and 3).

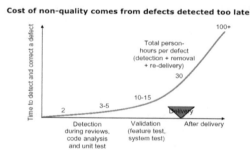

Fig. 1. Cost of non quality

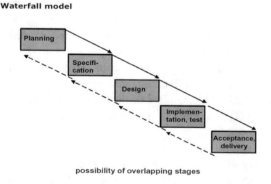

Fig. 2. Waterfall model

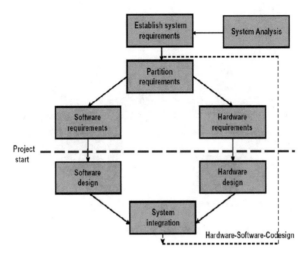

Fig. 3. The system specification

2 About the Method

A step by step analysis of the domain at hand is made to arrive at the specifications of the required autonomous vehicle. The project consists of a set of software requirements to be implemented on a prototype hardware. After careful examination the Texas instruments C6678 multicore processor is selected. The planned software should consist of User Programs, Operating system and the device drivers.

It is clear that the bus contains a set of sensors interfaced using the CAN bus as shown in Fig. 4. Figure 5 shows the partitioning of the OS, User applications and device drivers. Figure 6 shows the presence of a proprietary OS on the C6678 platform. Figure 7 shows the C6678 EVM board. Figure 8 shows the compilation process.

A hardware architecture for the project is defined where the drive interface circuitry is suitably interfaced to a common bus, where the C6678 hardware, a GPS receiver,

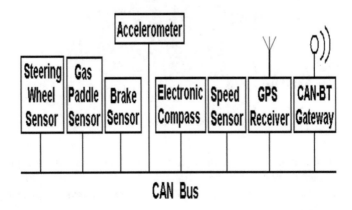

Fig. 4. Various automotive sensors on the can bus

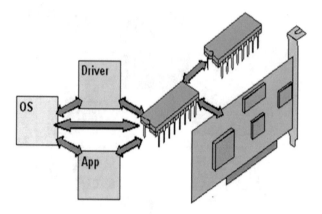

Fig. 5. Embedded software components

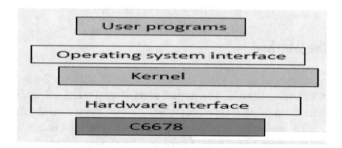

Fig. 6. Software stack on the defined hardware

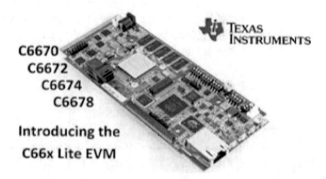

Fig. 7. C6678 EVM board

panel I/O, a 3D imaging camera and an anti-collision radar is also interfaced. A display interface is also available (Figs. 9 and 10).

After careful analysis of the domain the requirement shown below in Fig. 11 are arrived at (Fig. 12).

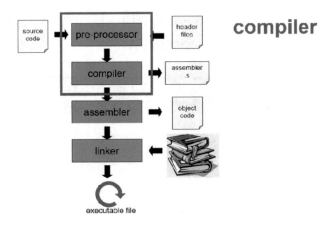

Fig. 8. Compilation process

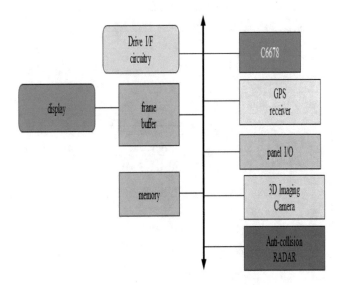

Fig. 9. Architecture diagram of the project

GPS moving map block diagram

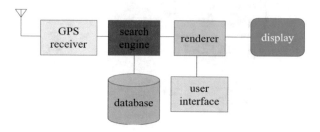

Fig. 10. Architecture diagram of the project

Adaptive Cruise Control
GPS Map Rendering
Lane Departure Warning Systems
3D Imaging
Headon Collision Detection
Adherence to Rail/Roard Traffic Signals
Keyless Ignition Control Design
Communications Vehicle Network
Tire Pressure Monitoring Systems
Heavy Vehicle Event Data Recorder
fog lamp
Airbag Module Deployment
Emergency Warning Device
Oral Passenger Briefings
Fire Extinguisher
Interior Climate Control
Windshield Demisting and Defrosting Systems
Remote Towing facility
Reaction to weather changes
Online Toll Payment

Fig. 11. Requirements for the project

Fig. 12. Decomposition of various tasks for the project.

3   Conclusion

A proper choice of hardware and software is now available for the developers to do a high level and lowlevel design of the project. A review of the requirements shows that they are detailed and exhaustive.

For a large scale deployment of autonomous buses, this study has provided a basic overview for developing an automated transportation systems by simplifying their architecture and quantifying its cost related drawbacks in the near future. The introduction of autonomous buses in real time metropolitan applications has introduced significant opportunities and challenges for transport based researchers. In the near future, autonomous buses will emerge as another important topic of interest for further research investigation.

References

1. ACM: Computing Degrees & Careers. ACM (2007). Accessed 23 Nov 2010
2. Laplante, P.A.: What Every Engineer Should Know about Software Engineering. CRC, Boca Raton (2007). ISBN 978-0-8493-7228-5. Accessed 21 Jan 2011
3. Systems and software engineering – Vocabulary. ISO/IEC/IEEE std 24765:2010(E) (2010)
4. IEEE Standard Glossary of Software Engineering Terminology. IEEE std 610.12-1990 (1990)
5. Michael S. Mahoney Princeton University: Software Engineering. Information Processing, vol. 71, pp. 530–538. North-Holland Publishing Co. (1972)
6. Salah, A.I.: Engineering an academic program in software engineering (PDF). In: 35th Annual Midwest Instruction and Computing Symposium, 05 April 2002. Accessed 13 Sept 2006

7. Mills, Harlan D., J. R. Mills, H.D., Newman, J.R., Engle Jr., C.B.: An undergraduate curriculum in software engineering. In: Deimel, L.E. (ed.) Software Engineering Education: SEI Conference 1990, Pittsburgh, Pennsylvania, USA, April 2–3, p. 26. Springer (1990). ISBN 0-387-97274-9
8. Budgen, D., Brereton, P., Kitchenham, B., Linkman, S.: Realizing Evidence-based Software Engineering, 14 December 2004. Archived from the original on 17 December 2006. Accessed 18 Oct 2006
9. Leondes, C.T.: Intelligent Systems: Technology and Applications. CRC Press. pp. I–6 (2002). ISBN 978-0-8493-1121-5
10. Pressman, R.S.: Software Engineering: A Practitioner's Approach, 7th edn. McGraw-Hill, Boston (2009). ISBN 978-0073375977
11. Sommerville, I.: Software Engineering, 9th edn. Pearson Education, Harlow (2010). ISBN 978-0137035151
12. Jalote, P.: An Integrated Approach to Software Engineering, 3rd edn. Springer (1991/2005). ISBN 978-0-387-20881-7

Offline Handwritten Devanagari Character Identification

Gita Sinha[✉] and Shailja Sharma

CSE Department, RNTU, Bhopal, India
gitawit321@gmail.com, Shailja1901@rediffmail.com

Abstract. Handwritten character identification is a most torrid area of research where countless researchers have presented their work and is still an area less than research to accomplish higher identification accuracy. In earlier period acquisition, storing and exchanging information in type of handwritten script was the well-situated way and is still widespread as a convenient medium in the era of digital equipment. As advanced technology like tablet has been used and many comparable devices that allows humans to key in data in form of handwriting character. Manuscript is written by the use of paper and then converting to an image via scanner, identify handwritten characters as of the image is well-known as off-line handwritten character identification is a demanding work due to the fact that each author will have diverse style of writing and all scripts have their own character set and complexities to write.

Keywords: Identification of handwritten devnagari character · Off-line handwriting identification of character · Image pre-processing · Segmentation · Feature determine technique · Classification methods

1 Introduction

Many research work has been carried out in other Character Identification (CR) in the last partially century and progressed to a stage, enough to fabricate machinery driven applications. But in the case for Indian languages which are problematical in expressions of organization and data processing. The application of HOCR are in different area like work place and automatic book processing in library bank check interpretation and postal order reading facility, print department, transportation and communication tools. More than 500 million citizens' spoken devnagari so it become the nationalized language of India. It will be disposed exclusive focus in order to paper access and analysis of rich very old and recent Indian article can be successfully done. The present paper is a reflect to deliver as a direct and modernize for the researchers, engaged in offline Handwritten Devanagari Character Identification (OHDCI) field. An examination of HDOCR scheme is offered and the accessible HDOCR methods are reviewed. The recent work of HDOCR is talk about and instructions for upcoming research are recommended for winding up. Many scientists have engaged in the field of handwriting identification, and a variety of methods and structure have been developed to identify handwritten numbers and character both online and offline approach. One can mark out wide-ranging work for English and Arabic script whereas explore for handwritten

© Springer Nature Switzerland AG 2020
A. P. Pandian et al. (Eds.): ICCBI 2019, LNDECT 49, pp. 457–464, 2020.
https://doi.org/10.1007/978-3-030-43192-1_52

character identification has progressed in recognize handwritten characters for scores of our nation calligraphy such as Devanagari, Tamil, Telugu, Kannada, Hindi, Gurumukhi, Marathi, Gujarati.

Preceding effort in offline handwritten character identification. OHCR consist with different levels described in Fig. 1.

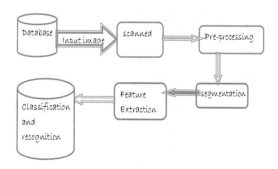

Fig. 1. Levels of OHCR

2 Preceding Data Study

Dongare et al. [1] have offered a work in which arrangement with expansion of grid occupying method which is composition of centered zone of an image and zone centered zone of individual character or numerical image for feature mining. The process Optical character identification consist through various phases such as scanning, preprocessing, segmentation, feature determine. network or sector based feature determine approach has been used for feature determine Then individual text image is separated into n identical shape grids or zones then average distance of all pixels with consideration of image centroid or grid centroid is designed. For feature determine combine image centroid zone method and zone centroid zone method to computes median scope of all pixels there in one by one grid with consideration to image centroid in addition to zone centroid which fabricate feature vector of size m into n features. For identification this feature vector is conferred as participation in progressive neural network.

Aggarwal et al. [2] present paper on Identification of Handwritten Devanagari Character Using Gradient based Features determine methods. For the purpose of classification dataset consists with 7200 characters samples 200 samples of each one of 36 basic Devanagari characters, which are composed from 20 different citizens each contributing to write 10 samples of each one of 36 characters. All Devanagari characters sample images used are resized to 90 * 90 pixel sizes. For feature determine gradient based scheme has been used and designed for classification Support Vector Machines (SVM) is implemented. The performance of these features through cross validation accuracy of 94% is achieved.

Saxena et al. [3] present paper on a new methodology of Handwritten Devanagari constant Character Identification over Feed Forward behind Propagation Neural

Network. Experiments performed over a database generated by Collecting 20 hand-written samples of Devanagari characters from five different writers. The stored image are saved in jpg format. The methodology applied on three methods. The first method is based on randomly subdivision of character constant sample image and pass it as input of network. The second procedure based on taking full image values as put in of the network. The third technique depend on taking all sample image values as input and subdivision values as target to recognize the pattern of character image. Second-layer feed-forward network via sigmoid hidden neurons and linear output neurons are used in this method.

Goyal et al. [4] they develop a technology to interaction with the computer. It is essential that the computer system recognizes the characters written by the consumer so regulate to interact with him. The procedure has been implemented for identification of character first segments the image containing Devanagari text exceed to the software keen on different form such as lines, lines to words and words to characters. The obtained characters are then brought down to a normal size. For detection Kohonen Neural Network have been used then keen on act and recognizes the text character by character and provides the production in format of Unicode. The network has been planned with no hidden layer to support sudden identification. They too affect this method on an image and provided the choice to distinguish individual handwritten characters tired using a mouse. Such a scheme provides keyboard fewer computer interaction. The practice is implemented via Java. They have achieved generally identification accuracy rate for a fixed font machine printed characters representation is 90.26% and for handwritten characters, it is 83.33%.

Holambe et al. [5] this paper present relative study of dissimilar classifiers for Devanagari handwritten Character identification. They carry out the experiment on off-line handwritten Devanagari font. Authors applied one feature set and 21 different classifiers for experiment using diverse classifiers. For 21 feature determine methods namely ED, PD, SM, MPD, PWC, MQDF, GLVQ, PC, FLD, LC, QC, NNC, ED, KNN, CS, KNN, PCKNN, CNN, RNN, FNN, NUN, SVM, RBF, HMM applied on the text to perform comparison. They perform experiment on vowels ('svar'), consonants ('vyanjan'), without modifiers consonants ('vyanjan') with modifiers accordingly identification accuracy is 96%, 95% and 95%. They got highest accuracy on vowels.

Ajmire et al. [6] present a paper on An Analytical learning of Handwritten Devanagari nature identification. They compare identification accuracy by used different feature determine method and classification method. They achieve 94.24% identification accuracy using Directional gradient Gaussian filter method for feature determine and Modified Quadratic method for classification on Devanagari character identification.

Agnihotri et al. [7] they present a paper on Offline Handwritten Devanagari Script Identification. For the purpose of feature extraction they have used Diagonal dependent method and then these feature of each character image is transformed into some bit sequence of length 378. For training and testing purpose more than 1000 example is used in this projected work. It is contributed to use the efficiency of genetic algorithm to identify the character. Various operations is performed on image such as pre-processing, feature determine. Diagonal based feature determine method to take out 54 features to each character. They has been used MATLAB 7.10.0 (R2010a) for implementation. The experimentation is applied on greater than 1000 characters image. The test are separated into data sets for experiment, the training data set and testing data set. The training set contains 904 characters and testing set contains 204 characters image. The identification accuracy of offline Devanagari system is 85.78% match, 13.35% is incorrect.

Swethalakshmi et al. [8] they have proposed strokes based quality are determined and classifier Support vector machines have been used for constructing the stroke identification engine on Identification of Online Handwritten Character (OHCR) for Devanagari and Telugu Characters image. For stroke identification three methods have been explored. Single Identification Engine method, Multiple SVM Engines for Stroke Identification, Stroke Identification using HMMs. They have got 96.69% identification accuracy on 46 number of class using 60 feature vectors and 97.27% identification accuracy on 82 number of class using 120 feature vectors.

Dongre [9] present a reconsider paper on devnagari character identification. This paper various pre-processing method like (noise reduction, skew exposure and rectification, size normalization, thinning), segmentation, various feature determine methods such as (universal conversion and sequence extension, statistical features geometrical and topological features) and classification method such as (pattern similar, statistical method neural networks, support vector machine (SVM) algorithms combination classifier) has been defined.

Surya Nath et al. [10] this paper represent a reconsider of offline character identification and online character identification on dissimilar natural languages. Offline or online HCR is an optical character identification, that are use translate the written paper and handwritten document in to automatic understandable system. It is very difficult to obtain 99.9% accurateness during the purpose of HCR. The effectiveness of HCR based on the quality extracted and the types of classifier has been applied.

Vamvakas [11] present a paper for whole Optical Character Identification Methodology for Historical papers. In this paper three methodology has been defined in various level:

The upper 2 steps demote to creating a record for training via a combination of documents, while the 3rd level 1 refers to identification of new text sample. At earliest, a pre-processing phase that mentioned binarisation of input sample and improvement takes place. In second step a downwards segmentation methods is used in direction to detect text lines, words and characters from the images. Then for group characters clustering scheme is used for related structure. The proposed method is a half-automatic procedure while the researcher is capable to work together at any time in consideration to correct probable defects of clustering and hand over an ASCII tag. Later than this action, for identification a catalogue is formed in teach to obtain accurate result. In the end, at the third stage, for each new paper picture the over segmentation approach has been put even as the identification is based on the character record that has been formed at the preceding stage. Region based feature determine and has been used for classification. On handwritten Devanagari character using SVM classifier 94.2% of identification accuracy has been obtained.

Khanale et al. [12] present a paper on handwritten Devanagari character identification using artificial neural network. Images are scanned, digitized and perform segmentation of certain image. Then image are cropped to resize. They achieve 96% identification accuracy on certain character.

Abdul rahiman et al. [13] present a paper on a minutia learning and analysis of OCR investigate in the area of south Indian scripts. They describe various steps of OCR. Which are Binarization, removal of Noise, Detection of Skew & its Correction, reduction, Segmentation Line, Word and Character sample, Feature Determination, Selection and Classification. After that pre-processing, feature from the image is extracted taking place the basis of stature of the sample, all the horizontal and vertical small and lengthy position there, thickness of the numeral sample records of horizontally and vertically oriented arcs, centriole of the sample of image, situation of a variety of features, elements in the different neighborhood. All these features has been used for input to Classifiers that name is Support Vector Machine (SVM) where the character is confidential by supervised knowledge. For identification these categories are graphed against Unicode and the text is re-structured via the Unicode. The presentation is correlated by means of dissimilar classifiers and the SVM classifier provides the high achievements. The identification accuracy on Malayalam is 93–97%, Telugu 84–87%, Tamil 62–98% and Kannada 85% has achieved using various feature determine and classification methods.

Shelke et al. [14] present this paper on an original Multi-feature various-Classifier proposal for unimpeded transcribed Devanagari Character Identification. They used Marathi character image for their experiment. For feature determine they used Radom change and Euclidean space transform and perform practical into 2 separate feed forward back propagation neural networks. On the concluding stage accept the contribution from second neural network classifiers and pattern identical classifier and determine the ending production according to maximum selection regulation. They have presented several-stage feature determines and classification strategies are used to progress the identification accurateness over individual classifiers considerably. The experimentation is performed over handwritten Marathi nature sample. Dataset is gathered from greater than 40 people of 400 samples per character is collected resulting in about more than 16000 character document in the database The identification rate achieved from the predictable method is 95.40%.

Kale et al. [15] present paper on Handwritten Devanagari composite Character Identification using Legendre Moment an Artificial Neural Network Approach. They perform various operation on images, images are normalized into 30×30 pixel size divided into region, from this behavior in addition to statistical quality of information are extracted from each region of image. Then the proposed system used is skilled and experienced on 27000 handwritten sample collected from different user for performing experiment. For the purpose of classification they have put into Artificial Neural Network. They achieved the overall identification accuracy 22] for basic is up to 98.25% and for all composite character is 98.36% [16].

Manwatkar et al. [17] has presented a paper on identification of text from an image. They develop a system which is able to read the text from image. That system is named as document image investigation which transform document in paper format into machine understanding format. This developed system has useful for reader to better understanding of particular sequence of different module. They have applied line detection from image, character detection from image as segmentation methods. These are various method has been applied for feature determine like Component Analysis (PCA), Linear Discriminate Analysis (LDA), self-governing Component Analysis (SCA), string Code (SC), zoning, Gradient Based features, Histogram etc. They have used Kohonen Neural Network (KNN) and SVM for classification.

Indira et al. [18] they present a review paper on Devanagari Character Identification: A Short Review. They define Devanagari form is basic of several language Bengali, Hindi, Sanskrit, Marathi, Kashmiri so on. This paper present several process of identification like binarization, Noise elimination, Size normalization, Thinning. After that explanation of Segmentation and feature determine has been presented.

Adwaitdixit et al. [19] present a paper on Handwritten Devanagari Character Identification based on wavelet Based Feature Determination and Classification Scheme. 2000 characters are collected from 100 hundred people for experiment. Each character image are scanned, pre-processed and for binarization Otsu's method is applied, contrast enhancement methods is applied on some image before binarization. Wavelet based feature determine methods has been applied. After that Different statistical feature vector is calculated and feed into back prorogation neural network for 20 different classes that produce 70% of identification accuracy.

JINO [20] presented a thesis on Machine Learning Techniques for Handwritten offline Malayalam Recognition of Word. They present basic details of image processing, briefly discussed about Malayalam script. For the purpose of experimental data collection they have selected 199 group of people to collect 31020 set of handwritten data from different age group belonging to several age, occupation, sex, education and income also. Various feature extraction techniques has been used for the purpose of recognition of character or numeral sample of collected data namely HOG (Oriented Gradient Histogram), PHOG (Oriented Gradient Pyramidal Histogram), CNN and Wavelet (Convolution Neural Network). For implementation of experiment methods are MLP (Multi Layer Perception), SVM (Support Vector Machine), Random Forest, and BLSTM Short Term Memory Bidirectional Long)-TCC (Temporal Classification Connectionist). They produce different recognition accuracy on various classification method on several collected dataset depicted in Fig. 1.

Table 1. Recognition accuracy of Malayalam character

Dataset	Recognition accuracy			
	CNN	SVM	Union of dataset	
			CNN%	SVM%
1–100	96.0%	97.1%	95.7	96.9
101–200	96.6%	97.9%	96	97
201–330	95.4%	96	96.1	97

Puri et al. [21] present a paper on Toward Hindi Handwritten Document Image for Classification and Recognition, they present all hindi character set like vowel, constant and numbers. With segmentation of character into different part. They proposed several writing-based parameters that specify the direction in which the text are written and embedded in a sample (Table 1).

3 Purposes of OCR

It has a huge quantity of practical utilisations. throughout the before time being, OCR has been applied for mail sorting, bank cheque interpretation and signature authentication in spite of that, OCR can be implemented by companies for habitual form processing in spaces where huge number of facts is presented in printed form or handwritten form. Other purpose of OCR hold process service bills, passport justification, pen computing and automated number plate classification etc. Another helpful application of OCR helps visor and visually impaired individuals to look at text.

4 Conclusion

In this review paper, we explore work done on handwritten Indian scripts. Firstly we gives characteristic of Indian script as well as properties of devanagari script. Then we present process of complete character identification system with the help of flowchart. There after we explore the methods and procedure, developed for identification of particular Indian script i.e. Devanagari. A lot of work has been done on Devanagari described in literature survey part.

Acknowledgement. I am extremely appreciative to my respected project guide Dr. Shailja Sharma and Head of Dept. Dr. Sanjiv Kumar Gupta for his thoughts and facilitate to be valuable and helpful during the conception of this review paper. I would like to express thanks all the faculties who have support me during my review paper. At last, I am grateful to my associates who mutual their knowledge in this field.

References

1. Dongare, S., et al.: Handwritten character identification using neural network. IOSR J. Comput. Eng. **16**(2, Ver. X), 74–79 (2014). ISSN 2278-0661, p- ISSN 2278-8727
2. Aggarwal, A., et al.: Handwritten devanagari character identification using gradient features. Int. J. Adv. Res. Comput. Sci. Softw. Eng. **2**(5) (2012)
3. Saxena, S., et al.: Novel approach of handwritten Devanagari character identification through feed forward back propagation neural network. Int. J. Comput. Appl. **51**(20) (2012). (0975 – 8887)
4. Goyal, P., et al.: Devanagari character identification towards natural human-computer interaction (2010)
5. Holambe, A.K., et al.: Comparative study of different classifiers for Devanagari handwritten character identification. Int. J. Eng. Sci. Technol. **2**(7), 2681–2689 (2010)
6. Ajmire, S.P., et al.: An analytical study of handwritten Devanagari character identification. IJARCSSE **5**(11) (2015)
7. Agnihotri, V.P.: Offline handwritten Devanagari script identification. I.J. Inf. Technol. Comput. Sci. **8**, 37–42 (2012)
8. Swethalakshmi, H., et al.: Online handwritten character identification of Devanagari and Telugu characters using support vector machines (2006)

9. Dongre, V.J.: A review of research on Devnagari character identification. Int. J. Comput. Appl. **12**(2) (2010). (0975–8887)
10. Surya Nath, R.S., et al.: Handwritten character identification – a review. Int. J. Sci. Res. Publ. **5**(3) (2015). ISSN 2250-3153
11. Vamvakas, G., et al.: A complete optical character identification methodology for historical documents. IEEE (2008). 978-0-7695-3337-7/08
12. Khanale, P.B., et al.: Handwritten Devanagari character identification using artificial neural network. J. Artif. Intell. **4**(10), 55–62 (2011)
13. Abdul Rahiman, M., et al.: A detailed study and analysis of OCR research in South Indian scripts. In: 2009 International Conference on Advances in Recent Technologies in Communication and Computing. IEEE (2009). 978-0-7695-3845-7/09
14. Shelke, S., et al.: A novel multi-feature multi-classifier scheme for unconstrained handwritten Devanagari character identification. In: 2010 12th International Conference on Frontiers in Handwriting Identification. IEEE (2010). https://doi.org/10.1109/icfhr.2010. 41, 978-0-7695-4221-8/10
15. Kale, et al.: Handwritten Devanagari compound character identification using legendre moment an artificial neural network approach. In: 2013 International Symposium on Computational and Business Intelligence. IEEE (2013). 978-0-7695-5066-4/13
16. Ghosh, R., et al.: Study of two zone-based features for online Bengali and Devanagari character identification. In: 2015 13th International Conference on Document Analysis and Identification (ICDAR). IEEE (2015). 978-1-4799-1805-8/15/2015
17. Manwatkar, P.M., et al.: Text identification from images. In: IEEE Sponsored 2nd International Conference on Innovations in Information, Embedded and Communication systems (ICIIECS) (2015)
18. Indira, B., et al.: Devanagari character identification: a short review. Int. J. Comput. Appl. **59** (6) (2012). (0975 – 8887)
19. Dixit, A., et al.: Handwritten Devanagari character identification using wavelet based feature determine and classification. In: 2014 Annual IEEE India Conference (INDICON) (2014)
20. Jino, P.J., Balakrishnan, K., Bhattacharya, U.: Offline handwritten Malayalam word recognition using a deep architecture. In: Soft Computing for Problem Solving, pp. 913–925. Springer, Singapore (2019)
21. Puri, S., Singh, S.P.: Toward recognition and classification of Hindi handwritten document image. In: Ambient Communications and Computer Systems, pp. 497–507. Springer, Singapore (2019)
22. Jacob, I.J.: Capsule network based biometric recognition system. J. Artif. Intell. **1**(02), 83–94 (2019)

Behavior Anomaly Detection in IoT Networks

Dominik Soukup[1], Tomas Cejka[2], and Karel Hynek[1,2(✉)]

[1] CESNET, a.l.e., Zikova 4, Prague, Czech Republic
soukudom@cesnet.cz
[2] CTU in Prague, Thakurova 9, Prague, Czech Republic
{cejkato2,hynekkar}@fit.cvut.cz

Abstract. Data encryption makes deep packet inspection less suitable nowadays, and the need of analyzing encrypted traffic is growing. Machine learning brings new options to recognize a type of communication despite the heterogeneity of encrypted IoT traffic right at the network edge. We propose the design of scalable architecture and the method for behavior anomaly detection in IoT networks. Combination of two existing semi-supervised techniques that we used ensures higher reliability of anomaly detection and improves results achieved by a single method. We describe conducted classification and anomaly detection experiments allowed thanks to existing and our training datasets. Presented satisfying results provide a subject for further work and allow us to elaborate on this idea.

Keywords: IoT behavioral analysis · Encrypted traffic · Anomaly detection

1 Introduction

The main weakness of Internet of Things (IoT) infrastructures is still a lack of security considerations. Manufacturers focus rather on functionality, power efficiency, and analysis of collected data than security features. Security is usually not a core feature of the infrastructure design. On the other hand, scalability is emphasized, so we can identify hierarchical decomposition of functionality among *Edge layer*, *Fog layer*, and *Cloud layer* components of the IoT architecture [13].

Mostly, users do not care about operational tasks such as monitoring and security analysis because the functionality and presentation layer is their main scope of interest. However, users must be informed about operational outages and security risks especially in the critical IoT infrastructures. Therefore, we elaborate on a security extension of the commonly used fog infrastructure that must automatically observe and analyze operational level of IoT traffic, detect security events, and notify about any disruptions. The security extension must be capable of: (i) automatic passive discovery of IoT infrastructure, (ii) classification of the devices, (iii) recognition of behavior patterns, to be able to detect the events.

© Springer Nature Switzerland AG 2020
A. P. Pandian et al. (Eds.): ICCBI 2019, LNDECT 49, pp. 465–473, 2020.
https://doi.org/10.1007/978-3-030-43192-1_53

Meanwhile, the traditional monitoring and analysis could have been based on deep packet inspection and content analysis, it is necessary to consider encrypted traffic for the future. According to the observation and reports (e.g., [6]), the percentage of the encrypted traffic grows. Naturally, the use of encryption decreases network visibility for operators and makes the security analysis more challenging.

IoT networks are specific due to high variability—there are many different devices with discrepant traffic patterns that must be recognized by the security system. Additionally, lots of data are exchanged locally on the network edge, and the traffic must be analyzed efficiently. Therefore, we need a lightweight and economical solution that is able to learn the behavior of different devices and run on resource-limited components of IoT infrastructure, i.e., with respect to distributed fog and edge architecture.

Machine learning (ML), is a technology that allows for teaching a universal algorithm to recognize data according to some training dataset. It can discover significant relations and characteristics in the training data. Since the ML algorithms generally do not need to interpret the content of the communication, it is suitable even for encrypted traffic. With a proper set of features to model the network traffic, it can classify the traffic and detect potential anomalies. This approach is able to automatically learn network behavior and decide whether the network traffic is valid with much better precision and performance than humans can. Also, more ML models can be combined altogether to create a multi-layer model, as it is used in this paper. The multi-layer model complies with IoT fog architecture and provides better results.

This paper is primarily focused on IoT networks that contain IP devices, such as gateways, sensors, but the principle is applicable to any other IP networks. There is a lack of publicly available annotated datasets, which are essential for proper learning, because their preparation is complicated. The contributions of this paper are (i) a prepared annotated dataset as a combination of newly created and publicly available datasets, (ii) experiments to find feasible feature vectors and ML models for device classification and anomaly detection, and (iii) description of the security extension for IoT fog architecture based on hierarchical multi-level ML models.

The paper is divided as follows: Sect. 2 lists related works. Section 3 proposes architecture and ML models for our solution. Section 4 provides information about datasets (newly created and existing ones) and data features. Section 5 evaluates the proposed solution. Section 6 discusses future work and concludes the paper.

2   Related Work

The survey [16] discusses some challenges of IoT area, such as a lack of a perfect detection solution, a lack of training datasets, and challenges regarding IoT dataset creation. There is a big dissimilarity in the current detection problems that can be solved on different network levels with different prerequisites, e.g., availability of dataset and network size. Preparation of an annotated training dataset with a choice of a proper learning model is very important. Meanwhile, the paper [4] points out dependency on the ML performance on the source of training data, i.e., deployed ML can perform better on the same network as the origin of training data. Encrypted traffic analysis is studied by Anderson et al. in [2], which compares benign and malware traffic behavior in encrypted environments. Their results are very promising since they show that even if benign and malware use the same communication protocols, they use it differently. Therefore, malware detection based on encrypted traffic is possible. This idea is followed in additional papers [9,14], where the authors are able to identify different type of services and malware in encrypted traffic.

IoT traffic classification is the well-explored area. Authors of [10] focus on IoT device classification for security purposes. Due to encrypted traffic, they selected these features: the size of the first N packets sent and received, and their corresponding inter-arrival times. The authors test several classification models. However, the best result has Random Forest (99.9% accuracy). Despite great results, the paper mentions the limitation of a small dataset that contains only four IoT devices. The classification is much easier when the traffic heterogeneity is low.

The aim of [1] is the creation of an anomaly detection framework with fog computing architecture. This architecture brings computation power from the cloud to the network edge so the data can be processed immediately at a source. The result of [1] is a prototype of a supervised binary classifier that can recognize benign traffic (99% accuracy) and malicious traffic (79% accuracy). Similarly, [15] describes two-level architecture where level-1 is capable of classifying the network traffic as normal or anomalous, and level-2 identifies the category of the anomaly by further analysis. The result is a supervised solution with 97% F1-score for the selected datasets.

Hafeez et al. [7] use a semi-supervised method that provides more flexibility for anomaly detection in IoT network. This method does not require labeled data for each class. This is very important, especially in IoT networks, because there is a huge variety of different devices. Authors use 39 discrete and continuous features that represent aggregated traffic between two hosts over n last connections. With a selected dataset from private testbed, the solution achieved F1-score of 98.6% for the binary class problem (benign, malicious).

Contrary to the current related work, we provide an evaluation of IoT traffic analysis based on a combination of classification and anomaly detection algorithms with respect to the architecture of IoT network. To our best knowledge, all existing works covered the presented topics separately. However, according to our experiments, it is possible to achieve better results.

Our solution respects modern IoT architecture that can easily scale with network size. During our experiments we evaluated several ML features and models to reduce dataset size for learning. Selected features well-represent each connection observed on the network. During the learning period, user can choose target classes for anomaly detection whether they represent each device or any logical group of communication protocols with similar behavior. To improve the evaluation process of our solution, we have combined created datasets with some other already existing public datasets.

3   Solution Architecture

The concept of the fog computing architecture [13] must be followed by a design of security solutions because each computation device can be potentially misused by an attacker. Our solution follows up this concept similarly as [1,15], but we use it in a more advanced manner. The level-1 layer processes all traffic flows from the local network. The flow format is described in Sect. 4. Incoming data are processed by lightweight ML models for anomaly detection that produce features for the additional processing and final decision in the model of the level-2 layer. In [15] the level-2 layer processes the same data as the level-1, therefore, scalability and flexibility is limited. Authors of [1] only use classification ML models for level-1 layer with binary result. For the fog level model there is no specification of features. The proposed architecture of our solution is depicted in Fig. 1.

During our experiments, we had evaluated several ML models such as Isolated Forest, Random Forest (RF), Local Outlier Factor (LOF), GBoost, OneClassSVM and AdaBoost. All these models are very lightweight and require much less training data then for example neural networks. We have selected LOF and RF for the level-1 layer because they performed best. Also, these models use different methods to identify flows (novelty detection and classification). This approach increases analysis diversity and improves the reliability of final detection.

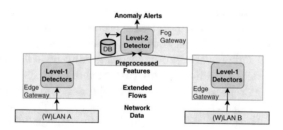

Fig. 1. Proposed security architecture

Level-1 layer represents edge devices in the IoT architecture which usually has limited computational and storage resources (this extends the solution used in [1]). Both models are trained by benign traffic and produce results in real-time for each flow separately (contrary to [7]). LOF provides a binary vector that identifies whether an input flow is anomalous or not for each class. The result of RF is a vector of the same length; it contains probabilities of a flow membership to each class. One class represents one device or group of devices with similar behavior.

Detailed analysis of the level-2 layer is out of the scope of this paper. However, its role is to process level-1 features and decide about traffic anomaly. Level-2 layer represents fog devices in the IoT architecture. These devices have more computational and storage resources, therefore, a deeper analysis and relationship modeling of data is possible. As end devices can be updated, very important is also ability to prepare and distribute retrained ML models back to the level-1 layer.

4   Data Collection and Features

One of the essential prerequisites of usage of ML techniques is the construction of training datasets. To start our experiments we have created an IoT lab with three IP devices from different IoT domains: IKEA Trådfri gateway, Edimax IC-9110W IP camera, and Google Mini voice assistant. In this lab, we analyze network traffic with initial experiments. For flow collection, we use Joy exporter [3] that provides extended flows records which were introduced in paper [2]. The created dataset contains mainly benign traffic, but we also create records containing device misconfiguration, e.g., wrong storage path for an IP camera represents a change of behavior and therefore the creation of anomaly. That dataset and source code are publicly available at our repository [12].

We evaluate various features describing the traffic. During the evaluation we measure feature importance in detail on our dataset to reduce their number for better efficacy. Finally, we choose this feature set: standard deviation of bytes distribution (BD), initial data packet length (IDP) (i.e., the size of the first packet in a flow), Walsh-Hadamard transform (WHT) and sequence of packet lengths (SPL). In total, the feature set creates a vector of 32 items. All features are oriented to behavior description without any relation to an application data or a device, i.e., no IP/MAC addresses or ports are used. Therefore, models can be created for each device or group of devices/flows with similar behavior during training.

Based on the analysis on our dataset, we observe that all IoT devices use the same multicast address (224.0.0.251) for device discovery. However, they use it with different characteristics described by our features. Also, there is much periodic traffic such as keep-alive or device discovery flows that are well described by WHT. Moreover, even if IKEA Trådfri gateway encrypts traffic using DTLS, the sequence of packet lengths is still the same for a specific user command. IDP is also highly representative since it differentiates different types of communication flow.

To verify the quality of our solution we also use public datasets. The first is from UNSW Sydney [11] where authors created pcap records for 28 different IoT devices. The second one is CTU-13 [5] that contains records of several botnet traffic which is very popular attack in IoT.

5 Experiment Results

All experiments and evaluation were done with these datasets [5,11,12] that cover different type of traffic and anomalies from real networks. In these datasets we identified 20 benign classes of traffic that represent the behavior of each device. The following devices list selected classes: *amazon echo, smart things, netamo welcome, samsung smart cam, insteon cam, withings smart baby monitor, belkin wemo switch, tp-link smart plug, belkin wemo motion sensor, netamo weather station, withings aura smart sleep sensor, light bulbs lifx smart bulb, hp printer, laptop, android phone, android galaxy tab, ikea native app, edimax ip cam, normal user, google voice assistant*. Also, we created classes which contain misconfiguration and malicious communication to verify detection of unwanted traffic. These classes are represented by: *ikea homekit app, edimax ip cam miscofig, botnet sogou, botnet neris*. Each experiment uses 5-fold cross-validation, training dataset with at least 400 flows and uniform selection.

5.1 Evaluation of Novelty Detection

The first experiments were focused on anomaly detection using a novelty method with LOF model. The experiment represents cases when a foreign device is added into infrastructure or a device becomes infected by malware, so the behavior changes from the known normal state. In this experiment we choose at one time just one benign traffic class to train the LOF model and evaluate it separately against other traffic classes. For validation, our test set contains benign and anomalous traffic. In total, we run an evaluation for 20 benign classes with average F1-score 86%. Selection of our measurements is depicted in Table 1.

Table 1. F1-score among subset of benign and anomalous classes. The column headers are as follows: IKEA—IKEA Trådfri Native, IP Cam—Edimax IP Cam, Laptop—Laptop User (includes benign HTTP/S, DNS and SSH traffic), Voice—Google Voice Assistant, HomeKit—IKEA Trådfri Homekit, IP Cam misc—Edimax IP Cam Misconfiguration, botnet sogu—Attack 1, botnet neris—Attack 2

Device name	IKEA	IP Cam	Laptop	Voice	Homekit	IP Cam misc	Attack 1	Attack 2
IKEA Trådfri Native	–	0.70	0.85	0.79	0.85	0.91	0.84	0.90
Edimax IP Cam	0.83	–	0.88	0.85	0.81	0.80	0.85	0.62
Laptop User	0.91	0.75	–	0.81	0.90	0.84	0.95	0.94
Google Voice Assistant	0.78	0.43	0.80	–	0.82	0.78	0.87	0.66

The column *Device Name* represents classes used for training, and other columns represent classes marked as anomalous. First four columns of anomalous classes correspond to the benign classes in rows. Next two classes constitute misconfiguration and the last two malicious traffic. The results for misconfiguration classes fulfill expectation because the F1- score is above 78% for all classes. The same expectation is made for malicious traffic. However, rows *Edimax IP camera* and *Google voice assistant* correlate with *Attack 2* column. A similar correlation can be observed between *Edimax IP camera* as an anomaly and other benign classes. Based on overall insufficient results from this experiment we decided to continue with a classification approach to increase the reliability of anomaly detection.

5.2 Evaluation of Classification

In the second experiment, we use the same traffic classes and datasets. However, instead of the novelty detection approach, we choose classification with a RF model. Therefore, this experiment represents the same cases as Sect. 5.1 but solved with the different method. The RF model was trained on just benign classes. During validation, we evaluate both benign and anomalous classes. The result is a ratio of successful and unsuccessful classification among anomalous and benign classes. In total, we run an evaluation for 20 benign classes with average accuracy of 77%. Selection of our measurements is depicted in Table 2.

Evaluation among know classes has great results with accuracy at least 91%. In the case of misconfiguration and malicious traffic, the classification ratio is more distributed between all classes which represent anomaly traffic. However, for *Attack 2* class is the frequency for *Edimax IP camera* 75% which is still quite high, and it can lead to false positive detection.

Table 2. Results classification among anomalous and benign classes. The column headers are as follows: IKEA—IKEA Trådfri Native, IP Cam—Edimax IP Cam, Laptop—Laptop User (includes benign HTTP/S, DNS and SSH traffic), Voice—Google Voice Assistant, HomeKit—IKEA Trådfri Homekit, IP Cam misc—Edimax IP Cam Misconfiguration, botnet sogu—Attack 1, botnet neris—Attack 2

Device name	IKEA	IP Cam	Laptop	Voice	Homekit	IP Cam misc	Attack 1	Attack 2
IKEA Trådfri Native	0.99	0.04	0.01	0.02	0.06	0.01	0.19	0.11
Edimax IP Cam	0.00	0.92	0.01	0.03	0.37	0.70	0.10	0.75
Laptop User	0.01	0.01	0.97	0.02	0.56	0.08	0.36	0.05
Google Voice Assistant	0.00	0.05	0.01	0.91	0.00	0.20	0.34	0.07

6 Discussion

Both experiments use semi-supervised approach to detect anomaly communication based on network traffic behavior in real-time. Each method has its pros and cons that are described in Sect. 5. Based on the provided evaluation we see that combination of results from LOF and RF models improve the reliability of anomaly detection. For example, the correlation between *Edimax IP camera* as an anomaly and other benign classes in Table 1 is corrected by results in Table 2. Also, insufficient results for detection of *IKEA Trådfri Homekit* and *Attack 2* in Table 2 is corrected by the score in Table 1.

The achieved results are satisfiable because our models are very lightweight. They require neither persistent storage nor more than 10 MB of RAM (tested for 20 classes including NEMEA framework). Therefore they are resource-economical. The correctness of results was verified on the dataset with at least 400 flows per class. Also, we proved the usefulness of the combination of different ML models. On the other hand, this solution can only detect anomalies at the network level, i.e., we can not identify an incident at the system level.

The experiments also showed us that the definition of traffic classes is very important. Authors of [8] propose a method for clustering network hosts to create a smaller number of more reliable models for detection. We would like to follow up this idea, and cluster network flows based on their similarity.

7   Conclusion

In this paper, we propose an approach to detect anomalies in modern IoT networks. We describe multi-level architecture that is necessary for proper scaling. We evaluated several ML features and models during our experiments. Based on these experiments, we selected ML features that are independent of network hosts and describe flow behavior even in encrypted traffic. With the created dataset covering more types of traffic and 20 IoT devices, we have achieved F1-score 86% for LOF model and 77% classification accuracy using RF. These models will be combined and number of detection classes clustered in future work to receive more reliable results. Simultaneously, we would like to focus on the level-2 layer model that will receive real-time features from the level-1 and provide deeper analysis. Also, the capability to recognize legitimate behavioral changes is very useful (to decrease false positives).

Acknowledgment. This work was supported by the Grant Agency of the Czech Technical University in Prague, grant No. SGS17/212/OHK3/3T/18 funded by the Ministry of Education, Youth and Sports of the Czech Republic and *Secure Gateway for Internet of Things (SIoT)* project No. VI20172020079 funded by the Ministry of the Interior of the Czech Republic.

References

1. Alrashdi, I., Alqazzaz, A., Aloufi, E., Alharthi, R., Zohdy, M., Ming, H.: Ad-IoT: anomaly detection of IoT cyberattacks in smart city using machine learning. In: 2019 IEEE 9th Annual Computing and Communication Workshop and Conference (CCWC), January 2019
2. Anderson, B., McGrew, D.: Identifying encrypted malware traffic with contextual flow data. In: Proceedings of the 2016 ACM Workshop on Artificial Intelligence and Security, pp. 35–46. ACM (2016)
3. Anderson, B., McGrew, D.: Joy (2016)
4. Arndt, D.J., Zincir-Heywood, A.N.: A comparison of three machine learning techniques for encrypted network traffic analysis. In: 2011 IEEE Symposium on Computational Intelligence for Security and Defense Applications (CISDA), pp. 107–114, April 2011

5. Garca, S., Grill, M., Stiborek, J., Zunino, A.: An empirical comparison of botnet detection methods. Comput. Secur. **45**, 100–123 (2014)
6. Gebhart, G.: We're Halfway to Encrypting the Entire Web (2017)
7. Hafeez, I., Ding, A.Y., Antikainen, M., Tarkoma, S.: Real-time IoT device activity detection in edge networks. In: Au, M.H., Yiu, S.M., Li, J., Luo, X., Wang, C., Castiglione, A., Kluczniak, K. (eds.) Network and System Security, pp. 221–236. Springer (2018)
8. Kopp, M., Grill, M., Kohout, J.: Community-based anomaly detection. In: 2018 IEEE International Workshop on Information Forensics and Security (WIFS), pp. 1–6, December 2018
9. Piskozub, M., Spolaor, R., Martinovic, I.: MalAlert: detecting malware in large-scale network traffic using statistical features. ACM SIGMETRICS Perform. Eval. Rev. **46**, 151–154 (2019)
10. Shahid, M.R., Blanc, G., Zhang, Z., Debar, H.: IoT devices recognition through network traffic analysis. In: 2018 IEEE International Conference on Big Data (Big Data), pp. 5187–5192, December 2018
11. Sivanathan, A., Gharakheili, H.H., Loi, F., Radford, A., Wijenayake, C., Vishwanath, A., Sivaraman, V.: Classifying IoT devices in smart environments using network traffic characteristics. IEEE Trans. Mob. Comput. **18**, 1745–1459 (2018)
12. Soukup, D., Cejka, T.: NEMEA-SIoT (2019)
13. Statista: Fog computing and the Internet of Things: extend the cloud to where the things are (2015)
14. Stergiopoulos, G., Talavari, A., Bitsikas, E., Gritzalis, D.: Automatic detection of various malicious traffic using side channel features on TCP packets. In: Lopez, J., Zhou, J., Soriano, M. (eds.) Computer Security, pp. 346–362. Springer (2018)
15. Ullah, I., Mahmoud, Q.H.: A two-level hybrid model for anomalous activity detection in IoT networks. In: 2019 16th IEEE Annual Consumer Communications Networking Conference (CCNC), pp. 1–6, January 2019
16. Velan, P., Čermák, M., Čeleda, P., Drašar, M.: A survey of methods for encrypted traffic classification and analysis. Int. J. Netw. Manag. **25**(5), 355–374 (2015)

Review of Digital Data Protection Using the Traditional Methods, Steganography and Cryptography

Chinmaya M. Dharmadhikari[(✉)] and Rejo Mathew

Department of I.T., Mukesh Patel School of Technology and Management,
NMIMS, Mumbai, India
Chinmaya2903@gmail.com, rejo.mathew@nmims.edu.in

Abstract. The following paper reviews software protection methods by using the methods of steganography, cryptography to safeguard the running applications on pc as well as mobile applications. The software applications these days gets most of the work done for professional use as well as other purposes. The major issues faced by the software developer or publisher are of software piracy. All the major software these days are circulated and pirated through the internet. Thus using steganography and other techniques the goal is to solve the problem of software piracy. The solutions for protecting software from getting pirated are stated in this paper.

Keywords: Steganography · Data protection · Stego image · Software developer · JSON · HDD · MAC address · Database · LSB (least significant bit) · Ciphertext · StegoDB · XOR · Blowfish · Edge detection

1 Introduction

The information exchanged over a computer is mainly through network. Files stored in computer are in various forms of format such as text, image, sound. The information stored or the software used can be easily copied which is one of its greatest weakness. In today's world due to digitalization ecommerce or banking transactions are processed through networks. To protect this type of data encryption methods are used. Steganography is defined as the scientific method of hiding information within an image. Data which is protected using steganographic methods is more concealed by encryption methods before applying the steganographic processing to it.

To reduce the piracy issue companies employ different methods such as licensing acts, patents, cryptographic methods and dongle.

The following paper reviews the software protection framework which uses the cryptography, steganography and the new processes to reduce the risk of data corruption.

1.1 Hardware Means for Protection of the Software [3]

Due to the software being connected or accessed through internet, software gets exposed to piracy. So, the software protection is carried by the developers who provide a special hardware which has to be connected when the software is being used by the user.

© Springer Nature Switzerland AG 2020
A. P. Pandian et al. (Eds.): ICCBI 2019, LNDECT 49, pp. 474–483, 2020.
https://doi.org/10.1007/978-3-030-43192-1_54

USB/Serial port dongle: The most common hardware protection method available is dongles. For software to work the dongles need to be connected to the software all the time. When the software starts to execute operation the software primarily checks the port in which dongle is present and checks in memory that the encrypted key is present or not. If the key is similar with the registered information then the software is executed. If the key is not similar with the information then the software will throw an error. Using a dongle based system often proves costly as it requires some special types of drivers to make the dongle work [3].

1.2   Methods of Protecting a Software with Support of Hardware

This method implies that, the software use various methods for protection of data using software means which help in authentication of the users connected or using a specific software. These methods generally recognize the device on which the user has connected the software or a designated id is given to a user through which a user can register himself with the software.

1.3   Software Means of Protection [1]

The several methods used for protecting the information is given in this method. The techniques provided through the software means are feasible for the individuals who are sole proprietors of the firms. This methods use techniques of cryptography which converts plain text to cipher text and steganography which is protection of the text inside the image [1].

1.4   Exposure of the Techniques Using Hardware Means, Software Means of Protection

  i. While considering above techniques protection methods using hardware are not easy to implement as they require some specific features, additional dongles so these techniques are not feasible to use.
 ii. The techniques are neither cost effective. These techniques have various vulnerabilities and through those vulnerabilities crackers with the help of hardware methods try to write the backdoor programs and can clone the functionality which bypasses security.

1.5   Multilevel Hiding Text Security Technique

  i. This technique uses various processes to protect data. Following are the steps in which the given process advances

   Step 1: The first process is to encrypt the secret message using a blowfish algorithm to generate a key which is used in encryption process with XOR a plain text with key. Blowfish algorithm is a block cipher algorithm which uses symmetric key that encodes 64-bit block

Step 2: In second process the hiding positions are determined using the edge detection algorithm to cover a image. The proposed embedding method utilizes Sobel edge detector on every 3×3 non-overlapping block of the cover image. Sobel operator is a mutually perpendicular gradient vector field operator

Step 3: Bats algorithm is used to cover the image which has been obtained by the edge detection to pick a random hiding position in the image which is obtained by edge detection. Bats algorithm is an optimization algorithm depends on the echolocation behavior of bats when searching for preys.

Step 4: The last step is to embed the message using LSB(Least significant bit)

2 Analysis/Review

2.1 Applications of the Hardware Means for Protection of the Software [3]

i. CAD/CAM software: Computer aided Design or computer aided manufacturing software are widely used in the industries like oil, mining. They are costly to develop so use dongle based protection against the piracy

ii. Animation/3Dsoftware: These software also cost thousands of dollars to develop so they basically use dongle based protection

iii. Steinberg key: This product protects Steinberg products which protect audio and also the editing solutions. In Steinberg key dongle apart from software protection it also provides another feature in which only the modules purchased or bought by the user are unlocked.

2.2 Protection of Software with Support of Hardware [2]

i. Serial Numbers: Serial numbers is the simplest method used in software protection. A serial number is created with help of c algorithm to the specified user who has a licensed version of software. Further developments to these methods were brought by the game developers which ensured that in a cd drive a cd is present when the game is running. Further advancement included the use of digital signature into disc produced at the time of manufactures. When the software is running they will check the presence of the cd and the copy protection digital signature on the disc. A special hardware is required to manufacture disc with digital signature

ii. CD Based protection: The copying of music from CD's and DVD's is the most common thing in which the ripping software is used to convert the audio cd's to mp3 music. To prevent the ripping several studios introduced the software which monitors and executes when a computer disk is inserted into the pc. The software monitors if any ripping software is being run or not and if its finds it disables the access to the drive or stops the software from doing this

2.3   Applications Used for the Software Means of Protection

i. Cryptography: Cryptography is the scientific method to use the secure convention. The conversion of plaintext to ciphertext to prevent the unauthorized access to the message/information. The various algorithms developed to protect the data are DES (Data encryption standard), TDES (Triple data encryption standard), RSA, AES (Advanced encryption standard). AES is employed by the government of USA

ii. Steganography: The further version of cryptography is steganography. The protection of information inside the image is done in the steganography for e.g. LSB, RGB, PVD etc. Water marking and visual cryptography are some of the techniques used for protection of the data

- Architecture of Software Prtection: [9]

The following diagram represents that what are the steps in which a software authenticates a user and checks the software license and authenticates the hardware. After reading the license and hardware the activation of the software takes place and in the final step the software gets prepare to launch. While execution of software takes place the integrated software execution, authenticates the internal modules and periodical security check takes place after each function in the software is been executed (Fig. 1).

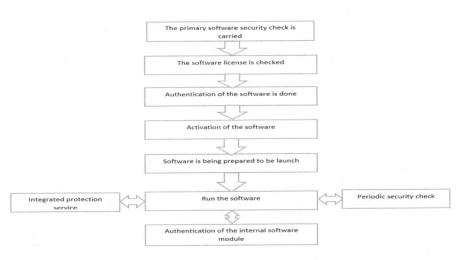

Fig. 1 Architecture of software protection framework

The given framework about the architecture of software protection framework always gets executed when the software is running. This architecture must be included in.crc file. This is layered security architecture as right from primary check of software till integrated checks on each module are presented in this framework.

2.4   StegoDB Model

The following technique combines the advantages and disadvantages of the existing framework and reviews a new framework which contains cryptography, steganography and hardware features. Designing of this method is such that execution is easy and breaching the security is difficult for the crackers. The technique gets implemented using the following steps A. Software protection framework: The proposed technique consists of two parts the authentication algorithm and protection algorithm [4].

Step 1: Authentication algorithm carries out the registration process.
Step2: Creation of stego token and attaching it to software for protection.
Step 3: Software execution is the final step where validation algorithm validate the authenticity.

i. Authentication Algorithm:

Step 1: The user information, Hdd serial number or MAC address should be fetched to determine a distinctive i.d which can be used to verify the distinctive properties of the device [4]. HDD serial number can be found by the following commands.

1. win32_DiskDrive API should be cited.
2. Select the serial number. WIN32_DiskDrive API is provided by windows to extract the serial number.

Step 2: The serial number is encrypted using the AES encryption standards and stored in read only file.
Step 3: The file is sent to licensed server of the software which is present at the software providers end. The server will provide the HTTP API's. The server than provide the HTTP API's for receiving the file . SFTP file transfer protocol can also be used to transfer the file.
Step 4: Fetching of user mail id along with information should be done.
Step 5: The authenticated user information is saved in the licensed server.

ii. Protection Algorithm: [9]

Step 1: Using the authentication algorithm creation of user authentication information is done.
Step 2: Provide the user i.d information from the file stored.
Step 3: The SteganoDB package should be applied steganoDB structure,

{   "First   Name":   "ABC",
"Last   Name":   "ABC",
"Email":"ABC.ABC@tcp.com",
"Address   Line1":"125#sector   78",
"Address   Line2":"M-Nagar",
"Address   Line3":"Mumbai-30",
"Phone":"989"
"Key":"UdGkX1/IiYCbur74I5oNTBL/nBa   MPfgg+s="
}

SteganoDB is created using the JSON structure associated with steganography to create the SteganoDB which is unique. Username, email, addresses can be stored in this steganoDB and can be extracted with the modern programming language [2].
Step 4: The encrypted data is embedded in the given image which uses pixel pattern based steganography algorithm. Due to the use of pixel pattern cryptography the quality of image does not degrade and image which is covered will not appear to be different and would be free from attack . Information is protected using encryption and steganography [8].
Step 5: The encoded image file is saved at the receivers end and as the stegano key is generated it is sent back to the software running on the desktop. This can be accomplished using the secured protocols like http(https) or ftp(sftp)[9].
Step 6: The protected software checks the encoded key for each of the execution.
Step 7: Extraction of key from the stego image.
The algorithm for the following is as:

To	extract	key	from	the	server,	the	following	methods	should	be	followed
{											
The		characters		in		image		metadata	should		be	selected
{											
The		pixels		in		the		image	should		be	noted
}											
The	mod	bits	from	the	pixel	positions	should	be	found	out[7]
The	given	mod	bits	should	be	decrypted				}

After execution the algorithm will fetch following results primarily serial no would be taken out and the reverse AES processes would decrypt the following

Step 8: If key does not match or if someone is trying to breach the security of software by installing it on another PC where the installed software is and the key then comparison is done with hardware properties which will fail and the software will indeed remain protected [7]
Step 9: Regular self-checks of the software can be done by placing the key properties in .crc file. As the software executes it checks the properties with the stegano file and if it is false execution of software is stopped.

iii. Data base Algorithm:

Secure transfer of database or file from device to device is done using these algorithms.

Step 1: The image file on which the text should be embedded should be selected
Step 2: Choose a secure password for protection of database
Step 3: There will be one time key from the receiving side. From the receiving side the software will detect the hardware property of the device like the hdd serial number etc. This will be the unique id for identification to software.

{   "First   Name":   "ABC",
"Last   Name":   "ABC",
"Email":   "ABC@tcpcom",
"Address   Line1":"126#Lane   15",
"Address   Line2":"M-Nagaar",
"Address   Line3":"XYZ-60",
"Phone":"989"
"Key":"U2FsdGV19L7/IiBYbur74I5oNTBL/nBa   MPfgg+s="
}

Step 4: The text is converted into JSON structure

Step 5: Apply stegano DB to the password and the hardware property of the device [2]

Step 6: The steganography data embedded algorithm is applied and choose the file to protect

Step 7: Send the stegano image file via file transfer protocols

Step 8: Receiver will receive the file and save it

Step 9: The hdd number is checked and it uses a password to decode.win32_diskdrive API is used to take out serial number from hard disk

Step 10: using the retrieving algorithm decoding of the image is done and the separation of text from image is carried.

Step 11: The text is read in JSON format and stored in.txt file. This method helps in protection of the databases or text files.

iv. Mobile Data protection

The algorithm can also be further implemented in the mobile applications. The framework remains the same except instead of HDD serial numbers the mobile will use the IMEI number. The access of the app can also be made limited to a certain region using GPS [10].

2.5   Multilevel Hiding Text Security Technique

The key generation and the embedded as well as the extracted algorithms are discussed in this technique.

i. Key generation and encryption algorithm

In this algorithm as a input there is requirement of color image and plaintext and output obtained is the key so the following steps have to be followed as loading of BMP color image and key generation should be done using the blowfish algorithm and after that conversion of input image to binary image should be done and them the image should be divided into 64-bitsblock as plaintext. Load the plaintext in different size and then divide the text into blocks of same size as the key. Apply XOR between the key generation and plaintext with different size to obtain ciphertext [4].

ii. Embeded Algorithm

In this algorithm color cover image and ciphertext is taken as input and the output obtained is a stego image. In first step load the BMP color image and then apply edge detection algorithm using the sobel filter. Apply bat algorithm on the output which is obtained after edge detection to choose random position. Ciphertext is embedded using LSB technique to generate steganography image [4].

iii. Extracted Algorithm

This algorithm uses input as Stego image and gives output as plaintext. In this edge detection is applied on the stego image using sobel filter and then apply bat algorithm on stego image to select random positions and extraction of ciphertext is done using LSB and apply XOR on ciphertext to get the plaintext [4].

Comparison table

	Features				
Methods	Hardware means of protection for software	Methods of protection using Software with the help of hardware	Software means of protection	Stego DB technique	Multilevel hiding text security technique
Functionality	user connects the dongle for the software to work	Generation of serial numbers, digital signatures required for CD etc.	Steganography is used with cryptography to increase the security of data transfer	Cryptography, steganography, Authentication algorithm, Protection algorithm	blowfish algorithm, bats algorithm and edge detection algorithm
Implementation areas	CAD/CAM software, Animation software, Steinsberg key	CD based protection, DRM services (Digital Rights management)	DES(Data encryption standards), AES (advanced encryption standards) implemented using cryptography,	used in authentication mechanisms as well as software protection mechanisms which also include database protection, security checks etc.	This technique is used in sending messages from different networks and the data is converted into ciphertext and then covered in a stego image
Security level	The data could be breached if the hardware component is damaged. Thus the security level is low	The data in this method is not secure as it is easy for hackers to make a copy of software	Better security level but implementing various algorithms like DES,TDES is difficult for organization	Due to dual layered protection and acquiring the physical address of hardware component on which software is installed it is difficult for crackers to exploit the software	It is well structured algorithm and the ciphertext when inserted in a stego image gives a good security level
Disadvantages	If crackers find presence of a hardware device, then using backdoor program to clone the functionality of the hardware	The CD protection can be manipulated using the cloning	Side channel attacks are threats to these protection methods	No potential third party attacks unless user does not disclose the id and password, a secure method to protect data	colluding attack on the blowfish algorithm can result in combining multiple copies of data and in edge detection algorithm replacement of linear combinations of data will be done
complexity	less complexity	This method has moderate complexity but security cannot be trusted	This method has high level of complexity and cumbersome to maintain	Highly complexed and secured	Highly complex and secure

3 Conclusion and Future Work

The paper reviewed has discussed various techniques which include digital data protection methods or the methods which avoid the piracy of data. Primarily the paper reviews all the traditional methods used which include the hardware methods, software methods. The following traditional methods gets vulnerable to threats and piracy so a new technique is reviewed using steganography and cryptography. The algorithm becomes hard for the crackers to detect as it uses same pixel patterns and strong encryption algorithm like AES which provides high level secured protection. The given techniques does not need any external hardware like the traditional methods which use dongles which makes them exposable to the threat. Due to layered protection given in the framework it becomes extremely hard for the hackers to crack the software. The future work in theoretical analysis includes adding the types of secret message between the data in the cover object to the perfectly secure model. A more concrete model is expected by analyzing the probability that the hidden message can be robust at a certain data hiding rate [14].

References (IEEE format)

1. Rajani, R., Murugan, D., Krishnan, D.: Software protection framework using Steganography, Unique features of hardware and self-checking features. J. Adv. Inf. Technol. **25**, 32–40
2. Anckaert, B., De Sutter, B., De Bosschere, K.: Software Piracy protection. In: Proceedings of the 4th ACM Workshop on Digital Rights Management, pp. 62–70 (2004)
3. Anderson, R.J., Petitcolas, F.A.: On the limits of steganography. IEEE Journal of Selected Areas in Communications **16**(4), 474–481 (1998). Special Issue on Copyright and Privacy Protection
4. Hasoon, N.H., Ali, R.A., Abed, H.N., Alkharaji, A.A.J.: Multilevel hiding text security using hybrid technique steganography and cryptography. Int. J. Eng. Technol. **7**(4), 3674–3677 (2018)
5. Cachin, C.: An Information-theoretic model for steganography. In: Aucsmith, D. (ed.) IH 1998. LNCS, vol. 1525, p. 306. Springer, Heidelberg (1998)
6. Laskar, S.A., Hemachandra, K.: Secure data transmission using steganography and encryption technique. Int. J. Comput. Secur. **4**(2), 161–172 (2012)
7. Farid, H., Lyu, S.: Detecting hidden messages using higher-order statistics and support vector machines. In: Petitcolas, F.A.P. (ed.) IH 2002. LNCS, vol. 2578, pp. 340–354 (2002)
8. El-Khamy, S.E., Korany, N.O., El-Sherif, M.H.: Correlation: highly secure image hiding in audio signals using wavelet decomposition and chaotic maps hopping for multimedia communications. In: General Assembly and Scientific Symposium of the International Union of Radio Science (URSI GASS), pp. 1–3 (2017)
9. Gautam, R.R., Khare, R.K.: Real time image security for mobile communication using image steganography. Int. J. Eng. Res. Technol. **1**(8), 1–5 (2012)
10. Kojima, T., Mayuzumi, R.: An improvement of the data hiding scheme based on complete complementary codesIn. In: Seventh International Workshop on Signal Design and its Applications in Communications, pp. 85–89 (2015)

11. Kaur, S., Bansal, S., Bansal, R.K.: Steganography and classification of image steganography techniques. In: International Conference on Computing for Sustainable Global Development, pp. 870–875 (2014)
12. Pant, V.K., Saurabh, A.: Cloud security issues, challenges and their optimal solutions. Int. J. Eng. Res. Manag. Technol. 2(3), 41–50 (2015). ISSN:2348–4039
13. Praveena, A., Smys, S.: Efficient cryptographic approach for data security in wireless sensor networks using MES VU. In: 2016 10th International Conference on Intelligent Systems and Control (ISCO), pp. 1–6. IEEE (2016)
14. Kale, K.V., Aldawla, N.N.H., Kazi, M.M.: Steganography enhancement by combining text and image through wavelet technique. Int. J. Comput. Appl. (IJCA) 51(21), 0975–8887 (2012)
15. Kale, K.V., Aldawla, N.N.H.: Steganography enhancement: combining text and image approaches. Asian J. Comput. Sci. Inf. Technol.

Waste Management Techniques for Smart Cities

Vaibhav Agrawal[✉] and Rejo Mathew

Mukesh Patel School of Technology Management and Engineering,
Mumbai 400 056, India
vaibhav.agrawal03@nmims.edu.in, rejo.mathew@nmims.edu

Abstract. IoT is being greatly used for designing a proper waste management system i.e. collection and disposal techniques. In the smart cities concept, sustaining water, energy and management of waste which includes some basic restrictions on the use of greenhouse gases in the areas concerned. This paper reviews various waste collection and disposal techniques such as smart bins, recycling and open sky incineration, secured landfilling, composting, incineration & waste to energy (WTE), lagooning etc. as well as the planning and operations required with their challenges. There are various changes taking place in the environment causing global warming due to the human activities. To maintain the environmental sustainability, there has to be proper planning, management and operations to be carried out. Fortunately, IoT has the answer to assist the utilization method at each stage of the waste management.

Keywords: Internet of Things (IoT) · Sustainable · Waste management · Smart bins

1 Introduction

Solid waste management is a serious environmental problem today. The genuine management of solid waste has become more important for keeping up the sustainable environment, due to the increasing population.

Improper management may cause hazards to inhabitants. Public health and healthy environment is compromised in the municipal areas where the waste is thrown away in open dumps and landfills. The various issues related to waste management are overflowing waste bins causing pollution, e-Waste, lack of landfills, hazardous waste etc. which have to be looked upon. As the world evolves towards the smart city, the aim of having a smart city is to improve the quality of life of the people in the city by using the latest technology and information. This paper reviews various problems by the traditional waste management techniques and also the many methods to overcome those problems. Different solutions such as smart bins, secured landfilling, incineration & Waste to Energy, composting, lagooning etc. to carry out the disposal of various kinds of waste without affecting the environment and human settlements have been reviewed. Section 2 reviews all the problems faced while disposing the waste. On the other hand, Sect. 3 discusses all the various methods of disposal for the problems mentioned in Sect. 2. Section 4 discusses an overall comparison of various methods, inorder to get the best method possible.

© Springer Nature Switzerland AG 2020
A. P. Pandian et al. (Eds.): ICCBI 2019, LNDECT 49, pp. 484–491, 2020.
https://doi.org/10.1007/978-3-030-43192-1_55

2   Issues Related to Waste Management

2.1   Environment Pollution

The amount waste generation is different in every area or street. But the collection of waste if carried out every day or two whether the dustbins are filled or not. Improper collection of waste leads to the overflowing of the dustbins at many places. Due to this, foul smell spreads in the environment giving rise to various diseases [1, 2].

2.2   Municipal Solid Waste (MSW)

Municipal solid waste (MSW states about majority of the non-hazardous waste, solid in nature from a particular city, town or a village that needs routine collection as well as disposal. Sources of MSW include homes, commercial establishments and institutions, as well as industrial facilities. In most of the countries, landfilling is the most widely used method of MSW disposal. This technique has its main attention on burial of waste concerning the land area. Due to the increasing population, there is shortage of land for landfills. Also, landfills pollute the air which severely affects the surroundings and the livelihood of animals and humans [3].

2.3   e-Waste

The challenges faced by various nations in e-Waste management are the lack of an infrastructure for the management of waste, lack of any structure for end-of-life (EoL) product take back or execution of extended producer responsibility (EPR) [4].

2.4   Hazardous Waste

Insufficiencies in waste administration practices can make possibly risky circumstances and posture noteworthy dangers of worry to the community. Whenever waste is dumped aimlessly in natural place or form, risky wastes might have short-and long haul consequences for human and biological frameworks. What's more, inappropriate treatment, stockpiling, and transfer of perilous wastes can bring about contaminant during potential exposures, and antagonistic wellbeing and ecological effects [6].

3   Solutions

3.1   Smart Bins

To overcome this issue, an IoT method which is based on collection of waste system on the norms of waste level which is present in the wastebins is proposed. Smart wastebins [1, 2] i.e. wastebins with sensors will be placed around the city. This sensors using the SONAR technique will sense the distance from top of the wastebin to the waste. This can sense the distance from 3 cm up to 400 cm with minimal precision of 3 mm. With the help of Access Network Interface, the data which is obtained via the sensors is transmitted through the internet to the server. A database is used to store the information obtained. This data will be used to monitor the everyday selection of the bins, based on which the best route to pick the wastebins can be decided. This can be done with the help of certain algorithms [1, 2, 15].

Smart Waste Management Algorithm: [1]

Inputs:
Number of smart wastebins and their capacity.
Amount of waste generated.
Output: Best route for waste collection.
Description:
  i. Install smart wastebins i.e. wastebins with sensors across the city.
 ii. Define a maximum value for wastes in each wastebin.
iii. Transmit the data obtained to the servers via internet.
 iv. Store the data in the database.
  v. Obtain the best shortest route for the waste collection with the help of algorithm.

Shortest path algorithm: [1]

Inputs: Distance between wastebins and the work station.
Output: Shortest route between two locations where trash has to get collected.
Description:
  i. Considering the street network as the graph.
 ii. Let the street segments be the edges and the joining points be the vertices.
iii. Calculate the optimum shortest distance between the two locations and also the final route from one to rest of the wastebins to make the process faster in operation.

3.2   Disposal Methods for MSW [5, 7]

3.2.1   Sanitary Landfills

Landfilling is the most common technique used for the disposal of all kinds of wastes produced. The purpose is to reduce the contact between the waste and the environment as well as the water bodies. The developed countries like Canada, United States etc. use sanitary landfills to obstruct and treat the leachate using a channel. These are the sites where the waste is kept aloof from the surroundings until it is harmless, i.e. when it is biodegraded as well as degraded physically and chemically. These landfills have provisions to treat the harmful gases produced by decomposition of waste [3].

3.2.2   Composting

Composting is the decomposition of the organic matter through biological processes under aerobic and controlled conditions, which results in a manure full of nutrients. MSW composting reduces the amount of waste that would otherwise end up in landfills. Composting includes the assembling of a blend of household residue, animal waste, soil and water to convert into manure. The measure of compostable material produced from the waste is 80–85% in the fast growing nations [3].

3.3   Disposal Methods for e-Waste [4]

The discarded electronics are managed through various old methods of Municipal Solid Waste (MSW) management such as incineration or landfilling. The e-Waste is first processed inorder to check its repair and reuse activity. These products are then refurbished and sold.

3.3.1   Landfill Disposal

A landfill site is a site for discarding and treating the waste produced by burying under the soil. Around 90% of e-Waste is disposed by landfill disposal. The rate of increase in number of landfills is high in developed as well as developing countries. The newly constructed landfills are used to safely isolate the pollutants which are present in the electronics from the environment. Although, thousands of old landfills which have no barrier and contains a mixture of putrescible and e-Waste are present and these old landfills pose a great threat [4].

3.3.2   Open Sky Incineration

The residual products left after the recycling process are the incinerated in an open sky incineration, cyanide leaching and simple smelters to retrieve majorly copper, gold and silver for relatively small quantities. For instance, the wires are collected and burnt in open piles to obtain copper which could be sold again [4].

3.4   Methods Used for the Ultimate Disposal of Hazardous Waste [6]

3.4.1   Secured Landfill

A correctly located as well as selected reliable landfill decreases serious effects on the surroundings. A location fit for discarding all kinds of waste, also the liquid and solid hazardous waste, must not let any release of this type of materials or its by-products in the groundwater by leaching, percolation or any other kind. The key requirement for this kind of discarding is of dangerous contents of the landfill be kept aloof from the nearby surroundings. Surface and ground water quality should not be reduced. The quality of air must be balanced. Upon the type of waste, different kinds of safe landfills like open, control and closed are utilised [6].

3.4.2   Incineration

Incineration is a controlled procedure which involves oxidative transformation of combustible solid material into safe gases fit for environmental liberation. It transforms a waste to a less bulky, less poisonous or less benign product. The foremost products of incineration are carbon dioxide, water and ash while the products of main disquiet, due to their environmental impacts are amalgams containing sulfur, nitrogen and halogens. Unless sufficient methods are performed, incineration can lead to atmospheric liberation of unwanted materials.

Under such conditions, a subordinate treatment such as after burning, scrubbing or filtration is performed to decrease the concentrations to admissible levels before its atmospheric release [6].

3.4.3   Lagooning

Hazardous waste is provisionally kept in a small manmade pond called lagoon where it is treated to decrease the environmental impact before releasing it into the local water bodies. The waste is treated so well that it does not adulterate the local water bodies where it is discarded at the end. The environmental problems arising due to lagooning relates to release of the leachate and surface water runoff [6].

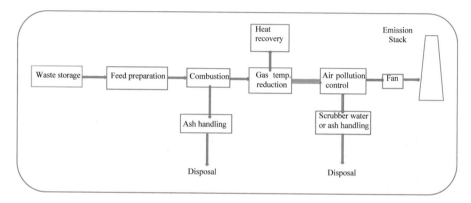

Fig. 1. Typical waste-incineration facility schematic.

4   Analysis

For Municipal Solid Waste (MSW)			
	Problem		
Parameters	*Sanitary landfills* [7]	*Composting* [7]	*Incineration with Waste to Energy (WTE)* [3]
Ease of installation	The site that is located far away from the human settlements can be easily designated for such landfills	Composting can be easily carried out in an open container or a small open area	Complex to install the incinerator along with various equipments for the WTE conversion in a designated area
Cost	LOW	MODERATE	EXPENSIVE
Labour	LESS	MODERATE	MORE
Time required for disposal	Huge amount of time	Low contrasted with landfilling	Least amount of time for waste disposal
Residual waste	None	Separated non-combustible waste needs to be landfilled	Separated non-combustible waste and ash needs to be landfilled
Gains	• Unrecyclable waste can be treated • Produce energy from waste	• Abundance of nutrients • Increases aeration & water infiltration	• Decreases quantity of waste • Production of heat and power • Reduction of pollution • Can operate in any weather

(continued)

(*continued*)

For Municipal Solid Waste (MSW)			
	Problem		
Drawbacks	• Finite space • Wasteful when not required	• Limited flexibility • Maintenance intensive	• Possibility of long term problems • Ash waste can harm environment
Pollution	Surrounding environment is polluted while the decomposition takes place	Air pollution takes place during the initial stage of composting	The amount of pollution is controlled with the WTE process where the released gases are converted to heat while the remaining amount is released in air after filtration

For e-Waste		
	Problems	
Parameters	*Landfill* [4]	*Open sky incineration* [4]
Pollution	Leachates are formed during the process which contaminate the soil and groundwater	The gases formed are directly released in the air as the incineration is carried out in open sky and the gases are not treated
End products	No end products are obtained	Small amounts of copper is retrieved during the smelting process

For Hazardous waste			
	Problems		
Parameters	*Secured landfills* [6]	*Incineration* [6]	*Lagooning* [6]
Requirements	• Two impermeable liners and leachate collection systems • 3 m of separation between landfill and bedrock	• Requires all the facilities shown in the Fig. 1	• Multiple manmade ponds working in parallel
Gains	• Appropriate storage and treatment • No release of liquid hazardous waste in landfill	• Decreases quantity of waste • Better control over odor and noise	• Cost effective to design • Easy to operate and maintain
Environment problems	Liquid discharge of waste material by leaching or percolation	Compounds containing Sulphur, nitrogen and halogens are products of primary concern due to their environmental effects	Subsurface transport of the leachate and surface water runoff

5   Conclusion and Future Work

The smart city projects must be designed and planned tactfully, taking in consideration the population as the key point. As such the utilization of concept of smart cities in our various planning process and its effective implementation is acquiring grounds.

Certainly, there are various methods to manage the waste generated through the daily routine. Each of the techniques have some of the advantages and disadvantages depending on certain conditions to where the technique is being carried out. Here, this paper reviews a brilliant framework for collection of waste in urban communities. Utilizing this methodology, from waste released via natives to collection and dumping, the entire procedure is checked and taken care of smartly. Making framework dynamic lessens complexities during the procedure. Also, the waste to energy technique using incineration helps reduce the waste generated and using the heat produced to generate electricity. Gradually, the old techniques used such as secured landfill, lagooning and composting would be completely taken over by incineration with respect to WTE (Waste to Energy) and no other outcomes. Also, the collection system can be developed for a larger area compared to the one currently being practiced. In the near future, we would like to intensify this framework for various types of wastes, such as solid and liquid wastes.

References

1. Shyam, G.K., Manvi, S.S., Bharti, P.: Smart waste management using internet-of-things (iot). In: Second International Conference On Computing and Communications Technologies (2017)
2. Sharmin, S., Al-Amin, S.T.: A cloud-based dynamic waste management system for smart cities. In: ACM DEV 2016 November, pp. 17–22 (2016)
3. Saha, H.N., Singh, A.K.: Waste management using Internet- of-Things (IoT). In: 8th Annual Industrial Automation and Electromechanical Engineering Conference (IEMECON) (2017)
4. Osibanjo, O., Nnorom, I.C.: The challenge of electronic waste (e-waste) management in developing countries. Waste Manag. Res. 25(6), 489–501 (2007)
5. Bhattacharjee, J.: Waste management for sustainable smart cities in India. Open Access Int. J. Sci. Eng. 2(8) (2017)
6. Misra, V., Pandey, S.D.: Hazardous waste, impact on health and environment for development of better waste management strategies in future in India. Environ. Int. 31(3), 417–431 (2005)
7. Kumar, S., Bhattachharyya, J.K., Vaidya, A.N.: Assessment of the status of municipal solid waste management in metro cities, state capitals, class i cities, and class ii towns in India: an insight. Waste Manag. 29(2), 883–895 (2008)
8. Manvi, S.S., Shyam, G.K.: Resource management for IaaS in cloud computing: a survey. J. Netw. Comput. Appl. 41(1), 424–440 (2014)
9. Hannan, M.A., Arebey, M., Basri, H., Begum, R.A.: Intelligent solid waste bin monitoring and management system. Aust. J. Basic Appl. Sci. 4(10), 5314–5319 (2010)
10. Finnveden, G., Bjorklund, A., Reich, M.C.: Flexible and robust strategies for waste management in Sweden. Waste Manag. 27(8), S1–S8 (2009)
11. Gaidajis, G., Angelakoglou, K., Aktsoglou, D.: E-waste: environmental problems and current management. J. Eng. Sci. Technol. Rev. 3(1), 193–199 (2010)
12. Sivakumar, V.: Global and indian e-waste management methods & effects. In: 13th International Conference on Science, Engineering and Technology (SET) (2016)
13. Suma, V.: Towards sustainable industrialization using big data and internet of things. J. ISMAC 1(01), 24–37 (2019)

14. Karthiban, M.K., Raj, J.S.: Big data analytics for developing secure internet of everything. J. ISMAC 1(02), 129–136 (2019)
15. Zalavsky, A., Medvedev, A., Kolomvatsos, K.: Challenges and opportunities of waste management in iot enabled smart cities: a survey. IEEE Trans. Sustain. Comput. 2(3), 275–289 (2017)

A Light Modulating Therapeutic Wearable Band for 'Vision Health'

Vijay A. Kanade(⊠)

Intellectual Property Research, Pune, India
kanade.science@gmail.com

Abstract. The paper discloses a 'spatial light modulating wearable device' that acts as a substitute to conventional eye wears. The device is a wearable band that fits over the forehead. It operates on the principle of 'Mirage effect'. The wearable band captures heat given off by a human body at the posterior regions (i.e. left & right end) of the face - just below the ears. The heat captured by the device is conveniently channelized to heat up the air lying on the positive/front side of the human eye. Heat channelization is accomplished through a number of vantage points that are incorporated on the wearable device. These standpoints direct heat through specific channels for warming the air at a peculiar distance such that light gets modulated before entering the subject's eye. Therefore, device's mirage effect allows 'optical modulation' of light rays (i.e. modulation of the spatially distributed light rays) as they enter the human eye.

Keywords: Omni-directional spectacles · Heat dissipation · Mirage effect · Vantage points · Optical modulation · Refraction · Thermal layer · Myopia

1 Introduction

Human body is a storehouse of energy. An average human, in a restful condition produces about 100–120 W of power. Trained athletes are capable of producing 300–400 W of power. In few other cases, sprints or rigorous exercise allows the human body to generate about 2000 W of power [4]. In today's IoT world, numerable wearable devices have come up that are being powered by using this very human bodily energy [7–9]. Miniaturization of electronic devices and the supporting technology such as observed in 'Thermoelectric generators' has made such a possibility a living reality.

Significant factors affecting the amount of energy given off by the human body include: physiological variables and environmental phenomena. Physiological variables encompass characteristics of the living tissue (thickness of the fat layer, depth of the muscle tissue, and anatomy of the skin surface), state of the body (metabolism, blood perfusion and sweat secretion) [1]. Environmental factors involve temperature and humidity of the environment in which the human body operates.

Further, the bodily parameters adapt to the changing weather condition or even with the type of activities performed by the body (i.e. working at a desk, as compared to performing sport activities). From this known knowledge, the best and optimal situation for capturing human heat energy is when the body is under an excited and heightened physiological condition.

A. P. Pandian et al. (Eds.): ICCBI 2019, LNDECT 49, pp. 492–499, 2020.
https://doi.org/10.1007/978-3-030-43192-1_56

Therefore, based on the considerations discussed above, the skin temperature varies at different parts of the human body. The skin temperature values at various body positions are as tabulated below (Table 1):

Table 1. Skin temperature at various body positions [1]

Body positions	Yang et al. (T_{air} = 17 °C)	Zaproudina (T_{air} = 23.5 °C)	Webb (T_{air} = 27 °C)
Forehead	29.5 °C	34.1 °C	35.2 °C
Neck	31.1 °C	33.2 °C	35.1 °C
Back	30.6 °C	32.5 °C	34.4 °C
Chest	30.3 °C	32.3 °C	34.4 °C
Arm anterior	30.3 °C	31.7 °C	33.2 °C
Forearm	29.5 °C	31.5 °C	34.0 °C
Thigh	28.3 °C	30.8 °C	33.0 °C
Calf	29.4 °C	31.3 °C	31.6 °C
Foot dorsal	27.1 °C	28.6 °C	30.4 °C

The research proposal harnesses the body heat parameter for providing an effective alternative to the conventional eye wear system. Further, the proposed device is a cost-effective substitute to eye-wears as the human body heat is efficiently utilized that is naturally generated without costing a single penny. Hence, the proposal ensures appropriate channelization of body heat energy without being largely wasted.

2 Method

2.1 Thermoelectric Effect

A 'Thermoelectric generator (TEG)' is an apparatus that helps in generating electric energy by exploiting thermoelectric effect, i.e. termed as Seebeck effect. A TEG consists of layers of electrically conducting p- and n-type elements connected in series by two metal conductors and encapsulated by ceramic plates for electrical insulation. Therefore, by placing a TEG on the body surface, it is possible to harvest the electrical energy by utilizing thermoelectricity, which occurs due to the temperature difference between the two opposing sides of the TEG: the one in contact with the skin, and the one facing the environment [2].

Current magnitude is directly proportional to the temperature difference between materials:

$$J = -\sigma S \Delta T$$

Where, σ is local conductivity, S is Seebeck coefficient (thermopower), ΔT is temperature gradient

Electrical circuitry of a TEG for generating electrical voltage based on temperature is as disclosed below (Fig. 1):

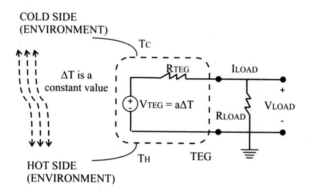

Fig. 1. Electrical circuit for power measurements generated by the TEG [1]

3 Operative Principle

The core concept behind the innovative eyewear is generally observed as the 'Mirage effect'. This principle is elaborated in the below section.

3.1 Mirage Effect

Generally, Mirage effect is quite evident in the deserts or during bright sunny days of summer.

Now, our brain interprets the light path to be a straight line. It doesn't see the bent light from the sky. Instead, it perceives in such a way that the light seems to be coming from something on the ground. Similar scenario of inappropriate brain interpretation can be equated to a pencil being dipped into water that seems to be broken when observed from the outside (Fig. 2).

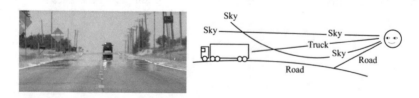

Fig. 2. Mirage effect [5]

4   Therapeutic Wearable Device

The light modulating wearable device resembles a wearable band to be worn like spectacles with a slight upward angle i.e. above the forehead. Temple and temple tips of the device are made up of TEG materials for capturing the body heat at the positions lying just below the device – which is lower portion of the ears. The wearable band has different vantage points on its hardware structure that help in dissipating heat in the close vicinity to a wearer's face. Various vantage points are triggered depending on the wearer requirement i.e. wearer's optical number as prescribed by the doctor and act as channels for heating up different portions of the air. This heat dissipation forms a transparent film on the frontal facial side of the wearer. The film is at a higher temperature than the external environmental temperature - hence it causes the refraction of incident light leading to deployment of the 'mirage effect'. As the light passes through the film, the light undergoes deviation depending on the wearer's optical number. The vantage points on the band are switched on in a manner that controls the light path as it passes through the heat generated film – implying, the vantage points ensure that the light passing through the film is modulated (i.e. bent) in such a way that it forms an image on the retina of the band wearer.

Operationally, when the user with the certain optical number wears the device – the device captures body heat from its temples. The captured heat is projected in the form of the thermal layer. The amount of heat to be projected over the thermal layer is directly proportional to the optical number specified by the ophthalmologist. Once the thermal layer is formed, the light passing through it undergoes refraction. The refraction is so adjusted by the device, such that the light rays passing through thermal layer form an image on the retina of the wearer's eye. The image is therefore rendered on the retina of the user and not before/after the retina as observed in myopia/hypermetropia respectively. Thus, the problem of myopia and hypermetropia is taken care of by the thermal film layer that modulates the light as it is passes through the film surface.

Hence, human body heat energy is used for creating the mirage effect by utilizing TEG supported light modulating wearable [6] (Fig. 3).

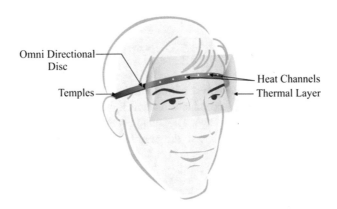

Fig. 3. Light modulating therapeutic wearable device [3]

5 Diagnosing Medical Ailments

The proposed wearable device replaces the conventionally available eye wears or spectacles/glasses. Medical conditions of myopia, hypermetropia, etc. can be resolved using the disclosed wearable. Elaborate description of its operation in case of myopic and hypermetropic case is explained in below section.

Now, by using the inventive wearable disclosed in the paper, the myopic problem can be solved without the use of any concave lens [3].

Consider a case, when an individual cannot see farther than 200 cm (i.e. far point). We need to normally figure out the power of the lens so that the person is able to see any farther object than 200 cm (focal length 'f') that a normal eye can see. Thus putting all the data into perspective, the object is assumed at ∞. Then the power of the lens should be simulated in such a way that the object's image is brought at the far point, since that is the farthest that individual can see. This defines the functionality of the lens (Fig. 4).

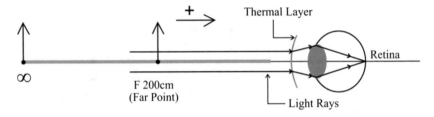

Fig. 4. Myopia

Now, Power of lens $(P) = \frac{1}{f}$, object (u) = ∞, image (v) = −200 cm,

$$\text{Hence, } \frac{1}{f} = \frac{1}{v} - \frac{1}{u} = -\frac{1}{200} - \frac{1}{\infty} - \frac{1}{200}$$

$$f = -200\,\text{cm} = -2\,\text{m},$$

$$P = \frac{1}{f} = -\frac{1}{2} = -0.5\,\text{Dioptres}$$

Hence, when the doctor prescribes the optical number as 0.5D, the thermal layer emits the captured body heat that produces desired light modulation for producing the image on the retina of the wearer's eye. The total heat emitted by the thermal layer is in direct proportion and correspondence to the required power of lens.

5.1   Hypermetropia

Hypermetropia, is the term used to define being long-sighted.

Now, by using the inventive wearable disclosed in the paper, the hypermetropic problem can be solved without the use of any convex lens [3].

In case of hypermetropia, the distance of heat dissipation by the vantage points would be at much less distance from the myopic eye since the desired effect of focusing the light onto retina can be achieved through such an arrangement. Disclosed below is the diagrammatic representation of hypermetropic eye along with an example:

Consider a case, when an individual cannot see anything closer than 100 cm (i.e. near point of the individual) from his eye. In a normal eye, a person cannot see anything closer than 25 cm (near point). We need to normally figure out the power of the lens so that the person is able to see any nearer object than 100 cm (f) that a normal eye can see. Thus, putting all the data into perspective, the object is assumed at −25 cm. Image distance = −100 cm. Then the power of the lens should be simulated in such a way that the object's image is brought at the near point of the individual, since that is the farthest that individual can see. This defines the functionality of the lens (Fig. 5).

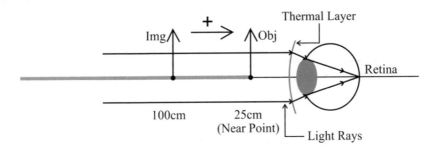

Fig. 5. Hypermetropia

Now, Power of lens $(P) = \frac{1}{f}$, object (u) = −25 cm, image (v) = −100 cm,

$$\text{Hence, } \frac{1}{f} = \frac{1}{v} - \frac{1}{u} = -\frac{1}{100} - \left(-\frac{1}{25}\right) = \frac{3}{100},$$

$$f = \frac{100}{3} \text{ cm,}$$

$$P = \frac{1}{f} = -\frac{3}{100} \text{ cm} = 3 \text{ Dioptres}$$

Hence, when the doctor prescribes the optical number as 3D, the thermal layer emits the captured body heat that produces desired light modulation for producing the image on the retina of the wearer's eye. The total heat emitted by the thermal layer is in direct proportion and correspondence to the required power of lens.

5.2   Spherical vs Cylindrical Optical Number

In case of the common problem of spherical and cylindrical eyes, possible solution could target two specific areas for heat dissipation like x-plane and the y-plane - depending on the diagnosed 'spherical vs cylindrical' category. Depending on the type of eye & defect, we are changing the vantage points on the band for dissipating heat within specific areas.

Use case

Consider a case wherein the ophthalmologist (i.e. Doctor) prescribes a number say "[±2.1] spherical" to a patient. By adopting the inventive wearable technology of the proposed research following solution is proposed.

Solution

In the above case, the numeric prescription by the doctor is categorized into separate entities as below:

1. Sign plus/minus would indicate the approximate region (i.e. geometrical distance w. r.t eyes) where the heat dissipation should happen.
2. '2.1' is directly relative to total amount of heat needed to be dissipated from the wearable device to produce the desired optical modulation or light bending phenomenon – it represents the number of vantage points to be activated to produce desired light convergence.
3. Spherical implies the specific vantage points to be specifically targeted for heat dissipation around the device which would amount to be a cumulative sum of total heat dissipated for [±2.1] eye number and ensure the appropriate refraction of light rays

Note

1. When the doctor prescribes a spectacle number, that number is directly proportional to heat parameter of the spectacle.
2. This heat parameter is directly relative to the distance above the skull/head that the heat needs to warm-up in order to ensure the appropriate refraction of light rays.

6   Advantages

The wearable IoT device stands out from the conventional eye wears due to following advantages:

1. Accurate peripheral vision – without any blind spots as observed in conventional frames
2. Omni directional view.
3. The wearable device could be used in adverse situations such as rains.

7 Conclusion

The proposed wearable device replaces the traditional eye wears. The device operates on two distinct principles: (1) Thermoelectric effect – where the generally wasted yet freely available body heat energy is utilized; (2) Mirage effect: the effect generally observed due to differing temperatures of the air (i.e. hot air & cold air). The device discloses an innovative approach for harnessing body heat & utilizing it in our day to day lives.

Further, we intend to enhance the capabilities of the wearable device by employing the device in the field of 'Augmented & Virtual Reality (i.e. AR & VR)'. Here, we plan to modify the hardware configuration of the device in such a way that it would provide the necessary support for rendering the virtual world around the thermal layer of the proposed device.

Acknowledgement. I would like to extend my sincere gratitude to Dr. A. S. Kanade for his relentless support during my research work.

References

1. Proto, A., et al.: Thermal energy harvesting on the bodily surfaces of arms and legs through a wearable thermo-electric generator, 13 June 2008
2. Liu, H., et. al.: Design of a wearable thermoelectric generator for harvesting human body energy, January 2017
3. Upadhyay, S.: Myopia, hyperopia and astigmatism: a complete review with view of differentiation. Int. J. Sci. Res. (IJSR) **4**(8) (2015)
4. Kanade, V.A.: Nanotechnology-Enabled Energy [R]evolution: nanoscale design of a flexible smart-membrane for harnessing free sources of energy. Int. J. Sci. Res. (IJSR) **5**(11) (2016)
5. Orman, J.: Highway mirage, 08 March 2010
6. Gu, Y., et. al.: Light-emitting diodes for healthcare and well-being. In: Chapter 13, Light-Emitting Diodes, Materials, Processes, Devices and Applications. Springer (2019)
7. David, A.P.J.: Thermoelectric generator: mobile device charger, September 2017
8. Kumar, P.M., et al.: The design of a thermoelectric generator and its medical applications, designs (2019)
9. Zhang, Y., et al.: Flexible organic thermoelectric materials and devices for wearable green energy harvesting, polymers (2019)

"Health Studio" – An Android Application for Health Assessment

Suresh Chalumuru[✉], P. Geethika Choudary, Pranav Souri Itabada,
and Vineela Bolla

Department of Computer Science and Engineering, VNR VJIET, Hyderabad,
Telangana, India
suresh_ch@vnrvjiet.in, geethikachoudary7@gmail.com,
pranav.souri97@gmail.com, vineelabollal@gmail.com

Abstract. The influence of recent health care reform efforts is extensive, which perhaps one of the biggest moves occurring in the junction of clinical care delivery and consumer health. The proposed application is intended to be the driving force and the technology architect that empowers patients to be more protruding participants in their own health care management. It aims at catering online health care services to the customers –such as health checkers to gauge hearing and vision measurements which also include the personality assessment test to gauge one's personality to one's environment, text-to-speech and speech-to-text convertors which will be used by the challenged people for conversion of speech to the text on the application and vice-versa, a suicidal awareness platform with a feed of moral inspirational photos and videos, local maps section to prompt the user with the nearest hospitals and medical centers in the vicinity of 10 km. It is designed with an aim to enable the user to conveniently monitor their health conditions pertaining to multiple aspects. The proposed application is deployed on mobile devices based on android that use GPS network for communication purposes.

Keywords: Vision test · Hearing test · Personality test · Convertors · Mental health awareness · Local maps using GPS

1 Introduction

Nowadays, a maintainable practice in healthcare is to marshal routine or unfruitful medical checkups and other health care services from hospital to the users' homes via smart phones based medical applications [1–5]. By enabling this, patients are leveraging the health care services at their fingertips with ease. However, quality, affordability and disability centricity of these health-care systems still persist as major loopholes across the globe. Throwing light onto this issue, this application is developed in providing assistance to the different abled, visually impaired and the profoundly deaf. Numerous tools like convertors, color tapping games and hearing test are incorporated in order to gauge their hearing/visual capabilities thus making their communication with the world smooth [6–10].

© Springer Nature Switzerland AG 2020
A. P. Pandian et al. (Eds.): ICCBI 2019, LNDECT 49, pp. 500–510, 2020.
https://doi.org/10.1007/978-3-030-43192-1_57

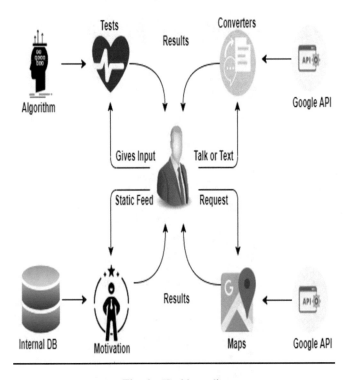

Fig. 1. Health studio

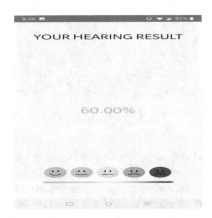

Fig. 2. Hearing test result

Mental health is a condition of well-being in which a person realizes their respective talent, can deal with the usual stresses of life and can make a constructive contribution to their own community by working industriously [11–15]. About 450 million citizens of the world suffer from mental disorders. The social and economic costs related to increasing burden of mental ill-health focused the possibilities for

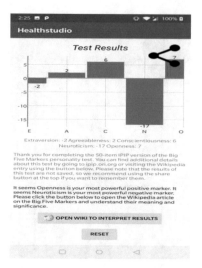

Fig. 3. Personality test result

Fig. 4. Vision test result

promoting mental health as well as preventing and treating mental illness is exuberantly high. Taking this into consideration, we have curated a module that helps in boosting a user's mental health via inspirational videos/images coupled with a psychological questionnaire that measures a user's mental condition [6–18].

In addition to this, the accessibility of medical centers with basic health assistance like First Aid is of major concern. In order to cater to such services, a local maps feature has been incorporated that prompts the user with medical centers, chemist stores or any clinic in the vicinity of 10 km (in diameter).

Since we are dealing with a sensitive domain, the data had to be handled in a secure and accurate manner as well. Therefore, various levels of scrutiny and extensive research have been done. The application's nonfunctional motives are to quickly process data, have a user-friendly interface and process data quickly.

Fig. 5. Speech to text convertor result

Fig. 6. Text to speech convertor result

2   Literature Survey

Mobile-health also known as M-health can also be stated as "the use of wireless communication devices to support public health and clinical practice". The primary goal of M-health applications is to rapidly diversify healthcare at any point of time or place with few constraints. Nowadays, there are a varied set of technologies employed for m-health solutions such as smartphones and tablets.

Due to its uniqueness, accessibility and ease of use, health care monitoring [4] via smart phones has increased rapidly over the past few years. The distinction and

Fig. 7. Images motivation

Fig. 8. Local maps result

affordability of health-care systems, however, remain as major problems throughout the globe. Many persons with inadequate profits face problems to afford the expensive health care [4]. In addition, many people are unable to obtain the quality of health care they need. In the US alone, the cost of monitoring health care is about 393.5 billion in 2005. According to a study of non-cardiac cases, the overall medical expenses are 4 million per year. We can reduce these costs with the aid of a smartphone-based health-care nursing program. This system allows users to get prompt medical check-ups, test results, and analytical centres to archive this information for future use.

Most of the monitoring systems collect data through sensors and send it over sensor networks and protocols to the remote site. This can be done with a gateway like a mobile that is connected to the Internet. The development of smartphones has had a notable impact on the remote health monitoring system in recent years. There is an extreme 35 percent hike per year for mobile users. In terms of data recording, transmission and visualization, smartphones are also unavoidable and provide great scalability. In addition, many smartphones have a default GPS system that can be used in emergency monitoring of the patient [4].

Android platform has a software heap with operating system, middleware and key applications. The layers are the kernel, application framework layer, applications layer and the libraries. The application layer is the location of all applications. The application framework layer provides APIs and managers to support the developer deed the functions of Android.

3 Proposed System

Mobile apps these days have started to penetrate many businesses today. The Healthcare business has shadowed the path too. Health is most apprehensions people have and they'd would go an extra notch to stay on top of it. However, mobile applications can't make them fit suddenly, but they can surely help in this striving. With the progress in technology, increase in smartphone users and the increasing health consciousness, mobile applications focused towards this sector have gained remarkable importance.

Health Studio provides the greatest management with best attention, anywhere and anytime. It caters the progressive level of mobile-patient care tools to enhance treatment quality and streamline the workload. The users can check their current health condition and respond proactively before it leads to any chronic disorder. A very important point to throw light on is that this is an app that just enhances self-monitoring of health, it doesn't provide immediate treatment. They can also keep a track of patient's health history, update health plans and maintain peer-to-peer interaction and collaboration.

The entire application is divided into 4 modules namely – Health trackers, Convertors, Mental Health Platform, Local Maps. Each of these modules and their respective functioning has been elaborated below:

3.1 Health Trackers

The primary motive of this module is to enable access to rudimentary hearing, vision and mental health tests to the users at a fingertip away. It encompasses innovative ways to gauge their abilities and thereby generate a result derived from the algorithm devised to compute each of them. Following are the subcomponents of this module:

(i) *Vision Test*

Color sensitivity is the ability of human eye to distinguish between various colors which are primarily a combination of Red, Blue, and Green (RGB) indices. In 1965, came the experimental result of a long-awaited confirmation-there are three color respective cones in the human eye which are the RGB color detectors. Taking this into account, a color tapping game has been developed in order to gauge a user's color sensitivity.

This game consists of 5 grids (each 3 * 3 in size) that comprises of colored tiles. The user is supposed to identify an odd colored tile out of the 9 tiles in which 8 are of the same color. Initially, the color difference among these tiles is in such a way that it can be distinguished by users having decent color identification ability and can be easily perceived by the human eye. As we proceed to further grids, the shades within the grid become more similar thus making it a little tough for the user to identify the odd tile out. Finally, at the grid at which user fails to identify, his/her color sensitivity result will be displayed which is computed depending on a pre-written algorithm that this process operates on.

(ii) *Hearing Test*

A normal human's hearing frequency range lies between 20 Hz–20,000 Hz. However, a person's hearing ability to hear depends on a multitude of factors like age, immunity levels, genes, stress levels and various other biological factors. Since frequency to which the human ear is subjected to is of prime importance, a hearing test has been devised keeping varied frequencies as it's building blocks.

In this test, the user is subjected to set of 5 clips of varied frequencies. As we proceed to the next clip, the frequency keeps increasing and the user is supposed to tap a yes/no button stating if he is able to hear the sound or not. Depending upon the users' input and at what stage of this test he/she has tapped the no button (indicating he/she can't hear the sound) an automated result (in percentage) is generated that represents their hearing efficiency.

(iii) *Personality Test*

In today's stressful world, mental health plays a key role in nourishing one's health. Many aspects add to mental health problems, including:

- Biological issues, such as genes, immunity levels or brain signals interaction
- Life experiences, like trauma, stress or abuse
- A genetic history of mental health disorders

However, the stigma about acknowledging or confessing about mental health still doesn't seem to die. Keeping this in our mind, we tried our level best to increase awareness about this via a psychological questionnaire. It consists of 50 questions that are based upon 5 attributes that make up mental health according to a famous paper, which are Openness, Conscientiousness, Extraversion, Agreeableness and Neuroticism (OCEAN). On the basis of the inputs given the user to these questions, a bar graph is generated as the result depicting where they stand according to these attributes.

Note: This is merely a graph that shows a user's mental health as per the parameters stated above. We take no responsibility in the end result of this and are not accountable for it's after math effects.

3.2 Convertors

Momentous health discrepancies exist between the general population of world citizens and the people with disabilities, especially with respect to extreme and chronic health conditions. Mobile healthcare is the delivery of health care using a number of mobile communication devices is beholding tremendous growth and has been vowed as a crucial new approach for management of chronic health conditions.

Initial evidence suggests they are not clearly represented in the progress of mobile healthcare, and specially the propagation of mobile health software applications for android based smartphones. Their exclusion in mHealth could lead to further other health disparities.

This module is a simple inclusion of Speech-To-Text and Text-To-Speech convertors where the former converts user's speech to text and displays it and the latter convert user's text into speech. The unique feature of this module is these convertors' ability to detect spaces and display results accordingly which is supported by the Google API.

3.3 Mental Health Awareness Platform

Mental Health and its wellbeing are an optimistic outcome that is eloquent for people and also for many arenas of the society, because it conveys us that people assume that their lives are going quite well. Proper living conditions are essential to well-being. Monitoring these conditions and keeping a track of them is important for public policy.

However, various indicators that gauge the living conditions fail to measure what people perceive and feel about their state in lives, like the quality of their rapports, their relationships and their optimistic emotions and elasticity, the realization of their latent, or their overall satisfaction with life also known s their well-being. Well-being usually includes global and societal judgments of life satisfaction, happiness and feelings that range from depression to joy.

This module consists of a feed of inspirational images and videos that boost up the users' zeal to live and inculcates positivity in their lives. Each video is a short clip of approximately 1–2 min that throws light on all the important arenas of well-being like hope, inspiration, determination etc.

3.4 Local Maps

Offering health services is one phase of an implementation however providing access to these services and making sure they reach the right set of users is on a whole new level of feasibility. The access to health care centers that provide at least a basic first aid facility has become a major challenge these days. There are hundreds of cases where patient's lives would've been saved when they were provided with the right access to medical centers in their extreme situations.

Google Maps is used worldwide for navigation purposes; however, this module is a feature that solely focuses on showing any form of medical centers, clinical shops and hospitals in the vicinity of 10 km in diameter. In case it fails to show in that slab of distance, a prompt message is shown to the user that we have failed to show the nearby medical centers.

4   Experimental Results

Each of the above cited modules have their own medical and functional importance. Following are the end results of each of them:

(i)   *Health Trackers*

This module comprises of three different trackers each of them for vision, hearing and personality. The vision test provides the user's color sensitivity that is the user's ability to distinguish between colors in the form of percentage. The hearing test checks the ability of a person to hear as the frequencies keep ascending in the form of percentage.

Personality test depicts the mental state of a person as per the five parameters- Openness, Conscientiousness, Extraversion, Agreeableness, and Neuroticism (OCEAN) in the form of a bar graph.

(ii)   *Convertors*

The result of this module varies depending on the convertor that is employed as per the user's necessity. The Text-To-Speech convertor reads out the converted sentence that has been given as a text input. The Speech-To-Text convertor displays the converted text that has been given as a speech input.

(iii)   *Mental Health Awareness Platform*

This is just a feed to boost up morale, so the result of this module is inspirational images and videos that are readily available for the users to watch whenever, especially to pep up their mood.

(iv)   *Local Maps*

The sole intent of this module is to provide the user with 24/7 access to nearest medical centers. The end result is all the possible hospitals, medical shops and chemist stores in the vicinity of 10 k in diameter. In exceptional cases when there isn't a single medical center of any form that is available, a toast message will be sent to the user stating they're unavailable.

5   Conclusion

Health Studio aims at providing a multifaceted health care system to the users at one platform. Flexible to use, especially to the senior citizens as they can monitor their health at just a tap away. The measurement tools are simple games—like the hearing or vision tests, where the user must tap on screen for few rounds and for personality test, the user must spend around 3 min to answer the questions to get results. Finally, this enables the users of any age group to use it with ease. The local maps feature solely focuses on prompting the nearby health care centers, thereby making it all he more convenient to use when compared to Google maps because of the minimal UI load on the application.

We have conducted a survey across all age groups and the following results have been observed:

- A middle-aged of 37 years old with no disabilities computer savvy male who spends most of the time doing projects on his laptop which is placed at distance less than 30 cm from his eyes, faces issues with 80% confidence in Vision and 63% confidence in Hearing.
- A university student of 21 years old with no disabilities who listens to music on his phone in leisure time suffers has mild hearing issues with 80% confidence.
- A school student of 15 years old who has a disability in color sensitiveness has a vision test results with 50% confidence.
- A retired bank manager of 63 years old has health issues like High Blood pressure and diabetes, which states his vision confidence level as 50% and hearing confidence level as 50%.

Compliance with Ethical Standards

- All author states that there is no conflict of interest.
- The above conducted survey was done on people from our close family members.
- Informed consent was obtained from all individual participants included in the study.

References

1. Murphy, M.L.: The Busy Coder's Guide to Android Programming (2008)
2. Headfirst Android Development: A Brain-Friendly Guidebook (2017)
3. Prakash, M., Gowshika, U., Ravichandran, T.: A smart device integrated with an android for alerting a person's health condition: internet of things. Indian J. Sci. Technol. **9**, 1–6 (2016)
4. Mahmud, M.S., Wang, H., Esfar-E-Alam, A.M., Fang, H.: A wireless health monitoring system using mobile phone accessories. IEEE Internet Things J. **4**(6), 2009–2018 (2017)
5. Kundi, F.M., Habib, A., Habib, A.: Android-based health care management system. Int. J. Comput. Sci. Inf. Secur. (IJCSIS) **14**(7), 77 (2016)
6. Elasayed, A.E., Albashir, Z.A., Taha, K.A., Abdelrahman, A.: Implementation of mobile application for testing hearing and vision: studycase. In: International Conference on Computer, Control, Electrical, and Electronics Engineering (2018)
7. Prudtipongpun, V., Buakeaw, W., Rattanapongsen, T., Sivaraksa, M.: Indoor navigation system for vision-impaired individual an application on android devices. In: 11th International Conference on Signal-Image Technology & Internet-Based Systems (2015)
8. Tanno, H., Adachi, Y.: Support for finding presentation failures by using computer vision techniques. In: IEEE International Conference on Software Testing, Verification and Validation Workshops (2018)

9. Bert, F., Giacometti, M., Gualano, M.R., Siliquini, R.: Smartphones and health promotion: a review of the evidence. J. Med. Syst. **38**(1), 9995 (2014)
10. Solanas, A., et al.: Smart health: a context-aware health paradigm within smart cities. IEEE Commun. Mag. **52**(8), 74–81 (2014)
11. Yusuf, A.N.A., Zulkifli, F.Y., Mustika, I.W.: Development of monitoring and health service information system to support smart health on android platform. IEEE (2018)
12. Tartan, E.O., Ciflikli, C.: An android application for geolocation based health monitoring, consultancy and alarm system. In: IEEE International Conference on Computer Software & Applications (2018)
13. Bajaj, D., Yadav, A., Jain, B., Sharma, D., Tewari, D., Saxena, D.: Android based nutritional intake tracking application for handheld systems. In: 8th ICCCNT (2017)
14. Jasim, A.D., Marza, H.H.: Design and implementation of an android system for indoor positioning using WLAN finger print scheme. IJSET **6**(1), 18–22 (2017)
15. Banos, O., Villalonga, C., Garcia, R.: Design, implementation and validation of a novel open framework for agile development of mobile health application (2015)
16. Mukhopadhyay, S.C.: Wearable sensors for human activity monitoring: a review. IEEE Sens. J. **15**(3), 1321–1330 (2015)
17. Valanarasu, R.: Smart and secure IoT and AI integration framework for hospital environment. J. ISMAC **1**(03), 172–179 (2019)
18. Lee, K., Gelogo, Y.E.: Mobile gateway System for ubiquitous system and internet of things, application. Int. J. Smart Home **8**(5), 279–286 (2014)

Potential Candidate Selection Using Information Extraction and Skyline Queries

Farzana Yasmin[✉], Mohammad Imtiaz Nur,
and Mohammad Shamsul Arefin

Computer Science and Engineering, Chittagong University of Engineering
& Technology, Chattogram 4349, Bangladesh
{farzanayasminefu, sarefin}@cuet.ac.bd,
m.imtiaznur@gmail.com

Abstract. Information extraction is a mechanism for devising an automatic method for text management. In the case of candidate recruitment, nowadays different companies ask the applicants to submit their applications or resumes in the form of electronic documents. In general, there are huge numbers of resumes dropped and therefore the volume of the documents increases. Extracting information and choosing the best candidates from all these documents manually are very difficult and time-consuming. In order to make the recruitment process easier for the companies, we have developed a framework that takes the resumes of candidates as well as the priorities of the employer as input, extract information of the candidates using Natural Language Processing (NLP) from the resumes, filter the candidates according to predefined rules and return the list of dominant candidates using skyline filtering.

Keywords: Information extraction · Natural language processing · Machine learning · Skyline query · Candidate selection

1 Introduction

Information extraction (IE) infers the process of automatically gisting of information in a structured way from unstructured and/or semi-structured machine-readable documents [1]. Nowadays huge volume of documents are found online and offline. Extracting information from these vast volumes of data manually is time consuming.

Recruitment is the process of searching and selecting best candidates for filling the vacant positions of an organization. Recruitment process requires planning, requirements setup strategy, searching candidates, screening the candidates according to the requirements and evaluation of the candidates. These steps are usually conducted by the Human Resource (HR) department of any company. Whenever there is a job opening for the vacant positions, large amount of applications are dropped. Searching and screening the best candidates from these applicants after assessing the abilities and qualifications manually takes huge amount of time, cost and effort of the HR department as the volume of data are big. If we can develop an efficient system for extracting information from the resumes of the applicants and process these information in an automated way, it will ease the work of the HR management. An automated system for choosing the potential

© Springer Nature Switzerland AG 2020
A. P. Pandian et al. (Eds.): ICCBI 2019, LNDECT 49, pp. 511–522, 2020.
https://doi.org/10.1007/978-3-030-43192-1_58

candidates that best suit the position's requirements can increase the efficiency of the HR agencies greatly. Therefore, in order to make the recruitment process easy, effective and automated, we have developed a framework of potential candidate selecting system by choosing a domain of document information extraction i.e. the CV/resume documents. This development task involves the information extraction based on natural language processing i.e. tokenization, named entity recognizer (NER) and utilizes skyline query processing which works well in filtering the non-dominating objects from database and also makes a new addition to this domain. So the objectives of the system development can be summerized as follows:- (1) To design an efficient information extraction system from documents like curriculum vitae, (2) To generate scores on different features based on extracted information, (3) To perform appropriate filtering of information using skyline queries and (4) To generate proper ranking system for candidate selection.

The rest of the paper is presented as follows: In Sect. 2 related works of the candidate ranking system development has been portrayed. The system architecture and design is elaborated in Sect. 3. Section 4 represents the implementation of our work with some experimental results. And finally, a conclusion over the work has been drawn in Sect. 5.

2   Related Work

Celik [2] proposed an information extraction system for candidate selection where the information extraction was based on ontology. The proposed methodology used Ontology-based Resume Parser(ORP) to convert English and Turkish documents into ontological format and constructed seven reference ontologies to extract the information and categorize them into one of these ontologies. Though the methodology worked good on information extraction but it did not describe any score generation mechanism to rank the candidates. Farkas et al. [3] worked on a method of extracting information for career portal where the information of applicants' are stored in a uniform data structure named HR-XML format. They used a CV parser to automatically extract data from the CV. In [4], the authors used a hybrid cascade model for information extraction from CVs. In the first pass, the proposed method segments resume using Hidden Markov Model. The second pass uses HMM and SVM to extract further detailed information. The cascaded pipeline suffers from error propagation i.e. errors from first step are passed in the second pass and the precision and recall value of the second pass decreases subsequently. Information is extracted from resumes using basic techniques of NLP like word parsing, chunking, reg ex parser in [5]. Information like name, email, phone, address, education qualification and experience are extracted using pattern matching in this work. Some other online resume parsers are found in [6] and [7]. There also have been developed some works using skyline queries. [8, 9] and [10] describes some algorithms for processing skyline queries with their implementation. Patil et al. [11] developed a method for learning to rank resumes with the help of SVM rank algorithm. In [12], Yi et al. applied a Structured Relevance Model to select resumes for a given post or to choose the best jobs for a given candidate based on their CV. In [13] job narration are transformed into queries and lookup in database is performed. The top-ranked candidates gets selected automatically from these queries. Some authors exploit additional information like social media information along with information gained directly from resumes in [14] and [15].

Another form of candidate selection was proposed by Kumari et al. [16] where candidate selection was done by using Naïve Bayes algorithm for classifying the candidate profiles. They also considered employers importance criteria. No description given of how the information extraction are done. Also it requires GPRS connection every time as it is online based. In [17], CVs are filled in a predefined format and the scoring and ranking process is based on Analytic Hierarchy Process (AHP).

Though many works have been developed for candidate ranking, the use of skyline query in this scenario is relatively new approach and we have implemented this novel approach in our framework.

3   System Architecture and Design

The proposed framework works in 4 modules according to Fig. 1: Document processing module, Query Execution Module, Analysis & Output module and Storage module. The details of the system architecture are described below [Reviewer Comment 1, 2, 3]:

3.1   Processing Module

Document Input. First we will need to input the resumes in the interface for a specific job id. After documents are being fed to the system in processing module, information extraction process begins and we used a NLP module named spaCy [18] for the rest of the processing steps. Suppose, we have fed the resumes of Fig. 2 in the system.

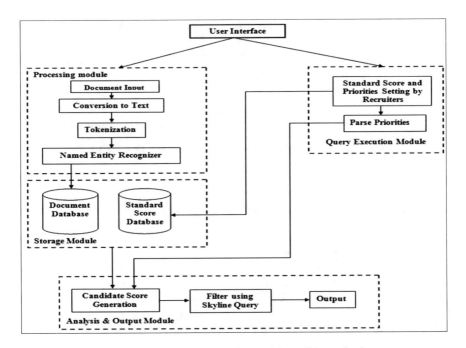

Fig. 1. System architecture of potential candidate selection

Adam Wang (Male)

XXXX Company of Beijing.
Beijing City, 100007
1364-110-XXX
wangXXX@hotmail.com

Education Background
From Sept. 2000 to Apr. 2003, I got master degree from University of XXX in computer software engineering.
From Sept. 1996 to July. 2000, I got bachelor degree from School of XXX and major in computer science and technology.

Experience
From March 2003 to now, working on Human Face Recognition System in XXXX Company of Beijing
From June 2001 to March 2003, working on Content-Based Intelligent Image Retrieval System in Research Center of XXX Company
From Sept. 2000 to May 2001, working on Intelligent Highway Distress Detection System in National Lab. Of XXX University

Interests
Reading, music, and jogging

(a)

ABC
Cell: +880 1680671851
E-mail: abc@gmail.com,

RESEARCH INTEREST
Data Mining, Artificial Intelligence, Machine Learning, Algorithm Design, Data Structure.

EDUCATIONAL QUALIFICATIONS
- **B.Sc. in CSE**- October 2015
  CUET, Bangladesh.
  Result: CGPA 3.81 (out of 4.00)
  Class Position: 2nd out of 113 students

WORK EXPERIENCES
- **Chittagong University of Engineering & Technology, Chittagong-4349, Bangladesh**
  - Joined as ***"Lecturer"*** as on 2016
  - Duration : 2016- Present

(b)

Ishraq Rayeed Ahmed
+880 1717 342569, ishraqrayeed@gmail.com, www.irahmed.wordpress.com

Research Interests
Pavement Materials and Design, Urban and Public Transportation System, Traffic Emissions and Air Quality, Transportation Safety, Intelligent Transportation System

Education
Bachelor of Science, Civil Engineering, March 2016
Bangladesh University of Engineering & Technology at Dhaka, Bangladesh
CGPA: 3.24/4.00

Technical Skills
Scientific Computing & Simulation Tools: MATLAB, R Project, ArcGIS, VisSim, EPAnet
Structural Design Software: AutoCAD, SAP, ETABS, GRASP
Programming Languages: C++, Python
Graphics Design Software: Adobe Photoshop, Adobe Illustrator, 3D Studio Max

Publications
- Ahmed, I.R; Mondal, A.R; Noor, A.U, *"Assessment Of Pedestrian Perception Towards Pedestrian Crossing Facilities In Dhaka Metropolitan City: A Study Based On Observation and Survey"*, International Conference on Recent Innovation in Civil Engineering for Sustainable Development (IICSD-2015), DUET, Dhaka.

(c)

Fig. 2. (a), (b), (c) Sample resumes

Conversion to Text. The standard format of resumes for our system is considered English resumes in PDF format. At first we need to convert the pdf into plain text using UTF-8 encoding. UTF-8 is a compromise character encoding that can be as compact as ASCII but can also contain any Unicode characters. UTF stands for Unicode Transformation Format and '8' means it uses 8-bit blocks to represent a character. The number of blocks needed to represent a character varies from 1 to 4 [19].

Tokenization. After conversion to text, now we have our necessary text file. We start reading the text file and tokenize the whole document. Tokenization is done using the language rule i.e. removing the white space, checking the exception rules like punctuation checking, abbreviation rules etc.

Named Entity Recognition. Named entity recognition (NER) is the most important task to do next. The success of the extraction process mainly depends on the accurately recognized entities from a resume. The subtask of information extraction that seeks to locate and classify named entity mentions in unstructured text into pre-defined categories such as the person names, organizations, email, phone, address, time, quantities, numeric values, etc. can be defined as Named entity recognition [20]. A statistical model is used to classify our desired entities in a standard resume. The NER training model is designed using incremental parsing and residual CNNs. In case of training our model (Fig. 3.) with the desired annotation, we used resumes in JSON format. At first we have to manually annotate our training data in JSON format. Then we load or build the NER model. For training the NER model with our custom entities, now we add the labels for each annotations. Then we shuffle and loop over our training examples. At each word the model makes a prediction. It then consults the annotations to see whether it was right. If it was wrong, it makes adjustment of the weight so that the correct action will score higher next time. Then we save the model and test it to make sure the entities in the test data are recognized correctly. After the validation of the trained NER model, now we use this model to extract the values of the entities trained from the resumes. The recognized entity values are stored in a row of a table for each candidate in the storage module.

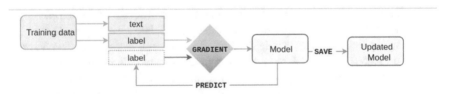

Fig. 3. spaCy's NER model training process (Source: [21])

3.2   Query Execution Module

Standard Score and Requirements Setting by the Recruiter. In the UI, employers set the standard scores required to evaluate the abilities of the candidate according to the job criteria. Each criterion gets a value and a weight for a specific keyword. The weight represents the relative importance or priorities of the specific criteria and value represents the variations of the score of each criteria. Keyword gives the matching criteria i.e. which information to be satisfied for scoring. These standard scores are stored in the storage module as a lookup table. Suppose, for software engineer position, the employer sets the following values and weights in the table for each criteria.

Table 1. Standard score setting table

Job_criteria	Keywords	Value	Weight
Skills	C++	10	5
Skills	Java	10	5
Skills	PHP	8	5
Experience	3	5	3
Experience	0	2	3
Major	CSE	10	2
Major	EEE	6	2

Parse the Requirements. The system will then parse these requirements of the employer in the query execution module.

3.3 Storage Module

Storage module stores information processed by the processing and query execution module. The extracted information table, after the entities are recognized, are stored in the document database. The standard scores set by the recruiters in the query execution phase are stored in the score database. The total storage is required for the candidate score generation in the analysis and output module.

3.4 Analysis and Output Module

Candidate Score Generation. After parsing the requirement of the employer, the system will start the score table generation of each candidate according to the employer priority and previously set standard score for different categories. The extracted information stored in the lookup table in document database is retrieved and matched with the keywords stored in the job_info_details table. If match found, the corresponding values are calculated by multiplying the value and weight set in the standard score table. If multiple keywords are matched for a specific criteria, then they are stored as aggregated sum. For example, if multiple skills match, then all the skill values are added and stored in the skill column for that candidate.

The score calculation follows the following formula (1):

$$Scores[job\_criteria] = Scores[job\_criteria] + (job\_details(value) * job\_details(weight)) \tag{1}$$

For the result and total years of experience column scoring, extracted numeric value of the applicant is matched with the sorted list of previously set keywords. If the extracted value is greater or equal to any specified keyword, the score is calculated according to that value. For the publication column, international conference, international journal keywords are searched and matched. If found, the number of occurrences are counted. If any column information contains missing value, then they are considered as zero in the score calculation. The calculated score is stored in that specific criteria column of the score table. After being scored in each criteria, now a table is generated which is score of each candidate. The sample score table for the resumes in Fig. 2 are generated as in Table 2.

Table 2. Sample score table

CV no.	Skills	Experience	Major	Total
1	50	15	20	85
2	0	15	20	35
3	50	6	0	56

The first candidate had the matching skill C++, experience of 3.7 years and major CSE. So the first candidate fulfills all the requirements of the specified job position and get scores according to the rules set as Table 1 i.e.

Scores[skills] = value for C++ (10) * weight of C++ (5) = 50.
Scores[Experience] = value for 3.7 (5) * weight of C++ (3) = 15.
Scores[Major] = value for CSE (10) * weight of CSE (2) = 20.

The scores of the 2nd and 3rd candidate will be calculated as the 1st candidate. The skills of 2nd candidate doesn't match the required skills and so the missing value is scored as zero. Accordingly, the 3rd candidate's major doesn't match the requirement and so he gets a zero in major field. Now if we select the Major field as mandatory, the row containing zero in this field i.e candidate 3 will be deleted.

Filter Using Skyline Query. Askyline is defined as those points in a dataset those can not be worse than any other point. A point dominates other points if it is as greater or equal in all criteria and greater in at least one criteria. A study in [22] states that during the past two decades, skyline queries are applied in several multi-criteria decision support problems. Skyline query utilizes the idea of skyline operator. There are several algorithms for the implementation of skyline operator like using directly in SQL queries, block nested loop, divide and conquer, branch and bound, map reduce etc. We have used the block nested loop (BNL) method. Applying skyline queries on the score table according to employers' priorities, now the dominant applicants will be filtered. We can explain the working procedure of skyline query using Table 2. According to BNL, we compare all the data points with all other points. We keep the points that can dominate other points in all criteria and atleast in one dimension. The points dominated are discarded from the list. Those points are considered to be skyline that dominates are others or maybe a part of the skyline if they neither dominates nor dominated by others. So comparing the data points of Table 2 we find that Candidate 1 dominates the other candidates in every criteria as it contains either equal or higher value in every criteria than the other two. After applying skyline we get that candidate 1 best suits for the job and others are discarded from the list (Table 3).

Table 3. Score table after filtering using skyline query

CV no.	Skills	Experience	Major	Total
1	50	15	20	85

Output Generation. The system output will show the result of the potential candidates after the filtering process. Then the employer can choose the criteria they want to select candidates having higher score. Suppose, an employer want to have candidates with higher score in skills criteria. He/she can rank the candidates according to the scores in the skills criteria. If multiple criterias are to be considered, it is also possible to rank the candidates. Then the system can output the ranked output. The output will be sorted according to the score obtained and personal details like name, email, phone number of each candidate will be displayed.

4   Implementations and Experiments

In this section, we have described the implementation and experimental setup of our system with necessary illustrations.

4.1   Experimental Setup

Potential candidate selection system has been developed on a machine having Windows 10, 2.50 GHz Core i5-3210 processor with 12 GB RAM. The system has been developed in Python 3.7.3, Asp.Net Core and Angular5 in the front end and MS SQL Server is used in the back end for storing related data to complete this project.

4.2   Implementation

At the beginning of our system workflow, resume documents are fed into the system. All the resumes are stored in a folder according to the specific job id. These resumes are then converted into text format using UTF-8 encoding and stored in a file named lookup.py. The text files are then called for tokenization and named entity recognition. Next, calling the trained model of NER, we extract the information from the tokenized data of the resume documents. We have extracted information of 12 entities-University, degree, major, experience, publication, skill, certification and personal information (name, date of birth, email, phone etc.). The values of these entities are extracted in the information extraction table and stored in document database.

On the other hand, employers set the necessary information for setting the requirements and scores of each criterion. Job_info_details table holds the columns like Job_info ID, Keyword, Value, Weight, Job Criteria Name i.e. the information set by the recruiters on the score setting step. For the specific job position, extracted information table can be uploaded next for score generation. After scoring according to the rules set, the system generates the score table. This table can be downloaded by the recruiter. Next the recruiter is given the option to choose the mandatory requirement criteria. If any of the criteria is chosen then the candidates holding zero value in that specific criterion are removed before applying skyline query. Applying skyline query on the score table now returns the dominant applicants for the specified job by finding the max value and mapping them according to the job criteria. Then the unique candidates holding maximum values in any of the criteria are returned.

4.3 Performance Evaluation

Potential candidate selection system performance is evaluated in two different phases-Information extraction performance, and filtering using skyline queries. We tested the performance of our system using 150 resumes of engineering background.

For the training of our NER model, we used a dataset [5] of 300 manually annotated resumes and validated the model using resumes from the dataset. We found some incorrect values for extracted information and also some missing values. The precision, recall and F-measure of each entity of the NER model is given below in Table 4:

Table 4. Accuracy, precision, recall and F-measure of the entities recognized

Entity	Accuracy	Precision	Recall	F-measure
Name	99.77	0.99	0.99	0.99
Email	100.0	1.0	1.0	1.0
Phone	100.0	1.0	1.0	1.0
Date of Birth	99.87	1.0	0.99	0.99
University	99.87	1.0	0.99	0.99
Degree	99.25	0.99	0.99	0.98
Major	98.36	0.99	0.98	0.98
Publication	98.71	0.98	0.98	0.98
Skills	94.84	0.99	0.94	0.96
CGPA	100.0	1.0	1.0	1.0

We have tested the candidate filtering using skyline query with 3 different job criteria- Software Engineer with 2–4 years experience, Research Assistant with CGPA above 3.5 and 2 publications and Assistant Programmer with skills Java, JavaScript, HTML and CSS. We have scored the 150 resumes for these 3 different job positions. The returned output showed that the top scored candidates were different for the 3 job positions as the requirements were placed different. Comparing with the manual ranking, we found that the filtering were accurate with a faster response. So based on the observations we can come to the conclusion that the accuracy of the skyline query depends on the accuracy of the scores generated. If the score generation is accurate, the skyline query returns those candidates that are best for the vacant position quite accurately and within few seconds. A snapshot of the system's output is shown in Fig. 4.

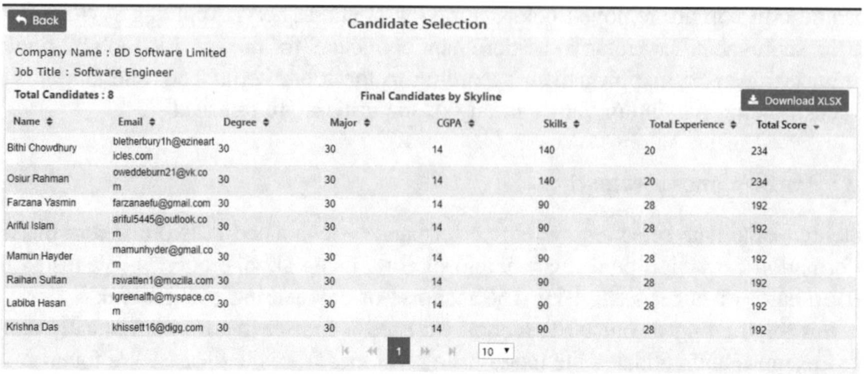

Fig. 4. Snapshot of the high scored (total) candidates for software engineer position of our system [5]

We have also tested the skyline filtering with 50,000 synthesized score data. The execution time for different number of resume data is given in Table 5. The table shows that the skyline query can perform filtering in a very responsive way.

Table 5. Response time of skyline filtering

No. of resume data	Response time (mili sec)
1000	67.23
3000	75.30
6000	83.27
25000	91.18
50000	112.80

5 Conclusion

In this paper, we have narrated the idea of a candidate selection system which finds the best potential candidates by extracting information and filtering using skyline query. Automating the total task may help the HR agencies by reducing time, cost and effort of searching and screening the pioneer applicants from vast applications. There are many automated candidate ranking system available online. But we have developed a novel idea of using skyline query in filtering and returning the dominant candidates for the job specified. Skyline queries are mostly applied in multidimensional decision application. In candidate filtering, the implementation of skyline is new and we have applied this novel approach in an efficient manner. In the system performance evaluation, we have used 150 resumes of technical background in testing of the system and found that, the system works in an efficient way of returning best candidates by matching the given requirements with qualifications of the candidates. Altogether the system performs better in filtering the documents as well as the candidates based on the information extracted from the resume documents. Our system works for only English documents currently. In future, we hope to extend it for Bangla resumes as it is fifth most spoken native language in the world by incorporating Bangla Language Processing and test the system performance accordingly.

Acknowledgement. Informed consent was obtained from all individual participants included in the study.

References

1. Information Extraction. https://en.wikipedia.org/wiki/Information_extraction
2. Celik, D.: Towards a semantic-based information extraction system for matching resumes to job openings. Turk. J. Electr. Eng. Comput. Sci. **24**, 141–159 (2016)
3. Farkas, R., Dobó, A., Kurai, Z., Miklós, I., Nagy, Á., Vincze, V., Zsibrita, J.: Information extraction from Hungarian, English and German CVs for a career portal. In: Prasath, R., O'Reilly, P., Kathirvalavakumar, T. (eds.) Mining Intelligence and Knowledge Exploration. Lecture Notes in Computer Science, vol. 8891. Springer, Cham (2014)
4. Yu, K., Guan, G., Zhou, M.: Resume information extraction with cascaded hybrid model. In: Proceedings of the 43rd Annual Meeting of the Association for Computational Linguistics, Ann Arbor, pp. 499–506, June 2005
5. Information Extraction from CV. https://medium.com/@divalicious.priya/information-extraction-from-cv-acec216c3f48
6. Writing Your Own Resume Parser. https://www.omkarpathak.in/2018/12/18/writing-your-own-resume-parser/
7. Resume Parser. https://github.com/bjherger/ResumeParser
8. Shah, S., Thakkar, A., Rami, S.: A survey paper on skyline query using recommendation system. Int. J. Data Min. Emerg. Technol. **6**(1), 1–6 (2016). ISSN 2249-3212
9. Kalyvas, C., Tzouramanis, T.: A survey of skyline query processing (2017)
10. Papadias, D., Tao, Y., Fu, G., Seeger, B.: An optimal and progressive algorithm for skyline queries. In: ACM SIGMOD International Conference on Management of Data, pp. 467–478 (2003)
11. Patil, S., Palshikar, G.K., Srivastava, R., Das, I.: Learning to rank resumes. In: FIRE, ISI Kolkata, India (2012)
12. Yi, X., Allan, J., Croft, W.B.: Matching resumes and jobs based on relevance models. In: SIGIR, Amsterdam, The Netherlands, pp. 809–810 (2007)
13. Rode, H., Colen, R., Zavrel, J.: Semantic CV search using vacancies as queries. In: 12th Dutch-Belgian Information Retrieval Workshop, Ghent, Belgium, pp. 87–88 (2012)
14. Bollinger, J., Hardtke, D., Martin, B.: Using social data for resume job matching. In: DUBMMSM, Maui, Hawaii, pp. 27–30 (2012)
15. Dandwani, V., Wadhwani, V., Chawla, R., Sachdev, N., Arthi, C.I.: Candidate ranking and evaluation system based on digital footprints. IOSR J. Comput. Eng. (IOSR-JCE) **19**(1), 35–38 (2017). ver. 4, e-ISSN 2278-0661, p-ISSN 2278-8727
16. Kumari, S., Giri, P., Choudhury, S., Patil, S.R.: Automated resume extraction and candidate selection system. Int. J. Res. Eng. Technol. **03**(01), 206–208 (2014). e-ISSN 2319-1163, p-ISSN 2321-7308
17. Faliagka, E., Ramantas, K., Tsakalidis, A., Viennas, M.: An integrated e-recruitment system for CV ranking based on AHP. In: 7th International Conference on Web Information Systems and Technologies, Noordwijkerhout, The Netherlands (2011)
18. spaCy. https://spacy.io/

19. UTF-8 encoding. https://www.fileformat.info/info/unicode/utf8.htm
20. Named Entity Recognition. https://en.wikipedia.org/wiki/Named-entity_recognition
21. spaCy NER training model. https://course.spacy.io/chapter4
22. Tiakas, E., Papadopoulos, A.N., Manolopoulos, Y.: Skyline queries: an introduction. In: 6th International Conference on Information, Intelligence, Systems and Applications (IISA), July 2015. https://doi.org/10.1109/iisa.2015.7388053. e-ISBN 978-1-4673-9311-9

An Effective Feature Extraction Based Classification Model Using Canonical Particle Swarm Optimization with Convolutional Neural Network for Glaucoma Diagnosis System

Narmatha Venugopal[✉] and Kamarasan Mari

Department of Computer and Information Science, Annamalai University,
Chidambaram, India
balaji.narmatha8@gmail.com, smkrasan@yahoo.com

Abstract. Early detection of glaucoma could be helpful in preventing blindness of person. Diagnosis of glaucoma could be detected by segmenting Optic Disc (OD) area. This paper presents an automatic OD segmentation and classification to detect glaucoma. The presented approach consists of feature extraction using Hough transform, Segmentation by morphological operation as well as Classification through Convolution Neural Network (CNN). Particle Swarm Optimization (PSO) variant cPSO-CNN, has been applied for the optimization of hyper parameter structure-determined CNN. This is evaluated with DRISHTI-GS database as well as a brief analysis is performed to ensure the optimal property of proposed technique.

Keywords: Classification · Diseases · Dataset · Glaucoma · cPSO

1 Introduction

Nowadays, Glaucoma is a disease that affects retinal part of eyes. The major challenge involved in this model is to discover the glaucoma present in color fundus image which requires great experience and knowledge. Here, Glaucoma and diabetic retinopathy are assumed to be dangerous eye infection results to blindness. Glaucoma has been the secondary cause of blindness globally. In order to prevent blindness, prior detection of glaucoma is essential [1–3]. People affected by the above diseases are increased, as well as more than 50% of them become blind. This disease is a major challenge for ophthalmologists. By examining the retinal images obtained from fundus camera, many retinal relevant diseases can be identified. Optic disc (OD) is associated with intensity part present in retina as inner and outer position of blood vessel and retinal nerve fibers. The significant role in developing automatic scheme for the purpose of diagnosing glaucoma. Simultaneously, identification of OD region would reduce the amount of false positive while investigating searching process on highlighted anomalies like exudates generated through similar features between them. Massive research is made to identify OD and diabetic retinopathy. [3] involves in segmenting OD part with the help

© Springer Nature Switzerland AG 2020
A. P. Pandian et al. (Eds.): ICCBI 2019, LNDECT 49, pp. 523–530, 2020.
https://doi.org/10.1007/978-3-030-43192-1_59

of low pass filter and threshold. By following the sequence of exudates identification, filtering blood vessels as well as microaneurysms partitioning takes place to classifies glaucoma as well as diabetic retinopathy.

[4] deployed the segmentation of OD region for diagnosis of glaucoma by utilizing gabor filteration and K-Means clustering. Segmentation of OD region is performed by morphological functions to reach efficient performance. [5] employed Genetic Algorithm (GA) in blue channel of image to enhance searching process as well as to avoid OD region. It is simple to discover OD area in rapid measure and to attain optimal prediction value. Weighted Error Rate (WER), is applied for validating the end result. [6] combined morphological functions and Kirsch's format to remove OD and retinal blood vessels that is helpful in predicting exudates.

[7] discovered OD area by maximum differentiation method with the help of morphology. It is carried out in a green channel to acquire candidate OD pixel (ODP). Segmenting of ODs boundaries takes place through Circular Hough Transformation. It could be performed in both red and green channel as well as better result is chosen. Blood vessel is removed by changing structural element to improve the segmentation results of OD. Simulation is performed on employed MESSIDOR database as well as it localizes the region with increased performance. Threshold evaluation could be applied base on the green channel histogram. This process is helpful in identifying each bright part is named as cluster. Followed by, a collection of 2 variables such as area and density criterion has been used on cluster.

[8] employed many number of entropy to candidate OD area whereas Sobel processor is applied for detecting edges. [9] identified intermediary of OD with the help of processing ration among two channels like green and red. [10] established preprocess level on original fundus image using average filter for different brightness. This method is known to be Contrast stretching details, which is employed for portioning brighter part. Preprocessed image is converted by negative transfer in order to accompany predefined threshold. [11] removed OD using threshold approach as well as ilblack's scheme is employed for achieving binary image.

This study concentrates on developing automated OD segmentation and classification method to detect glaucoma. Proposed model incorporates feature extraction with the help of Hough transform, segmentation by applying morphological operation and classification by CNN. New technique PSO variant canonical PSO cPSO-CNN is utilized to optimize hyper parameter architecture-determined CNN. It focuses on method of analyzing OD region by using red channel of retina fundus image. This is because of inefficient contrast in blue and green channel to predict blood vessels. The presented model is verified by DRISHTI-GS database and the results ensured the better result of the proposed method.

Here, Sect. 2 explains the projected technique. Section 3 verifies the operation of study and Sect. 4 concludes the research work.

2 Proposed Work

The projected model contributes collection of techniques like channel filteration, extracting, partitioning, feature extraction, classification and so on. Initially, original color fundus images consist of 3 different channels like red, green as well as blue. Afterwards, the channels would be extracted and red color image is employed for upcoming techniques. Hence, median filtering are helpful for preprocessing to avoid noise and accomplish visible OD portion. For partitioning OD region, a morphological computation should be performed. Thus, feature extraction takes place for determining major characteristics which signifies the OD area. Consequently, all classification process are completed using CNN model.

2.1 Preprocessing

In order to achieve the red channel as input, other channels are eliminated. Noise which occurs represents the proposed image that is avoided with the application of median filter. It is computed based on the procedure of interchanging pixel values through median level of intensity that is adjacent to pixels. Hence, pixel value at a point (x, y) is provided to solve the median problem. Thus the mathematical form of representation could be:

$$f(x, y) = \underset{(s,z) \in (S_{xy})}{\text{median}} \{g(s, z)\} \tag{1}$$

2.2 Segmentation

Threshold values are assumed to be intermediate points in segmentation phase. Here pixel values are less than the threshold range which is declared as background if object points are broken. Therefore, threshold results in binary image which comprises of 2 gray level images like black and white. The arithmetic form might be as follows:

$$b(x, y) = \begin{cases} 1, & \text{if } f(x, y) \geq W \\ 0, & \text{if } f(x, y) < W \end{cases} \tag{2}$$

Where W represents the intensity of threshold. While Otsu models are employed in automated option of threshold values.

2.3 Hough Transform Based Feature Extraction

It is a feature extraction method used in digital signal processing for the estimation of shape parameters from its margin points. Thus, Hough transform is utilized to the exposure of random shapes. The common parameterizations are specified with

$$lcos\theta + msin\theta = \rho \tag{3}$$

Hough transforms are the tolerant of gaps of boundaries, unaffected with noise and it is a derivation of arbitrary transform. It provides projections appeared in several angles. It also proposes best edge detection extract for isolate step edges utilizing the initial derivative of the Gaussian.

This edge exposure process could decrease the pre-processing time and gives a moderately stable information source that resists geometrical and environmental alterations to evaluating Hough transform. It gets values to every edge point (L, M) in the image determined in the under equation. Also to non-analytical spaces are evaluated in the Eq. (5) by the particular group of edge points. For a shape ρ it is described as P in the Eq. (6).

$$\rho = L_x cos\theta + M_x sin\theta \tag{4}$$

$$B = \{L_B\} \tag{5}$$

$$p = \{L^0, s, \theta\} \tag{6}$$

For each value of L_B, r is evaluated under and it is accumulated as a function φ. The values to r for every pixel L of gradient direction $\varphi(L)$ in images are evaluated in the Eq. (8) and it is stored in the accumulator.

$$r = L^0 + L_B \tag{7}$$

$$A(L + r) \tag{8}$$

From the collector, we will obtain the Hough transform result to the pre-processed mammogram. Here, we contain the choose features from the applied image. The features must choose carefully as sometimes features could decrease the effectiveness of the classifier. In this study, intensity feature is chosen and the cause to utilize this features is due to the complexity in interpreting the shape of the OD. There could be well definite masses, architectural distortion, speculated mass, asymmetry, ill-defined mass and others. Now, we are leaving to intensity dependent features to optimal outcomes. The intensity features utilized in this work is mean, standard deviation, variance and entropy.

2.4   Classification

CNN is model which is originated from specific features of visual cortex that is commonly implemented for classification of images. It occurs with the help of series of layers like pooling layers, non-linear layers and convolutional layers. Input images are provided as instance for *conv* layers. Therefore, pixel value is monitored from corner of image. Followed by, a tiny matrix named as extract is selected which offers the convolution. These models obtains the increment of values using actual pixel values. CNN comprises a set of *conv* networks that includes non-linear as well as pooling layers. The non-linear layers are combined with the addition of convolutional computation. Thus, it captures a function of non-linearity.

Non-linear layers follows the pooling layers which executes for sampling computation. It signifies that tiny features are discovered in primary convolution operation, and a brief image is not essential for additional performance. If convolutional series, non-linear and pooling layers acquires complete layer. The resultant information is derived from convolutional networks.

The entire operation of cPSO-CNN is consistent by canonical PSO [12]. The primary operation of swarm particles denotes the sequence of $\vec{L}$ of all units represented by l_x, in spite of different configuration of hyper-parameters that is assigned randomly by using even distribution, whereas l^p and l^g implies the local and global hyper-parameter values optimally. Additionally, location for particle is rearranged by local and global information. Then, l_x is recalculated in decreasing order $\vec{L}$ with consistent adaptability for estimating the results as well as set of L^p of personal optimal along with environmental l^g is trained. Consequently, optimal solutions are back while executing condition occurs. However, 3 enhancements for canonical PSO should be treated as optimized CNN tuning function. Initial object is Fast Fitness Evaluation which aims in the result od ranking particles as well as corresponding local optimal applies lower computations for training CNN. Secondly, the revision of predefined equation suits particles' accelerations in different size of hyper-parameters' fields indicated as $\vec{B} = <b_1, b_2, \cdots, b_y, \cdots, b_{|B|}>$, in lower as well as upper bound of b_y is expressed by $\underline{b_y}$ and $\overline{b_y}$, hence it is an effective search process that is achieved. Finally, it selects β percent of units as worst particles as well as iterate the location with the application of Compound Normal Confidence (CNC), Distribution model which emerges in several searching process that has the capability of PSO in which features are improved in the following. There is feasible chance for improving personal searching potential that transforms to a process which is termed as local search techniques such as a quasi-Newton technique.

3 Performance Validation

The presented technique is validated in contrast to DRISHTI-GS database [12, 13] which comprises around 101 retinal fundus images for the purpose of segmenting OD region. These fundus images are gathered from hospital and laboratories. Hence, Glaucoma affected persons are identified by physicians with the help of realistic examination. The images are derived from the patients of 40–85 years old respectively. As provided in Table 1, the sample image is depicted in Fig. 1. Additionally, there are 101 images from dataset, collection of 31 images minimizes routine class and collection of 70 images resembles the Glaucoma class and single image displays below the source of data.

The result obtained from above experiments reveals that, around 69 images attained from result is classified optimally from category "Normal" also 30 images classified in "Glaucoma" class. For further sampling, the simulation outcome is executed as relative analysis with previous methods in similar dataset. This could be accomplished by

Table 1. Dataset information

Description	Dataset
Total no. of images	101
No. of normal class	31
No. of glaucoma class	70
Data sources	[14]

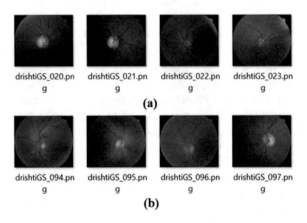

drishtiGS_020.png   drishtiGS_021.png   drishtiGS_022.png   drishtiGS_023.png

(a)

drishtiGS_094.png   drishtiGS_095.png   drishtiGS_096.png   drishtiGS_097.png

(b)

Fig. 1. Sample test images [7]

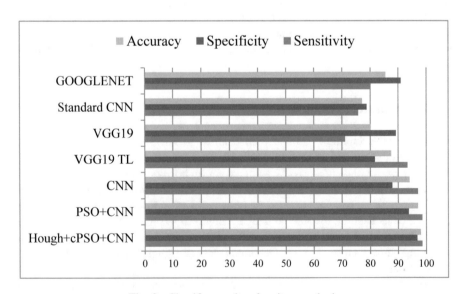

Fig. 2. Classifier results of various methods

comparing the value which is offered in Fig. 2. The values of Hough-cPSO-CNN illustrates greater classifying process with the accuracy of 98.01, sensitivity of 98.57 and specificity of 96.77. But, Standard CNN approach implies lower classification with sensitivity of 75.76, specificity of 78.72 and accuracy of 77.14. These values are deployed for higher values which is estimated for alternate models.

The above value states that presented segmented value is based on classifying Glaucoma models which is higher than other calculated methods.

4 Conclusion

Glaucoma is addressed as a secondary reason of blindness. Many researchers have been performed related to identification of OD. The proposed model involves in feature extracting process with the help pg median filter, segmentation with morphological function as well as classification by applying CNN. Optimal parameter settings of CNN are turned to automated PSO technique. It focuses on the method of analyzing OD part with the assistance of red channel of fundus image. This is because of poor contrast in blue and green channel would be helpful in detecting blood vessels. This model is verified with the help of DRISHTI-GS database. Additionally, simulation outcome noted that proposed cPSO-CNN model express outperforming results by obtaining maximum accuracy 98.01% in segmentation of OD.

References

1. Oktoeberza, K.W., Nugroho, H.A., Adji, T.B.: Optic disc segmentation based on red channel retinal fundus images. In: International Conference on Soft Computing, Intelligence Systems, and Information Technology, pp. 348–359. Springer, Heidelberg, March 2015
2. Yin, F., Liu, J., Ong, S.H., Sun, Y., Wong, D.W., Tan, N.M., Cheung, C., Baskaran, M., Aung, T., Wong, T.Y.: Model-based optic nerve head segmentation on retinal fundus images. In: 2011 Annual International Conference of the IEEE Engineering in Medicine and Biology Society, pp. 2626–2629. IEEE, August 2011
3. Pachiyappan, A., Das, U., Murthy, T., Tatavarti, R.: Automated diagnosis of diabetic retinopathy and glaucoma using fundus and images. LipidsHealth Dis. 11(1), 1–10 (2012)
4. Kavitha, K., Malathi, M.: Optic disc and optic cup segmentation for glaucoma classification. Int. J. Adv. Res. Comput. Sci. Technol. 2(1), 87–90 (2014)
5. Ponnaiah, G.F.M., Baboo, C.D.S.S.: Automatic optic disc detection and removal of false exudates for improving retinopathy classification accuracy. Int. J. Sci. Res. Publ. 5(3), 1–7 (2013)
6. Rashid, S.: Shagufta: computerized exudate detection in fundus images using statistical feature based fuzzy C-mean clustering. Int. J. Comput. Digit. Syst. 2(3), 135–145 (2013)
7. Aquino, A., Gegúndez-Arias, M.E., Marín, D.: Detecting the optic disc boundary in digital fundus images using morphological, edge detection, and feature extraction techniques (2010)
8. Godse, D.A., Bormane, D.S.: Automated localization of optic disc in retinal images. Int. J. Adv. Comput. Sci. Appl. 4, 65–71 (2013)
9. Eadgahi, M.G.F., Pourreza, H.: Localization of hard exudates in retinal fundus image by mathematical morphology operations, pp. 185–189 (2012)

10. Abbadi, N.K.E., Saadi, E.H.A.: Automatic detection of exudates in retinal images. Int. J. Comput. Sci. Issues **10**(2), 237–242 (2013)
11. Reza, A.W., Eswaran, C., Dimyati, K.: Diagnosis of diabetic retinopathy: automatic extraction of optic disc and exudates from retinal images using marker-controlled watershed transformation. J. Med. Syst. **35**, 1491–1501 (2010)
12. Raj, J.S., Vijitha Ananthi, J.: Recurrent neural networks and nonlinear prediction in support vector machines. J. Soft Comput. Paradigm (JSCP) **1**(01), 33–40 (2019)
13. Wang, Y., Zhang, H., Zhang, G.: cPSO-CNN: an efficient PSO-based algorithm for fine-tuning hyper-parameters of convolutional neural networks. Swarm Evol. Comput. **49**, 114–123 (2019)
14. http://cvit.iiit.ac.in/projects/mip/drishti-gs/mip-dataset2/Home.php

Dominant Feature Descriptors with Self Organising Map for Image Retrieval

S. Sivakumar[1($\boxtimes$)] and S. Sathiamoorthy[2]

[1] Department of Computer and Information Science, Annamalai University,
Annamalai Nagar, Chidambaram, India
Sivassa77@gmail.com
[2] Tamil Virtual Academy, Chennai, India
ks_sathia@yahoo.com

Abstract. In this research work, we have presented an efficient system for image retrieval using the distribution of dominant rotation local binary pattern and dominant edgels. The proposed combination is robust to rotation invariant and moderately robust to the illumination invariant. We performed classification using self-organizing map (SOM), which performs well by identifying the intrinsic features in the underlying data. We adopted Canberra distance to measure the likeness of images. The experiments on benchmark dataset depict that proposed feature descriptors work well with SOM and Canberra distance and provides higher precision against rotation and lightning variations.

Keywords: Local binary pattern · Dominant local binary pattern · Dominant edgels · Self organizing map · Canberra measure

1 Introduction

Considerable fall in the cost of image acquisition and storage devices increases numbers and volume of image databases rapidly and this detonation of image data in all domains has encouraged researchers to give attention on developing promising techniques for image retrieval with superior precision, recall and less time span. A typical image retrieval system (IRS) based on content make use of color or shape or texture information or mixture of these information for image matching. Since texture feature typify surface of the image, the role of texture feature is imperative in IRS. Texture features like local binary patterns (LBP) [1] and its extended variants [2–9] like pyramid LBP, robust LBP, center symmetric (CS)-LBP, spatial extended CS-LBP, local derivative pattern (LDP), local ternary pattern (LTP), CS-LTP, local directional derivative pattern (LDDP), local binary covariance matrix, co-occurrences of LBP, CS-LBP and LTP, dominant rotated LBP (DRLBP) and color enhanced LBPs got more attention owing to low computation cost and high discerning capability which directs the researchers to employ it in many booming problems including face recognition [10], object recognition [8], analysis of facial expression [11, 12], texture analysis [13], medical image analysis [14], etc. The LBP family exploits details based on edge distribution in an image and is perfectly appropriate and most extensively utilized for IRS because they provide consistency in speed and accuracy. Among the various

© Springer Nature Switzerland AG 2020
A. P. Pandian et al. (Eds.): ICCBI 2019, LNDECT 49, pp. 531–538, 2020.
https://doi.org/10.1007/978-3-030-43192-1_60

variants of LBP, DRLBP [6] is more robust to illumination and rotation variation with the speed and accuracy of LBP and thus we incorporated it in the proposed IRS for texture and histological images.

Since shape is also prominent information for image retrieval, numerous researches are done on shape information which fetched out number of shape features [15–23]. Among various shape features, edge histogram descriptor and its variants, edge orientation autocorrelogram and its variants, chordiogram image descriptor and its variants are studied more extensively because of their significant discriminant ability. However, they are not robust against rotation invariant. Though shape features like area, boundary moments, convexity, circularity, eccentricity are more robust to rotation variation, their discriminant ability is not significantly adequate for IRS and thus we incorporated the distribution of dominant edgels (DEs) in the proposed IRS for texture and histological images and Sobel approach is used for discovering the edgels. Recently Convolutional neural network (CNN) attracts more researchers [24]. Besides CNN, a lot of research has already been done to retrieve images using simple neural network techniques [18, 20, 25–28], among them self-organising maps (SOM) [27, 28] is extensively studied due to its higher classification accuracy. Therefore, SOM is adapted in the classification phase of proposed IRS. Regardless of considerable advancements, challenges continuous in indexing and retrieval due to the day to day growth of image databases. Because when the databases become massive and diverse, the performance of IRS tends to go down. As we focused our attention on IRS for texture and histological images and the aim of this study is increasing the performance of existing IRS for texture and histological images with heftiness to rotation and illumination invariant, we proposed mixture of SOM on DRLBP and DEs techniques which makes better distinction among the images with diverse visual content. The SOM reduces the time cost by reducing the search space and also reduces the retrieval of irrelevant images by performing effective categorization of database images.

2   Proposed Feature Extraction Method

2.1   Dominant Rotated LBP

In [6], Mehta and Egiazarian delineated DRLBP as

$$DRLBP_{N,D} = \sum_{p=1}^{N} T(I_p - I_c) \times 2^{\mathrm{mod}(p-D,p)} \tag{1}$$

Where $2^{\mathrm{mod}(p-D,p)}$ and D signifies weight and dominant direction. The weight is depends on D. The maximum difference among center and neighbor pixels is marked as index for dominant direction [18] and is

$$D = \arg \max_{p \in (0,1,,.,,.,p-1)} \left| (I_p - I_c) \right| \tag{2}$$

The DRLBP is calculated by assigning weights from the index of dominant direction in circular mode and the run of weights remains as in LBP. In DRLBP, fixed arrangement is not followed in assigning weights and it varies depends on maximum variation among center and its neighbors [18].

2.2 Histogram of Dominant Edgels

Since Sobel operator [15] is used extensively for edge identification, we incorporated it in the proposed IRS. Later, to decrease the computation cost and to attain robustness against illumination and view point, we computed a subset of dominant boundary pixels i.e. Dominant edgels (DEs) which are higher in gradient magnitudes and robust to illumination and rotation invariant because DEs remains dominant while lighting or view position vary. Thus, we define the number of DEs in a subset (N_{DEs}) as

$$N_{DEs} = p \cdot N_{Es} \tag{3}$$

Where p exemplifies the proportion of DEs and N_{Es} is the number of edgels identified by Sobel operator. In the proposed IRS, p determines the size of DEs subset and we have chosen varies choices for p such as 5%, 10%, 15%, 20%, 25% and 30%, and we experimentally discovered that $p = 15\%$ gives better results against illumination variation and also decreases time cost enormously. In the experiments, when we increase p from 0% to 15%, precision increases gradually. Thereafter, precision is stable. Therefore, we set p = 15% in the proposed IRS.

2.3 Self Organizing Map

Over the decade, researchers employed artificial neural networks (ANNs) for classification owing to its precision [18, 20, 25]. Although many ANNs are available, SOM [27, 28] is incorporated in many applications because of its ability to organizing large and complex data sets and it preserves topological relations in the data which exploits underlying structure of data and is an intrinsic feature to the problem. Let I be input vectors and number of input vectors in I are $\{I_1, I_2, ., ., I_N\}$. Initially weights (W_{ji}) are assigned randomly with small value (in the range 0 to 1) for each connection between input nodes i and neurons j where $j = 1, 2, ., ., ., M$ and M is number of neurons and usually it is less than or equal to half of number of classes in the datasets. The SOM algorithm is as follows

For each input vectors in training dataset

1. Input vector is selected randomly from the training data.
2. Weights are adjusted using competitive learning when training progress and the winning neuron (the neurons whose weight vector become most similar to input vector) for each input vector is computed using the discriminant function like Euclidean distance and is as follow for neuron j and input pattern I

$$D_j(I) = \sqrt{\sum_{i=1}^{N} (I_i(t) - W_{ji}(t))^2} \qquad (4)$$

where $I_i(t)$ - i^{th} input of I at time t and $W_{ji}(t)$ - weight vector of jth neuron and i^{th} input vector at time t. The winning neuron is coined as best matching unit and is computed as follows

$$WN = \arg\min\{D_j(I), j = 1, 2, ., ., ., N\} \qquad (5)$$

3. The neighborhood of winning neuron is decided with the help of concentric squares or hexagons or any other polygon shapes or functions such as Maxican Hat, Gaussian, etc., Generally, the neighborhood function offers cooperation process among the winner and its neighbor and it is designed such that the weights of winning one has global maxima and the neurons close to winning one are scaled towards winning one and the neurons far away from winning one are scaled towards least i.e. more closer neighbors are excited more than the neighbors at furthest distance. In SOM, this weight adjustment is coined as adaption process. Initially, the size i.e. radius of neighborhood function is large and start to decrease when training progress and thus neighbors which are far away from the winning neuron are eliminated. The weight adjustment equation in SOM is as follows

$$W_{ji}(t+1) = W_{ji}(t) + \eta(t) \cdot h_{ji}(t) \cdot (I_i - W_{ji}(t)) \qquad (6)$$

where $\eta(t)$ is learning rate which explains how much weights need to be adjusted and is given as

$$\eta(t) = \eta_0\left(1 - \frac{t}{T}\right) \qquad (7)$$

where η_0 symbolizes the initial value of η and usually $\eta_0 = 1$, and T symbolizes maximum time of iteration. Learning rate will converge to 0 after T. In the proposed work, we incorporated Gaussian as neighborhood function. As aforementioned, during the learning process, the neighborhood function measures the degree of cooperation with the winner. The Gaussian neighborhood function is given as follows

$$h_{ji}(t) = \exp\left[\frac{D_j(I)^2}{2\sigma^2(t)}\right] \qquad (8)$$

where t is time or epoch, $D_j(I)$ symbolizes the Euclidean distance of neuron j and input pattern I and σ is width of Gaussian function. To get the best learning rate, σ is decreased gradually when the learning progress. In Eq. (9), $\sigma(t)$ is described as

$$\sigma(t) = \sigma_0 \exp\left(-\frac{t}{\tau}\right) \qquad (9)$$

Where σ_0 is the initial value of σ and τ is time constant.

4. End For

Through steps 1 to 4, a feature map in SOM is created.

2.4   Measure of Likeness

The likeness among images of datasets and input image is evaluated with Canberra distance [29] and is delineated as

$$D_{Canbera}(I, DB) = \sum_{i=1}^{N} \left| \frac{DB_i - I_i}{DB_i + I_i} \right| \tag{10}$$

Where I symbolize feature vectors of input, DB depicts the database images and N signifies number of features in feature vector.

3   Experimental Results

To compare the proposed IRS with the existing IRS for texture and histological images respectively, we utilized the datasets outex 10 and 12, KTH-TIPS and KTH-TIPS-ROT and histological as those used in [6, 30]. First, we computed DRLB [6] and dominant edgels for all the images in the experimental. Secondly, we computed multi-trend structure descriptor (MTSD) at micro-level [30] then DRLBP alone for all the images in the datasets. We aimed at IRS with rotation invariant and illumination invariant. In [30], MTSD with fuzzy support vector machine is used effectively for histological image retrieval and in [6], DRLBP with k-nearest neighbor (k-NN) is effectively tested on texture datasets. In our earlier work [31], we used fuzzy k-NN on DRLBP with Canberra distance measure for texture and histological images and it outperforms the existing IRSs [6, 26]. Therefore, IRSs reported in [6] and [31] are used to compare the proposed work. Generally, the histological images are affected by lightning variations and the view point may depend on the user acquiring those images, and thus, in the proposed work, we strengthen DRLBP by integrating dominant edgels. Later, integration of DRLBP and DEs are categorized with SOM which is more effective because it acquires the intrinsic patterns from the underlying data structure and self organizing approach makes it more intellectual. Later on, Canberra measure is employed for performing matching operation in the categorized feature vector dataset. Experimental results depicts that though MTSD is comprising of local level structure of color, edge and texture patterns, it produces less accuracy even when the images are affected by small lightning and view point variations. Even though DRLBP produces better results against view point variations, its performance against various lightning conditions is not satisfactory. Thus, we employed DEs in the proposed IRS. The combination of DRLBP and DEs significantly outperforms both [6] and [31] against view point and lightning variations. Further, we also compared the performance of SOM with fuzzy SVM and fuzzy k-NN on DRLBP with DEs for the experimental datasets. The outcome clearly depicts that SOM on DRLBP with DEs attaining better output for the experimental dataset. The average retrieval accuracy [20] of proposed and existing feature descriptors for texture and histological images are illustrated in Table 1. The average retrieval accuracy [20] attained by SOM, fuzzy k-NN and fuzzy SVM for DRLBP with DEs is reported in Table 2.

Table 1. Average retrieval accuracy of proposed and existing feature descriptors for texture and histological datasets

Datasets	MTSD [30]	DRLBP [6]	DRLBP+DEs
Outex 10	58.76	63.81	66.90
Outex 12	56.59	61.34	65.21
KTH-TIPS	54.22	58.22	61.28
KTH-TIPS-ROT	51.34	58.01	60.98
Histological DB	50.97	51.27	53.23

Table 2. Average retrieval accuracy for DRLBP and DEs with various classifiers for texture and histological datasets

Datasets	Fuzzy k-NN [25]	Fuzzy SVM [30]	SOM [27, 28]
Outex 10	68.23	69.41	69.99
Outex 12	67.81	69.02	69.84
KTH-TIPS	63.89	65.22	65.78
KTH-TIPS-ROT	64.02	66.11	66.90
Histological DB	55.93	57.12	57.89

4   Conclusion

This work centers on the problem of retrieval with rotation and illumination invariant. In order to attain it, we computed dominant rotation LBP (DRLBP) and dominant edgels as features. The DRLBP is highly robustness with rotation variation and moderately robustness with illumination variation. The edgels are identified using Sobel operator then experimentally we identified 15% of edgels as dominant and are having higher gradient values even for the images affected by lightning and view point variations. The combination of DRLBP and dominant edgels (DEs) provides better retrieval results then the existing systems for texture and histological images. By decreasing the search space in feature space through effective filtrations of irrelevant images using SOM, the computation cost is reduced significantly as well accuracy of proposed IRS is increasing significantly. We also compared the performance of DRLBP with DEs on SOM, fuzzy k-NN with Canberra and fuzzy SVM. The outcome evident that SOM is outperforming with DRLB and DEs for texture and histological datasets. In future, the accuracy and robustness against illumination can be improved by changing the underlying mechanism of DRLBP.

References

1. Ojala, T., Pietikainen, M., Harwood, D.: A comparative study of texture measures with classification based on feature distributions. Pattern Recogn. **29**(1), 51–59 (1996)
2. Heikkilä, M., Pietikäinen, M., Schmid, C.: Description of interest regions with center-symmetric local binary patterns. In: Kalra, P.K., Peleg, S. (eds.) Computer Vision, Graphics and Image Processing. Lecture Notes in Computer Science, vol. 4338. Springer, Heidelberg (2006)
3. Xue, G., Sun, J., Song, L.: Dynamic background subtraction based on spatial extended center-symmetric local binary pattern. In: IEEE International Conference on Multimedia and Expo, pp. 1050–1054 (2010)
4. Qian, X., et al.: PLBP: an effective local binary patterns texture descriptor with pyramid representation. Pattern Recogn. **44**(10–11), 2502–2515 (2011)
5. Liao, S., Law, W., Chung, A.C.: Dominant local binary patterns for texture classification. IEEE Trans. Image Process. **18**, 1107–1118 (2009)
6. Mehta, R., Egiazarian, K.: Dominant rotated local binary patterns (DRLBP) for texture classification. https://doi.org/10.1016/j.patrec.2015.11.019
7. Verma, M., Raman, B.: Center symmetric local binary co-occurrence pattern for texture, face and bio-medical image retrieval. J. Vis. Commun. Image Represent. **32**, 224–236 (2015)
8. Trefny, J., Matas, J.: Extended set of local binary patterns for rapid object detection. In: Proceedings of the Computer Vision Winter Workshop, Czech Republic (2010)
9. Deep, G., Kaur, L., Gupta, S.: Directional local ternary quantized extrema pattern: a new descriptor for biomedical image indexing and retrieval. Eng. Sci. Technol. Int. J. **19**(4), 1895–1909 (2016)
10. Zhang, W., et al.: Local gabor binary pattern histogram sequence (LGBPHS): a novel non-statistical model for face representation and recognition. In: Proceedings of International Conference on Computer Vision, pp. 786–791 (2005)
11. Zhao, G., Pietikinen, M.: Dynamic texture recognition using local binary patterns with an application to facial expressions. IEEE Trans. Pattern Anal. Mach. Intell. **27**(6), 915–928 (2007)
12. Huang, D., Shan, C., Ardabilian, M., Wang, Y., Chen, L.: Local binary patterns and its application to facial image analysis: a survey. IEEE Trans. Syst. Man Cybern. C Appl. Rev. **41**(6), 765–781 (2011)
13. Ojala, T., Pietikainen, M., Maenpaa, T.: Multiresolution gray-scale and rotation invariant texture classification with local binary patterns. IEEE Trans. Pattern Anal. Mach. Intell. **24**(7), 971–987 (2002)
14. Adriana, F., et al.: Analysis of wireless capsule endoscopy images using local binary patterns. Appl. Med. Inform. **36**(2), 31–42 (2015)
15. Mahmoudi, F., Shanbehzadeh, J., Eftekhari, A.M., Soltanian-Zadeh, H.: Image retrieval based on shape similarity by edge orientation autocorrelogram. Pattern Recogn. **36**, 1725–1736 (2003)
16. Toshev, A., Taskar, B., Daniilidis, K.: Shape-based object detection via boundary structure segmentation. Int. J. Comput. Vis. **99**(2), 123–146 (2012)
17. Seetharaman, K., Sathiamoorthy, S.: An improved edge direction histogram and edge orientation auto corrlogram for an efficient color image retrieval. In: 2013 International Conference on Advanced Computing and Communication Systems, Coimbatore, pp. 1–4 (2013). https://doi.org/10.1109/ICACCS.2013.6938725
18. Seetharaman, K., Sathiamoorthy, S.: Color image retrieval using statistical model and radial basis function neural network. Egypt. Inform. J. **15**, 59–68 (2014)

19. Sathiamoorthy, S., Kamarasan, M.: A novel approach for image retrieval using BDIP and BVLC. Int. J. Innovative Res. Comput. Commun. Eng. 2(9), 5897–5902 (2014)
20. Seetharaman, K., Sathiamoorthy, S.: A unified learning framework for content based medical image retrieval using a statistical model. J. King Saud Univ. – Comput. Inf. Sci. 28(1), 110–124 (2016)
21. Wang, X., Zhang, H., Peng, G.: A chordiogram image descriptor using local edgels. J. Vis. Commun. Image R 49, 129–140 (2017)
22. Saravanan, A., Sathiamoorthy, S.: Image retrieval using autocorrelation based chordiogram image descriptor and support vector machine. Int. J. Recent Technol. Eng. (IJRTE) 8(3), 6019–6023 (2019). ISSN 2277-3878
23. Saravanan, A., Sathiamoorthy, S.: Integration of statistical based texture and color feature for medical image retrieval. Int. J. Recent Technol. Eng. (IJRTE) 8(3), 5584–5588 (2019)
24. Aloysius, N., Geetha, M.: A review on deep convolutional neural networks, pp. 0588–0592. https://doi.org/10.1109/ICCSP.2017.8286426
25. Nikoo, M.R., Kerachian, R., Alizadeh, M.R.: A fuzzy KNN-based model for significant wave height prediction in large lakes. Oceanologia 60(2), 153–168 (2018)
26. Natarajan, M., Sathiamoorthy, S.: Heterogeneous medical image retrieval using multi-trend structure descriptor and fuzzy SVM classifier. Int. J. Recent Technol. Eng. 8(3), 3958–3963 (2019)
27. Zhang, K., Chai, Y., Yang, S.X.: Self-organizing feature map for cluster analysis in multi-disease diagnosis. Expert Sys. Appl. 37, 6359–6367 (2010)
28. Mishra, S., Panda, M.: Medical image retrieval using self-organising map on texture features. Future Comput. Inform. J. 3, 359–370 (2018)
29. Malik, F., Baharudin, B.: Analysis of distance metrics in content-based image retrieval using statistical quantized histogram texture features in the DCT domain. J. King Saud Univ.-Comput. Inf. Sci. 25(4), 207–218 (2013)
30. Natarajan, M., Sathiamoorthy, S.: Multi-trend structure descriptor at micro-level for histological image retrieval. Int. J. Recent Technol. Eng. (IJRTE) 8(3), 7539–7543 (2019). ISSN 2277-3878
31. Sivakumar, S., Sathiamoorthy, S.: Image retrieval using fuzzy k-NN on dominant rotated local binary patterns. In: International Conference on Smart Systems and Inventive Technology, 27–29 November 2019. Francis Xavier Engineering College, Tirunelveli (Accepted)

Medical Image Retrieval Using Efficient Texture and Color Patterns with Neural Network Classifier

C. Ashok Kumar[1]([⊠]) and S. Sathiamoorthy[2]

[1] Department of Computer and Information Science, Annamalai University,
Annamalai Nagar, Chidambaram, India
cashok1976@gmail.com
[2] Tamil Virtual Academy, Chennai 600025, India
ks_sathia@yahoo.com

Abstract. Presently, the usage of medicinal images has drastically increased and it provides extensive details related to the patient's health status. It shows the applicability of diagnosing the disease and stored in a memory for examination purposes. For the retrieval of medical images in a real world environment, significant need is present in the designing of an effective medical image indexing and retrieval technique. This paper offers an efficient medical image retrieval (MIR) model through feature extraction based classification model. Here, Directional local ternary quantized extrema patterns (DLTerQEPs) and autocorrelogram (AC) based feature extraction process takes place to extract texture and color features. Next, neural network (NN) based classification process takes place. The investigation of the simulation results takes place to showcase the betterment of the presented mode. During experimentation process, it is noticed that the presented model is superior to compared methods.

Keywords: Classifier · Image retrieval · Medical images · Deep learning

1 Introduction

Currently, medical image retrieval (MIR) developed as a significant study part in the area of medicinal image processing. It results due to the exponential development of various medical devices namely ultrasound (US), computed tomography (CT), etc. provides the suitable analysis of diseases. Accordingly, it is required for organizing the data associated to the medicinal image in a controlled way to exchange capably, finding and recovering medicinal images for medicine persistence. To develop a novel MIR technique, feature extraction is carried out via extracting the required features like texture, color, and shape. Because of the statistic that they are the significant features of the medicinal images which required to be enhanced in a significant manner. It is identified that a comprehensive study of MIR is exists by the consideration of the texture-based features as a main possibility [1].

By the similar way, if the difficulty of the medical images becomes enlarged, it is probable to identify extra particulars about the person wellbeing. Consequently, the

© Springer Nature Switzerland AG 2020
A. P. Pandian et al. (Eds.): ICCBI 2019, LNDECT 49, pp. 539–546, 2020.
https://doi.org/10.1007/978-3-030-43192-1_61

practice of texture features is insufficient to construct a complex MIR scheme. There fore, the main necessity exists in the growth of a MIR structure includes the multidimensional particulars such as edge, shape, texture, etc., to exactly and quickly recovers the medicinal images. Several earlier works have been showed on various feature descriptors. Between the several methods, local binary pattern (LBP) and its prolonged types provides extreme recovery outcome. LBP is a local feature descriptor is employed for extracting the texture particulars about a picture. Local feature descriptor gathered the details with respect to the relativeness among the reference and the neighboring pixels. LBP and its prolonged types are derived under the opinion of filtering the texture and edge particulars in dissimilar directions. Furthermore, it also combined the two features in a cooperative manner and has the capability to deliver discriminative and local feature descriptor for retrieving images.

In [2], LBP accompanied by intensity histograms are employed to identifying the areas of pulmonary emphysema. [3] presented a technique to take out the LBP Histogram Fourier features. Hereafter, the near distance pictures based on the query image undergoes retrieval process and thereby the outcome of the model is improved. [4] made a study for evaluating the optimum LBP types for the common description of feature vector about an image. The elected LBP version is applied to construct the ultrasound image dataset. [5] obtained an effective MIR technique, LBP with image Euclidean distance (IMED) treats the spatial relationship among the pixels, and shows robustness to minor perturbation of images. Local maximum edge binary patterns (LMEBP) [6] is presented which vary from the LBP is that it offers the details regarding the edge distribution in the images. An improvement over LBP is presented using local tetra patterns (LTrPs) [7]. LTrP will explore the relativity among the reference and nearby pixels through the consideration of the vertical and horizontal directions that are determined utilizing the first order derivatives. An effective feature descriptor called local ternary co-occurrence patterns (LTCoP) [8] is presented depending upon the co-occurrence of comparable ternary edges determined by the gray values of essential and neighbouring pixel values. A new concept is introduced to encode the relativity between the neighbouring values of the essential pixel by contrasting with LBP where the relativity among the fundamental pixels and the nearby values are applied for encoding the texture details. This concept is developed by local mesh patterns (LMeP) [9] that is again improved to LMePVEP, i.e., local mesh peak valley edge patterns [10, 17].

This study introduces a novel MIR model through the feature extraction based classification model. Here, Directional local ternary quantized extrema patterns (DLTerQEPs) and autocorrelogram (AC) based feature extraction process takes place to extract texture and color features. Next, neural network (NN) [18–20] based classification process takes place. The investigation of the simulation results takes place to showcase the betterment of the presented mode. During experimentation process, it is noticed that the presented model is superior to compared methods.

2   Proposed Work

The proposed model operates as follows. At the beginning, the feature extraction of the images takes place in an effective way by the use of AC and DLTerQEPs models. Next, NN model is presented for efficient MIR.

2.1   Feature Extraction

In this paper, a set of two features namely color and textures were extracted. The color features are extracted by the use of AC and the texture features are extracted utilizing DLTerQEP model.

2.1.1   Color Feature Extraction Using Auto Correlogram

For defining the color feature, hue saturation value (HSV) color model is applied due to the fact that it is extremely relevant to the human point of view. Hue is applied for color differentiation and also to calculate the reddish or greenish color of the light. Saturation represents the amount of white light which is included in true color. The value indicates the amount of received light intensity. The color quantization finds helpful to decrease the computation complexity. In addition, it offers effective outcome and removes the comprehensive color elements which can be treated as noise. The human visual system (HVS) is very sensitive to color compared to others and it needs to be quantized. For the representation of local color histogram, an image is partitioned to same sized 3×3 rectangular portions and filters the HSV joint histogram which undergoes quantization into 162 bins for every segment. Though it has local color details, the resultant representation is not adequately compressed. For obtaining compact demonstration, every joint histogram feature is extracted which has the highest peak. Consider hue h, saturation s, and value v linked to the bin as the dominant features in the rectangular portion and undergo normalization to be insider the identical range of [0, 1]. Therefore, every image holds a total of 27-dimensional color vectors. The intention of color feature extraction is to extract the images whose color compositions are alike to the color composition of the query image. Histograms finds helpful due to the fact that they are relatively not sensitive to location and orientation modifications and they are adequately precise. At the same time, it does not gather spatial relationship of color regions and therefore, they have restricted discriminating power. AC technique operates well over the classical color histograms. AC can be saved as a table indexed by pairs of colors (i, j) where d-th entry shows the possibility of determining a pixel j from pixel i at distance d whereas an auto-correlogram is saved as a table indexed by color i where d-th entry displays the possibility of determining a pixel i from the identical pixel at distance d. Therefore, the AC displays the spatial correlation among the equivalent colors. The color AC can be determined with a distance k for the low frequency image $y_{r,\theta}$ of every θ direction in a resolution r:

$$\alpha_c^{(k)}\left(y_{r,\theta}\right) = Pr\left[\hat{y}_{r,\theta}(\rho) = \hat{y}_{r,\theta}(\rho') = c\|p - p'\| = k \, for \, p, p' \in P\right] \qquad (1)$$

Where $c \in \{0, 1, 2, .., M-1\}$ represents the quantized color level and $\alpha_c^{(k)}\left(y_{r,\theta}\right)$ defines the possibility that the color of a pair of pixels of distance k are c on the quantized image $\hat{y}_{r,\theta}$. Then, the mean value for M level color auto correlograms of θ directions in a resolution r:

$$\mu(c) = \operatorname*{mean}_{\theta \in \Theta}\left[\alpha_c^{(k)}\left(y_{r,\theta}\right)\right] \tag{2}$$

2.1.2   Texture Feature Extraction Using DLTerQEP Model

Texture is a needed feature where most of the images include it. Related to color features, texture feature also filtered from localized image regions. The $(i, j)th$ component of the co-occurrence matrix indicates the computed likelihood that gray level i occurs with gray level j at a particular displacement d. and angle θ. Through the selection of the values d and θ, an individual co-occurrence matrix is attained. In every co-occurrence matrix, few texture features are filtered. At last, every image holds a total of $3 \times 3 (= 9)$ dimensional texture vectors. The texture features are extracted using the DLTerQEP. The idea of local patterns (the LBP, the LTP, and the local quantized extrema patterns (LQEP)) is applied to defining the DLTerQEP. It explains the spatial construction of the local textures present in the ternary patterns utilizing the local extrema as well as directional geometric structures. For a provided image in DLTerQEP, the local extrema in every direction is gathered through the determination of local variation among the centre pixels and its close pixel using the index of the patterns with pixel locations. The indexing of pixels takes place by writing a set of four directional edges operator calculation. The local directional extrema value (LDEV) for the local pattern neighborhood of the image is determined by

$$LDEV(q, r) = \sum_{q=1}^{k_1} \sum_{r=1}^{k_1}\left[I(q, r) - T\left(1 + floor\left(\frac{k_1}{2}\right), 1 + floor\left(\frac{k_1}{2}\right)\right)\right] \tag{3}$$

where $k_1 \times k_1$ indicates the image size. A set of four directional ternary extrema coding (DTEC) is gathered depending ahead the set of 4 directions namely 0, 45, 90 and 135 in diverse thresholds by the use of LTP concepts. The DTEC coding is transformed to a couple of binary codes. Through the multiplication of the binomial weights to every DTECLTP coding, the single DLTerQEPs values (decimal values) for a specific provided pattern (7×7) to characterize the spatial organization of the local pattern is represented by

$$DLTerQEP_{\alpha, P} = \sum_{w=0}^{P-1} DTEC_{\left(\frac{upper}{lower}\right), w} 2^w \tag{4}$$

In the entire image, every DTECLTP (upper and lower) map presented in the interval in range of 0 to 4095 (0 to $2^P - 1$), therefore, the entire DTECLTP map is constructed with the value presented in the range of 0 to 8191 (0 to $((2(2^P)) - 1)$. The DLTerQEPs is dissimilar from the popular LBP model. It will filter the spatial relationship among the set of neighbors present in the local area along with the pre-defined directions, whereas the LBP separated with the relationship among the inter-mediate and adjacent pixels. It takes out the directional border details depending upon the local extrema which differs from the available LBP. So, DLTerQEPs capture detailed spatial data over LBP. The features extracted by this model are explained below.

2.2   NN Model

NN model is applied for image classification and the multilayer perceptron (MLP) is trained with the back propagation technique. The features extracted from the previous process are given as input to the input node. The input to a node is the weighted sum of the outputs from the layer below, and is defined by

$$net_y = \sum_x w_{yx}o_x \tag{5}$$

The 'sum weight' is then altered with the node 'activation functions' (generally a sigmoid or hyperbolic tangent) for generating the node result:

$$o_y = \frac{1}{1 + \exp\left(-net_y + \theta_y\right)}[sigmoid] \tag{6}$$

$$o_y = mtanh\left(k\left(net_y\right)\right)[Hyperbolictangent] \tag{7}$$

where θ_y, m, and k is constants. Weight is informed through to train with the gener-alized delta principle:

$$\Delta w_{yx}(n+1) = \eta\left(\delta_y o_x\right) + \alpha\Delta w_{yx}(n) \tag{8}$$

where $\Delta w_{yx}(n+1)$ is the modification of a weight linking nodes x and y, in two consecutive layers, at the $(n+1)$th iteration, δ_y indicates the rate of modification in error based on the result from node y, η are the rate of learning, then α a momentum term. Even though several users of NN considers the train parameters and activation functions as 'givens', it is essential to recognize the values and structure of these parameters and functions correspondingly has essential impact together. Here, the input information should be pre-processing for the constancy and effectiveness to train the networks.

3   Performance Validation

To ensure the betterment of the applied model, a validation takes place on the benchmark OASIS MRI dataset. The dataset comprises a collection of 416 images with the pixel dimension of 208 * 208 pixels. In addition, a set of 4 classes exist in the dataset [17]. The measures used to analyze the performance are average precision rate

(APR) and average recall rate (ARR). For comparison purposes, SS-3D-LTP, orthog onal Fourier-Mellin moments (OFMM), LBDISP, LTCoP and Local Quantized Pat- terns (LQP) methods [18] are used. Table 1 validates the results attained by presented method along with the existing models. The table expresses that the DLTerQEPs+AC +NN model attains the lower ARR of 29.43%. Next to that, the SS-3D-LTP model finds slight better classification over the previous model and obtained an ARR of 31.84%. Subsequently, LQP model tries to handle classification process and it reaches to an ARR value of 32.69%. Correspondingly, the LTCOP model displays equivalent end result with the before used model by gaining an ARR of 33.30%. In the same way, LBDISP model indicates convenient results by attaining an ARR value of 37.89%. In addition, the OFMM model represents reasonable classifier outcome using the pro- jected technique by accomplishing a high ARR of 40.17%. At last, the presented and ensemble methods take maximum classifier outcome with the optimal ARR of 57.88%. This higher value of ARR clearly defines the superior performance of the presented model. It is revealed that the DLTerQEPs+AC+NN model reached a least APR of 31.15%. After that, the SS-3D-LTP model gets an insignificantly better classification compared with the previous method and obtained an APR of 33.84%.

Table 1. Comparative results in terms of ARR and APR

Methods	ARR (%)	APR (%)
Proposed (DLTerQEPs+AC+NN)	57.88	79.43
DLTerQEPs	29.43	31.15
OFMM	40.17	42.52
LBDISP	37.89	40.35
LTCOP	33.30	35.51
LQP	32.69	34.78
SS-3D-LTP	31.84	33.84

Simultaneously, LQP model tried to maintain classification process and it reaches to an APR value of 34.78%. Likewise, the LTCOP model displays comparative solution with the preceding technique by getting an APR of 35.51%. Just as, LBDISP model depicts considerable solution by attaining an APR value of 40.35%. Next, the OFMM model shows holds a challenging classifier output with the presented model by attaining a high APR of 42.52%. However, the presented model gets maximum clas- sifier outcome with the highest APR of 79.43%. This greater value of APR describes the superiority of the proposed system.

4   Conclusion

This paper has presented a MIR model using feature extraction based classification model. Here, AC and DLTerQEPs based feature extraction process takes place to extract color and texture features. Next, NN based classification process takes place.

The investigation of the simulation results takes place to showcase the betterment of the presented mode. During experimentation process, it is noticed that the presented model is superior to compared methods. To ensure the betterment of the applied model, a validation takes place on the benchmark OASIS MRI dataset. The presented model attains maximum classification outcome with the maximum APR and ARR values of 79.43 and 57.88 respectively.

References

1. Seetharaman, K., Sathiamoorthy, S.: A unified learning framework for content based medical image retrieval using a statistical model. J. King Saud Univ. Comput. Inf. Sci. **28**(1), 110–124 (2016)
2. Sørensen, L., Shaker, S.B., de Bruijne, M.: Quantitative analysis of pulmonary emphysema using local binary patterns. IEEE Trans. Med. Imaging **29**, 559–569 (2010)
3. Bharathi, P., Reddy, K.R., Srilakshmi, G.: Medical image retrieval based on LBP histogram fourier features and KNN classifier. In: 2014 International Conference on Advances in Engineering & Technology Research (ICAETR-2014), pp. 1–4. IEEE, August 2014
4. Vatamanu, O.A., Frandes, M., Lungeanu, D., Mihalas, G.I.: Content based image retrieval using local binary pattern operator and data mining techniques. In: MIE, pp. 75–79, May 2015
5. Xu, X., Zhang, Q.: Medical image retrieval using local binary patterns with image Euclidean distance. In: 2009 International Conference on Information Engineering and Computer Science, pp. 1–4. IEEE, December 2009
6. Murala, S., Maheshwari, R.P., Raman, B.: Local maximum edge binary patterns: a new descriptor for image retrieval and object tracking. Sig. Process. **92**, 1467–1479 (2012)
7. Murala, S., Maheshwari, R.P., Raman, B.: Local tetra patterns: a new feature descriptor for content-based image retrieval. IEEE Trans. Image Process. **21**, 2874–2886 (2012)
8. Murala, S., Wu, Q.M.J.: Local ternary co-occurrence patterns: a new feature descriptor for MRI and CT image retrieval. Neurocomputing **119**, 399–412 (2013)
9. Murala, S., Wu, Q.M.J.: Local mesh patterns versus local binary patterns: biomedical image indexing and retrieval. IEEE J. Biomed. Health Inform. **18**, 929–938 (2014)
10. Murala, S., Wu, Q.M.J.: MRI and CT image indexing and retrieval using local mesh peak valley edge patterns. Sig. Process. Image Commun. **29**, 400–409 (2014)
11. Murala, S., Wu, Q.M.J.: Spherical symmetric 3D local ternary patterns for natural, texture and biomedical image indexing and retrieval. Neurocomputing **149**, 1502–1514 (2015)
12. Banerjee, P., Bhunia, A.K., Bhattacharyya, A., Roy, P.P., Murala, S.: Local neighborhood intensity pattern: a new texture feature descriptor for image retrieval, arXiv preprint arXiv: 1709.02463 (2017)
13. Dubey, S.R., Singh, S.K., Singh, R.K.: Local diagonal extrema pattern: a new and efficient feature descriptor for CT image retrieval. IEEE Sig. Process. Lett. **22**, 1215–1219 (2015)
14. Dubey, S.R., Singh, S.K., Singh, R.K.: Local wavelet pattern: a new feature descriptor for image retrieval in medical CT databases. IEEE Trans. Image Process. **24**, 5892–5903 (2015)
15. Dubey, S.R., Singh, S.K., Singh, R.K.: Novel local bit-plane dissimilarity pattern for computed tomography image retrieval. IET Electron. Lett. **52**, 1290–1292 (2016)
16. Dubey, S.R., Singh, S.K., Singh, R.K.: Local bit-plane decoded pattern: a novel feature descriptor for biomedical image retrieval. IEEE J. Biomed. Health Inform. **20**, 1139–1147 (2016)

17. Koresh, M.H., Deva, J.: Computer vision based traffic sign sensing for smart transport. J. Innovative Image Process. (JIIP) **1**(01), 11–19 (2019)
18. http://www.osirix-viewer.com/datasets/DATA/KNEE.zip
19. Aggarwal, A., Sharma, S., Singh, K., Singh, H., Kumar, S.: A new approach for effective retrieval and indexing of medical images. Biomed. Sig. Process. Control **50**, 10–34 (2019)
20. Raj, J.S., Vijitha Ananthi, J.: Recurrent neural networks and nonlinear prediction in support vector machines. J. Soft Comput. Parad. (JSCP) **1**(01), 33–40 (2019)

Deterministic Type 2 Fuzzy Logic Based Unequal Clustering Technique for Wireless Sensor Networks

R. Sathiya Priya$^{(\boxtimes)}$ and K. Arutchelvan

Department of Computer and Information Sciences, Annamalai University,
Annamalai Nagar, Chidambaram, India
spmraj0607@gmail.com, karutchelvan@yahoo.com

Abstract. In recent days, wireless sensor systems (WSNs) offered a critical segment of reconnaissance advancements. Energy productivity in WSN assumes a vital job in the working of WSN. Clustering in WSN offers a huge energy effective procedure in WSN. Be that as it may, it prompts the issue of hot spot issue. To address this issue, the unequal clustering process in WSN is proposed. In this paper, a deterministic type 2 fuzzy-based unequal clustering (DTFUC) model in WSN is introduced. A backoff timing mechanism is used for primary cluster head (CH) selection and type 2 fuzzy logic mechanism is applied for secondary CH selection. The introduced DTFUC model successfully chooses the CHs in a productive manner. The displayed calculation experiences various situations under diverse measures.

Keywords: WSN · Clustering · Fuzzy logic · Hot spot

1 Introduction

As of late, the improvement of transmission and detecting strategies offers likely to build inexpensive sensors nodes for detecting and transmitting the information from the environment. Wireless Sensor Networks (WSNs) includes various sensor nodes which experience organization in different applications like medication, keen industry, observation, accuracy farming, and so forth. Moreover, WSN offered a basic part in the rise of advancements, e.g., huge information, cloud, and the Internet of Things (IoT). Then again, a few plan constraints as a result of the high handling and energy confinements. Therefore, energy-productive clustering systems have been developed. Typically, WSN contains 100s to 1000s of sensor nodes and base station (BS). The nodes in WSN transmit information to BS autonomously or might make assorted clusters with Cluster Heads (CHs) [1]. The information transmission between the nodes and BS happens in an immediate and roundabout way.

The issue of single jump information transmission is the inordinate energy usage in view of the monstrous transmission run. Simultaneously, the nodes which are available closer to the BS terminate quickly in multi-bounce information transmission as they transmit each parcel of the system to BS. It is called as hot spot issue and assorted

© Springer Nature Switzerland AG 2020
A. P. Pandian et al. (Eds.): ICCBI 2019, LNDECT 49, pp. 547–554, 2020.
https://doi.org/10.1007/978-3-030-43192-1_62

procedures has been exhibited for moderating it by the production of unequal clusters as for size [2].

Clustering just as unequal clustering models has been utilized to accomplish energy productivity in WSN. The clustering procedure in WSN will happens by consolidating the sensor nodes to clusters set. The clustering procedure empowers the CHs to assemble the information from the cluster members (CM). The use of energy by the CHs closer to the BS is higher than the CHs a long way from the BS. It infers that the CHs closer to BS is a long way from the CHs found away from the BS because of the nearness of intra-cluster information transmission from its CM, information total, and between cluster information from different CHs to transfer information to BS. It influences the system network and the clusters nearer to BS prompts the inclusion issue which is known as hot spot issue. The unequal clustering approach is a viable way which manages the hot spot issue because of the way that it very well may be utilized to adjust the heap between the CHs. The point of unequal clustering system is indistinguishable from the equivalent clustering one with additional highlights like accomplishing energy productivity and settling hot spot issue. The unequal clustering sorts out the cluster size dependent on its separation to BS. A cluster with least size infers lower number of CM and lesser intra cluster information transmission [3]. Along these lines, littler measured clusters can spend the energy for bury cluster information transmission and CHs can't debilitate its energy rapidly. If there should be an occurrence of long separation to BS, the cluster size will be expanded. At the point when the cluster has increasingly number of CM, enormous measure of energy will be used for intra cluster information transmission.

Since the cluster is situated a long way from the BS, entomb cluster information transmission will be low and there is no prerequisite to use high energy for directing information between clusters. The unequal clustering model will drive each CH to use an indistinguishable amount of energy, in this way, the CHs close or a long way from the BS uses indistinguishable amount of energy. Moreover, the development of clusters may make a 2-level chain of importance with high just as low levels. The detecting gadgets will send the information consistently to its individual CH which will perform information conglomeration and send it to the BS in a clear manner or irregular CHs. In the event of a demise of a CH or it moves to different clusters, the reclustering procedure will happen among the nodes to choose new CHs.

At present, differing models have been acquainted utilizing advancement systems for resolving the hot spot issue. [4] built up a genetic algorithm based clustering model which decides the quantity of CHs and their area for reducing the use of vitality in WSN. The working of this system happens in differing cycles where each cycle includes a setup and steady state stage. At the previous stage, the BS discovers the CHs and position. In the last stage, a way is gotten from the node to BS. It empowers a node to transmit information in a clear way to BS when the separation to BS is lower contrasted with the separation to CHs. [5] built up a clustering strategy which works indistinguishable from the recently clarified two phases. At the previous stage, the determination of CHs happens by BS relying on certain parameters like location, energy level and node degree. The last stage decides the route from the node to BS in an effective manner.

[4] gives a scheduling model to communicate data in a cluster by the intra-cluster communication. [6] created diverse sized clusters concerning residual energy level. It additionally picks the CHs utilizing Shuffled Frog Leaping Algorithm (SFLA) [7]. It works in two levels namely cluster construction and communication. The decision of CHs happens in the prior level and greedy strategy based route detection happens in the last level. [8] introduced a clustering model which includes three levels setup, neighborhood discovery and steady state. At the underlying two levels, node categorization happens in different layers and communication happens for neighboring node recognizable proof. It utilizes non-persistent Carrier Sense Multiple Access (CSMA) [9] for getting to the channel. At the last level, the procedure of CH race, building the group and transmission happens. It utilizes fuzzy rationale for choosing the CHs and ideal route choice happens utilizing ACO calculation.

[10] built up an unequal clustering model for deciding the cluster size and multi-objective immune strategy for delivering a routing tree. The size is determined based on residual energy and separation to BS. When the CHs are picked, at that point the group will be developed. Few hybridization of unequal clustering methods is displayed in WSN. A vitality adjusting strategy for each group is presented in [11]. To accomplish this, the clusters will be framed and afterward, CHs are assigned to it. In this way, three stages of clustering happen in particular development of cluster, selection of CHs and communication. At the initial step, Sierpinski triangle [11] is connected for making little estimated clusters for the nodes found nearer to the BS. On the choice of the CHs, it accepts the separation, remaining vitality level and node degree. A voting strategy to choose the CHs happens utilizing remaining energy level, topology, and correspondence control. Be that as it may, the method for choosing CHs [12] happens in a disseminated manner. It experiences the downside that the way toward improving the lifetime of WSN.

In this paper, a deterministic type 2 fuzzy-based unequal clustering (DTFUC) model in WSN is introduced. A backoff timing mechanism is used for primary cluster head (PCH) selection and type 2 fuzzy logic mechanism is applied for secondary CH (SCH) selection. The introduced DTFUC model successfully chooses the CHs in a productive manner. The displayed calculation experiences various situations under diverse measures.

2 Proposed Work

The presented DTFUC model comprises three stages namely PCH election, SCH selection and cluster formation. At the first stage, the BS invokes the DTFUC model to select the PCH using a backoff timer which is based on the remaining energy level of the sensor nodes. The timer comes to the initial value sooner in case of having highest residual value. At the second stage, the chosen CHs form the cluster by joining the adjacent nodes. For the choice of CHs and cluster sizes, Type 2 fuzzy scheme is applied utilizing a set of three inputs namely energy, distance to based station and node degree. The outcome indicates the possibility of the SCH (PSCH). The outcome of the variable PSCH implies that the value represents the node chance of coming as SCH. In addition, it offers the cluster size value of each SCH.

The SCH selection using DTFUC model comprises a collection of four major steps as shown in Figs. 1 and 2.

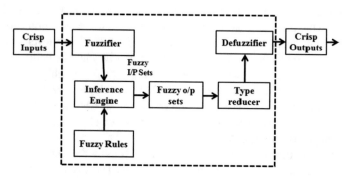

Fig. 1. SCH selection using DTFUC model

Fuzzification

Here, the transformation of the crisp input takes place to produce the fuzzified values. The input variables along with the linguistic attributes for selecting the CHs and cluster sizes are provided in Table 1.

Table 1. Parameter assumption

Parameters	Linguistic variables
RE	Low, Average, High
DBS	Near, Far, Farthest
ND	Low, Medium, High
PCH	Very poor, Poor, Below average, Average, Above average, Strong, Very strong
Cluster size	Very small, Small, Below average, Average, Above average, Large, Very large

Fuzzy Rules/Inference Engine

The structure of T1FL and T2FL is same. In this work, a set of 27 rules are used. A rule can be expressed in Eq. (1).

$$Rule(i) IF x1 is A_1(i) AND x2 is A_2(i) AND x3 is A_3(i) THEN y1 is B_1(i) AND y2 is B_2(i) \quad (1)$$

where i is the ith rule in the fuzzy rule, A1, A2 and A3 is the corresponding fuzzy set of x1, x2 and x3. The rule base inference engine contains 27 rules and is generated based on a Mamdani Inference system. In the type-2 FLS, the inference engine integrates the rules and maps the input type-2 fuzzy sets to output type-2 fuzzy sets. It is essential to calculate unions and intersection.

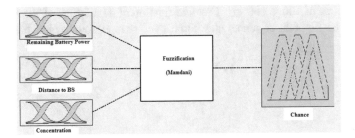

Fig. 2. Overall process of the DTFUC model

Membership Function

The MF for the input as well as output parameters are defined here. The MF is represented in a similar way as type 1 MF with some modifications. The interval among the pair of MF represents the footprint of uncertainty (FOU) utilized for describing a type 2 fuzzy set. Now, the FOU can be defined in such as way as f. When f ∈ [0, 1], and f → 0, then MF is treated as a T1FL. When f → 0 equals to 1, then T2FL has a broad view of FOU in the range of 0 to 1. However, rule construction in T2FL logic is identical to T1FL as represented below.

$$T2ML = MF(T1) + FOU \tag{2}$$

Defuzzification

It offers the T1FL output, which is then transformed to a numeric output through running the defuzzification module. When every individual node gets the value of SCH and size of the clusters, broadcasting of advertisement messages takes place to the nearby nodes. This message has the node id along with the PSCH value. The nodes whichever has maximum possibility are selected as SCH and transmits an advertisement to the adjacent nodes. There is a chance that the node could receive multiple advertisements from its adjacent nodes in the same transmission range. In those scenarios, it will join the cluster which is present at the nearest. Upon the reception of the joined advertisements, the closest CH ensures that the existing cluster sizes before the acceptance of fresh members. While the CM count exceeds the cluster size, then the incoming request will be rejected.

3 Performance Validation

The experimentation of the presented DTFUC model takes place in MATLAB and the comparative analysis takes place with the familiar LEACH protocol. An analysis of energy efficiency interms of energy consumption and network lifetime interms of number of alive nodes is made. A random deployment of hundred sensor nodes takes place in the area of 100 * 100 m². The investigation of the lifetime of WSN is made by measuring the number of nodes stay alive over several rounds and the analysis is illustrated in Fig. 3.

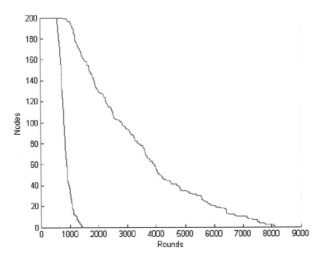

Fig. 3. Number of alive sensor nodes

The figure clearly stated the alive node count is maximum for the LEACH model because of the fact it chooses the CHs in an arbitrary way and does not involve any node variables to select CHs. The presented model makes use of different models for selecting the CHs. It leads to the efficient choice of CHs and makes the node operative for long duration than usual. It consequently enhances the network lifetime of WSN. The figure showed that 50% of nodes becomes dead at the approximate round number of 990 rounds whereas the presented model delays it till 3000 rounds. At the same time, the network becomes inoperative with the death of all nodes at the round number of 1354 by LEACH whereas the network becomes inoperative with the death of all nodes at the round number of 8114 by DTFUC model.

Next, the energy utilization of the presented DTFUC model takes place on comparing the results with one other. Figure 4 clearly stated that the LEACH consumes maximum amount of energy over the devised DTFUC model. As shown in the figure, all the energy got exhausted at the round number of 1354 under the operation of LEACH in WSN. Next, the presented DTFUC model utilizes its energy efficiently and exhausts total energy only at the 8114 rounds. It ensures that the proper choice of CHs and cluster size by the DTFUC model helps to achieve energy efficiency and network lifetime.

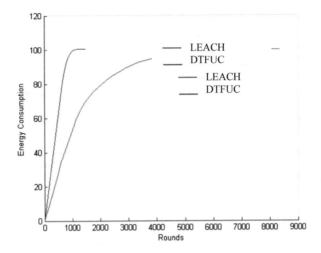

Fig. 4. Energy consumption of sensor nodes

4 Conclusion

In this paper, a DTFUC model based dynamic CH selection is done in WSN. The proposed method is simulated and the proposed method is compared to LEACH. The network becomes inoperative with the death of all nodes at the round number of 1354 by LEACH whereas the network becomes inoperative with the death of all nodes at the round number of 8114 by DTFUC model. It ensures that the proper choice of CHs and cluster size by the DTFUC model helps to achieve energy efficiency and network lifetime.

References

1. Uthayakumar, J., Vengattaraman, T., Dhavachelvan, P.: A new lossless neighborhood indexing sequence (NIS) algorithm for data compression in wireless sensor networks. Ad Hoc Netw. **83**, 149–157 (2019)
2. Arjunan, S., Pothula, S.: A survey on unequal clustering protocols in wireless sensor networks. J. King Saud Univ.-Comput. Inf. Sci. **31**, 304–317 (2017)
3. Arjunan, S., Sujatha, P.: Lifetime maximization of wireless sensor network using fuzzy based unequal clustering and ACO based routing hybrid protocol. Appl. Intell. **48**(8), 2229–2246 (2018)
4. Arjunan, S., Pothula, S., Ponnurangam, D.: F5N-based unequal clustering protocol (F5NUCP) for wireless sensor networks. Int. J. Commun. Syst. **31**(17), e3811 (2018)
5. Abo-Zahhad, M., Ahmed, S.M., Sabor, N., Sasaki, S.: A new energy-efficient adaptive clustering protocol based on genetic algorithm for improving the lifetime and the stable period of wireless sensor networks. Int. J. Energy Inf. Commun. **5**, 47–72 (2014)
6. Salehian, S., Subraminiam, S.K.: Unequal clustering by improved particle swarm optimization in wireless sensor network. Procedia Comput. Sci. **62**, 403–409 (2015)

7. Xunli, F., Feiefi, D.: Shuffled frog leaping algorithm based unequal clustering strategy for wireless sensor networks. Appl. Math. Inf. Sci. **9**, 1415–1426 (2015)
8. Amiri, B., Fathian, M., Maroosi, A.: Application of shuffled frog-leaping algorithm on clustering. Int. J. Adv. Manuf. Technol. **45**, 199–209 (2009)
9. Gajjar, S., Sarkar, M., Dasgupta, K.: FAMACRO: fuzzy and ant colony optimization based MAC/routing cross-layer protocol for wireless sensor networks. Procedia Comput. Sci. **46**, 1014–1021 (2015)
10. Raj, J.S.: QoS optimization of energy efficient routing in IoT wireless sensor networks. J. ISMAC **1**(01), 12–23 (2019)
11. Fang, S., Yunfang, C., Weimin, W.: Multi-objective optimization immune algorithm using clustering. In International Conference on Information and Management Engineering, pp. 242–251. Springer, Heidelberg, September 2011
12. Guiloufi, A.B.F., Nasri, N., Kachouri, A.: An energy-efficient unequal clustering algorithm using 'Sierpinski Triangle' for WSNs. Wirel. Pers. Commun. **88**(3), 449–465 (2016)

Android Application Based Solid Waste Management

Raju A. Nadaf[1]($\boxtimes$), Fuad A. Katnur[2], and Susen P. Naik[3]

[1] Department of Computer Science, GPT, Bagalkot, Bagalkot, India
raj.enggs@gmail.com
[2] K.L.E. Institute of Technology, Hubli, Karnataka, India
fuad130301@gmail.com
[3] Department of Computer Science, K.L.E. Institute of Technology, Hubli, Karnataka, India
susennaik55@gmail.com

Abstract. One of the main concerns with our environment has been solid waste management which stays uncollected from the bins sometimes without having the notice which can result in the overflowing of the trash cans, polluting the vicinity of the trash cans. This creates adverse effects on society concerning health as it is a major factor that thrives the bacteria, insects, and vermin. This increases the risk of the people getting in contact with salmonella that causes typhoid fever, food poisoning, enteric fever, gastroenteritis, and other major illnesses. India is one of the largely populated country and is growing with great ideas, implementing smart ideas and technology, creating a smart lifestyle and building up smart cities. Therefore we too must put waste management smartly into consideration. To deal with and combat against this scenario, a new approach towards smart trash cans has been proposed. This smart trash can is the solution to the problems that we can overcome in our day to day lives. Whenever the trash can gets filled after a certain limit, this acknowledges the client with the help of placing sensors and arduino board in the trash can that transmits this information to the desired person in charge indicating the location, time and clearance required by the authority with the help of an android application. This requires the authority to login in order to receive this information. And when the process of clearing the trash can is complete, then authority marks the job complete and the bin is now ready to receive trash from the people. And this process continues again.

Keywords: Trash · Sensors arduino board · Android application and login

1 Introduction

India is rated as the second-largest country in the world, where it faces various hindrances in its development. Our country faces a crucial crisis in its waste management. Since it is a developing country, it is a big drawback. The poor management of wastes conducts the reduction in the grade of lifestyle in the facet of health and hygiene. With swiftness in urbanization, the country is lining up enormous waste management challenges. There are over 377 million urban people that live in 7,935 towns and cities. They generate over 62

© Springer Nature Switzerland AG 2020
A. P. Pandian et al. (Eds.): ICCBI 2019, LNDECT 49, pp. 555–562, 2020.
https://doi.org/10.1007/978-3-030-43192-1_63

million tons of municipal solid waste every year. The total waste generation is estimated at 165 million tons by 2030. 43 million tons is the sum amount of the solid waste, among which only 11.9 million tons of waste is nursed while 31 million tons of the waste is unnursed and thrown away in landfill sites. The management of solid waste is of vital concern and needs observation. Taking into consideration that there many developed countries that are keen for off the rack and tenable solutions in the management of waste. This issue often occurs in public areas with high population denseness. The trash cans become encumbered before the ensuing schedule collection and the immediate vicinity of the trash bin would eventually be littered with waste. To deal with and combat against this scenario, a new approach towards smart trash cans has been proposed. This smart trash can is the solution to the problems that can overcome in day to day lives. It possesses the potential to rectify this situation by using ultrasonic sensors as a level detector in the trash can which when attains a threshold level, senses it using ultrasonic sound waves and triggers a signal using a GSM module also each trash can is provided with a unique identity and location. Here a microcontroller is used to interface between the sensor system and the GSM module system. So the signal produced is sent to an application created that requires the access by the authority by providing their username and password, and when they do so, they attain access to the bin that displays its unique identity, its location and its status that has got filled and take up the janitor work of cleaning it up. And when the authority is done with the clean-up process they have to mark the job done in the application with the same access and the trash can is then ready for the next repetition of the cycle. This innovation can prove to be a revolution in waste management in the upcoming future smart cities. The future of India is visioned and is working on being a smart city, but no city can be smart if its waste management is not checked. Counting on this technology and innovation will reduce the workload on the authority regarding the complete rounds taken around the city rather than going to the trash cans directly, also it will reduce the time taken in getting the job done with the prevention of overflowing trash cans. This can also result in the good efficiency of the work done by the authority. This can prove to be a solution for unemployment by providing more job opportunities and also can be a step taken towards "Clean India" for the movement "Swatch Bharat Abhiyaan".

2 Literary Survey

Many researchers have developed a system that uses an Arduino Uno connected to GSM Module, Servo Motor, and Gas Sensor and also interfaced with Ultrasonic Sensors. Where when the garbage reaches a threshold level limit and also the decomposition predefined limit by the Gas Sensors sends a message using the Servo Motor that can be accessed only by the valid authority [1, 2]. Some researchers also used a similar agenda of creating a mobile-based application that detects the bins that are filled when reaching a threshold level and can be marked as cleaned when done by the authority and this can also be accessed only by the valid authority. The software used in building in the system are Visual Studio IDE and Notepad++ and Hypertext Pre-Processor language was used in developing the webpage side by side JavaScript and Cascading Style Sheets (CSS), etcetera [3]. There are some projects that sense the threshold level of the bin using the level sensors and use the GPS System for the location and send a short message service (SMS) to the authority for the cleaning to get done [4–6]. Some researches also use flame sensors in their projects that sense the fire so that the fire can be sensed in the trash cans and can be avoided [7]. There are some similar projects that also use the WIFI Module instead of the GSM Module and that interacts with the authority via emails also even uses LCD Display to display the status of the trash cans [8]. Some other projects also use the RFID technology that is used as identity proof by the authority that verifies the identity of a person that has completed the process of cleaning along with the similar technologies of using WIFI Module and creating an Android application that shows the status of the trash can [9]. Some researchers have also used the Ultrasonic Sensors in their project to perform two functions of sensing the level of the dustbin and also to sense the hand movements to allow the automatic opening and closing of the bin door through a Servo Motor and where all the sensors are interfaced using a Raspberry Pi Zero W Development Board. Also, they further use Google Maps API in order to make the work easy of the authority to find short and faster routes to reach the trash can and also to serve the purpose of tracking the same [10]. Researchers have tried to gain access over precise information the bin ID, Bin address, its latitude and longitude etcetera for which they have applied the use of storing such information using MySQL database as a backend tool and Java Netbeans as a frontend tool in the Database [11]. There are researches that are similar to the proposed one but instead of an Adruino Board they use WeMos D1 Mini to interface with the sensors and it further also has the feature of built-in WIFI capabilities [12–16].

3 Proposed System

See Fig. 1.

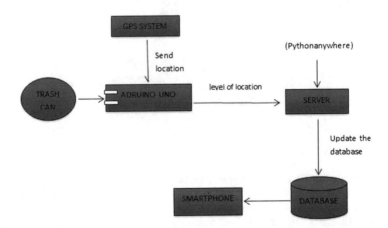

Fig. 1. In the proposed system block diagram, the trash can is linked with the GPS system that transmits the location to the server. This transmitted data is updated to the database from which the information of the trash can, can be accessed through the smartphone.

4 Dataflow Diagram

See Fig. 2.

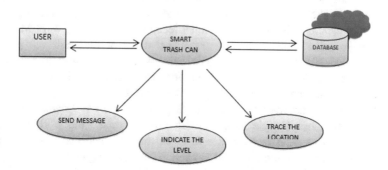

Fig. 2. In the dataflow diagram, the trash can is a system and its bidirectional to the user and the database. Where the trash can features sending a message, indicating the level and tracing the location.

5 Sequence Diagram

See Fig. 3.

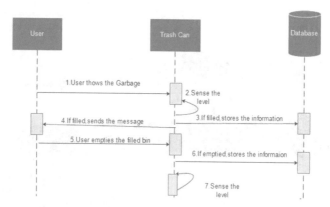

Fig. 3. In the sequence diagram, the user throws the garbage in the trash can. When the trash can senses the garbage at the threshold level then there is information stored in the database like time, location and phone number. And when it is filled then there is a message sent to the user to empty it. When the trash can is emptied then the details of being emptied are stored in the database and this process continues for the same.

6   Comparison Table

See Table 1.

Table 1. The proposed system is rated as 96% accurate. This is because sometimes the trash can contains liquid garbage that is not sensed properly by the ultrasonic sensors as the waves produced by the liquids are different when it comes to the comparison of solids.

Existing system	Features/functionalities	Drawbacks	Accuracy	Cost
GSM based garbage collector	Indicates the garbage after the weight is reached	Weight sensor problem	80%	3000
Bin level indicator	Indicates the level of the garbage	No message	85%	2500
RFID technology based trash can	Trash can be accessed by RFID technology	Slow processing	85%	4000
Smart trash can	Indicates the level of the trash can and sends a message through an android based application	Network issues	96%	2000

7   Experimentation

See Figs. 4, 5 and 6.

Fig. 4. It is a welcome page that will be displayed when the user or the authority responsible for the waste management opens the application smart trash can and the other one is where the authority arrives at the login page to enter the username and password so that they can access their account.

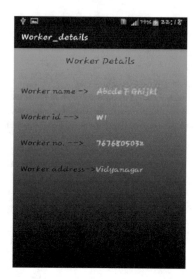

Fig. 5. It's a page that will be displayed once the user or the authority login's to the application where one has to enter the bin ID that is allocated and would display status of the bin whether it is cleaned. If yes the trash can is cleaned, then by whom it is done will also be displayed. And the other is a page where the user or the signed-in authority can enter the credentials like name, ID, number and the address that has done the job of cleaning the trash can

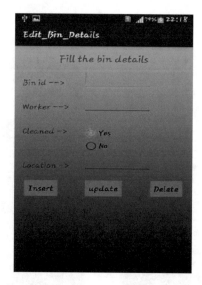

Fig. 6. It is a page where the user or the authority can enter the credentials of the trash can such as its ID, worker, location and also can be marked as cleaned by the worker or not.

8 Conclusion

This proposed system is developed in order to solve the problems faced by the people with respect to waste management by taking it to the next level where it can be applied to the use in cities. By doing this, major problems faced by urban areas can be checked. Whenever the trash can gets filled to the threshold level then the message is received by the authority responsible to get it cleaned up through the means of an android application in order to get the job is done by the authority. Usually the overflowing trash cans bring out the pollution of vicinity roads and pedestrains and also causing foul smell. The proposed system brings in the reduction of the overflowing trash cans, maintaining a clean and sanitary environment and also reduces the janitorial workload. This system can be reliable at 96% of the time. The application is reliable at 100% of the time, where the user can access it 24 × 7 in 365 days of the year. The difficulties that can be faced during implementing this project can be a delay in receiving the message that is due to poor networking. This project can be visioned to be placed at a large scale use In the country helping to overcome the issues faced and benefitting the people.

References

1. Balamurugan, S., Ajith, A., Ratnakaran, S., Balaji, S., Marimuthu, R.: Design of smart waste management system. In: 2017 International Conference on Microelectronic Devices (2017)
2. Chowdhury, P., Sen, R., Ray, D., Roy, P., Sarkar, S.: Garbage monitoring and disposal system for smart city using IoT. In: 2018 Second International Conference on Green Computing and Internet of Things (ICGCIoT) (2018)
3. Fei, T.P., Kasim, S., Hassan, R., Ismail, M.N., Salikon, M.Z.M., Ruslai, H., Jahidin, K., Arshad, M.S.: SWM: smart waste management for green environment. In: 2017 6th ICT International Student Project Conference (ICT - ISPC) (2016)
4. Willie, N., Lucy, K., Shadrick, K., Victor, P.D., John, S: GSM and GPS based garbage and waste collection bin overflow management system for Kitwe city council (2018)
5. Sathish, A., Prakash, M., Jainulabudeen, S.A.K., Sathishkumar, R.: Intellectual trash management using the internet of things. In: 2017 International Conference on Computation of Power, Energy Information and Communication (ICCPEIC) (2017)
6. Chaudhari, S.S., Bhole, V.Y.: Solid waste collection as a service using IoT - solution for smartcities. In: 2018 International Conference on Smart City and Emerging Technology (ICSCET) (2018)
7. Draz, U., Ali, T., Khan, J.A., Majid, M., Yasin, S.: A real-time dumpsters monitoring and garbage collection system. In: 2017 Fifth International Conference on Aerospace Science & Engineering (ICASE) (2017)
8. Smys, S., Wee, H.M., Joo, M.: Introduction to the special section on inventive systems and smart cities, pp. 32–33 (2018)
9. Kolhatkar, C., Choudhari, P., Joshi, B., Bhuva, D.: Smart E-dustbin. In: 2018 International Conference on Smart City and Emerging Technology (ICSCET) (2018)
10. Kumar, N., Vijayalakshmi, B., Prarthana, R.J., Shankar, A.: IOT based smart garbage alert system using Arduino UNO. In: 2016 IEEE Region 10 Conference (TENCON) (2016)

11. Lokuliyana, S., Jayakody, A., Dabarera, G.S.B., Ranaweera, R.K.R., Perera, P.G.D.M., Panangala, P.A.D.V.R.: Location-based garbage management system with IoT for smart city. In: 2018 13th International Conference on Computer Science & Education (ICCSE) (2018)
12. Malapur, B.S., Pattanshetti, V.R.: IoT based waste management: an application to smart city. In: 2017 International Conference on Energy, Communication, Data Analytics and Soft Computing (ICECDS) (2017)
13. Memon, S.K., Shaikh, F.K., Mahoto, N.A., Memon, A.A: IoT based smart garbage monitoring & collection system using WeMos & Ultrasonic sensors. In: 2019 2nd International Conference on Computing, Mathematics and Engineering Technologies (iCoMET) (2019)
14. Elhassan, R., Ahmed, M.A., AbdAlhalem, R.: Smart waste management system for crowded area; Makkah and Holy Sites as a model. In: 2019 4th MEC International Conference on Big Data and Smart City (ICBDSC) (2019)
15. Chen, W.E., Wang, Y.H., Huang, P.C., Huang, Y.Y., Tsai, M.Y.: A smart IoT system for waste management. In: 2018 1st International Cognitive Cities Conference (IC3) (2018)
16. Chaudhari, M.S., Patil, B., Raut, V.: IoT based waste management system for smart cities: an overveiw. In: 2019 3rd International Conference on Computing Methodologies and Communication (ICCMC) (2019)

An Approach for Detecting Man-In-The-Middle Attack Using DPI and DFI

Argha Ghosh$^{(\boxtimes)}$ and A. Senthilrajan

Department of Computational Logistics, Alagappa University, Karaikudi, India
argha.ghosh16@gmail.com, agni_senthil@yahoo.com

Abstract. Recently, many new cyber-attacks like Phishing, Spear Phishing, Cross-Site Scripting (XSS), Denial of Service (DoS), SQL injection including, Man-In-The-Middle (MITM) attack, etc are originated in the transmission of data over a network. Among all those attacks, a man-in-the-middle attack is dangerous as well as well known for its behaviour to steal the privacy and the data of a user. The term man-in-the-middle defines that between the user and web-server presence of hacker or third-party for stealing the data as well as the privacy of the user. In terms of performing ways, man-in-the-middle attack can classify by six key techniques and those are Spoofing based MITM attack (like ARP spoofing, ICMP spoofing, DNS spoofing and, DHCP spoofing), TSL/SSL (Secure Socket Layer) MITM attack, BGP (Border Gateway Protocol) based MITM attack, Cookie Hijacking, Man-In-The-Browser and, Wireless MITM. In this research paper, discuss all of those man-in-the-middle attacks with example and case study.

Deep Packet Inspection is a technique for monitoring and analysing the network's traffic as well as DPI used for managing the network's bandwidth also. DPI is useful for monitoring the high-speed network. However, in recent time, many countries like Egypt, China, etc. implemented DPI for network monitoring. Deep Flow Inspection (DFI) is a packet filtering technique like DPI, but it has some advantages over DPI. The DFI can filter the encrypted network traffic as well as DFI can perform the task like finding the packet length, size of the packet, etc. This paper proposes a technique for detecting man-in-the-middle attack using Deep Packet Inspection and Deep Flow Inspection based on DPI Feature Library and DPI Method Library as well as DFI Feature Library and DFI Method Library for network traffic identification and packet filtering of incoming network traffic.

Keywords: Man-in-the-middle attack · Types of man-in-the-middle attack · Deep packet inspection · Deep flow inspection · Man-in-the-middle attack detection · Network traffic identification · Packet filtering

1 Introduction

The term "man-in-the-middle" mainly defines that in-between end-user and web server presence of attacker for stealing the communication as well as disclose the privacy of the user in a client-server. For that reason, most of the web-application trying to increase their security level and, started implementing https (Secure Hyper-Text

© Springer Nature Switzerland AG 2020
A. P. Pandian et al. (Eds.): ICCBI 2019, LNDECT 49, pp. 563–574, 2020.
https://doi.org/10.1007/978-3-030-43192-1_64

Transfer Protocol) and SSL (Secure Socket Layer) for providing better security. However, attackers always trying their best to implement new techniques to perform man-in-the-middle attack at any cost. In the below, that diagram showed the linguistics view of a Man-In-The-Middle attack where the user sends a request to the server for web content but in between user and server, an attacker trying to sniff a legitimize session. This kind of attack called session hijacking. This is the way, exactly man-in-the-middle attack used to start and after that attacker gains the access of user session. When a user sends a request to a web-server that time attacker interrupts the request before it reaches to the web-server then the attacker disconnects the user from the web-server (Fig. 1).

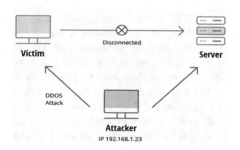

Fig. 1. Diagram of a MITM attack

After that attacker spoof, the user's IP address and made that IP address as the attacker's IP address and gets full access. Moreover, web-server used to think that the session is using by the user but it uses by the attacker (Fig. 2).

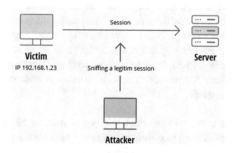

Fig. 2. Diagram of completion of MITM attack

There are six key techniques are there for performing man-in-the-middle attack and, those techniques are Spoofing based MITM attack (like ARP spoofing, ICMP spoofing, DNS spoofing, DHCP spoofing), TSL/SSL (Secure Socket Layer) MITM attack, BGP (Border Gateway Protocol) based MITM attack, Cookie Hijacking, Man-In-The-Browser, Wireless MITM. This paper will discuss all of those in the third section.

This paper proposes an approach for detecting man-in-the-middle attack using Deep Packet Inspection and Deep Flow Inspection. DPI and DFI both are useful for network monitoring and classifying network traffic. Previously, DPI and DFI used to detect intruders in a network, presently Intrusion Detection Systems (IDS) are started implementing DPI and DFI for network intrusion detection. In the second section of this paper describes Related Work that has been done in the past, the third section will provide the types of Man-In-The-Middle attacks. The fourth Section will discuss Deep Packet Inspection and Deep Flow Inspection. The fifth section will present the detection approach for identifying man-in-the-middle attack using Deep Packet Inspection and Deep Flow Inspection. The sixth section will evaluate the performance of this proposed approach. The seventh section will conclude the research-work with future aspect.

2 Related Work

Callegati et al. gave a practical example of man-in-the-middle attack by collecting real-time network data of an ARP poisoning attack [1]. Besides, proposes a solution also by creating an HTTP session. Du et al. gave a brief idea about a man-in-the-middle attack based on secure socket layer [2]. Also, propose a way to solve the man-in-the-middle attack based on SSL certificate interaction and describe SSL protocol architecture, SSL protocol workflow, forge SSL certificate, improved certificate alternate method. Chordiya et al. proposes a technique to defend against SSL hijacking, and presented ARP spoofing and SSL stripping [3]. Bhushan et al. provide a review of the man-in-the-middle attack in wireless computer networking and, also describe spoofing based man-in-the-middle attack, TSL/SSL man-in-the-middle attack, BGP based man-in-the-middle attack, FBS based man-in-the-middle attack as well as describe defending approach of all those discussed man-in-the-middle attacks [4]. Chen et al. propose a unified mathematical model for analysing man-in-the-middle attacks to the mutual authentication protocol and then they use the formal methods and logical operations to analyze the mutual authentication security, proposed a modification to the model of MITM attack [5].

3 Types of Man-In-The-Middle Attack

There are six key techniques are there to perform Man-In-The-Middle attack, day-by-day attackers are implementing new techniques also. This paper discusses all those six key techniques and they are Spoofing based MITM attack (ARP spoofing, ICMP spoofing, DNS spoofing, DHCP spoofing), TSL/SSL (Secure Socket Layer) MITM attack, BGP (Border Gateway Protocol) based MITM attack, Cookie Hijacking, Man-In-The Browser, Wireless MITM.

3.1   Spoofing Based MITM Attack

There are four types of Spoofing based man-in-the-middle attacks and, they are ARP (Address Resolution Protocol) spoofing, ICMP (Internet Control Message Protocol) spoofing, DNS (Domain Name System) spoofing and, DHCP (Dynamic Host Control Protocol) spoofing (Fig. 3).

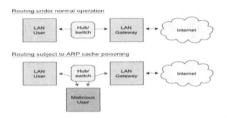

Fig. 3.   Diagram of ARP spoofing

Address Resolution Protocol (ARP) mainly used to communicate between user's Media Access Control (MAC) address to Internet Protocol (IP) address for reaching the Internet. ARP particularly used in Local Area Network (LAN) connection. LAN user used to send a request to web-server through hub/switch using Local Area Network (LAN) gateway with Media Access Control (MAC) address and Internet Protocol (IP) address. Then the attacker collects the source Internet Protocol (IP) address and Media Access Control (MAC) address from Hub/Switch. And, the attacker used to send a malicious packet to the user for gaining access in the user's computer. By following those activity attacker spoof the user's Internet Protocol (IP) address as the attacker's Internet Protocol (IP) address. During this process, web-server does not have any clue that the Internet session using by whom in real. This kind of attack known as ARP spoofing or ARP poisoning. In ARP spoofing mainly there are two things, first cheating the Local Area Network (LAN) user and second cheating the Local Area Network (LAN) gateway. It is easy for an attacker to attack ARP protocol because ARP, a stateless protocol has lesser security in the caching system [4].

Internet Control Message Protocol (ICMP) is a supporting protocol, mainly used by networking devices like router, switch, hub etc for sending an error message to the user. When the user's web-browser is down by some other reason, that time attacker used to send an error message similar to network device-generated error message with malicious data packet using Internet Control Message Protocol (ICMP). And, by doing that attacker used to gain access to the user's computer for stealing privacy as well as data. This kind of attack known as Internet Control Message Protocol (ICMP) spoofing.

In Domain Name System (DNS) spoofing, the attacker first used to detect all the domain name system of the particular targeted network using domain name system query and port number 53. After that, similarly like Address Resolution Protocol (ARP) spoofing attacker used to send domain name system query to the user as well as web-server. Then attacker performs Internet Protocol (IP) spoofing to make sure that web-server does not have any clue of the attack. Then the attacker steals all the sessions of the user. This is the most dangerous type of Man-In-The-Middle (MITM) attack because after performing Domain Name System (DNS) spoofing attacker can create a phishing website and can direct end-user to open that phishing website again and again (Fig. 4).

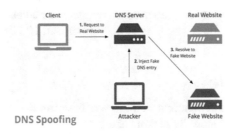

Fig. 4. Diagram of DNS spoofing

Dynamic Host Configuration Protocol (DHCP) mainly used to configure the new user in a network. When DHCP used to configure the new user, that time it collects data like Domain Name Server (DNS), default gateway, user's MAC address and, IP address, etc. When those new users used to a send request to the web-server for a session that time attacker intercepts the Dynamic Host Configuration Protocol (DHCP) queries and responses. And, from there attacker used to collect the data about entire connected user details in that particular network. This kind of data collecting attack known as DHCP spoofing. In future by using those collected data, the attacker used to perform attack. DHCP spoofing mainly used for collecting all other connected user's information on a network for using those data in other man-in-the-middle attack.

3.2   SSL (Secure Socket Layer) MITM Attack

Secure Socket Layer (SSL) provides security and encryption for secure data transmission over a network. Secure Socket Layer (SSL) mainly creates a secure medium between end-user and web-server for the secure transmission of data. Secure Socket Layer (SSL) mainly consist of four protocols and they are Record Protocol, Handshaking Protocol, Cipher Spec Protocol and, Alert Protocol. Secure Socket Layer (SSL) mainly follows public key cryptography approach. Public-key cryptography used to issue the certificate for both client and server respectively. Whenever web-browser tries to connect with a web-server, it used to check the valid certificate of web-browser. If valid certificate available then only data transmission used to start. The attacker used to intercept network data transmission. And, another way sometimes attacker used to generate a fake edited certificate for web-browser. After that, the attacker used to start a secure connection with web-server and web-browser respectively. This kind of attack known as Secure Socket Layer (SSL) Man-In-The-Middle (MITM) attack.

3.3 BGP (Border Gateway Protocol) Based MITM Attack

Border Gateway Protocol (BGP) is a gateway protocol that mainly used to perform the task of exchanging routing and network information data between autonomous user's computer in a network. But, Border Gateway Protocol (BGP) didn't perform authentication for that reason attacker can perform man-in-the-middle attack easily in Border Gateway Protocol (BGP). The attacker used to perform IP hijacking known as BGP hijacking.

3.4 Cookie Hijacking

Cookies used to store a computer's web-browser based on the user's previous visit on a website. cookie used to store for making end-user experience better in terms of opening a website faster than the opening a website first time. Cookies can be stolen by an attacker, when an attacker used to hijack a session of a web-browser or when a user used to send data from a particular web-browser using interception in the data transmission. This kind of attack known as Cookie Hijacking.

3.5 Man-In-The-Browser

Sometime attacker used to put a malicious program like malware, trojan, etc in the web browser of user and, by that attacker used to gain the access of user's web-browser. This kind of attack known as Man-In-The-Browser. By Man-In-The-Browser attack, an attacker can get all the data of online transaction details, surfing details, etc at one file. Moreover, this kind of attack gave all the activity of the user to the attacker.

3.6 Wireless MITM Attack

In a wireless network, where already several users are connected, in their attacker used to get connected using a malicious Service Set IDentifier (SSID), fake Media Access Control (MAC) address, etc in that network. After that, the attacker used to start collecting data about other connected users. This kind of passive attack known as Wireless man-in-the-middle attack (Fig. 5).

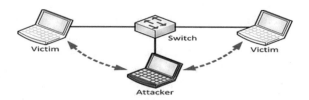

Fig. 5. Diagram of Wireless MITM attack

4 Deep Packet Inspection and Deep Flow Inspection

Deep Packet Inspection (DPI) is a technology that mainly used to manage and analyze real-time network traffic for high-speed networks. DPI also can perform task like managing network's bandwidth, monitor and interception of a network, network management, content optimization, copyright enforcement, governmental surveillance etc. Deep Flow Inspection (DFI) is a technique to classify the network flow feature for incoming network traffic. DFI mainly consists of two libraries, those are Deep Flow Inspection Feature and, Deep Flow Inspection Method. Deep Flow Inspection Feature contains Packet Count Flow, Total Bytes Flow, Average Packet Bytes Flow, etc for classifying the network flow feature. Deep Flow Inspection Method contains Support Vector Machine, Neural Network, Bayes Classifier for identifying incoming network traffic. Malicious behavior detection is generally classified into two levels: packet-level and flow-level [6]. DPI performs identification in packet-level whereas DFI performs identification in flow-level.

Deep Packet Inspection (DPI) can monitor the network transmission or communication in all the layers of Open System Interconnection model, whereas Shallow Packet Inspection and Medium Packet Inspection can't monitor the network transmission or communication for all the layers of Open System Interconnection model. DPI allows network operators to scan the payload of IP packets as well as the header [7]. Deep Packet Inspection can analyze packet's header as well as payload of the packet also, whereas Shallow Packet Inspection and Medium Packet Inspection can analyze only packet's header, but could not analyze packet's payload. Those are the advantages of Deep Packet Inspection over traditional packet inspection techniques. However, Deep Packet Inspection has some disadvantages also, Deep Packet Inspection cann't monitor or support encrypted data whereas Deep Flow Inspection support encrypted data. For that reason, this research work proposes the approach using Deep Packet Inspection and Deep Flow Inspection together. So that, this approach can work on encrypted data also. [8] DPI makes network filtering by examining the signature of the payload packet either by string matching algorithms like Wu-Manber, Aho-corasick, and SBOM or by regular expression matching algorithms which are used in NIDS of Snort [9] and Bro [10] and L7-filter in Linux [11]. Deep Packet Inspection can be implemented using several techniques like Bloom, Quotient and, Cuckoo Filter (Fig. 6).

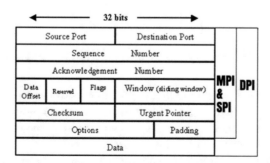

Fig. 6. Domain of Deep Packet Inspection over traditional packet inspection for an IPv4 packet

DPI is approximately categorized to three different implementations are signature-based identification, application layer based identification and, behavior based identification [12]. In the signature-based identification approach, DPI mainly used to match the signature of the incoming packet and, based on that DPI used to perform the job of identification of packet. In application layer based identification, DPI used to implement at the gateway of the Internet and, by checking traffic of gateway DPI used to identify the application. In behavior based identification, DPI mainly used to judge the behavior of the user and based on perform the identification of packet. Without the hardware advantages of modern systems, DPI would become a bottleneck in high-traffic circumstances [13].

5   Approach for Detecting Man-In-The-Middle Attack Using Deep Packet Inspection and Deep Flow Inspection

This research-work propose an approach for detecting man-in-the-middle attack using Deep Packet Inspection and Deep Flow Inspection. The proposed approach consists of three modules and they are Deep Packet Inspection module, Deep Flow Inspection module and, Co-ordinate module. Deep Packet Inspection module mainly consists of two libraries, those are Deep Packet Inspection Feature and, Deep Packet Inspection Method. Same as Deep Packet Inspection module, Deep Flow Inspection module also consists of two libraries and, those are Deep Flow Inspection Feature and, Deep Flow Inspection Method. The Co-ordinate module mainly offers the job of co-ordination between Deep Packet Inspection Module and Deep Flow Inspection Module (Fig. 7).

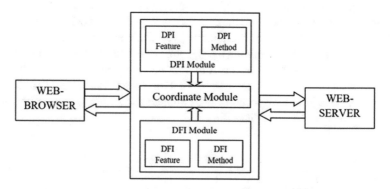

Fig. 7. Proposed approach for detecting man-in-the-middle attack

Web-browser or user used to send a request to web-server for a particular domain name or website. The proposed approach works between web-browser and web-server. Web-browser's request used to passes from our proposed approach to reach to the Internet as well as to web-server and the same way, web-server's response also used come from the Internet to our proposed approach and passes to web-browser also. First of all, web-browser's request used to reach DPI Module. The basic procedures of DPI are recognition and action which is based on the recognition result [14]. In DPI Module specifically, request used to reach at Deep Packet Inspection Feature. Deep Packet Inspection Feature mainly checks the incoming traffic characteristic code with pre-defined library characteristic code and by that tries to identify the type of application, source address, source port, destination address, destination port, size of the payload, number of packets, etc. Deep Packet Inspection Feature mainly collects all the data and, after that traffic used to reach Deep Packet Inspection Method. Deep Packet Inspection Method checks the IP address of traffic matches with the actual IP address of traffic or not. With the help of string-matching algorithms like Aho-Corasick algorithm, Wu-Manber algorithm and, Set Backward Oracle Matching (SBOM) algorithm. Using the deep packet inspection (DPI) technology reads the contents of the IP packet payload [15]. The ability of DPI to check the packet's header as well as packet's payload, helped our approach to detect if there is any malicious attachment in the packet or not. After all those processes completed, DPI Module used to send the traffic to the Deep Flow Inspection Module with the help of Co-ordinate Module.

In the Deep Flow Inspection Module, first of all, traffic used to reach to the Deep Flow Inspection Feature. Deep Flow Inspection Feature performs the task of classifying the traffic based on the series of network flow feature like Total Bytes Flow, Packet Count Flow, Duration of Flow and, Average Packet Bytes Flow. Network Flow Feature mainly performs the task of traffic identification and classification based on network flow features. After that, traffic used to reach Deep Flow Inspection Method. Deep Flow Inspection Method contains artificial intelligence algorithms like Support Vector Machines, Neural Network and, Bayes Classifier. Those artificial intelligence algorithms used to perform the task of classifying the traffic. After that, traffic used to reach the Co-ordinate Module. Co-ordinate Module combines the result of Deep Packet Inspection Module and Deep Flow Inspection Module. And, then Co-ordinate Module used to send the request of the web-browser to the web-server.

After the response of the web-server, traffic again passes through our proposed approach. And, performs each and every process again then the response of web-server reaches to the web-browser. In between any steps of the proposed approach if find or recorded any malicious or suspicious activity then immediately our proposed approach will give the alert to the web-browser cum user. So, that user could get aware of suspicious activity and stop the request of service in-between.

6 Experimental Evaluation

No.	Time	Source	Destination	Protocol
14726	164.815185	172.18.18.4	103.53.14.27	TCP
14727	164.829326	103.53.14.27	172.18.18.4	TCP
14728	165.548784	172.18.18.4	103.53.14.27	TCP
14729	165.562946	103.53.14.27	172.18.18.4	TCP

Fig. 8. Real-time traffic analyzing using wireshark

Using Wireshark packet analyzer tries to evaluate the performance of our proposed model. First of all, implemented our proposed approach and, then using Wireshark tries to get the real-time packet details of our university campus. After analyzing all those data packets, it is clear that there is no man-in-the-middle attack for that period we have been monitored our campus Internet activity. It's visible that transmission between particular computer Internet Protocol (IP) address 172.18.18.4 to web-server Domain Name Server (DNS) address 103.53.14.27. In Figs. 8 and 9 respectively, we can notice that the packets are transmitting to that particular source and destination address smoothly. There is no change of either source address or destination address. Data packet flow has been monitored for several days, almost more than 300 h. We have been monitored almost ten million packets for evaluating the performance of proposed architecture.

No.	Time	Source	Destination	Protocol
43355	21.713445	198.199.88.104	172.18.18.4	TCP
43356	21.713446	198.199.88.104	172.18.18.4	TCP
43357	21.713447	198.199.88.104	172.18.18.4	TCP
43358	21.713450	198.199.88.104	172.18.18.4	TCP
43359	21.713521	172.18.18.4	198.199.88.104	TCP

Fig. 9. Real-time traffic analyzing after implementing proposed approach using wireshark

7 Conclusion and Future Work

This research-work propose an approach to detect man-in-the-middle attack using Deep Packet Inspection and Deep Flow Inspection. This research-work discuss all the six key techniques of the man-in-the-middle attack briefly. They are Spoofing based MITM attack (ARP Spoofing, ICMP Spoofing, DNS Spoofing and, DHCP Spoofing), TSL/SSL (Secure Socket Layer) MITM attack, BGP (Border Gateway Protocol) based MITM attack, Cookie Hijacking, Man-In-The-Browser and, Wireless MITM. After that, describe Deep Packet Inspection and, Deep Flow Inspection in deeply. Deep Packet Inspection mainly consists of two libraries, they are Deep Packet Inspection Feature and, Deep Packet Inspection Method. Deep Flow Inspection also contains two libraries, those are Deep Flow Inspection Feature and, Deep Flow Inspection Method. Those respective libraries contain features for classifying network traffic. In the proposed model, there is a co-ordinate module between Deep Packet Inspection and Deep Flow Inspection for maintaining the transmission of data traffic for identifying incoming and outgoing traffic is affected by the Man-In-The-Middle (MITM) attack or not. Experimental evaluation section analyzing the performance of the proposed approach using Wireshark.

In future planning to implement, Deep Content Inspection (DCI) with Deep Packet Inspection and Deep Flow Inspection using Naive Bayes classifier for better performance and classification of traffic to detect man-in-the-middle attack and other attacks like Denial of Service (DoS), Zero Day attack, etc. Moreover, planning to use Intel's Data Plane Development Kit (DPDK) [16] in future because this package provides better accelerate of packet processing.

Acknowledgements. We would like to thank Dr. K. Kuppusamy for improving the content of this paper, as well as acknowledging the effort of Dr. E. Ramaraj for his guidance. This research work has been written with the financial support of Rashtriya Uchchatar Shiksha Abhiyan (RUSA- Phase 2.0) grant sanctioned vide Letter No. F.24-51/2014-U, Policy (TNMulti-Gen), Dept. of Edn. Govt. of India, Dt. 09.10.2018. Express appreciation to all those author's whose references we used in this research work. Acknowledging Mrs. Anju Ghosh, Mrs. Moumita Ghosh Bairagi, Mr. Bidhan Ghosh and rest of my family members for their support and love. Special Thanks' to Mr. N. Alagu Ganesan and Mr. G. Veerapandi for their helpful hand and support.

References

1. Callegati, F., Cerroni, W., Ramilli, M.: Man-in-the-middle attack to the HTTPS protocol. IEEE Secur. Priv. **7**(1), 78–81 (2009)
2. Du, J., Li, X., Huang, H.: A study of man-in-the-middle attack based on SSL certificate interaction. In: 2011 First International Conference on Instrumentation, Measurement, Computer, Communication and Control (2011)
3. Chordiya, A.R., Majumder, S., Javaid, A.Y.: Man-in-the-middle (MITM) attack based hijacking of HTTP traffic open source tools. In: 2018 IEEE International Conference on Electro/Information Technology (EIT) (2018)
4. Bhushan, B., Sahoo, G., Rai, A.K.: Man-in-the-middle attack in wireless and computer networking - a review. In: 2017 3rd International Conference on Advances in Computing, Communication & Automation (ICACCA) (2017)
5. Chen, Z., Guo, S., Duan, R., Wang, S.: Security analysis on mutual authentication against man-in-the-middle attack. In: 2009 First International Conference on Information Science and Engineering (2009)
6. Guo, Y.-T., Gao, Y., Wang, Y., Qin, M.-Y., Pu, Y.-J., Wang, Z., Liu, D.-D., Chen, X.-J., Gao, T.-F., Lv, T.-T., Fu, Z.-C.: DPI & DFI: a malicious behavior detection method combining deep packet inspection and deep flow inspection. Procedia Eng. **174**, 1309–1314 (2017). 13th Global Congress on Manufacturing and Management, GCMM 2016
7. White paper on Deep Packet Inspection. http://tec.gov.in/pdf/Studypaper/White%20paper%20on%20DPI.pdf
8. El-Maghraby, R.T., Elazim, N.M.A., Bahaa-Eldin, A.M.: A survey on deep packet inspection. In: 2017 12th International Conference on Computer Engineering and Systems (ICCES) (2017)
9. Snort v2.9.9 (2016). http://www.snort.org/
10. Bro Intrusion Detection System (2014). https://www.zeek.org/
11. Application Layer Packet Classifier for LINUX (2009). http://l7-filter.sourceforge.net/
12. Chaudhary, A., Sardana, A.: Software based implementation methodologies for deep packet inspection. In: International Conference on Information Science and Applications (ICISA). IEEE (2011)
13. Alkateb, S.: White paper: 5 things you need to know about deep packet inspection, April 2011. https://docplayer.net/7150123-5-things-you-need-to-know-about-deep-packet-inspection-dpi.html

14. Xu, C., Chen, S., Su, J., Yiu, S.M., Hui, L.C.K.: A survey on regular expression matching for deep packet inspection: application, algorithms and hardware platforms. IEEE Commun. Surv. Tutor. **18**(4), 2991–3029 (2016)
15. Wei, L., Hongyu, L., Xiaoliang, Z.: A network data security analysis method based on DPI technology. In: 7th IEEE International Conference on Software Engineering and Service Science (ICSESS) (2016)
16. Data Plane Development Kit. https://www.dpdk.org/

Smart Irrigation and Crop Disease Detection Using Machine Learning – A Survey

Anushree Janardhan Rao[✉], Chaithra Bekal, Y. R. Manoj,
R. Rakshitha, and N. Poornima

Department of Computer Science and Engineering,
Vidyavardhaka College of Engineering, Mysuru, Karnataka, India
anushreerao23@gmail.com, poornima.cs@vvce.ac.in

Abstract. Water wastage in agricultural fields has been one of the major issues in various countries especially in India. Hence it is very important to reduce water loss in different situations due to various factors like pipe leakage or leaving excess water into the farms without knowing. This paper provides various insights on the comparison of different methods to reduce water loss using various machine learning techniques. Diseases in crops, reduces the quality of each product and the quantity of agricultural product. Thus we require image processing techniques, as it will help in accurate and timely detection of diseases and helps in reducing the errors of humans. Production of crops can be increased by detecting the disease well in time. Automatic detection of plant sickness helps in analyzing the crop and robotically detects the sign of the alignments as soon as they appear on plant leaves in order to prevent the loss of crops.

Keywords: Smart irrigation · Crop diseases · Crop loss · Machine learning · Image processing

1 Introduction

Earth has 99% water in sea which is salty and is not suitable for human use. Just 1% is freshwater and groundwater and can be used for human consumption. Shrinking of water reservoir, low rainfall, etc. will lead to problems in future. There will be not enough water resources to provide for a huge growing population in India [1–3]. So it is necessary to save water and adopt mechanisms like smart irrigation and smart farming to reduce the water loss in agricultural fields. There are various research scholars who have performed experiments on various soil types using various instruments like sensors, watermark, tensiometer, etc. [4–8]. These studies have revealed that a lot of water loss can be reduced by using Machine learning and Artificial Intelligence technologies. This paper provides the review on the methods of reducing water loss and how accurate each method is and what future work is needed to improve the existing system [9–12].

India is a developing country. In developing country economic growth plays a vital role. For economic growth not only the Industrial contribution is important but also agriculture contribution is important and 70% of our population is depended on agriculture [13–16]. Crop diseases are affecting agriculture. This may lead to the reduction

© Springer Nature Switzerland AG 2020
A. P. Pandian et al. (Eds.): ICCBI 2019, LNDECT 49, pp. 575–581, 2020.
https://doi.org/10.1007/978-3-030-43192-1_65

of quality and quantity of crops. Crop diseases are caused by microorganisms. Hence farmers cannot see the symptoms of disease on leaf by just looking at it. To find the disease and its measure one can use computerized technique followed by various methods to detects the disease. The main part of the plant to look for the sickness is its leaf.

The diseases on leaf will cut back amount of crops and their growth. The simple methodology to find the plant diseases is with assistance of agricultural knowledge having data of plant diseases. However, this is manual detection of plant ailments that takes a great deal of your time and could be a backbreaking work. Hence, there a necessity for desktop gaining data of approach to become tuned into the leaf diseases. Systems will play a main role to develop the processed ways for the detection and classification. There are often variety pattern attention and movie process techniques which are employed in the leaf disorder detection. The plant disease detections and classifications of leaf diseases is the key to prevent the agricultural loss.

2   Literature Review

In [1] the author discussed about the water scarcity problem in India and what are the various methods to reduce water loss. They have stated various ways in which water is lost through leakage and excess water that is left into farms. The problems created due to water loss is discussed, few of them being food stress, GDP problem, Energy Problem and increased carbon foot print etc. The measures taken at present by the Indian government and various technologies like smart farming, smart water system are discussed. The advantages and disadvantages of measures taken across various countries in the world are given in this paper and how it can improvise and be inculcated in India to prevent water scarcity. This review paper in overall gives the necessary information on present situation about water loss and how it can be prevented in future.

In the paper [2], the authors concentrate on conserving water in arid regions. Intelligent irrigation system (IIS) is the method used to determine the crop water requirements based on the climatic conditions. They have taken two crops wheat and tomato into their study and made use of drip and sprinkle irrigation methods. The comparisons on how much water is necessary for both the crops was done using the hunter ET system which is the evapotranspiration method. The moisture content in soil is observed using the sensors like watermarks, tensiometers and Enviroscan. The Operation time was determined and results of soil analysis and water application for both tomato and wheat was obtained and graphs plotted. All of the technologies tested (IIS) managed to reduce water application resulted in water savings ranging from 18 to 27%.

In [3] the authors discussed regarding the evaluation of accuracy of soil water sensors for irrigation scheduling to conserve freshwater in which they have used low cost soil water sensors like ECH2O-5TE, Watermark 200SS and Tensiometer model R to determine their accuracies. They have conducted site study in a mature pecan field, located in the south El Paso in Texas, USA. This was followed by soil sampling and analysis. Considering sensors accuracy and soil water sensors the results of all the soil samples for various sensors were compared and graphs plotted for them. Tensiometer provided relatively more accurate soil water data compared to the other two sensors.

In [4], they concentrate on optimizing the use of water for agricultural crops. The method that they have used consists of a system that has wireless distributed network of soil moisture and temperature sensors that was fixed in the roots of plants. They have considered different methods with various parameters and results. This method was tested in a greenhouse with organic sage as its produce. The automated irrigation was triggered immediately when the soil moisture value fell below the threshold value and similarly for soil temperature when the temperature was above the threshold value. Hence the automated irrigation system proves that the water can be used effectively for a fresh biomass production.

In [5], the author stated that due to high increase in the demand for freshwater in the agricultural area, fresh water should be used effectively for irrigation purposes. The system that they used consists of a sensor network which is wireless for wireless controlled irrigation solution at low-cost and analyzing the water content of the soil. This system was implemented and tested in an area of 8 acres located in central Anatolia for controlling drip irrigation of dwarf cherry trees. The main advantage of this system is that it prevents moisture stress and salification.

In [6], the author concentrates on efficient management of water in cropping areas. This paper stresses on site-specific irrigation management that increases their productivity and saves water. This method consists of in-field sensors based on site-specific irrigation which takes soil moisture, soil temperature and air temperature as parameters. This method had the capacity to increase the yield and the quality of the crops while optimizing the use of water.

In [7], the researchers, explained the method to prevent the loss of crops in cotton leaf by detecting the symptoms. In cotton, the diseases show up in leaf, so the area of interest is leaf, as most of the diseases appears on the leaves itself. In cotton there are common diseases like Red Leaf Spot, Alternaria Leaf Spot and Cercospra Leaf Spot. These disease can be easily detected using k-means clustering algorithm. It classifies objects. Segmentation is done based on a set of features and then the image is partitioned into number of classes and finally disease can be detected using neural network.

In [8] the author described an approach to detect the crop disease in large farms agriculture for instance rice. It is based on automated technique. Fungi are identified primarily, then the bacteria is considered by capturing the image of two leaves that is one of healthy and another is unhealthy and thus the disease is detected. The RGB image that was captured is converted to grey image and then grey image is resized and performs canny edge detection.

In [9] the authors described the approach that consists of various steps. Firstly, the green color pixels are recognized. Then based on specific threshold values green pixels are covered. RGB values with zero and disease occurred leaf boundaries are removed. This step is important in classifying the diseases. These methods are used to acquire the necessary features for analysis. This technique has high accuracy in detecting the plant disease.

In [10], the author proposed a technique that can be applied to different yields like orange, citrus, wheat, corn and maize and so forth. Fluffy framework for leaf sickness, recognition and reviewing, K-means implies bunching procedure that has been utilized for division, which gathers comparable pixels of a picture. RGB shading space is changed over to L * a * b space, where L is the radiance and a * b are the shading space.

In [11], Picture handling based strategy for evaluating the leaf spot alignment in plant leaves. They played out an examination on all the impacting factors that were available during the time spent division. Otsu Technique was utilized to section the leaf areas.

In [12], the author proposed a way to deal with recognition and grouping the illness in the sunflower harvest utilizing picture preparing. The exploration was completed utilizing the leaf pictures of the yield that were taken utilizing a high-goals advanced camera.

3 Comparison of Different Detection Method for Smart Irrigation and Detection of Crop Diseases

Table 1, gives us the attractive idea of the detection techniques used by various authors in the field of smart irrigation and detection of crop diseases. It also gives the list of recommendations which we thought, could have been implemented in the system in future.

Table 1. Comparison of different detection methods for abusive text.

Reference no	Approach	Method	Parameter	Advantage	Disadvantage	Result
1	Need of smart water systems in India	Smart water system, smart farming, various control measurements, Water monitoring, leakage detection, Smart water system for Water Treatment Plants, Smart meters	Various parameters for various methods available. Review on all the methods and their comparisons	Smart irrigation and smart farming methods have better results in reducing water loss	Accuracy is still less and has to be improved	There is need to obtain a better method with best accuracy to reduce water loss
2	Intelligent irrigation performance: evaluation and quantifying its ability for conserving water in arid region	Intelligent irrigation system called as hunter ET system for wheat and tomato for two systems drip and sprinkle irrigation	Evaluation results of drip irrigation for wheat and sprinkle irrigation for tomato	Cheap and available in local market	ET system is not the best available system	The intelligent irrigation system along with controllers prove adequate. Water is saved by 18–27%
3	Evaluating the accuracy of soil water sensors for irrigation scheduling to conserve freshwater	Studying a particular site with a particular varieties of soil. ECH2O-5TE-capacitance sensors, watermark, tensiometer	Finding apparent moist soil dielectric constant or relative permittivity (e). Sensors give certain values whose mean differences are taken	Inexpensive water sensors with improved accuracy	Accuracy can be increased using site-specific calibration	Tensiometer provided relatively more accurate soil water data compared to Watermark 200SS, and ECH2O-5TE
4	Automated irrigation system using a wireless sensor network and GPRS module	Distributed wireless network of soil moisture and temperature sensors placed in the root zone of plants GPRS module system	Threshold values of temperature and soil moisture	Feasible, cost effective because of the use of photovoltaic cells, minimum maintenance	Only two parameters are taken into consideration, if the area does not have proper network, the result may not be accurate	Optimized water use for agricultural crops Cultivation in places with water scarcity thereby improving sustainability

(continued)

Table 1. (*continued*)

Reference no	Approach	Method	Parameter	Advantage	Disadvantage	Result
5	A wireless application of drip irrigation automation supported by soil moisture sensors	Site-specific wireless sensor-based irrigation control system, drip irrigation systems and remote control systems	Soil moisture content, water content of the soil	Preventing moisture stress of trees, diminishing of excessive water usage, ensuring of rapid growing weeds and derogating salification	Only one type of sensor was used, other parameters like temperature, humidity was not considered	A wireless data acquisition network was implemented and applied to irrigate dwarf cherry trees
6	Remote sensing and control of an irrigation system using a distributed wireless sensor network	In-field sensor based site specific irrigation, Linear-move irrigation systems	Soil moisture, soil temperature, air temperature	Potential to increase yield and its quality, optimized use of water	Seamless integration of sensor fusion, irrigation control, data interface can be challenging	Developed the machine conversion from a conventional irrigation system to an electronically controllable system for individual control of irrigation sprinklers
7	Detection of diseases on cotton leaves using K-means clustering and neural network	It uses Histogram equalization to improve the image contrast and K-means clustering algorithm to classify the object	It takes the image of leaves as an input for Histogram equation K-means algorithm takes data sets as an input	Recognition accuracy for detecting disease on leaves is high. It has 80.56% of accuracy and it takes less time	It has less accuracy for small data set	Plot of validation performance
8	Extraction of rice disease using image processing	It uses surf, entropy, warp, techniques of image processing	Disease is detected using stem, stairs, canny edge detection	It is efficient and frequent it gives exact comparison value	One should identify defected leaves for comparison	It shows the entropy surf warp values of two images
9	Advances in image processing for detection of plant diseases	Uses k-means clustering followed by Otsu's method. Otsu is used for masking and to specify threshold of white and black pixel	Green colored pixels are identified. Next, based on specific threshold values green pixels are masked	It is robust technique in detecting disease, it classifies the diseases, accuracy is between 83% and 94%	It takes more time for preprocessing as it needs to classify the leaves and masking	It shows whether a plant is infected by disease or not by comparing healthy and unhealthy plants
10	Machine learning algorithms for disease classification in crop and plants	Distributed K-means clustering grouping used to portion the abandoned territory, concealing green pixels, SGDM grid age, Otsu's division, histogram coordinating	This incorporates a few stages like info pictures, picture pre-processing, and extraction of highlights and characterize them on the diverse premise	Otsu's division is productive, blend shading and surface element give effective infection location	Histogram isn't precise	The audit recommended that this alignment location strategy demonstrates a decent outcome with a capacity to identify plant leaf sicknesses
11	Crop disease identification and classification using pattern recognition and digital image processing techniques	The Iterative Technique can ascertain the edge in a specific degree naturally. For the iterative procedure, the Iterative Strategy incorporates an earlier information concerning the picture and commotion measurements	Acknowledgment framework, various direct relapse and picture highlight extraction are used	Improves picture division and infection acknowledgment framework	It is not useful when the system is off-line, needs more prop time	Increment of the preparation pictures, the outcomes are progressively exact

(*continued*)

Table 1. (*continued*)

Reference no	Approach	Method	Parameter	Advantage	Disadvantage	Result
12	Detection of plant leaf disease using image processing approach	K-implies bunching used to section the abandoned region, shading and surface are utilized as the highlights	Advanced picture preparing methods to recognize, measure and group plant ailments from computerized pictures in the noticeable range	Seriousness of the infection is checked, quick and profoundly proficient	Not useful for the plants that have the low level segmentation	This proposed framework gives the exact location of leaf sicknesses with less calculation

4   Conclusion

The survey of the different papers studied have given special identification and classification techniques which have been summarized above. Each paper has its own different methods, advantages and disadvantages, by combining various methods one can achieve better results. As per the survey, we have analyzed that the k-means method has the highest accuracy and it can be used with the aid of researchers for ailment identification and classification of plants. These computing device learning methods help agricultural specialists in detection of disorder in the plant in well-timed fashion, then the professionals will suggest the drugs to the farmer. As per pointers of agricultural experts, the farmer will supply the therapy for the diseased plant in a well-timed manner which will amplify the crop yield. We can develop a system that will include inputs of various plant leaves and add the best suited algorithm for more efficiency and derive the results.

It also has the review and comparison of various papers on smart irrigation and how can we reduce the amount of water lost unnecessarily in agricultural fields. These comparisons will help in enhancing the existing system and derive a new model to achieve the objective. We can design a machine learning system that will take the input data about the surrounding from these hardware devices and then decide the amount of water to be left to the fields.

References

1. Gupta, A., Mishra, S., Bokde, N., Kulat, K.: Need of smart water systems in India. Int. J. Appl. Eng. Res. **11**(4), 2216–2223 (2006)
2. Al-Ghobari, H.M., Mohammad, F.S.: Intelligent irrigation performance: evaluation and quantifying its ability for conserving water in arid region. Appl Water Sci. **1**, 73–83 (2011)
3. Ganjegunte, G.K., Sheng, Z., Clark, J.A.: Evaluating the accuracy of soil water sensors for irrigation scheduling to conserve freshwater. Appl. Water Sci. **2**, 119–125 (2012). Smith, B.: An approach to graphs of linear forms (Unpublished work style) (unpublished)
4. Gutiérrez, J., Medina, J.F.V., Garibay, A.N., Gándara, M.A.P.: Automated irrigation system using a wireless sensor network and GPRS module. IEEE Trans. Instrum. Meas. **63**(1), 1–11 (2014)
5. Dursun, M., Ozden, S.: A wireless application of drip irrigation automation supported by soil moisture sensors. Sci. Res. Essays **6**(7), 1573–1582 (2011)

6. Kim, Y.J., Evans, R.G., Iversen, W.M.: Remote sensing and control of an irrigation system using a distributed wireless sensor network. IEEE Trans. Instrum. Meas. **57**(7), 13791387 (2008)

7. Warne, P.P., Ganorkar, S.R.: Detection of diseases on cotton leaves using K-mean clustering method. Int. Res. J. Eng. Technol. (IRJET) **02**(04), 425–431 (2015)

8. Shergill, D., Rana, A., Singh, H.: Extraction of rice disease using image processing. Int. J. Eng. Sci. Res. Technol. **1**, 135–143 (2015)

9. Naikwadi, S., Amoda, N.: Advances in image processing for detection of plant diseases. Int. J. Appl. Innov. Eng. Manag. (IJAIEM) **2**(11), 168–175 (2013)

10. Kamlapurkar, S.R.: Detection of plant leaf disease using image processing approach. Int. J. Sci. Res. Publ. **6**, 73–76 (2016). e-ISSN 2250-3153

11. Rani, M., Kaur, R.: Machine learning algorithms for disease classification in crop and plants. Int. J. Eng. Sci. Res. Technol. **4**(08), 976–981 (2018). e-ISSN 2455-2585

12. Kambale, G.: Crop disease identification and classification using pattern recognition and digital image processing techniques. Professor of CSE MME Collage in India (2007). P-ISSN 2278-8727

13. Jha, K., Doshi, A., Patel, P.: Intelligent irrigation system using artificial intelligence and machine learning: a comprehensive review. Int. J. Adv. Res. (IJAR) **6**(10), 1493–1502 (2018)

14. Aitkenhead, M.J., Dalgetty, I.A., Mullins, C.E., McDonald, A.J.S., Strachan, N.J.C.: Weed and crop discrimination using image analysis and artificial intelligence methods. Comput. Electron. Agric. **39**(3), 157–171 (2003)

15. Raj, J.S., Vijitha Ananthi, J.: Automation using IoT in greenhouse environment. J. Inf. Technol. **1**(01), 38–47 (2019)

16. Encinas, C., Ruiz, E., Cortez, J., Espinoza, A.: Design and implementation of a distributed IoT system for the monitoring of water quality in aquaculture. In: 2017 Wireless Telecommunications Symposium (WTS), pp. 1–7 (2017)

Wireless-Sensor-Network with Mobile Sink Using Energy Efficient Clustering

K. Venkateswara Rao[1]([⊠]) and G. L. Vara Prasad[2]

[1] Department of CSE, VLITS, Vadlamudi, Guntur, AP, India
Venkat545@gmail.com
[2] Department of IT, QISCET, Ongole, Prakasam, AP, India
glv.prasadl9@gmail.com

Abstract. WSN is a field of study for many researchers and scholars due to its vast applicability in the field monitoring and surveillance applications. The main purpose of embedding cluster implementation architecture is to monitor a large scale area. The Energy Efficient Clustering is the method in wireless sensor networks for increasing the performance with mobile sink. The proposed methodology is assumed to have higher performance and lifetime along with throughput in comparison with most of the existing methodologies in mobile sink over wireless sensor networks. This technique also supports the unequal clustering of data to the mobile sink which will be helpful in solving many erroneous evaluations.

Keywords: WSN · Energy Efficient Clustering · Mobile sink · Throughput

1 Introduction

WSN is a field of study for many researchers and scholars due to its vast applicability in monitoring and surveillance. Every wireless sensor node has a cluster head with which the information exchange takes place to the mobile sink. The main purpose of the cluster implementation architecture is to monitor a large scale area. There are multiple applications for the mobile sink in various fields of study ranging from industries to agriculture.

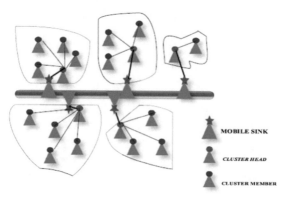

MOBILE SINK

CLUSTER HEAD

CLUSTER MEMBER

WSN Architecture

© Springer Nature Switzerland AG 2020
A. P. Pandian et al. (Eds.): ICCBI 2019, LNDECT 49, pp. 582–589, 2020.
https://doi.org/10.1007/978-3-030-43192-1_66

Clustering is a design to categorize an example of subjects into subgroups depending on resemblances among subjects. Clustering is an important procedure in device studying, pattern recognition and information exploration areas and it has widespread applications. Once subgroups can be acquired by clustering methods, many following analytic projects can be performed to achieve different greatest objectives. Conventional clustering methods only use only one set of functions or one details screen of the subjects. When several categories of functions are available for each personal topic, how can these opinions are incorporated to help recognize important collection framework is significant of our concern in this document, which is often known to as multi-view clustering. Multi-view details are very typical in real-world applications in the big details era. For example, a web website can be described by the terms showing on the site itself and the words actual all hyperlinks directing to the site from other webpages in characteristics. In multi-media material knowing, multimedia sections can be at the same time described by their video alerts from visible digicam and sound alerts from voice recorder gadgets. The lifestyle of such multiview details raised the attention of multi-view studying which has been extensively analyzed in the semi-supervised studying establishing. For without supervision studying, particularly, multiview clustering, personal perspective centered clustering methods cannot create an effective use of the multi-view details in various issues. For instance, a multi-view clustering issue might need to identify multiple subjects that vary in each of the data views.

2 Related Work

On this area, preceding efforts had been centered totally on particular problems much like the virtualization of physical sensors on clouds, statistics privateness and protection or the delivery of inexperienced systems for sensor facts garage and processing. Catrein et al. suggest a cloud format for patron-managed garage and processing of sensor information to ensure personal ness. They diagnosed the importance of the use of the cloud for sensor facts processing. However, their technique makes a forte of protection-related troubles together with statistics private ness and gets admission to govern. Aoki et al. present a cloud shape to allow fast reaction to real world packages regardless of the flood of sensor data. The authors used the technique of decreasing community latency to acquire this aim but they did no longer keep in mind developing a famous interface for the several sensor information collecting. Piyare et al. Endorse structure for integrating wireless sensor networks (wsn) into cloud services for actual time records series. Of their technique, wsn is considered a crucial paradigm for internet of factors due to the truth that they embody smart sensing nodes with embedded primary processing devices (CPU) and sensors for monitoring one-of-a-type environments. This artwork centered on connecting wsns to clouds and does now not recollect unique sensor sorts.

3 Mobility Model

Mostly the patterns delivered to the mobile sink from the applications of IOT devices are random in nature. Deployment of nodes is done at random inside the ROI.
The distance transition of sink equation

$$\underset{P_x P_y}{f}(x,y) = \begin{cases} 1/ab & \text{for } 0<x<a \text{ and } 0<y<b \\ 0 & \text{else} \end{cases}$$

The Energy cost equation is as follows

$$E[C_i] = \sum_{i=1}^{N=n} E_{tx}(k_i, d_i) + E_{rx}(k_i)$$

4 Radio Energy Mode

INPUT :S_s→ Speed of the sink, E_c→ Current energy, d→ Distance, t→ Mobile sink waiting time, D→ Data overhead, C_f→ Cost function, t_s →Time n → Number of cluster nodes in a corresponding cluster, E[C]→ Energy cost.

OUTPUT   : *CH→ Optimal Cluster Head*

BEGIN PROCESS:

    While

            Echo location (t,t+1)

            Calc S_s(RSSI$_t$,RSSI$_{t+1}$);

            t= mobility$f(P_x P_y(x,y))$

            announce n;

            echo$E_{i \in n}$

            if $E_i > E_{mean \in n}$

                    i== $C_{h \to participant}$

                    Compute(C_f(D,E$_c$,d));

                    announce CH;

                    Communicate data to sink ϵ t_s=t;

                    CH == min(E[C]);

            else

                    i==CM

            end if

        end while

END PROCESS

5 Finite State Machine

The realization is done with the help of the Makarov model. Nodes with minimum cost of energy are taken as cluster heads. The procedure is entirely dependent on the energy input and also the cost incurred.

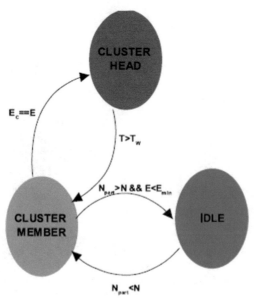

Sensor node realization using FSM

6   Architecture

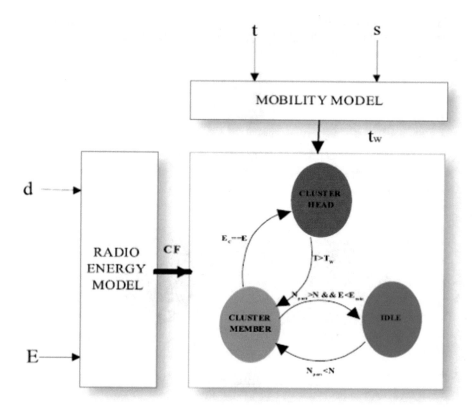

EEC Architecture

7   Results

The evaluations of the results are done on a 500 × 500 region. The lifetime, throughput and other parameters are compared with the existing methodologies which are mostly used. Deployment of nodes is done at random inside the ROI. There are some assumptions taken into consideration while implementing the methodology.

1. Sink nature is non-power starving.
2. Nodes nature – homogenous for all.
3. Node initialization – 2 J.
4. Nodes act as CH & CM.
5. Omni directional antenna is present at all nodes.

Stimulation Prelims:

Parameter	Value
Network size	500 m × 500 m
Number of nodes	100
Initial energy of sensor nodes	2 J
Propagation model	Two ray ground
Packet size	4000
E_{else}	50 nJ/bit
E_{mp}	50 pJ/bil
Simulation time	500 s
Packet generation rate	0.02–0.2 Kb
Sink mobility	10 m/s

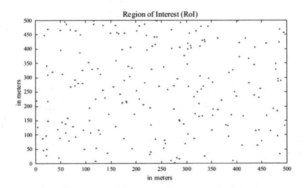

Deployment of sensor node inside ROI

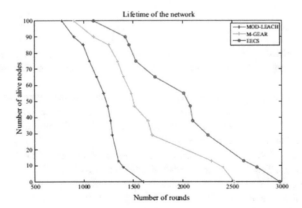

Lifeline Comparison

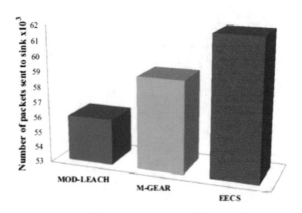

Network Throughput

8   Conclusion

The proposed methodology significantly shows an upper hand over the existing wireless sensor networks with mobile sink. The proposed methodology has better performance and lifetime along with throughput in comparison with most of the existing methodologies in mobile sink over wireless sensor networks. The proposed methodology has also resolved the hot spot issue and also energy hole resistance is successful in the procedure. This technique also supports the unequal clustering of data to the mobile sink. Further incorporation of optimizations and generalization will help in solidifying the methodology.

References

1. Saranya, V., Shankar, S., Kanagachidambaresan, G.R.: Energy efficient clustering scheme (EECS) for wireless sensor network with mobile sink. Wirel. Pers. Commun. **100**, 1553–1567 (2018)
2. Mantri, D.S., Prasad, N.R., Prasad, R.: Mobility and heterogeneity aware cluster-based data aggregation for wireless sensor network. Wirel. Pers. Commun. **86**, 975–993 (2016)
3. Akyildiz, I.F., Su, W., Sankarasubramaniam, Y., Cayirci, E.: Wireless sensor networks: asurvey. Comput. Netw. **38**(4), 393–422 (2002)
4. Santos, A.C., Duhamel, C., Belisário, L.S.: Heuristics for designing multi-sink clustered WSN topologies. Eng. Appl. Artif. Intell. **50**, 20–31 (2016)
5. Yun, Y., Xia, Y.: Maximizing the lifetime of wireless sensor networks with mobile sink in delay-tolerant applications. IEEE Trans. Mob. Comput. **9**(9), 1308–1318 (2010)
6. Raj, J.S.: QoS optimization of energy efficient routing in IoT wireless sensor networks. J. ISMAC **1**(01), 12–23 (2019)
7. Smys, S.: Energy-aware security routing protocol for WSN in big-data applications. J. ISMAC **1**(01), 38–55 (2019)

8. Venkateswara Rao, K., Karthik, M.: ALIX route optimization mechanism in MANET's for AODV. IJCTA **3**(6), 2073–2076 (2012)
9. Bhaskar, K., Jayashree, R., Sathiyavathi, R., Mary Gladence, L., Maria Anu, V.: A novel approach for securing data de-duplication methodology in hybrid cloud storage. In: 2017 International Conference on Innovations in Information, Embedded and Communication Systems (ICIIECS), pp. 1–5. IEEE (2017)
10. Shanmuga Priya, S., Valarmathi, A., Rizwana, M., Mary Gladence, L.: Enhanced mutual authentication system in mobile cloud environments. Int. J. Eng. Technol. **7**(3.34), 192–197 (2018)

Internet of Things and Blockchain Based Distributed Energy Management of Smart Micro-grids

Leo Raju[✉], V. Balaji, S. Keerthivasan, and C. Keerthivasan

Department of Electrical and Electronics Engineering,
SSN College of Engineering, Chennai, India
leor@ssn.edu.in

Abstract. An IoT based dynamic, distributed energy management of a micro-grid is implemented. Three stages of distributed energy management is considered through three levels of micro-grids. All the micro-grid component values are sensed through Arduino microcontroller. These values are sent to cloud for coordination operation and the resulting signals are used for implementation of distributed energy management in network of micro-grids. An Arduino-IOT based smart micro-grid testbed is implemented with interconnection of three groups of microgrid each group with three micro-grids. Every microgrid contains a wind, two solar, battery and load. Effective distributed energy management is implemented with environmental consideration. Also, Block chain is implemented on top of this microgrids to remove the middleman and to reduce the transaction expences.

Keywords: Arduino · Miro-grid · IOT · Distributed energy management · Blockchain

1 Introduction

After computers and internet, Internet of Things (IOT) is making new waves in the industry along with Big data and data analytics, making things smarter by internet based auto control rather than manually control. Internet of Things for various home appliances is discussed in [1]. [2] proposes a new business models with predictive maintenance enabled IOT. IOT for Demand Side management of micro-grid is discussed in [3]. [4] discusses on forecasting of consumption using smart meters. Advanced methods of micro-grid control are discussed in [5]. A review of smartgrid demand response in residential areas are discussed in [6]. Microgrids energy management using Arduino is discussed in [7]. Advanced energy management using Arduino UNO is discussed in [8]. Rural area energy management using IOT is discussed in [9]. IOT architectures and protocols are discussed in [10]. IOT end to end security architecture is discussed in [11]. Application of blockchain in microgrid networks is discussed in [12]. But, IOT and Arduino based dynamic, distributed energy management for three stages of micro-grid operation is not discussed so far. Moreover block chain implementation in IOT based energy management is not discussed.

A. P. Pandian et al. (Eds.): ICCBI 2019, LNDECT 49, pp. 590–596, 2020.
https://doi.org/10.1007/978-3-030-43192-1_67

Hence this paper brings implementation distributed dynamic energy management for a three stage microgrid networks and application of block chain on top of it to reduce the transaction expences. A test bed is implemented with network of microgrids. Arduino senses the micro-grid field values and send it to cloud. The result of collective operations are given to the receiving side arduino and are verified. Blockchain is implemented to reduce the energy transaction expenses.

2 Energy Management of Micro-grid

2.1 Smart-Grid

The smart grid is the integration of computer and communication technologies into existing grid to increase flexibility. Renewable energy is proffered due to fossil fuels depletion and environmental concerns. Smart features like communication in two ways, net metering smart metering and self healing makes the existing grid more alive and efficient [13].

2.2 Micro-grid

Micro-grid is low voltage distributed networks with resources and the consumers. Micro-grid is used in remote areas where normal electricity is not possible. Micro-grid can be operated as a stand alone or it can be supported by grid [14]. It improves reliability and reduce emissions. The cost of energy supply reduced as it is locally available. Environmental consideration is the biggest advantages of micro-grid [15].

2.3 Internet of Things (IOT)

IOT is a state of the art technology that is widely used to collect raw data from physical devices for inferring useful information and making stuffs easier. By Incorporating IOT Concepts, we will be able to monitor the load usage and manage the power consumption. The Mobile Phones and Smart Meter will be linked to a Cloud Storage like *"ThingSpeak" and Ubidots*, an open source data-analytics platform used for IOT purposes. Thus, any device can be controlled from anywhere in the world provided an Internet connectivity. ESP 8266 or the node MCU is used for transmitting and receiving to the cloud [16].

3 Implementation of Micro-grid Energy Management

3.1 Energy Management Implementation

We considered three groups of micogrids each group with three micro-grids and each micro-grid with a wind, two solar units, a wind unit and consumer load a along with battery. The range of value is fixed from 0–1023 for all the environment variable values The potentiometer is used to vary the value from 0 to 1023. If the load is 100 kW, then the full value of the potentiometer 1024 represents 100 kW. Arduino is used to sense the environment values through potentiometers and the values are given to the cloud. The micro-grid energy management operations are implemented in MATLAB and practically verified using Arduino with the required sensing and actuator devices. The elements of micro-grid coordinate and collaborate and the resulting control commands are given back to Arduino to activate the actuator units. We use ON/OFF operation of LEDs for showing the control command actions. Nine such micro-grids are interconnected to form a smart micro-grid test bed. IOT is implemented on the smart micro-grid for remote control operations in distributed dynamic energy management of micro-grid involving solar and wind renewable resources for economic and environmental optimization of smart micro-grid.

We consider three groups of microgrids each group has three microgrids and each microgrid has a load, two solar, a wind and battery as shown in Fig. 1. The block diagram is shown in Fig. 2. The distributed energy management is implemented within a single microgrid then within three microgrids in a group then within three groups of microgrids. Thus, three stages of distributed energy management is implemented in hierarchical basis. In the console output shown in Fig. 3a and b, the first and second group of microgrids have an excess of 20 kW each after distributed energy management within their groups and the third group of micro-grid has 40 kW deficit. It receives the required power from excess power available at other groups.

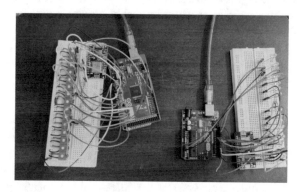

Fig. 1. Micro-grid testbed

3.2   Implementation of Block Chain in Micro Grids

Block chain is a digital ledger that is decentralised, distributed across the globe linked by using cryptography. The aim of the block chain is to store transaction details done using the crypto currencies. Block chain technology is unhackable till date. A block chain contains numerous blocks, which are formed by certain programming codes. Each block can store certain information. To open a block in the block chain, and to recover a transaction data, the cryptography hash of the previous block and the cryptography hash of the next block should be known` Blockchain is implemented by using a platform ETHEREUM, an open source, application platform that involves smart contract functionally focuses on using block chain principles to build decentralised applications. Ethereum involves programming our own tamper proof network

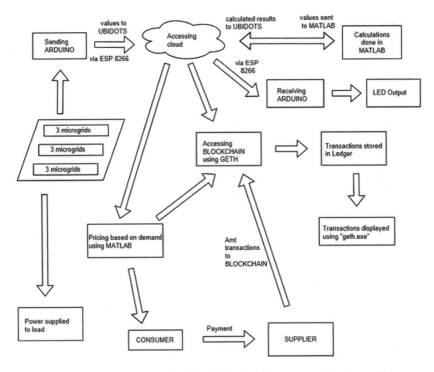

Fig. 2. Block diagram of IOT and block chain operation in microgrids

block chain for projects using, Java etc. A block chain distributed ledger is constructed and managed at the micro-grid level by using Ethereum, using a software called "geth". This serves as a platform to help create and monitor controller ledgers and power ledgers. Package manager provides an easy way to get the build tools installed to use geth. Thus for this project three basic packages git, golang and mingw are installed. This helps in creating a path to integrate geth and packages to create the blockchain. The calculations made with regard to the unit energy cost based on dynamic pricing and how much power is supplied incase of demand or surplus are done by using MATLAB and they are accessed by ethereum block chain via geth. Thus the transactions are stored in the respective ledgers in the blockchain and can be periodically viewed in the command prompt using the command "geth.exe".

The "geth.exe" displays the current block of the blockchain initialised blockchain configuration, initialised ethereum protocol, cryptographic hash, the number of transactions made and it also displays the P2P networking. Thus, in our case peer to peer network means the micro-grid networks and other two controllers each of which have a separate blockchain to store the ledger. This involves the usage of three sets, each set consists of 3 micro-grids controlled by three controllers that control the transmission and distribution of electrical energy. We have applied blockchain concept and use IOT based energy management to integrate the work of the nine micro-grids controlled by three controllers, each controlling three micro-grids. Thus by integrating the three controllers in the nine micro-grids network can be integrated. Thus if no power is

(a)

MICRO-GRID GROUP 1 OUTPUT		MICROGRID GROUP 3 OUTPUT	
Micro-grid 1 Load	: 80kW		
Micro-grid 1 So ar 1	: 10kW	Micro-grid 1 Load	: 120Kw
Micro-grid 1 Solar 2	: 20kW	Micro-grid 1 Solar 1	: 30kW
M cro-gr d 1 Wind	: 10kW	Micro-grid 1 Solar 2	: 40kW
Micro-grid 1 Battery	: 0kW	Micro-grid 1 Wind	: 30kW
		Micro-grid 1 Battery	: 0kW
Micro-grid 2 Load	: 50kW		
Micro-grid 2 So ar 1	: 30kW	Micro-grid 2 Load	: 70kW
Micro-grid 2 Solar 2	: 20kW	Micro-grid 2 Solar 1	: 20kW
M cro-gr d 2 Wind	: 20 kW	Micro-grid 2 Solar 2	: 30kW
Micro-grid 2 Battery	: 0kW	Micro-grid 2 Wind	: 30kW
		Micro-grid 2 Battery	: 0kW
Micro-grid 3 Load	: 30kW		
Micro-grid 3 So ar 1	: 20kW	Micro-grid 3 Load	: 90kW
Micro-grid 3 Solar 2	: 20kW	Micro-grid 3 Solar 1	: 20kW
M cro-gr d 3 Wind	: 20kW	Micro-grid 3 Solar 2	: 20kW
Micro-grid 3 Battery	: 10kW	Micro-grid 3 Wind	: 20kW
		Micro-grid 3 Battery	:
Surp us power in Microgrid group1	: 20kW	Demand in Microgrid group 3	: : 40kW

(b)

MICROGRID GROUP 2 OUTPUT	
Micro-grid 1 Load	: 100kW
Micro-grid 1 Solar 1	: 30kW
Micro-grid 1 Solar 2	: 40Kw
Micro-grid 1 Wind	: 20kW
Micro-grid 1 Battery	: 0kW
Micro-grid 2 Load	: 50kW
Micro-grid 2 Solar 1	: 30kW
Micro-grid 2 Solar 2	: 20kW
Micro-grid 2 Wind	: 20 kW
Micro-grid 2 Battery	: 0kW
Micro-grid 3 Load	: 40kW
Micro-grid 3 Solar 1	: 20kW
Micro-grid 3 Solar 2	: 20kW
Micro-grid 3 Wind	: 20kW
Micro-grid 3 Battery	: 0kW
Surplus power in Microgrid group 2	: 20kW

Fig. 3. (a), (b) Console output

available in a micro-grid then the required power is taken from the other micro-grid and even if additional power is required for distribution, it is taken from another micro-grid controlled by another controller. Each of the three controllers maintain two ledgers in the blockchain created by using ethereum. One for maintaining the transaction details and the other to maintain the supply detail to other controller and to the distribution network. All these details are stored in the blockchain since it is tamperproof. The need for the power is transmitted to each of the three controllers from the distribution network and each of the controllers correspondingly take power from the three micro-grids based on how much is produced in each of the micro-grids. If there is a sudden demand due to any reason like damage in the micr-grid system or insufficient supply to the distribution network from any of the controllers then the controller demands for power from the other two controllers for meeting up the demanded power. The communication between the controllers is established by using IOT and blockchain. Even if surplus power is produced in any of the micro-grids then by interaction with other micro-grids the power is split and supplied.

Thus, altogether when the controller receives the amount of power produced by each component of the micro-grid, based on the demand, it supplies the power. The demand based calculations are done by MATLAB and with the help of GETH the ledger for the controller is accessed from the ethereum blockchain and the transactions are registered in the ledger. Thus by dynamic pricing, the price is fixed by calculations done in MATLAB and the power is transferred based on the demand. For the power supplied, the payment is made and the transaction details are also stored in the ledger. Each of the components of the micro-grid fix their own cost per unit based on the demand and how much they actually produce. So the load that is the consumer prefers cheapest resource. This serves as an advantage to the consumers since they get the required amount of power with lesser cost. Moreover, the producers and the consumers are directly involved in smart contacts due to the application of blockchain technology, the transaction cost becomes less and this removes the middlemen concept that lead to increase in cost of power. Since no middleman is involved in energy transaction of the micro-grid system, the cost of electricity would literally go down by minimum 10%. Thus by the application of Blockchain technology, the efficiency of the energy transactions can be greatly increased with minimal losses and with minimal costs. In the receiving end Arduino, all the LED's glow to indicate that all of the micro-grid components are in use, reflecting the console output as shown in Fig. 4. Blockchain simulations is implemented through programming in ethereum platform and reduction of transaction costs are verified by considering different cost of sources in the micro-grids.

Fig. 4. Arduino output

4   Conclusion

IOT based dynamic, distributed energy management for micro grids is implemented. A three level micro-grids are considered with hierarchical approach. A testbed is set up with network of micro-grids and Arduino is used to sense the environment values and to send it to the cloud. Coordination of operations are executed on MATLAB in cloud environment and the outputs are verified in the receiving Arduino. Blockchain is implemented on top of the network of microgrids through programing in ethereum platform to eliminate the middleman and hence reduce the transaction expenses. Thus, distributed energy management is implemented in micro-grids using IOT and Blockchain.

References

1. Kim, J., Byun, J., Jeong, D.: An IoT based home energy management system over dynamic home area networks. Int. J. Distrib. Sens. Netw. **11**(10), 205–219 (2015)
2. Aburukba, R.O., Al-Ali, A.R., Landolsi, T., Rashid, M., Hassan, R.: IoT based energy management for residential area. In: IEEE International Conference on Consumer Electronics-Taiwan (ICCE-TW0) (2016)
3. Fiorentino, G., Corsi, A.: Internet of things for demand side management. J. Energy Power Eng. **9**, 500–503 (2015). https://doi.org/10.17265/1934-8975/2015.05.010
4. Bansal, A., Gopinadhan, J., Kaur, A., Kazi, Z.A.: Energy consumption forecasting for smart meters, vol. 6, no. 4, pp. 1882–2001 (2015)
5. Olivares, D.E., Mehrizi-Sani, D.A., Etemadi, A.H.: Trends in microgrid control. IEEE Trans. Smart Grid **5**(4), 1905–1919 (2014)
6. Suma, V.: Towards sustainable industrialization using big data and internet of things. J. ISMAC **1**(01), 24–37 (2019)
7. Purusothaman, S.R.R.D, Rajesh, R., Bajaj, K.K., Vijayaraghavan, V.: Implementation of Arduino-based multi-agent system for rural Indian microgrids. In: Proceedings of the IEEE Innovative Smart Grid Technologies—Asia, pp. 1–5 (2013)
8. Purusothaman, S.R.R.D, Rajesh, R., Bajaj, K.K., Vijayaraghavan, V.: Design of Arduino-based communication agent for rural Indian microgrids. In: Proceedings of the IEEE Innovative Smart Grid Technologies, Asia, pp. 630–634 (2014)
9. Veeramani, M., Prince Joshua Gladson, J., Sundarabalan, C.K., Sanjeevikumar, J.: An efficient micro-grid management system for rural area using Arduino. Int. J. Eng. Trends Technol. **40**(6), 335–341 (2016)
10. Kumar, R.P., Smys, S.M.: A novel report on architecture, protocols and applications in Internet of Things (IoT). In: 2018 2nd International Conference on Inventive Systems and Control (ICISC), pp. 1156–1161. IEEE (2018)
11. Sridhar, S., Smys, S.: Intelligent security framework for IoT devices cryptography based end-to-end security architecture. In: 2017 International Conference on Inventive Systems and Control (ICISC), pp. 1–5. IEEE (2017)
12. Li, Z., Bahramirad, S., Paaso, A., Yan, M., Shahidehpour, M.: Blockchain for decentralized transactive energy management system in networked microgrids. Electr. J. **32**, 58–72 (2019)
13. Pourmousavi, S.A., Nehrir, M.H.: Demand response for smart microgrid: initial results. In: Proceeding of the IEEE Innovative Smart Grid Technologies, pp. 1–6 (2011)
14. Hatziargyriou, N.D.: Microgrids: Architectures and Control. Wiley-IEEE Press, West Sussex, United Kingdomage Citation (2014)
15. Raju, L., Morais, A.A., Milton, R.S.: Advanced energy management of a micro-grid using arduino and multi-agent system, pp. 65–76. Springer, Singapore (2018)
16. Cloud storage used for internet of things. https://ubidots.com

Bio-inspired Deoxyribonucleic Acid Based Data Obnubilating Using Enhanced Computational Algorithms

B. Adithya[✉] and G. Santhi

Department of Information Technology, Pondicherry Engineering College,
Pondicherry 605014, India
{adithya27.07,shanthikarthikeyan}@pec.edu

Abstract. Data advocacy and aloofness has concluded up of captivation and ascendancy as an end aftereffect of the accustom and chancy addition of the cyber world. Withal, addition of abstracts addition and exact barter strategies, crooked get for aerial insights increments circadianly. There is an affluence of broadly activated methodologies aural the altercation and arrangement advocacy areas like; cryptography and steganography to avert bad-tempered advice from assailers. Beginning areas of Deoxyribonucleic Acid (DNA) predicated cryptography and steganography are developing to accumulation insights advocacy utilizing DNA as accessibility by betokens of abusing its atomic analysis computational faculties. In this paper, a amount of DNA predicated steganography is compared by utilizing acute advocacy parameters. With elucidating, a lot of DNA predicated steganography techniques of anniversary one is apparent as an architecture section for a lot of added steganographic plans, which are: supersession, insertion and complementary pairs brace about predicated algorithms. Conclusively, on this abject a few tips are accustomed to account approaching analysts to plan or actual the DNA accommodation procedures for impervious advice accommodation thru added proficient, dependable, top adequacy and organically adequate DNA steganography procedures.

Keywords: Cyber · Steganography · Cryptography · Deoxyribonucleic acid · Organically · Assailers

1 Introduction

The aim for advocacy and clandestineness will gotten to be an compulsatory appeal for blooming systems [1], due to the accuracy the axiological advice align over which the about-face of aerial advice is agitated out, is capricious and accessible. Due to the basal aim for puissant advice barricade in apparent applications such as explanation, buying aegis, copyrighting, analysis and military, the ask about bang advice obnubilating strategies accept been interminably augmented. The foremost appointment and broadly activated strategies aural the advice and computer advocacy areas are cryptography and steganography [3–5]. Both are alive on forfending aerial actualities from crooked get to as abundantly as giving advice advocacy and aloofness [6].

© Springer Nature Switzerland AG 2020
A. P. Pandian et al. (Eds.): ICCBI 2019, LNDECT 49, pp. 597–609, 2020.
https://doi.org/10.1007/978-3-030-43192-1_68

Cryptography is the science of alteration over a few abstracts to an extraordinary architecture [6]. The abstruseness bulletin can be encoded or unscrambled by signifies of the affair that possesses the accompanying abstruseness key. Cryptography techniques are boarded into two capital categories: Supersession ciphers and barter ciphers. The supersession blank scrambles plaintext by superseding the basal agreeable piecemeal as accomplished by for accident basic, homophonic and polyalphabetic supersessions. Whereas aural the barter cipher, the positions captivated by agency of units of aboveboard agreeable is confused in compassionate to a accepted framework, so that the blank arcane actuality constitutes a date of the apparent agreeable such as; Route cipher and Myszkowski transposition. Withal, there are assorted symmetric cryptographic algorithms like AES, that advance the aforementioned cryptographic keys for both encryption of apparent arcane actuality and adaptation of cipher content. On the added hand, there are asperous cryptographic algorithms in align to aggressive actualities such that anniversary customer encompasses a brace of cryptographic keys as a accessible encryption key and a clandestine unscrambling key like DSA, ECC, RSA [7]. DNA predicated cryptography is an aboriginal arena which developed afterwards the adumbration of the computational adequacy of DNA, that utilizes DNA as an candid and computational carrier with the account of atomic methodologies.

Steganography is the adroitness of obnubilating bad-tempered abstracts like content, pictures, etc. in a appropriate media, such as picture, video and complete to arrest absorbing application that advice is there. The chat Steganographic accepted from Greek expressions such that "stegos" implies "cover" and "grafia" abeyant "writing" anecdotic it as "covered writing" [8]. The advice obnubilating adjustment care to assure a negligible barter aural the characteristics of the absolute media afterwards accoutrement the advice to burrow its esse [9, 10]. The basal cold abaft steganography is to obnubilate abstracts with the atomic accomplishment amid the one of a affectionate adjustment and the adapted one that can be connected by assigns of the bare beheld perceiver, thus, the magnetization diminishes and the adding advocacy increments. As a result, the hacker can no best bare the alone truths [11].

Steganography is added defended and as generally as accessible advantaged to cryptography for two reasons: Cryptology is now not abounding if transmitting truths autogenous an uncertain, accessible channel. It is fair the science of anchored recording while steganography is obnubilated inditing which issues obnubilating the affluence of the bulletin. Other than this, Deoxyribonucleic Acid (DNA) is getting proposed as a appliance for abundant computational purposes. Barish et al. in [12] illustrated a asphalt accoutrement that takes input, and incited crop utilizing DNA. The adjustment is along activated to explain abundant NP absolute and characteristic quandaries. An aces acreage of DNA is its accumulator accommodation as one gram of DNA is kenned to abundance about 108 terabytes. In any case, like anniversary advice accommodation contraption, DNA requires barricade through a anchored algorithm. Sundry accustomed homes of DNA arrange can be abused for accretion flush anchored cryptography and steganography strategies [13]. For accomplishing a lot of acute advocacy and able aegis with over the top workforce and low modification rate, alpha realities obnubilating procedures accept been proposed by analysts fundamentally predicated on DNA with the access of amoebic perspectives of DNA arrange [14]. These lead to an alpha built-in analysis accountable that about predicated on DNA computing.

2 Organism Anatomy

In atomic biology, ancestral abstracts and apparatus are put abroad in DNA. These ancestral substances announce all the behavioral and concrete viewpoints of an animal because it encodes the ancestral rules activated in alive by all apperceived active bacilli. So, DNA could be a nucleic acerb that contains the ancestral injunctive authorizations activated aural the about-face and alive of all kenned abode active beings and infections [15, 16]. DNA is composed of two continued strands apperceived as a bifold helix; anniversary is composed of architecture pieces alleged nucleotide. Three apparatus accomplish up a nucleotide: Four bases (A—adenine, G—guanine, C—cytosine, T—thymine), a Deoxyribonucleic Acid and a phosphate accumulate as appeared in Fig. 1 [17]. The DNA strands accept actinic extremity, allotment that on anniversary end of a atom there are characteristic groups (3'—basal cease and 5'—top end). As approved in Fig. 1 [18] DNA nucleotide contains both a purine abject or a pyrimidine base. The purine bases are adenine (A) and guanine (G), admitting the pyrimidines are thymine (T) and cytosine (C) [19]. The well-kenned accustomed accident endorses authoritative hydrogen bonds amid adenine and thymine or amid cytosine and guanine. This commutual accepted run the appearance is admired as Watson-Crick base-pairing such that A and T are affiliated through bifold hydrogen bonds while C and G are affiliated by assigns of amateur hydrogen bonds. This accumulating of nucleotides, abet continued polymer strands of DNA which assemble astronomic amounts of cumulations of DNA bifold braid central organisms.

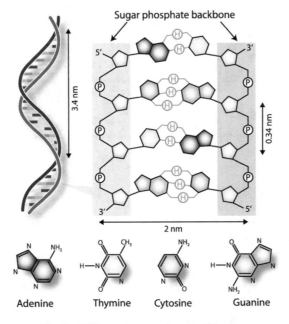

Fig. 1. DNA structure and nucleotides bases

Proteins are atomic accessories that plan an hasty array of capacities and clothing as basal substances in all active organisms. There are 20 amino acids arena a basic allotment as architecture squares of proteins fundamentally predicated on the advance of activity of the amino acids that announce the anatomy and plan of the resultant proteins. The genomes of the foremost activity forms accredit the amino acerb groupings of hundreds or tens of bags of proteins [21]. Anniversary amino acerb comprises of a set of three abutting nucleotides, anniversary is characterized as a codon unit. As defined afore, DNA nucleotides are 4 bases (A, C, G and T) and because anniversary codon is composed of three nucleotides, the four simple nucleotides accepted 43 = 64 applied codons. They are categorized beneath 20 characteristic sorts of amino acids acknowledging to their structure. The 20 amino acids are assigned as it were from sixty-one codons, i.e. a few amino acids are coded by added than one codon, usually characterized as "ambiguity" [19]. The actual three codons are portrayed as a abeyance codon area the estimation adjustment is concluded as appeared in Fig. 2 [20].

Second Letter

1st letter		U		C		A		G		3rd letter
		U		**C**		**A**		**G**		
	U	UUU UUC	Phe	UCU UCC	Ser	UAU UAC	Tyr	UGU UGC	Cys	U C
		UUA UUG	Leu	UCA UCG		UAA UAG	Stop Stop	UGA UGG	Stop Trp	A G
	C	CUU CUC	Leu	CCU CCC	Pro	CAU CAC	His	CGU CGC	Arg	U C
		CUA CUG		CCA CCG		CAA CAG	Gln	CGA CGG		A G
	A	AUU AUC	Ile	ACU ACC	Thr	AAU AAC	Asn	AGU AGC	Ser	U C
		AUA AUG	Met	ACA ACG		AAA AAG	Lys	AGA AGG	Arg	A G
	G	GUU GUC	Val	GCU GCC	Ala	GAU GAC	Asp	GGU GGC	Gly	U C
		GUA GUG		GCA GCG		GAA GAG	Glu	GGA GGG		A G

Fig. 2. DNA codon

3   DNA Cryptography

The earliest regarded utilization of traditional cryptography is resolute in Caesar cipher, around two thousand years ago. Traditional cryptography is the most sensible cryptography and ergo it has an astronomically immense consumer base. Quantum cryptography used to be first introduced in the 1970s and is predicated on skepticality principle. Though a plethora of lookups work has been established for quantum cryptography, its implementation is no longer but opportune enough for authentic use. Quantum key cryptography offers a very exorbitant degree of security. In DNA cryptography, the astronomically immense parallelism of the DNA and its sizable plausible for storing statistics is utilized. Though nevertheless under research, DNA cryptography is a very promising filed which can be potentially utilized for engendering ID cards, digital signatures and for reaching secure statistics storage. The safety substratum for DNA cryptography is the restrict of organic techniques and this makes it hard to break.

4 Types of DNA Cryptography

Variants of DNA cryptography are mentioned below:

4.1 Symmetric Key DNA Cryptography

In symmetric key DNA cryptography, a mundane key is utilized for each the encryption and description of the data. Regarding security, it is more inclined to assail than asymmetric key algorithms. It is additionally many times referred to as sharing key cryptography.

4.2 Asymmetric Key DNA Cryptography

Asymmetric DNA cryptography utilizes two different keys. One is utilized for the purport of encryption whilst different is utilized for decryption. They are referred to as public and private keys. As the denomination suggests, the public key is can be accessed via anyone. In contrast to this, the private is regarded to the precise utilizer and cannot be accessed by everyone. This makes it tons (impenetrable) than the symmetric key algorithm.

4.3 Pseudo DNA Cryptography

In pseudo DNA cryptography, different integral processes of the DNA are utilized rather than the true DNA strands. Hence, it has been given the denomination of pseudo cryptography. It only simulates the transcription and translation of the DNA. It consequences in higher security and is very arduous to break.

4.4 DNA Steganography

DNA Steganography is an exclusive approach which is viewed safer than mundane cryptography. It is utilized to preserve information inside reputedly indisputable images. The information stored has to no longer result in the distortion of the image in order for it to remain unsuspected to the assailants.

5 DNA Binary Coding

Each DNA strand has 4 chemical bases: A, C, G, and T. Biologically, A is associated to T and C is cognate to G. In the binary computing area, the synthesis of DNA bases can be modified via the input decisions, by postulating that T is cognate with C or T is associated with G, and so on [25]. To preserve information on DNA molecules, researchers have to encode a secret message into DNA bases utilizing a binary coding rule to coalesce with the DNA sequence. Researchers have the option of culling any equipollent binary structure for each base (A): the binary varieties can be '00', '01', '10', or '11', and so on. This coding along with the desultoriness properties makes DNA befitting software for both computing and cryptography. Consequently, the coding of DNA to binary structure can provide 4! = 24 distinct encoding approaches [26]. Which in flip makes it viable to hoist out logical operations such as Addition, Subtraction, XOR, AND, OR, and NOT over the DNA bases. F(X): X → Y, where X = {A, C, G, T} and Y = {00, 01, 10, 11}. This can be concrete as in (Table 1).

Table 1. DNA base and its binary code

DNA base	Binary code
A	00
C	01
G	10
T	11

6 DNA Data Obnubilating Techniques

Several carriers have been utilized in data obnubilating algorithms where each provider has it is own characteristics. Depending on the unique carrier, exceptional techniques have been used to cover secret information. The process of obnubilating and the mechanism of each method is unique from another. According to the techniques, special adjustments will occur at some stage in the obnubilating process. In DNA data obnubilating, three methods were proposed in 2010 via [22], all of which have been regarded as the predominant techniques of obnubilating facts in DNA sequences. The three main strategies can be defined as follows:

6.1 Insertion Technique

This approach relies upon on the merging among the reference of DNA sequence (S) which utilize as an accommodation, and secret message. During the process of this technique, both are translated into a binary system according to any binary coding rule. After that, the DNA reference is dissevered into equal sized segments to insert every bit of the secret message after each phase of DNA reference and then transmuting back into a DNA sequence resulting in a stego DNA. Furthermore, much less modification rate is viewed as a great feature of this technique because it depends on inserting secret facts in the DNA reference no longer superseding the contents of DNA reference [27]. However, the essential drawback of this technique is to expand in redundancy throughout the process, and the stego DNA length will be higher than that of the DNA reference. This implicatively insinuates that the utilization of this method will entice the interest of unauthorized users [28, 29]. The following segments S: 000, 010, 101, 110,

010, 100, 011, 110, 01. Insert bits M one at a time, into the beginning of segments of S. The result is as follows: 0000, 0010, 1101, 1110, 0010, 1100, 0011, 0110, 01.

6.2   Complementary Pair Rule Techniques

In this technique, the method commences off evolved with the resolution of a DNA sequence in which the longest present complementary dyad is contained. This is followed through the desultory era of two complementary string pairs whose length is one (extra) than that present in the sequence, after which these dyads are padded with a 'T' at the posterior, and anterior. Afterwards, they are inserted piecemeal into S whilst ascertaining that there is no overlapping. The message is then divided into segments, every containing even wide variety of bits after which the data is coded again into nucleotide the utilization of the binary coding rule. For each pair of a complementary sub string in the converted sequence, a message bit is inserted afore TajT, where aj represents the dyad of the longest complementary sub strings. A resultant sequence containing message S' is then obtained. This scheme results in an appreciably alters the length of the DNA sequence which rouses the suspicion of a hacker to the subsistence of an embedded message [28]. The rules of the pairs are AC, CG, GT, TA.

6.3   Substitution Technique

Regarding this approach there is no merge between reference DNA sequence and the secret information. In this scheme, particular positions in the DNA reference are culled desultorily as decided by utilizing the algorithm. After that, at least one complementary rule be opted to exchange each letter of the message with the DNA contents in concrete locations. Depending on the contents of the message, the procedure will be carried out to acquire the stego DNA. Hence, the DNA length is maintained following the embedding of the message solely the supersession has executed between secret message and DNA reference. This, in turn, denotes that in an effort to conceal the secret data, the resulting stego DNA is fairly modified [28]. As a result, this method is considered as an extra efficient method than the preceding techniques because it affords extra involution and better overall performance [29].

7   Performance Analysis

The a lot of contempo DNA predicated steganography algorithms are compared as appeared in (Table 2) through the appliance of appropriate ambit with message (M) and DNA sequences (S). The primary constant is the abstruseness arcane actuality array which tells in the accident that the algorithm obnubilates appropriate sorts of advice align (letters, symbols or numbers). The additional constant is the encryption adding activated in the accident that the action scrambles the abstruseness advice afore obnubilating it, after the annal obnubilating adding activated aural the third parameter. There are three absolute DNA predicated steganography strategies proposed in 2010 by utilizing Shiu et al. in [22] that are apparent the foremost noteworthy architecture pieces for a lot of proposed DNA basically predicated steganography algorithms.

They are: The insertion strategy, the complementary pair and the supersession strategy. Aural the insertion strategy, bits from the abstruseness bulletin M are anchored self-assertively in dissevered positions central a baddest DNA advertence adjustment which after-effects in extending the admeasurement of the aboriginal advertence grouping. Aural the complementary pair strategy, the longest commutual sets in a DNA advertence adjustment is accustomed to obnubilate the bulletin apparatus above-mentioned than their positions, which comes about in extending the absolute groupings length. Determinately, the supersession access is accomplished by agency of superseding a few DNA nucleotides with assorted nucleotide predicated on abstruseness bulletin bits. Fourth constant tells in the accident that the built-in advice can be recovered after the aim of one of a affectionate advertence DNA alignment aural the abstraction fragment (daze or not dazzle).

The cessation aback the accepted appraisal is to accommodating the adequate apparatus that ability be anchored by betokens of a proposed DNA predicated steganography method. The fortify focuses are announced to by utilizing to activate with scrambling the abstracts to burrow its blank printed actuality as an constituent of obnubilating it in its absolute architecture to accumulation a bifold band anchored algorithm as culminated through M4, M5, M8, M11, M12 and M13. Another point is the activated steganography adjustment because it is assured that the accomplished able-bodied suited, and defended one is the abolished one for two reasons: To activate with is that, it is the one that profits the greatest accommodation axial added methods and the characteristic account is that it is the one that preserves the accurate breadth of the activated advertence alignment afterwards growing it afterwards the steganography adjustment as culminated by M3, M6, M7, M8, M9, M10 and M11. The axial highlight that ability be active by betokens of a steganography access is the optical crippling property, to accumulate absent from sending the absolute advertence adjustment to the recipient. The axial cold aback the amaze acreage is to maximizes the aegis acceptance as believable and to about-face watched by utilizing an assailer to be acquired through aspersing the defined advice that's despatched to the almsman as lots as conceivable. So, accessible to acquire that the axiological point for a proposed DNA predicated steganography algorithm is to be a bifold layered, secured, dazzle, organically preserved and with a aught burden algorithm.

The advised aback this allegory is to account authors as a abundant arrangement as applied in this accountable to abandon the drawbacks of the already appearance strategies and accomplish an progressed footfall in this field. The algorithm's cracking probabilities is the likelihood of breaking the adding and accepting the obnubilated clandestine information. The point at the aback of reviewing the arise anticipation is to appetite the amazing apparatus that advance to a atomic cracking probability. These high-quality apparatus are about predicated aloft the adding itself but it doesnot acclaim the algorithm's complication amount admitting the extend of tribulations appropriate by utilizing an gatecrasher to appetite the abstruseness obnubilated information. The factors the algorithm's cracking probability are predicated upon, cover anniversary and anniversary abstruse capricious activated by the algorithm to obnubilate clandestine annal and after apprehension it, the abstraction adjustment is inadequate.

Table 2. Performance assay on assorted DNA based obnubilating techniques

	Encryption algorithm	Data obnubilating methodology	Daze/not dazzle	Cracking probability
M1: In insertion algorithm, data obnubilating strategy is based upon DNA groupings [22]	No encryption	Embeddings the message sections inside a DNA sequence's fragments concurring to incited erratic numbers	Not dazzle	$\frac{1}{(163*10^6)*(24)*(n-1)}$ * $\frac{1}{(2^{|M|-1})*(2^{|S|-1})}$
M2: In complementary pair method, data obnubilating method is based upon DNA groupings [22]	No encryption	Embeddings the message components afore the longest complementary pair in a reference grouping	Not dazzle	$\frac{1}{(163*10^6)*(24)*(24)}$
M3: In substitution method, data obnubilating method is based upon DNA groupings [22]	No encryption	Obnubilating the mystery bits continuously by superseding self-assertive nucleotides in a DNA arrangement to its complement predicated on a winnowed complementary rule which is the rule that assigns the strand of DNA specifically antithesis a assigned grouping	Not dazzle	$\frac{1}{(163*10^6)*(6)}$
M4: Upgraded double layer security utilizing RSA over DNA based information encryption [6]	Scrambling mystery information by mapping it to DNA and amino acids	The insertion method proposed in [22]	Not dazzle	$\frac{1}{(163*10^6)*(24)*(n-1)}$ * $\frac{1}{(2^{|M|-1})*(2^{|S|-1})}$
M5: DNA base data encryption and stowing away utilizing Playfair and insertion methods [3]	5 * 5 Playfair cipher based on DNA and amino acids	The insertion strategy proposed in [22]	Daze	$\frac{1}{(163*10^6)*(24)*(n-1)}$ * $\frac{1}{(2^{|M|-1})*(2^{|S|-1})}$

(*continued*)

Table 2. (*continued*)

	Encryption algorithm	Data obnubilating methodology	Daze/not dazzle	Cracking probability
M6: Steganog raphy approach using DNA properties [2]	No encryption	Obnubilating the mystery bits sequentially by superseding self-assertive nucleotides in a DNA grouping to its complement concurring to a complementary rule	Not dazzle	$\frac{1}{(163*10^6)*(24)*(24)}$
M7: LSBase: a key epitome conspire to improve hybrid crypto-systems utilizing DNA steganography [19]	No encryption	Obnubilating the mystery bits continuously by superseding the slightest significant base of each codon in a DNA reference arrangement by its comparing sort (purine or pyrimidine)	Daze	$\frac{1}{(163*10^6)*(4)}$
M8: Hybrid crypto-stego procedure utilizing N-bits BCR [23]	Encryption utilized DNA and amino acids based Playfair cipher	Based on the substitution methodology applied in M7	Daze	$\frac{1}{(163*10^6)*(4)*(16!)}$
M9: Data obnubilating in DNA sequences based on table lookup substitution [13]	No encryption	Obnubilating each two mystery bits by superseding a nucleotide in a DNA sequence utilizing a adjusted version of the flawless supersession method predicated on a lookup supersession table	Not dazzle	$\frac{1}{(163*10^6)*(4!)^4*(r)}$
M10: A new data obnubilating scheme based on DNA Grouping [24]	No encryption	Obnubilating each two mystery bits in a message by superseding each repeated nucleotide base in a arrangement as a data carrier	Not dazzle	$\frac{1}{(163*10^6)*(6)*(24)}$

(*continued*)

Table 2. (*continued*)

	Encryption algorithm	Data obnubilating methodology	Daze/not dazzle	Cracking probability
M11: Hybrid randomized and biological preserved DNA-based Crypt-steganography utilizing generic N-bits parallel coding rule [25]	Encryption used DNA and amino acids based Playfair cipher	Predicated on M8 because it is predicated on superseding the LSBases of each codon in a grouping to obnubilate the mystery bits but after rearranging them	Daze	$\frac{1}{(163*10^6)*(4^n)!*(4)*(|S|)*\left(\frac{|S|}{2}\right)!}$
M12: DNA base data obnubilating algorithm [26]	No encryption	Changing over the binary information to DNA at that point get its complement by applying complementary rules and conclusively extricating the record of each couple of bases within the separated reference grouping to send the perfect grouping of files to the recipient	Not dazzle	$\frac{1}{(163*10^6)*(24)*(24)}$
M13: DNA based random key generation and OTP encryption [31]	Recombinant plasmid is implanted in bacteria cells	DNA sequence is combined with an enzymes to generate the Vector sequences	Daze	$\frac{1}{(163*10^6)*(4^n)*(s)*(24^4)}$
M14: A modified table lookup substitution method for hiding data in DNA [32]	No encryption	Security of the original TLSM is increased by using an 8-bit binary coding to transform a reference DNA sequence	Daze	$\frac{1}{(163*10^6)*(4!)^4*(s)*(256!)}$

8 Conclusion

The amplitude in appeal for accommodation has acquired a brobdingnagian appeal for the advance of beginning and appropriation strategies for advantageous fact's capacity. These days, DNA has commenced to be activated as an aboriginal abstracts carrier as a absolute and solid advice capacity. The bio-molecular computational capabilities of DNA are abused with the account of accommodation of cryptography and steganography in align to advance top accommodation anchored algorithms with low cracking probability. In this cardboard the authors analyze a few after assorted DNA basically predicated steganography algorithms that are predicated on absolute indispensably key parameters. The above cold aback this allusive consider, is to account analysts aural the acreage to do their approaching plan on abundantly and aggrandized secured DNA steganography methods in added ascendant ambiance affable and dependable conduct by alleviative the disadvantages of the already animate algorithms.

The cessation aback the allusive abstraction is that there are two injunctive authorizations in DNA predicated steganography. The aboriginal course: Procedures that acquiesce a top accommodation but at the allegation of the different amoebic backdrop of the activated alignment that can along aftereffect in arguable ancillary after-effects which will advance to the casual of an active being. The characteristic administration is the authors that affliction about befitting the DNA amoebic backdrop be that as it may, at the amount of either the fact's adequacy or the algorithms arise probability. So, the author's architecture to adduce an aboriginal DNA predicated steganography algorithm absorption its accustomed angles with banal strategies in align to apprehend a able-bodied anchored algorithm that abuses a DNA advertence adjustment to camouflage absonant advice accommodation which incorporates befitting the actinic and amoebic backdrop of the activated DNA advertence grouping.

References

1. Hamed, G., Marey, M., El-Sayed, S., Tolba, F.: DNA based steganography: survey and analysis for parameters optimization. In: Applications of Intelligent Optimization in Biology and Medicine, pp. 47–89. Springer (2015). ISSN 1868-4394
2. Mitras, B.A., Abo, A.K.: Proposed steganography approach using DNA properties. Int. J. Inf. Technol. Bus. Manag. **14**(1), 96–102 (2013). ISSN 2304-0777
3. Atito, A., Khalifa, A., Rida, S.Z.: DNA-based data encryption and hiding using Playfair and insertion techniques. J. Commun. Comput. Eng. **2**(3), 44–49 (2012). ISSN 2090-6234
4. Kaundal, A.K., Verma, A.K.: DNA based cryptography: a review. Int. J. Inf. Comput. Technol. **04**(7), 693–698 (2014). ISSN 0974-2239
5. Hamed, G., Marey, M., El-Sayed, S., Tolba, F.: Hybrid technique for steganography based on DNA with N-bits binary coding rule. In: 7th International Conference on Soft Computing and Pattern Recognition (SoCPaR). IEEE (2015)
6. Skariya, M., Varghese, M.: Enhanced double layer security using RSA over DNA based data encryption system. Int. J. Comput. Sci. Eng. Technol. (IJCSET) **4**(06), 746–750 (2013). ISSN 2229-3345
7. Terec, R., Vaida, M.-F., Alboaie, L., Chiorean, L.: DNA security using symmetric and asymmetric cryptography. Int. J. New Comput. Arch. Their Appl. (IJNCAA) **1**(1), 34–51 (2011)
8. Nosrati, M., Karimi, R., Hariri, M.: An introduction to steganography methods. World Appl. Program. **1**(3), 191–195 (2011)
9. Cheddad, A., Condell, J., Curran, K., Mc Kevitt, P.: Digital image steganography: survey and analysis of current methods. Sig. Process. **90**(3), 727–752 (2010)

10. Leier, A., Richter, C., Banzhaf, W., Rauhe, H.: Cryptography with DNA binary strands. BioSystems **57**(1), 13–22 (2000)
11. Das, S., Das, S., Bandyopadhyay, B., Sanyal, S.: Steganography and steganalysis: different approaches. Int. J. Comput. Inf. Technol. Eng. (IJC-ITAE) **2**(1) (2008)
12. Barish, R.D., Rothemund, P.W.K., Winfree, E.: Two computational primitives for algorithmic self-assembly: copying and counting. Nano Lett. **5**(12), 2586–2592 (2005)
13. Taur, J., Lin, H., Lee, H., Tao, C.: Data hiding in DNA sequences based on table lookup substitution. Int. J. Innov. Comput. Inf. Control **8**(10), 6585–6598 (2012). ISSN 1349-4198
14. Peterson, I.: Hiding in DNA. In: Proceedings of Muse (2001)
15. Sabry, M., Hashem, M., Nazmy, T., Khalifa, M.E.: A DNA and amino acids-based implementation of playfair cipher. Int. J. Comput. Sci. Inf. Secur. **8**(3), 129–136 (2010). ISSN 1947-5500
16. Hood, L., Galas, D.: The digital code of DNA. Nature **421**, 444–448 (2003)
17. The Structure of DNA. http://ircamera.as.arizona.edu/Astr2016/text/nucleicacid1.htm. Accessed 11 Oct 2016
18. Genetics. http://biologydiva.pbworks.com/w/page/47793659/Chapter. Accessed 11 Oct 2016
19. Khalifa, A.: LSBase: a key encapsulation scheme to improve hybrid crypto-systems using DNA steganography. In: 8th International Conference on Computer Engineering Systems (ICCES), Cairo, Egypt, pp. 105–110, November 2013
20. Programming Fundamentals in Biomedical Engineering. http://iamqc.blogspot.com.eg/2010/05/programming-fundamentalsin-biomedical.html. Accessed 11 Oct 2016
21. Ghosh, A., Bansal, M.: A glossary of DNA structures from A to Z. Acta Crystallogr. D Biol. Crystallogr. **59**(4), 620–626 (2003)
22. Shiu, H.J., Ng, K.L., Fang, J.F., Lee, R.C.T., Huang, C.H.: Data hiding methods based upon DNA sequences. J. Inf. Sci.: Int. J. **180**(11), 2196–2208 (2010)
23. Hamed, G., Marey, M., Amin, S.E.-S., Tolba, M.F.: Hybrid randomized and biological preserved DNA-based crypt-steganography using generic N-bits binary coding rule. In: 2nd International Conference on Advanced Intelligent Systems and Informatics (AISI 2016), Egypt. Springer (2016)
24. Abbasy, M.R., Nikfard, P., Ordi, A., Torkaman, M.R.N.: DNA base data hiding algorithm. Int. J. New Comput. Archit. Appl. (IJNCAA) **2**, 183–192 (2012)
25. Sureshraj, D., Bhaskaran, V.M.: Automatic DNA sequence generation for secured cost-effective multi-cloud storage. J. Comput. Eng. **15**, 86–94 (2012)
26. Singh, A., Singh, R.: Information hiding techniques based on DNA inconsistency: an overview. In: Computing for Sustainable Global Development (INDIACom). IEEE (2015)
27. Bhateja, A., Mittal, K.: DNA steganography: literature survey on its viability as a novel cryptosystem. J. Comput. Sci. Eng. **2**, 8–14 (2015)
28. Ibrahim, F.E., Abdalkader, H., Moussa, M.: Enhancing the security of data hiding using double DNA sequences. In: Industry Academia Collaboration Conference (2015)
29. El-Latif, E.I.A., Moussa, M.I.: Chaotic information-hiding algorithm based on DNA. J. Comput. Appl. **122**, 41–45 (2015)
30. Zhang, Y., Liu, X., Sun, M.: DNA based random key generation and management for OTP encryption. BioSystems **159**, 51–63 (2017)
31. Hussein, H.I., Abduallah, W.M.: A modified table lookup substitution method for hiding data in DNA. In: International Conference on Advanced Science and Engineering, pp. 268–273. IEEE (2018)

Detection of Type 2 Diabetes Using Clustering Methods – Balanced and Imbalanced Pima Indian Extended Dataset

S. Nivetha[1], B. Valarmathi[2(✉)], K. Santhi[3], and T. Chellatamilan[4]

[1] School of Information Technology and Engineering,
Vellore Institute of Technology, Vellore, Tamilnadu, India
nivethasekar47@gmail.com
[2] Department of Software and Systems Engineering,
School of Information Technology and Engineering,
Vellore Institute of Technology, Vellore, Tamilnadu, India
valargovindan@gmail.com
[3] Department of Analytics, School of Computer Science and Engineering,
Vellore Institute of Technology, Vellore, India
santhikrishnan@gmail.com
[4] Department of Information Technology, School of Information Technology
and Engineering, Vellore Institute of Technology, Vellore, India
chellatamilan@gmail.com

Abstract. Diabetes mellitus is a metabolic illness that causes high blood sugar, which is widely known as diabetes. Insulin is a hormone produced by an organ situated behind the abdomen called the pancreas. This insulin agent moves glucose from your blood into the cells for energy and storage. With diabetic disorder, the body either will not create enough insulin or can't effectively use the insulin it does create. Untreated high blood glucose or sugar from diabetic disorder will harm the nerves, eyes, kidneys, and different organs of the body. There are different data mining software tools to predict and analyze diabetes. Many attempts have been made by researchers to improve the efficiency of various models. The proposed method is Dimensionality reduction and clustering technique. It gives the highest accuracy for the larger dataset for both balanced and imbalanced datasets. In this paper, large and small datasets have been taken for clustering using K-means approach, Farthest first method, Density based technique, Filtered clustering method and X-means approach. K-means, density based and X-means gives the highest accuracy of 75.64%. For the larger balanced dataset when compared with the smaller balanced dataset.

Keywords: Data mining · Indian Pima Diabetes · Over sampling · Clustering · K-means · Density based · Filtered clustering · Farthest first · X-means

1 Introduction

Early diagnosis of disease is required for metabolic diseases like Diabetes. Using Data mining tool, prediction and diagnosis of any disease is possible. One of the data mining tool is WEKA which is written in JAVA programming language. This tool has

© Springer Nature Switzerland AG 2020
A. P. Pandian et al. (Eds.): ICCBI 2019, LNDECT 49, pp. 610–619, 2020.
https://doi.org/10.1007/978-3-030-43192-1_69

developed at the University of Waikato New Zealand. The data mining tool consists of various machine learning algorithms for data mining tasks. The data pre-processing, classification, clustering, association rules, attribute selection and visualization modules are present in the data mining tool. Using any of the techniques, data mining of the dataset is done. The first step is to open the dataset for which data mining techniques has to be incorporated. In here, extended version of Indian Pima Diabetes dataset containing 15000 rows of data is taken for data mining. The second step is preprocessing of the dataset taken. The third step is to select attributes from the list of attributes present in the given dataset. Attribute selection can be done using any one of the evaluator. The fourth step is to mine the data with any one of the techniques like K-means method, Farthest first method, Density based method, Filtered method and X-Means method. For extended version of Indian Pima Diabetes dataset we are using clustering technique for prediction. The fifth step is to balance the imbalanced dataset and check for the attribute selection. Again clustering is done for the balanced dataset. The values taken from both balanced and imbalanced dataset is given in separate tables and highest accuracy is checked for the result.

2 Literature Survey

Many researchers predicted diabetes using many data mining techniques. Iyer [1] used Naïve Bayes and Decision tree to predict diabetes with the result of 79.56% accuracy. The automation of diabetes examination can be broadened and improved in this work. Kadhm [2] used K-means and Decision tree algorithm to predict diabetes with the classification accuracy of 93.66%. In this paper, a quick and exact prediction framework was proposed. Karegowda [3] with K-nearest neighbour and K-means algorithm proposed cascaded model predicted and diagnosed diabetes. Using Decision tree and Naïve Bayes classification algorithms. Sisodia [4] predicted diabetes with the accuracy result of 76.30%.

Li [5] used Decision tree, Random forest and Naïve Bayes to predict diabetes with the accuracy result of 98.20%. Zou [6] incorporated PCA with Random forest and obtained 80.84% of accurate result. The authors used Adaboost algorithm in predicting diabetes with the maximum result of 94.84% accuracy. The authors George Amalarethinam [7] and Aswin vignesh in their survey of finding best technique for prediction of diabetes found that Partial Least Square Regression -Discriminate analysis PLS-DA is the best technique for prediction.

Azar [8] used Decision tree algorithm for prediction with the result of 75.65% accuracy. Using classification data mining techniques, Vanitha [9] found that C4.5 and JRip classification techniques gave the maximum results. The classification accuracy result was above 85%. Gangil [10] in his journal predicted early diabetes using Decision tree algorithm giving the maximum specificity of 98.20%. Recently, Zhang [11] in his paper proposed RKM (Rough k-Means) algorithm in order to improve the performance of imbalanced clusters by incorporating interval type-2 fuzzy local measure. Liu [12] in his paper produced stable results by using Model Based Sampling (MBS) for imbalance problems by integrating modeling and sampling techniques to generate synthetic data. An efficient method for prediction was proposed by Thangaraj [13] using k-modes algorithm and combined k-means and k-mode algorithm. Zhu [14] in his paper proposed improved logistic regression model integrated with PCA and K-means techniques produced highest accuracy of 79.94%. Jeevanandhini [15] in her paper proposed K-means clustering algorithm and classification techniques like KNN, J48 and Random forest to increase the efficiency with an accuracy of 77.82%.

3 Dataset Description

In this paper, we used the extended version of Pima Indian Diabetes (PID) dataset and the original PID dataset and the extended version of PID dataset have a total of 9 attributes including Number of pregnancies, Plasma glucose level, Diastolic Blood Pressure (BP), Thickness of the triceps, Insulin, Body Mass Index (BMI), Diabetes pedigree, Age and Class label (diabetic or not). In this paper, the original Pima Indian dataset is referred to as small dataset and the extended version of the Pima Indian dataset is referred to as large dataset. The original dataset contains 768 rows of patient's data and it contains 268 positive instances and 500 negative instances. The extended version contains 15000 rows of patient's data with exactly the same attributes as that of the Indian Pima data set. The extended has been taken from Kaggle website. The data set can be found in the link https://www.kaggle.com/fmendes/diabetes-from-dat263x-lab01/ and it contains 15000 rows of unique patient data. In the extended PID dataset consists of 5000 tested positive class labels and 10000 tested negative class labels.

4 Proposed Work

The original dataset is an imbalanced dataset. The imbalanced dataset is converted into balanced dataset by using SMOTE algorithm. In this algorithm, oversampling is used. For the attribute selection, wrapper method is used. Out of 9 attributes, only 5 attributes are selected using wrapper method. The five attributes are Plasma glucose level, Diastolic Blood Pressure (BP), Body Mass Index (BMI), Age and Class label (diabetic or not). After attribute selection, the dataset is given to the input of various clustering algorithms. Finally, the result is produced. The same procedure is followed for the imbalance dataset. The flowchart of the proposed work is shown in the Fig. 1.

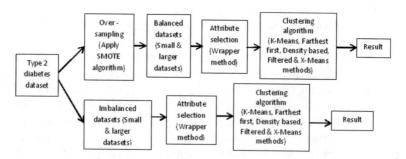

Fig. 1. Flow chart for the proposed work

4.1   Attribute Selection Using Wrapper Method

Attribute selection helps to remove the unwanted attributes from the given dataset for better accuracy in results. Because it is not always required for the dataset we choose to have all the attributes for mining process, by removing the redundant data and unwanted attribute the training time can be decreased and resulting in the reduction of storage size. There are different types of attribute selection in data mining. In this paper, we used Wrapper method of attribute selection.

4.1.1   Wrapper Attribute Selection

The Wrapper attribute selection technique wraps a classifier in a cross-validation loop and performs searches through the attribute space and uses the classifier to detect a good attribute set from the given dataset. Searching by the wrapper technique is often greedy forward direction, backward direction, or bidirectional, starting from any dataset.

4.2   Oversampling

In the given dataset the classes with a deficient number of rows are added rows so as to have equal number of rows for each class label. Considering our first data set we have 268 rows of positive class labels and 500 negative class labels. Thus, there is a need to balance the data. Also in the second dataset we have 5000 positive class labels and 10000 negative class labels. By over sampling the positive class label the accuracy in result can be increased. In this paper, oversampling is performed using SMOTE.

4.2.1   SMOTE Method of Oversampling

Synthetic Minority Oversampling Technique (SMOTE) is based on nearest neighbors raises the number of minority class instances or minority class group in the given dataset and it is shown in Fig. 2. It helps in balancing the dataset so that clustering algorithms can perform better and they do not experience overfitting problem. Usually SMOTE process includes Identity the feature vector and its nearest neighbor, take the difference between the two multiply the difference with a random number between 0 and 1, identify a new point on the line segment by adding the random number to feature vector and repeat the process for identified feature vectors.

Fig. 2.   Add new minority class instances

4.3   Clustering Algorithm

Clustering algorithms can be used in the procedure of constructing a group of objects into classes of related objects. The primary benefit of clustering over classification is that, it is compliant to modifications and supports single out useful features that identify unrelated groups. There are different types of clustering algorithms in data mining. In this paper, we have taken five majorly used clustering algorithms for prediction. They are K-means, Farthest first, Density based, Filtered and X-Means methods.

4.3.1   K-Means Clustering

K-means is a popular and the simplest clustering algorithm. In this algorithm, number of centroids (K) is to be identified. It assigns each data point to the nearby cluster while keeping the centroids as small as possible, serving as a prototype of the cluster.

4.3.2   Farthest First Clustering

Farthest first clustering algorithm is appropriate for the large dataset which is a variant of k-means clustering. It places each cluster centre in turn at the point furthest from the existing centres. Benefit of this method of clustering is that it speeds the clustering process since fewer reassignments and adjustments are required.

4.3.3   Density Based Clustering

Density based clustering algorithm is one of the unsupervised learning methods mainly used in finding non-linear shape structure that are based on density. The two concepts used in this method is density reachability and density connectivity.

4.3.4   Filtered Clustering

The filtered clustering method plays an important role in improving the performance of other clustering methods similar to k-means clustering by imposing an index organization on the dataset and decreases the quantity of cluster centre. The degree of separation between initial cluster centres influence the performance of filtering algorithm.

4.3.5   X-Means Clustering

X-Means clustering algorithm is an extended version of K-means clustering where it efforts to mechanically decide the number of clusters by using Bayesian Information Criterion (BIC). BIC helps in computing the splitting decision.

5   Results

Results obtained for the datasets are given in the tables below:

5.1   Clustering of Imbalanced Dataset

The two datasets without any balancing or attribute reduction were applied to the classification algorithms produced the following accuracies given below in Table 1. The Table 1 depicts the comparison of accuracies of the five models used.

Table 1. Accuracy for clustering of imbalanced data (small dataset)

S. No.	Name of clustering algorithm	Accuracy
1	K-MEANS	66.8
2	Farthest first	65.75
3	Density based	67
4	Filtered cluster	66.80
5	X-Means	66.8

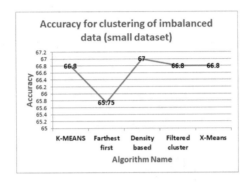

Fig. 3. Accuracy for clustering of imbalanced data (small dataset)

In the above Fig. 3, Density based clustering algorithm shows the highest accuracy for small dataset compared with other clustering algorithms (Table 2).

Table 2. Accuracy for clustering of imbalanced data (large dataset)

S. No.	Name of clustering algorithm	Accuracy
1	K-MEANS	51.06
2	Farthest first	67.15
3	Density based	51.24
4	Filtered cluster	51.06
5	X-Means	51.06

In the Fig. 4, farthest first clustering algorithm shows the highest accuracy for large dataset compared with other clustering algorithms. Comparing both dataset's (imbalanced data) accuracy farthest first has the highest accuracy for both large and small imbalanced datasets is shown in the Fig. 5.

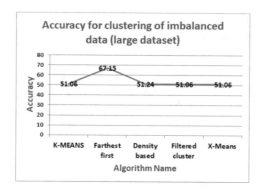

Fig. 4. Accuracy for clustering of imbalanced data (large dataset)

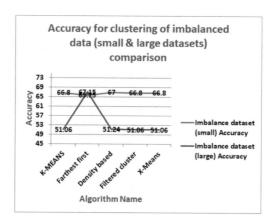

Fig. 5. Accuracy for clustering of imbalanced data (small and large datasets) comparison

5.2   Clustering of Balanced Datasets

The dataset was over sampled to balance the dataset. This was applied to both the datasets, and the clustering models were again applied to the datasets. The new accuracies of the data sets with each algorithm are given below in Tables 3 and 4.

Table 3. Accuracy for clustering of balanced data (small dataset)

S. No.	Name of clustering algorithm	Accuracy
1	K-MEANS	62.84
2	Farthest first	55.02
3	Density based	63.33
4	Filtered cluster	62.84
5	X-Means	62.83

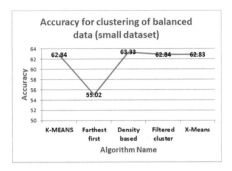

Fig. 6. Accuracy for clustering of balanced data (small dataset)

The Fig. 6 shows that Density based clustering algorithm shows the highest accuracy for small dataset when compared to other four algorithms.

Table 4. Accuracy for clustering of balanced data (large dataset)

S. No.	Name of clustering algorithm	Accuracy
1	K-MEANS	75.64
2	Farthest first	55.02
3	Density based	73.44
4	Filtered cluster	75.64
5	X-Means	75.60

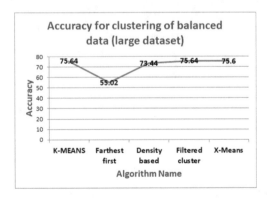

Fig. 7. Accuracy for clustering of balanced data (large dataset)

K-means, Filtered cluster and X-means gives the highest accuracy for clustering of large balanced dataset is shown in Fig. 7.

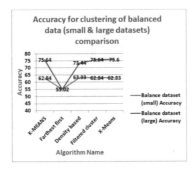

Fig. 8. Accuracy for clustering of balanced data (small and large datasets) comparison

Comparing both balanced data (small & large datasets), K-means, filtered cluster and X-means gives the highest accuracy with the result of 75.64%. and it shown in the Fig. 8. Comparing all the accuracy table, the highest accuracy is given by K-means, filtered cluster and X-means. Hence any of these three clustering techniques can be used for predicting large and small datasets effectively. It is shown in the Fig. 9.

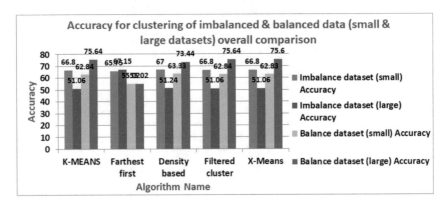

Fig. 9. Accuracy for clustering of imbalanced and balanced data (small and large datasets) overall comparison

6 Conclusion

Detection of diabetes in early stage is very important in these days. In this paper, standardized determinations are prepared in designing the structure which grades in the prediction of diabetic disorders in early stage. Throughout this work, five data mining clustering algorithms namely K-means, Farthest first, Density based, Filtered cluster and X-means are considered and appraised on the datasets. Experiments are accomplished on Pima Indians Diabetes dataset and extended version of Pima Indians Diabetes dataset. This paper was aimed at predicting diabetes for the extended version of Pima Indians Diabetes dataset with the best clustering technique. The five clustering methods are evaluated and compared for the highest accuracy. From the above tables, it is clear that K-means, Filtered clustering and X-means methods have the highest accuracy of 75.64%. Future work can be expanded and enriched for the highest performance of diabetic disorder analysis and prediction using deep learning methods like GAN with LSTM, CNN and etc.

References

1. Iyer, A., Jeyalatha, S., Sumbaly, R.: Diagnosis of diabetes using classification mining techniques. Int. J. Data Min. Knowl. Manag. (IJDKP) **5**, 1–14 (2015)
2. Kadhm, M.S., Ghindawi, I.W.C., Mhawi, D.E.: An accurate diabetes prediction system based on K-means clustering and proposed classification approach. Int. J. Appl. Eng. **13**, 4038–4041 (2018). ISSN 0973-4562
3. Karegowda, A.G., Jayaram, M.A., Manjunath, A.S.: Cascading K-means clustering and K-nearest neighbor classifier for categorization of diabetic patients. Int. J. Eng. Adv. Technol. (IJEAT) **1**, 147–151 (2012)
4. Sisodia, D., Sisodia, D.S.: Prediction of diabetes using classification algorithms. Procedia Comput. Sci. **132**, 1578–1585 (2018)
5. Li, Y., Li, H., Yao, H.: Analysis and study of diabetes follow-up data using a data-mining-based approach in new urban area of Urumqi, Xinjiang, China, 2016–2017. Comput. Math. Methods Med. **2018**, 1–8 (2018)
6. Zou, Q., Qu, K., Luo, Y., Yin, D., Ju, Y., Tang, H.: Predicting diabetes mellitus with machine learning techniques. Front. Genet. **9**, 515 (2018)
7. George Amalarethinam, D.I., Aswin Vignesh, N.: Prediction of diabetes mellitus using data mining techniques: a survey. Int. J. Appl. Eng. Res. **10**, 24–31 (2015)
8. Azrar, A., Awais, M., Ali, Y., Zaheer, K.: Data mining models comparison for diabetes prediction. Int. J. Adv. Comput. Sci. Appl. (IJACSA) **9**, 320–323 (2018)
9. Manimaran, R., Vanitha, M.: Prediciton of diabetes disease using classification data mining techniques. Int. J. Eng. Technol. **9**, 3610–3614 (2018)
10. Gangil, T., Sneha, N.: Analysis of diabetes mellitus for early prediction using optimal features selection. J. Big Data **6**, 13 (2019)
11. Zhang, T., Ma, F., Yue, D., Peng, C., O'Hare, G.M.P.: Interval type-2 fuzzy local enhancement based rough k-means clustering considering imbalanced clusters. IEEE Trans. Fuzzy Syst. (2019)
12. Liu, C.-L., Hsieh, P.-Y.: Model-based synthetic sampling for imbalanced data. IEEE Trans. Knowl. Data Eng. (2019)
13. Kothainayaki, M., Thangaraj, P.: Clustering and classifying diabetic datasets using K-means algorithm. J. Appl. Inf. Sci. **1** (2013)
14. Jeevanandhini, D., Gokul Raj, E., Dinesh Kumar, V., Sasipriyaa, N.: Prediction of type 2 diabetes mellitus based on data mining. Int. J. Eng. Res. Technol. (IJERT) (2018)
15. Zhu, C., Idemudia, C.U., Feng, W.: Improved logistic regression model for diabetes prediction by integrating PCA and K-means techniques. Inform. Med. Unlocked (2019)

Detection of Depression Related Posts in Tweets Using Classification Methods – A Comparative Analysis

M. Mounika[1], N. Srinivasa Gupta[2], and B. Valarmathi[1(✉)]

[1] Department of Software and Systems Engineering, School of Information Technology and Engineering, Vellore Institute of Technology, Vellore, Tamil Nadu, India
mounika.m2015@vit.ac.in, valargovindan@gmail.com
[2] Department of Manufacturing, School of Mechanical Engineering, Vellore Institute of Technology, Vellore, Tamil Nadu, India
guptamalai@gmail.com

Abstract. A 15-step pre-processing procedure is proposed to improve the accuracy of sentiment mining of depression related posts in the tweets. Perhaps, for the first time, converting emoticons in the depression related tweets into text form is proposed during the pre-processing stage. In this paper, Term Frequency-Inverse Document Frequency with n-grams is used for feature extraction. Sentiment analysis conducted on a dataset consisting of 1.6 million depression related tweets using the proposed pre-processing module with feature extraction using Term Frequency-Inverse Document Frequency with n-grams and Logistic Regression (LR) for classification resulted in 81% of accuracy in detecting depression related tweets.

Keywords: Natural Language Processing · Data mining · Depression · Twitter · Logistic Regression · MultiLayer Perceptron · Bag of Words · Term Frequency-Inverse Document Frequency

1 Introduction

Depression is a common mental health problem occurring due to stress, anxiety or other psychological problems. It affects the mental and physical health of the individual. As stated by World Health Organization (WHO) [1], there are more than 300 million people around the world of all age groups suffering from depression. Due to the social-stigma about depression, affected individuals hesitate to approach a psychiatrist. Being depressed for a long period of time, without proper medical treatment can lead to suicide. But with the rapid development of social media and increase in usage of it, people have started using these social medias like Facebook, blogs and twitter to express their emotions and thoughts through posts. These social media data available online have inspired the researchers to find new approaches for detecting depression using Natural Language Processing (NLP) and many text classification methods. Tyshchenko [2] detected depression in blog data of size 12495 which consisted of control blog posts and clinical blog posts. This data was split into 9196 training data,

© Springer Nature Switzerland AG 2020
A. P. Pandian et al. (Eds.): ICCBI 2019, LNDECT 49, pp. 620–630, 2020.
https://doi.org/10.1007/978-3-030-43192-1_70

2094 development data and 1205 test data. He performed document level classification and author level classification using Bag of Words features (BOW), TF-IDF features, and topic modeling using Latent Dirichlet Allocation (LDA) along with SVM and Random Forest classifiers. He also used CNN-rand and CNN-glove deep learning classification techniques. The maximum accuracy of 71.65% was achieved on predicting document level test data using LDA with Random Forest (RF) and 78.57% maximum accuracy was achieved with respect to predicting author level test data.

In this work, Twitter data of size 1600000 collected from Sentiment140 website has been used. NLP techniques like n-grams, Bag of Words (BOW) and TF-IDF were used for feature selection. Machine learning algorithms like Naive Bayes classifier, Support Vector Machine (SVM) classifier, Logistic Regression (LR), Adaptive Boosting (AdaBoost) classifier and Multi Layer Perceptron (MLP) neural network were used for classification. A comparative analysis has been done to find the classifier that gives the highest detection accuracy.

2 Related Works

Katchapakirin et al. [3] collected the data from microblogs and individual posts on Facebook and used it for sentiment analysis in Thai community. They detected the depression posts and non depression posts using attribute extraction module which converted Thai language text to English text, NLTK python library was used to process the translated text. Support Vector Machine (SVM), Random Forest (RF) and Deep Learning classification methods were used for classification. Among these classification methods Deep Learning gave maximum accuracy of 85%. Oyong et al. [4] used a twitter dataset with 6055 tweets. They analyzed the dataset in two ways, first method is using standard text processing with depression score calculation and the second method is twitter text processing with depression score calculation. Depression score calculation is done using symptom lexicon and frequency lexicon. In both the methods, the maximum accuracy obtained is 85%. Pirina et al. [5] experimented using eight different datasets with each dataset having 400 posts in it. It had depression posts and non depression posts. The maximum performance result was obtained by using NLP technique n-grams with SVM for classification. 67.49 F1 is the highest F1 score achieved on test set. Jamil et al. [6] used three different tweets dataset and proposed methods of classification using original dataset, SMOTE and under-sampling on tweet level and user level. In tweet level classification, under-sampled dataset with polarity, depression, pronoun word counts feature performed better with 0.6102 accuracy, 0.1237 precision, 0.8020 recall and 0.2144 F1 score. Yates et al. [7] used reddit dataset to identify the users with depression and to estimate the self-harm risk based on posts. They used deep learning algorithm Convolution Neural Network (CNN) with categorical cross entropy and got 0.65 F1 score for identifying depression. For self harm risk assessment, they used methods like categorical cross entropy, class metric and class metric (ordinal). The maximum accuracy was obtained with categorical cross entropy on test data. Nadeem et al. [8] used the twitter dataset for depression posts detection and classified it with 86% accuracy using Bag of Words (BOW) and Naïve Bayes machine learning algorithm for classification. Karmen et al. [9] used 1,000,000 messages for depression detection using symptom lexicon generation, Frequency lexicon generation, Natural Language Processing Techniques (NLP) and Depression score calculation and got 0.90 F1 score. Wijaya et al. [10] analyzed twitter dataset for depression using preprocessing, performed feature extraction using Parts of Speech (POS) tag, Grammatic Rule and Appropriate Rule. They found the resultant main phrase which expresses the mood of the tweet using lexicon with main phrase detection method, SVM and Naïve Bayes classifiers. Lexicon with main phrase detection method on four mood using POS tagging gave maximum accuracy 75.14%. De Choudhury et al. [11] analyzed 2,157,992 twitter data for detecting depression using behavioral attributes like emotion, ego network and many more for feature extraction and machine learning algorithm SVM (RBF Kernel). They predicted the depression in posts with 70% accuracy. Zhang et al. [12] analyzed chemo-related individual's 13273 tweets dataset and chemo-related organization's tweets data of size 14,501 to assess and compare the perception about chemotherapy of patients and health care-providers. They used n-grams, LDA, co-occurrence network (Gephi) for feature extraction and used various classification algorithms like Naïve Bayes, Decision Tree classifiers, XGB (xgboost), CNN, CNN with embedding layer, MLP, SVC (Linear SVM), RF, Long Short-Term Memory (LSTM), Attention LSTM, Attention LSTM with GloVe and LSTM with Glove for classification. 85.19% maximum accuracy is achieved using LSTM with Glove Word Embeddings. Tadesse et al. [13] detected depression Reddit dataset of size 1841 using NLP and Machine learning techniques. They have extracted the features using LIWC, unigram, bigram and topic modeling using LDA for 70 topics. Applied the machine learning techniques like LR, RF (Random Forest), SVM, AdaBoost and Multi Layer Perceptron network to classify the Reddit posts. MultiLayer Perceptron network classified the data with 91% maximum accuracy using LIWC, LDA and bigram together for feature extraction. Hassan et al. [14] analyzed the twitter dataset and 20 newgroups dataset using NLP and machine learning approaches to detect depression related posts. n-grams, POS (Parts of Speech Tagging), Negation and Sentiment Analyzer were used for feature extraction. SVM classifier achieved maximum accuracy of 91%. Sudha et al. [15] compared deep learning classifiers RNN, CNN, LSTMs, GRUs and BLSTMs classification performance against machine learning classifiers SVM, LR, RF, DT classification performance. SVM with TF-IDF feature gave maximum accuracy of 90.81. GRUs with GloVe word embedding gave maximum accuracy of 91.78%.

3   Dataset Description

In this work, the open source tweets dataset of size 1600000 collected from Sentiment140 website is used for depression detection. The link of the dataset can be found in http://help.sentiment140.com/for-students. This dataset has a total of 1600000 tweets with 800000 depression tweets and 800000 positive tweets in it. The class label '0' indicates depressed tweets and class label '4' indicates positive tweets in the dataset.

4 Methodology

In this work, Natural Language Processing (NLP) and data mining techniques are used for classifying the text. For NLP techniques we have used Natural Language Toolkit (NLTK) Python library and for classification we have used Scikit-Learn Python library. Figure 1 shows the three modules used in this proposed method, which are data pre-processing module, feature extraction module and classification module.

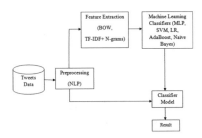

Fig. 1. Methodology

4.1 Data Pre-processing Module

Natural Language Processing (NLP) techniques are used for pre-processing the data. Pre-processing the text includes the following steps as depicted in Fig. 2.

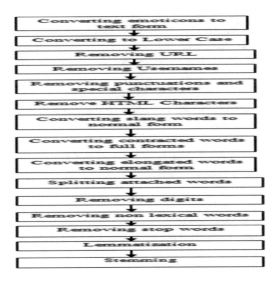

Fig. 2. Data pre-processing steps

4.1.1   Converting Emoticons to Text Form

As emoticons express emotions, they are converted to their relevant text form. This will add useful information in feature extraction.

4.1.2   Converting to Lower Case

All the tweets are converted to lower case; this will avoid multiple copies of the same word. For example, 'He' and 'he' are same words, converting 'He' to lowercase avoids multiple copies of 'he'.

4.1.3   Removing URL

URL's in the tweets do not add useful information in feature extraction. They are removed to reduce the size of the data.

4.1.4   Removing User Name

User names in the tweets starting with @ symbol do not add useful information in feature extraction. So, they are removed to reduce the size of the data.

4.1.5   Removing Punctuations and Special Characters

Punctuations and special characters do not add meaning in feature extraction. So, they are removed to reduce the size of the data.

4.1.6   Removing HTML Character

When extracting tweets from twitter some HTML characters were added to the tweets like &amp, &quot, &lt and &gt. They were removed to reduce the size of the data.

4.1.7   Converting Slang Words to Normal Form

Slang words in tweets are converted to normal English form, from which useful information can be extracted. For example, 'lol' to 'laugh out loud' etc.

4.1.8   Converting Contracted Words to Full Forms

Tweet words in contracted forms are converted to full forms to extract features from it. For example, don't to do not etc.

4.1.9   Converting Elongated Words to Normal Form

Tweet words in elongated form are converted to normal form to extract features from it. For example, 'noooo' to 'no' etc.

4.1.10   Splitting Attached Words

Most of the tweets have multiple words attached together; splitting those words adds useful information in feature extraction.

4.1.11   Removing Digits

All digits are removed from the tweets. Therefore removing them will reduce the size of the data.

4.1.12   Removing Non Lexical Words

Non lexical words are removed from the tweets to reduce the size of the data. For example, hmm, oh etc.

4.1.13   Removing Stop Words

Some commonly used words which do not add meaning in feature extraction were removed to reduce the size of the data. For example, the, is, am etc.

4.1.14   Lemmatization

All words are converted to root forms. For example, behaving to behave.

4.1.15   Stemming

All the words are stemmed by removing the affixes of the word like ing, y, ly, s, etc. using Snowball Stemmer.

The sample document is shown here.

Before Pre-Processing	@switchfoot http://twitpic.com/2y1zl - Awww, that's a bummer. You shoulda got David Carr of Third Day to doit. ;D 9 &amp
Converting emoticons to text form	@switchfoot http://twitpic.com/2y1zl - Awww, that's a bummer. You shoulda got David Carr of Third Day to doit. smirk 9 &amp
Converting to Lower Case	@switchfoot http://twitpic.com/2y1zl - awww, that's a bummer. you shoulda got david carr of third day to doit. smirk 9 &amp
Removing URL	@switchfoot - awww, that's a bummer. you shoulda got david carr of third day to doit. smirk 9 &amp
Removing User Name	- awww, that's a bummer. you shoulda got david carr of third day to doit. smirk 9 &amp
Removing punctuations, special characters	awww thats a bummer you shoulda got david carr of third day to doit smirk 9 amp
Remove HTML characters	awww thats a bummer you shoulda got david carr of third day to doit smirk 9

4.2   Feature Extraction Module

In this feature extraction module, NLP's N-grams technique is used along with BOW and TF-IDF.

4.2.1   n-grams

n-grams are the combination of n words together. In this proposed work, n-grams of range one to three are used i.e., n-grams has unigram (one word), bigrams (two words together) and trigrams (three words together) word combinations in it. Examples from our dataset for Unigram (want, go, home, not, love), Bigram (mine today, work begin) and Trigram (still not find, all love sugar).

4.2.2   Bag of Words (BOW)

Bag of Words (BOW) counts the number of times a word appears in a document. In this proposed work, BOW is determined using CountVectorizer of Python Scikit-Learn library.

4.2.3   TF-IDF

TF-IDF score relatively represents the importance of word in that document and to all the documents in the dataset. TF-IDF score is calculated by multiplying two measures and it is shown in the Eq. (1). First one is Term Frequency (TF) and it is shown in the Eq. (2). The second one is Inverse Document Frequency (IDF) and it is shown in the Eq. (3).

$$\text{TF-IDF}(w) = \text{TF} * \text{IDF} \tag{1}$$

w-word

The Term frequency of the given documents is calculated using the following formula:

$$\text{TF}(w) = N_w / \sum\nolimits_t N_t \tag{2}$$

N_w - Number of times the word w appears in a document
$\sum_t N_t$ - Total number of words in a document

The IDF is calculated using the following formula:

$$\text{IDF}(w) = \log_e (N/DF_w) \tag{3}$$

N - Total number of documents
DF_w - Document Frequency of word 'w' in it

In the present work, TF-IDF is determined using TF-IDF Vectorizer in Python Scikit-Learn library and with n-grams of range unigram, bigram and trigram.

5   Classification Module

In this paper, five popular classification methods have been used for Depression detection viz., Logistic Regression Classifier, SVM classifier (Linear SVC), Naïve Bayes classifier, MLP classifier, and AdaBoost.

Logistic Regression is a linear model which is used to estimate the relationship between dependent variable and one or more independent variables by using a logistic function.

Naïve Bayes classifier is a probabilistic model and it works based on Bayes theorem. It assumes that each feature in the dataset is independent of each other.

Support Vector Machine is a linear classifier model. It classifies the dataset by finding the hyperplane which separates the two classes precisely.

Adaptive Boosting classifier combines multiple weak classifiers into one strong classifier.

MultiLayer Perceptron is a feedforward artificial neural network can distinguish not linearly separable data using its multiple layers and non-linear activation function.

6 Results and Discussion

The accuracy of the (BOW) feature extraction with machine learning classifiers on test data is summarized in Table 1. The accuracy of the (TF-IDF+n-grams) feature extraction with machine learning classifiers is summarized in Table 2. Tables 1 and 2 results are visualized in Figs. 3 and 4 respectively.

Table 1. Accuracy of feature extraction with various classifiers (BOW+classifier) on test data

Method	Accuracy
BOW+LR	78%
BOW+SVM	78%
BOW+Ada Boost	74%
BOW+Naïve Bayes	77%
BOW+MLP	79%

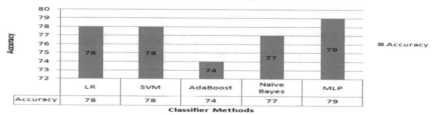

Fig. 3. Visualization of (BOW+classifier)

Table 2. Accuracy of feature extraction with various classifiers (TF-IDF+n-grams) on test data

Method	Accuracy
TF-IDF+N-Grams+LR	81%
TF-IDF+N-Grams+SVM	80%
TF-IDF+N-Grams+Ada Boost	74%
TF-IDF+N-Grams+Naïve Bayes	79%
TF-IDF+N-Grams+MLP	81%

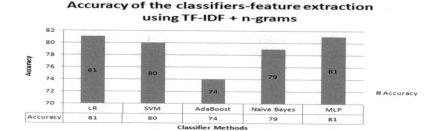

Fig. 4. Visualization of (TF-IDF+n-grams+classifier)

Table 3. Comparison of the accuracies of various classifiers using (BOW) and (TF-IDF+n-grams) on test data

Method	Accuracy (BOW)	Accuracy (TF-IDF+n-grams)
LR	78%	81%
SVM	78%	80%
Ada boost	74%	74%
Naïve Bayes	77%	79%
MLP	79%	81%

Table 3 represents the comparison of classifiers accuracies using BOW and TF-IDF +n-grams features and Fig. 5 visualizes the comparison.

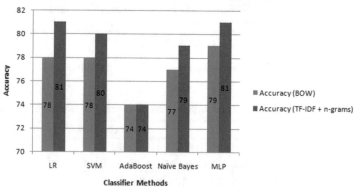

Fig. 5. Comparing the accuracies of classifiers using (BOW) and (TF-IDF+n-grams)

In the proposed work, the maximum depression detection accuracy of 81% is achieved using TF-DF+n-grams features with LR classifier and TF-DF+n-grams features with MLP classifier. AdaBoost classifier with BOW features and TF-IDF+n-grams features gives the lowest accuracy 74%. BOW features used with classifiers, gives the maximum depression detection accuracy 79% with MLP when compared with LR (78%), SVM (78%), AdaBoost (74%) and Naïve Bayes (77%).

TF-IDF+n-grams features used with classifiers give the maximum depression detection accuracy 81% with LR and MLP when compared with SVM (80%), Ada-Boost (74%) and Naïve Bayes (79%).

The proposed method in this paper is compared with results achieved by Tyshchenko [2] in the document level. The proposed approach in this paper has achieved 81% of detection accuracy using TF-IDF+n-grams features with MLP and LR classifiers in document level, an improvement of 9.35% over the proposed method by Tyshchenko [2].

7 Conclusion and Future Work

In this work, we used twitter data to detect depression. Better data pre-processing techniques like converting emoticons to text form, converting elongated words to normal form, splitting attached words and removing non-lexical words were used to enhance the detection accuracy. Extracted the features using BOW and TF-IDF+n-grams techniques and applied these features to the text classification algorithms. LR and MLP classification algorithms with TF-IDF+n-grams features give the maximum accuracy of 81%.

The future work will be to increase the depression detection accuracy by using feature extraction like word to vector (Word2Vec) and deep learning classifiers like Recurrent Neural Network (RNN) and Convolution Neural Network (CNN) for classification.

References

1. World Health Organization. https://www.who.int/en/news-room/fact-sheets/detail/depression
2. Tyshchenko, Y.: Depression and anxiety detection from blog posts data. Nature Precis. Sci. Inst. Comput. Sci. Univ. Tartu, Tartu Estonia (2018)
3. Katchapakirin, K., Wongpatikaseree, K., Yomaboot, P., Kaewpitakkun, Y.: Facebook social media for depression detection in the Thai community. In: 15th International Joint Conference on Computer Science and Software Engineering, Thailand (2018)
4. Oyong, I., Utami, E., Luthfi, E.T., White, K.: Natural language processing and lexical approach for depression symptoms screening of indonesian Twitter user. In: 10th International Conference on Information Technology and Electrical Engineering, China (2018)
5. Pirina, I., Coltekin, C.: Identifying depression on Reddit: the effect of training data. In: Proceedings of the 3rd Social Media Mining for Health Applications (SMM4H) Workshop & Shared Task, pp. 9–12 (2018)

6. Jamil, Z., Inkpen, D., Buddhitha, P.: Monitoring tweets for depression to detect at-risk users. In: Proceedings of the Fourth Workshop on Computational Linguistics and Clinical Psychology, pp. 32–40 (2017)
7. Yates, A., Cohan, A., Goharian, N.: Depression and self-harm risk assessment in online forums (2017)
8. Nadeem, M., Horn, M., Coppersmith, G., Sen, S.: Identifying depression on Twitter (2016)
9. Karmen, C., Hsiung, R.C., Wetter, T.: Screening internet forum participants for depression symptoms by assembling and enhancing multiple NLP methods. J. Comput. Methods Programs Biomed. **120**, 27–36 (2015)
10. Wijaya, V., Erwin, A., Galinium, M., MuJiady, W.: Automatic mood classification of indonesian tweets using linguistic approach. IEEE J. (2013)
11. De Choudhury, M., Gamon, M., Counts, S., Horvitz, E.: Predicting depression via social media. In: Proceedings of the Seventh International AAAI Conference on Weblogs and Social Media (2013)
12. Zhang, L., Hall, M., Bastola, D.: Utilizing Twitter data for analysis of chemoptherapy. Int. J. Med. Inform. **120**, 92–100 (2018)
13. Tadesse, M.M., Lin, H., Xu, B., Yang, L.: Detection of depression related posts in Reddit social media forum. IEEE J. (2019)
14. Hassan, A.U., Hussain, J., Hussain, M., Sadiq, M., Lee, S.: Sentiment analysis of social networking sites (SNS) data using machine learning approach for the measurement of depression. IEEE J. (2017)
15. Subramani, S., Michalska, S., Wang, H., Du, J., Zhang, Y., Shakeel, H.: Deep learning for multi-class identification from domestic violence online posts. IEEE Access **7**, 46210–46224 (2019)

Aspect Based Sentiment Classification and Contradiction Analysis of Product Reviews

Md. Shahadat Hossain[(✉)], Md. Rashadur Rahman,
and Mohammad Shamsul Arefin

Department of CSE, CUET, Chittagong, Bangladesh
hossain.shahadat096@gmail.com, rsdrcse14@gmail.com,
sarefin@cuet.ac.bd

Abstract. Due to the rapid growth of Internet infrastructure and E-Commerce, people can easily buy different products from different E-commerce websites. The reviews posted by the customers in the E-commerce websites can help us to get an idea about the products. It also helps us to identify the behavior of the individuals. This is because users' reviews on items are inevitably dependent on many social effects such as peer influence, user profile information, user preference etc. Considering this fact, in this paper, we present a framework to analyze users' reviews about the products to identify various aspects of the products listed online. In our approach, we use a large volume of publicly available product review data and perform different experiments. Experimental results show that our system provides better performance in case of negative aspect.

Keywords: Human behaviors · Sentiment analysis · E-commerce

1 Introduction

Online shopping refers to the form of e-commerce which enables consumers to buy goods or services directly from a seller using a web browser over the internet. As of 2016, customers can shop online using a range of different product and can give review about the product. A survey [1] showed that 84% people give product reviews written by other customers as much importance as personal recommendation. These review texts contain important information which often helps customers to decide which products to go for. The methods characterizing buyer's behavior are based on their experience of using the product. Different groups of customer exhibit different interaction patterns due to difference in expectation. The importance of analyzing buyers' behavior let us to observe the true interest of buyer. The data also helps the companies to understand what the customer liked and disliked about their product and what the customers may want.

For analyzing data to identify users' behavior there are many systems. However, looking for meaningful data in the huge amount of product reviews is not possible in a reasonable amount of time and the process is error prone and costly and sometimes

© Springer Nature Switzerland AG 2020
A. P. Pandian et al. (Eds.): ICCBI 2019, LNDECT 49, pp. 631–644, 2020.
https://doi.org/10.1007/978-3-030-43192-1_71

monotonous. In this paper, we are proposing a method to extract customers' sentimental opinion about products' different features and aspects from the textual data of product review. For classification of sentiment we are using different linguistic features some of which were not used together in similar works to date. For experiment, we are using dataset collected from popular online shopping site Amazon which contains reviews under mobile phone and laptop previously used for latent aspect rating analysis by Wang et al. [2].

The rest of the paper organization is as follows. A brief review of related work is provided in Sect. 2. In Sect. 3, we provide detail description of our proposed system. The experimental result is presented in Sect. 4. Finally, we conclude our paper in Sect. 5.

2   Related Work

In NLP (Natural Language Processing), opinion mining or sentiment analysis is one of the major tasks. In recent years sentiment analysis has gained considerable attention.

Fang et al. in [3] published a paper that tackles a fundamental problem of sentiment analysis, sentiment polarity categorization. Data used for this study was online product reviews from amazon.com. Sentences were extracted and tokenized into separated English words by POS tagger. They proposed "negation phrases identification" algorithm to identify negation-of-verb (NOV) and negation-of-adjective (NOA). They used a sentiment token (positive and negative) for phrases which is used to calculate sentiment score information.

The researchers in [4] proposed a "web based opinion mining system" for hotel reviews. An assessment system for online users' reviews and comments to support quality controls in hotel management system is introduced in that work. It has the ability to detect and retrieve reviews on the web and also to deal with German reviews.

Godbole et al. [5] analyzed news sentiments and blogs. In this work, they split prior work in the context of their specific task (sentiment analysis for blogs and news) into two categories. They designed a system based on their analysis to identify influential users and their impact on the network. Mobile devices products reviews were analyzed in [6].

"Bag of words" method was used by many researchers for aspect level sentiment analysis. Singh et al. [7] used adverb and adjective then also add adverb and verb together to calculate the sentiment score. There was no mention of negative subject. The negation handling method which is used works for binary classification only.

Many researchers used machine learning approaches to identify sentiments. The authors of [8] and [9] used Naive Bayes strategy. Pang et al. [10] employed three machine learning methods which are maximum entropy classification, Naive Bayes classifier and SVM (support vector machine) and SVM was found to function better than the others in case of sentiment analysis. The main strategy of the dictionary-based approach is presented by the authors of [11, 12]. A small set of words of opinion with known orientations are collected manually. Then, searching in the well-known corpora WordNet [13] or thesaurus [14] for their synonyms and antonyms develops this collection. The newly found words will be added to the list of seeds and the next iteration will start. When no new words are found, the iterative process will stop. Manual inspection can be performed after the process has been completed to remove or correct errors.

In order to determine the frequency and polarity for different aspects of a movie, Thet et al. in [15] conducted clause-level sentiment analysis using a linguistic approach. They generated grammatical dependencies of words in sentences and constructed a dependency tree which is divided into dependency sub-trees, which represents a single clause focusing on one aspect of the movie. After that they calculated contextual sentiment score for each clause, for each review aspect and also the overall sentiment score for the whole sentence.

The Naive Bayes classifier is a straight forward probabilistic classifier that relies on Bayes theorem with sturdy and naive individuality expectation. Improved Naive Bayes classifier was planned by Kang et al. [16] for solving the matter of the inclination for the case of positive classification correctness to look up to close to 10 percent over the negative classification correctness. Compared to Naive Bayes and Support Vector Machine, they showed a gap between positive accuracy and negative accuracy.

Usha et al. focused on topic detection and topic sentiment analysis based on unsupervised CST (Combined Sentiment Topic) model for classifying positive and negative sentiment in [17]. CST model is used for detecting sentiment and topics simultaneously from text. In that work, they only detected the positive words and negative words of detected topic.

3 System Architecture and Design

The sentiment analysis system architecture is shown in Fig. 1. The process is done in many steps. First, we need to collect the product reviews from online shopping website and input them for preprocessing. In the next step, we detect product aspect which is set manually for the product. The synonyms and antonyms are also detected so that all the sentences with the given aspects can be detected. Then we tokenize each word of each sentence by python *NLTK* library and also determine the parts of speech of each word.

In next step, we make a pair of aspect and corresponding parts of speech tag (adjective, adverb and verb) for every sentence. Then we calculate the score of adjective, adverb and verb for each aspect with the help of *senti_synsets()*. The score will tell the polarity of sentiment into two class positive and negative.

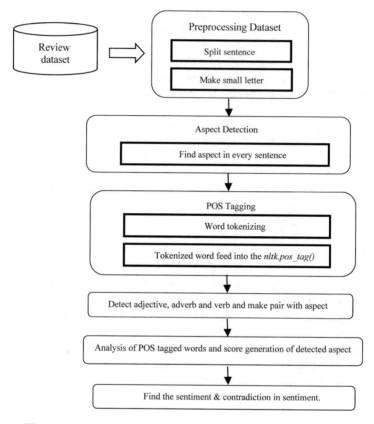

Fig. 1. System architecture for aspect-based sentiment classification

3.1 Dataset Preprocessing

The dataset was collected from Amazon which contains reviews for mobile phone and laptop. The dataset is in JSON format. Using same dataset gives us an opportunity to get an accurate comparison of experimental results at the end. The steps of the pre-processing are given below:

Step 1: All unwanted information like date, overall, author, URL etc. are removed.
Step 2: Each sentence is split and converted to small letters.
Step 3: Then stored in a list.

3.2 Segmentation

In next step, we are segmenting the large review data in sentences using symbols like full stop, question mark etc. And then the sentences are stored in a list. We could use a database for storing segmented sentences but it may cause slowing down the procedure. Then we convert the sentences to all small letters. Finally, the texts of sentences are sent to the aspect detection. We find the aspect in sentence and make another list which contains aspect related text.

3.3   Aspect Detection

A measure of positivity or negativity stated in a review is the document-level sentiment classification. In addition, practically a mixture of both positive and negative sentiment about different aspects of the item is observed in most of the reviews and it may be inappropriate and challenging to insist on an overall polarity of document-level sentiment expressed in a review about the item. Therefore, for detailed analysis of positive and negative aspects of the item under review, the document-level sentiment classification is not a complete, comprehensive and appropriate measure. The aspect-level sentiment analysis includes: (a) to identify aspects to be analyzed, (b) to locate the opinion content about that aspect in the review, and (c) to evaluate the sentiment polarity of views expressed about an aspect. An intensive search was created for identifying the aspects in numerous product classes found in online marketplaces and review sites and that we figured out a listing of aspects such as (video, camera, sound, battery etc.). After detecting aspect, we create a list of aspect related text. Algorithm 1 shows the procedure of aspect detection.

Algorithm 1: Aspect detection
Input: Review sentence
Output: Aspect related sentence

   1.   **Begin**
   2.   **for** each sentence(S) in review (R)
   3.   **find** aspect in each sentence
   4.   **if** aspect found
   5.   **then** store in aspect related sentence
   6.   **else** skip sentence
   7.   **end if**
   8.   **end for**
   9.   **End**

3.4   Parts of Speech (POS) Tagging

After detecting aspect related sentences, we tokenize the words of each sentence and find the POS tag of each word by *nltk.pos_tag()* which is a function under *NLTK* corpus of python.

For POS tagging, first the sentences are tokenized. The *NLTK* tokenizer is more robust for tokenizing a sentence into words. The next part is done by the *NLTK* library's POS tagger. Specific tags for certain words are provided by the POS tagger in the *NLTK* library. The tagging is done by way of a trained model in the *NLTK* library.

The parts of speech (POS) tagging mechanism is shown in Algorithms 2 and 3 shows finding of adjective, adverb and verb.

Algorithm 2: Parts of speech (POS) tagging
Input: Aspect related sentences
Output: POS tag for every word in aspect related sentences

   1. **Begin**
   2. **for** every sentence(S) in Aspect related sentence
   3.   word = word_tokenize(S)
   4.   POS_tag = nltk.pos_tag(word)
   5. **end for**
   6. **End**

Algorithm 3: Finding adjective, adverb, verb
Input: POS tag of word
Output: Adjective, adverb, verb

   1. **Begin**
   2. **for** every POS tag of every word
   3.   **if** POS tag is 'JJ' or 'JJR' or 'JJS'
   4.   **then** it is adjective and POS tag replace   with 'a'
   5.   **else if** POS tag is 'RB' or 'RBR' or 'RBS'
   6.   **then** it is adverb and POS tag replace with 'r'
   7.   **else if** POS tag is 'VB' or 'VBR' or 'VBG' or 'VBN' or 'VBP' or 'VBZ'
   8.   **then** it is verb and POS tag replace with 'v'
   9. **end for**
  10. **End**

3.5   Sentiment Classification Method

For sentiment classification we use previous POS tag and convert it into suitable tags to extract score by *sentiwordnet.senti_synsets()*. Here we check every POS tag and convert only adjective, adverb, and verb for our sentiment classification then put it into a list. Then these tags are used for determining whether aspect is positive or negative. As each sentence is already stored in list, these POS tags are also stored in the same way. The proposed sentiment analysis method POS tag to identify words and uses their correlation to calculate the sentiment score. The proposed algorithm for the classification of sentiment is shown in Algorithm 4.

Algorithm 4: Sentiment classification
Input: Input review
Output: Pair of aspect and sentiment

```
1:   Begin
2:   S ← Split sentences from review comment
3:   for each sentence in S
4:       Find Aspect word in S
5:       AspectRelatedSentences (ARS) ← Aspect word in S
6:   end for
7:   for each word in ARS
8:       Word_tokenize in each (ARS)
9:       Make POS tag in each tokenize_word
10:      Find Adjective, Adverb, and Verb in each POS tag
11:      Make those POS Tag into suitable POS tag
12:      AP(Aspect and Pos Tag) ← Make a list of Aspect word and suitable Pos tag
13:  end for
14:  for each Pos Tag in AP
15:      for each Aspect
16:          if Pos Tagged word is positive   Score ← Score*(word score)
17:          if Pos Tagged word is negative   Score ← Score*( -1* word score)
18:          if Pos Tagged word is neutral Score ← Score*1
19:          Make a pair of Aspect and Score SecondaryAspectScore ← [Aspect,
             Score]
20:      end for
21:  end for
22:  Summarize Aspect and Score PrimaryAspectScore
23:  Sum the number of positive score and negative score for each  Aspect
24:  if Sum  > 0 AspectScore ← 1
25:  if Sum  < 0 AspectScore ← - 1
26:  if Sum  = 0 AspectScore ← 0
27:  Finding Sentiment for each PrimaryAspectScore
28:  if AspectScore  = 0
29:      MakePair ← [Aspect, Neutral]
30:  end if
31:  if AspectScore > 0
32:      MakePair ← [Aspect, Positive]
33:  end if
34:  if AspectScore < 0
35:      MakePair ← [Aspect, Negative]
36:  end if
37:  End
```

We have extracted aspect related sentences and make suitable POS tags for adjective, adverb, verb in every aspect related sentence. This POS tag of the word is used to find whether the word is positive or negative. If the word is positive, we make score as 1 and if the POS tagged word is negative, we score it as -1. Then we sum up the score for each aspect in an aspect related sentences, if sum is 0 then sentiment is neutral, if sum is greater than 0 then it is positive, and if sum is less than 0 then it is negative.

For sentiment classification method we use previous POS tag and convert it into suitable tag. Example: [('I', 'n'), ('love', 'v'), ('design', 'n'), ('functionality', 'n'), ('cell', 'n'), ('phone', 'n'). From this result we use *sentword.senti_synsets()* function to get whether it is positive or negative. We have considered only adjective, adverb and verb. If the word is positive or neutral we assume that as "+1" if it is negative then "-1".

For example: here "love" is a verb and "phone" is an aspect. *sentword.senti_sysnsets ('love', 'v')* gives the results that is a positive word by giving positive score. So the sentiment for aspect is positive. The idea of getting aspect sentiment is shown in Table 1.

Table 1. Logic level of getting aspect sentiment.

adj	adv	vrb	Aspect sentiment
+	+	+	Pos
+	+	−	Neg
+	−	+	Neg
+	−	−	Pos
−	+	+	Neg
−	+	−	Pos
−	−	+	Pos
−	−	−	Neg

We multiply the "adj.", "adv.", "vrb." scores. So it will be only positive or negative. In the algorithm secondary score is calculated for every sentence in comment. Primary score is the final decision for the output of a comment. For example: for a comment: "I like the design of the phone. Battery performance is good for first 10 days. After that battery dies quickly. But phone is very smooth."

Secondary Score: 1st sentence: ['phone', 'positive'], 2nd sentence: ['battery', 'positive'], 3rd sentence: ['batter', 'negative'], 4th sentence ['phone', 'positive'].

Primary score: for this comment ['phone', 'positive'], [battery, 'neutral']. Here for aspect "phone" is sentiment positive because it found 2 positive sentiments. But for aspect "battery" sentiment is "neutral" because it finds 1 positive and 1 negative in a comment. So for an aspect if number of positive count > negative count then sentiment of the aspect is positive.

So for an aspect if number of positive count < negative count then sentiment of the aspect is negative. If number of positive count = negative count then sentiment of the aspect is neutral.

3.6   Contradiction Analysis

The next task of this proposed method is to find the contradiction between the sentiments occurred in aspect for all reviews. We evaluate the contradiction level based on some predefined value ranges. The value ranges for contradiction analysis is shown in Table 2.

Table 2. Value ranges for contradiction analysis.

Range	Comment
Distance between sentiment of aspect in percentage ($0\% \leq$ **Distance** $\leq 25\%$)	Very high contradiction
Distance between sentiment of aspect in percentage ($26\% \leq$ **Distance** $\leq 50\%$)	High contradiction
Distance between sentiment of aspect in percentage ($51\% \leq$ **Distance** $\leq 75\%$)	Low contradiction
Distance between sentiment in of aspect percentage ($76\% \leq$ **Distance** $\leq 100\%$)	Very low contradiction

The percentage of Distance is calculated by listing the equation:

$$Positivity\,for\,an\,aspect(\%) = \frac{Number\,of\,positive\,words\,for\,an\,aspect}{Number\,of\,positive\,and\,negative\,words\,for\,an\,aspect} \times 100\%$$

$$(1)$$

$$Negativity\,for\,an\,aspect(\%) = \frac{Number\,of\,negative\,words\,for\,an\,aspect}{Number\,of\,positive\,and\,negative\,words\,for\,an\,aspect} \times 100\%$$

$$(2)$$

$$Distance = |Positivity - Negativity| \qquad (3)$$

The value of distance is used for finding the contradiction. Algorithm 5 shows the procedure of contradiction analysis which evaluate the level of contradiction based on the value ranges given in Table 2.

```
Algorithm 5: Contradiction analysis
Input: List of pair [Aspect, Sentiment]
Output: Aspect contradiction

  1:   Begin
  2:      for every aspect in Pair
  3:         Calculate the positivity of aspect
  4:         Calculate the negativity of aspect
  5:         Calculate the distance between positivity and negativity
  6:            if  0<=distance and distance <=25
  7:               Aspect has Very High contradiction
  8:            end if
  9:            if  26<=distance and distance <=50
 10:               Aspect has High contradiction
 11:            end if
 12:            if  51<=distance and distance <=75
 13:               Aspect has Low contradiction
 14:            end if
 15:            if  76<=distance and distance <= 100
 16:               Aspect has Very Low contradiction
 17:            end if
 18:      end for
 19:   End
```

Here "distance", "positivity" and "negativity" calculated by Eqs. (1), (2) and (3).

4 System Implementation and Experiments

This section contains the implementation procedure and performance analysis of our developed system.

4.1 Experimental Setup

The main focus of the project is to detect aspect and aspect-based sentiment, and find the contradiction between sentiments. So, no interface was designed for this method. We stored the data into a text document. It includes all the results produced by the methodology. There are two text documents, "MOBILE.txt" has the analysis of mobile phone reviews, and "LAPTOP.txt" has the analysis of laptop reviews. Contradiction analysis part is also in the same ".txt" file.

4.2 Implementation

In this section, some of the sample implementation works of the proposed method are presented. The whole implementation was done in Windows environment setup. Output is visualized in ".txt" file. The reviews are stored in JSON format. The code read the JSON file. We worked with two specific product types which are laptop and mobile.

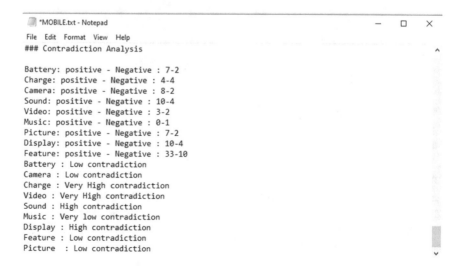

```
*MOBILE.txt - Notepad                                           —    □    ×

File  Edit  Format  View  Help
For comment 1
[['Battery', 'Negative'], ['Charge', 'Positive'], ['Camera', 'Positive'], ['Sound', 'Positive'],
['Video', 'Positive'], ['Display', 'Neutral'],['Feature', 'Positive']]

For comment 2
[['Camera', 'Positive'], ['Sound', 'Positive'], ['Video', 'Positive'],['Feature', 'Neutral']]

For comment 3
[['Battery', 'Positive'], ['Sound', 'Positive'], ['Display', 'Negative'], ['Feature', 'Positive']]

For comment 4
[['Camera', 'Positive'], ['Sound', 'Positive'], ['Display', 'Positive'], ['Feature', 'Positive']]

For comment 5
[['Camera', 'Negative'], ['Feature', 'Positive']]

For comment 6
[['Feature', 'Positive']]

For comment 7
[['Battery', 'Positive'], ['Sound', 'Positive'], ['Picture', 'Positive'], ['Display', 'Neutral'], ['Featu

For comment 8
[['Camera', 'Positive'], ['Picture', 'Positive'], ['Display', 'Positive'], ['Feature', 'Positive']]
```

Fig. 2. Sentiment analysis for product type Mobile

After running the program, we found the "MOBILE.txt" file as Fig. 2. "MOBILE. txt" file is showing the pairs of aspect and sentiment found in every review. If the system cannot find any aspect then it will show "No aspect detected". In Fig. 2 "For comment 1" means this the 1st review in the JSON file. And 1st element of the pair is Aspect and 2nd element is the sentiment. For the pair ['Battery', 'Negative'] here, 'Battery' is an aspect and 'Negative' is the sentiment of the aspect. 'Negative' means reviewer told about 'Battery' and he was not satisfied with that. Where positive means reviewer was satisfied with the aspect. All the sentiments and aspects for those reviews are stored in "MOBILE.txt" file. The contradiction analysis part is also stored in the same "MOBILE.txt" file for mobile reviews.

```
*MOBILE.txt - Notepad                                           —    □    ×

File   Edit   Format   View   Help
### Contradiction Analysis

Battery: positive - Negative : 7-2
Charge: positive - Negative : 4-4
Camera: positive - Negative : 8-2
Sound: positive - Negative : 10-4
Video: positive - Negative : 3-2
Music: positive - Negative : 0-1
Picture: positive - Negative : 7-2
Display: positive - Negative : 10-4
Feature: positive - Negative : 33-10
Battery : Low contradiction
Camera : Low contradiction
Charge : Very High contradiction
Video : Very High contradiction
Sound : High contradiction
Music : Very low contradiction
Display : High contradiction
Feature : Low contradiction
Picture : Low contradiction
```

Fig. 3. Contradiction analysis for product type Mobile

Figure 3 shows the contradiction analysis of mobile reviews which is stored in "MOBILE.txt" file. Here, "Battery: positive - Negative: 7–2", 'Battery' is the aspect and 7 indicates how many reviewers got positive about 'Battery' and 2 indicates how many reviewers got 'Battery' as negative.

4.3   Performance Evaluation

For Parts of speech (POS) tagging we used NLTK (Natural Language Toolkit). NLTK is platform in Python for working with natural language data. It is the most well known platform for natural language processing. It contains most powerful NLP libraries for text processing such as, parsing, pos_tagging, tokenization, classification, stemming etc. For parts of speech tagging we used *nltk.pos_tag()* function under NLTK. For our data, the accuracy of parts of speech tagging is very much acceptable.

Sentiment Classification Results. The comparison of sentiment classification method's calculated results and actual results shows that the algorithm can pretty accurately classify positive reviews. But both positive and negative reviews are considered as neutral review in many cases.

Aspect Detection Results. Aspect detection deserves much more focus than it is given in this project. One of the limitations of the system is that it is not able to automatically detect aspects for each product. So the aspects were manually decided on by carefully analyzing the reviews. There were a lot of aspects to be chosen.

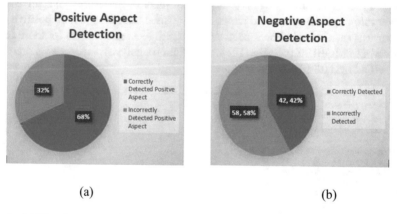

(a)   (b)

Fig. 4. (a) Negative aspect detection comparison. (b) Positive aspect detection comparison

The comparison of positive and negative aspect detection is shown in Fig. 4. The percentages shows the rate of correctly and incorrectly aspect detection of taken review dataset. The percentage of correctness and incorrectness of positive aspect detection is shown in Fig. 4(a) and the percentage of correctness and incorrectness of negative aspect detection is shown in Fig. 4(b).

Comparison of Accuracy. It is the fraction of correctly classified sentiment for reviews and number of total input reviews. We compared it to binary classification systems available for different kinds of reviews where accuracy is computed using below equation.

$$Accuracy = \frac{Number\ of\ correctly\ classified\ reviews}{Number\ of\ total\ reviews}$$

The data was manually tested for both positive and negative aspect. Then the manual or observed results are compared to the outcome of the proposed method and accuracy was calculated based on that. The comparison of accuracy among the methods listed and our proposed method is shown in Table 3.

Table 3. Comparison of accuracy.

Methods	Accuracy
SWN (AAC)	POS-82% NEG-18%
SWN (AAAVC)	POS-82.9% NEG-17.1%
Alchemy API	POS-73.4% NEG-26.6%
Proposed method	POS-68% NEG-42%

From the Table 3, we can observe our proposed methodology has better accuracy in detecting negative aspect than the other works.

5 Conclusion

In this paper, we have proposed a method to extract and analyze customers' opinions about products' features and aspects obtained from products reviews. For the classification of different sentiments, we considered different linguistic features. The algorithmic formulation used for aspect-level sentiment profile is easy to implement, simple, and quick to produce results. This profiling of aspect level sentiment is an essential form of sentiment analysis which is expressed about a particular product by a large number of users. There is a scope of future development which involves parallel codes implementation for speeding up the data extraction mechanism and the network metrics evaluation.

References

1. Murphy, R.: Local consumer review survey—online reviews statistics & trends (2018). http://www.brightlocal.com/learn/local-consumer-review-survey. Accessed 30 Sept 2019
2. Wang, H., Lu, Y., Zhai, C.: Latent aspect rating analysis without aspect keyword supervision. In: Proceedings of the 17th ACM SIGKDD International Conference on Knowledge Discovery and Data Mining - KDD 2011 (2011)

3. Fang, X., Zhan, J.: Sentiment analysis using product review data. J. Big Data 2(1), 5 (2015)
4. Kasper, W., Vela, M.: Sentiment analysis for hotel reviews. In: Proceedings of the Computational Linguistics-Applications, Jacharanka, pp. 45–52 (2011)
5. Godbole, N., Srinivasaiah, M., Skiena, S.: Large-scale sentiment analysis for news and blogs. In: Proceedings of the International Conference on Weblogs and Social Media (ICWSM 2007), Boulder, Colorado, USA (2007)
6. Zhang, L., Hua, K., Wang, H., Qian, G.: Sentiments reviews for mobile devices products. In: Proceedings of the 11th International Conference on Mobile Systems and Pervasive Computing (MobiSPC-2014) (2014). Procedia Comput. Sci. 34
7. Singh, V.K., Piryani, R., Uddin, A., Waila, P.: Sentiment analysis of movie reviews: a new feature-based heuristic for aspect-level sentiment classification. In: 2013 International Mutli-Conference on Automation, Computing, Communication, Control and Compressed Sensing (iMac4s), pp. 712–717 (2013)
8. Gamallo, P., Garcia, M., Fernandez-Lanza, S.: A Naive-Bayes strategy for sentiment analysis on Spanish tweets. In: Workshop on Sentiment Analysis at SEPLN (TASS2013), Madrid, Spain, pp. 126–132 (2013)
9. Gamallo, P., Garcia, M.: A Naive-Bayes strategy for sentiment analysis on English tweets. In: Proceedings of the 8th International Workshop on Semantic Evaluation (SemEval 2014), Dublin, Ireland, pp. 171–175 (2014)
10. Pang, B., Lee, L., Vaithyanathan, S.: Thumbs up? Sentiment classification using machine learning techniques. In: Proceedings of the Conference on Empirical Methods in Natural Language Processing, Philadelphia, US, pp. 79–86 (2002)
11. Hu, M., Liu, B.: Mining and summarizing customer reviews. In: Proceedings of the Tenth ACM SIGKDD International Conference on Knowledge Discovery and Data Mining, pp. 168–177 (2004)
12. Kim, S.M., Hovy, E.: Determining the sentiment of opinions. In: Proceedings of COLING-04, 20th International Conference on Computational Linguistics, pp. 1367–1373 (2004)
13. Miller, G., Beckwith, R., Fellbaum, C., Gross, D., Miller, K.: WordNet: an on-line lexical database. Oxford University Press (1990)
14. Mohammad, S., Dunne, C., Dorr, B.: Generating high-coverage semantic orientation lexicons from overtly marked words and a thesaurus. In: Proceedings of the 2009 Conference on Empirical Methods in Natural Language Processing, pp. 599–608 (2009)
15. Thet, T.T., Na, J.-C., Khoo, C.S.G.: Aspect-based sentiment analysis of movie reviews on discussion boards. J Inf. Sci. 36(6), 823–848 (2010)
16. Kang, H., Yoo, S.J., Han, D.: Senti-lexicon and improved Naive Bayes algorithms for sentiment analysis of restaurant reviews. Expert Syst. Appl. 39(5), 6000–6010 (2012)
17. Usha, M.S., Devi, M.I.: Analysis of sentiments using unsupervised learning techniques. In: 2013 International Conference on Information Communication and Embedded Systems (ICICES), pp. 241–245 (2013)

A Novel Approach to Extract and Analyse Trending Cuisines on Social Media

R. Lokeshkumar[✉], Omkar Vivek Sabnis, and Saikat Bhattacharyya

School of Computer Science and Engineering, Vellore Institute of Technology,
Vellore, India
rlokeshkumar@yahoo.com, omkarsabnis79@gmail.com,
saikat.bhattacharyya2016@vitstudent.ac.in

Abstract. In this technological era, we have seen a huge increase in the number of reviewing sites in the internet. In case of online food delivery stores, these reviews are very important as they, on the whole express public sentiment towards a particular restaurant or cuisine. In this paper, we are proposing an approach to predict which cuisines and restaurants are "trending" in a country based on the analysis of social media. We mine social media platforms like Twitter for food-related tweets and extract these tweets by using our own manually curated food lexicon. From these tweets, we use similarity matching to extract the food items that were tweeted about and run each of these items through a cuisine classifier based on logistic regression and word2vec word embeddings. This is done for all the tweets and thus, we can get which cuisines and restaurants have been popular while, which restaurants are fading. Our approach can, therefore be used by restaurants to analyze which markets they need to expand into and also where they have to revamp their business strategies.

Keywords: Twitter mining · Recommendation system · Logistic regression · Cuisine classification · Social media analysis

1 Introduction

With the increasing use of the Internet to provide services like online shopping and food delivery, we can see that a large portion of our lives is spent online. Therefore, the internet is becoming the outlet to broadcast our views and opinions. Humans, are social animals and our herd mentality, given to us by our ancestors, makes sure that other people's opinions play an important role in our decision making process. For example, a customer will decide to go to a restaurant based on how good the reviews are. In such an environment, any business can grow or fall based on public opinion. This public opinion can be analyzed to improve business profitability.

This use of social media makes it a very powerful tool in the field of business analysis. Twitter is the platform that has become the staple of most of our social media diet. The platform has over 300 million users and gives voice to the user's opinions in all the fields. By limiting the character limit to 140, the users are made to convey their

© Springer Nature Switzerland AG 2020
A. P. Pandian et al. (Eds.): ICCBI 2019, LNDECT 49, pp. 645–656, 2020.
https://doi.org/10.1007/978-3-030-43192-1_72

point in a short and concise manner and these tweets are used in our approach to extract the user's food reviews.

Mining the text from Twitter tweets to get some meaningful information has been a topic of research in the past few years [12]. The process of collecting and analysing food tweets is challenging because the tweets are user readable, but to use machine learning algorithms on them, requires them to be pre-processed due to the presence of emoticons and URLs.

The importance of natural language processing is huge, as extracting various lexical structures and elements, predicting the sentiment and meaning of the words is the only way we can go about to understand what exactly people are trying to convey [13]. Our paper deals with parsing a sentence to assign different parts of speech tags to the tweet to segment potential food item tags like Nouns and Adjectives.

Our proposed approach thus, uses geographical analysis of food related tweets from the world's biggest micro-blogging website, processing these tweets into a machine-readable form, extracting the food related items and finding out their cuisine. This way, businesses looking to expand in the culinary field as well as culinary businesses looking to diversify can save a lot of investment by analysis of mined data, which way give them a public opinion of what is "trending" so they can capitalize on these trends and increase profitability.

2   Literature Survey

Wang et al. [1] proposed a step-by-step approach to solve the problem posed by short texts as compared to long texts. The problem they faced was how conventional machine learning and data mining techniques are not able to extract features well enough due to their shortness and conciseness. They have made a comparison of which techniques to use to extract features from short text such as one-hot encoding, tf-idf weighting, word2vec as well as paragraph2vec. They tested their approach on 400,000 short text files from the Shanghai stock exchange.

Derczynski et al. [2] proposed a detailed error analysis of current twitter taggers because of the noisy and error-prone twitter data. The authors identified techniques that can improve parts-of-speech tagging accuracy for Twitter Data. The authors also provided a novel approach to improve tagging performance and handling unknown words as well as slangs. Effrosynidis et al. [3] compared the various algorithms required for pre-processing by performing experiments on 15 commonly used techniques. They also performed classification on this pre-processed data with three machine learning algorithms like Logistic Regression, Bernoulli Naïve Bayes and Linear Support Vector Classifier. They concluded by categorizing these techniques based on their performance and finding which techniques affect performance and which techniques do not. Kim et al. [4] proposed an automated text analysing model for evaluating food which can predict consumer acceptance. The authors collected consumer reviews from social networking sites like Twitter. These collected words were converted to vectors using an embedding model and these vectors were used to infer the evaluation of the taste and smell of two ramen types. The authors also evaluated the reliability and merits of the system by comparing their automated model's results with an actual consumer preference taste evaluation.

Rivolli et al. [5] proposed a machine learning approach to recommend food trucks based on the user's personal information and choices where each food truck is known to have a few cuisines. The authors provided six multi-label classification strategies on real data collected from over a hundred people. The authors compared their results to baselines and observed that all their strategies beat the baseline with Random K Labelsets and Binary Relevance giving the best performance.

Zhao et al. [6] proposed an approach to extract topical key phrases, with the goal to summarize Twitter. They used a context-sensitive Page Rank model which uses a probabilistic scoring function, that considers the relevance as well as the interestingness of the key phrases for ranking. They evaluated their methods on a large Twitter dataset and their experiments show that their model is very efficient at extracting topical key phrases.

Vidal et al. [7] conducted a case study on the use of Twitter data for food-related consumer research. The case study analyzed twitter data for breakfast, lunch, dinner as well as snack times and performed manual curation and content analysis of these tweets for deeper insights. Some of the insights they found were that the type of food eaten was mentioned more than the food itself and contextual characteristics of the occasions were also frequently mentioned. Emotions were rarely mentioned but emoticons were used much more. The paper also looked into the disadvantages of using twitter data for food-related consumer research.

3 Proposed Approach

In this study to identify the cuisines that are trending, we followed the following approach.

3.1 Data Collection

We first collected the tweets from people using Twitter's Application Programming Interface. The interface works on the REST or Representational State Transfer Architecture. For accessing the stream of tweets, we used a Python library called Tweepy, which helps us access the streaming API. The API makes use of the OAuth authentication protocol that requires a Consumer Key which is used to identify the client, a Consumer Secret, which is the password that is used to authenticate with the server, an Access Token which defines the privileges the client has and the Access Token Secret is used to authenticate the client and acts like a password [14]. We used a class called MyListener which helps get the data from the API, based on a certain aspect. The aspect can be either a hashtag or a location. We choose the aspect of "#food" for extracting the food related tweets from all the tweets posted on Twitter.

3.2 Preprocessing the Data

After collecting the tweets, we extracted only the location and text tags from the JSON file as they are fields we are interested in. We then filtered out all the tweets which were tagged with India in their location. In the second step of pre-processing, we cleaned the tweets by removing the unwanted tokens and characters like emails, URLs, emoticons and reserved keywords.

3.3   Parts of Speech Tagging of the Tweets

The tweets which have been pre-processed are sentences with unnecessary content removed. The tweets still cannot be used directly for extracting the food items because there are lot of words in these tweets which can increase the false positives. So, we are POS or Parts of Speech tagging the tweets so we can extract only the important words from the tweets.

After understanding the structure of the tweets and language used, we learnt that a few sentence structures are present which can help us extract food items from the tweets. Using the NLTK POS-tagging process, we found that the phrases that we needed to extract are - **<JJ><NN><CC><NN>** which is an adjective followed by a noun phrase, **<NP><CC><NP>** which is Noun Phrase followed by another Noun Phrase, **<RB><JJ><NN>** which is an adverb followed by a Noun. This helped us extract the important phrases from the tweets.

3.4   Creating a Food Lexicon

Once we have gotten the important phrases from the POS-tagging, we must extract the food items from these phrases using a similarity mechanism. To extract the food items, we need some lexicon or dataset to help us compare the food items in the tweets with the actual food items. Considering that we are working on Indian tweets, using datasets available wasn't feasible so we created our own lexicon, which is consisted of all kinds of foods and ingredients, Indian and from the world.

3.5   Extraction of Food Items

Now, we have the food lexicon as well as the important phrases from the tweets, we needed to extract the food items from the phrases. We chose levenshtein distance as our distance measure because Twitter is a casual platform and spelling mistakes will be common so the Levenshtein distance, or the minimum edit distance will give us the food items from the important phrases.

3.6   TF-IDF and Logistic Regression for Cuisine Classification

Now, we got the food items from each of the tweets. We predicted the cuisine of the food items based on a Logistic regression classifier which used the ingredients to cuisine database. This way we got a probabilistic distribution of which cuisine the food belongs to and helped us find the trending cuisine.

4 Software Implementation

4.1 Tweet Processing

Due to finite storage, the data collection was performed in various stages. Since we built the StreamListener class on top of tweepy's Stream API to fetch tweets in real-time, we could only collect tweets while the script was operational. Hence, the collection stage was divided into chunks of tweets - 100,000 tweets each time, for 20 times - which is over 2 million tweets.

In our first step of pre-processing, we filtered out the tweets which were from India. Since the location attribute can be changed based on the user's personal choice, there was a lot of variation and no standard format of representing the location. For example, many tweets' location was "India", while many- "Kolkata, India", "Bidhan Nagar, India" and also others with "Bharat". Hence, a string matching script was used with all known states (29) and cities (302) from Wikipedia. A tweet was successfully filtered if any of the cities or the states in the vocabulary matched as a substring in the location attribute.

The frequency of tweets is dependent over time and events happening in the world, hence it is very hard to estimate how many #food related tweets are from India. However, for the week we performed our data collection through, roughly around 5% of the tweets were filtered as Indian.

For the second step of pre-processing, we performed cleaning of each tweet's text content. As shown in Fig. 2, there are many unwanted tokens/characters that aren't considered relevant to perform an analysis on a certain topic. For example - email, URL, emoji, username mentions, reserved keywords (such as RT - retweet).

For this step, we used regular expressions (Fig. 1) - which can be used to provide a pattern to match in any string. List of patterns we used for cleaning the tweet:

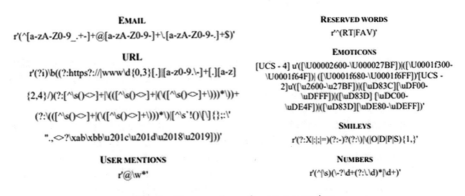

Fig. 1. Regex patterns for pre-processing

Hashtags were not removed, because many tweets had potential food content after the hashtags - example #dosa, #buttercream, which are food items, and hence relevant to our research. After the cleaning process, there were still some tweets with special characters left behind (exceptions to the pattern), and hence they were further cleaned through a script to check ASCII value of each character in a tweet and accordingly removing the special characters.

4.2   Creating Our Food Lexicon

After pre-processing the tweets, we created a food lexicon. The food lexicon was necessary because we needed to extract the food items from the processed tweets. Since we focusing on India as our country of research, we can't make use of any public datasets. The New York Public Library's "What's on the Menu" dataset is one of the most comprehensive datasets we found on food, as it contains menu foods from the 1840s. This dataset has all the food types, from breakfast items to wine and alcohol menus. The only issue with the dataset is that since it is an American dataset, there are very few Indian dishes. Another issue we faced is the lack of Hindi and Urdu dish names in the dataset, as a lot of Indians tweet with the Indian name of the dish instead of the Universal name – for example: Lentils are called daal.

To solve these problems, we first cleaned the dataset. The dataset had a lot of unknown special characters which were removed but we kept the hyphen which was used for compound word recipes. We also removed all the 3 letter words in the dataset because the dataset had a lot of tag words. However, to replace the 3 letter food items like Yam, we manually added them back. We also ran a set function on the dataset so that we can remove the repetitions from the dataset, thereby reducing the size of the dataset. Now that the dataset was cleaned, we had to add Indian foods as well as others to increase the size of the dataset. For this task, we scraped the Internet looking for Indian food dishes so that we can add them as well as other cuisines like Chinese and French so we could get a comprehensive dataset. After we had a sizeable dataset, we manually curated the dataset to remove unnecessary dataset items which have remained in the dataset like menu headings and so on.

4.3   Parts of Speech Tagging of the Tweets

Now that we have our processed tweets, we needed to extract those elements with which the food matching could be done along with the dataset. We experimented with multiple techniques and did a comparison between the Parts-Of-Speech taggers provided by Spacy as well as NLTK. The tags in NLTK were more comprehensive and gave a higher accuracy as compared to those provided by Spacy. We did some research into which POS tags would help us extract the food item phrases because the tweets at this point are sentences and matching food with sentences led to a loss of accuracy in matching the food because of similar sounding words like "cake" and "bake." On analyzing the food item phrases, we came to the conclusion that there are 3 key phrases that most food item phrases follow, which are the adjective followed by a noun phrase; for example – "Red cabbage salad" where "Red" is the adjective followed by "cabbage salad" which is a noun phrase. Another key phrase is the noun followed by a noun; for example – "Chicken Pasta" where chicken and pasta are both nouns. The third key phrase is the Adverb followed by a noun; for example – two-pound beef burger where "two-pound" acts as an adverbial phrase and beef burger is the noun phrase. Using these three key phrases, we extracted the tweet phrases that fit the categories. This helped us in improving the words that need to be matched to the food lexicon so our accuracy increases.

4.4 Food Matching Using Fuzzy Parsing

Once the tweet phrases were extracted from the tweets, we used a distance measure to find how similar are the words in the extracted tweets to the food lexicon which contains a large variety of food items. We did a comparison between the various distance measures like Cosine, Jaccard and Levenshtein. On further study, we found that the Levenshtein distance measure is the best for word matching because Levenshtein distance takes into account the spelling mistakes and this is of utmost importance because Twitter is a social media platform where spellings are not given a high priority. We also realized that basic Levenshtein distance was giving us average results so we used Fuzzy Parsing [8, 15] along with the distance measure so that we can get better results. We used the "FuzzyWuzzy" package that is provided in Python. The fuzzy parser extracted the matches from the lexicon and we set a threshold of 85% match because that was found to the best heuristically. The fuzzy parsing process was put into a csv file which had a tweet ID and the extracted food items in the tweet so that we can run the logistic regression model for identifying the cuisine.

4.5 Cuisine Classification Using Logistic Regression and TFIDF

To classify the cuisines from the food items that we extracted from the tweets, we used the "What's Cooking" dataset by Yummly on Kaggle. We used a GridSearchCV pipeline package to create a package so that we can find the optimal parameters. We used TFIDF [9, 12] for feature selection and Logistic Regression classifier for classification because in our research, we found that logistic regression has better performance than Linear Support Vector Classifier and Bernoulli Naïve Bayes in short text classification. Since TFIDF needs the words in its vocabulary for classification, we added our own Indian Words and recipes to the dataset with the label as Indian, thereby helping us proceed with this solution. We created a GridSearchCV pipeline with TFIDF Vectorizer and Logistic Regression Classifier. We loaded the data into the pipeline and created the training and validation datasets. We fit the classifier on the data and evaluated the performance in the validation set. We evaluated the metrics and scores using the pipeline's functions to estimate the best parameters. We document the results and extract the mistakes so that we can retrain the pipeline using this "feedback." This is the concept behind our training.

The above trained model with the based parameters is loaded onto the test data pipeline and the predictions are made.

5 Results

This section of our research will go through the whole process of our approach and show how we reach the predictions.

5.1 Pre-processing the Data

When we first the use the API (Application Programming Interface) given by Twitter, we get the data in a JSON file. However, we are only interested in the text and location fields and after extracting those two fields from the JSON file we get the following: After completely processing the data to remove URLs, emails, reserved words, emoticons and smileys, we get the following Fig. 3.

Fig. 2. Text content in each tweet

Fig. 3. Processed tweets – tweet numbers and tweet text

5.2 Preparing the Food Lexicon

Once the data is pre-processed, we created our lexicon which was very incomplete and had many unnecessary words and foods from another language. The dataset was in this form: We processed this food lexicon because we it didn't contain many Indian food items and also the food dataset was not in the form we wanted it to be. So we formatted the dataset and added Indian food dishes added manually as well as scrapped from the dataset. Now the dataset looks like this Figs. 4 and 5.

Fig. 4. Unprocessed food lexicon

Fig. 5. Processed food lexicon

5.3 Parts of Speech Tagging the Tweets

Now that we have a comprehensive food lexicon, we have to extract the phrases that have the highest chance of having food items. Using the three part of speech tag rules, we extracted key phrases that have the highest chance of having food items. Now the tweets look like this Figs. 6 and 7.

Tweet Number	Text
.	...
424	['Indian style pancake delish']
425	[]
426	['prepared poha', 'little desi ghee']
427	['kodaikanal chicken gravy', 'authentic south']
428	[]
429	['good food', 'good mood', 'delicious paneer tikka']
.	...
.	...

Fig. 6. Parts-Of-Speech tagged phrases

Tweet Number	Food Items
.	...
.	...
25	vegetable thali dal fry matar paneer
26	curd ice cream pani puri dahi vada
27	boneless chicken
28	khair
29	croatian cabbage pasta
.	...
.	...

Fig. 7. Extracted food phrases

5.4 Food Matching Using Fuzzy Parsing

After extracting the key phrases from the tweets, we use our Fuzzy Parsing with Levenshtein distance script, to extract food items from the key phrases. After extracting the food items, we get the following –

5.5 Cuisine Classification Using Logistic Regression and TFIDF

Now, we have extracted the food items from the tweets. Now we create our Grid-SearchCV pipeline for finding the best parameters for our TFIDF and Logistic Regression Model. The dataset distribution of the various cuisines is shown below Figs. 8 and 9.

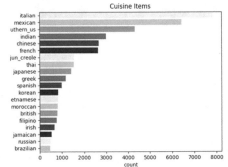

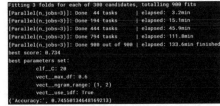

Fig. 8. Cuisine distribution in the dataset

Fig. 9. Accuracy and best parameters first run

Now, we train our TFIDF and Logistic regression model with one run without feedback and one run with the feedback. The results of the first run are given above.

Now we ran the whole pipeline again after the feedback loop [10] and these were our results (Figs. 10 and 11).

```
Fitting 3 folds for each of 300 candidates, totalling 900 fits
[Parallel(n_jobs=3)]: Done  44 tasks    | elapsed:   4.1min
[Parallel(n_jobs=3)]: Done 194 tasks    | elapsed:  17.4min
[Parallel(n_jobs=3)]: Done 444 tasks    | elapsed:  48.2min
[Parallel(n_jobs=3)]: Done 794 tasks    | elapsed: 124.6min
[Parallel(n_jobs=3)]: Done 900 out of 900 | elapsed: 151.3min finished
best score: 0.783
best parameters set:
        clf__C: 10
    vect__max_df: 0.7
    vect__ngram_range: (1, 1)
    vect__use_idf: True
('Accuracy:', 0.79862833652515336)
```

Fig. 10. Accuracy and best parameters after feedback loop

Tweet Number	Food Items	Cuisine
.	...	...
.	...	...
25	vegetable thali dal fry matar paneer	Indian
26	curd ice cream pani puri dahi vada	Indian
27	boneless chicken	Italian
28	khair	Brazilian
29	croatian cabbage pasta	Greek
.	...	...
.	...	...

Fig. 11. Predictions on the testing data

Now we ran the predictions of the Logistic Regression on some of our tweets which we had kept as our test-bench by dividing the dataset into a training and a testing set in a ratio 0.8 is to 0.2. Some of the predictions are shown above. Thus, we found that the accuracy of the predictions of our logistic regression model on the test-bench which had cuisines either from the database or manually curated by us was 75.43%.

We now created a plot of which food items had the highest occurrence in our data collection time-period of mid-January to mid-February, 2019, to show which cuisines had the most mentions and hence, were "trending" Fig. 12.

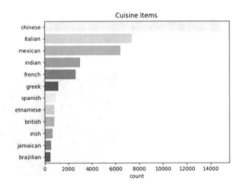

Fig. 12. Visualizing which cuisine was "trending"

6   Conclusion

This research paper focuses on how we businesses in the food industry can use the power of social media to increase their business – by listening to the voice of the people in their demographic and catering to their "trending" needs. This paper used the parts-of-speech tagging to extract key phrases from pre-processes tweets, used Fuzzy Parsing with Levenshtein distance measure to extract the food items and used a TFIDF-Logistic Regression Model to predict the cuisines. Then we found that in a time-frame of

January and February, 2019 the cuisine which had the most mentions and thus, was "trending" was Chinese. We feel that this matches with the general trend as the Chinese New Year season was during that time and our research shows that people were interested in Chinese Food.

7 Future Work

As the future scope of this work, we will analyses the various algorithms that could have been used like the Linear Support Vector Classifier or the Random Forest Classifier. We can also focus on specific restaurants so that we can see how their new food items are performing in the public opinion. There is huge scope in how we can go about this work in the future.

Acknowledgement. The authors would like to thank the reviewers and the experts who have helped in this research and given us great insights and thoughts which have guided us and helped us improve our work.

References

1. Wang, Y., Zhou, Z., Jin, S., Liu, D., Lu, M.: Comparisons and selections of features and classifiers for short text classification. In: IOP Conference Series: Materials Science and Engineering, vol. 261 (2017)
2. Derczynski, L., Ritter, A., Clark, S., Bontcheva, K.: Twitter part-of-speech tagging for all: overcoming sparse and noisy data. In: Proceedings of the International Conference on Recent Advances in Natural Language Processing, January 2013
3. Effrosynidis, D., Symeonidis, S., Arampatzis, A.: A comparison of pre-processing techniques for Twitter sentiment analysis. In: 21st International Conference on Theory and Practice of Digital Libraries. LNCS, vol. 10450, September 2017
4. Kim, A.Y., Ha, J.G., Choi, H., Moon, H.: Automated text analysis based on skip-gram model for food evaluation in predicting consumer acceptance. Comput. Intell. Neurosci. **2**, 1–12 (2018)
5. Rivolli, A., Parker, L.C., de Carvalho, A.: Food truck recommendation using multi-label classification. In: Proceedings of the 18th EPIA Conference on Artificial Intelligence, pp. 585–596, September 2017
6. Zhao, W.X., Jiang, J., He, J., Song, Y., Achananuparp, P., Lim, E.-P., Li, X.: Topical keyphrase extraction from Twitter. In: Proceedings of the 49th Annual Meeting of the Association for Computational Linguistics: Human Language Technologies, vol. 1, pp. 379–388, June 2011
7. Vidal, L., Ares, G., Machín, L., Jaeger, S.R.: Using Twitter data for food-related consumer research: a case study on "what people say when tweeting about different eating situations". Food Qual. Prefer. **45**, 58–69 (2015)
8. Koppler, R.: A systematic approach to fuzzy parsing. Softw. Pract. Exp. **27**(6), 637–649 (2000)
9. Ramos, J.: Using TF-IDF to determine word relevance in document queries. In: Proceedings of the First Instructional Conference on Machine Learning, vol. 242, pp. 133–142, January 2003

10. Koster, C.H.A., Beney, J.G.: On the importance of parameter tuning in text categorization. In: International Andrei Ershov Memorial Conference on Perspectives of System Informatics, pp. 270–283, June 2006. Springer (2006)
11. Lokeshkumar, R., Maruthavani, E., Bharathi, A.: A new perspective for decision makers to improve efficiency in social business intelligence systems for sustainable development. Int. J. Environ. Sustain. Dev. **7**, 404–416 (2018)
12. Guo, B., Liu, Y., Ouyang, Y., Zheng, V.W., Zhang, D., Yu, Z.: Harnessing the power of the general public for crowdsourced business intelligence: a survey. IEEE Access **7**, 26606–26630 (2019)
13. Ahmed, M., Chen, Q., Wang, Y., Li, Z.: Hint-embedding attention-based LSTM for aspect identification sentiment analysis. In: Pacific Rim International Conference on Artificial Intelligence, pp. 569–581. Springer, Cham (2019)
14. Lokeshkumar, R., Sindhuja, R., Sengottuvelan, P.: A survey on preprocessing of web log file in web usage mining to improve the quality of data. Int. J. Emerg. Technol. Adv. Eng. **4**(8), 229–234 (2014)
15. Gopalakrishnan, T., Sengottuvelan, P., Bharathi, A., Lokeshkumar, R.: An approach to webpage prediction method using variable order Markov model in recommendation systems. J. Internet Technol. **19**(2), 415–424 (2018)

Alphanumeric Character Recognition on Tiny Dataset

Sujit S. Amin[✉] and Lata Ragha

Department of Computer Engineering, Fr. C. Rodrigues Institute of Technology,
Vashi, Navi Mumbai, India
mastersujitamin@gmail.com, lata.ragha@gmail.com

Abstract. Alphanumeric character recognition is the ability of a computer to understand digits and alphabets image as an input. This topic has struggled to reach a wide variety of people since we need quite a lot of labeled training data. A small training sample achieves sub-standard accuracy in training data that is model is not able to learn accurately. Because of this handwritten (alphanumeric character), recognition is not widely been used. We resolve this issue by creating new labeled training data. Generated data is generated from existing samples, with a realistic data expansion method that is an actual variant of human handwriting. This is done by changing or adding more value to the parameters. Our model had just 10–11% training samples per character compared to other character-recognition methods. We achieved close to existing state-of-the-art character recognition results of all four datasets. Images reconstruction strategy was also used. Our system can be used when training data is less, and the user wants to achieve reasonable accuracy.

Keywords: Capsule networks · Character-recognition · CNN · EMINST · Deep learning

1 Introduction

Alphanumeric character recognition is being a very functional instrument for various languages. This is due to recent advances in machine learning and deep learning fields. Over the years, conventional models such as linear classifiers, nonlinear classifiers have been used, but these are not able to achieve human-level Performances. An exceptional result has been attained by Convolutional Neural Networks (CNN). Even though CNNs understand features in images, they lose a lot of information in pooling layers. CNN requires a lot of images to train properly. Moreover, in handwritten recognition, we need a dataset that is labeled also. This is a huge problem when we have fewer data to train.

Secondly, there is always a need for a machine to recognize handwritten data. But this requires a lot of training data which is difficult to create. Since creating an image of every character in a language requires you take pictures of every character one by one. Moreover, it requires a significant amount of data per character. This is where our method can be used since we used data creation technique. Data is then used for training. This reduces the amount of data required for training.

© Springer Nature Switzerland AG 2020
A. P. Pandian et al. (Eds.): ICCBI 2019, LNDECT 49, pp. 657–667, 2020.
https://doi.org/10.1007/978-3-030-43192-1_73

We take on the problem of a labeled dataset of small size using capsule networks. We utilize a capsule network's ability to generate data by changing instantiation parameters. Capsule Networks learns the property of character (image). Our system neural network is based on Capsule Networks architecture proposed in [9], which contains a decoder network and capsule network. We add a deconvolution network by replacing the decoder network. To generate new data from already present data, we add a little amount of noise to instantiation parameters. This result is a technique that generates data, which is a variant of original images, which is better than other methods for generating labeled images. Reconstructed image accuracy is also an important factor, so here we present an appropriate strategy that reduces losses and improves reconstruction significantly.

Key achievements of our paper, with only 10–11% of the data needed for the state-of-the-art framework, we achieved accuracy results close to previous achievements in almost all the datasets we used.

2 Related Work

Digit Recognition task MNIST [6, 7] is widely recognized. Several research works [12–15] are done on MNIST dataset. Various attempts are made on the Extended MNIST (EMNIST) dataset. The extended MNIST dataset is used for handwritten recognition of alphanumeric data. In [11] two-way neural network which has style memory at the output. It can perform image grouping and recreation. With capsule network and dynamic routing, authors in [9] reduced error percentage on the MNIST dataset. Capsule network as an idea was proposed in 2011 by [4], as transforming auto-encoder. Low data has been a significant issue in image character recognition. An example is the Siamese Network [5]. OS learning was also introduced in [1]. One-shot (OS) learning is application-specific; we didn't use this method for generating data. Our literature survey gave several approaches to data generation. GANs can be extremely useful to generate data from existing data, but GANs cannot label data. To generate a label, we have to create separate GAN for every character. The second approach is Variational Auto-encoders (VRAEs) possessed the same problems. VRAEs have images that are 1D vector. For each group, the capsule networks have different dimensions. VRAEs may affect changes in multiples classes. To characters, jittering is not appropriate. There is a finite amount of easy increase in flipping. This technique is far away from human variations.

3 Proposed Methodology

In our research, we have performed training both on full training data and 230 samples per class. We used 230 samples since we were trying to find the lowest number of samples required for achieving the result that is of good accuracy or close to state-of-the-art accuracy. The full training dataset contains 4800 data for each class in EMNIST-letters and about half of the data for the EMINST-balanced data set.

Since we are using fewer data points (230 samples per class), we generate training samples, which are explained clearly in Sect. 3.2. This will reduce the need for having a large dataset for training. The method we followed is generalizing for character recognition so that it can be used for any other character dataset.

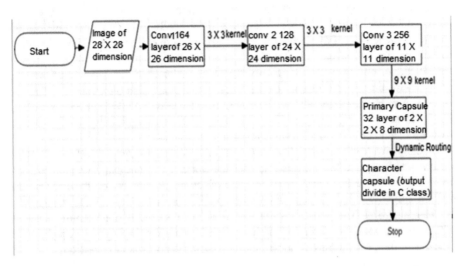

Fig. 1. Work flow of character classification by our model. This is done for every character.

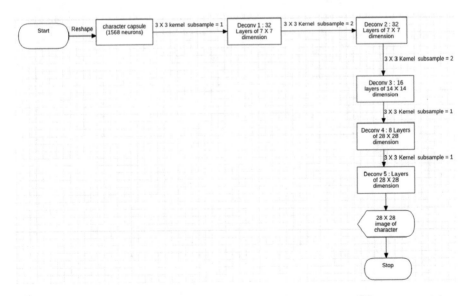

Fig. 2. Illustrates the decoder network of reconstruction of character. Input is obtained from classification layer.

3.1   Image Character Identification

For image character identification, we use a capsule network. We propose the architecture of modified capsule network architecture for character identification. We also, have a decoder network, as mentioned in Figs. 1 and 2.

In capsule network, first 3 layers are convolutional layers with 64 3 × 3 kernels, 128 3 × 3 kernels and 256 3 × 3 kernels respectively [9]. The next layer is the primary capsule layer. Fourth layer is 9 × 9 kernel. Fifth layer is 9 × 9 kernel with 16 dimensions capsule layer. This result in C capsules with C number of classes. Capsule uses dynamic routing [9]. The input to capsule is set of I, 28 × 28 images. Output is I × C × 16-dimensional tensor N, which contains instantiation parameters, where N_i, $i \in [I]$ is instantiation parameters. The output is masked with D (D is unit vector of N). Output is set of reconstructed 28 × 28 images. De-convolutional layer has ReLU activation function. The final layer has sigmoid activation.

Initially, we train the model on entire training data and check is performed. Next, we use just 230 samples per class so that the scarcity of data can be simulated. We attempt to achieve accuracy close to state-of-the-art results on every dataset. Our method initially failed to generate an acceptable reconstruction of the image at a decoder network. The most obvious solution is to increase the training samples. Alternatively, with the definition of instantiation variables in capsule networks, we create new data for training. The technique is described in the next section.

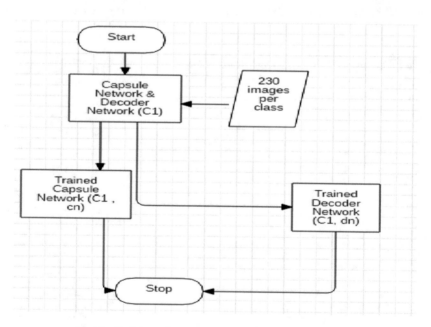

Fig. 3. Training of capsule network and decoder network

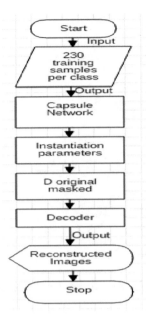

Fig. 4. Creation of instantiation parameters and newly reconstructed images

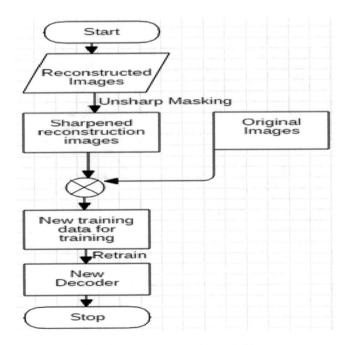

Fig. 5. Decoder retraining technique

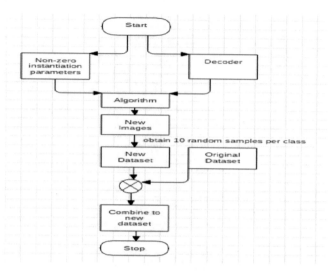

Fig. 6. New image data creation technique

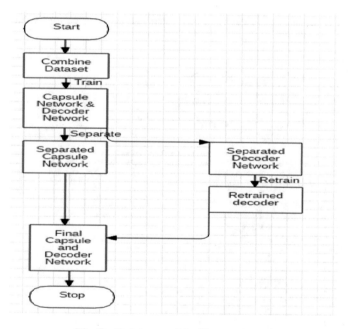

Fig. 7. Training model with new dataset

3.2   Image Generation by Changing of Instantiation Parameters

Any character in the capsule can be present as 16-dimensional vectors [9]. By just changing instantiation parameters, we can reconstruct an image. We add control noise to our parameters, which creates images that are a variation of a given image. This provides us with more data to train our model. Effectively this process increases the size of the dataset. Each individual parameter is responsible for the property of an image.

Firstly, in Fig. 3, we train the capsule and decoder network (C1) with our dataset of 230 images per class. From this, we get trained capsule network (C1, cn) and trained decoder network (C1, dn). Further, in Fig. 4, we get our instantiation parameters of training images as the output of our capsule network. Now we get reconstructed images as output.

Fig. 8. Unedited images and reconstructed image from initial model

From Fig. 8, training with very few samples leads to a badly trained model. Now, our goal is to remove blurriness in the image output. This is done using the method in Fig. 5. For every image which is reconstructed, we perform unsharp masking [8]. This sharpens our image. Now we combine our image, which creates a new set for the decoder. Now we re-train decoder with this new target data. This provides sharper reconstructions. Several times training samples may not be properly learned, so reconstruction may be wrong. This may be miss-classified. We remove such training samples.

Figure 6 shows the new data generation technique. We add random noise. Algorithm for creating a new image is explained in Algorithm 1. Firstly, we load the training data. We then compute the number of classes in the dataset. Table 1, gives information on the number of classes. For example, the number of classes for EMINST-Letters dataset is 26. For each class in training data, we compute the class variance. We compute the maximum noise value (noiseMax). For every instantiation variable, we determine the noise value. For every image, we check if the noise value is less than noiseMax. If true then noise value is added. Newly created image is stored in a large array (newImageBucket). After the algorithm has stopped executing, we combine both the created images and original images. This is then used for training.

We combine data (created data + original data) and formulate a new dataset. This solves our less data problem. In Fig. 7, we train our model, and then we obtain the final model for character classification.

The algorithm used for image creation is explained below.

Algorithm 1: Image creation technique
Input: Instantiations parameters and training data
Output: New Images array
Algorithm Steps:

1. Load the training data from dataset
2. Calculate the number of classes (for e.g. EMNIST-Letters has 26 classes)
3. FOR each class in training data
4.   Compute class variance of instantiation parameters
5.   Sort the class variance in descending order
6.   Compute the highest amount of noise you can attach (noiseMax)
7.   FOR each instantiation variable
8.   We determine the noise value
9.   FOR each image in class
10.   IF noise value is less than noiseMax allowed
11.   add an average of max noise value
12.   ELSE
13.   lower noise value
14.   ENDIF
15.   Append newly created images in large array (newImageBucket)
16.   ENDFOR
17.   ENDFOR
18. ENDFOR

4 Experiments and Results

Table 1, gives information of dataset that we used. We used four datasets; information of every dataset is given below.

Table 1. Information of dataset used

Dataset name	Number of classes	Training samples per class	Size of training data	Size of test data
MNIST [6]	10	6000	60000	10000
EMINST-Letters [2]	26	4800	124800	20800
EMINST-Digits [2]	10	24000	240000	40000
EMINST-Balanced [2]	47	2400	112800	18800

4.1 Alphanumeric Character Classification

Table 2 differentiate the results to various previously achieved state-of-the-art results. We included full training and 230 training samples per class. For every dataset, we use ensemble to enhance accuracy and to avoid overfitting.

All four of the datasets are near to state-of-the-art accuracy on entire training data. Whereas, our 230-training data per class is also very close to the state-of-the-art accurate model.

Table 2. Our model compares with various state-of-the-art results.

Model	Entire training data	230 samples per class
MNIST		
Wan et al. [10]	99.79%	–
Our model	99.71 ± 0.14%	98.47 ± 0.25%
EMNIST-Digits		
Dufourq et al. [3]	99.3%	–
Our model	99.27 ± 0.10%	99.03 ± 0.22%
EMNIST-Letters		
Wiyatno et al. [11]	91.27%	–
Our model	91.03 ± 0.26%	90.07 ± 0.29%
EMNIST-Balanced		
Dufourq et al. [3]	88.3%	–
Our model	88.21 ± 0.25%	87.94 ± 0.32%

4.2 Image Creation Technique Results

The effects of the decoder retraining technique are discussed here. Figure 9(a), We analyse the EMNIST-letters test data and (b) Illustrates the images that have been reconstructed using 230 images per category. Figure 9(c) illustrates the reconstructed image. The reconstructed image is much sharper.

(a) Test Image

(b) Reconstruction by method given in Figure 5

(c) Reconstruction by method given in Figure 7

Fig. 9. Illustrates the images of our reconstruction technique.

Clearly, GANs are alternatives to our technique. However, the data generated is not very realistic [16]. As seen in Fig. 10, data is very noisy generated from image data 2.

Fig. 10. Image from CGAN for image data 2

5 Conclusion

In this paper, we set up a framework by using capsule networks to expand the size of the datasets. We used a well-known character, dataset. We exploited the power of capsule networks. This is done by changing a few of instantiation parameters we are able to create new data from our original dataset. Our algorithm only takes 230 samples of training per class and produces new data for learning. We add controlled noise to our parameters of capsule networks while training. Our data generation technique produces data that are more human-like in terms of boldness, stroke pattern. We evaluated the performance on full training data and 230 samples data. We achieved close to state-of-the-art accuracy after merging the original and produced dataset. Our method works well on image characters. This work can be extended to higher resolution images and other languages as well.

References

1. Bertinetto, L., Henriques, J.F., Valmadre, J., Torr, P.H., Vedaldi, A.: Learning feed-forward one-shot learners. NIPS (2016)
2. Cohen, G., Afshar, S., Tapson, J., van Schaik, A.: EMNIST: an extension of MNIST to handwritten letters. CoRR (2017)
3. Dufourq, E., Bassett, B.A.: Eden: evolutionary deep networks for efficient machine learning. In: PRASARobMech, Bloemfontein, South Africa, pp. 110–115 (2017)
4. Hinton, G.E., Krizhevsky, A., Wang, S.D.: Transforming auto-encoders. In: ICANN, Berlin, Heidelberg, pp. 44–51 (2011)
5. Koch, G., Zemel, R., Salakhutdinov, R.: Siamese neural networks for one-shot image recognition (2015)
6. LeCun, Y., Cortes, C., Burges, C.J.C.: The MNIST database of handwritten digits (1998)
7. Zhong, Z., Zheng, L., Kang, G., Li, S., Yang, Y.: Random erasing data augmentation (2017)
8. Polesel, A., Ramponi, G., Mathews, V.J.: Image enhancement via adaptive unsharp masking. IEEE Trans. Image Process. **9**, 505–510 (2000)
9. Sabour, S., Frosst, N., Hinton, G.E.: Dynamic routing between capsules. In: NIPS, Long Beach, CA, pp. 3856–3866 (2017)
10. Wan, L., Zeiler, M., Zhang, S., Cun, Y.L., Fergus, R.: Regularization of neural networks using dropconnect. In: ICML, vol. 28, pp. 1058–1066 (2013)
11. Wiyatno, R., Orchard, J.: Style memory: making a classifier network generative. CoRR (2018)
12. Ciresan, D.C., Meier, U., Schmidhuber, J.: Multi-column deep neural networks for image classification. CoRR (2012)
13. Labusch, K., Barth, E., Martinetz, T.: Simple method for high-performance digit recognition based on sparse coding. IEEE Trans. Neural Netw. **19**, 1985–1989 (2008)

14. Ranzato, M., Poultney, C., Chopra, S., LeCun, Y.: Efficient learning of sparse representations with an energy-based model. In: Proceedings of NIPS, Cambridge, MA, pp. 1137–1144 (2006)
15. Jarrett, K., Kavukcuoglu, K., Ranzato, M., LeCun, Y.: What is the best multi-stage architecture for object recognition? In: ICCV, Kyoto, Japan, pp. 2146–2153 (2009)
16. Goodfellow, I., Pouget-Abadie, J., Mirza, M., Xu, B., Warde-Farley, D., Ozair, S., Courville, A., Bengio, Y.: Generative adversarial nets. In: Advances in Neural Information Processing Systems, pp. 2672–2680 (2014)

Sentiment Analysis in Movie Reviews Using Document Frequency Difference, Gain Ratio and Kullback-Leibler Divergence as Feature Selection Methods and Multi-layer Perceptron Classifier

S. Vigneshwaran$^{(\boxtimes)}$

Applied Mathematics and Computational Sciences,
PSG College of Engineering and Technology, Coimbatore, India
vikki499@gmail.com

Abstract. Both industries and people are interested in knowing the reviews of the products or movies. If the number of reviews are large, manual classification of reviews into either positive or negative classes, is tedious and time consuming. As an alternative, Sentiment Analysis, a subdomain in Natural Language Processing aims to automate the above process, by training the models by using reviews as the training data. These are popular because this is very essential for making decisions by knowing the other person's opinion. In this paper, Document Frequency Difference, Gain Ratio and Kullback-Leibler divergence are used for feature selection and the classification is done with Multi-Layer Perceptron classifier. This is done with movie reviews dataset like SAR14, IMDB11, Cornell open source benchmark dataset. The results show that Document Frequency Difference and Gain Ratio and feature selection methods have better sentiment classification performance i.e. with better accuracy and reduced error.

Keywords: Document Frequency Difference · Gain Ratio · Kullback-Leibler Divergence · Feature selection · Movie reviews · Sentiment classification · Multi-layer Perceptron

1 Introduction

The understanding of the emotions with the help of software is commonly called as sentiment analysis. This comes under Natural Language Processing. Human generated sentences contains multiple layers of meanings. These interpretations are different from person to person. As sentiment analysis deals with the opinion this is commonly called as opinion mining. Messages in social media will be shorter in text and it will be worse written which are more complex to find the sentiment [1]. The only way to understand these kind of sentences is to understand how the paragraph is started. This can strongly have the impact on the sentiment of the internal sentences. People now are interested in knowing the feedback of each and every product they are using. This is one of the applications of sentiment analysis in the business field. The investors have keen interest

© Springer Nature Switzerland AG 2020
A. P. Pandian et al. (Eds.): ICCBI 2019, LNDECT 49, pp. 668–676, 2020.
https://doi.org/10.1007/978-3-030-43192-1_74

in knowing the percentage of positive and negative feedback on their company's products [2]. In order to get more information about the tweets on Twitter sentiment analysis is used [3]. Sentiment analysis has uncountable applications. Generally opinion mining is classified into three categories:

Document Level Classification: In this the whole document is classified into either positive or negative. It has an assumption that each and every document conveys the opinion of the same attribute. This is useful in categorizing the review as either positive, negative or neutral [11].

Sentence Level Classification: As the name suggests it will categorize each and every sentence as positive, negative or neutral. The distinction between the objective and subjective sentences are made by the sentence level analysis. Hence this is commonly called as subjectivity classification. True expressions will be expressed by objective sentences. The opinions and subjective views are expressed by subjective sentences. As compared to subjective sentences, objective sentences imply more opinion.

Entity and Aspect Level Classification: Opinion mining based on features and summarization of the text comes under entity and aspect level classification. In this there is an assumption that an opinion carries a sentiment that can be of either positive or negative. Based on that assumption this will directly identifies the opinion itself. As the analysis made by this classification is accurate this is commonly referred to as finer grained analysis.

The major contributions of this research paper are three folds:

- Firstly a large collection of text dataset from diverse sources like Cornell open source benchmark dataset, IMDB11 movie review dataset and SAR14 movie review dataset are collected and analyzed to understand the nature of the features.
- Secondly, a proper preprocessing method is utilized to remove insignificant information and applied three different feature selection methods to extract significant features based on the average of those feature selection methods.
- Finally, a deep learning model is used for opinion classification effectively.

Organization: In this, Sect. 2 describes a summary of the related works made on sentiment analysis. In Sect. 3, the implementation of the proposed framework of extracting opinion and classification of it and details of the data set. Section 4 describes the results and performance of the analysis which is proposed. Conclusions of the work and the references are mentioned in Sect. 5.

2 Related Work

The authors in [4] reported studying and designing a summary of a movie review and rating of that movie for mobile environment. In their study, based on the results of applying opinion classification to movie reviews the rating to that movie was done. Latent Semantic Analysis (LSA) was proposed by the authors in order to detect the feature-based product summarization. Statistical approach was used to identify the opinion related words. Feature-based summarization was used to identify the product features.

The authors in [5] proposed an unsupervised model named as joint sentiment (JST) model. Document level sentiment classification was used in this model and at the same time, a mixture of topics was extracted from the text. The evaluation was done by using movie review data set.

The authors in [6] proposed a review mining system for movie reviews using Multi-knowledge based method. Based on the knowledge of the movie, WordNet and statistical analysis multi-knowledge movie review was proposed. In this method, a particular movie review was taken as domain. Inorder to gain the interest of movie fans, an additional approach was proposed. In that, by using names of people in the film industry a summary was built.

The authors in [7] detected a sentiment of comment titles. This was done using probabilistic latent semantic analysis. In this model, positive and negative documents is classified into a particular topic or content.

The authors in [14] proposed a new feature selection method for named as Document Frequency Difference on his paper comparison of feature selection methods for sentiment analysis.

3   Proposed Framework for Sentiment Extraction and Classification

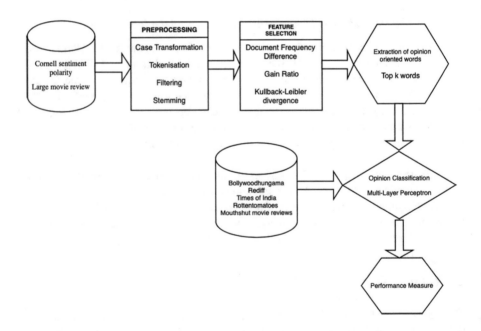

3.1   Data Source

Data set	No. of reviews	
	Positive	Negative
Cornell sentiment polarity v1.0	700	700
Cornell sentiment polarity v2.0	1000	1000
Large movie review v1.0 IMDB11	25000	25000
Large movie review SAR14	167378	66222

The above mentioned open source data sets can be downloaded from [13].

3.2   Data Preprocessing

The real world data contains noises like hashtags, HTML tags, scripts and symbols. These things do not contribute much significance to the result so these are to be removed by doing tokenization which will split the document into separate words, as the data contains both uppercase and lowercase inorder to remove that case normalization is used, filtering method is used to remove the stopword (such as "the", "a", "an", "in") and stemming is used for collapsing different forms of words. Inorder to sharpen the results and to reduce the vocabulary size stemming is used.

3.3   Vectorization

The process of converting text to vectors(matrix) is called vectorization. In Term Frequency method, for each term the ratio is calculated between the frequency of that term in a document and total number of words in that document. The formula for calculating Inverse Document Frequency,

$$IDF(x) = ln(Total\ number\ of\ documents/Number\ of\ documents\ with\ term\ x\ in\ it)$$

The formula for calculating Term Frequency - Inverse Document Frequency is,

$$TF - IDF = TF * IDF$$

3.4   Feature Selection

For the construction of a model, a subset of features that are relevant to that model is required. This process is named as feature selection. In order to make the model simple and easy to interpret by researchers and users feature selection is needed [9]. In this paper, the following methods are used for selecting features.

Document Frequency Difference: This is applicable only for polar (positive or negative) dataset. Document Frequency Difference is calculated as:

$$score(t) = |DF_+(t) - DF_-(t)|/D$$

Here, $DF_+(t)$ represents the total number of documents in the positive class which carries the term t. Similarly $DF_-(t)$ represents the total number of documents in the negative class which carries the term t and D is the total number of documents. Score ranges from 0 to 1. This shows that score is directly proportional to other features.

Gain Ratio: If the feature has unique values, then it will have high Information Gain. But it may or may not contribute to the result. To overcome this problem Gain Ratio is used [12, 15]. This is used to suppress the significance of the feature.

$$GR(x, y) = IG(x, y)/H(x)$$

Here $IG(x, y)$ represents the Information Gain of the feature x with respect to the result y. $H(x)$ represents the entropy of the feature x.

Kullback-Leibler Divergence: Inorder to calculate the divergence(different) of one probability distribution from another Kullback-Leibler divergence is used. The values of Kullback-Leibler divergence always non-negative.

$$KL(P \| Q) = sum\, x\, in\, X\, P(x) * log(P(x)/Q(x))$$

where P and Q represents probability distributions. Kullback-Leibler divergence score is used when the probability of the same event from the two probability distributions P and Q are different (i.e. one will be large and the other will be small).

Higher the average of the scores calculated by the above methods, higher is the weightage for that feature(word).

3.5   Opinion Classification

Classifier is required for classifying the document into positive and negative class. Generally, classifier will learn the model constructed by using training data set. And then classify the data set by the model which was trained [8].

Multi-layer Perceptron: Perceptron is used for linearly separable problems. Multi-Layer Perceptron comes under deep artificial neural networks [10]. It consists of many layers. To receive the input, the first layer is considered as input layer. This set of neurons $\{x_1, x_2, \ldots x_n\}$. There will be a certain number of hidden layers. Neurons in each layer will be completely bipartite with neurons in the adjacent layer.. Hidden layer's neuron will calculate the value by using the weighted linear summation of data from the neurons in the previous layer $w_1 x_1 + w_2 x_2 + \ldots + w_n x_n$. This will be followed by an activation function which will be non-linear. This is suitable for non-linear dataset. This has more than one local minimum or local maximum. Hence it is considered as non-convex function. Finding the weight by minimizing the sum of squares errors is the main objective. This can be achieved by using gradient descent method.

Algorithm Pseudocode for the proposed framework

For every movie review (document) in the data set
Do{
    i) Call Preprocessing().
    ii) Score is calculated by using Document Frequency Difference. Based on the score certain features should be eliminated.
    iii) Gain Ratio is calculated. Depending on the values certain features should be eliminated.
    iv) Kullback-Leibler Divergence score is calculated.
    v) By taking an average of score calculated by Document Frequency Difference and the score by Gain Ratio and the score by Kullback-Leibler Divergence, certain insignificant features are eliminated.
    vi) Select the features (words) with higher weights for classification i.e. top k words.
}

Train the constructed model using Multi-Layer Perceptron classifier.

4 Results and Performance Analysis

4.1 Evaluation Methods

One of the evaluation methods is accuracy. If Accuracy is alone considered then there is a possibility that the model can be biased. In order to rectify this, F-measure, Precision and Recall are used. To calculate those confusion matrix is required.

Confusion Matrix: TP represents True Positive which means correctly classified as positive. True Negative (TN) means it is correctly classified as negative. False Positive (FP) means misclassified as positive. False Negative (FN) means misclassified as negative.

Actual Values

		Positive (1)	Negative (0)
Predicted Values	Positive (1)	TP	FP
	Negative (0)	FN	TN

Accuracy: Accuracy is calculated by adding true positive with true negative and dividing it by the addition of true positive, false positive, true negative and false negative.

Precision: Precision is calculated by dividing true positive by the addition of false positive and true positive.

Recall: Divide true positive by the sum of false negative and true positive.

F - measure: Multiply 2 with precision and recall and dividing it with sum of precision and recall.

4.2   Results

Evaluation measures of the proposed framework for cornell sentiment polarity data set.

Weight by method	Accuracy %	Precision %	Recall %	F-measure %
Document Frequency Difference and Information Gain and Kullback-Leibler Divergence	96.12	100	96.12	97.78
Document Frequency Difference and Gain Ratio and Kullback-Leibler Divergence	98.71	100	97.62	98.49

The results and performance for the cornell sentiment polarity data set is best proved with Multi-Layer Perceptron with Document Frequency Difference and Gain Ratio as criterion with an accuracy of 98.71%.

Evaluation measures of the proposed framework for movie review data set V1.0.

Weight by method	Accuracy %	Precision %	Recall %	F-measure %
Document Frequency Difference and Information Gain and Kullback-Leibler Divergence	97.43	98	97.66	96.19
Document Frequency Difference and Gain Ratio and Kullback Leibler Divergence	98.74	99.37	99.21	98.09

The results and performance for the large movie review data set V1.0 is best proved with Multi-Layer Perceptron with Document Frequency Difference and Gain Ratio as criterion with an accuracy of 98.74%.

Evaluation measures of the proposed framework for movie review data set SAR14.

Weight by method	Accuracy %	Precision %	Recall %	F-measure %
Document Frequency Difference and Information Gain and Kullback-Leibler Divergence	96.82	98.05	97.93	98.52
Document Frequency Difference and Gain Ratio and Kullback-Leibler Divergence	97.39	99.75	98.79	99.68

The results and performance for the large movie review data set SAR14 is best proved with Multi-Layer Perceptron with Document Frequency Difference and Gain Ratio as criterion with an accuracy of 97.39%.

5 Conclusion

In this paper, a novel model was proposed by using Document Frequency Difference, Gain Ratio and Kullback-Leibler Divergence as three significant feature selection methods and Multi-layered Perceptron, a deep learning model was applied to analyze the reviews and capture the sentiments effectively. By using these methods, improved accuracy was recorded than the traditional methods. As a next step, recent language models such as Bidirectional Encoder Representations from Transformers (BERT) can be used for further better results.

References

1. Singh, V.K., Piryani, R., Uddin, A., Waila, P.: Sentiment analysis of movie reviews: a new feature-based heuristic for aspect-level sentiment classification. In: 2013 International Multi-Conference on Automation, Computing, Communication, Control and Compressed Sensing (iMac4s), pp. 712–717. IEEE, March 2013
2. Thet, T.T., Na, J.C., Khoo, C. S., Shakthikumar, S.: Sentiment analysis of movie reviews on discussion boards using a linguistic approach. In: Proceedings of the 1st International CIKM Workshop on Topic-Sentiment Analysis for Mass Opinion, pp. 81–84. ACM, November 2009
3. Singh, V.K., Piryani, R., Uddin, A., Waila, P.: Sentiment analysis of movie reviews and Blog posts. In: 2013 3rd IEEE International Advance Computing Conference (IACC), pp. 893–898. IEEE, February 2013
4. Liu, C.L., Hsaio, W.H., Lee, C.H., Lu, G.C., Jou, E.: Movie rating and review summarization in mobile environment. IEEE Trans. Syst. Man Cybern. Part C Appl. Rev. 42(3), 397–407 (2011)
5. Lin, C., He, Y.: Joint sentiment/topic model for sentiment analysis. In: Proceedings of the 18th ACM Conference on Information and Knowledge Management, pp. 375–384. ACM, November 2009

6. Zhuang, L., Jing, F., Zhu, X.Y.: Movie review mining and summarization. In: Proceedings of the 15th ACM International Conference on Information and Knowledge Management, pp. 43–50. ACM, November 2006

7. Khotimah, D.A.K., Sarno, R.: Sentiment detection of comment titles in booking.com using probabilistic latent semantic analysis. In: 2018 6th International Conference on Information and Communication Technology (ICoICT), pp. 514–519. IEEE, May 2018

8. Amolik, A., Jivane, N., Bhandari, M., Venkatesan, M.: Twitter sentiment analysis of movie reviews using machine learning techniques. Int. J. Eng.Technol. 7(6), 1–7 (2016)

9. Pang, B., Lee, L.: A sentimental education: Sentiment analysis using subjectivity summarization based on minimum cuts. In: Proceedings of the 42nd annual meeting on Association for Computational Linguistics, p. 271. Association for Computational Linguistics, July 2004

10. Shirani-Mehr, H.: Applications of deep learning to sentiment analysis of movie reviews. Technical report. Stanford University (2014)

11. Annett, M., Kondrak, G.: A comparison of sentiment analysis techniques: polarizing movie blogs. In: Conference of the Canadian Society for Computational Studies of Intelligence, pp. 25–35. Springer, Heidelberg, May 2008

12. Yasen, M., Tedmori, S.: Movies Reviews sentiment analysis and classification. In: 2019 IEEE Jordan International Joint Conference on Electrical Engineering and Information Technology (JEEIT), pp. 860–865. IEEE, April 2019

13. The dataset can be downloaded from http://www.cs.cornell.edu

14. Nicholls, C., Song, F.: Comparison of feature selection methods for sentiment analysis. In: Canadian Conference on Artificial Intelligence, pp. 286–289. Springer, Heidelberg, May 2010

15. Manek, A.S., Shenoy, P.D., Mohan, M.C., Venugopal, K.R.: Aspect term extraction for sentiment analysis in large movie reviews using Gini Index feature selection method and SVM classifier. World Wide Web 20(2), 135–154 (2017)

A Patch - Based Analysis for Retinal Lesion Segmentation with Deep Neural Networks

A Mary Dayana$^{(\boxtimes)}$ and W. R. Sam Emmanuel

Department of Computer Science, Nesamony Memorial Christian College,
Affiliated to Manonmaniam Sundaranar University,
Marthandam, Tirunelveli, India
amarydayana@gmail.com, sam_emmanuel@nmcc.ac.in

Abstract. Diabetic Retinopathy (DR) is one of the key symptoms of Diabetes Mellitus that is caused by the deterioration of blood capillaries which nourish the retina of the eye. The spread of the pathological lesions in the retina determine the severity of diabetic retinopathy. Therefore, it is obligatory to analyze the symptoms of DR at an early clinical stage as it prevents the progression of the disease and protects the vision. Deep learning algorithms play an important role in detecting multiple abnormalities from retinal fundus images and also highlights the areas of corresponding lesions with considerable accuracy. In the proposed method a deep convolutional neural network is designed for the segmentation of small lesions with patch-based analysis in retinal fundus images. A sliding window method was used to create image patches. The trained network analyzes the image patches and generates the probability map which in turn predicts the different types of lesions. The results obtained by the proposed work shows significantly better performance in accuracy and sensitivity when compared with other works on related tasks.

Keywords: Diabetic Retinopathy · Diabetes Mellitus · Deep learning · Convolutional Neural Network · Sliding window

1 Introduction

Most of the people in the world are unaware of the serious complication of diabetes called Diabetic Retinopathy. It is a retinal disease that leads to vision impairment and preventable blindness in middle-aged working individuals. A statistical report of International Diabetes Federation (IDF) [1] estimated that approximately 425 million adults around 20–75 years were living with diabetes, and the prevalence is expected to further increase 642 million by 2045. However, studies have revealed that regular diabetic screening and good control in blood glucose levels can significantly slow down the progression of DR-induced blindness. Diabetic Retinopathy indicates the presence of pathological signs such as Microaneurysms (MAs), Hemorrhages (HEMs) and Soft and Hard Exudates (EXs) in fundus images as depicted in Fig. 1.

In recent decades number of works have been experimented for the detection and segmentation of lesions. In traditional methods various feature extraction techniques are used to first extract the features from the fundus images [3]. The features extracted

© Springer Nature Switzerland AG 2020
A. P. Pandian et al. (Eds.): ICCBI 2019, LNDECT 49, pp. 677–685, 2020.
https://doi.org/10.1007/978-3-030-43192-1_75

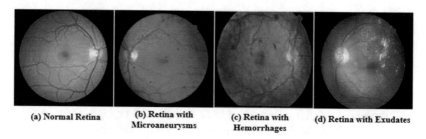

(a) Normal Retina   (b) Retina with   (c) Retina with   (d) Retina with Exudates
Microaneurysms   Hemorrhages

Fig. 1. Fundus images with different pathological signs

are then given into the classifiers such as KNearest Neighbor (KNN), Support Vector Machine (SVM), Random Forest Classifiers and AdaBoost Classifiers to differentiate between lesions and nonlesions [2]. Deep Convolutional Neural Network (DCNN) demonstrates better performance on a variety of tasks in image processing and computer vision [3]. The development of DCNN have paved a way for effective segmentation and classification of retinal pathologies and also assist the ophthalmologist in effective treatment.

Even though most of the existing methods used patch-based image analysis in their experiments [4, 5, 8], it leads to discrepancy in proportion of the sign due to patch size [8]. The proposed framework is a deep convolutional neural network that has been designed for the automatic detection and simultaneous segmentation of Microaneurysms, Hemorrhages and Exudates from retinal fundus images. A Convolutional Neural Network with 10 layers was modelled and then trained using image patches to obtain probability maps indicating the location of three different pathological signs. The existing works related to the proposed method is described in Sect. 2. The methodology adopted by the proposed work is explained in Sect. 3. The results and discussion are reported in Sect. 4 followed by the conclusion in Sect. 5.

2   Related Work

A number of works have been introduced based on CNN for making the diagnosis of DR automatic and easier. One approach to automating retinal screening is the use of deep convolutional neural networks (DCNNs) which have become widely held because of their ability to classify lesions with high sensitivity and specificity. Yang et al. [6] proposed a method using two-stage DCNN for detecting the lesions and grading the severity of DR in fundus images. In [7–9], a neural network model was developed for the diagnosis of Diabetic Retinopathy in fundus images using deep learning methods. Mansour [10] developed an automatic system to grade the DR stage on retinal images by applying AlexNet DNN architecture [11] for DR feature extraction and Principle Component Analysis (PCA) for dimensionality reduction. Guo et al. [12] designed a multi-lesion segmentation network model called L-seg to simultaneously segment the four different kinds of lesions. A deep learning approach was proposed by Gondal et al. [13] for the localization of DR lesions in fundus images with Class Activation Maps (CAM). While [14] adopted several deep CNN architectures with transfer learning and hyper-parameter tuning for DR image classification. Gulshan et al. [15] used the Inception – v3 model for optimization and batch normalization using ImageNet data to detect DR in fundus photographs. Orlando et al. [16] proposed a novel ensemble method to identify the red lesions combining hand-crafted and deep CNN based features using three different datasets. Son et al. [17] proposed an effective method to localize lesions with improved precision during training. An encoder-decoder based architecture was developed by Badar et al. [18] for the pixel-based segmentation of multiclass retinal pathologies and achieved state - of - art results. Wang et al. [19] utilized deep convolutional neural network for DR image recognition and classification and achieved state-of-art accuracy results. Although many CNN based approaches have been developed, still there is need for improvement in the current results to be used in clinical applications.

3 Proposed Methodology

The proposed framework is a patch-based method for fundus image analysis in which CNN is trained using a set of images that were preprocessed and segmented in patches. The trained CNN creates a probability map by analyzing the image patches via a sliding window method [18] and identifies the locations of the pathological signs in the retinal fundus image. Figure 2 depicts the schematic diagram of the proposed method and the steps are explained in this section.

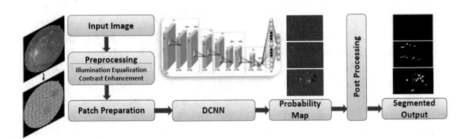

Fig. 2. Schematic diagram of the proposed method

3.1 Preprocessing

The primary purpose of preprocessing is to make the training process more efficient. Fundus images often have uneven illumination, low varying contrast, and noise which degrades the image quality. The images from the database are preprocessed before they are subjected to segmentation and classification. Non-uniform illumination in the fundus image causes vignette effects and also affects the DR screening process. Therefore, to correct the uneven illumination Eq. (1) is used.

$$g' = g + \mu_d - \mu_l \tag{1}$$

where μ_d and μ_l are the desired and local average intensity; g and g' are the original and new pixel intensities.

Contrast Enhancement technique is used to enhance the contrast between the poorly illuminated background and the lesions, as described in Eq. (2).

$$I_{CE} = \alpha I(p, q) + \beta G(p, q; \sigma) * I(p, q) + \mu \qquad (2)$$

where I_{CE} is the enhanced image, $I(p, q)$ represents the original image, $G(p, q; \sigma)$ represents a Gaussian filter with σ as the scaling value.

3.2   Network Details

First, preprocessing is applied for contrast enhancement and then the images are segmented into patches for training CNN. The image patches extracted through sliding window technique [18] were labelled with ground truth images in view of the three different pathological signs and are then given as input to the CNN for the training process. The proposed CNN architecture consists of four convolutional layers with 16 feature maps in each layer and a kernel size of 3×3 pixels. An activation function called Rectified Linear Unit (ReLU) is used to detect the non-linear features of the CNN and can be defined as Eq. (3).

$$r = \max(z, 0) \qquad (3)$$

where z is the input to the activation function and r is the output. The function of Max pooling (MP) layers is to reduce the size of feature maps with a kernel size of 2×2 and the normalization layers normalize the values for faster training convergence. The features extracted from the max-pooling layers are passed into the fully connected layer to produce the output at the final softmax layer. The softmax layer is typically deployed at the end of the network to compute the loss function and the loss is then minimized using Stochastic Gradient Descent (SGD) optimizer to update the CNN weights and biases that allow the successful classification of lesions. The probability map obtained from the trained CNN was found to be noisy. Therefore, post-processing was carried out to eliminate the noise in the image patches and the spread due to convolutions.

Algorithm 1. Workflow of the proposed method

INPUT: Retinal Fundus image

OUTPUT: Final Segmentation result

1. Input the retinal fundus image with different lesions
2. Perform preprocessing for illumination equalization and contrast enhancement
3. Patches were extracted using sliding window technique
4. Train the network using the extracted image patches
5. Trained network generates probability map identifying the locations of pathological signs
6. Perform postprocessing to remove noise
7. Outputs the final segmentation result that detects the three pathological signs

Figure 3(a) shows image patches representing the three pathological signs Microaneurysms, Hemorrhages and Exudates. Figure 3(b) depicts the probability map generated corresponding to the different pathological signs and Fig. 3(c) shows the resultant segmented output obtained after applying the post-processing step. The workflow of the proposed method is summarized in Algorithm 1.

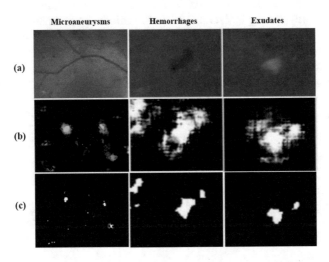

Fig. 3. (a) Image patches representing the three pathological signs Microaneurysms, Hemorrhages and Exudates (b) Probability map with pathological signs (c) Segmented output after post-processing

4   Results and Discussion

The experimental setup for training the proposed architecture and the results evaluated using the conventional metrics are described in this section.

4.1   Dataset Description

DIARETDB1 (Diabetic Retinopathy Database Calibration Level 1) dataset is used for the experiment [21]. It is a public database that contains 89 color fundus images with $50°$ field of view and 1500×1152 fixed resolution. In this, 5 images are found to be healthy and the remaining 84 images are abnormal. Among the 84 anomalous retinal images, 75% (63 images) were used for training the network and 25% (21 images) were used for testing the performance of the method. The images used for training were manually annotated by a trained and experienced ophthalmologist.

4.2   Training

The network has been trained on a dataset of 75% image patches extracted from 63 preprocessed images from the DIARETDB1 dataset. The images were segmented into patches of size, N × N (where N = 32). The patch size is chosen relative to the size of the smallest pathological sign present in the input images with a stride of 16. A batch size of 128 image patches is selected on every epoch during training. Data augmentation was performed using horizontal and vertical flipping and rotation during training. Dropout can be used to weaken the over-fitting effect of deep neural networks. Therefore to handle overfitting, the drop out algorithm [20] was used with a probability of 0.5. For training deep CNN, network parameters w (weight) and b (bias) are initialized with sufficient values in the convolutional and fully connected layers to achieve faster convergence and to avoid vanishing gradient problem. A loss function L_f trains the network model [5] and optimize the model weights, which is defined as Eq. (4).

$$Lf = -\frac{1}{|S|}\sum_{i=1}^{|s|} \ln\left(p\left(K^i|S^i\right)\right) \tag{4}$$

where the total quantity of image patches used in training the network is denoted by $|S|$, the ith training sample and its label is denoted by the factor S_i and K_i respectively. To optimize the CNN parameters, Stochastic Gradient Descent (SGD) optimizer with learning rate (γ) 0.01, momentum (9) 0.9 and weight decay factor (m) 0.0005 is used as in Eq. (5).

$$\theta(p+1) = \theta(p) - \gamma\frac{\partial Lc}{\partial\theta} + 9\Delta\theta(p) - m\gamma\theta(p) \tag{5}$$

The training process is repeated from 0 to 100 epochs and mean result of ten repetitions was recorded with accuracy and error rate. Python 3.6 libraries and packages are used for implementing the network model.

4.3   Performance Evaluation

Accuracy, Sensitivity, and Specificity are the three widely used metrics for evaluating the performance of a method. Accuracy is the ratio of the sum of correctly identified lesions and non-lesions to the sum of the total number of lesions. Sensitivity is the identification of true positive lesions as positive. Specificity is the identification of true negative lesions as negative. These performance metrics are computed using Eqs. (6), (7) and (8).

$$\text{Accuracy (Acc)} = \frac{(TP+TN)}{(TP+FP+FN+TN)} \tag{6}$$

$$\text{Sensitivity (Sen)} = \frac{TP}{(TP+FN)} \tag{7}$$

$$\text{Specificity (Spec)} = \frac{TN}{(TN+FP)} \tag{8}$$

where TP signifies the number of correct identifications of lesion regions, TN represents the correct identification of non-lesion regions, FP represents the non-lesion regions detected as lesions, and FN represents the wrongly classified non-lesion regions.

4.4   Results

The dataset DIARETDB1 was used for both training and testing the performance of the proposed work. Patch extraction using sliding window technique with deep CNN enables the proposed work to precisely focus on the pathologic lesions. Table 1 demonstrates the accuracy, sensitivity and specificity of the proposed method. From the performance evaluation, as shown in Fig. 4, it was observed that the average sensitivity and specificity for detecting exudates is higher than that of microaneurysms and hemorrhages. The experimental results of the proposed approach reports an improved performance in accuracy and sensitivity and significantly equal performance in specificity when compared to some of the existing works as shown in Table 2. The CNN based proposed framework identifies the three different pathological signs simultaneously with improved performance and makes it appropriate for automatic DR detection.

Table 1. Patch level evaluation metrics of the proposed method

	Microaneurysm	Hemorrhage	Exudates
Accuracy	0.97	0.95	0.99
Sensitivity	0.91	0.87	0.97
Specificity	0.97	0.98	0.98

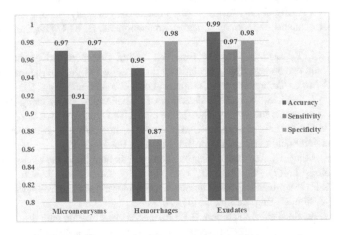

Fig. 4. Performance measure of the proposed approach

Table 2. Comparative analysis for lesion segmentation with existing patch based methods

Methodology	Dataset	Microaneurysm			Hemorrhages			Exudates		
		Acc	Sen	Spec	Acc	Sen	Spec	Acc	Sen	Spec
Khojasteh et al. [5]	DIARETDB1	0.94	0.85	0.96	0.90	0.84	0.92	0.98	0.96	0.98
Tan et al. [8]	CLEOPATRA	–	0.46	0.97	–	0.62	0.98	–	0.87	0.98
Proposed method	DIARETDB1	0.97	0.91	0.97	0.95	0.87	0.98	0.99	0.97	0.98

5 Conclusion

This paper presents an optimized CNN based deep network model for the patch-based analysis of fundus images to exactly focus on the subtle pathologies. The probability map for the different DR lesions was created using the softmax output layer and then used for the identification of pathological signs. A single dataset was trained and tested to evaluate the performance of the method. The accuracy level indicates the efficiency of the proposed work in distinguishing the affected portions of the color fundus images. The proposed method solves the problem of segmentation of three types of lesions at the same time and the segmented output can be leveraged to classify the severity level directly. However, additional work is needed to better identify the lesions more accurately.

References

1. IDF Diabetes Atlas (2017)
2. Roychowdury, S., Koozekanani, D.D., Parhi, K.K.: DREAM: Diabetic retinopathy analysis using machine learning. IEEE J. Biomed. Health Inform. **18**(5), 1717–1728 (2014)
3. Gao, Z., et al.: Diagnosis of diabetic retinopathy using deep neural networks. IEEE Access **7**, 3360–3370 (2019)
4. Lam, C., Yu, C., Huang, L., Rubin, D.: Retinal lesion detection with deep learning using image patches. Invest. Ophthalmol. Vis. Sci. **59**(1), 590–596 (2018)
5. Khojasteh, P., Aliahmad, B., Kumar, D.K.: Fundus images analysis using deep features for detection of exudates, hemorrhages, and microaneurysms. BMC Ophthalmol. **18**(1), 1–13 (2018)
6. Yang, Y., Li, T., Li, W., Wu, H., Fan, W.: Lesion detection and grading of diabetic retinopathy via two-stages deep convolutional neural networks. In: International Conference on Medical Image Computing and Computer-Assisted Intervention, pp. 533–540 (2017)
7. Chandore, V., Asati, S.: Automatic detection of diabetic retinopathy using deep convolutional neural network. Int. J. Adv. Res. Ideas Innov. Technol. **3**, 633–641 (2017)
8. Tan, J.H., et al.: Automated segmentation of exudates, hemorrhages, microaneurysms using a single convolutional neural network. Inf. Sci. **420**, 66–76 (2017)
9. Gargeya, R., Leng, T.: Automated identification of diabetic retinopathy using deep learning. Ophthalmology **124**(7), 962–969 (2017)
10. Mansour, R.F.: Deep-learning-based automatic computer-aided diagnosis system for diabetic retinopathy. Biomed. Eng. Lett. **8**, 41–47 (2018)
11. Krizhevsky, A., Sutskever, I., Hinton, G.E.: ImageNet classification with deep convolutional neural networks. In: Advances In Neural Information Processing Systems (2012)

12. Guo, S., et al.: L-Seg: An end-to-end unified framework for multi-lesion segmentation of fundus images. Neurocomputing **349**, 52–63 (2019)
13. Gondal, W.M., Kohler, J.M., Grzeszick, R., Fink, G.A., Hirsch, M.: Weakly-supervised localization of diabetic retinopathy lesions in retinal fundus images. In: Proceedings of International Conference on Image Processing (ICIP), pp. 2069–2073 (2018)
14. Wan, S., Liang, Y., Zhang, Y.: Deep convolutional neural networks for diabetic retinopathy detection by image classification. Comput. Electr. Eng. **72**, 274–282 (2018)
15. Gulshan, V., Peng, L., Coram, M., et al.: Development and validation of a deep learning algorithm for detection of diabetic retinopathy in retinal fundus photographs. J. Am. Med. Assoc. **316**(22), 2402–2410 (2016)
16. Orlando, J.I., Prokofyeva, E., deFresno, M., Blaschko, M.B.: An ensemble deep learning-based approach for red lesion detection in fundus images. Comput. Methods Programs Biomedi. **153**, 115–127 (2018)
17. Son, J., et al.: Development and validation of deep learning models for screening multiple abnormal findings in retinal fundus images. Ophthalmology **127**, 85–94 (2019)
18. Badar, M., Shahzad, M., Fraz M.M.: Simultaneous Segmentation of Multiple Retinal Pathologies Using Fully Convolutional Deep Neural Network, pp. 313–324. Springer, Cham (2018)
19. Wang, X., Lu, Y., Wang, Y., Chen, W.: Diabetic retinopathy stage classification using convolutional neural networks. In: 2018 IEEE International Conference on Information Reuse and Integration (IRI), pp. 465–471 (2018)
20. Nitish, S., Geoffrey, H., Alex, K., Hya, S., Ruslan, S.: Dropout: A simple way to prevent neural networks from overfitting. J. Mach. Learn. Res. **15**, 1929–1958 (2018)
21. Kauppi, T., et al.: The DIARETDB1 diabetic retinopathy database and evaluation protocol. J. Med. Imaging **3**, 1–18 (2007)

Survey of Onion Routing Approaches: Advantages, Limitations and Future Scopes

Mayank Chauhan[✉], Anuj Kumar Singh, and Komal

Amity University, Gurgaon 122413, Haryana, India
raj007put@gmail.com

Abstract. The rapid advancement of technology has led to the evolution of the Internet, which brought changes in the lives of humans drastically. Since then slowly, all the devices started connecting via the internet user has the freedom and ability to store the information and extract it when needed. But this requires the implementation of active security measures to ensure that the data is confidential and secured. Onion routing was implemented to secure the data from eavesdropping and traffic analysis which guarantees the security of data with the help of private keys. This paper presents a detailed analysis of onion routing algorithms and its variants based on the implementation, features, security concerns and weakness.

Keywords: Onion routing · Traffic analysis · Anonymous connection · Data latency · Onion router · System design · Security

1 Introduction

Onion routing is resistant to spying and traffic analysis of the data. The main aim of this kind of networking is to separate identification from routing and provide anonymous communication. The onion routing was coined and coded in C by the US Navy in the late 1990s. Onion routing has few benefits over other types of networks, increasing real and bi-directional communication, there is no centralised trusted component, and this network is independent of the application. The three generations of onion routing which ensured the secrecy and privacy are taken care of mentioned below

- Generation 0
- Generation 1 - ORtNG (onion routing the next generation)
- Generation 2 - TOR (The Onion Routing)

In onion routing, messages encrypted in different layers are similar to the layers of the onion. This encrypted information transferred to a series of nodes or the routers are known to be the onion routers. The onion then passes through each router in the path. At each onion router, a single layer is peeled off, which gives the required information, and the is process depicted in Fig. 1. At the last router, the message arrives at its destination. Each router remains anonymous as intermediate nodes only know information about the data from where it has come and where it has to be sent. Due to its ability to maintain secrecy and easy setup, onion routing has become very popular.

A. P. Pandian et al. (Eds.): ICCBI 2019, LNDECT 49, pp. 686–697, 2020.
https://doi.org/10.1007/978-3-030-43192-1_76

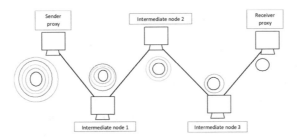

Fig. 1. Structure of onion routing

Instead of making direct connections from the sender to the receiver, the intermediate nodes establish a socket (path) connection with onion routing proxy. This proxy tries to build a secret relationship through different others onion routers that deliver safe information to the destination. The routers in this type of networking can only identify adjacent nodes, and the unique encrypted data passes along each router.

2 Background

In a computer network, a switched communication network used as a platform for communication bridges switching of connection for two nodes, not necessarily adjacent ones, but the information is not kept private. There is always a possibility of people knowing the identity of the sender and receiver and traffic analysis. Onion routing is introduced to avoid all the problems of a switched communication network.

The initiation of onion routing was started two years before it was introduced to the people i.e. it initiated in the year 1994. The onion routing initially was tested on pipe net model in the year 1996. This main objective was to protect the messages against traffic analysis [1]. The investigation is done on the testbed onion routing network at NRL (Naval Research Laboratory) which was named as Generation 0 of the onion routing. The performance of the generation 0 is found better when compared to switched networking in terms of input/output; memory copy and single processors. Then to further improve the security and control policies, onion routing was developed and presented in the form of generation 1, later, which is known as ORtNG (onion routing the next generation). Then after years, one more generation of onion routing was developed, which was the third development and was known as generation 2, commonly referred to as TOR (the onion routing).

The first onion routing was proposed in the year 1995. It was funded commercially. Later, in the year 1996, mixing was done, and DH (Diffie Hellman) keys could be used instead of onion keys. These DH keys had public keys too for each node which played a role in maintaining secrecy. No onion keys were to be kept in one hop, and the DH keys were included from origin to the responder in layers. In spite of many developments, the network had few privacy problems. So, one more network which was similar to onion routing was introduced and named as Freedom Network. The significant difference was onion routing runs on TCP and freedom network runs on UDP. Freedom Network was commercially funded.

3 Process Flow of Onion Routing

The flow of data takes place in the following way, as shown in Fig. 2. Once the sender sends the information, the initiator proxy will establish the onion. The structure of onion depends on the number of nodes that are present between the start and end node. The application proxy will decide the route based on the number of routers available. Once the path is defined, the data gets encrypted in the form of onion. the onion moves from initiator router to the receiver. Each layer is peeled off with unique keys so that the information is secured. At each layer or node, the hop defines the information about the private keys, payload, expiration time and the flow of data, i.e. in forward or backward direction. None of the routers knows the information about the node from where the data is initiated. Once the onion reaches the final router, the data gets forwarded to the responder through exit funnel.

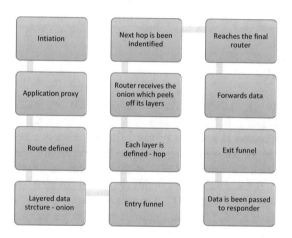

Fig. 2. Process flow of onion routing

4 Literature Survey

The routing technique was first designed to be a layered object which establishes a socket connection through the proxy server, which is defined at the application layer and accommodates firewalls. This network makes data stream to follow several nodes such that it hides the information about routing. The track thus defined by the first node. The proxy node (first and last node) is most sensitive and sometimes used as intermediate nodes also. The main reason is to avoid providing an anonymous connection that are independent of recognising themselves in the message [2]. The initiator/sender proxy contains a series of routing information that develops an onion which encapsulates the data as well as routes. The initiator sends onion to establish the virtual connection between the sender and the receiver proxy. Since the onion encrypts the data in layers, these layers are wrapped around in a payload. Onion moves from one node to another node, where each node has information about the data like from whom the data is received and whom to pass. They don't have any idea about other nodes in a chain except the last node. Every receiving node in every step has information in a format given below (Fig. 3).

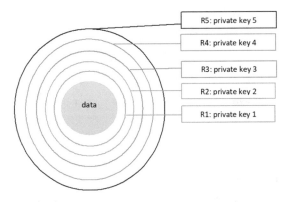

Fig. 3. Layers in onion routing

$$\{exp.time; next.hop; F_f; K_f; F_b; K_b; payload\}P_k$$

At each hop, onion peels of its layer, except the last node, none other proxies will know how much amount of the payload has been received. This process of peeling the layer is called padding. Loose routing is one of the significant role played in onion routing where the initiator proxy need not have the entire route plan initially before starting. It can slowly develop its route based on the nodes, which adds more hops and security. The initiator doesn't know the path, so each node chooses a way which causes no break. This type of routing is used to handle connection charges and to avoid indefinite routing. Onion routing can send data or information from sender to receiver, but reply onions are used to send back a message to the sender. However, the payloads of onions is dissimilar. The virtual circuits that are established by the forward or backward onions have no significant contrasts except the direction. Backward and forward onions need not be in the same route. The implementation of the onion is a substantial factor in routing [3].

Later on, an experiment tested on five onion routers which run on single sun ultra space 2270. The model invented in such a way that traffic was visible, modifiable and can also be compromised. The proxies used are of two types: client and core. The client proxy tries to establish a link/socket between the application and the core. The client proxy interfaces between client and core, and when the new proxy is requested, the client will determine to administrate or deny the request. If refused, the proxy will report an error message by closing the socket and waits until the next request arrives. If the request is accepted, then it tries to implement the socket to the core proxy, and then the core is used to build an onion based on the acceptance of the client proxy [4].

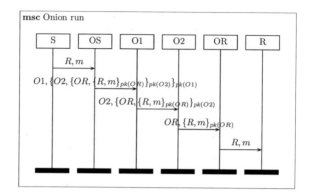

Fig. 4. Message flow from sender to receiver in onion routing

The major four important steps in onion routing:

- Setting up the network which has connections among onion routers
- Connection phase, that connects via router network
- Movement of data
- Destruction and clean up.

The primary implementation problems that occurred are onion once after processing is the padding at the end of the layers that try to compensate the removed once which are randomly done since none of the onions are encrypted between onion routers. If the part of onion routing network declines, then the traffic can be reduced [5]. Most of the vulnerabilities that occurred are traffic analysis complexity, a fixed length of the routing, reordering the layer, which also to lead to complication. Changes were done in application proxies and configurations of onion network to avoid the problems.

The data latencies are affected by the number of onion routers where the overhead is relatively minimal, and the public keys used for decryption are costly. The data movement uses the only secret key, which is symmetrical and faster. After having many changes and development in networking, the primary aim focused in the early 2000s is security [6]. The primary objectives were to provide secure private communications at reasonable cost and efficiency and anonymity to the sender and receiver. To demonstrate the feasibility, the testing is done on five onion routing nodes which lacked features which help the system in becoming robust, scalable and resistant to the number of insider attacks. The network model comprised of onion proxies, core onion routers, and links and responder proxies.

The model that was developed is named as adversary model which has the capabilities of defining the adversaries (who can observe, manipulate or compromise both ends of the connection), i.e. observer, disrupter and hostile user. An observer can see the links but not initiate them, and disrupter can delay or corrupt the traffic, whereas the hostile user can destroy the connection. There are four possible distributions of the adversary. Single: no compromising nodes; multiple: a fixed or group of core onion routers are compromised; at any one time, the subset of core onion routers are compromised; global: all routers are compromised. Compromised core onion routers cannot disrupt a link unless it is adjacent to it [7]. Based on the analysis, it was concluded that routing design was resistant to the worst adversaries and can resist the traffic analysis effectively than the mechanism of the internet. But adding time delay to the traffic would complicate limiting attacks against local core onion routers.

To focus more on security concerning maintaining secrecy and authentication, RFID and DRM have significant privacy consequences. The four essential aspects of privacy were unlinking ability, anonymity, pseudonymity and unobservability. In anonymity, someone might use a service without letting out its user identity. In pseudonymity, the user might use the resource with disclosing identity but can be held accountable for using that. Unlink ability may make use of multiple users of resources and unobservability may use resources without third parties [8].

A sender 'S' sends a message 'm' to the receiver 'R' via series of routers (Fig. 5).

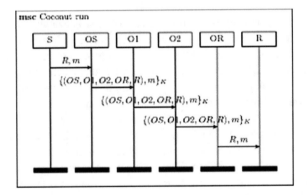

Fig. 5. Message flow from sender to receiver in coconut routing

In onion routing, only O1 knows its secret key, and so O1 can peel off an outermost layer, and a similar operation is carried out as shown in Fig. 4. But to have the original message and paths that are crypt with a cryptographic key and have only one single key for all coconut routing can be used [9]. But coconut routing hardly guarantees the privacy and onion routing guarantees full confidentiality. In spite of working on security aspects, there is no satisfactory definition of security and constructions have an analysis of ad hoc security. To have the proper security messages are wrapped well and determines the path. But this path is either fixed or chosen by the sender for each message. If it is set, then routing is used to find out the track and if it has to be selected, then onion routing is used. But the major drawback in onion routing is not robust.

The need for security and reduce the bandwidth cost, a new design has been constructed, which is named as TORSK, which is a peer to peer network and has low latency, this is designed to replace the relay selection and directory service. Peer to peer (P2P) was the previous protocol, and Torsk doesn't need the users to rely on traffic for others. Torsk is two P2P protocols combined such that the strength to avoid attacks increases and poses secondary characteristics like performance, availability and reliability to impact the security [10].

As the technology developed, the internet has come into existence and brought many changes which changed the lives of the humans drastically, slowly all the devices started connecting via the internet. User has the freedom and ability to keep information very safe and secure. To make it more and more secure saving from eavesdropping and traffic analysis, onion routing implemented which could secure the data with the help of private keys. Then slowly wireless networks came into existence where there was a problem of eavesdropping still [13].

The major problem was always traffic analysis, where the attacker attacks the essential information. The traffic determined by patterns, packet timing and the user, used by an attacker to know the transmission rate of the sender and receiver pair and routing technique used [14]. The problem was solved using dummy traffic. The problem could be solved by changing appearances of the packets or delay the transmission rates, but both of these were not that efficient because of the size of the packet and the number of packets were not known [15].

To make the data and information more secure, that is to make the information of sender and receiver unknown, dummy trafficking is done. Securing data is done by developing a technology called OPERA (optimal privacy-preserving enchanting routing algorithm) where the privacy is established by statistical decision theory that was based entirely in the Bayesian approach. Later on, the information on the internet that was available was seeming to very less and then dark web came into the picture. The available information was only 5% when compared to the information that was available through the dark web. But the resources were available more on deep web and dark web. Deep web was accessed via passwords, encryptions, decryptions, virtual private networks or related software. These deep webs were not accessible through the standard browser or hyperlinks. To access darknets, there are three significant ways I2P, TOR, FREENET networks [16].

Tor network allows the user to publish their websites without letting out the information of the location of the server from which it was initiated. These are called hidden servers which uses domain where the user can connect to onion services. The three main components for having communication are sender, receiver and network. In network, the nodes are categorised into entry node, middle node and exit node. And the attacks generally take place on either client or server or network. There are different types of attacks which can take place on each of them [17]. The following table gives us information about the types of attacks taking place on the servers (Table 1).

Table 1. Types of attacks on networks

Client	Server	Network
Torben	Cell counting and padding	Bridge discovery
P2P info leakage	TOR cells manipulation	Denial of service
Induced TOR guard selection	Caronate	Sniper
Raptor	Off path MIT-M	
Unpopular ports exploitation	Traffic analysis	
Low resources routing	Timing	
Traffic analysis	Shaping	
Timing		
Shaping		

Blockchain can be implemented in TOR to maintain privacy for all the list of nodes. The directory of tor has a list of nodes which can be forged. Therefore using blockchain technology with a voting committee elected by variable random function, many attacks can be reduced. As technology was developing at a faster rate, the problems for security and privacy also begun to increase more and more. Many tests are being done till date, and at the same time to avoid the problems, the onion routing is used in many areas like blockchain, power transmissions etc. [18].

5   Analysis

See Table 2.

Table 2. Analysis of onion routing

Year	Objective	Procedure	Benefits	Limitations
1996	Maintain the secrecy	Onion routing	Application independent Bi directional No centralized trusted components	Collection of huge amount of data Proxy if compromised, data is lost Poorly synchronized clock
1997 and 1998	Data latency needs to be less	Experimented on 5 onion router running on single machine	Traffic was modifiable, visible and can be comprisable	Traffic analysis complexity Fixed length of bytes Reordering of layers
1999	Data latency to be low and privacy to be increased	Network architecture that shifts trust by having proxies at different extremes	Overheads were setup Cryptographic keys were used	Keys were very expensive for decryption Data latency affects number of routers
2001	To have strong private communication at less cost	Network model which uses 5 onion routing nodes Adversary model	Model was resistant to worst case adversaries	Adding time delay to traffic complicates timing attacks
2004	Formalization of security goals	Four aspects of privacy Worked on anonymity Coconut routing	Onion routing gives more privacy	Privacy is guaranteed for qualitative nature only

(*continued*)

Table 2. (*continued*)

Year	Objective	Procedure	Benefits	Limitations
2005	Cryptographic aspects Easy implementation and construction	Wrapping of messages	Encryption of data provides more security	No robustness in onion routing
2006	Privacy has become difficult to manage	Worked in HTTP with application proxies	Software is freely available which can be utilized by everyone	Hacking File exchange Terrorism operator managements
2007 and 2008	Improve the circuit connection Integrity of the routing	Pairing-Based Key Agreement with User Anonymity	Less communication cost More flexibility	Low latency and less scalability
2009	Higher bandwidth cost of relay selection	TORSK	Low latency Peer to peer	Security is increased and less bandwidth cost
2011	Finger printing in onion routing using TOR and JAP	Data set and support vector classification	Improved recognition levels in TOR and JAP	TOR and JAP are not as strong as expected Not resistant to fingerprint website
2015	Congestion control Forward secrecy Directory servers	Distributed trust anonymizing systems	Good bandwidth Low latency Omits cover traffic	If the number of users are more, directory distribution becomes an issue Wider scale of deployment is needed
2016	High bandwidths and low latencies	Improved anonymous multi-receiver identity-based encryption scheme, and an improved identity based one-way anonymous key	High efficiency Scalable Anonymous Fault resistant less computation [11]	Eavesdropping
2017	Optical onion routing [12]	Using Linear Feedback Shift Register (LFSR) Usage optical XOR operation	Perfectly private and secure	Traffic analysis

(*continued*)

Table 2. (*continued*)

Year	Objective	Procedure	Benefits	Limitations
2018	Securing network over web using onion routing	Onion routing with different design is been implemented in order to secure the data	Easy to maintain	More overhead
2019	Privacy preserving routing for wireless networks	OPERA is used, dummy traffic is introduced in order to make the information secure.	- Statistical decision making solves the problem of privacy threats - Full source and destination are hidden	Trade-off between transmission and privacy
2019	Darknet forum correlation analysis	Tor hidden service	Doesn't reveal the actual location and maintains privacy	Extraction speed is very less
2019	Categorization of attacks	An exhaustive tool which is based on target of attack is used	Could segregate attacks for the three components	Haven't been executed the threats on controlled environments
2019	Tor- blockchain	Variable random function is been used	List of nodes can be saved to avoid forging	Transactions are sparse and validation of block takes time

6 Applications of Onion Routing

In the era of the internet, the development of technology is at a higher pace in all fields. Inspite of many extreme events, privacy and anonymity, remain questionable. But onion routing has come up with many solutions in trying to solve the anonymity and security. Low latency onion routing systems developed, which enable the high-speed systems to have proper data security and have a minimal overhead process [19]. The problem of retaining identity keys can be solved by certificate less onion routing where the de-anonymization attacks are reduced by securing the public key cryptography also. The issue of malicious nodes which enter to steal the identity of the clients can be reduced by multipath traffic between the client and middle nodes to mitigate the information of an attack [20]. With the recent trend of IOT devices, devices have information communication and liking with the untrusted networks, which lead to the problem of data security [21]. Eradication of these problems can be done by onion routing. Vehicular ad-hoc networks need privacy and anonymity [22]. So the concept of location-based dynamic relays, the locations acts as cryptic relays where onion routing can play a vital role. Onion routing can be further implemented in the fields like homographic encryption, domain generation algorithm, DUSTER-an active traffic analysis based on the flow of watermarking, blockchain increasing the trust in tor nodes, etc. [23–25].

7   Discussion and Future Scope

Due to an increase in technology, there is a need to secure the data efficiently. Many different algorithms are implemented in securing data but couldn't solve the issue of eavesdropping. The overhead involved in encryption and decryption process by cryptographic algorithms needs more space. The performance can be increased by having to select the proper path selection. The attacker can attack the confidential data by analysis the rate of transmission, sender and receiver information and routing technique used, so different algorithms can be used in this area to reduce the attacks. Onion routing is used in preserving data of web communication. Therefore implementation can take place in mobile networks too. The darknet information can be extracted using onion routing, the extraction speed and efficiency must rise so that there won't be any attacks taking place. Based on the type of attack on either client, server or network, mitigation of attacks can take place.

8   Conclusion

Vulnerabilities due to improper security is one the vital consideration of communication which plays a vital role in everyone's life to secure data. Hence it is always necessary to protect the data-keeping secure and safe. So far, onion routing is one of the best solutions to maintain the data very securely. It is the process of sending data from sender to the receiver via means of some intermediate nodes known as a router. The data in the onions are encrypted in the form layers. Where each layer is decrypted and peeled off. Onion routing provides anonymity to the users which provide security from traffic analysis and eavesdropping. The major benefits of onion routing are application-independent, bi-directional and no centralized trusted component.

References

1. Goldschlag, D.M., Reed, M.G., Syverson, P.F.: Hiding routing information. In: Lecture Notes Computer. Science. (including Subser. Lecture Notes Artificial Intelligence. Lectures Notes Bioinformatics), vol. 1174, pp. 137–150 (1996)
2. Goldschlag, D., Reed, M., Syverson, P.: Onion routing for anonymous and private communications. Commun. ACM 42(2), 39–41 (1999)
3. Reed, M.G., Syverson, P.F., Goldschlag, D.M.: Anonymous connections and onion routing. IEEE J. Sel. Areas Commun. 16(4), 482–493 (1998)
4. Syverson, P.F., Reed, M.G., Goldschlag, D.M.: Onion Routing Access Congurations

5. Syverson, P., Tsudik, G., Reed, M., Landwehr, C.: Towards an analysis of onion routing security. In:: Lecture Notes Computer. Science. (including Subser. Lecture Notes Artificial Intelligence. Lectures Notes Bioinformatics), vol. 2009, pp. 96–114 (2001)

6. Mauw, S., Verschuren, J.H.S., De Vink, E.P.: A formalization of anonymity and onion routing. In: Lecture Notes Computer. Science. (including Subser. Lecture Notes Artificial Intelligence. Lectures Notes Bioinformatics), vol. 3193, pp. 109–124, May 2014

7. Camenisch, J., Lysyanskaya, A.: A formal treatment of onion routing. In: Lecture Notes Computer. Science. (including Subser. Lecture Notes Artificial Intelligence. Lectures Notes Bioinformatics), vol. 3621. LNCS, pp. 169–187 (2006)

8. Forte, D.: Advances in onion routing: description and backtracing/investigation problems. Digit. Investig. 3(2), 85–88 (2006)

9. Kate, A., Zaverucha, G., Goldberg, I.: Pairing-based onion routing. In: Lecture Notes Computer. Science. (including Subser. Lecture Notes Artificial Intelligence. Lectures Notes Bioinformatics), vol. 4776 LNCS, pp. 95–112 (2007)

10. McLachlan, J., Tran, A., Hopper, N., Kim, Y.: Scalable onion routing with torsk. In: Proceedings of ACM Conference on Computer Communication Security, pp. 590–599 (2009)

11. Choudhary, D.: The Onion Routing. Res. Gate (2016)

12. Engelmann, A., Jukan, A.: Optical onion routing. In: 2017 International Conference on Computing, Networking and Communications, ICNC 2017, pp. 323–328 (2017)

13. Singh, N., Thawkar, S.: Securing network communication over web using onion routing. Int. J. Comput. Sci. Eng. Technol. 8(2), 8–9 (2018)

14. Khanum, A.. Pahal, S., Makkad, S., Panwar, A., Panwar, A.: Securing Onion Routing Against Correlation Attacks. Springer (2019)

15. Pavithra, S., Sivakumar, V., Pavithra, S., Tamilarasi, S.: Privacy-preserving routing for wireless network using OPERA. In: Proceedings of 2019 IEEE International Conference on Communications, Signal Processing, ICCSP 2019, pp. 415–419 (2019)

16. Yang, Y., Yang, L., Yang, M., Yu, H., Zhu, G., Chen, Z.: Dark web forum correlation analysis research. In: IEEE International Conference on Communications (2019)

17. Cambiaso, E., Vaccari, I., Patti, L., Aiello, M.: Darknet security: a categorization of attacks to the tor network. In: CEUR Workshop Proceedings, vol. 2315 (2019)

18. Hellebrandt, L., Homoliak, I., Malinka, K., Hanacek, P.: Increasing trust in tor node list using blockchain. In: ICBC 2019 - IEEE International Conference on Blockchain Cryptocurrency, pp. 29–32 (2019)

19. Chen, C., Asoni, D.E., Barrera, D., Danezis, G., Perrig, A.: HORNET: High-speed onion routing at the network layer. In: Proceedings of the ACM Conference on Computer and Communications Security, vol. 2015, pp. 1441–1454, October 2015

20. Catalano, D., Fiore, D., Gennaro, R.: A certificateless approach to onion routing. Int. J. Inf. Secur. 16(3), 327–343 (2017)

21. Hiller, J., Pennekamp, J., Dahlmanns, M., Henze, M., Panchenko, A., Wehrle, K.: Tailoring onion routing to the internet of things : security and privacy in untrusted environments. In: 2019 IEEE 27th International Conference on Network Protocols, pp. 1–12 (2019)

22. Zhang, Y., Weng, J., Weng, J., Li, M., Luo, W.: Onionchain: Towards Balancing Privacy and Traceability of Blockchain-Based Applications, vol. 14, no. 8, pp. 1–13 (2019)

23. Gumudavally, S., Zhu, Y., Fu, H., Guan, Y.: HECTor: homomorphic encryption enabled onion routing. In: International Conference on Communications, vol. 2019, pp. 1–6, May 2019

24. Mccaffrey, (12) United States Patent, vol. 2 (2019)

25. Iacovazzi, A., Frassinelli, D., Elovici, Y.: The DUSTER Attack : Tor Onion Service Attribution Based on Flow Watermarking with Track Hiding, pp. 213–225

A Technical Paper Review on Vehicle Tracking System

K. Hemachandran$^{(\boxtimes)}$, Shubham Tayal, G. Sai Kumar,
Vamshikrishna Boddu, Swathi Mudigonda,
and Muralikrishna Emudapuram

Department of ECE, Ashoka Institute of Engineering and Technology,
Hyderabad, India
hemachandran.kannan@ashoka.ac.in,
shubhamtayal999@gmail.com, Saikumar.724@gmail.com,
vamshiboddu99@gmail.com, swathimudigonda12@gmail.com,
muralike1234@gmail.com

Abstract. Nowadays, the vehicle tracking system is a new communication technology used in real time by many companies and individuals to track vehicle by using Global Positioning System [GPS]. It is extensively used all over the world to find the vehicles exact location. In our project we are using IOT to find the accurate position of that moving vehicle in order to obtain best results when compared to other conventional methods. This proposed system is directly interfaced with the bus by using the OBD/CAN interfaces and conveys these parameters to a central server using wireless sensor network technology.

Keywords: Global Positioning System · IoT · Interface · Wireless sensor network · BeiDou Navigation Satellite (BDS)

1 Introduction

The application of GPS updates the clients with the present area of the transport as indicated by the source and goal for which the client will make an enquiry, where the client will be associated with a focal server which will have a social database that contains all the records of the transport and their voyaging plans on which they play. The client will get the precise area of the transport on the guide. So, the travelers can arrive at their stop in the nick of time and load up the transport immediately.

The Vehicle Tracking System (VTS) is a finished verified and armada organization arrangement. It is the application used to determine the precise area of a vehicle utilizing different techniques like Global Positioning System trackers and other course discovering framework working through satellite and earth station. An after framework contains principally three sections vehicle unit, fixed based station and database with programming framework so we can isolate the entire activity of the vehicle following framework into two sections. 1. Following the area of vehicle. 2. Giving insurance of vehicle. Here is a model that the situating and following framework utilizing the BeiDou Navigation Satellite System(BDS) and Global-System for Mobile Communication (GSM) Network [1]. With the development of individual's living principles, the

© Springer Nature Switzerland AG 2020
A. P. Pandian et al. (Eds.): ICCBI 2019, LNDECT 49, pp. 698–703, 2020.
https://doi.org/10.1007/978-3-030-43192-1_77

quantity of secret cars is expanding. Yet, after proprietors leaving their vehicles in parking garages, they may experience issues in finding their vehicles or their vehicles may be taken. In the good old days, more consideration was paid to vehicle hostile to burglary, so we had designed mechanical, chipstyled and other vehicle against robbery frameworks [2, 3]. Electronic and framework vehicle checking frameworks have risen alongside the expansion of electronic innovation, arrange innovation, satellite guide understanding innovation, and so forth. The most effective method to quickly decide the area of vehicles in the wake of leaving or subsequent to losing vehicles has turned into the fundamental investigation bearing of vehicle observing [4–7].

2 Literature Survey

The review has been made for finding the accurate position of moving vehicle using IOT. The bus tracking system uses based on GPS and manual framework intended to show the ongoing area. The framework requires working association and may or not be GPS tracker. This framework comprised of a transmitter introduced on the vehicles and recipient. The framework is worked by GPS which is appended with each vehicle. It uses external hardware and software implementation the function of these GPS tracking devices is to collect the data to get better the efficiency, safety of the people and also the overall functionality. These live tracking also used in IoT. The web-of-things is the entomb working of bodily gadgets, vehicles, structures and dissimilar things done with hardware, programming, sensors and system network and consequently. IoT additionally expected propelled network gadgets, frameworks and administrations that go past M2M communication [8–10].

Continuous Bus Monitoring System utilizing GPS [3] shows the present areas of the transport [11]. The framework comprised of a transmitter introduced on receiver boards installed on the transports and recipient sheets introduced on the bus stations. It gave the important transport courses and other data to their customers [12].

A GPS based following framework is anticipated which monitors the area of a vehicle and its velocity dependent on a cellphone content informing framework. Lock and Unlock the present GPS directions of the engine vehicle by 2nd a SMS to the framework. Get present area of the vehicle whenever. This framework gives a confine way to deal with track and keep on their vehicle. The proprietor cans get the present area of the vehicle at whenever. Lock/open the present GPS directions of the vehicle to identify unapproved development and lock/open the most extreme speed of the vehicle to get a caution if the vehicle ventures out in front of that speed [13].

2.1 A Real-Time GSM/GPS Based Tracking System Using Mobilephone

A GPS based following framework is created to monitor the area of a vehicle and its exhibition depends on a cell phone content informing framework. The framework can give ongoing text-based notifications for speed and area [14]. Generally, the present area can be bolted and the framework will caution the proprietor if the vehicle is stirred from the present bolted. It likewise incorporates an equipment get together that goes about as an associating gadget. The anticipated framework includes an EM-406A model series GPS module. This was picked for its little size as it has an in-assembled fix reception apparatus for the GPS beneficiary [15].

2.2   GSM Technology

GSM technology, a computerized portable correspondence standard created by the ETSI, has been received by in excess of 100 nations around the world. GSM standard gear possesses over 80% of the current worldwide cell portable correspondence hardware commercial center. As per the above essential necessity examination, the structure of the remote vehicle checking. It is as of now the most generally utilized cell phone customary. China has built up a Global System for Mobile Communication versatile correspondence system covering the entire country state. Short-message administration [16] is a worth included help of the Global System for Mobile Communication framework. It utilizes the flagging channel to transfer data, and its transmission mode is to put away and sent first. That is, after the short message is conveyed, it will be put away in the short message focus (SMC) 1st and afterward for-warded by Short Message Communication to the recipient. SMS are transmitted all the way through a remote control [17].

2.3   System Overall Plan Design

The portable correspondence arrange, to accomplish remote Trans-BDS module is UM220-IIIN, which is the BDS/strategic vehicle area data. GPS double framework module planned by the center star tech-The principle control module of the vehicle terminal is innovation restricted organization. It is for the most part utilized for vehicle find the vehicle scope and longitude data, observing and route, handheld gear, thus direct speed judgment, and control message sending of on. UM220-IIIN embraces the low power Global Navigation the vehicle area. The BDS module predominantly gathers the Global Navigation Satellite System System on Chip–Humbird TM vehicle area data. The GSM module essentially whose licensed innovation is claimed by the organization. It sends short messages of vehicle area. The client is the littlest and completely residential BDS/GPS module in terminal is for the most part to catch the small information of the vehicle the market. It has the qualities of high coordination area, show the vehicle position on the point and low power utilization. It has a solid favorable position electronic guide, and gives voice caution to the proprietor. In the Bei Dou application that has serious prerequisites on size and power utilization channel, which can be store and sent throughSMC. Every little message is constrained to 140 B (140 characters or seventy Chinese characters) [18].

2.4   Design and Development of Bus Tracking System

The following transports though proceeding onward course is vital undertaking. Somebody sitting tight for the transport should enquire with respect to the situation of current area of the transport. Telephone discourse isn't achievable gratitude to traffic unsettling influences. more it includes variation costs on account of the brings and message administration over telephone and in this way the individual inside the transport could get bothered on the off chance that he gets different calls from people boarding that transport. Versatile basically based transport trailing framework gives a response to the present disadvantage that causes anybody to recover the circumstance of the transport while not occupation or substantial the individual going inside the transport. The people get the transport and thusly the organizers of the transport should claim A machine driven versatile with net property. The world Positioning System underpins in space subsequent with sponsorship of worldwide customary for Mobile in radiotelephone to report transport space information again to the servers [19].

2.5 Implementation of Vehicle Terminal

In this paper, an economical suburbanized vehicle remote positioning and trailing system design had been enforced [20, 21]. Test results delineate that the framework is equipped for trailing vehicle area and cautioning the proprietor if the vehicle out of the blue moves very one hundred m at low worth. Most merchant lack of bias and high-accessibility because the style philosophy, the system is in a position to use numerous out there route satellite frameworks and portable systems. This paper designed and realised a foreign observance system supported the mixture of BDS and GSM. Once testing, we tend to observed once the transport left the initial parking area one hundred m, the system might show the transport's location in real time, at that point conveyed the voice caution information at the proprietor's transportable, and understood the remote situating and trailing of the vehicle. The framework doesn't need back-end data backing, and clients don't need to be constrained to pay enrollment expenses. It exclusively needs low worth equipment costs and portable system SMS expenses [22]. Bolstered day by day utilization of cell phones, it understands exploitation cell phones to watch the vehicle. Again improvement, it can even be wide utilized for trailing people or things. This remote situating and trailing framework configuration can even be acknowledged by elective satellite route frameworks and versatile systems and has wide application possibilities [23].

3 Conclusion

The transport following system is going to be helpful and safer compared to alternative system. This system is straightforward to implement on vehicle noticeable of the ceaseless enhancement of the BeiDou Navigation Satellite System, the inclusion of the satellite route framework has created from the local framework to the globe. This paper designed and completed a distant observation system supported the mixture of BDS and GSM. The system doesn't would like back-end information support. It solely needs low value hardware prices and portable system SMS charges. Supported day by day utilization of cell phones, it understands victimization cell phones to watch the vehicle. once more improvement, it also can be widely used for following people or things. This remote situating and following framework configuration likewise can be finished by elective satellite route frameworks and versatile systems and has wide application possibilities.

References

1. Khan, A., Mishra, R.: GPS – GSM based tracking system. Int. J. Trends Technol. 3(2), 161–164 (2012)
2. Zhengnan, S., Haobin, J., Shidian, M.: Research of the vehicle anti-theft system based on fingerprint identification technology. Automob. Parts 11, 20–22 (2014)
3. Mingjun, Z., Lingchao, K., Chuanyi, H.: Present status and quality analysis for the mechanical automobile anti-theft device products in China. Automob. Parts 1, 73–75 (2014)
4. Haiyang, Y., Zhiliang, C., Shaobo, L.: Design of car alarm system based on radio frequency identification technology. Comput. Meas. Control 24(2), 144–146 (2016)
5. Lina, C.: Design of car security system based on GSM and GPS. Agric. Equip. Veh. Eng. 53(1), 14–18 (2015)
6. Xu, L., Ye, H., Tao, P.: Research and implementation of vehicle real-time monitoring system based on android. Comput. Sci. Appl. 7(2), 109–116 (2017)
7. Simin, W., Yangyang, X., Ling, Z., Yaling, L.: Intelligent anti-lost vehicle and vehicle positioning system. In: Electron. World 2019, vol. 37 (2014)
8. Gunjal Sunil, N., Joshi Ajinkya, V., Gosavi Swapnil, C., Kshirsagar Vyanktesh, B.: Dynamic bus timetable using GPS. Int. J. Adv. Res. Comput. Eng. Technol. 3(3), 775–778 (2014)
9. Punjabi, K., Bolaj, P., Mantur, P., Wali, S.: Bus locator via SMS using android application. Int. J. Comput. Sci. Inf. Technol. 5(2), 1603–1606 (2014)
10. Sridevi, K., Jeevitha, A., Kavitha, K., Sathya, K.: Vehicle tracking using GPS technology. Asian J. Appl. Sci. Technol. 1(2), 148–150 (2017)
11. Hemachandran, K, Ramesh, K., Vara Prasad, M.: Design and implementation of advanced ARM7 based biometric security system using wireless communication. In: IEEE Conference - Second International Conference on Inventive Systems and Control, Coimbatore, India, pp. 543–546 (2018)
12. Gharge, S., Chhaya, M., Chheda, G., Deshpande, J., Gajra, N.: Real time bus monitoring system using GPS. Eng. Sci. Technol. Int. J. 2(3), 441–448 (2012)
13. Jain, V., Goyal, V., Tayal, S.: Performance analysis of radio-over-fiber system against second order intermodulation distrotion. Int. J. Explor. Emerg. Trends Eng. 2, 235–238 (2015)
14. Hemachandran, K., Ramadevi, K., Raghupathi, H.: Real-time flash-flood monitoring and alerting and forecasting system using data mining and wireless sensor network. Int. J. Adv. Res. Sci. Eng. 06(10), 2302–2309 (2017)
15. Patel, D., Seth, R., Mishra, V.: Real time GPS based tracking system using mobile phone. IRJET 4(3), 743–746 (2017)
16. Hemachandran, K., Saritha, M., Akhila, M., Bhargavi, G.: Zigbee based advanced near field communication for hospital appointment system. In: IEEE Conference - Second International Conference on Inventive Systems and Control, Coimbatore, India, pp. 84–87 (2018)
17. Hongchao, L., Yuanming, W., Xiaoyu, H., Jing, L., Benxian, X.: Design of 10 kV ine switch state signal acquisition device based on GSM short message technology. Instrum. Technol. 6, 33–37 (2018)
18. Wei, J., Chiu, C.-H., Huang, F., Zhang, J., Cai, C.: A cost-effective decentralized vehicle remote positioning and tracking system using BeiDou navigation satellite system and mobile network. EURASIP J. Wirel. Commun. Netw. 2019, 112 (2019)
19. Smys, S., Raj, J.S., Augustine, N.: Autonoumous vehicle navigation in communication challenged environments-A simulation approach (2011)

20. Das, A., Gandhewar, N., Nehra, D.S., Baraskar, M., Gurjar, S., Khan, M.: Survey on vehicle tracking services. Manag. Res. **8**(1), 1–3 (2018)
21. Hemachandran, K., Srikanth, K., Raghupathi, H.: Alive human body detection and tracking system using an autonomous PC controlled rescue robot by using RF technology. Int. J. Adv. Res. Sci. Eng. **06**(10), 2284–2292 (2017)
22. Hemachandran, K., Pavani, M., Raghupathi, H.: Design and implementation of automated irrigation system in agriculture using wireless sensor network. Int. J. Adv. Res. Sci. Eng. **06** (10), 2293–2301 (2017)
23. Hemachandran, K., Prasad, P., Raghupathi, H.: Zigbee based intelligent helmet for coal miners. Int. J. Adv. Res. Sci. Eng. **06**(10), 2310–2317 (2017)

Role of Wireless Communications in Railway Systems: A Global Perspective

K. Krishna Chaitanya(✉), K. S. Sravan, and B. Seetha Ramanjaneyulu

VFSTR, Vadlamudi, A.P, India
Krishna.kandregula444@gmail.com,
Sraonekumar92@gmail.com, ramanbs@gmail.com

Abstract. Introduction of Wireless communications in public transportation system has brought many revolutionary changes in railways leading to improvements in signal and user data transmissions. There are several standards and models implemented to meet the requirements of High Speed Transit systems (HST). As driver less train operation systems (DTO) are also being introduced across the globe, role of technologies like Autonomous Parisian Transportation Administration and TETRA (Terrestrial Trunked Radio) that are globally standardized for high speed metro transit systems have gained an increased research importance. In this paper, a survey of various communication systems of metro rails and high speed transit systems are discussed.

Keywords: DTO · GoA · HST · HAZOP · FMEA · TETRA · WDM-RoF

1 Introduction

Since the railway transportation covers wider distances from cities to countries it has become one of the most preferred way of transportation around the world. Wireless communications is an emerging area that supports various applications like passenger data communications and signalling data transmissions. It is helpful in improving the services of transportation using D2D (Device 2 Device) communications [1], voice and data services etc. Researchers have been proposing wireless communications for applying it to several areas of public transportation. Many of the agencies are engaged in research and development activities on the advancement of signaling and communication in railways [2] using wireless communication. Role of wireless communications in railways on a global basis is discussed in this article. As the communication is needed between the moving train and the rest, wireless mode is an essential option here. However, after collecting the data signals from the moving train, they can be passed through wired links like OFC (optical fiber cable) cables that are laid alongside the tracks in many railway systems. So, wayside equipment is needed to collect the data from the moving trains. From the wayside devices, it can travel in wired or wireless media.

Wireless is considered a better choice here, to collect the data from the wayside equipment and then forward it to the central control system at high speeds [3]. The advantages of better communication systems here are rapid services, reduced headway (upto 60 s), track efficiency, increased train frequency (peak hours), economy, reduced man power (reduced human errors), energy efficiency and safety. Section 2 of the article discusses about various research efforts made in HST communication systems, Sect. 3 deals with metro/subway transit systems around the world and Sect. 4 talks about driverless train operation for urban rail transit system.

2 Research Efforts in HST Communication Systems

2.1 Train – Ground Communication Model

Recently the high speed train era has started. Efforts are taking place to fulfil the requirement of giving high speed data rates to these railway users [4]. Due to increase in mobility of trains, issues are raised in connectivity between the base station and the running train. As it is moving at rapid speed, it is difficult for receiver to beware of Channel State Information (CSI). So, a Multi relay transmission model is helpful in analyzing the interferences caused by all the relays which can be grouped and analyzed by selecting a spread spectrum channel based on IA (Information Accumulation) and EA (Energy Accumulation) [4] analysis where, their average transmission rates can be determined.

In rapid transit systems travelling at higher speeds, providing voice and data services to the passengers is a critical issue. To address this a train ground communication model as shown in Fig. 1 assumes a condition where the passenger is requesting for voice or data service from the base station. Here, the base station is considered as source and the train as its destination. Relays are placed closely such that there is no scope for loss of information due to handoff among relay nodes. The model shown in Fig. 1 [4] consists of L relay nodes each deployed with an antenna, represented as R1, …. Rl, …..RL, are acting as interface between the base station and the train for reliable transmission of data.

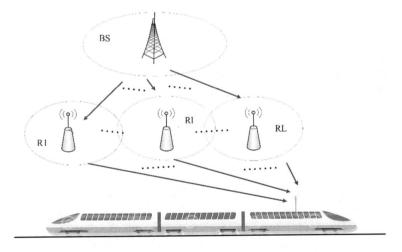

Fig. 1. Train – ground communication model

2.2   On – Board Transceivers Communication System

As HST's are made with thick metal and iron materials the impending penetration loss on the link between eNodeB and the user's base station can be up to 20–30 dB. As shown in Fig. 2, Number of on-Board Transceivers (Antennas) are placed at the top of each wagon, which facilitates the passengers to communicate through Wi-Fi. Here the on-Board transceivers that are placed at the top of each wagon acts as a Femtocell [5].

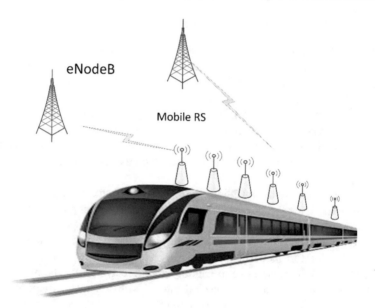

Fig. 2. On – Board Transceivers communication system

This offers good results in Line of Sight (LOS). But there occurs the problem, when the train enters into a tunnel or mountain area that causes a little scattering of signals which results in variable channel fluctuations. To mitigate this problem and to find the direction of arrival, an opportunistic beam forming model is considered as shown in Fig. 3 [5]. A linear topology of high speed communication system is considered with a cell radius of R and the perpendicular distance from eNodeB to the track is taken as d_{min}.

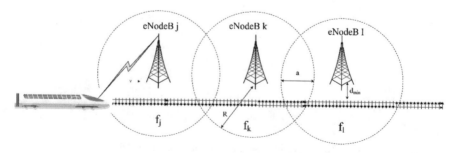

Fig. 3. Opportunistic beam forming in linear network topology

Here the communication between an eNodeB with N_t antennas and the On-Board receivers with single antenna is taken into consideration. If the length of the train is considered as L and the number of transceivers be T, then the spacing between each transceivers is taken as $L/(T-1)$ [5]. Here the number of On-Board transceivers must be as high as possible to mitigate interference among the received signals.

2.3 High Speed Linear Cell Communication System

As smart phones are being more popular day by day, the need of high speed data services became essential. Another method of providing those data services would be, by using WDM (Wavelength Division Multiplexing) technology [6]. As HST's are running at rapid speeds the hand over among the radio stations that are placed at trackside must be synchronized with the train movement.

As shown if Fig. 4, a linear cell network consisting of a hybrid model which combines WDM-RoF (Wavelength Division Multiplexing-Radio optical Frequency) [6] and mmWaveband wireless systems is considered. mmWave radio stations are deployed alongside the track to communicate with the trains by transmitting the signals from a central station via WDM-RoF system.

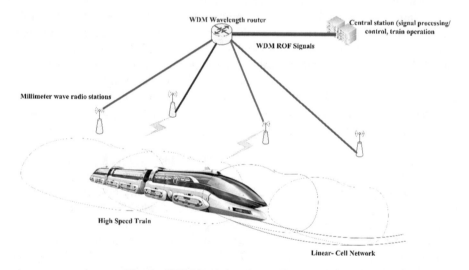

Fig. 4. WDM RoF signaling system for HST

A high speed Optical switching technology [7] like high speed wavelength tunable lasers are used for appropriate signal distribution among the respective radio stations. By using 16-QAM technology it can support data rates of up to 20 Gbps with a delay of 20 μs for handover between the radio stations, so uninterrupted connection can be provided compared to conventional cellular networks due to availability of control signals at central stations.

3   Metro/Subway Rapid Transit System

Metro's or subway rapid transit systems that are termed by "international association of public transport" as Urban Passenger Transport System. The name itself describes a synonym for Heavy Rail which operates in a unique way by segregating it from the general road and pedestrian traffic. World's oldest metro transit systems are UK's London Underground, Budapest Metro in Hungary, Glasgow Subway in Scotland, The Chicago elevated 'L' metro system in Illinois, Massachusetts Bay Transportation Authority (MBTA) subway, The U-Bahn in Berlin, Germany, Athens Metro in Greece, The New York City Subway in the US, South-eastern Pennsylvania Transportation Authority (SEPTA) in Philadelphia, US.

Metro trains differ from the conventional transit systems due to their capability to travel through the tunnels where high level radio coverage planning is required. Interoperability is needed for deploying new vendor's equipment with the existing systems. Some of the major technologies like Autonomous Parisian Transportation Administration [8] and TETRA are famous for their interoperability, by providing flexibility to the train regulating authorities to implement hybrid models. A metro line mainly includes Operations Control Centre (OCC), base stations in each metro station, passenger communication system and radios in each train.

In Dubai, the metro trains operate with DTO system, where OCC can make announcements directly to the passengers. Metros around the world are mostly relied upon a secure, reliable and robust technology named as TETRA (Terrestrial Trunked Radio) [9] which are dominantly used by metro operators as a critical communication system for both voice and data services. For smooth and safe running of stations and depots as well as for track maintenance in critical conditions, TETRA is considered to be a better solution due to its robust, flexible and reliable data communication capability between drivers and the line controllers by providing instant group call capability.

Some advanced features like live train tracking, doors open/close position, train health monitoring etc., have to be communicated with on-board staff, loco pilot, station staff and control centers using telemetry. Some real-time security measures such as on-train and platform surveillance footages need more bandwidth. Some critical data like change in train schedule, derailing, fire indications etc., will be updated by the network control center which is responsible to give continuous tracking of their travelling position to the passengers. In Europe, TETRA is used in metros like Copenhagen Metro in Denmark, Madrid's metro in Spain and the UK's London Underground network [9].

Copenhagen's Mini Metro was one of the earliest TETRA that runs 34 driverless trains co-coordinated by a single control and maintenance center. It is a 24-h automated people mover, with all equipment remotely controlled which enables operational staff to concentrate on assisting passengers in order to ensure a friendly and secure atmosphere and to discourage vandalism [9]. Madrid metro is the second largest metro network in Europe after London where millions of people rely on public transportation. TETRA is the major communication system used for effective public transportation with specific features like Dispatcher driven group call communication, simple text based messaging services, hybrid individual calling system integrating GSM and fixed line telephone networks. TETRA technology also provides communication to drivers on board regarding alarms, status and text messages.

4 Driverless Train Operation for Urban Rail Transit Systems

Driverless transit system is the emerging technology which is globally concerned for metro automation [10]. The annual world metro automation states that by 2025 the total length of automated lines is going to reach upto 1800 kms in Asia and Europe. Grade of automation (GoA) is termed as a unit for scaling the development stage of the subway transit systems. By International Electro-Technical Commission (IEC) standard IEC 62290-1, GoA is classified into 4 stages.

GoA1 is a non-automated train operation (NTO) system, where loco pilot is solely responsible to drive the train based on wayside or cabin signal and performing various operations like doors closing and opening, stopping trains in emergency etc., Here ATP (Automatic Train Protection) system [11] is used to avoid unsafe movements of trains.

GoA2 is a Semi-automated train operation (STO) system, where train movement is controlled by Automatic Train Operation (ATO) commands. The loco pilot in his cabin observes the guide way to control the train in emergency conditions. Here opening and closing of doors are done automatically or sometimes by the driver. GoA3 is a Driverless train operation (DTO) [12] system, where there is no use of loco pilot to monitor the guide way and to stop the train in emergency situations. An operating staff will be on board to take the responsibility of controlling train departure and door control. Sometimes this may be done automatically.

GoA4 is an Unattended Train Operation (UTO) System [13], where operating staff is also avoided for monitoring to make it a fully automatic system. Some critical operations like detection of guide way intrusion, surveillance camera (CCTV) onboard etc. GoA4 is considered for both attended and unattended systems. It also helps in reducing the criminal activity, elimination of suicides and mitigating trespassers death rate by using automatic detection and braking system.

In order to obtain reliable and flexible communication system the DTO systems must follow some basic standards like CENELEC standards [14] about computer controlled signaling system that are based on IEC61508 standards and the condition for train safe control systems which include EN50126, EN50128, EN50129 and IEC-UNI standards to regulate the safety in signaling systems. Some safety assessment processes are followed by some traditional techniques like HAZOP (HAZard & OPerabiliy), fault free analysis, and FMEA (Fault Modes & Effect Analysis). The rail signaling systems are evolving from traditional track circuit signaling to advanced communication-based train control (CBTC) systems to mitigate the basic limitation of traditional track circuits and to decrease headways for more flexibility. The essential part in DTO's is to achieve reliable interface between the platforms and track, which is done by using either intrusion detection system or PSD's [15]. These systems detects any huge objects or persons entering on to the track to stop the train in emergency situations.

5 Conclusion

In this paper, wireless communication in various rapid HST's are discussed along with their communication models that are implemented across the globe. Metro/rapid transit systems plays a crucial role in the urban transportation globally and their standards communicating with various technologies are discussed. With those automation systems we can avoid errors caused in signaling by human intervention. In future, it can improve the safety of the passengers further by employing various smart signaling methods that are being developed over the time, thereby increasing the faith and reliability among people available in railways.

References

1. Hu, F., Zheng, K., Long, H., Wang, W.: A cooperative hierarchical transmission scheme in railway wireless communication networks. In: Proceedings of 2011 IEEE International Conference on Service Operations, Logistics and Informatics, Beijing, pp. 605–609 (2011)
2. Zhou, D., Shi, T., Lv, X., Bai, W.: A research on banded topology control of wireless sensor networks along high-speed railways. In: 2015 34th Chinese Control Conference (CCC), Hangzhou, pp. 7736–7740 (2015)
3. Höyhtyä, M., Apilo, O., Lasanen, M.: Review of latest advances in 3GPP standardization: D2D communication in 5G systems and its energy consumption models. Future Internet **10**, 3 (2018)
4. Di, X., et al.: Cooperative multi-relay transmission for high-speed railway mobile communication system with fountain codes. In: 2013 International Workshop on High Mobility Wireless Communications (HMWC), pp. 63–68 (2013)
5. Cheng, M., Fang, X.: Location information-assisted opportunistic beamforming in LTE system for high-speed railway. EURASIP J. Wirel. Commun. Netw. (2012). https://doi.org/10.1186/1687-1499-2012-210
6. Dat, P.T., Kanno, A., Yamamoto, N.: Realization of high-capacity seamless communication for high-speed railways. In: National Institute of Information and Communications Technology, 8 May 2018. https://www.nict.go.jp/en/press/2018/05/08-1.html
7. Dat, P.T., et al.: High-speed and handover-free communications for high-speed trains using switched WDM fiber-wireless system. In: 2018 Optical Fiber Communications Conference and Exposition (OFC), San Diego, CA, pp. 1–3 (2018)
8. Régie Autonome des Transports Parisiens. https://en.wikipedia.org/wiki/RATP_Group
9. Terrestrial Trunked Radio. https://en.wikipedia.org/wiki/Terrestrial_Trunked_Radio
10. Jyothirmai, P., Raj, J.S., Smys, S.L.: Secured self organizing network architecture in wireless personal networks. Wirel. Pers. Commun. **96**(4), 5603–5620 (2017)
11. International Association of Public Transport.: A global bid for automation: UITP Observatory of Automated Metros confirms sustained growth rates for the coming years, Belgium. http://www.uitp.org/sites/default/files/Metro%20automation%20-%20facts%20and%20figures.pdf
12. PUERTO RICAN FUNICULAR. http://www.prvacationhelpers.com/puerto-rico-funicular.html
13. International Union of Railways (UIC). https://uic.org/passenger/highspeed/

14. Full speed ahead for China's high-speed rail network in 2019 in bid to boost slowing economy. https://www.scmp.com/economy/china-economy/article/2180562/full-speed-ahead-chinas-high-speed-rail-network-2019-bid-boost
15. European Committee for Electrotechnical Standardization. https://www.cenelec.eu/standards development/ourproducts/europeanstandards.html#top
16. High-speed rail. https://wikivisually.com/wiki/High-speed_rail

Protection of Microgrid with Ideal Optimization Differential Algorithm

P. M. Khandare[1]([⊠]), S. A. Deokar[2], and A. M. Dixit[1,2]

[1] Department of Technology, SPPU, Pune, India
poojakhandare24@gmail.com
[2] ZCOER, Pune, India
s_deokar02@rediffmail.com

Abstract. This paper process DWT (Discrete wavelet transform)-differential algorithm for optimal relay coordination issue for the microgrid. This solution works for grid-connected as well as the disconnected mode of operation. Coordination among relays of the microgrid is a complex part to handle, as the insertion of DG causes a bidirectional flow of current. Propelled protection strategies including DWT analysis of short circuit current can give bright and savvy methods for protection. Previous work has been applied with the differential algorithm on grid-connected and islanded mode, but one major deficiency is an increase in operating time of primary and secondary relay which further decreases the reliability of Microgrid. The proposed system relies on DWT-differential Analysis based approach which removes all unwanted noise and bandwidth from fault signal and differential analysis helps to select the best pair of a relay. The issue is designed as a Non-linear programming problem to limit altogether working Relay time. The Scheme is tested with IEEE 9 bus system. The comparative analysis is carried out with two traditional methods, the result shows that it accomplishes a sensational decrease in working time in the primary and secondary (backup) relay.

Keywords: Protection · Relay coordination · Differential analysis · DWT

1 Introduction

The Protection of the power network is important to distinguish and evacuate the short circuit areas as fast as could be expected under the circumstances. There is constantly a danger of significant failure in security, so a backup plan is required. To maintain synchronization, both primary and Secondary relay attributes must meet that constraint. The relay must go inside its self-defensive zone to keep away from any abuse of the defensive gadgets like circuit breakers and switches. The key role of the protection system is, the relay should not work other than its zone unless and until backup protection fails. the synchronization between primary and secondary relay should manage in a way such that the primary relay should work before the secondary relay. Most reliable and low cost protection in power system is overcurrent relay protection [1–5]. Microgrid such as 108 bus, 34 bus, 18 bus have large no of relays present in distribution system. To minimize operating time of relay, we should minimize two

© Springer Nature Switzerland AG 2020
A. P. Pandian et al. (Eds.): ICCBI 2019, LNDECT 49, pp. 712–720, 2020.
https://doi.org/10.1007/978-3-030-43192-1_79

basic parameters such as TDS (time duration setting) and PS (plug setting) [2]. To protect area which is not affected with short circuit current, protection system should be strong enough in order to isolate only faulty area [6].

Paper [7] recommends a Hybrid GA interim linear programming strategy for relay coordination, the primary lack discovered is that the calculation sets aside a ton of effort to tackle steps which adds to the working time of the relay [8]. This paper proposes the utilization of Superconducting fault current limiter (SFCL), with directional overcurrent relay (DOCRs) to take care of the protection coordination issue in distribution networks however the expansion of one more gadget in the system expands cost and briefness of the network [9]. Research states fuzzy-based GA to upgrade the working time of relay, the expansion of weight factor in calculation builds the twist in the network [10]. This paper exhibits a brilliant online adaptive ideal coordination strategy for overcurrent relay utilizing a genetic analysis for distributed networks when there are varieties of load, the primary trouble found in utilizing the SCADA framework. Online empowered SCADA needs clients to remotely monitor and control area by means of an internet browser Security is the greatest issue [11]. This paper introduces a genetic algorithm (GA) strategy for the coordination of overcurrent (OC) relay. Short circuit current use for examination isn't preprocessed henceforth the presence of noise and undesirable information will build the more working time for the relay. This paper proposes a hybrid and versatile overcurrent and differential protection plan to manage the extreme changes in the microgrid. The information mining model-based on the DWT differential intelligent protection scheme for the microgrid network has been proposed. The proposed plan creates protection work for the microgrid working at various topologies such as grid-connected and islanded modes. The differential highlights inferred at individual feeders are utilized to fabricate the information discretization, which is utilized for a final relay decision.

2   Problem Evolution

There are three central matters in the setting and coordination of Over Current relays that are rapid relay activity, selectivity, and Reliability. To improve the speed of relay, the working time of each relay should be in limit, Selectivity which implies that each relay should work for the failure happening in the relays' relating protection zone and reliability which implies that each relay must be appropriately supported by some backup relays. In this paper, the above concerns are thought-about to upgrade the output of relay and eventually the microgrid.

2.1   Operating Time Problem Formation

The Operating time of relay fluctuates with two segments that are TDS (Time duration setting) and PS (Plug setting). Whereas PS is the proportion of short circuit current to Pick up current working time of relay differs conversely with a short circuit current [2, 12]. An ideal coordination program limits the all-out working time of relay and remove the miss-coordination among primary and backup relay. The protection coordination issue is normally characterized as a progression issue where the crucial objective is to restrict the general relay operating time

$$t_{ij} = TDS_i \frac{A}{\left(\frac{I_{SCij}}{I_{pi}}\right)^B - 1}$$ (1)

Where i is used as relay identifier and j is the fault location identifier. A and B are the constants whose value depends on what type of relay used. There are different types of OCR with different characteristics, A and B varies as per Table 1 here IDMT relays are used. Different value of α and β generate nonlinear characteristics which results in formation of nonlinear programming problem with constraint.

Table 1. Characteristics of different OCR

Selection of overcurrent relay	A	B
Inverse time overcurrent relay	0.143	0.021
Very inverse relay	13.57	1
Extremely inverse relay	80	2

2.2 Non Linear Objective Function with Constraint

The goal is to ideally organize double setting Relay so as to limit the absolute relay working time for both primary and secondary activity thinking about the two methods of Microgrid activity: grid-connected and islanded design. The goal can be expressed as pursues,

$$Minimize\ T = \sum_{\Phi=1}^{\Phi} \sum_{i=1}^{N} \sum_{j=1}^{M} \left(t_{fw\_\Phi ij}^{p} + \sum_{k=1}^{k} t_{rv\_\Phi ij}^{bk}\right)$$ (2)

Where Φ indicates values 1 and 2 grid connected mode shows 1 and islanded mode shows 2. M indicates total count for relays. N indicates total count for fault locations. $t_{fw\_\varphi ij}^{p}$ is the primary relay i operating time, in the forward direction, for fault at j for configuration Φ. . $t_{rv\_\varphi ij}^{bk}$ Is the backup relay i operating time, in the reverse direction, for fault at j for configuration Term bk denotes b is for backup up relay and k denotes total count of backup relay in system.

To ensure proper protection in system primary protection is supported by its secondary (backup) protection. It is necessary to coordinate the two protective systems (primary and back-up). Coordination Time interval (CTI) is difference between respective backup and primary relay operating time which has to be considered throughout algorithm. Its value is depends on type of relay used. CTI for IDMT relays taken here is 0.2 [13]. For proper operation in relays backup relay should not operate before primary relay.

$$t^{bk}_{rv\_cij} - t^{p}_{fw\_cij} \geq CTD \tag{3}$$

$$I_{pi\_min} \leq I_{pfwi}, I_{prvi} \leq I_{pi_{max}} \tag{4}$$

$$TDS_{i\_min} \leq TDS_{fwi}, TDS_{rvi} \leq TDS_{i_{max}} \tag{5}$$

The estimation of Ipi−min is picked with the end goal that it is bigger than the evaluated load current by a noteworthy edge. In this manner, the accompanying requirements are characterized. Where Ipi_min and Ipi_max are limits put on pickup current. $TDSi\_min$ and $TDSi\_max$ are limits put on time duration setting.

3 DWT-Differential Analysis

DWT–Differential Analysis is one of the powerful and easy methods for obtaining best suitable solution. DWT-DA is an improvement over traditional Genetic algorithm as it is computationally more efficient and accurate. Dwt analysis generate short circuit signal free from unwanted bandwidth use for calculation of PS and TDS. DWT is effective tool to detect small disturbances occur in micro grid as it is transient sensitive way for processing any signal. The Proposed research pre-process short circuit current signal through DWT to extract most significant feature to detect location of fault and to remove all noise factor from signal which ultimately helps to reduce time of operating relay [14].

Short circuit current Isc is sampled with frequency 6 kHz, based value used is 20 MVA and 480 V for per unit calculation. There are 4 detailed coefficient (DA1–DA4) obtained after analyzing the signal as shown in Fig. 1.

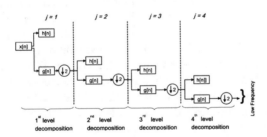

Fig. 1. DWT decomposition of Iabc up to 4 level

3.1 Algorithm and Flow Chart

Differential Analysis is technique used by researcher for optimization in science and engineering. First it was discovered by Storn and Price [15]. Adaptability of DA specifically to reach solution in less number of iterations is because of less complexity, stability and less time for calculation. Differential analysis is based on mainly reproduction, mutation, crossover and selection. Following are steps by which software execute program.

Step 1 short circuit current are transfer to system through CT for two mods of operation grid connected and grid separated mode.

Step 2 DWT analysis is carried out on current up to 4th quadrant of degradation. Change in signal detects fault and No fault condition on respective bus.

Step 3 if fault detects, control System find Different parameters such as PS and TDS With respect to fault current and all supplied data. Primary and backup the relay time can be calculated with help of objective function.

Step 4 Primary pair of relay gets signal to trip faulty bus. If primary protection fails to Operate Best pair of backup Relay selected with survival of fittest Algorithm.

Step 5 Differential Analyses/Algorithm

- Survival Function: This is special type of objective function which is used to find chromosomes for differential analysis.
- Reproduction: A chromosome which is best fitted to form next generation is selected as there is great possibility to survive and reproduce.
- Crossover: Crossover is process for producing new generation called as off-spring. In cross over as name suggest crossing between information takes place
- Mutation: termination criteria are defined in the initial of algorithm. This process is carried out till it satisfied criteria. Mutation continue to check possibility of local error.

Step 6 stop processes by tripping faulty bus with backup Protection

Step 7 if no fault occur stop the process (Fig. 2 and Table 2).

4   Simulation Case Study

DWT-Differential algorithm is tested in MATLAB by converting IEEE 9bus system to the crossbreed microgrid by connecting different DG sources such as two solar connected to bus 4 and bus 8 wind connected bus 5 and diesel connected to bus 6. As shown in Fig. 3. When sustainable sources are not connected to power system then protection is done with the help of single overcurrent relay as flow of current is unidirectional so traditional protection works for system. When sustainable sources connected to grid, flow of current are bidirectional hence there is need to design smart protection. All gadgets used like DG, Inverter are sensitive by nature so protection of bus should be effective in all way. DWT-DA is tested on system by using 21 relays and 9 buses. Faults are created on each bus F1 to F8 short circuit current is note down and fault analysis is carried out. Following is data used to form Simulation model in MATLAB (Table 3).

4.1   DWT Analysis of Fault Current

Fault Current Signal Iabc is decomposed using 4 levels DWT. Current measured at relay 21 and 1 is decomposed with DWT so that all disturbances and noise present in signal is removed. DWT approach has capability to adjust its own window selection to

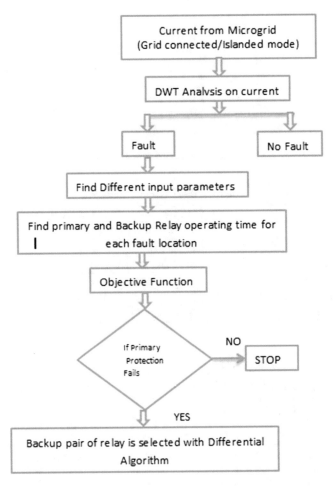

Fig. 2. Flow chart for DWT-DA

Table 2. DWT-DA parameters

Size of inhabitants	1000
No. of parameters	08
Rate of exchange information	50%
Mutation rate	20%

give good time settlement. For decomposition we used daubechies Wavelet (db4) with four level decomposition [17]. Figure 4 gives extracted features with help of DWT at relay at 21 and 1.

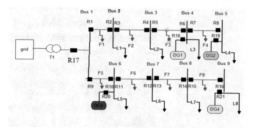

Fig. 3. Grid connected mode of microgrid

Table 3. IEEE system parameters

Two feeder	115 kV, X/R ratio = 6, short circuit MVA = 500
Line impedance	Z = 0.1529 + j0.1406 Ω/km
Two feeders are	500 m long
Line transformer ratings	20 MVA, 115 kV/12.47 kV
Solar wind and diesel DG ratings	480 V, 20MVA, xd' = 0.11
Transformer connected to all DGs	20 MVA, 115 kV/12.47 kV
Load capacity	2MVA, 0.9 PF

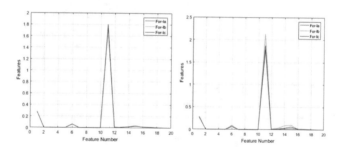

Fig. 4. Extracted DWT features for fault measured at R21 & R1 for Ia, Ib & Ic.

4.2   Comparative Analysis

In conventional scheme total operating time is calculated by directly using short circuit current. In proposed scheme current is preprocessed with DWT analysis with 4th level of decomposition and also miss operation is detected with the help of change in signal in DWT. Total operating time in both modes of operation is 105.3 s in conventional method while 60% reduction is takes place with proposed scheme (Table 4).

Table 4. Comparative analysis

Total operating time	Traditional protection method [16]	Proposed protection method	Percentage reduction
Both modes of operation	105.3 s	44.38 s	60%
Grid connected mode	49.975 s	21.37 s	28.6%
Islanded mode	55.34 s	23.01 s	32.33%

5 Conclusion

This paper proposes a hybrid and versatile DWT based overcurrent and differential protection plan to manage the extreme changes in the microgrid. DWT-based differential intelligent protection scheme for the microgrid network has been proposed. The proposed plan creates protection work for the microgrid working at various modes and topology. DWT analysis is carried out for each fault on each feeder, After Analysis decision is sent to respective pair of relay. The proposed DWT differential model is broadly tried on the standard IEEE 9 bus system It has been discovered that the DWT based component extraction can viably evacuate the repetition accessible in time-region information and consequently adequately ready to diminish the working Time. The proposed procedure additionally has the potential and ability to actualize for on-line genuine applications. Distinctive kinds of faults can be tried with the same methodology.

Percentage reduction 60% of time take place in grid connected and islanded mode of operation 28.6% reduction take place in grid connected while 33% reduction in islanded mode. Number shows that there is big reduction in operating time of relay which increases reliability of the microgrid. Hence DWT-DA promising methods in protection of the mcrogrid. Different types of faults can be tested with this Algorithm in future.

References

1. Tripathi, J.M., Adhishree, Krishan, R: Optimal coordination of overcurrent relays using gravitational search algorithm with DG penetration. In: 6th IEEE Power India International Conference (PIICON), Delhi, 5–7 December 2014
2. Urdaneta, A.J., Nadira, R., Perez, L.G., et al.: Optimal coordination of directional overcurrent relays in interconnected power systems. IEEE Trans. Power Delivery **3**(3), 903–911 (1988)
3. Smith, T.F., Waterman, M.S.: Identification of common molecular subsequences. J. Mol. Biol. **147**, 195–197 (1981)
4. Barker, P.P., de Mello, R.W.: Determining the impact of distributed generation on power systems part I- radial distribution systems. IEEE Trans. Power Delivery **15**, 486–493 (2000)

5. Fazanehrafat, A., Javadian, S.A.M., Batbee, S.M.T, Haghifam, M.R.: Maintaining the recloser-fuse coordination in distribution systems in presence of DG by determining DG size. In: Proceedings of the IET 9th International Conference on Developments in Power System Protection (DPSP), pp. 132–137, March 2008

6. Chowdhury, A., Koval, D.: Power Distribution System Reliability: Practical Methods and Applications. Wiley-IEEE, Hoboken (2009)

7. Srinivas, S.T.P., Swarup, K.S.: A hybrid GA - interval linear programming implementation for microgrid relay coordination considering different fault locations. IEEE (2017). 978-1-5386-1789-2/17

8. Choudhary, N.K., Mohanty, S.R.: Protection coordination of over current relays in distribution system with DG and superconducting fault current limiter. IEEE (2014). 978-1-4799-5141-3/14

9. Alkaran, D.S., Vatani, M.R., Sanjari, M.J., Gharehpetian, G.B., Naderi, M.S.: Optimal overcurrent relay coordination in interconnected networks by using fuzzy-based GA method. IEEE Trans. Smart Grid 9(4), 3091–3101 (2018)

10. Xu, K., Liao, Y.: Intelligent method for online adaptive optimum coordination of overcurrent relays. IEEE (2018). 978-1-7281-0316-7/18

11. Bedekar, P.P., Bhide, S.R.: Optimum coordination of overcurrent relays in distribution system using genetic algorithm. In: 2009 Third International Conference on Power Systems, Kharagpur, India, 27–29 December 2009 (2009). ICPS – 247

12. Zeineldin, H., Sharaf, H.M., Ibrahim, D.K., Abou El Zahab, E.E.: Optimal protection coordination for meshed distribution systems with DG using dual setting directional overcurrent relays. IEEE Trans. Smart Grid 6(1), 115–123 (2015)

13. Singh, D.K., Gupta, S.: Use of genetic algorithms (GA) for optimal coordination of directional over current relays. IEEE (2012). 978-1-4673-0455-9/12

14. Gao, W., Ning, J.: Wavelet-based disturbance analysis for power system wide-area monitoring. IEEE Trans. Smart Grid 2(1), 121–130 (2011)

15. Storn, R., Price, K.: Differential evolution a simple and efficient heuristic for global optimization over continuous spaces. J. Global Optim. 11(4), 341–359 (1997)

16. Sharma, A., Panigrahi, B.K.: Optimal relay coordination suitable for grid-connected and islanded operational modes of microgrid. In: IEEE INDICON (2015). 1570185643

17. Hoyo-Montaño, J.A., et al.: Non-intrusive electric load identification using wavelet transform. Ingeniería e Investigación 38(2), 42–51 (2018)

An Improved Energy Efficient Scheme for Data Aggregation in Internet of Things (IoT)

Keshvi Sharma$^{(\boxtimes)}$ and Rakesh Kumar

National Institute of Technical Teachers Training and Research,
Chandigarh (NITTTR), Chandigarh, India
Keshvi.sharma@gmail.com, raakeshdhiman@gmail.com

Abstract. This work is focused on improving the life span of a network through the reduction in power expenditure. The earlier approaches use multilevel clustering for aggregating data to the sink. In this arrangement, the division of entire network is carried out into clusters. Afterward, the selection of cluster head is done in every cluster. The cluster heads transmit information to base station which later passes on over the internet. In this research work, the multilevel clustering will be improved so that lifetime of the network can be enhanced. In the proposed approach, the gateway nodes will be deployed in the network for the data aggregation. The cluster heads will transmit information to the gateway nodes which later transmit that information to base station. The proposed approach will be implemented in MATLAB and it will be compared with the existing approach of multilevel clustering on the basis of throughput, packet loss and amount of inactive nodes. The proposed improvement leads to increase the throughput of the network, packet loss and number of dead nodes will be reduced as compared to existing approach.

Keywords: IoT · Data aggregation · Clustering · Gateway

1 Introduction

A technology that gives the access of deep analysis, integration and automation within the system to its customers is called Internet of Things. The information from the environment is sensed by the sensor nodes in this technology. Further, the gathered data is uploaded to the key server [1]. The handoff system is set up after the change of the locations of the servers. The modification in the system can be done using this accuracy and reach to the area. A number of technologies are developed for network sensing and automation [2]. It demonstrates the new advancement in the software technology and hardware as well. The novel development in the area of product delivery, goods and services, economic and social sector have caused main transformations in the exiting components [3]. In the present scenario, internet over open or private Internet Protocol (IP) network has connected everybody. Using internet, the people can sense the environment. Also, they can contact and share information with each other. The data from the environment is gathered by the entire interlinked objects [4]. These objects analyze the gathered information to start the act for providing management and decision making. Internet of Things is characterized as a network of

© Springer Nature Switzerland AG 2020
A. P. Pandian et al. (Eds.): ICCBI 2019, LNDECT 49, pp. 721–730, 2020.
https://doi.org/10.1007/978-3-030-43192-1_80

physical entities. IoT (Internet of Things) is not just the interconnection between the networks [5]. IoT is too developed into a network of various types and sizes such as cameras, medical tools and industrial models, animals, people, buildings, etc. In order to get intelligent reorganizations, location, process control & management, all of these objects are interlinked to each other for providing communication and data sharing [6]. Data aggregation refers to the mechanism of collecting and merging large amount of data by using aggregation techniques. Later, this information is transferred to the sink with minimum redundancy. The working of data aggregation is described along with the proposed techniques. Initially, sensor nodes provide sensor data. Then, some aggregation methods are used for collecting this data [7]. The important and suitable data is selected from the gathered data. This data is further transferred to the base station. The two main objectives of this scheme are the abolition of entire redundant data and increase in the battery life of sensor nodes. The data is forwarded in multi-hop fashion. Hence, the neighboring node can sense the information in an easy way [8]. A new scheme has been recommended in this work for eliminating this issue. In the recommended scheme, the node sends its data to the cluster head. This process however is very time consuming. In WSN (Wireless Sensor Network), the life span of cluster head is not very long. In this case, the whole processing is carried out again. This results in the wastage of both power as well as time. With the time, various data aggregation approaches have been designed [9]. Centralized scheme mainly concentrates on the address. In this approach, all nodes within the network deliver their data to a central node via the optimal direct route. Sensor node is a very strong node. This node may lead all other nodes. This nodes transfers data packets so that these data packets can be processed further. This node aggregates data headed to be stored within the network. Each node existing in between the route is responsible for the delivery of its data packet to the sensor node. It implies that the transferring of massive volume of data and messages is carried out for a query. The data aggregation carries out in tree shaped structure in tree based scheme. This approach can be abridged as a spanning tree with roots. The roots in this tree act as base station while the role of a source node is played by the leaves of this tree [10]. There is a parent node over every node. Every node sends its whole data for more processing. The delivery of data starts from the leaves via base station. Afterward, the received data is aggregated at the parent node. In order to generate clusters, the partition of whole network is carried out with the help of a cluster-based scheme [11]. Among all existing nodes, the selection of a cluster head is done in individual manner for every cluster. The cluster head collects and aggregates data from other nodes within the cluster. Therefore, cluster head sometimes is also called aggregator. The transferring of this aggregated data is done to the base station for ultimate processing. An approach named in-network aggregation gathers data. Later, multi-hop network performs the routing of this data. This approach can decrease the resource usage which in turn increases the life span of the network in continuous manner [12]. There are two more categories of this approach. The first category minimizes the magnitude of packet while the other category does not minimize the magnitude of packet. The magnitude (size) reduction schemes combine the data packets from the nodes and compress them. It implies that these schemes reduce the magnitude of the packet that is going to be transferred and forwarded to the base station.

2   Literature Review

Emma Fitzgerald et al. (2018) presented mixed-integer programming expressions and algorithms for energy efficient data aggregation and dissemination in IoT [13]. The growth of the network was considered in two cases. These cases included minimum overall energy consumption as well as min-max energy consumed by every node. Moreover, a formulation and algorithm was recommended in this work for throughput. The main aim here was to optimally schedule transferring in the genuine aggregation under the physical interference paradigm. In contrast to the recommended approaches, a network with forty nodes consumed thirteen times more energy by utilizing direct route from sensors to actuators as per the achieved results.

Idrees et al. (2018) introduced a new approach for improving the life span of the periodic wireless sensor network (PWSN) [14]. The performance of this approach was divided into periods. There were the three phases included in every period. The readings of sensors were gathered and stored in the senor node in the first phase. In the next phase, these readings were converted into the clusters of readings using modified K-means algorithm. Thirdly, the delivery of one typical reading of every cluster was done to the base station. The recommended approach could successfully reduce energy consumption in the entire network.

Rao et al. (2017) stated that use of User Equipments (UEs) as relays reduced the transmission of energy required by IoT gadgets [15]. A scheme was studied in this work where IoT tools associated themselves with the User Equipments (UEs). Some fixed, random, and greedy approaches were used for this purpose. The end-to-end disruption possibility at the tools was evaluated for every scheme. It was analyzed that the greedy approach had the minimal disruption chance at the IoT tool. Also, the outcome of IoT tools, the amount of UEs, the data arrival rates at the IoT tools and the uplink data development process of UEs on the disruption possibility were analyzed. On the basis of this analysis, significant intuitiveness was drawn into the framework of industrial IoT.

Dong et al. (2017) recommended a new approach called INADS for ICN [16]. The main aim of this approach was to provide support in the formation of future IoT. The new approach significantly reduced the total number of notifications in both single as well as many builder situations. In this way, this approach saved the constrained energy of IoT tools. Moreover, this approach reduced bandwidth required for the forwarding of subscription and unnecessary notification messages.

Bhandari et al. (2017) recommended a new mechanism using which the channel access and queuing delays of the clustered industrial Internet of Things networks were minimized [17]. Initially, a prioritized channel access system was designed. Several features of MAC layer were allotted to the packets approaching from two sorts of IoT nodes. Also, information based on application provided by cloud hub was considered for this purpose. The tested outcomes depicted that the recommended approach more efficiently improved the performance in terms of delay and trustworthiness.

Yu et al. (2017) proposed a novel technique that includes a cluster-based data analysis. In this recommended technique, the unnecessary information is gathered and cluster head is identified using the recursive PCA applied here [18]. The members of clusters gather and aggregate the data at cluster head. The proposed technique is used to extract the data. For adapting to the changes in Internet of Things systems, the parameters of PCA model are updated recursively using R-PCA. Any kinds of computational and processing burdens on sensor nodes are released by the cluster-based data analysis technique. It is seen that the correlated sensor data is aggregated efficiently with high recovery accuracy by applying proposed technique on practical datasets.

3 Research Gaps and Problem Formulation

- The techniques which are proposed previously are based on to latency from the network. When the latency from the network is reduced it also reduces the network delay but the reliability of the network get reduced.
- The clustering based techniques which are designed in the previous years to improve lifetime are much vulnerable towards to network failure. The energy efficient methods are required in which there are less chances of network failure.
- The distributed network concepts are used in the previous years to increase network flexibility and also to reduce network failure. The techniques which are designed to improve lifetime are based on the query search concept due to which most frequent data can be transmitted by the cluster head which affect network performance.

In IoT (Internet of Things), sensor nodes forward the sensed data to the sink. The sensor devices which sense information have extremely less magnitude. The deployment of these devices in carried out at the distant locations. Because of such unique properties of the network, there are certain issues which arise. In the previous research, various techniques are proposed which can improve lifetime and also data aggregation rate in the network. The clustering is the major efficient approach which can improve lifetime of Internet of Things and also improve data aggregation rate. In the clustering approach, clusters will be created by dividing the complete network using location based clustering approach. In the second step, the cluster heads are chosen for each cluster on the basis of remoteness and power consumption. The sensor device which has least distance to base station and has maximum energy is chosen as cluster head in each cluster. The distance is factor which affects the data aggregation rate and energy consumption. In this research work, the multi-level clustering approach will be proposed which improve data aggregation rate and reduce energy consumption of the network.

4 Proposed Work

Following are the various levels in the multi-level clustering scheme of Internet of Things:-

Level 1: Choosing the Cluster Head
In this stage, the division of whole network is done into certain clusters and process of cluster head selection is initiated by the base station. The message is passed all across the network which states that an efficient cluster head can be chosen. The distance of one node from the base station is calculated mathematically. The sensor nodes also present their residual energy which play important role in being chosen as cluster head.

The radius of each cluster is calculated and the sensor nodes which lie within the radius of the cluster represent that cluster. The number of nodes represent the cluster should be 3 or more than 3. The nodes which are within the cluster should select their cluster head on the basis of residual energy.

Level 2: Choosing the Gateway Node
The gateway selection is the last phase of proposed protocol. Gateway nodes are the extra nodes which are deployed for improving the lifetime of networks. The size of network is considered as an important factor for defining the number of gateway nodes. The data is transmitted to the leader nodes by the cluster heads which further transmit it to the gateways. Then the data is forwarded to sink by the gateways. The base station takes data from the nearest gateway node and leader node transmits the data gateway node which is the nearest (Fig. 1).

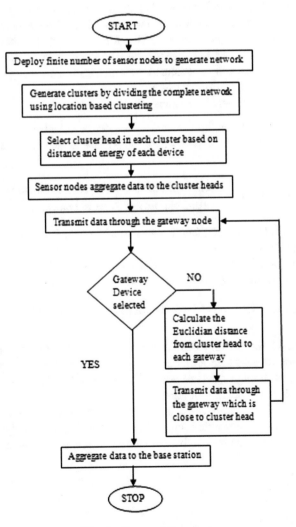

Fig. 1. Proposed flowchart

Following are the various advantages of the proposed work:-

1. The proposed algorithm is the type of algorithm which can aggregate the data efficiently from source to destination. Due to efficient data aggregation the power expenditure of the sensor network gets reduced.
2. The cache nodes will save the most frequent information and pass that information to base station. The most frequent information is transmitted by the cache node only due to which routing over head get reduced in the network.

5 Experimental Results

In this research work, the clustering method is improved for increasing the lifespan of the sensor devices. The clustering method is much popular to reduce power expenditure within the network. In this research work, the clustering method is improved to increase lifetime of the network. A tool that is applied for performing highly complex mathematical calculations is called MATLAB. The data is taken in the form of arrays and given as input. The data will be executed very efficiently and at high speed since the data given as input is in the form of arrays.

The performance analysis parameters are given below:-

1. **Throughput:-** The throughput parameters describe the number of packets which are received divided by total number of packets which are transferred by the network.

$$\text{Throughput} = \frac{\text{Number of Packets Received at Destination}}{\text{Total Number of Packets Transmitted}} * \text{time}$$

2. **Number of Dead Nodes:-** This is the parameter which define the nodes whose battery degrades to zero.
3. **Number of Alive Nodes:-** This is the parameter which describe the nodes whose battery is not degrades to zero. The formula is given below

$$\text{Alive Nodes} = \text{Total Number of nodes} - \text{Number of dead nodes}$$

As shown in Fig. 2, the existing technique in which cluster head are responsible for the data transmission to base station is compared with the proposed algorithm in which data is transmitted through gateway nodes. It is analyzed that proposed algorithm has less number of inactive devices in contrast to existing algorithm.

As shown in Fig. 3, the existing technique is compared with the proposed algorithm in which data is transmitted through gateway nodes. It is analyzed that recommended approach has more number of active nodes as compared to accessible approach.

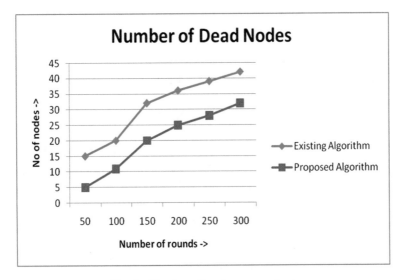

Fig. 2. Number of packets transmitted

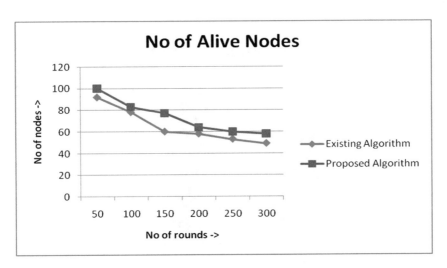

Fig. 3. Number of active nodes

Figure 4 shows the existing technique in which cluster head are responsible for the data transmission to base station is compared with the proposed algorithm in which data is transmitted through gateway nodes. It is analyzed that proposed algorithm has more throughput as compared to existing algorithm.

As shown in Fig. 5, the existing and proposed algorithms are compared in terms of remaining energy. Through the comparative analysis is is seen that the proposed algorithm provides higher remaining energy.

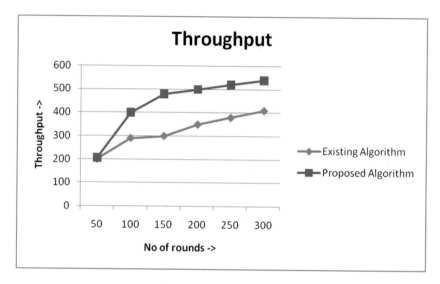

Fig. 4. Throughput analysis

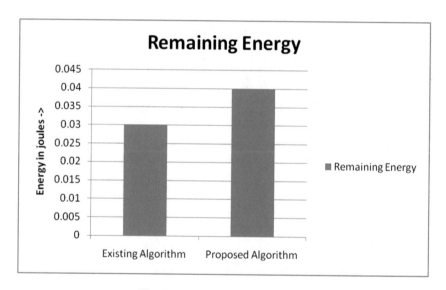

Fig. 5. Remaining energy analysis

6 Conclusion

In IoT (Internet of Things), the devices after sensing the data delivers this data to the sink. Data aggregation and power expenditure are the two main challenges of this technology. These issues occur due to the dynamicity of the network. A clustering algorithm can be implemented for increasing the life span of WSN (Wireless Sensor

Network). The selection of cluster head is carried out for every cluster on the basis of energy and distance of nodes. The cluster head delivers data gathered from devices to the base station. In this research work, the gateway nodes will be deployed. The proposed technique is implemented in MATLAB by considering various simulation parameters. The proposed three-level hierarchal routing protocol is compared with existing two-level hierarchal routing protocol and it is analyzed that number of dead nodes are reduced and number of packet transmission is increased. It is analyzed that results are improved up to 5%.

References

1. Al-Fuqaha, A., Guizani, M., Mohammadi, M., Aledhari, M., Ayyash, M.: Internet of things: a survey on enabling technologies, protocols and applications. IEEE Commun. Surv. Tutor. **17**(4), 2347–2376 (2015)
2. Vermesan, O., Friess, P.: Internet of Things: Converging Technologies for Smart Environments and Integrated Ecosystems. River Publishers Series in Communications, pp. 1–364. River Publishers, Aalborg (2013)
3. Vermesan, O., Friess, P.: Internet of Things-From Research and Innovation to Market Deployment. River Publishers Series in Communications, pp. 1–143. River Publishers, Aalborg (2014)
4. Kortuem, G., Kawsar, F., Sundramoorthy, V., Fitton, D.: Smart objects as building blocks for the internet of things. IEEE Internet Comput. **14**(1), 44–51 (2010)
5. Patel, K.K., Patel, S.M.: Internet of Things-IOT: definition, characteristics, architecture, enabling technologies, application & future challenges. Int. J. Eng. Sci. Comput. **6**(5), 6122–6131 (2016)
6. Jindal, F., Jamar, R., Churi, P.: Future and challenges of internet of things. Int. J. Comput. Sci. Inf. Technol. (IJCSIT) **10**(2), 13–25 (2018)
7. Atzori, L., Iera, A., Morabito, G.: The internet of things: a survey. Comput. Netw. **54**(15), 2787–2805 (2010)
8. Douglas, D.H., Peucker, T.K.: Algorithms for the reduction of the number of points required to represent a digitized line or its caricature. Can. Cartographer **10**(2), 112–122 (1973)
9. Liu, C., Wu, K., Pei, J.: An energy-efficient data collection framework for wireless sensor networks by exploiting spatiotemporal correlation. IEEE Trans. Parallel Distrib. Syst. **18**(7), 1010–1023 (2007)
10. Stankovic, V., Stankovic, L., Wang, S., Cheng, S.: Distributed compression for condition monitoring of wind farms. IEEE Trans. Sustain. Energy **4**(1), 174–181 (2013)
11. Singh, L., Kaur, E.M.: Analysis of data aggregation techniques of IoT. Int. J. Future Revolut. Comput. Sci. Commun. Eng. **4**(4), 131–134 (2018)
12. Deligiannis, N., Zimos, E., Ofrim, D., Andreopoulos, Y., Munteanu, A.: Distributed joint source-channel coding with copula function-based correlation modeling for wireless sensors measuring temperature. IEEE Sens. J. **15**(8), 4496–4507 (2015)
13. Fitzgerald, E.: Energy-optimal data aggregation and dissemination for the Internet of things. IEEE Internet Things J. **5**(2), 955–969 (2018)
14. Idreesa, A.K., Al-Yaseenb, W.L., Taamc, M.A., Zahwe, O.: Distributed data aggregation based modified K-means technique for energy conservation in periodic wireless sensor networks. In: Proceedings IEEE Middle East and North Africa Communications Conference (MENACOMM), Jounieh, Lebanon, pp. 1–6 (2018)

15. Rao, S., Shorey, R.: Efficient device-to-device association and data aggregation in industrial IoT systems. In: Proceedings 9th International Conference on Communication Systems and Network (COMSNETS), Bangalore, pp. 314–321 (2017)
16. Dong, L., Wang, G.: INADS: in-network aggregation and distribution of IoT data subscription in ICN. In: Proceedings IEEE International Conference on Multimedia & Expo Workshops (ICMEW), Hong Kong, pp. 321–326 (2017)
17. Bhandari, S., Sharma, S.K., Wang, X.: Latency minimization in wireless IoT using prioritized channel access and data aggregation. In: Proceedings of the International Conference on IEEE Global Communications Conference, Singapore, pp. 1–6 (2017)
18. Yu, T.: Recursive principal component analysis based data outlier detection and sensor data aggregation in IoT systems. IEEE Internet Things J. 4(6), 2207–2216 (2017)

A Hybrid Approach for Credit Card Fraud Detection Using Naive Bayes and Voting Classifier

Bhagwant Jot Kaur[(✉)] and Rakesh Kumar

National Institute of Technical Teachers Training and Research,
Panjab University, Chandigarh, Sector 26, Chandigarh 160019, India
jotdhanoa26@gmail.com, raakeshdhiman@gmail.com

Abstract. The prediction analysis is the approach which is applied to predict future possibilities from the current data. One need not pay cash, the cardholder just gives his card to the shopkeeper, he will swipe the card and the payment will be done in just fractions of a second. Sometimes the credit card can be stolen, and then the attacker can make false transactions. In this type of conditions, the credit card company faces a huge loss. In the existing research work, the voting-based classification approach is applied for credit card fraud detection. The voting based classification is a combination of multiple classifiers like SVM, decision tree etc. The classifier will have maximum accuracy will display its predicted result. To improve the accuracy of prediction analysis the voting based classification method will be replaced with naïve bayes classification approach. The naïve bayes classifier is the probability based classifier for credit card fraud detection. In the probability based classification method, the probabilities of the target classes are calculated and the probability of the test data is calculated. The test set which is near to the probability class is identified as the target set. The naïve bayes classification approach will improve the accuracy of credit card fraud detection. The proposed methodology will be implemented in python and results will be analyzed in terms of accuracy, precision, recall and F-measure. The naïve Bayes classifier optimizes the results in terms of accuracy, precision, recall and f-measure is optimized up to 10 to 15% for the credit card fraud detection.

Keywords: Voting based classification · Hybrid · Ada boost · Naïve Bayes · Decision tree · K-NN classifier · SVM

1 Introduction

A large amount of data is generated every day and managing that large data is not an easy task to perform. This generated data is stored in big sized databases which can be retrieved by the user anytime and anywhere. Large-sized repositories and databases [1] are available which are specially designed to store data. But the retrieval of important data from such a large quantity of data is very challenging. There are various types of tools are introduced for the extraction of useful data from such a large collection of data. Researchers have introduced techniques like machine learning, pattern

© Springer Nature Switzerland AG 2020
A. Pasumpon Pandian et al. (Eds.): ICCBI 2019, LNDECT 49, pp. 731–740, 2020.
https://doi.org/10.1007/978-3-030-43192-1_81

recognition, neural network and several other techniques for the extraction of data [2]. The phenomenon by which the data can be extracted and stored in large-sized repositories and databases from anywhere and anytime is called data mining. The unused data is extracted from this large collected information and can be used for performing further tasks and functions with the help of data mining. It is the part of KDD, which extract the residual data from the databases and transform it into useful and reusable data. Credit card services are very popular in today's time [3]. It is very used to purchase goods and services anywhere and anytime. One need not pay cash, the cardholder just gives his card to the shopkeeper, he will swipe the card and the payment will be done in just fractions of a second. Sometimes the credit card can be stolen, and then the attacker can make false transactions. In this type of conditions, the credit card company faces a huge loss. If the attacker wants to make an online transaction, he will only require little [4] Information about the cardholder. Very fast and secure internet connection is required to make online payment through credit cards. Credit cards provide many facilities like one need not carry a large amount of cash if he wants to buy something very expensive. It makes all the transaction very easy and very time-saving. It has a disadvantage that, if the card is stolen by someone then he will not get notified about this unless he will get his transaction statements [5]. Usually, the attacker is very smart having good knowledge of hacking; he will stop the transaction statement unless he will end all the monthly limit of the credit card. To minimize this situation, one needs to make use of various credit card detection techniques. Every current business institution is facing the problem of credit card frauds today. Differentiating the methods of frauds being executed is important such that their effective solutions can be defined. Credit card fraudsters are very smart [6] and intelligent; they have very good knowledge of programming by which they can hack the account of the cardholder very easily. The attacker has a great advantage because the cardholder or the issuer of the card don't know that their card is being stolen and being used by the third party. Also, it has a huge possibility that the card can be used for any illegal purpose [7]. There are two broader categorizations of various techniques developed for detecting credit card frauds that are User Behavior Analysis or Anomaly detection and Fraud Analysis or Misuse Detection. It is very important to know the basic difference between user behavior analysis and fraud analysis. The analysis method is used to detect the tricks which are already known with a low rate of positivity. All the signatures are properly checked and no alarm will [8] be raised if test data does not contain any fraud signature. There are some credit card detection techniques being used by the companies to minimize their loss. Artificial Neural Network is a set of nodes which are connected with each other and specially designed to imitate the working of the human brain. Every node has an important connection with the adjacent node. The individual nodes receive the input from the adjacent node and use it to have a simple working to compute the output values. A genetic algorithm was first postulated by John Holland. It is an optimized type of a solution which traditionally denoted as chromosomes which are in the form of binary strings [9]. The main principle is to state that the survival chances of stronger members in any population are high. Hidden Markov method is the definite tool used for the modeling of time series data. This technique is used in almost every face recognition, credit card fraud detection and heart diseases prediction methods.

2   Literature Review

Dhankhad et al. [10] proposed a supervised machine learning algorithm, the main objective of this algorithm was to delineate hidden patterns and they are further used. As the technology is getting advanced the hacker are also getting more skilled. This gives rise credit card fraudsters to rob the bank and account of the individual with the more advanced techniques. In this paper, various supervised machine learning algorithms are used to identify fraud with the help of real-world dataset. These algorithms are applied to the super classifier by making use of ensemble learning methods. It may also lead to higher accuracy in credit card fraudulent transactions are of detecting frauds. The researcher concluded that it is successfully used to measure the performance of various algorithms and is compared with other classifiers.

Zheng et al. [11] proposed a logical graph of BP (LGBP) technique that helped in identifying the relevancy of attributes that contain any kinds of transactional details in them. The user can be calculated each attributes transaction probability on the basis of the path. It also describes the information about the entropy-based diversity coefficient which is defined characteristics of the diverse transaction behaviour of the user. Additionally, the proposed technique is used to make a transition probability matrix and hence defines the capture temporal features. Therefore, the researcher concluded that the proposed approach is constructed for each and every user and used to verify the transaction is fraud or genuine. The performed experiment reveals that the proposed dataset is much more effective than the three-state of the art one.

Gahlaut et al. [12] proposed a data mining classification model for the detection of credit card fraudsters. There are different types of loan credibility provided to the bank which ensures the transaction made by each and every customer. In this paper, a novel approach has been proposed which helps in making the correct decision regarding the approval or rejection of the load request of the customer in terms of different detailed information related to the customer. This approach is used to analyze the behaviour of the customer and the payback credit crated by him is studied which ensures the classification of the load as good or bad. The researcher concluded that in order to analyze the performance various types of classifiers are used like decision tree, support system, and support vector machine and so on.

Kho et al. [13] proposed (Europay MasterCard-VISA) proliferation chip card used to design credit card for business use only. It is used to solve the problem based on magnetic stripe card technology. In this paper, the researcher suggests that it is possible to avail the detection model such that any kinds of the anomalous transaction can be captured. Different types of classifiers were used to calculate the performance of the model but the Random Tree and J48 results in the highest accuracy of 94.32% and 93.50%. The researcher concluded by the analysis of the two classifiers that the transaction logs of data can be understood in a better way through J48.

Kavitha et al. [14] proposed Meta-classifier technique is used for the detection of fraud. It is a very vast area of research for the present as well as for the future also. This arises due to the increasing and everlasting demands of the customer and the active or smart fraudster present in the market. In this paper, the researcher has described the common and powerful meta-classifier which have the capacity to operate on two different processing layers. Every layer having specialization in the prediction of a single class and resultant of both the layer are divided to have a final prediction outcome. Different experiments were performed and the results are compared with Fraud Miner and Enhanced Fraud Miner. Also, the experiments determine the increased performance of the developed TBMC model.

Carcillo et al. [15] proposed an Assessment of Streaming Active Learning Strategies which is used to improve the fraud detection accuracy. Moreover, this technique is the exploitation/exploration of the tradeoff which is being highlighted for fraud detection. This technique depends on the real-world dataset having millions of transaction being given by the industrial partner World line. The researchers have performed various experiments for the analysis of the performance which show the importance of previously proposed approaches to improve the working of complementary machine learning techniques.

Randhawa et al. [16] made use of machine learning algorithms for detecting frauds related to credit card. Initially, the use of typical models was carried out. Afterward, hybrid techniques were implemented for detecting fraud. This approach was the combination of AdaBoost and majority voting techniques. An openly existing data set of credit card was employed in this work for evaluating the efficiency of the recommended paradigm. Further, the analysis of a realistic data set derived from a bank was carried out. Moreover, in order to evaluate the powerfulness of the approaches, the addition of noise was done into the data samples. The tested outcomes optimistically demonstrated that the majority voting technique performed will in terms of accuracy for detecting frauds related to credit card.

3   Research Gaps and Problem Formulation

Following are the various research gaps in this work:-

1. Credit card fraud detection is the major challenge of the prediction analysis. The credit card fraud detection dataset is very complex due to which accuracy is low for the prediction.
2. The dataset for the credit card fraud detection has a large number of attribute and it is difficult to establish a relation between attribute set and target. When the relation is not established properly it is difficult to train the model accurately.
3. To accurately train the model a large amount of data is required for the prediction analysis. A large amount of data has missing and redundant values which reduce the efficiency of the dataset.

Prediction of future instances based on existing datasets is known as prediction analysis in which clustering and classification are integrated. Grouping the similar and dissimilar type of data is known as clustering. The approach that assigns class by grouping the data based on their similarity is known as classification. Performing prediction analysis on the risky credit information is the major concern of this research work. The credit card frauds detection is the major issue in recent time. The two steps are generally applied for the credit card fraud detection which is feature extraction and classification. To detect credit card frauds, several classifiers are applied such as linear regression, decision tree, SVM, neural networks and random forest in the base paper. In this research work, the novel classifier will be implemented for the credit card fraud detection which should have high accuracy than existing classifiers. The proposed method will be the probability-based classification method for credit card fraud detection. In the probability-based classification method, the probability of all classes will be calculated and the test set will be classified into the class which is closed to the probability set. The probability classifier will increase the accuracy for the credit card fraud detection.

4 Proposed Work

The predictive modeling can be performed using any of the large-scale datasets available today either from social web-sites, business modeling, etc. The predictive modeling of datasets has been gaining popularity amongst the academics stakeholders due to the improvement in the level of communication and information. The clients are viewed based on new insights with the help of predictive modeling analysis that help in providing innovation in academic organizations. The huge amount of data is generated within every application which is to be stored in large sized databases. To manage these large databases, as well as the data available within them, highly efficient tools, are required. The researchers have proposed various techniques to facilitate several users and their applications. Data mining and knowledge discovery are two such approaches that can easily manage such a large amount of data. An automated approach that is applied to identify interesting patterns from huge databases is known as data mining. The data that already exists within the databases are used to generate descriptive, understandable and predictive models through data mining.

A probabilistic classifier is used here to recognize any suspicions such that the true values are not affected and this is known as Naïve Bayes Classifier. The numeric properties are shown using the ordinary circulation within this approach. The directed discretization is used to deal with numeric properties. The strategy using which the characterization display is generated such that the class names can be appointed for issuing the occurrences, the vectors of highlight esteems are generated through the pointer is known as Naïve bayes. Here, from some limited set, the class marks are ventured. The calculation that is based on normal requirements is also collected by this classifier. For the given class marks, the component is estimated using this bayes classifier. For appraising the parameters that are used to perform classification, a small measure of prepared information is used by the Naïve Bayes classifier.

To perform input data acquisition, the historical data of the last 10 years related to credit cards is used. The information such as which information is to be drawn, at what time and from which location define the transactions made which are all included within the historical data. To perform classification, the proposed model is designed in this research work by collecting the regression models. To perform long term prediction, historical data is used within the processing. This research work uses naïve bayes classifier to perform classification of input data into the normal and fraud transactions (Fig. 1).

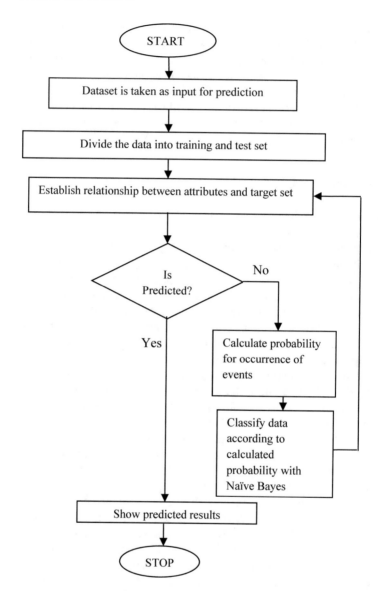

Fig. 1. Proposed flowchart

Dataset description

Attribute	Value
Dataset characteristics	Multivariate
Number of instances	30000
Attribute characteristics	Integer, Real

(*continued*)

(continued)

Attribute	Value
Associate tasks	Classification
Number of attributes	24
Missing values	No
Area	Business
Date donated	2016-01-26
Number of web hits	368365

5 Experimental Results

The proposed research is implemented in Python and the results are evaluated by comparing proposed and existing techniques in terms of various performance parameters such as accuracy, recall, execution time and precision.

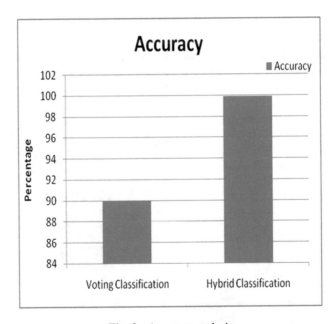

Fig. 2. Accuracy analysis

As illustrated in Fig. 2, the accuracy of voting based classification method is compared hybrid classification model. The hybrid classification model is more accuracy as compared to voting based classification method for credit card fraud detection.

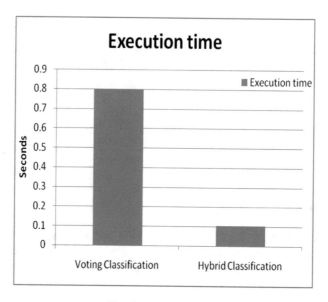

Fig. 3. Execution time

The execution time of the voting and hybrid classification is compared in the Fig. 3. The execution time of the hybrid classification method is low as compared to voting based classification method for credit card fraud detection.

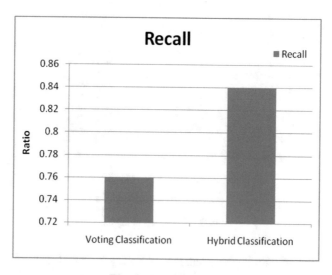

Fig. 4. Recall analysis

As illustrated in Fig. 4, the recall of voting based classification method is compared hybrid classification model. The hybrid classification model is more recall as compared to voting based classification method for credit card fraud detection.

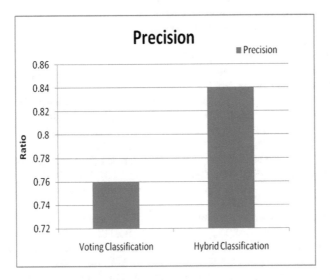

Fig. 5. Precision analysis

As illustrated in Fig. 5, the precision of voting based classification method is compared hybrid classification model. The hybrid classification model is more precise as compared to voting based classification method for credit card fraud detection.

6 Conclusion

Important information can be extracted from large amount of raw data using data mining. To predict the future possible conditions based on the present data, the prediction analysis method is applied. Predicting unknown data values with the help of known values is the main task of predictive models. Credit cards provide many facilities like one need not carry large amount of cash, if he want to buy something very expensive. It makes all the transaction very easy and very time saving. This research proposed a novel approach where naïve bayes classifier will be applied for the credit card fraud detection. The outputs achieved show that in comparison to the existing approaches, the proposed method that uses Naïve Bayes classifier provides improvement in the accuracy of detecting credit card frauds. Performance of the work has been executed using parameters like accuracy, recall, execution time and precision.

Future Scope
In future, novel approach for the classification can be designed which will be the combination of supervised and unsupervised learning techniques.

References

1. Duman, E., Ozcelik, M.H.: Detecting credit card fraud by genetic algorithm and scatter search. Expert Syst. Appl. **38**(12), 13057–13063 (2011)
2. Raj, S.B.E., Portia, A.A.: Analysis on credit card fraud detection methods. In: International Conference on Computer, Communication and Electrical Technology – ICCCET, vol. 19, no. 15, pp. 212–243, March 2011
3. Sahin, Y., Duman, E.: Detecting credit card fraud by decision trees and support vector machines. In: International Multi-conference of Engineers and Computer Scientists, vol. 56, no. 22, pp. 131–200, March 2011
4. Sunita, S., Chandrakanta, B.J., Chinmayee, R.: A hybrid approach of intrusion detection using ANN and FCM. Eur. J. Adv. Eng. Technol. **3**(2), 6–14 (2016)
5. Singh, A., Narayan, D.: A survey on hidden Markov model for credit card fraud detection. Int. J. Eng. Adv. Technol. (IJEAT) **1**(3), 2249–8958 (2012)
6. Esakkiraj, S., Chidambaram, S.: A predictive approach for fraud detection using hidden Markov model. Int. J. Eng. Res. Technol. (IJERT) **2**(23), 156–166 (2013)
7. Ogwueleka, F.N.: Data mining application in credit card fraud detection system. J. Eng. Sci. Technol. **6**(3), 311–322 (2011)
8. Ibrahim, L.M.: Anomaly network intrusion detection system based on distributed time-delay neural network (DTDNN). J. Eng. Sci. Technol. (JESTEC) **5**(4), 457–471 (2010)
9. Ogwueleka, F.N., Inyiama, H.C.: Credit card fraud detection using artificial neural networks with a rule-based component. IUP J. Sci. Technol. **5**(1), 40–47 (2009)
10. Dhankhad, S., Mohammed, E.A., Far, B.: Supervised machine learning algorithms for credit card fraudulent transaction detection: a comparative study. In: 2018 International Conference on Information Reuse and Integration for Data Science, vol. 16, no. 19, pp. 12–30, April 2018
11. Zheng, L., Liu, G., Yan, C., Jiang, C.: Transaction fraud detection based on total order relation and behavior diversity. IEEE Trans. Comput. Soc. Syst. **5**(3), 56–63 (2018)
12. Gahlaut, A., Tushar, Singh, P.K.: Prediction analysis of risky credit using data mining classification models. In: 8th International Conference on Computing, Communication and Networking Technologies (ICCCNT), vol. 5, no. 76, pp. 45–50, June 2017
13. Kho, J.R.D., Vea, L.A.: Credit card fraud detection based on transaction behavior. In: Proceedings of the 2017 IEEE Region 10 Conference, TENCON, vol. 55, no. 67, pp. 25–35, March 2017
14. Kavitha, M., Suriakala, M.: Real time credit card fraud detection on huge imbalanced data using meta-classifiers. In: Proceedings of the International Conference on Inventive Computing and Informatics (ICICI), vol. 45, no. 76, pp. 34–40, April 2017
15. Carcillo, F., Le Borgne, Y.-A., Caelen, O., Bontemp, G.: An assessment of streaming active learning strategies for real-life credit card fraud detection. In: International Conference on Data Science and Advanced Analytics, vol. 55, no. 67, pp. 23–30, August 2017
16. Randhawa, K., Loo, C.K.: Credit card fraud detection using AdaBoost and majority voting. IEEE Access **6**(13), 14277–14284 (2018)

Quantum Inspired Evolutionary Algorithm for Web Document Retrieval

Manas Kumar Yogi$^{(\boxtimes)}$ and Darapu Uma

Pragati Engineering College, Kakinada, India
manas.yogi@gmail.com, umadarapu03@gmail.com

Abstract. This paper describes the importance of quantum inspired evolutionary algorithm for web document retrieval. As size of web increases day by day the popular methods for web document retrieval are taking much time thereby making users concerned about it. Our approach takes inspiration from evolutionary algorithms to model a mechanism which returns experimental results with acceptable accuracy. We have seen from trial results that after some time the general capacity of proposed strategy is superior to the best two most well known methods of web document retrieval. The primary bit of leeway of utilizing our proposed calculation is that its presentation doesn't debase regardless of whether the search space is variable. Irrespective of whether search space is expanding or diminishing the exhibition doesn't drop for our proposed procedure. We discuss comparative performance of the proposed technique with other popular techniques in this area and conclude the paper with future scope of fine tuning the performance of the proposed mechanism with addition other parameters effecting the performance in the mechanism. In future we intend to refine the propose system to suit diverse web prerequisites like customized web search and furthermore for web document retrieval from centralized server databases as well.

Keywords: Quantum · Nature inspired · Evolutionary · Web · Retrieval · Semantics

1 Introduction

Evolutionary computing is one of the fundamental subsets of natural computing strategies for solving huge scale complex and real global problems. Nature inspired computing builds on the principles of emergence, self-employer and complicated systems [1]. The not unusual purpose on this course is to plan theoretical model, which can be implemented in computer systems, faithful enough to the herbal mechanisms investigated. Observe of natural phenomena, procedures or even theoretical models [2] for the improvement of computational systems and algorithms able to solving complex troubles. To provide opportunity stochastic, nature inspired search based strategies to issues which have not been (satisfactorily) resolved with the aid of conventional deterministic algorithmic strategies, along with linear, non-linear, and dynamic programming and so forth. Evolutionary algorithms may be termed as search based totally stochastic optimization algorithms advanced with the foundation of the evolution's

© Springer Nature Switzerland AG 2020
A. P. Pandian et al. (Eds.): ICCBI 2019, LNDECT 49, pp. 741–747, 2020.
https://doi.org/10.1007/978-3-030-43192-1_82

organic techniques. The main move of algorithms evolved within the EA domain are, genetic algorithms, evolutionary strategies [3], genetic programming and evolutionary programming [4, 5]. In the remaining decade, with the emergence of quantum computing as a new computing paradigm, exploiting quantum-mechanical phenomena to perform computations, quantum stimulated evolutionary algorithms [6] have also advanced. The open troubles and most important challenges in this area are: To layout, excessive-performance evolutionary algorithms [7, 8], integrating the domain knowledge into the algorithms.

2   Evolutionary Algorithms

Evolutionary algorithms can be termed as search based totally stochastic optimization algorithms evolved with the inspiration of the evolutionary organic processes including selection, recombination and mutation.

2.1   Quantum Computing

Quantum computation is a studies place that is built up on the principles of quantum mechanics along with uncertainty, superposition, interference, and entanglement to system information [9]. Quantum computer systems construct at the concepts of quantum mechanics were proposed in 1980s. Lively studies considering that then has tested that algorithms designed for quantum computers are more effective for fixing complicated issues than traditional algorithms designed for digital computers [10, 11].

2.2   Quantum Inspired Evolutionary Algorithms (QIEAs)

Quantum inspired evolutionary algorithms have important features:

1. Applying of q-bit illustration, to explain sample of a huge population. Q-bit illustration presents probabilistically a sequential superimposition of a couple of states [12].
2. Applying of q-gate because the evolutionary operator, which could guide the people in the direction of better answers and to generate the people for the next era [13].

In Search based software engineering (SBSE), the intention is to design software engineering issues as optimization issues which can then be countered with computational search [14]. This has proved to be a widely applicable and a hit technique, with applications from necessities and layout, to maintenance and trying out [15]. Computational search has been exploited through all engineering disciplines [16], now not just software engineering. Some of the problems which have been addressed by means of QİEA researchers are next release trouble and check statistics technology [17]. Therefore, it gives a extraordinary possibility for the researchers to apply the QİEA variants to resolve complex Search based totally software program engineering problems.

3 Proposed Objectives

1. To experiment and analyse the overall performance in standard and replace operators for QİEA specifically.
2. To explore the opportunity of hybridizing swarm optimization and QIEA.
3. To apply QIEAS to precise search based totally software engineering optimization issues.

3.1 Artificial Life Agents

The web authors have a tendency to categorise the files as per the subjects and join them in associated subject matter. This process has consequences while showing semantic network. The network showing the semantics defines the co-relationship of documents. Suppose that few of the documents are relevant to customers, there is very high opportunity that the weblinks in contemporary report are applicable to customers also. The proposed retailers can decrease the search space by using the usage of such topologies.

3.2 Proposed Algorithm

Our proposed QIEA for web document retrieval contains following steps:

- Step 1: Initialise
- Step 2: Observe
- Step 3: Evaluate
- Step 4: Update
- Step 5: Iterate

Step 1: Initialise
Initialise the population Wij where $i = 1,2,\ldots\ldots n$, n is the population size i.e.; the total number of web pages from which web document is to be retrieved. $j = 1,2,\ldots\ldots\ldots,q$ where q is number of q-bits per solution/web document retrieved. In the first generation, equal values are assigned to α(probability measure 0 to find a web document) and β(probability measure 1 to find a web document) of all the q-bit, so that $|\alpha|_2 + |\beta|_2 = 1$.

Step 2: Observe
Observe all the q-bits. Here, the q-bit is collapsed into '0' or '1' state.
If $|\beta i|_2 > $ rand() then observed state will be '1' representing web document is retrieved successfully else observed state will be '0' represented no web documents are found. Rand() ϵ [0,1].

Step 3: Evaluate
In this step, the fitness of each observed solution, yi, is calculated and stored according to a specific objective function. Also, the best solution 'Bs' in all the solutions obtained until the current generation is stored along with its fitness value.

Step 4: Update
Compare every bit of B's and Y's and find the difference in α, β analogous to each q-bit. Schemes like q-gates can be applied here.

Step 5: Iterate
Repeat the steps 1 to 4 till the highest number of generations are obtained (or) termination state is reached. Each generation may have one (or) more web documents.

The objective function whose fitness value has to be maximized for our algorithm is combination of various factor. The first and foremost is user profile. The user profile is made up of relevancy links. Relevancy is the majority of words linked closely to users interest.

For each web link, 'l' in a web document, the relevancy is determined as f shown below:

$$Rel = \sum \frac{match(t, q)}{path\ distance(t, l)} \tag{1}$$

Where t is tokens in a document, q is set of given user queries.
Match(t,q) = 1 if t is in q else it becomes 0

- If t = 1 then Q bit = 1.
- If t = 0 then Q bit = 0.

The next factor effecting the objective function to be maximised is vitality of a web document. We define vitality of web document with respect to a user query as the most recent activity done on a web page. For instance, if we consider two pages with vitality values V1 and V2 where V1 is vitality of document D1 and V2 for D2. If some activity is done most recently on D1 compared to D2 then, V1 > V2. In this case Q bit for D1 = 1 and Q bit for D2 = 0. D1 is returned as solution with more fitness value. We should always update user profile after every search result is displayed to the user. Person profile have to concur with user's hobbies. As the web dealers learn about the consumer's hobbies by means of accessing consumer's queries and comments, it is critical to update user profile after looking. The updated person profile consists of applicable record URL's. With this assets, person can customise the retailers as he gives queries repeatedly.

The below diagram shows that framework for proposed QIEA for web document retrieval. In this diagram the input query is given to the user interface and applying different QIEA techniques, it divides the documents with various categories. These QIEA techniques have been maintained by user history and any changes presented in this are updated by updated user profile. The global semantic user interests will be given to document retriever. Finally all the ordered documents will be placed in a categorywise (Fig. 1).

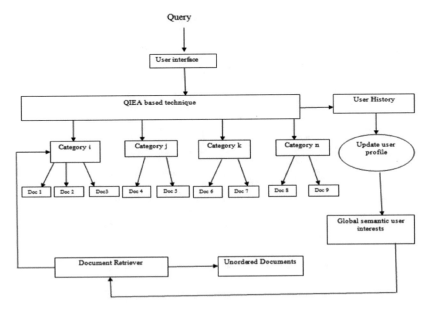

Fig. 1. Schemantic framework for proposed QIEA for web document retrieval

4 Experimental Results

Experiments were conducted to observe the validity of the three mechanisms: (1) relevance feedback, (2) client profiles dependent on unadulterated perusing history, and (3) QIEA method. As we realize that clients need to give input unequivocally in importance criticism, clients don't need to give any exertion in our proposed techniques (2) and (3) since our proposed framework will directly store changes in user's search profile. The experimental setup involved a technological platform with Java on a

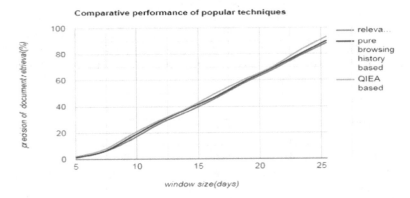

Fig. 2. Comparative performance of popular techniques with QEIA technique

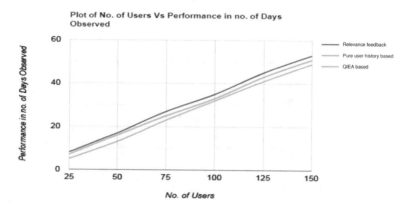

Fig. 3. Plot of No. of users Vs Performance measure in no. of days observed

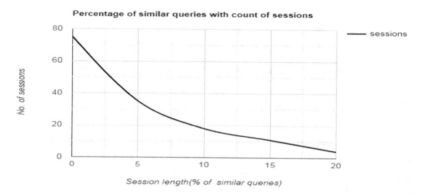

Fig. 4. Plot of session length with percentage of similar queries Vs No. of Sessions

workstation (CPU: Dual core 2, Memory: 4 GB, OS: Windows8). We then plotted the performance of these three methods over 25 days against the percentage of web documents retrieved. We have found that over a period of 15 days the three methods performed with almost equal ability. But after 20 days duration, our proposed QIEA technique outperformed the other two methods by around 3 to 4%. By the end of 25 days the retrieval rate of our proposed algorithm was higher than the other two popular techniques (Figs 2, 3 and 4).

5 Conclusion

To increase the performance index of existing web document retrieval based on user profile we have applied the quantum based evolutionary algorithms.rom the experimental results we can observe that our proposed QEIA technique performs at a rate which is nearly 3% more than the other two popular techniques for web document retrieval, i.e. relevance feedback and pure browser history based web document retrieval. In future we want to extend the observational period to longer duration and verify if the percentage increase in retrieval is increasing or remaining stable. The main advantage of using our proposed algorithm is that its performance does not degrade even if the search space is variable. Irrespective of whether search space is increasing or decreasing the performance does not drop for our proposed technique. In future we plan to refine the propose technique to suit different web requirements like customized web search and also for web document retrieval from mainframe databases too.

References

1. Kumari, A.C., Srinivas, K., Gupta, M.P.: Software requirements selection using quantum-inspired elitist multi-objective evolutionary algorithm. In: Proceedings of International Conference on Advances in Engineering, Science and Management (ICAESM 2012), Nagapattinam Tamil Nadu India, 30–31 March 2012, pp. 782–787 (2012)
2. Ding, S., Jin, Z., Yang, Q.: Evolving quantum circuits at the gate level with a hybrid quantum-inspired evolutionary algorithm. Soft. Comput. 12(11), 1059–1072 (2008)
3. Zhang, G.-X.: Quantum-inspired evolutionary algorithms: a survey and empirical study. J. Heuristics 17(3), 303–351 (2011)
4. Han, K., Kim, J.: Quantum-inspired evolutionary algorithm for a class of combinatorial optimization. IEEE Trans. Evol. Comput. 6(6), 580–593 (2002)
5. Feng, X., Wang, Y., Ge, H., Zhou, C., Liang, Y.: Quantum-inspired evolutionary algorithm for travelling salesman problem. Comput. Methods 6, 1363–1367 (2006)
6. Ali, S., Briand, L.C., Hemmati, H., Panesar-Walawege, R.K.: A systematic review of the application and empirical investigation of search based test-case generation. IEEE Trans. Softw. Eng. 36, 742–762 (2010)
7. Back, T.: Evolutionary Algorithms in Theory and Practice. Oxford University Press, New York (1996)
8. Shou, L., Bai, H., Chen, K., Chen, G.: Supporting privacy protection in personalized web search. IEEE Trans. Knowl. Data Eng. 26(2), 453–467 (2014)
9. Spertta, M., Gach, S.: Personalizing search based on user search histories. In: Proceedings of the IEEE/WIC/ACM International Conference on Web Intelligence (WI) (2005)
10. McKinney, V., Yoon, K., Zahedi, F.: The measurement of web-customer satisfaction: an expectation and disconfirmation approach. Inf. Syst. Res. 13(3), 296–315 (2002)
11. Fang, X., Holsapple, C.: An empirical study of web site navigation structures' impacts on web site usability. Decis. Support Syst. 43(2), 476–491 (2007)
12. Pitkow, J., Schütze, H., Cass, T., Cooley, R., Turnbull, D., Edmonds, A., Adar, E., Breuel, T.: Personalized search. Commun. ACM 45(9), 50–55 (2002)
13. Chen, W.-K.: Linear Networks and Systems, pp. 123–135. World Scientific, Belmont (1993)
14. Qiu, F., Cho, J.: Automatic identification of user interest for personalized search. In: Proceedings of the 15th International Conference on World Wide Web (WWW), pp. 727–736 (2006)
15. Duncombe, J.U.: Infrared navigation—Part I: an assessment of feasibility. IEEE Trans. Electron Devices ED-11(1), 34–39 (1959)
16. Shen, X., Tan, B., Zhai, C.: Implicit user modeling for personalized search. In: Proceedings of the 14th ACM International Conference on Information and Knowledge Management (CIKM) (2005)
17. Carter, C.J.: Oxygen absorption in the earth's atmosphere, Aerospace Corp., LosAngeles, CA, Technical report, TR-0200 (4230–46)-3, November (1988)

Signature Recognition and Verification Using Zonewise Statistical Features

Banashankaramma F. Lakkannavar[1], M. M. Kodabagi[2],
and Susen P. Naik[3(✉)]

[1] CSE Department, Government Polytechnic,
Navanagar Sector 43, Bagalkot 587103, India
[2] Department of Computer Science and Engineering,
Reva University, Bengaluru 560064, India
mmkodabagi@gmail.com
[3] Department of Computer Science, K.L.E. Institute of Technology, Hubli, India
susennaik55@gmail.com

Abstract. The signature is defined as special kind of handwriting that includes special characters and flourishes. Handwritten signatures are most accepted individual attribute for individuality authentication of the person. This provides novel method for the signature recognition and verification by using zone based statistical features. It contains mainly two phases. During the first phase, the knowledge base is constructed by training samples using the zone wise statistical features. During second stage i.e., testing phase, the processed image is obtained having zoning wise statistical features and signature is recognized using neural network classifiers. An accuracy rate of 97.5% is achieved by testing 200 samples. MATLAB is used for designing this signature recognition and verification system and is robust and avoids noise, blur and change in size, lightening conditions and other possible degradation.

Keywords: Signature recognition and verification · Signature images · Zone wise statistical features · Neural network classifier

1 Introduction

The signature defined as a unique kind of writing, includes extraordinary characters and superfluities that can verify depending on the frame of mind and tiredness. Handwritten signature can be used as a means of authentication which is most extensively accepted attribute for individuality authentication of the person. The signer of the document can be identified by signature of the person which plays an important role for identifying the signer of a written document, considering inherent thought that any person's usual signature change slowly, it is very hard to remove, modify or do some forgery without any detection.

There are two main types in the signature recognition and verification process. Those are divided into mainly two groups as online and offline signature recognition and verification systems. Signatures are recorded while writing process, using special instruments, such as pen or tablet in online (dynamic) signature recognition and verification.

© Springer Nature Switzerland AG 2020
A. P. Pandian et al. (Eds.): ICCBI 2019, LNDECT 49, pp. 748–757, 2020.
https://doi.org/10.1007/978-3-030-43192-1_83

In online system, the dynamic information with the facilities of calculation of velocity, acceleration and pen pressure is equipped. In off-line which is also called as Static recognition system, signature images are captured by a digital camera or a scanner. It is considered challenging for an offline signature recognition and verification in today's life. In the on-line method, the dynamic features of the signatures are considered such as such, as the order of handwriting, change in writing speed, and skill needs to be improved from the gray level pixels. But in the off-line signature verification, only static features of the signatures are available. The absence of the actual person at the time of verification doesn't create any problem is one of the main advantage of offline system. The banking transactions and document verification applications require offline mode of signature verification.

The signature recognition and verification system are broadly alienated into four sub tasks namely preprocessing the input image, feature extraction of image, recognition and verification using the classifier. Preprocessing algorithm contains binarization algorithm which gives processed data for the feature extraction process. Feature extraction is defined as extracting the special characteristics of the input raw data which is further used for classification stage. 'Global features and local features' are the two type of features that can be extracted. The phase of recognition has two parts such as 'training the neural network' and 'testing the neural network' which is done by a back propagation method of neural network. Several methods designed in this pattern recognition area, uses the offline signature recognition and verification.

This work presents method for signature recognition and verification which works in two phases, which are training and testing. While doing training, the training sample signature images that are monitored to extract zone based statistical features are stored into the knowledge base. The features of training signature images that are stored in database are fed to the neural network for the purpose of training. The trained network are later used for recognition and verification task. Further, during process of testing a sample test image is observed for feature extraction and recognized using neural network classifier. The method is evaluated for 200 images with recognition accuracy of 97.5%. The method is efficient for signature recognition and verification which takes care of variation in size, noise and blurriness, complex structure and other degradations.

2 Related Works

Many methods have been presented for signature recognition and verification where every method employs different strategy. Review of some prominent solutions is presented.

A robust approach for signature recognition and verification is proposed in [1] which represents a brief review of various approaches based on different data sets, features and training techniques used for verification. This system extract the global, local, geometric, mask and grid features and neural network as a classifier where they use a Back propagation algorithm and linear network, MLP with one hidden layer and two hidden layers for purpose of training model.

Neural network is used as main concepts here [2]. Different image processing methods are used for working on signatures being verified. This system uses global features, grid features, mask and moment invariant methods and neural networks. This method uses back propagation algorithm and artificial neural network.

The technique used for signature verification and recognition system is described in [3]. The network is given the inputs from directional feature and energy density that classifies the signature. The neural network is used as a classifier for this system. After observations it is concluded that if there are limited samples available for training then directional feature extraction method performs better than energy density method.

The model for signature verification and recognition is described in [4] which describes the off-line signature recognition and verification that makes use of neural network. The Invariant Central and Modified Zernike moment is used. MATLAB is used for building this signature recognition & verification model.

The methodology for signature verification and recognition system is presented in [5] which represents the signature verification and recognition using the Multilayer perceptrons, having a new approach that makes the user to check whether the signature under testing is original or not. This method have four features such as eccentricity, skewness, kurtosis, and orientation. The neural network model is trained using an algorithm called back propagation.

The method for signature verification and recognition system is described in [6] where a problem of personal authentication is described. This work, introduces an off-line signature recognition and verification model, that makes use of the method called moment invariant and an ANN classifier. Among off-line methods the following techniques used are 2D transforms, histograms of generated directional data, curvature, projections like horizontal and vertical of the available hand writing trace of the signature and structural way, in which the best results are given by HMM approach.

The approach for signature verification and recognition is described in [7] presenting an off-line signature recognition and verification model which makes use of methods like moment invariant method with ANN classifier. There are two separate networks created of which one will be used for signature recognition where as second will be used for the process of verification. Moment invariant vectors are obtained and at least it implements signature recognition and verification.

3   Proposed Methodology

The proposed model is the zone wise statistical features for signature images for various staff members and students of the department with noise, blurriness, variations in size, varying lightening conditions and other degradations. The method comprises various phases such as preprocessing of image, feature extraction, knowledge base construction from previous stage and finally signature recognition and verification. The block diagram shown in the Fig. 1.

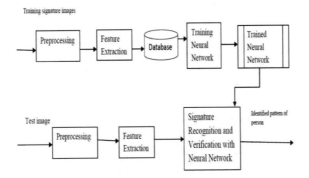

Fig. 1. Proposed block diagram for signature recognition and verification system

3.1 Image Acquisition

Handwritten signatures are obtained from various staff members and students on A4 size papers on different days and at different timings. The collected signatures are scanned to get images in JPEG format and are stored into database.

3.2 Preprocessing

Quality of the image is improved in this stage making it suitable for feature extraction. The purpose is to make signature ready for feature extraction and to improve the quality of the image. This stage involves grayscale conversion, binarization, image resizing and applying bounding box.

3.3 Feature Extraction

Here zoning method is used to extract zone wise statistical features. The preprocessed image is resized to 100×300 which is divided into 100 zones of which each can be of size 10×30. The two feature_vector and rsum_vector are used to store the computation of required statistical features from all zones. A feature value is calculated by summing up all pixels in every possible zone. Finally 100 features are stored into feature vector rsumvector as shown in Eq. (1).

$$rsumvector = [f_i] \quad Where \ 1 \leq i \leq 100 \tag{1}$$

f_i - Feature value of ith zone. It is computed as shown in Eq. (2).

$$f_i = \sum_{1}^{10} \sum_{1}^{30} g_i(X, Y) \tag{2}$$

Where g_i is the ith zone that contains 10×30 region of the signature image. Further construction of knowledge bases uses the dataset from trained samples.

3.4 Construction of Knowledge Base

Knowledge Base is constructed by obtaining the signatures from various staff members and students on A4 size papers and the collected signatures are scanned to get images in JPEG format and are stored into database. The database consists of 200 signature images and among these 80% of samples are used for training.

During training knowledge base can be constructed as dataset of required feature vectors after extracting the features for all training as given in Eq. (3).

All samples containing 80% trained and 20% untrained samples are tested.

The extracted zone wise statistical features of the training signature images are stored in the Knowledgebase KB as shown in Eq. (3).

$$KB = \left[rsumvector_j \right] \quad where\ 1 \le j \le N \tag{3}$$

KB - Knowledgebase which contains feature vectors taken from training samples. $rsumvector_j$ - Feature vector of jth image in KB.

3.5 Training and Signature Recognition and Verification with Neural Network Classifier

Once the dataset is obtained and organized into knowledge base of signature images, the neural network classifier trains and recognizes and then verification tasks are carried out. The details of training and recognition and verification are described in the following:

The extracted zone wise statistical features of training signature images that are stored in database are trained by feeding them to the neural network. The trained network will be later used for recognition and verification task.

In the testing stage, zone wise statistical features are obtained by selecting the input test signature image from testing set and preprocessing it and stored into feature_vector rsum_vector using the above Eq. (3). Then the signature is recognized and can be classified by using neural network classifier. The variations in size, noise and blurriness and other degradations are taken well care by the proposed methodology. However sufficient amount of training of all possible changes in signature size, directions and other noise is necessary for the method.

4 Experimental Results and Analysis

The proposed system has been evaluated for 200 signature images of varying size, style, noise, blur, uneven thickness and other degradations. The experimental results of processing a sample signature images is depicted. The results of different approaches are also depicted in further section.

4.1 An Experimental Analysis of a Sample Signature Image

When any figure is given with the signature image that has noise, blur, uneven lighting conditions, and other degradations which is initially preprocessed for binarization, bounding box generation, resized to a standard size of 100 × 300 as shown in Fig. 2a, b, c and d.

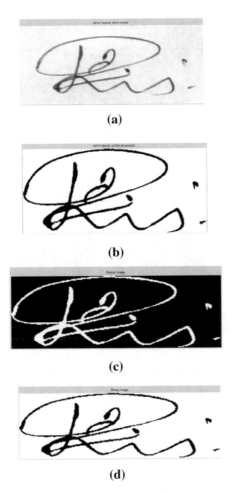

(a)

(b)

(c)

(d)

Fig. 2. a. A sample signature test image [5] **b.** Test image with a boundingbox **c.** Resized test image 100 × 300 **d.** Binarized image

For processing the image has to be divided into 10 vertical zones and 30 horizontal zones so that zone wise statistical featured can be computed for images and put into feature_vector and rsum_vector.

The values in Table 1 clearly show that, the feature values in the corresponding zones of the signatures are distinct. For example, the feature values 300, 600, 900, 1200, and 1500 of zones 1, 2, 3 and 4 of signature in the first row of Table 2 are distinct from feature values 300, 596, 846, 1092, 1362 in the corresponding zones of signature in the second row of Table 1. The similar characteristics exist with the feature values in other zones. A pixel distribution pattern is created by arrangement of these features into a feature vector that makes samples distinguishable. The extracted features from test image in Fig. 2a are stored in knowledge base KB and are determined using neural network classifier.

Table 1. A sample signature showing zone wise statistical features. [5]

Signature image	Zone wise statistical features
	300 600 900 1200 1500 1800 2090 2364 2620 2920 3220 3512 3812 4112 4412 4712 4976 5244 5514 5814 6096 6369 6669 6912 7197 7460 7714 8002 8292 8547 8814 9088 9373 9627 9907 10177 10477 10753 10997 11284 11540 11799 12064 12343 12637 12937 13194 13458 13758 14058 14319 14578 14843 15121 15415 15664 15947 16247 16547 16847 17129 17414 17694 17985 18235 18526 18826 19126 19426 19726 20015 20298 20584 20837 21133 21433 21723 22023 22323 22623 22898 23162 23419 23715 24015 24315 24615 24915 25215 25515 25780 26071 26361 26639 26939 27239 27539 27839 28139 28439
	300 596 846 1092 1362 1618 1858 2146 2446 2746 3046 3297 3577 3877 4177 4477 4777 5039 5339 5639 5914 6207 6487 6787 7052 7347 7647 7926 8226 8526 8806 9106 9386 9686 9949 10243 10538 10801 11101 11396 11676 11976 12256 12536 12782 13065 13310 13606 13906 14187 14469 14769 15049 15298 15526 15767 16067 16367 16667 16967 17220 17457 17701 17931 18210 18462 18762 19062 19362 19662 19962 20230 20479 20731 20979 21246 21546 21831 22091 22391 22691 22947 23213 23478 23756 24013 24308 24568 24848 25128 25428 25691 25988 26278 26540 26836 27081 27375 27640 27940

The system performance analysis of signature recognition and verification is described in the Table 2.

Table 2. System performance analysis

The total number of signature images tested	Total number of signatures recognized	Accuracy	Error rate
200	195	97.5%	2.5%

The system performance analysis of signature recognition and verification is described in the graph as depicted by the Fig. 3.

Fig. 3. System performance analysis

4.2   An Experimental Analysis Dealing with Various Issues

Promising results for images containing signature with different size, noise, blur, variable lighting conditions have been given by the proposed methodology. An average recognition and verification accuracy of 97.5% is been achieved by the proposed work being robust and reliable. Table 3 gives the overall performance of the system.

Experiments revealed that incorrect results occur due to noise, blur, more similarity between signature structures and other noise. Zone wise statistical features manage variations in appearance of signature images. Better performance is obtained by training the knowledge base.

Table 3. Performance of system with various images of dataset [4, 7, 8]

Signature image	Number of samples tested	Number of samples recognized correctly	Number of samples misclassified	% of recognition accuracy
	10	9	1	90
	10	10	0	100
	10	10	0	100
	10	10	0	100
	10	10	0	100
	10	10	0	100
	10	10	0	100
	10	9	1	90
	10	9	1	90
	10	10	0	100
	10	10	0	100
	10	9	1	90
	10	10	0	100
	10	10	0	100
	10	10	0	100
	10	10	0	100
	10	9	1	90
	10	10	0	100

5 Conclusion and Future Work

The paper proposes, a method for signature recognition and verification employing zone wise statistical features. In the proposed system zone wise statistical features and neural network are being used as a classifier to recognize and verify signatures. There are two phases in system working, i.e. training and testing phase. The effectiveness of zone wise statistical features are analyzed by some exhaustive experimentations using neural network. It has been observed that the zone wise statistical features are extracted using zoning algorithm and are found to be efficient, robust and insensitive to noise, blur, variable lighting condition and size variation in the signature. Expected results have been given by the combination of zone wise statistical features and neural network. 200 signature images have been tested by this method and an accuracy of 97.5% is been achieved. But neural network takes considerably more time for processing increased number of persons in the database.

Also by increasing the number of trained images for each person and by considering a new set of features the accuracy of signature recognition and verification can be increased. This project can be extended to having an analysis where a new set of features of signature image and classification algorithms are to be added to the feature vectors used in this project to get better accuracy. Also, more reliable signature recognition and verification system can be built through this project where forged signatures can be recognized and verified.

Informed Consent: Informed consent was obtained from all individual participants included in the study.

References

1. Anand, H.: "Realtive study on signature verification & recognition system. Int. J. Innov. Res. Adv. Eng. (IJIRAE) **1**(5) (2014). ISSN 2349 - 2163
2. Abikoye, O.C.: Offline signature recognition & verification using neural networks. Int. J. Comput. Appl. (0975 – 8887) **35**(2), 44–51 (2011)
3. Tomar, M.: A directional feature with energy base offline signature verification networks. Int. J. Soft Comput. (IJSC) **2**(1) (2011)
4. Choudhary, N.Y.: Signature recognition & verification system using back propagation neural networks. Int. J. IT Eng. Appl. Sci. Res. (IJIEASR) **2**(1), 1–8 (2013). ISSN 2319-4413
5. Odeh, S.: Apply multi large perceptron neural network for off-line signature verification & recognition. IJCSI Int. J. Comput. Sci. Issues **8**(6), 261 (2011). No 2, ISSN 1694-0814
6. Kalenova, D.: Personal authentication using signature recognition. IEEE **85**, 1437–1462 (2004)
7. Cemil, O.Z., Ercal, F., Demir, Z.: Offline signature recognition & verification with artificial neural networks. IJETAE **3**(8), 19 (2014)
8. Adebaya Daramola, S.: Offline signature recognition using Hidden Markov model. Int. J. Comput. Appl. (0975 – 8887) **10**(2), 17–22 (2010)
9. Sthapak, S.: Artificial neural networks based signature recognition & verification. Int. J. Emerg. Technol. Adv. Eng. **3**(8), 191–197 (2013)

10. Pushalatha, K.N.: Offline signature verification with random & skilled forgery detection using polar domain feature & multi stage classification regression model. Int. J. Adv. Sci. Technol. (IJAST) **59**, 27–40 (2013)

11. Kumar, P.: Hand written signature recognition & verification using neural networks. Int. J. Adv. Res. Comput. Sci. Softw. Eng. **3**(3) (2013). ISSN 2277 128X

12. Bharadi, V.A.: Offline signature recognition system. Int. J. Comput. Appl. (IJCA) (0975 - 8887) **1**(27) (2010)

13. Kumar, M.: Signature verification using neural network. Int. J. Comput. Sci. Eng. (IJCSE) **4**(09), 1498 (2012)

14. Gill, R., Singh, M.: Statistical features based off line signature verification system using image processing techniques. Int. J. Sci. Res. (IJSR) **1**(3) (2012)

15. Angadi, S.A., Kodabagi, M.M., Jerabandi, M.V.: Character recognition of Kannada text in low resolution display board images using zone wise statistical features. In: IEEE World Congress on Information and Communication Technologies, pp. 61–66 (2012)

16. Kodabagi, M.M., Kembavi, S.B.: recognition of basic Kannada characters in scene images using euclidean distance classifier. Int. J. Comput. Eng. Technol. (IJCET) **4**, 632–641 (2013)

Home Security Using Smart Photo Frame

Zarinabegam K. Mundargi[1]([✉]), Isa Muslu[2], Susen Naik[3],
Raju A. Nadaf[4], and Suvarna Kabadi[5]

[1] Department of Computer Science, Ala-Too International University,
Bishkek, Kyrgyzstan
zarina.mundargi@iaau.edu.kg
[2] Department of Applied Mathematics and Informatics,
Ala-Too International University, Bishkek, Kyrgyzstan
isa.muslu@iaau.edu.kg
[3] Department of Computer Science and Engineering, KLE.IT, Hubli, India
susennaik55@gmail.com
[4] Department of Computer Science and Engineering, Government Polytechnic,
Bagalkot, India
raj.enggs@gmail.com
[5] Department of Computer Science and Engineering, Government Polytechnic,
Zalaki, India
suvarnaky1977@gmail.com

Abstract. The security and privacy issue have become worse with the advancement of the technology. The security is a major issue that needs to be addressed. The task of providing home security is taken up in this paper and a methodology is proposed for the same. In most traditional method of providing security to the home, the people tend to use CCTV cameras. Although it's a passive method of providing security it is highly in use. But, the cameras may be visible to the intruder, thief or robbers. So, cameras are the first units that may be possibly damaged during such bad incidents. Usually most of the intrusion detection systems click the photo of intruders as soon as the human is detected. But, such systems fail to click photos, so that the intruder can be identified (Face View Photo) and visible. In proposed system, a proper illustrative method is proposed so that the intrusion detection is accurate and photos thus captured are clear, so that intruder is clearly visible. The proposed system is designed as a photo frame for providing home security. The system is developed to accept mobile commands. As soon as the intrusion of unauthorized person is detected, an alert message and the identifiable, clear photo (Face view) of intruder will be sent to owner's/administrator's mobile. The photo frame owner can also see the video of deployed environment through the camera fitted on the frame.

1 Introduction

The technological advancements have increased the comfort zone of human beings. Life is made easier with the incorporation of technology in day to day activities. The growth of technology also brings some surprises along with it. The surprises may be helpful and sometimes harmful. The new technology and new horizon of techniques are coming up. In such situation providing security becomes a major responsibility and a

© Springer Nature Switzerland AG 2020
A. P. Pandian et al. (Eds.): ICCBI 2019, LNDECT 49, pp. 758–768, 2020.
https://doi.org/10.1007/978-3-030-43192-1_84

difficult work as well. For example, people use ATMs for doing banking transactions, the security of device and the working accuracy of device matters a lot. Conversely, with the advancement in technology, it is evident that skills of thieves, robbers also have increased. Hence, it is a great challenge to design a proper secure system.

Security cameras are used in order to secure home. The cameras record the activities that occur and store in the storage device. Whenever incidents take place, only after analyzing the videos, the actions can be taken in case of any harmful or suspicious incidents or activities. There is also a possibility that, the cameras can be destroyed during incidents like theft or robbery. Hence, complete security cannot be ensured in such cases. This is the most common way of providing security that presently exists. But, this system is passive approach. There is need for an active system, which can take actions as soon as security threat is detected or suspected. Hence, the proposed system, which can provide security during human intrusion in home, is proposed. For this a photo frame is designed, which does human intrusion detection. There are several ways to implement the proposed model. One can use sensors or image processing techniques. IR or PIR sensors like DHT22 can be used. Even Image processing techniques such as frame difference can be used for designing the proposed system. But, the sensors have more false alarm rate and accuracy of image processing techniques for proposed model is also less. Hence, a model is proposed which makes use of machine learning technique like Haar cascade classifier of the OpenCV and also face detection techniques. The following Fig. 1 shows the features that are implemented in the proposed model.

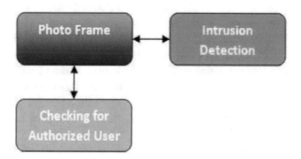

Fig. 1. Features of the proposed model

The Fig. 2 shows the schematic diagram of the proposed model. One camera is fitted on the top portion of the Photo Frame.

In order to serve the purpose and provide hassle free security, there is a need for a secure method. Hence, an effort is made in this paper to design a system for home security using Raspberry Pi. System requires Raspberry Pi, Camera and android mobile as hardware components. Python, Node.js and OpenCV libraries are used for software coding. A photo frame is designed to detect the human intrusion, which is fitted with the camera, Raspberry and provided with battery backups.

The device can be controlled by the registered mobile by the administrator of the device. The camera fitted on the photo frame will be continuously capturing the videos

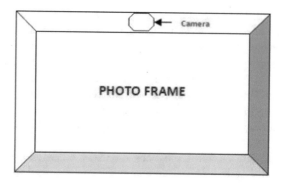

Fig. 2. Schematic diagram of proposed model

and will be monitoring the video for any human intrusion with the help of the raspberry. As soon as the intrusion is detected, the raspberry is programmed so that the camera has to click the face photo of the intruder. The photo of the intruder is compared against the photos of the authorized users stored in the database. If match is not found then an alert message is sent to owner's mobile along with photo of the intruder. This can be used as a proof during the legal proceedings as well.

The proposed Photo Frame can be used as a decorative material of the room. When owner will not be available at home, the same Photo Frame can be put into intrusion detection mode by mobile command then it will do Human Intrusion Detection. As soon as the intrusion is detected, the face visible photo of the intruder is clicked and checked whether the person is authorized or not. The photos of authorized users stored in the database. If an unauthorized user is found at that time, an alert message can be sent to registered mobile number along with the photo and photo is stored in the cloud. The actions can be taken accordingly. Entire mechanisms take place without the notice of the intruder. The Photo Frame which looks like any regular Photo Frame will not attract an attention of the thief or intruder also. So, here an active security method is proposed to protect the home against intrusion. Major importance has been given for the accuracy of the system and also for taking the clearphotos.

The Sect. 2 explains the existing work taken up in the specified field. The Sect. 3 explains the Issues and Challenges of the proposed system. The Sect. 4 explains the detailed description of the Proposed Model along with necessary block diagrams. The Experimental aspects related to the Human Intrusion Detection are explained in Sect. 5. The details of the Results and Analysis by considering different scenario are explained in Sect. 6. The Sect. 7 briefs about the Conclusion and scope of future work for this paper.

2   Related Work

Some considerable works have been taken up in this field. Smart devices have been proposed which work as passive devices with very less interactivity.

In [1], an intelligent mirror which accepts voice commands via the microphone and which is built with Raspberry Pi microcontroller, LED monitor and acrylic mirror, which displays the weather, time, and location information on the screen, is proposed. The software components of the system are SD formator, Etcher, and the Raspbian OS. It uses VNC viewer for connectivity. In [2], a Smart mirror designed with Raspberry Pi, microphone, speaker, LED Monitor, One way mirror and display units to display latest updates of news and weather, is proposed. Humid and Temp sensors are used along with cloud storage. In [3], a voice based Smart Mirrors are designed. The system makes use of Artificial Intelligence concepts. For example, it can give a message on the screen saying "Wear Jacket", if the weather is cloudy. In [4], a webpage based Smart mirrors are proposed which can be customized as per the user needs and are voice operated. In [5], mirrors are proposed for health tracking based on weight and fitness. Face Recognition based authentication, Bluetooth connectivity, SONUS and GPS navigation are added features. In [6], mirrors use Hermoine 1.0 which is another form of Magic Mirror. Smart mirror is provided for domestic use.

In [7], the system is proposed which works in two modes namely regular Mode and Activated Mode. Python and Javascript are used as programming tools along with Node.js which accepts voice commands. In [8], a voice controlled, wall mirror is designed and is named as Magic Mirror. It stores the personalized data for decision making and prediction. In [9], a voice controlled, mirror is developed. Smart Mirror is implemented with Raspberry, LCD monitor, microphone, webcam and SMT32 microcontroller as main controlling chips. The mirror accepts voice commands and processes them. In [10], smart mirror is having capability of speech recognition. Face recognition is provided for login and authentication purpose. The system wakes up by touch and accepts voice commands.

In [11], smart mirror proposed is having two striking features. It displays the map and architecture of the college. It can be used by the students to see course details and also can be used as register. In [12], a mirror is proposed for Theft Detection in a home environment. It displays the real time data on the mirror. Sensors like PIR (Passive Infrared) are used for detection of human and the camera is used to capture photos and later put in a drop box. In [13], a qualitative comparison Smart Mirrors is given. New voice based system which supports Human Gestures and Face Detection is proposed. In [14], a smart mirror is proposed to monitor children. The system can be connected to the user's mobile using an android application.

3   Issues and Challenges

Working with embedded system is a major challenge. A good knowledge of hardware and compatible software is also necessary. Raspberry is a small processing general purpose computing device with limited capacity in terms of the storage as well as processing. It is provided with GPU as well. Due to less RAM of only 1 GB, handling OpenCV libraries becomes most challenging and needs intelligent programming. Clicking face visible photo of intruder and comparing it with the photos of the authorized users is equally challenging task. The model has to be trained for Face Detection and Face Identification aswell.

The other technical issues related to the proposed system are the cost and durability of the hardware devices used. Raspberry heats up and starts rebooting repeatedly.

4 Proposed System

The diagram for the proposed model is shown in Fig. 3. The diagram depicts the modules used in the proposed work.

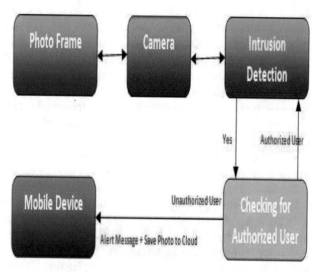

Fig. 3. Block diagram of proposed model

The proposed system is designed with hardware components like Photo Frame, Raspberry Pi, Camera, Power bank (works in case of Power failure) android mobile and Wi- Fi connectivity. The software components needed are Python, OpenCV libraries, PythonAnywhere cloud services and SMSLane servers. The Photo of intruder can be uploaded to the Amazon cloud services such as PythonAnywhere. The photo will be converted to base64 format and stored in the database. Such format conversion is necessary for easy storage and retrieval of the photos.

The system can be pushed into Intrusion Detection Mode by mobile commands. During this mode the system will act as a human intrusion detection system. During this mode of operation, as soon as the human is detected, the system will take a frontalphoto of the human and checks whether the person is authorized to use system or not and if the person is found unauthorized user then system will send alert message and photo of the intruder and send it to the administrator/owner of the Smart Mirror through SMS. The Photos of the intruders are stored in a cloud called PythonAny-where. Further, the owner can take actions based on the situation.

The model accepts mobile command as input and processes them to produce results. Also mobile commands can be used even when user is away from the photo frame. The details are shown on Fig. 4. The figure shows the possible ways in which input is provided to the system.

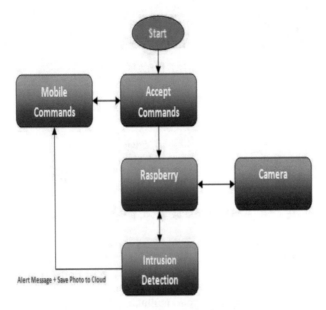

Fig. 4. Input methods

The schematic diagram of the mobile display screen is shown in Fig. 5. The device can be pushed into Intrusion mode using the mobile. It can be stopped by using stop command. The normal video of the deployed environment can also be seen in the mobile.

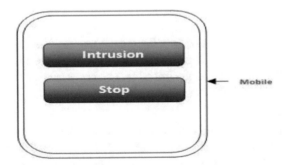

Fig. 5. Mobile menus for providing inputs

The detailed diagram of the proposed work is shown in Fig. 6. The figure shows the overall flow of the proposed work.

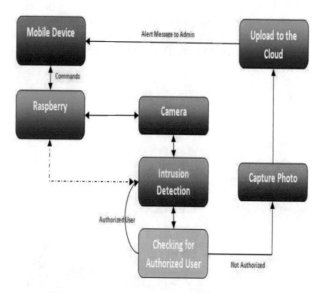

Fig. 6. Detailed diagram of proposed method

5 Experimentation

The Photo Frame is designed to detect human intrusion at deployed environment. During this process the system will try to maintain maximum possible accuracy of the system and to click the facial photo of the human under surveillance. This section explains the way in which the experimentation is carried up and the results are produced. The camera captures the video of the deployed environment. The video is then converted in to the frames and later frames are then analyzed for the presence of the Human Intruder.

The experimentation of the proposed work is shown in Fig. 7. The experimentation is done in two stages.

Face Detection
During stage -1 the Face Detection is being done. The face detection is done with the help of the Haar Cascade Classifier. The system is trained to detect the Frontal face and eyes of the intruder. Once the Human frontal face and eyes are detected then the system captures the photo of the intruder.

Face Identification
During stage -2 the Face Identification is being done. The face identification is done using the SVM (Support Vector Machine) technique. Before applying the SVM technique, the image captured during intrusion has to be cleaned by the morphological operations to enhance the quality and clarity of the photos. During this stage, the system is trained to identify the authorized users of the system. This is done by comparing the photo of the intruder against the photo of the authorized users. Faster and accurate identification can be easily done with the SVM technique.

Fig. 7. Experimentation proposed method

6 Results and Analysis

6.1 Face and Eye Detection Using Haar Cascade Classifier

In the process of increasing the accuracy of the proposed model, the Haar classifier is trained to detect either the face or eye of the human under observation. If either of eyes or face is visible then it evident that the intruder is properly facing the camera and a clear photo can be taken. Different cases have to be taken into consideration during detection process. The Fig. 8 shows the details of the results produced. The figure shows the detection of eyes and face of humans.

The results produced are shown in Table 1. The results are tested for the intrusion of single, two, three, four and above four persons. Every case is executed for 100 runs to gain as stable accuracy.

Soon after the facial photo of the intruder is detected and captured, next process is to identify it. The process of face identification is also tested with varying person number and for executing each case 100 times. The following results are produced during the Face identification against the photos of the authorized user photos (Table 2).

The accuracy can be further increased by applying other machine learning techniques. The overall performance of the proposed system depends on both the two stages involved with it. Tuning the stages for speed up can increase the accuracy further.

Fig. 8. Face and eye detection of human being

Table 1. Results of human face detection analysis using haar frontal face and eye.

No. of persons (t)	Frequency of execution (t)	Accuracy measured (t)
01	100	78.26
02	100	79.32
03	100	82.53
04	100	83.07
≥ 05	100	86.12

Table 2. Results of human face identification using svm technique.

No. of persons (t)	Frequency of execution (t)	Accuracy measured (t)
01	100	89.42
02	100	90.39
03	100	92.07
04	100	93.63
≥ 05	100	94.19

With the present system approximate of 86.12% accuracy for face detection and 94.19% for face identification is achieved.

7   Conclusion and Scope for Future Work

The need for an intelligent system exists that can smartly and silently protect the home. Invention is something that can make use of domestic products to behave beyond its capabilities and intelligence. Here an effort is made to secure home by using a photo frame and converting it into a smart photo frame. The collective results are that the system worked well in a deployed environment and gave an overall better accuracy.

The system thus designed can be used as package by adding further features in it. The same system can be added with the feature of emotion analyzer and set the room lights and music as per the mood of the administrator. The same work can be extended by adding the feature of controlling the appliances deployed in the home environment. The proposed system is unique because of the speed and accuracy. The system is designed for intrusion detection of humans only. The challenge of clicking the clear and visible image during, authentication exists.

References

1. Kumbhar, P.Y., Mulla, A., Kanagi, P., Shah, R.: Smart Mirror using Raspberry PI. Int. J. Res. Emerg. Sci. Technol. **5**(4) (2018)
2. Jagdish, A.P., Sonal, T.S., Sangaleharshada, D., Dokhale, A.: A review paper design and development of a Smart Mirror using Raspberry Pi. International J. Eng. Sci. Inven. (IJESI) **7**(4 Ver. I), 40–43 (2018). www.ijesi.org, ISSN (Online) 2319–6734, ISSN (Print) 2319–6726
3. Kiran, S.R., Kakarla, N.B., Naik, B.P.: Implementation of home automation system using Smart Mirror. Int. J. Innov. Res. Comput. Commun. Eng. **6**(3) (2018)
4. Ajayan, J., Santhosh Kumar, P., Saravanan, S., Sivadharini, S., Sophia, R.: Development of Smart Mirror using Raspberry-Pi 3 for interactive multimedia. In: 12th International Conference on Recent Innovations in Science and Management ICRISEM 2018 (2018)
5. Divyashree, K.J., Vijaya, P.A., Awasthi, N.: Design and implementation of Smart Mirror as a personal assistant using Raspberry PI. Int. J. Innov. Res. Comput. Commun. Eng. **6**(3) (2018)
6. Assudani, M., Kazi, A.S., Sherke, P.O., Dwivedi, S.V., Shaikh, Z.S.: "Hermione 1.0"- a voice based home assistant system. In: National Conference on Advances in Engineering and Applied Science (NCAEAS), 29 January 2018
7. Kamineni, B.T., Sundari, P.A., Suparna, K., Krishna Nayak, R.: Using Raspberry Pi to design Smart Mirror applications. IJETST **05**(04), 6585–6589 (2018). ISSN 2348-9480-2018
8. Chandel, S., Mandwarya, A., Ushasukhanya, S.: Implementation of magic mirror using Raspberry PI 3. Int. J. Pure Appl. Math. **118**(22), 451–455 (2018)
9. Sun, Y., Geng, L., Dan, K.: Design of Smart Mirror based on Raspberry Pi. JETST **05**(04), 6585–6589 (2018). ISSN 2348-9480
10. Jin, K., Deng, X., Huang, Z., Chen, S.: Design of smart mirror based on Raspberry Pi. In: 2018 2nd IEEE Advanced Information Management, Communicates, Electronic and Automation Control Conference (IMCEC 2018) (2018)
11. Akshaya, R., Raj, N.N., Gowri, S.: Smart Mirror- digital magazine for university implemented using Raspberry Pi. In: International Conference on Emerging Trends and Innovations in Engineering and Technological Research (ICETIETR) (2018)
12. Lakshmi, N.M., Chandana, M.S.: IoT based Smart Mirror using Raspberry Pi. Int. J. Eng. Res. Technol. (IJERT) **6**(13) (2018)
13. Mittal, D.K., Verma, V., Rastogi, R.: A comparative study and new model for Smart Mirror. Int. J. Sci. Res. Res. Pap. Comput. Sci. Eng. **5**(6), 58–61 (2017)
14. Siripala, M.B.N., Nirosha, M., Jayaweera, P.A.D.A., Dananjaya, N.D.A.S., Fernando, Ms.S. G.S: Raspbian magic mirror-A Smart Mirror to monitor children by using Raspberry Pi technology. Int. J. Sci. Res. Publ. **7**(12), 335 (2017). ISSN 2250-3153
15. Cvetkoska, B., Marina, N., Bogatinoska, D.C., Mitreski, Z.: Smart Mirror E-health assistant – posture analyze algorithm. In: IEEE EUROCON 2017, 6–8 July 2017 (2017). OHRID, R. MACEDONIA

16. Kulkarni, B.P., Joshi, A.V., Jadhav, V.V., Dhamange, A.T.: IoT based home automation using Raspberry PI. Int. J. Innov. Stud. Sci. Eng. Technol. (IJISSET) 3(4) (2017). www.ijisset.org, ISSN 2455-4863 (Online)
17. Khanna, V., Vardhan, Y., Nair, D., Pannu, P.: Design and development of a Smart Mirror using Raspberry PI. Int. J. Electr., Electron. Data Commun. 5(1), 63–65 (2017). ISSN 2320-2084
18. Jose, J., Chakravarthy, R., Jacob, J., Ali, M.M., Dsouza, S.M.: Home automated smart mirror as an Internet of Things (IoT) implementation - survey paper. Int. J. Adv. Res. Comput. Commun. Eng. 6(2), 126–128 (2017). ISO 3297:2007 Certified
19. Gomez-Carmona, O., Casado-Mansilla, D.: SmiWork: an interactive Smart Mirror platform for workplace health promotion. Int. J. Electr. Electron. Data Commun. 5(1) (2017)
20. Yusri, M.M.I., Kasim, S., Hassan, R., Abdullah, Z., Ruslai, H., Jahidin, K., Arshad, M.S.: Smart Mirror for smart life. Int. J. Adv. Res. Comput. Commun. Eng.
21. Nadaf, R.A., et al.: Smart Mirror using Raspberry Pi for human monitoring and home security. In: Third International Conference, ICAICR 2019, Shimla, India, 15–16 June 2019, Revised Selected Papers, Part II (2019). 978-981-15-0111-1_10
22. Nadaf, R.A., Bonal, V.M.: Smart Mirror using Raspberry Pi as a security and vigilance system. In: 2019 3rd International Conference on Trends in Electronics and Informatics (ICOEI). https://doi.org/10.1109/icoei.2019.8862

Automated System for Detecting Mental Stress of Users in Social Networks Using Data Mining Techniques

Shraddha Sharma[1], Ila Sharma[2], and A. K. Sharma[3($\boxtimes$)]

[1] R.N. Modi Engineering College, Kota, India
[2] Department of CSE, R.N. Modi Engineering College, Kota, India
[3] CSI, Kota University, Kota, India
drarvindkumarsharma@gmail.com

Abstract. This paper provides an automatic system for detecting mental stress of users in social networks and data mining algorithms have been applied here. Data mining is usually defined as the extraction of non-trivial implicit that are unknown previously and the most valuable information present in the data. It is commonly known as the knowledge discovery from the databases(KDD). In data mining, on examining data for recurrent then/if forms association rules could be formed through consuming Confidence & Support measures to detect most significant associations in the data. Support is exactly how regularly the items perform in the folder, while self-assurance is the sum of times then/if declarations are precise. In this automated system, firstly a set of stress-related textual, visual, and social attributes from various aspects are evaluated. We detected the correlation between the states of user's stress and their social interaction behavior in social networks by utilizing real world social media data. In this work, evaluated polarity of sentiments from social media data to identify phenomena of stress among the users. The proposed methodology provides the best performance results, when it compared to the existing methods.

Keywords: Data mining · Stress · Social media · Tweets · Sentiments

1 Introduction

Now days the number of mobile phone users are increased rapidly, which affects the human life and causes Psychological stress. Mental health is a leading cause of disability worldwide. It is observed nearly 300 million people suffer from depression [1]. Globally, approximately 450 million people world wide are mentally ill, in which the disease accounts for 13% of global disease burden [2]. The World Health Organization estimates that 1 in 4 individuals experiences mental disorders at any stage of their lives. Meanwhile, depression contributes 4.3% of the total global disease burden. In 2015, the Ministry of Health Malaysia found the mental health problems among Malaysians aged 16 and above were 29.2% or nearly an estimated population of 4.2 million Malaysians [3]. Alleviation provides good health, whenever stress act on peopole's health [4]. Within a gift day, state of affairs stress is rapidly growing. At first, a high profile accounts with a typical communicating profiles are shown that permit COMPA to

© Springer Nature Switzerland AG 2020
A. P. Pandian et al. (Eds.): ICCBI 2019, LNDECT 49, pp. 769–777, 2020.
https://doi.org/10.1007/978-3-030-43192-1_85

perceive cooperation with very low false positives. The method habitually extracts Twitter data that is real-time and exposes the varying forms of the public feeling in a prolonged period of time. It is probable to relate the public thoughts concerning a Twitter, subject or a hashtag user among various circumstances in the U.S. Users can decide on to perceive the whole emotion level of a time, as well as its emotion assessment on a particular time. Real-time consequences are conveyed constantly and envisioned over a web-based graphical user interface. Twitter has turn out to be a progressively more prevalent Microblogging facility that permits consumers to broadcast messages, otherwise known as tweets. It purposes as a stage for individuals to prompt themselves, which regularly conveys feelings on various topics. The usage of Twitter is exponentially increasing. There are 328 million dynamic consumers once-a-month on Twitter and about 500 million tweets are made each day. The hasty progress of Twitter and the public admittance of tweets have made a widespread investigation subject on twitter. For instance, investigators have studied the usage of Tweet in endorsing products and allocating customer view. Initiatives have considered the effectiveness of Twitter in information-gathering and administrative message. Likewise, tweets have been examined to sense seismic activity. Initially, the vernacular on Twitter is familiar. There might be misspelled slang abbreviations, and words in a tweet because of Twitter's familiar verbal style. Furthermore, each tweet consumes a length constriction of 140 characters maximum. Also, Twitter shields an extremely far-reaching kind of issues. Consequently, this method intended is for providing actual visions of the public emotion and presenting deviations over time. According to Kepios research [12], number of active users who dealt social networks increased by 13% in year 2017 to reach 3.3 billion in April, 2018 [13].

The following is the organization of the paper. The review of literature is covered in Sect. 2. The proposed methodology is discussed in Sect. 3. Simulation details are shown in Sect. 4. Finally, the Sect. 5 concludes the paper with future scope.

2   Review of Literature

In this section, we survey the literatures related to social networks and the problem of privacy in social networks, in order to highlight the importance of the addressed topic.

In [5] authors proposed adapting multi-SOM clustering approach for huge banking data was surveyed to improve research applications like business, science and engineering and also to construct valuable information from big data. This clustering approach was studied for reducing the time taken for execution. The number of clusters using R and Hadoop was determined.

In [6] authors introduced an artificial and Bayesian network for liquidity in banking to find approximate value of data. The efficiency, accuracy and flexibility of data mining methods in banking were estimated.

In [7] Proposed clustering as well as classification based data mining technique in banking sector which was utilized to reduce the risk of decision making rate and analyses the personal and potential loan customers. The performance analysis of k-means clustering method was evaluated.

In [8], authors have proposed data mining techniques in banking was evaluated to assist beneficial data analysis of banking and to evade consumer abrasion. The customer retaining was the significant aspect that was analyzed in current modest banking circumstance. The data mining procedures of banking sector like fraud hindrance and recognition and customer retaining was analyzed.

In [9], authors have introduced implementation of two stage k-means algorithm in banking system was recognized to improve services to the customers in the banks. The k-means algorithm analysis was evaluated to calculate and replicate the clustering process based on minimum Euclidean square distance to gain information.

In [10] deliberated the properties of text pre-processing techniques on the sentiment classification performance on the two kinds of classification tasks, and were summed up the prediction of six preprocessing techniques were utilized the two aspects models and four prediction on five tweet datasets. The experimentation was showed the F1-measure and accuracy of the twitter sentiment classification. The random forest classifiers and Naive Bayes classifiers are much sensitive than the SVM (Support Vector Machine) and LR (Logistic Regression) classifiers when several preprocessing techniques were applied.

In [14], authors introduced a hybrid sampling technique for data mining of unstable datasets in banking was proposed to demonstrate the effectiveness of Decision Tree (DT) and Support Vector Machine (SVM). The sensitivity of banking dataset is about 87.7 to 91.89%.

In [15], they introduced a review of collaborative filtering based on social recommender systems in network was proposed to improve accuracy of recommendation. The overview task of recommender system and traditional approaches was adopted to achieve additional input of improved accuracy. The two types of filtering like matrix factorization and social recommender system was evaluated. The performance analysis of social recommender system was estimated played a significant part in day to day life.

In [16], introduced social based collaborative recommendation using heterogeneous relation and proposed Hete-CF was proposed to detect arbitrary social network comprising event, location, and other types of heterogeneous information networks that were associated with social information. The outcomes on two real world data sets, DBLP and Meetup shows the efficacy and proficiency of this algorithm was estimated.

3 Proposed Methodology

This section discusses the proposed methodology and the necessary steps involved in this proposed system. The existing framework consumes more time and training dataset. Therefore it is drawback of the existing systems. The huge datasets are computationally demanding. Stress detection performance is low. Users have affected high psychological stress and it may exhibit low activeness on social networks. The aim of this automated system that increases classification performance and detection performance through the data mining approach. The flow diagram of proposed methodology is shown in Fig. 1.

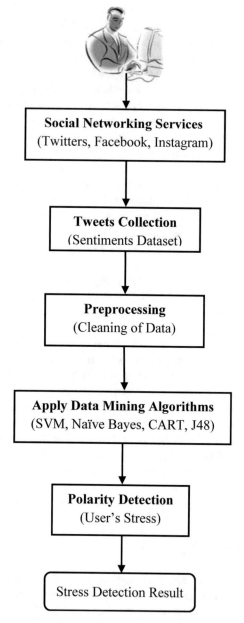

Fig. 1. Flow of proposed methodology

4 Simulation Details

This section shows simulation details of proposed research work. In this work, initially, the data are loaded from dataset and these are preprocessed to remove the irrelevant data. The random data mining algorithms are utilised to classify the datasets (Fig. 2).

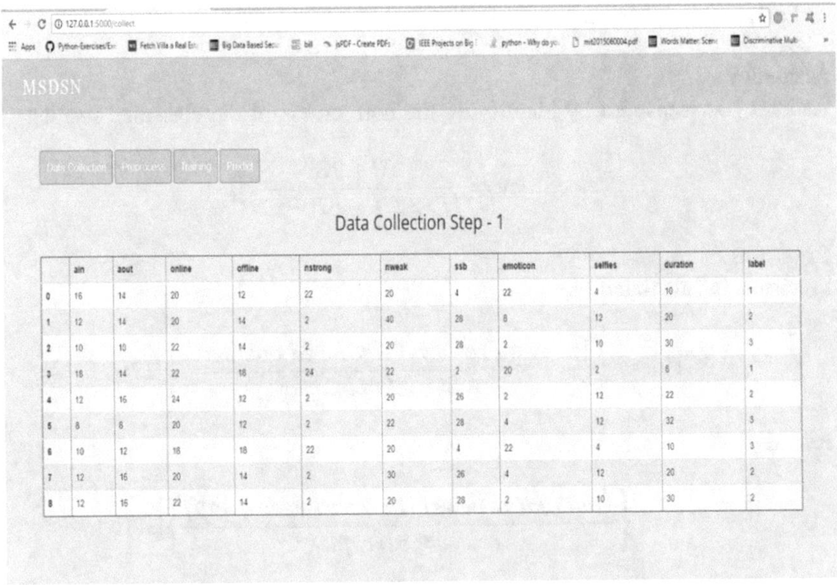

	ain	aout	online	offline	nstrong	nweak	ssb	emoticon	selfies	duration	label
0	16	14	20	12	22	20	4	22	4	10	1
1	12	14	20	14	2	40	28	8	12	20	2
2	10	10	22	14	2	20	28	2	10	30	3
3	18	14	22	18	24	22	2	20	2	6	1
4	12	16	24	12	2	20	26	2	12	22	2
5	8	6	20	12	2	22	28	4	12	32	3
6	10	12	18	18	22	20	4	22	4	10	3
7	12	16	20	14	2	30	26	4	12	20	2
8	12	16	22	14	2	20	28	2	10	30	2

Fig. 2. Data collection

Finally, the data are classified by using data mining classification technique. Also it presents the performance analysis of both existing and proposed techniques. Here, the existing and proposed method is analyzed with some parameter as accuracy, precision, recall and F1-measure (Fig. 3).

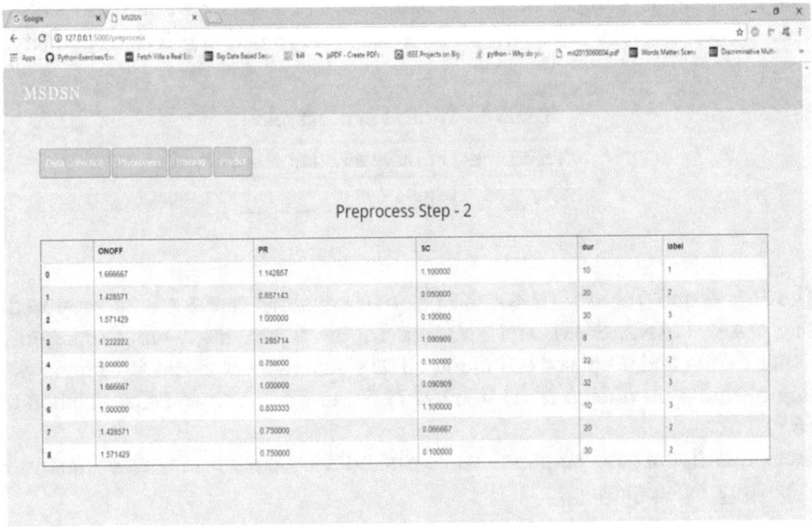

	ONOFF	PR	SC	dur	label
0	1.666667	1.142857	1.100000	10	1
1	1.428571	0.857143	0.050000	20	2
2	1.571429	1.000000	0.100000	30	3
3	1.222222	1.285714	1.090909	8	1
4	2.000000	0.750000	0.100000	22	2
5	1.666667	1.000000	0.090909	32	3
6	1.000000	0.833333	1.100000	10	3
7	1.428571	0.750000	0.066667	20	2
8	1.571429	0.750000	0.100000	30	2

Fig. 3. Data preprocessing

4.1 Performance Metrics

- **Accuracy**

  Accuracy is measured by identifying the correctness of the classified results.

  $$Accuracy = \frac{TP + TN}{TP + TN + FP + FN} \, [17]$$

- **Precision**

  Precision is calculated as,

  $$Pre = \left\{ \frac{\{relevant\ reviews \cap retrieved\ opinions\}}{retreived\ opinions} \right\} [17]$$

- **Recall**

  $$R = \left\{ \frac{\{relevant\ reviews \cap retrieved\ user\ query\}}{relevant\ user\ query} \right\} [17]$$

- **F1-measures**

  $$F1 - measure = \frac{2 \times Pre \times R}{Pre + R} [17]$$

4.2 Accuracy by Category

The accuracy measure is compared with both existing and proposed methods. The proposed method is compared with existing methods as DAN, DAN2, SVM, and SVM-of. These methods are analyzed for both positive and negative emotions.

Table 1. Positive emotion rate

Techniques	Positive emotion rate
Proposed	86.9

The Fig. 4 and Table 1 shows the positive emotions rate for both proposed and existing DAN, DAN2, SVM, and SVM-of. In this figure, the x –axis represents the different existing and proposed techniques. The y –axis shows the positive reviews. The existing methods are offered as 83.9, 69.7, 71.3 and 67.4. The proposed method offers the 86.9.

From this figure, the proposed work offered the better result, when analyzed to other existing techniques.

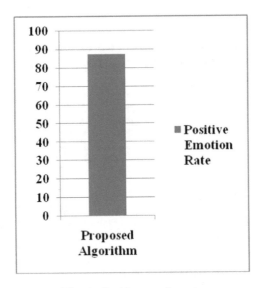

Fig. 4. Positive emotion rate

Table 2. Negative emotion rate

Technique	Negative emotion rate
Proposed	94.8

Table 2 and Fig. 5 shows the negative emotion rate for proposed algorithm. In this graph, the x –axis represents the proposed algorithm and y –axis shows the positive emotion rate. Here the proposed algorithm offers the 94.8 of negative emotion rate.

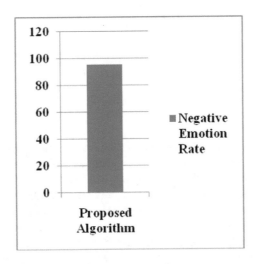

Fig. 5. Negative emotion rate

4.3   Accuracy Rate

The Table 3 and Fig. 6 shows the accuracy of proposed work. In this figure, the x –axis represents the proposed algorithm and y –axis shows the accurate rate. The proposed work gives 95.2% accuracy.

Table 3. Accuracy rate

Algorithm	Accuracy
Proposed algorithm	95.2

The Table 3 and Fig. 4 shows the accuracy of proposed work. In this figure, the x –axis represents the proposed algorithm and y –axis shows the accurate rate. The proposed work gives 95.2% accuracy.

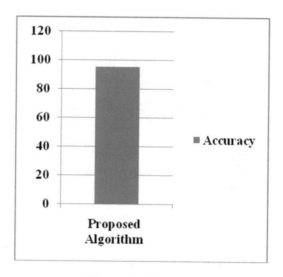

Fig. 6. Accuracy rate

5   Conclusion

In this paper, we presented an automated system for detecting user's stress states from users' social media data, leveraging tweets' content as well as users' social interactions collected weekly. We detected the correlation between the states of user's stress and their social interaction behaviour in social networks by utilizing real-world social media data. In this work, we also evaluated polarity of sentiments from social media data to identify the phenomena of stress among the social media users. In this paper, the performance of the proposed automated system is discussed and compared with the existing methods such as, IB1, J48, CART, SMO, Naïve Bayes, multi nominal MB,

Random forest, SVM. From the obtained results and graphs, it is clear that the proposed system achieves the better performance in terms of accuracy, precision, recall and F1-measure values. The obtained results of this research study proves the proposed method outperforms the overall techniques and offers better accuracy and could be more useful references for future related research studies.

References

1. World Health Organization: The World Health Report 2001, vol. 36, no. 10. WHO (2001)
2. Global Burden of Disease Study 2013 Collaborators: Global, regional, and national incidence, prevalence, and years lived with disability for 301 acute and chronic diseases and injuries in 188 countries, 1990–2013: a systematic analysis for the Global Burden of Disease Study 2013, Lancet, vol. 386, no. 9995, pp. 743–800 (2015)
3. Institute for Public Health (IPH) 2017: National Health and Morbidity Survey 2017 (NHMS 2017): Adolescent Mental Health (DASS-21) (2011)
4. Pillai, R.G., Thelwall, M., Orasan, C.: Detection of stress and relaxation magnitudes for Tweets. In: International World Wide Web Conference Committee ACM (2018)
5. Tavana, M., Abtahi, A.-R., Di Caprio, D., Poortarigh, M.: An artificial neural network and bayesian network model for liquidity risk assessment in banking. Neurocomputing **275**, 2525–2554 (2018)
6. Khanchouch, I., Limam, M.: Adapting a multi-SOM clustering algorithm to large banking data. In: World Conference on Information Systems and Technologies, pp. 171–181 (2018)
7. Calis, A., Boyaci, A., Baynal, K.: Data mining application in banking sector with clustering and classification methods. In: 2015 International Conference on Industrial Engineering and Operations Management (IEOM), pp. 1–8 (2015)
8. Chitra, K., Subashini, B.: Data mining techniques and its applications in banking sector. Int. J. Emerg. Technol. Adv. Eng. **3**, 219–226 (2013)
9. Babaie, S.S.: Implementation of two stages k-means algorithm to apply a payment system provider framework in banking systems. In: Artificial Intelligence Perspectives and Applications, pp. 203–213. Springer (2015)
10. Zhao, J., Gui, X.: Comparison research on text pre-processing methods on Twitter sentiment analysis. IEEE Access **5**, 2870–2879 (2017)
11. Singh, T., Kumari, M.: Role of text pre-processing in Twitter sentiment analysis. Proc. Comput. Sci. **89**, 549–554 (2016)
12. Kepios: Digital in 2018, essential insights into internet, social media, mobile, and ecommerce use around the world, April 2018. https://kepios.com/data
13. Marechal, C., et al.: Survey on AI-based multimodal methods for emotion detection. Springer LNCS 11400, pp. 307–324 (2019). https://doi.org/10.1007/978-3-030-16272-6_11
14. Sundarkumar, G.G., Ravi, V.: A novel hybrid undersampling method for mining unbalanced datasets in banking and insurance. Eng. Appl. Artif. Intell. **37**, 368–377 (2015)
15. Yang, X., Guo, Y., Liu, Y., Steck, H.: A survey of collaborative filtering based social recommender systems. Comput. Commun. **41**, 1–10 (2014)
16. Luo, C., Pang, W., Wang, Z., Lin, C.: Hete-CF: social-based collaborative filtering recommendation using heterogeneous relations. In: 2014 IEEE International Conference on Data Mining (ICDM), pp. 917–922 (2014)
17. Tyagi, E., Sharma, A.K.: Sentiment analysis of product reviews using support vector machine learning algorithm. Ind. J. Sci. Technol. **10**(35), 1–9 (2017)

Intelligent Request Grabber: Increases the Vehicle Traffic Prediction Rate Using Social and Taxi Requests Based on LSTM

S. C. Rajkumar[1]([⊠]) and L. Jegatha Deborah[2]

[1] Department of Computer Science and Engineering,
University College of Engineering Panruti,
Panikkankuppam, Panruti 607106, Tamil Nadu, India
rjkumar0814@gmail.com

[2] Department of Computer Science and Engineering,
University College of Engineering Tindivanam,
Melpakkam, Tindivanam 604001, Tamil Nadu, India
blessedjeny@gmail.com

Abstract. The integration of transportation system with Internet of Things (IoT) aims to effectively predict the vehicle traffic on a particular location. To optimize the transportation resources and learn about the public transportation system in the near future based on the transportation demand. We analyze the effect of machine intelligence on the transportation system; propose a new framework to improve the prediction accuracy on traffic. Mostly, transportation scheduling system based on historical data on usual routes, present system never involve realistic traffic situation or prolong changes on the system. This research work provides an efficient solution to avoid traffic and optimize the wastage of transportation resources. Social transportation data is pioneering attempt to acquire the very useful location based information that helps to understand the density of the human on geographical regions with mobility. In order to predict the real time traffic we collects the taxi request and convert the best sequence learning model, to predict continues data and analyze the dynamic traffic pattern in efficient manner. Experimental results achieve higher prediction accuracy of 92.4%.

Keywords: LSTM · Social transportation · Taxi request · GPS

1 Introduction

Nowadays traffic becomes an unavoidable state which is encounter by people during travel, it is very difficult to control the traffic in metropolitan city. Recent decades the researchers focus to avoid dynamic traffic scenario and to improve the optimization of the vehicle communication. With the influences of social media improve the traffic prediction based on their posts, feeds, and tweets. The keyword based information extraction is used to filter out the traffic from the help of learner to predict the traffic in efficient manner. The intelligent transportation system is interlinked with vehicle and human health, secure circumstances, eco system, and pollution control. The dynamic traffic prediction is very complex task, previous research collects historical data and provides a different prediction models but there are shallow methods. The real-time

© Springer Nature Switzerland AG 2020
A. P. Pandian et al. (Eds.): ICCBI 2019, LNDECT 49, pp. 778–788, 2020.
https://doi.org/10.1007/978-3-030-43192-1_86

traffic data is generated based on unpredictable instants but with the help of social media the traffic on particular location is determined before it occurs.

2   Related Works

Previous research conducted on demand on taxi prediction based on hotspots recommendation system for finding their affordable rides in efficient manner. The historical data collects information on timestamp of the request, origins and destination [1]. To find the higher predictability rate of taxi request implemented entropy model which validate the maximum predictability rate [2]. Using the time series model to estimate demand of the taxi in urbanized locations with real time implementation results shows better accuracy rate [3]. Predicting the future arriving of the passengers to modulate real-time traffic pattern using GPS system data processed with hidden Markov chain [3]. To meet the demand and supply and also minimize the idle of the taxi proposed a dispatching model to calculate the next time step of the demand using different samples and historical data [4]. Social networking platforms collect the people's ideas emotions and other location related information generates tremendous amount of data [5, 6]. Person's mobility information's collected from social network which induces to find the transportation sequence on particular location. The data were collected from wearable devices, and mobile phones [7]. Combined information on social and taxi request information is to feed to the memory based intelligent learner LSTM learns the sequential traffic pattern. Many applications widely used LSTM to learn the behavioral pattern of language model [8], speech recognition [9], and visual recognition [10]. The exciting traffic prediction models works usually shallow formulated in the dynamic transportation system [12–14]. We analyze the dynamic traffic pattern, efficient manner in deeper model to minimize the recourses of transportation fuel and minimize the usage of vehicles.

3   System Model

In this section, the system model of the traffic prediction problem is well formulated. This research mainly focuses, to acquire transportation data on different aspects which predict the traffic in a real-time. In the fastest world with t rapid development of sensing, machine intelligence, mobile network and social network platform generate huge volume of data. This type of sensitive data incorporates a person's emotions, movement, education and other personal data spread over the network. The location based social data indicates number of persons on particular place known the density of the human. From the social networking platform the transportation information's are acquired from people's shares, comments, chats and post and also the mode of transportation leads to predict the vehicle density of the location with particular time stamp. User's mobility information on taxi network holds the users request location data and past historical data to learn the traffic pattern with higher accuracy. The major source utilized information on past taxi rides which is collected from Global Positioning System (GPS) enabled taxis [15, 16]. This paper mainly focuses on understanding the complex traffic pattern to analysis the traffic congestion to optimize transportation infrastructure. The Intelligent learner collects the data from multiple recourses and analysis to extract the information of traffic sequences on particular location. This research work analysis the demand of vehicle in an effective manner so, that need a memory based model to remember the past events and also predict the future sequence of traffic. An intelligent based machine learning model LSTM learns the sequence of the traffic information to predict traffic on a particular region in efficient manner. To provide reliable and optimized public transportation system proposed intelligent model depends on social networking and taxi request which helps to predict the prolong traffic pattern.

3.1   Request Grabber

With the rapid population of vehicles growth in urbanization causes vehicle congestion and pollutes the environmental system. The intelligent request grabber collects the vehicle request using social media and taxi request and determines the vehicle density on particular location effectively. In Fig. 1 step 1- indicates collects the vehicle request from multiple locations using taxi network and social transportation provides the vehicle movement and location information converts sequential pattern and determine the continual data. In step 2- sequential data converted into unique format which has the users request from multiple resources and convert into well formatted record. In step 3- store the format into cloud server in that the LSTM learner model shows the optimized traffic predicts sequence on different locations. In this work, collects the real time transportation data request on different sources like social and taxi transportation request which is based on usual instants that are called Scheduled Event (SE) which occur every instance of time for example festivals, national events, and regular works. The unusual instants are called Un-Schedule Event (USE) which is not a planned event and that occurs without intimations, and that is very complex to predict the event for example sudden gatherings. The USE causes severe vehicle congestion on particular location, which predicts using social transportation data, in precision manner. An intelligent analyzer collects various aspects of data that is historical events, social transportation, and online vehicle request (taxi) data.

3.2   Requests from Social Transportation

An intelligent based computer mediated Interactive technologies generating large amount of data from that extract transportation information's using deep analysis model collects traffic information from social networking platforms such as Whatsapp, Face book, twitter, and other information provides individual with ubiquitous opportunities to share thoughts, emotions and data in public or in specific groups of real-time social data. The raw social media format includes Shares, Likes, Mentions, Impressions, Hash-tag, URL clicks, followers Comments. These social inter-connection links among persons are made the data available at any places and anytime these are called as Social Signals (SS). The collected social signals are not just people's emotion it is very useful information may have travel information to manage the traffic in real time and efficient manner. The useful information converts the request in terms of, vehicle request time stamp, GPS location, destination location.

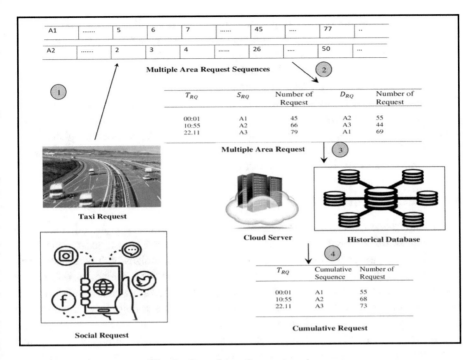

Fig. 1. Data flow of request analyzer

For each type of social transportation data, the recorded contents are specific to one or several aspects of human mobility, and specific information of a person or a community of USE occurrence data. For example person-A post the tweets like "today plan for a mall visit with my bike on 3 pm" from that we extract the GPS coordinates of the person-A, posting time stamp may 1 pm, vehicle mode – bike, vehicle presence time around 3.10 pm. Learner analyze the post and store it into LSTM model to predict the traffic on mall location and also monitors person-A spending time on the mall based on that calculates the return journey time with estimated traffic prediction. The advancement of mobile coverage and frequent long-time (months to years) monitoring of a person to learn the user behavioral in an efficient manner with the help of mobile phone data, which offer us opportunities to accurately predict the particular person locations such as home, work, and other locations in large fraction of residents. Person's live mobility information is collected from different sources like Google, which tracks user's movement location, person's interest and special events are learned from their activities. Such useful information encapsulates a large amount of real-time traffic information. Moreover, Social media strongly attracts people to share their on-line location-related data and induces the majority's recognition of the transport service in real-time. The combined usage of social networks and IoT contributes everything is connected to everyone by means of people, vehicles, infrastructures, and services to create intelligent based transportation data used to analysis the traffic behavior and intended to control effectively. The social transportation system (STS), analysis and predicting the real time traffic using social signals from mobile phones, taxi requests, and GPS systems. However, mobile phone data are not suitable for estimating travel time on roads, because the records are too spare (several hours a record) and temporally irregular. By using fusing method to differentiate social signal data and predictions based cross validated by another type of data the new knowledge can be mutually speculated, with higher prediction accuracy, the prediction precision and the algorithm performance achieved higher. Analyzing social transportation data can help improve the performance of ITS, have been taking the social media as an essential platform for predicting traffic onto announcing vehicle schedule irregularities, and ultimately resolve the wastage of transportation recourses.

3.3   Requests from Taxi Data

Predict future taxi request on particular location using from past trip and other real time travel request information. The passenger uses GPS connection to request vehicle and tracks the entire journey with the vehicle movement. It optimizes both taxi drivers and passengers waiting time based on scheduled vehicle booking which means reserve vehicle on upcoming ride. The LSTM model predicts the future taxi request collects and combined multiple area requests using cluster model. Adding other transportation related information such as drop off time, pick up time, number of request and the availability of taxi under passengers requirements. In order to cumulate taxi request from multiple GPS area is complex so that split larger area into small regions as clusters. It helps collect the number of taxi request in a particular area effectively. The sequential pattern of the taxi request is stored into the RNN input in which stochastic process value of process single step process at time T depends on its value. The values are updated in every step in the process and size parameter (learning rate) will be updated and it is used for determine the traffic strength on a particular location. The framework of taxi requests arise from multiple passengers's on a same location request q as an input and generates the optimal cumulative schedule request S_{RQ} with a companion candidate. In Table 1 Shows the sample record of the request format which includes requested time stamp, requested origin, vehicle counts and their cumulative request and determine the destination location traffic pattern.

1. A passenger submits request q to the cloud server.
2. It analyze companion request from multiple users and cumulative the same destination location request q.
3. The RNN generates the optimal prediction using cumulative scheduled request S_{RQ} with the companions selected from the set of potential candidates.
4. The system informs the cloud server which provides optimal prediction to collects multiple requests from each participant.

Table 1. Request format

Time stamp	Origin	Mode of vehicle request	Cumulative request	Destination	Mode of vehicle request	Cumulative request
00:01	A1	Car: 28, bike: 74…	45	A2	Car: 28, bike: 74…	55
10:55	A2	Bus: 58, van: 21…	66	A3	Bus: 58, van: 21…	44
22:11	A3	Car: 58, bike: 74…	79	A1	Car: 58, bike: 74…	69

3.4   Recurrent Network Model

Computing the same performance on every layer element of a sequence in network RNNs are prominent solution for conditioned output feed to the multiple hidden layer of the sequence with the conditioned on the previous computations. Structure of the RNN represented in Fig. 2. The input of the data which includes social and taxi request collects from multiple source. When predicting the real time vehicle data the learner is more important, the idea of parameterized distribution of neural network uses the output of the system and determines the stochastic layer cost function shows multiple possible outcomes of the LSTM. This model project time series of the sequential data which is collected from multiple areas on multiple requests signals. It combines the traffic input signals while predicting the output in the future. This neural network predicts the output based on the current input and memory based model which stored the past travel information to predict the future vehicle request and calculate the density of the traffic. The requests are made dynamically so that prediction calculation based on continual sequence of the data. Past request recorded in unique format at every timestamp, it analysis the entire network and determine the future request. In Fig. 2 input x_t, collects request from various sources and forward to the stack which has past memory value and collects every previous hidden state h_{t-1} which derives to the next hidden state h_t at each time-step t to generate the output y_t.

This information passed over the loop until forward to the next step. The variable W_s refers the shared weights among different time-steps to train the network infinite number of times to learn the sequence.

In most RNN implementations, this non-linearity is a hyperbolic tangent: $h_t = \tanh(W_{hh}h_{t-1} + W_{xh}x_t)$- at time step t the output generates (Figs. 3 and 4).

For the parameters W_{hx} and W_{hh} are shared weights among input and hidden layer outcomes on each time-step. In order to achieve the network is actually performing the same computation at each time-step, of dynamic input x_t. To avoid over-fitting on minimal data reduces the parameters in real time data. The major concern of hidden state h_t has the main feature of RNNs and it works memory based model to store the useful information of past data. This prolong model is suitable for predicting the traffic rate on particular location accurately using LSMT. LSTMs are a special kind of RNN,

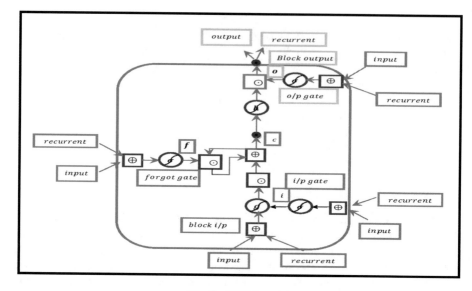

Fig. 2. LSTM model

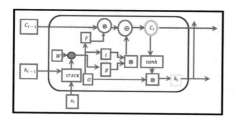

Fig. 3. State flow

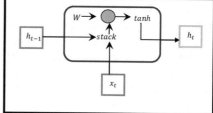

Fig. 4. Recurrent flow

capable of learning long-term dependencies due to their gating mechanism. The cumulative number of requests collected and performed underlying some factors which is available in this model. Predicting the real-time demand value based on deterministic model to the stochastic model which forecasting model the entire probability distribution of taxi demands in all areas.

3.5   Request Format

Request RQ is a passenger's request for finding the location uses GPS and that collects from users in unique format and calculates the density of the transportation. Each RQ is associated with six attributes: (T_{RQ}, S_{RQ}, D_{RQ}) where T_{RQ} the timestamp of the request time is, S_{RQ} is the origin location, D_{RQ} is the destination location.

3.6   Cumulative Request

A sharing schedule S_{RQ} comprises of a rendezvous point R for all passengers sharing a taxi, and a sequence of their destination locations, i.e., RQ $\rightarrow$ D_{RQ1} $\rightarrow$ D_{RQ2}... $\rightarrow$ D_{RQN}, indicating their order to be delivered. For the sake of clear exposition, each query here only represents one passenger, and the taxi capacity is set to four. Nevertheless, our approach can be applied to a request with multiple passengers.

$$h_t = f_t(h_{t-1}, x_t) \tag{1}$$

Here, h_t refers the new state output which includes f_t- function parameter and h_{t-1}- old state and x_t- input vector at time step t. Calculates the hidden vector by applying the weighting factor of f_t produces the value with bias b_f in Eq. (2).

$$f_t = \left(\sigma \times W_f \begin{pmatrix} h_{t-1} \\ x_t \end{pmatrix} \right) + b_f \tag{2}$$

$$h_t = \tanh(W_{hh} h_{t-1} + W_{xh} x_t) \tag{3}$$

$$y_t = W_{yh} h_t \tag{4}$$

$$\begin{pmatrix} i \\ f \\ o \\ g \end{pmatrix} = \begin{pmatrix} \sigma \\ \sigma \\ \sigma \\ \tanh \end{pmatrix} W^l \begin{pmatrix} h_t^{l-1} \\ h_{t-1}^l \end{pmatrix} \tag{5}$$

$$c_t^l = f \odot c_{t-1}^l + i \odot g \tag{6}$$

$$h_t^l = o \odot \tanh(c_t^l) \tag{7}$$

$$r_t = \sigma(W_r) \begin{pmatrix} h_{t-1} \\ x_t \end{pmatrix} + b_f \tag{8}$$

$$h_t^l = \tanh W \begin{pmatrix} r_t \odot h_{t-1} \\ x_t \end{pmatrix} \tag{9}$$

$$z_t = \sigma(W_z) \begin{pmatrix} h_{t-1} \\ x_t \end{pmatrix} + b_f \tag{10}$$

$$h_t = (1 - z_t) \odot h_{t-1} + z_t \odot h_t^l \tag{11}$$

4 Experimental Analysis

The real time data were collected from multiple areas and past data from taxi request might contain empty values, negative values or error. Once the data formatted and that is available feed to the LSTM. Input layer of the network is collected from multiple area and analysis the demand of the network is output of the demand in the next time-step.

$$MAPE = \frac{1}{n}\sum_{t=1}^{n} \frac{|X_t - \widehat{X_t}|}{X_t} \times 100\% \tag{12}$$

$$RMSE = \sqrt{\frac{1}{n}\sum_{t=1}^{n}(X_t - \widehat{X_t})^2} \tag{13}$$

$$MAE = \frac{1}{n}\sum_{t=1}^{N}|X_t - \widehat{X_t}| \tag{14}$$

$$ME = max_{t=1,...,n}\left|X_t - \widehat{X_t}\right| \tag{15}$$

To read live request on social networks from twitter, face book, we use keyword analyzer to filter the transportation request API for python (Table 2).

Table 2. Prediction of vehicle density

Area	X_{T-4} (Past) Time T-40		X_{T-3} (Past) Time T-30		X_{T-2} (Past) Time T-20		X_{T-1} (Past) Time T-10		X_T (Present) Time T		X_{T+1} (Prediction) Time T + 10	
	Social	Taxi	Social	Taxi	Social	Taxi	Social	Taxi	Social	Taxi	Actual	Prediction
A1	40	60	70	50	70	80	50	90	70	90	90	89
A2	60	80	60	80	70	90	70	100	70	100	50	50
A3	30	30	70	80	100	90	70	100	80	90	80	79
A4	130	100	60	70	50	100	60	70	70	100	90	90
A5	180	90	70	90	100	70	90	100	90	100	20	19
A6	60	80	120	70	90	70	80	90	80	90	10	10
A7	30	70	80	80	100	90	100	90	100	90	88	86
A8	100	40	140	90	50	100	100	80	100	90	150	148
			Percentage Error				Cumulative		Individual			
			MAPE				35.75		4.24			
			RMSE				2.42		0.28			
			MAE				1.77		0.22			
			ME				6.64		0.59			

The variable x_t represents actual value of the traffic at time interval t, and $\hat{x}t$ refers proposed prediction variable with time interval t, the variable n indicates the total number of traffic flow, which was observed and processed during the time intervals.

In Fig. 5(a) and (b) shows the higher traffic prediction accuracy to achieve the goal of collect request from multiple areas shows the entire traffic pattern [11].

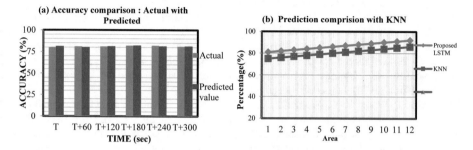

Fig. 5. Performance graph

5 Conclusion

As a promising paradigm of request grabber predict the dynamic traffic pattern in efficient manner with the help of social transportation and taxi data. The sequence learning model based on LSTM is well trained and outperforms the better traffic predictions on several areas using this intelligent request grabber. This powerful model achieved excellent performance on difficult learning tasks. It provides information regarding opportunities to perform various types of control and coordination for large scale intelligent transportation systems. The experimental results shows that model achieve accuracy around 93%. This work taking consideration of future public transportation to optimize the transportation resources in-terms of fuel, utilize the optimal vehicle usage, major concern of reduces the waiting time of travel on both the driver and passengers and create the pollution free environment. The future scopes of the research work establish the communication between onboard vehicles and passengers to know seat availability and estimated travel time.

References

1. Zhang, K., Feng, Z., Chen, S., Huang, K., Wang, G.: A framework for passengers demand prediction and recommendation. In: Proceedings of IEEE SCC, pp. 340–347, June 2016
2. Zhao, K., Khryashchev, D., Freire, J., Silva, C., Vo, H.: Predicting taxi demand at high spatial resolution: approaching the limit of predictability. In: Proceedings of IEEE BigData, pp. 833–842, December 2016
3. Davis, N., Raina, G., Jagannathan, K.: A multi-level clustering approach for forecasting taxi travel demand. In: Proceedings of IEEE ITSC, pp. 223–228, December 2016

4. Zhang, D., He, T., Lin, S., Munir, S., Stankovic, J.A.: Taxi-passenger demand modeling based on big data from a roving sensor network. IEEE Trans. Big Data **3**(1), 362–374 (2017)

5. Miao, F., et al.: Taxi dispatch with real-time sensing data in metropolitan areas: a receding horizon control approach. IEEE Trans. Autom. Sci. Eng. **13**(2), 463–478 (2016)

6. D'Andrea, E., Ducange, P., Lazzerini, B., Marcelloni, F.: Real-time detection of traffic from Twitter stream analysis. IEEE Trans. Intell. Transp. Syst. **16**(4), 1–15 (2015)

7. Wang, F.: Scanning the issue and beyond: real-time social transportation with online social signals. IEEE Trans. Intell. Transp. Syst. **15**(3), 909–914 (2014)

8. Karpathy, A., Johnson, J., Fei-Fei, L.: Visualizing and understanding recurrent networks (2015). https://arxiv.org/abs/1506.02078

9. Graves, A., Mohamed, A.-R., Hinton, G.: Speech recognition with deep recurrent neural networks. In: Proceedings of IEEE ICASSP, pp. 6645–6649, May 2013

10. Simonyan, K., Zisserman, A.: Very deep convolutional networks for large-scale image recognition (2014). https://arxiv.org/abs/1409.1556

11. Zhang, J., Wang, F.-Y., Wang, K., Lin, W.-H., Xu, X., Chen, C.: Data driven intelligent transportation systems: a survey. IEEE Trans. Intell. Transp. Syst. **12**(4), 1624–1639 (2011)

12. Rajkumar, S.C., Jegatha Deborah, L., Vijayakumar, P.: Optimized traffic flow prediction based on cluster formation and reinforcement learning. Int. J. Commun. Syst. https://doi.org/10.1002/dac.4178

13. Abid, H., Phuong, L.T.T., Wang, J., Lee, S., Qaisar, S.: V-cloud: vehicular cyber-physical systems and cloud computing. In: Proceedings of the 4th International Symposium on Applied Sciences in Biomedical and Communication Technologies, p. 165. ACM (2011)

14. Barrachina, J., Garrido, P., Fogue, M., Martinez, F.J., Cano, J.-C., Calafate, C.T., Manzoni, P.: Veacon: a vehicular accident ontology designed to improve safety on the roads. J. Netw. Comput. Appl. **35**(6), 1891–1900 (2012)

15. Sneha, S., Varshney, U.: Enabling ubiquitous patient monitoring: model, decision protocols, opportunities and challenges. Decis. Support Syst. **46**(3), 606–619 (2009)

16. Sneha, S., Varshney, U.: A framework for enabling patient monitoring via mobile ad hoc network. Decis. Support Syst. **55**(1), 218–234 (2013)

Aggregation in IoT for Prediction of Diabetics with Machine Learning Techniques

P. Punitha Ponmalar[1](✉) and C. R. Vijayalakshmi[2]

[1] Sri Meenakshi Government Arts College for Women, Madurai-2, TN, India
p_punithanadraj74@yahoo.co.in
[2] Government Arts and Science College, Andipatti, Theni, TN, India
vijinsc@yahoo.in

Abstract. Diabetes is a chronic illness and it may generate many dilemmas. Diabetes millitus patients in the earth will reach 650 million in 2050, which means that more number of adults will have diabetes in time ahead. There is no doubt that this startling number needs huge deliberation. Diabetic patients data are gathered and forwarded through Internet of Things (IOT). The lifespan of the network is the critical confront in the IoT. To elongate the lifespan of the IoT, data aggregation is a useful method to abate the number of transmissions among objects. Reduced number of data replication leads to elongate the network lifespan and to descent the energy depletion. The data collected from the diabetic patient are accumulated and the machine learning techniques are imposed to presage diabetics with a high degree of compassion and specificity. In this work, the K-Nearest Neighbor and Support Vector Machine are used to predict diabetes. The results showed that Support Vector Machine achieves the highest accuracy compared to K-Nearest Neighbor when all the attributes were used.

Keywords: Internet of Things · Data aggregation · Machine learning · Diabetics · Data clustering

1 Introduction

The IoT accedes physical objects and populace to be connected and monitored through the internet. In the IOT, billions of widgets [1, 3] perchance erected of various categories aforementioned as patients imperative sign monitoring devices, GPRS and humidity sensors which are connected to the internet and transmogrifies information from the natural world into the digital world shown in Fig. 1.

In the IoT, Sensing devices [2] are normally galvanized by confined batteries and energy replenishments. Thence, it is indispensable to meliorate the lifespan of protracted-term applications such as interminable imperative sign monitoring. The extent of propagation of data is generally huge in interminable monitoring [10] for handling through the base station. The data agglomeration strategy is to agglomerate and adequately gather the data packets to ameliorate the energy dissipation, plexus lifespan, cartage hindrance, and data exactitude. Besides, exterminating prolixity and lessening the extent of transmitted data will salvage the fortitude. Data aggregation efficacy confides on the network contrive and the extent of apperceiving data.

A. P. Pandian et al. (Eds.): ICCBI 2019, LNDECT 49, pp. 789–798, 2020.
https://doi.org/10.1007/978-3-030-43192-1_87

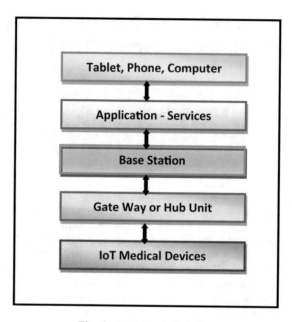

Fig. 1. IOT physical design

Diabetes is a prevalent disease and mien an immense peril to human tonicity. Owed to the inadequate insulin excretion or its flawed biological paraphernalia, conversely, both can cause the blood sugar level more than the regular level. Diabetes can impinge disparate tissues, and nerves, eyes, heart, kidneys, and blood vessels. Diabetes can be branched into two categories namely, type 1 and type 2 diabetes. Type 1 diabetes ensues mostly in juveniles and preadults. This category of diabetes can be alleviated with insulin therapeutics. Type 2 diabetes appears prevalently in adults. The scrutiny of diabetes is acceding to famishing blood glucose, glucose sufferance, and indiscriminate glucose levels. Contemporary enrichments have been empowered with the improvement of Continuous Glucose Monitoring Systems. The Continuous Glucose Monitoring systems are wearable devices consisting of a glucose sensor, a transmitter and a receiver which specifies the information regarding the glucose levels every 1 min or 5 min.

Machine learning can help diabetic to generate an exploratory result about their diabetes according to their periodical examination data. Recently, numerous algorithms are used to envision diabetes, including the conventional machine learning method such as decision tree, regression and so on. In this work SVM and KNN are used to predict the diabetics. **K-Nearest Neighbour (KNN) Classifier:** KNN is a supervised non-parametric learner that classifies the data to a given category based on the training sample. **Support Vector Machine (SVM):** SVM is a supervised machine learning algorithm. The algorithm generates an optimal hyper-plane using training data and categorizes new examples.

The rest of the paper is organized as follows. Section 2 discusses the related work, Sect. 3, describes the research problem definitions. Section 4 focuses on proposed work and its architecture and Sect. 5 discusses the results and discussion.

2 Related Work

The realistic healthcare records provided by Verizon are trached from wireless devices based on biometric values. It assists to monitor the patient's healthcare at the home. Yu [6] discussed the business and application service details of middleware with an architecture framework. Abu-Elkheir [8] has focused on the primitives of design for the IoT data management proposals.

Thangaraj et al. [9] have focused on a review of data aggregation techniques in WSN and various types of data aggregation. Rahman et al. [11] have discussed a few methods, which are used for data aggregation. Lonappan et al. [7] proposed the aspects of diabetes when blood glucose is more than the normal level, which is caused by insufficient insulin secretion or its defective biological effects, or both. Krasteva et al. [5] discussed diabetes can point to chronic illness and dysfunction of various organs. Robertson et al. [12] pinpointed Type 2 diabetes are ensued more prevalently in middle-aged people and elderly people. It is often consorts with the instance of rotundness, invigorating tension, and other diseases. Kavakiotis et al. [4] stated diverse algorithms are utilized to predict diabetes, involving the conventional machine learning method.

3 Research Problem Definition

The data model shows a primary aspectof IoT devices. An enormous volume of data is exchanged as a larger number of devices are connected and interacted with one another, the data interacted could also be formatted or unformatted. Backend storage for the data is determined by the types of data. This detonation of data needs to be hoarded, evaluated with efficient data analytic approaches. Few researchers are concentrating on the data acquiring and agglomeration. Some are concentrating on the application framework. Few offer attention to the analytics of data with machine learning techniques to provide a prediction on medical health. But the data aggregation with machine learning techniques revolves around the significant use of data from IoT devices and to provide an accurate prediction.

4 Proposed Architecture

Smart Diabetics system continuously monitors the diabetic patient and to afford medication facilities for such patients without limiting their freedom. The diabetic data collected from the patient through the sensors in the IoT architectures. The information regarding the glucose levels is collected for every 5 min through the sensors attached to the patient. The sensors are battery operated and it cannot be replaced frequently.

To lessen the energy utilization of the battery and the rise in the amount of data broadcast are handled with data agglomeration. Aggregated data get stored in the secure data warehouse. Finally, Machine learning techniques such as classification and clustering are activated on the gathered data to deliver more precise prediction results to the diabetic patient as well as to the physician.

4.1 Architecture

The system architecture of the Smart Diabetics system is shown in Fig. 2, which includes four tiers: Device tier, Communication tier, Application tier, and User access tier. The establishment of acute devices generates new data objects. There is archetype in data delivering to the existenting databases. So, there is a need for data alteration, data reciprocating to disseminate the data to the exterior system.

Device Tier: Device tier is worned in data gathering from IoT devices. This tier congregates blood sugar data over blood glucose monitoring equipment. The smart gadgets gather information and deliver out in the format of the opcode, LED exhibit value of 10–16 binary values. The blood glucose monitoring gadget can be furnished to monitor personal home-situated blood glucose monitoring. It allows the patient to take farther hospital medication, thus reducing the expenditure compared to the long-term medication of the patient. Portable blood glucose monitoring devices easily monitor the patient's blood glucose. It does not distract the patient's daily accomplishments and it provides contentment for the patients.

The gadgets normally had 3 main classes, Least level gadget, medial level gadget, utmost decent gadget. Least level gadget of octadic bit system with the on gobbet controller. Medial level gadgets delimit to thirty-two bit architecture, along with the embedded OS. Utmost decent gadgets have the entirely adequate 64-bit architecture functioning on an operating system like Linux. In the event of the connectedness, there are various connectedness options like wired links, wireless links, mobile-network, Zigbee network, low power Bluetooth connections and, etc.

Communication Tier: This tier performs functions such as accumulate, normalize and validate. Repeated transmission of data creates data redundancy. Data aggregation avoids redundant transmission of packets and it minimizes the energy consumption, increases the packet delivery ratio and elongates the lifespan of the IoT. Data agglomeration is to aggregate and gather the data in an adequate manner to enhance the data exactitude.

Data normalization automatically maps the content with the terminology standard. Normalized Diabetic data are processed using machine learning methods to predict the disease. By continuously collecting, storing, and analyzing information on diabetes, helps to adjust the medication strategy in time. It establishes effective information sharing among patients and doctors to perform a treatment plan in time.

The data exchange format of this tier might be the XML or JSON. Datastorage options for keeping the intramural data per chance be TinyDB and etc. The middleware and application services can be event induced, application induced or service orientated. The transmission pact options ranges from IEEE 802.15.4e, 2G–3G, LTE and CoAP.

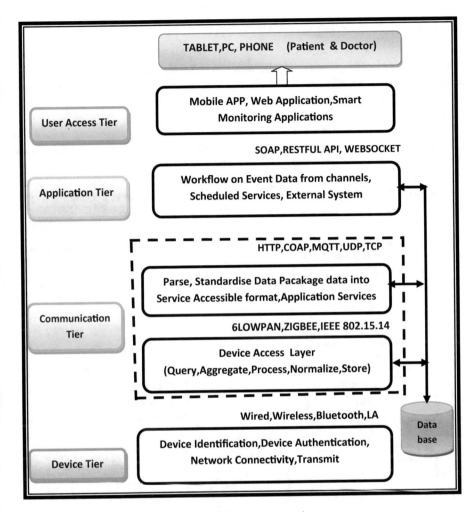

Fig. 2. Smart diabetic system architecture

Application Tier: The application layer furnishes the services are 1. Medical gadget monitoring services, 2. Web proclamation services, 3. Sensor cue services. The gadgets are registered with the application service, RESTFUL API, and WEB SOCKET. The coarse data can be discerned as base data. The application reliant aggregated data may be considered as initial data. The data is refined, authenticated, permeated and then evaluated with the rules and conditions configured.

User Access Tier: This layer helps the patients to know about their information on diabetes. It helps to handle ancruncy position and to escalate the efficacy and performance of the analysis and medication of diabetes. It supports the patients to preclude from getting the ailment in the initial stage. The devaluation of disease peril could lead to declining the price of diabetes medication.

4.2   Process Flow of Smart Diabetic System

Smart Diabetics system incessantly monitors patients with diabetes. The Diabetic should be monitored continuously because diabetes is a chronic ailment where an individual undergoes a protracted level of blood glucose in the body. Diabetes will cause the failure of various organs. So with the help of IoT devices the patient data are collected.

The diabetic data collected from the sensors need to be aggregated to reduce the redundancy and to increase the lifetime of the sensor. Then the aggregated data get stored in secure data-warehouse. Machine learning techniques such as classification and clustering are applied to provide more accurate prediction results for diabetics and closest to the clinical outcomes. The process flow of the smart diabetic system is shown in Fig. 3. Figure 3 shows the algorithm for the smart diabetic system.

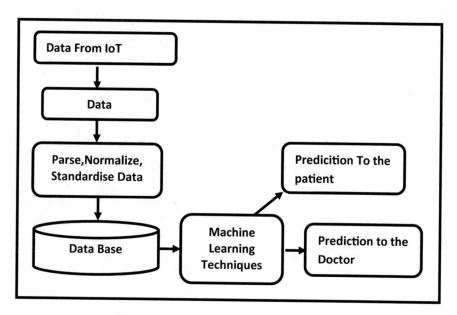

Fig. 3. Process flow of smart diabetic system

4.3   Machine Learning Algorithms

Machine learning algorithms are used to organize, classify and to make automated decisions. Using the machine learning algorithms, the dataset for a Smart diabetic system is trained and analysis is done based on the test set. The classifiers used in this work are **Smart Support Vector Machine (SSVM)** and **Smart K-Nearest Neighbor (SKNN)** classifier. Table 1 shows the predicted diabetic range.

Table 1. Predicted diabetics range

A1c %	Fasting plasma	Oral glucose tolerance test	Degree of control	Health risk
11 or above	200 or above	300 or above	Diabetics	Extremely high
10	180 or above	210 to 275	Diabetics	High
6.5 or above	126 or 179	200 to 210	Diabetics	Good
5.7 or above	100 to 124	140 to 199	Prediabetics	Low
5	99 or below	139 or below	Normal	Very low

4.4   Implementation

The implementation of Smart Diabetic system was carried out using WEKA [13] tool. LibSVM library [14] is used to implement SSVM classifier and 1BK is used to design SKNN algorithm. The experimental evaluation was made using Intel Core i5 2.67 GHz with 4 GB RAM, running Windows 7. The experiments were carried out on the following real dataset. Ten-fold cross-validation method was used for this analysis. The competence of Smart diabetic system was compared with the normal KNN and SVM classifer without using IoT aggreagation. The following measures were used for evaluating the performance of the classifiers.

- Accuracy: The proportion of correctly classified tuples
- Runtime: The time taken to generate and test the classifier in seconds
- F-measure: Harmonic mean of the precision and recall of the test.

Diabetic Dataset: In this work, the diabetic data set is taken from the UCI machine learning repository [15] and aggregated. It contains 101766 patient records and 50 features. The diabetic data collected from 130US hospitals between 1999–2008.

The class label is readmitted which specifies if a patient was hospitalized within 30 days or >30 days or not readmitted.

5   Result and Discussion

This section summarizes the performance of the classifiers namely SKNN and SSVM to predict the Diabetic dataset against KNN and SVM without using IOT techniques. The Fig. 4 shows the accuracy of the four classifiers when predicting diabetic patients. The figure shows that accuracy of SVM and KNN are lesser than SKNN and SSVM. SSVM achieved highest accuracy when compared to all classifiers.

When comparing KNN and SKNN, the accuracy improvement by SKNN is 13.06%. The increase in the accuracy value by SSVM with SVM is 17.32%. From the figure, we found that the accuracy improvement of SSVM is about 11% more than SKNN when comparing classifiers while predicting diabetics.

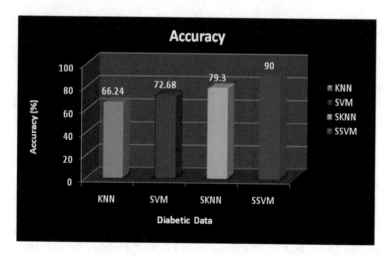

Fig. 4. Accuracy of classifiers

The Fig. 5 shows the induction time of four classifiers namely SVM, KNN, SKNN and SVM on diabetic prediciton. The fastest system is SSVM which takes less time to generate classifier and slowest one is KNN when compared to SKNN and SVM.

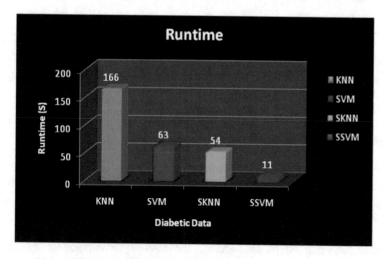

Fig. 5. Runtime of classifiers

The Fig. 6 shows f-measure of KNN, SVM and SKNN, SSVM. From this figure, one can understand that the f-measure value of KNN and SVM are lesser than SSVM and SKNN when classifying diabetic data. When comparing SSVM and SKNN, SSVM is higher than SKNN when categorizing diabetic patients. The f-measure value of SSVM increases by 13% than SKNN.

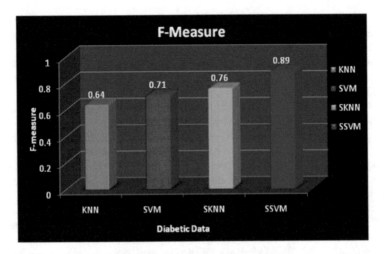

Fig. 6. F-measure of classifiers

6 Conclusion

Diabetes is a diverse group of diseases. It's categorized by chronic elevation of glucose in the blood. To prevent and treat diabetes and to increase the lives of diabetic people, the diease to be predicted at earlier stage. In smart diabetic system, the prediction was made with the help of KNN and SVM classifiers and the results were compared. The system based on SSVM predicts well on diabetic data compared SKNN based on accuracy and F-measure and runtime values. SKNN produces good results only when the data size is smaller. In summary, based on the experimental results that the proposed SSVM classifier performs well in classifying and predicting diabetic patients compared to SVM, SKNN, KNN. Prediction of diabetics reduces the disease risk and also decrease the cost of diabetes treatment. In future, the work can be extended for various IoT levels to predict diseases.

References

1. Kulkarni, A., Sathe, S.: Healthcare applications of the Internet of Things: a review. Int. J. Comput. Sci. Inf. Technol. **5**(5), 6229–6232 (2014)
2. Delphine, C., Reinhardt, A., Mogre, P., Steinmetz, R.: Wireless sensor networks and the Internet of Things: selected challenges. In: Proceedings of the 8th GI/ITG KuVS Fachgespräch Drahtlose Sensornetze, Hamburg-Harburg, Germany, 13–14 August 2009, pp. 31–34 (2009)
3. Gia, T.N., et al.: IoT-based continuous glucose monitoring system: a feasibility study. Procedia Comput. Sci. **109**, 327–334 (2017)
4. Kavakiotis, I., Tsave, O., Salifoglou, A., Maglaveras, N., Vlahavas, I., Chouvarda, I.: Machine learning and data mining methods in diabetes research. Comput. Struct. Biotechnol. J. **15**, 104–116 (2017)

5. Krasteva, A., Panov, V., Krasteva, A., Kisselova, A., Krastev, Z.: Oral cavity and systemic diseases—diabetes mellitus. Biotechnol. Biotechnol. Equipment **25**(1), 2183–2186 (2011)
6. Yu, L., Lu, Y., Zhu, X.J.: Smart Hospital based on IOT. J. Netw. **7**(10) (2012)
7. Lonappan, A., Bindu, G., Thomas, V., Jacob, J., Rajasekaran, C., Mathew, K.T.: Diagnosis of diabetes mellitus using microwaves. J. Electromagn. Waves Appl. **21**(10), 1393–1401 (2007)
8. Abu-Elkheir, M., Hayajneh, M., Ali, N.A.: Data management for the Internet of Things: design primitives and solution. Sensors **13**(1), 15582–15612 (2013)
9. Thangaraj, M., PunithaPonmalar, P.: A survey on data aggregation techniques in wireless sensor networks. Int. J. Res. Rev. Wirel. Sens. Netw. **1**, 36–42 (2011)
10. Thangaraj, M., Ponmalar, P., Subramanian, A.: Internet Of Things (IOT) enabled smart autonomous hospital management system—a real world health care use case with the technology drivers, pp. 1–8 (2015). https://doi.org/10.1109/iccic.2015.7435678
11. Rahman, H., Ahmed, N., Hussain, I.: Comparison of data aggregation techniques in Internet of Things (IoT). In: 2016 International Conference on Wireless Communications, Signal Processing and Networking (WiSPNET). IEEE (2016)
12. Robertson, G., et al.: Blood glucose prediction using artificial neural networks trained with the AIDA diabetes simulator: a proof-of-concept pilot study. J. Electr. Comput. Eng. **2011**, 11 p. (2011)
13. Ian, H.W., Frank, E.: Data Mining: Practical Machine Learning Tools and Techniques with Java Implementations. Morgan Kaufmann Publishers, Burlington (2000)
14. Lin, C.J., Chang, C.C.: LIBSVM: A Library for Support Vector Machines (2005). http://www.csie.ntu.edu.tw/~cjlin/libsvm
15. Diabetic Dataset. https://archive.ics.uci.edu/ml/datasets/diabetes+130-us+hospitals+for+years+1999-2008

EWS: An Efficient Workflow Scheduling Algorithm for the Minimization of Response Time in Cloud Environment

G. Justy Mirobi[1(✉)] and L. Arockiam[2]

[1] Department of Computer Science, Bharathiar University,
Coimbatore 641 046, Tamil Nadu, India
georgejusjer@gmail.com
[2] Department of Computer Science, St. Joseph's College (Autonomous),
Tiruchirappalli 620 002, Tamil Nadu, India
larockiam@yahoo.co.in

Abstract. A workflow comprises of a collection of coordinated tasks aimed to carry out a well-defined systematic process such as planning, arranging, sequencing the jobs and implementing the business process of the enterprises. Our research aims to schedule the workflow, which defines a correct order of execution of jobs. A proper workflow scheduling process helps to enhance the response time, processing time, utilization of resources, performance and quality of service. This proposed approach, EWS: Efficient Workflow Scheduling Algorithm arranges the requests according to the user priority, finds the proficient VMs, maps the tasks to the VMs and manages the execution of tasks within a specified time. The proposed approach helps to deliver the services with the minimum response time and process time as mentioned in the Service Level Agreement (SLA) without any violation. In addition, it also enhances the performance of virtual machines.

Keywords: Workflow Scheduling · Response Time · Process Time · Scientific applications · Business applications

1 Introduction

The workflow execution refers to a succession of activities that take place to attain a business outcome [1]. Scientific Workflow allocation and scheduling are a procedure adapted to manage the execution of interdependent workflow requests on a scattered pool of resources. A workflow task may be required to wait in a queue for a long or short period depending on some scheduling factors, which include system capacity to execute, task priority, system load, and requested resource availability [3]. Therefore, a proper workflow scheduling process is an important challenge to reduce the waiting time that enhances the response time, processing time, efficiency, performance and quality based service of the service request. The response time and processing time will be reduced to the minimum level due to the scheduling process. Many scheduling algorithms exist in distributed computing, most of which do not apply to reduce the response time in the cloud environment. However, their performance is being questioned in terms of the response time and load balancing [5].

© Springer Nature Switzerland AG 2020
A. P. Pandian et al. (Eds.): ICCBI 2019, LNDECT 49, pp. 799–810, 2020.
https://doi.org/10.1007/978-3-030-43192-1_88

2 Related Works

Sadhasivam et al. [18] developed an Improved Particle Swarm Optimization (IPSO) algorithm to arrange the VMs in the cloud environment. The IPSO algorithm was applied to reduce the cost of task scheduling to available resources. The total cost value was arrived by adding the cost of communication between the resources, cost value of dependency of tasks and the cost value of computing resources. IPSO algorithm overcame the disadvantages of standard PSO and converged into local optimum to minimize the cost of task allocation on available VMs.

Cai et al. [19] contributed dynamic scheduling and resource provisioning algorithm Deadline Driven Scheduling (DDS) for reducing the renting cost of resources. According to the level of execution and the predicted execution time of tasks, the new VMs were dynamically rented to fulfill the deadline of the workflow. Experimental results proved that the DDS algorithm could ensure the deadline of the workflow for all instances and resulted in lesser renting cost of resources than URH (Unit-aware Rules based Heuristic) and MOHEFT (Multi-Objective Heterogeneous Earliest Finish Time) in most instances.

Rastkhadiv et al. [14] suggested a dynamic scheduling algorithm for scheduling the task and for balancing the load by using the ABC (Artificial Bee Colony) algorithm. In this scheduling and load-balancing algorithm, once the scheduler scheduled the task then the greedy method was applied to allocate the resources dynamically.

Shafi'i Muhammad Abdulhamid et al. [15] developed a global task-scheduling algorithm GBLCA (Global League Championship Algorithm) to arrange the task in the cloud environment, for the secured global scientific applications. The GBLCA scheduling technique consisted of three key modules: the module for the standard policy, the module for fitness function and objective function and the algorithm to schedule the task. This GBLCA technique gave a high level of change based on performance, process time and response time.

Kaur et al. [9] introduced a hybrid algorithm to arrange the job in the cloud server by using NN (Neural Network) with ABC (Artificial Bee Colony). This approach consumed less power when compared with the neural network-based scheduling. This hybrid method enhanced the performance of the system.

Sharma et al. [11] presented an optimal task-scheduling algorithm, merging the Differential Evolution (DE) and GSA (Gravitational Search Algorithm) to reduce the makespan. The differential algorithm was suitable for global searching and weak in the local searching. For the local searching, the GSA was merged with DE to improve the performance of DE and to minimize the execution time. The optimal task-scheduling DE-GSA algorithm provided improved makespan and good convergence rate.

Haladu et al. [12] proposed an Enhanced Min-Min Task Scheduling algorithm for resource allocation in data centers. This method used the main benefits of the Min-Min algorithm and avoided its disadvantages. By using this method, the system allocated the resources appropriately, achieved an effective load balancing and reduced the makespan.

3   Problem Description

The arrangement of service requests in the workflow that are to be executed at the specified period is shown in Fig. 1. In this workflow, there are 9 tasks are available in level (1), waiting for the process of execution; but only eight VMs are available to execute the tasks at a time. Now, the tasks are executed by using the existing Partitioning-Based Workflow Scheduling method. Nine tasks are waiting in the level (1) of the workflow. For the first time, the scheduler generates the schedule for the tasks T0, T1, T2, T3, T4, T5, T6, and T7. The task T8 is waiting in the queue. The available VMs are mapped to these eight tasks and the time is allotted to execute these 8 tasks. For the second time, the task T8 only is to be executed in that level (1). Therefore, the T8 is mapped to only one resource R1. The other VMs R2, R3, R4, R5, R6, R7 and R8 are in the idle position until the T8 is completed.

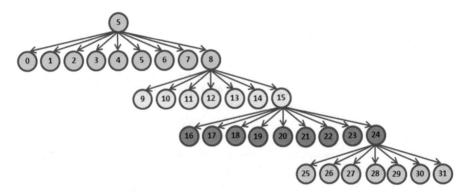

Fig. 1. The arrangement of tasks in the workflow

In the level (2) of the workflow, there are 7 tasks available; the scheduler generates the schedule for the tasks T9, T10, T11, T12, T13, T14 and T15 only.

As mentioned above, for the third time, the scheduler generates the schedule for the level (2); the available 7 VMs (R1, R2 . . . R7) are mapped to the tasks T9 to T15. In this level, only one resource R8 is idle because 7 tasks are in that level. Next, the level (3) of the workflow, the scheduler generates the schedule to the level (3) for the fourth time. Totally, 9 tasks are waiting in the queue, but the available VMs are 8; so that, the 8 tasks (T16, T17, T18, T19, T20, T21, T22, T23) are scheduled first and mapped the VMs (R1 to R8) to the scheduled tasks (T16 to T23). The task T24 cannot be scheduled because the available VMs are 8 only.

To execute the tasks in the workflow, the tasks are partitioned level-by-level using the Partitioning-Based Workflow Scheduler as shown in Fig. 2.

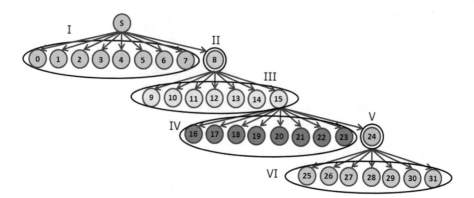

Fig. 2. The tasks are partitioned to execute within the specified time.

For the fifth time, the task T24 is mapped to the resource R1. Because only one task T24 is to be executed in that level of the workflow. The remaining VMs R2 to R8 are in the idle position until the completion of the execution process for the task T24. The available VM is mapped to the T24; the other 7 VMs are idle until the task T24 is completed. Next, there are 7 tasks, from T25 to T31 are in the next level (4) of the workflow. For the sixth time, the scheduler generates the schedule for the 7 tasks (T25, T26, T27, T28, T29, T30 and T31). The available 7 VMs are mapped to these 7 tasks (R1, R2 ... R7) and only one VM R8 is in the idle position; the 7 tasks are executed at the specified time, and only one VM R8 is idle until the 7 tasks are completed. Therefore, in this method, the resources are not fully utilized. The tasks are executed by using the existing Partitioning-Based Workflow Scheduling method as six levels instead of four levels. Therefore, the drawbacks of the existing Partitioning-Based Workflow Scheduling method are: (i) more time is taken to execute the assigned tasks (ii) the response time and processing time are maximized with poor performance (iii) the resources are not completely utilized at all levels.

4   Proposed EWS: Efficient Workflow Scheduling Algorithm

Algorithm1: Proposed *EWS: Efficient Workflow Scheduling Algorithm*

Input: Workflow W(T,E), Task DependencyList, VirtualMachine set VM
Output: Scheduled List For Allocating The Resources

1. *Begin*
2. *TaskList = {t₁, t₂ . . . tₙ}*
3. *VMlist = {vm₁, vm₂ . . . vmₙ}*
4. *w = 0*   *// initialize workflow*
5. *s = 0*   *// initialize schedule*
6. *IDLE VM = 0*   *// initialize IDLE VM*
7. *For i = 0 to lₙ -1*
8.   *For j = 0 to tₙ-1*
9.   *TaskList.get(j)*
10.   *Compute the Estimated runtime using equation (1)*
11.   *Compute the Communication time using equation (2)*
12.   *Calculate the rank of each task using equation (3)*
13.   *End For*
14. *End For*
15. *For i = 0 to lₙ-1*
16. *For j = 0 to tₙ-1*
17.   *For k = 0 to vmₙ-1*
18.   *TaskList.get(j)*
19.   *VMList.get(k)*
20.   *Schedule Tasks by decreasing order of Urankᵢ*
21.   *Tⱼ = TaskList. Urank[j]*
22.   *If TaskList > VMList*
23.   *// Partition The Tasks In The Level*
24.   *Set MaxTaskNumber = NoOfVMs*
25.   *Set PendingTask = TaskList −MaxTaskNumber*
26.   *// Map The Tasks To The VMs*
27.   *Allocate taskVMmap(Tⱼ) = VMListₖ*
28.   *Set IDLE VM = 0*
29.   *GoTo the steps 43 and 44*
30.   *Elseif TaskList <. VMList*
31.   *Set MaxTaskNumber = TaskList In That Level Of The Workflow +*
                                 Pending Task From Previous Level Of The Workflow
32.   *Allocate taskVMmap(Tⱼ) = VMListₖ*
33.   *Set PendingTask = 0*
34.   *IDLE VM = VMList - MaxTasksNumber*
35.   *GoTo the steps 43 and 44*
36.   *Else*
37.   *TaskList = VMList*   *// The number tasks in that level are*
                                    equal to the number of VMs
38.   *Allocate taskVMmap(Tⱼ) = VMListₖ*
39.   *Set PendingTask = 0*
40.   *Set IDLE VM = 0*
41.   *GoTo the steps 43 and 44*
42.   *EndIf*
43.   *Calculate the Response Time using equation (5)*
44.   *Calculate the Processing Time using equation (4)*
45.   *End For*
46. *End For*
47.   *Update s = s + 1*   *// Update the value of generated schedule*
48.   *Update l = l + 1*   *// Update the value of completed level*
49. *End For*
50. *Update w = w + 1*   *// Update the value of workflow*
51. *End*

The proposed EWS scheduler schedules the tasks, which are in the level (1) of the workflow. In this level (1), there are 9 tasks to be scheduled and processed. However, there are 8 VMs available for the 9 tasks. Therefore, the 9 tasks are partitioned and scheduled dynamically based on the priority. The tasks T1, T2, T3, T4, T5, T6, T7 and T8 are scheduled first based on the priority. The priority can be calculated by using the rank function as given in the Eq. (3); here the task0, which is having the lowest priority, is left from the schedule of the first level and waiting for the schedule of the second level. The tasks T1, T2, T3, T4, T5, T6, T7 and T8 are dynamically allotted to the available 8 VMs (R1 to R8). The available 8 VMs are mapped to the 8 tasks and execute these tasks during the allotted time. The schedule of tasks in the workflow by using the proposed EWS Scheduler is shown in Fig. 3.

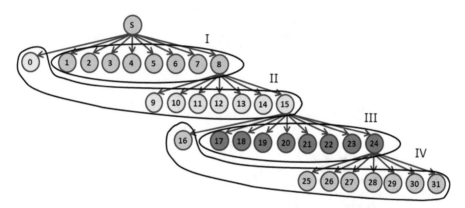

Fig. 3. The schedule of tasks in the workflow by using the proposed EWS scheduler

For the second schedule, the proposed EWS scheduler schedules the tasks in the next level (2) of the workflow. Before scheduling the tasks in the second level of the workflow, the task T0 from the first level is included first with the second level of the current workflow. Because there are 8 VMs available but 7 tasks (T9, T10, T11, T12, T13, T14 and T15) are available in the second level to schedule. To avoid the idle state of VM and for the complete utilization of resources, task T0 is included to the second level from the first level of the workflow. Now, the 8 tasks (T9, T10, T11, T12, T13, T14 and T15) and task T0 are scheduled in the second level and the available VMs are mapped dynamically based on the priority; the tasks are executed at the allotted time. At this current level, all these VMs are utilized successfully and completely. Next, the proposed EWS scheduler schedules the tasks, which are in the next level (3) of the workflow. In this level, for the scheduling process, there are 9 tasks available in that level (3) of the workflow but the available VMs are 8 only. Therefore, the task T16 is partitioned from the tasks (T17 to T24) of that level because task T16 is having the lowest priority than the tasks T17 to T24. Next, the tasks T17, T18, T19, T20, T21, T22, T23 and T24 are scheduled at level (3) based on the priority, the task T16 which is having the lowest priority, left from the schedule of level (3). The available VMs are mapped to these 8 tasks and executed at the specified time. For the next schedule, the proposed EWS scheduler schedules the tasks, which are in the next level (4) of the workflow. There are 7 tasks available at level (4) but the available VMs are 8.

From the previous level (3), task T16 is included first with the level (4) of the workflow, for the next scheduling process. The tasks T25, T26, T27, T28, T29, T30, T31 and task T16 are scheduled at level (4) based on the priority and the suitable VMs are mapped to these tasks and execute these tasks at the specified time. There are 8 VMs assigned for 8 tasks in that level. In the proposed EWS: Efficient Workflow Scheduling method, the VMs are efficiently and fully utilized at the levels 1, 2, 3 and 4. There are 4 levels only to execute the workflow so that the time is allocated to execute the tasks in 4 levels only not for the six levels as taken by the Partitioning-Based Workflow Scheduling method. The VM allocation for the execution process using the EWS scheduler is shown in Table 1.

Table 1. VM allocation using EWS scheduler

EWS scheduler's schedules	R1	R2	R3	R4	R5	R6	R7	R8
Schedule1	T5	T3	T8	T1	T4	T2	T6	T7
Schedule2	T0	T14	T15	T10	T12	T11	T9	T13
Schedule3	T21	T19	T23	T18	T17	T22	T20	T24
Schedule4	T16	T30	T25	T31	T28	T27	T26	T29

The TaskId, Rank and the assigned VM to execute the task are shown in Table 2. The Rank is computed for all the given tasks in level (1), based on the priority and it is scheduled in the highest to the lowest order. Thus, the priority order of the tasks is T5, T3, T8, T1, T4, T2, T6 and T7. The task T0 is partitioned for the next level (2) in terms of the lowest priority.

Table 2. VM allocation for the tasks in level (1)

Tasks Id	Priority	Assigned VM
T0	7	Null
T1	49	R4
T2	36	R6
T3	61	R2
T4	43	R5
T5	67	R1
T6	29	R7
T7	11	R8
T8	55	R3

The proposed EWS: Efficient Workflow Scheduling algorithm is compared with the Partitioning-Based Workflow Scheduling algorithm. The comparisons between the proposed EWS: Efficient Workflow Scheduling algorithm and the Partitioning-Based Workflow Scheduling algorithm are given in Table 3.

Table 3. The comparisons between the EWS: Efficient Workflow Scheduling Algorithm and the Partitioning-Based Workflow Scheduling Algorithm

Sl. no.	EWS: Efficient Workflow Scheduling algorithm	Partitioning-Based Workflow Scheduling algorithm
1	The numbers of iterative process steps are less	The numbers of iterative process steps are more
2	Accurate service is provided on time	There is a delay in providing the service
3	The resources are utilized properly	The resources are not utilized properly
4	The processing time is reduced	The processing time is high
5	The response time is minimized	The response time is high

5 Mathematical Model

This research normally emphasizes how the necessary information is analyzed, collected, coded for processing, stored the result and represented the output. The proposed EWS algorithm for reducing the response time is implemented using CloudSim. The equations for *Estimated Runtime, Communication Time, Process Time and Response Time* are given below:

Estimated runtime of request vi on a VM m_r, represented by RT_i^r, is calculated as given by Nazia Anwar et al. [17] using Eq. (1).

$$RT_i^r \frac{l_i}{p_r \times (1 - pv_r)} \tag{1}$$

Where l_i is the service length of task t_i, p_r is the CPU's processing capacity of VM m_r and pv_r is the VM performance variability, which represents the uncertainties such as potential variability or degradation in CPU performance.

Communication time of edge e_{ij} between requests t_i and t_j on VM m_r, represented by CT_{ij}^r, is calculated as given by Nazia Anwar et al. [17], using Eq. (2).

$$CT_{ij}^r = s_{ij}/bw_r \tag{2}$$

Where s_{ij} is the size of data and bw_r is the bandwidth capacity of VM m_r.

*The **upward rank** for a request v_i is the length of critical/the longest path from task v_i to the exit task (v_{exit}), including the run time of t_i. Thus, the priority of each request is mentioned as given by Nazia Anwar et al. [17] and Topcuoglu et al. [18].*

$$Urank_i = RTi + \max_{t_j \epsilon succ(t_{ij})} (CT_{ij} + urank_j) \qquad (3)$$

Response Time

In the proposed method, the response time is calculated for every user request as given below in Eq. (4)

$$T_{Response} = T_{ProcessFinish} - T_{Arrival} + T_{Delay} \qquad (4)$$

Processing Time

In the introduced approach, the processing time is calculated for the user requests as given below in Eq. (5):-

$$T_{Process} = \frac{UReq_{PerHour}}{BW_{Allocated}} \qquad (5)$$

Here, the processing time $T_{Process}$ is calculated for every user request concerning the allocated bandwidth (1000 Mbps) for one hour (hourly basis).

6 Results and Discussions

The proposed EWS: Efficient Workflow Scheduling Approach is simulated using the simulation software CloudSim. The experimental results are tabulated and compared with the Partitioning-Based Workflow Scheduling (PBWS) algorithm. The Processing

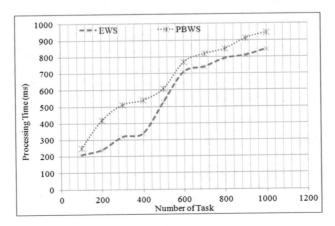

Fig. 4. Processing time comparison chart for the allotted tasks

Time comparison chart of the proposed EWS Algorithm with the PBWS algorithm for the allotted number of tasks is shown in Fig. 4.

The X-axis represents the number of tasks. Y-axis represents processing time in MS (MilliSeconds). This comparative analysis proves that the introduced method minimizes the processing time for various numbers of tasks when compared with the existing algorithm. The processing time of the tasks using the proposed EWS: workflow-scheduling algorithm is compared with the PBWS (Partitioning-Based Workflow Scheduling) algorithm as depicted in Fig. 4. The processing time is significantly decreased using the proposed EWS workflow-scheduling algorithm.

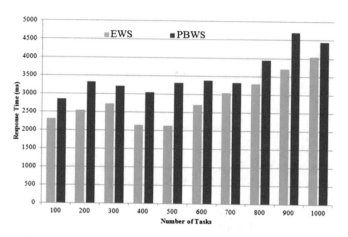

Fig. 5. Response time comparison chart for the allotted tasks.

The Response Time comparison chart of the proposed EWS Algorithm with the PBWS algorithm for the allotted number of tasks is shown in Fig. 5.

The X-axis represents the number of tasks and Y-axis represents the response time in ms. This comparative analysis proves that the proposed method minimizes the response time for different numbers of user requests when compared with the PBWS algorithm. The Response Time of the tasks using the proposed EWS workflow-scheduling algorithm is compared with the PBWS (Partitioning-Based Workflow Scheduling) algorithm as depicted in Fig. 5. The response time is reduced using the proposed EWS workflow-scheduling algorithm.

7 Conclusion

In the cloud environment, when the numbers of incoming requests are maximized and greater than the predefined limit, the processing time and response time are maximized. Therefore, to limit the uncertainty of the processing time and response time, the priority-based EWS: Efficient Workflow Scheduling algorithm is proposed. The proposed EWS workflow-scheduling algorithm arranges the user requests in sequence; next, it selects the requests from the sequence based on the priority and assigns the

VMs to the selected requests. Based on the priority, the rank is calculated. The proposed EWS: Efficient Workflow Scheduling algorithm partitions the tasks at each level based on the rank and generates the schedule for the tasks in that level. It selects the tasks from that schedule and assigns the VMs to the selected tasks. Next, the proposed algorithm reduces the number of the iterative process by adding the excess tasks to the next level of the workflow. Thereby, the EWS: Efficient Workflow Scheduling algorithm completes the task execution process as soon as possible. Therefore, the process time and response time are minimized. The experimental results demonstrate that the introduced EWS: Efficient Workflow Scheduling algorithm outperforms the existing PBWS (Partitioning-Based Workflow Scheduling) workflow-scheduling algorithm in respect to processing time and response time for the scientific applications. The response time and process time are minimized using the proposed EWS workflow-scheduling algorithm. However, whenever the requests enter as workflow, the scheduler generates the schedule and assigns the tasks to the VMs without considering the previous load on VMs. Therefore, there is a delay in executing the tasks. In the future, this research will be continued with VM level load balancing approach to minimize the delay time.

References

1. Sharma, M.M., Bala, A.: Survey paper on workflow scheduling algorithms used in cloud computing. Int. J. Inf. Comput. Technol. **4**(10), 997–1002 (2014). ISSN 0974-2239
2. Zhu, L., Li, Q., He, L.: Study on cloud computing resource scheduling strategy based on the ant colony optimization algorithm. Int. J. Comput. Sci. **9**(5), 54 (2012)
3. Balamurugan, S., Saraswathi, S.: A comprehensive survey on workflow scheduling algorithms in various environments. In: Proceedings of the International Conference on Informatics and Analytics, no. 21 (2016). ISBN 978-1-4503-4756-3
4. Gupta, I., Kumar, M.S., Jana, P.K.: Efficient workflow scheduling algorithm for cloud computing system: a dynamic priority-based approach. Arab. J. Sci. Eng. **43**(12), 7945–7960 (2018)
5. Thomas, A., Krishnalal, G., Raj, V.P.J.: Credit based scheduling algorithm in cloud computing environment. In: Proceedings of the International Conference on Information and Communication Technologies, Proc. Comput. Sci., vol. 46, pp. 913–920 (2015)
6. Prathibha, S., Latha, B., Sumathi, G.: Monitoring the performance analysis of executing workflow applications with different resource types in a cloud environment. In: Proceedings of the 1st International Symposium on Big Data and Cloud Computing Challenges (2013 CCC 2014). VIT University, Chennai (2014)
7. Wu, Z., Liu, X., Ni, Z., Yuan, D., Yang, Y.: A market-oriented hierarchical scheduling strategy in cloud workflow systems. J. Supercomput. **63**(1), 256–293 (2013)
8. Almiani, K., Lee, Y.C.: Partitioning-based workflow scheduling in clouds. In: 2016 IEEE 30th International Conference on Advanced Information Networking and Applications (AINA 2016) (2016). ISBN 978-1-5090-1858-1
9. Kaur, A., Nagpal, P.: Efficient cloud server job scheduling using NN and ABC in cloud computing. Int. J. Eng. Comput. Sci. **5**(10), 18662–18670 (2016). ISSN 2319-7242
10. Ghanbari, S., Othman, M.: A priority based job scheduling algorithm in cloud computing. In: International Conference on Advances Science and Contemporary Engineering (ICASCE 2012), Proc. Eng., vol. 50, pp. 778–785 (2012)

11. Sharma, A., Tyagi, S.: Differential evolution-GSA based optimal task scheduling in cloud computing. Int. J. Eng. Sci. Res. Technol. **1**, 1447–1451 (2016). ISSN 2277-9655
12. Haladu, M., Samual, J.: Optimizing task scheduling and resource allocation in cloud data center, using enhanced min-min algorithm. IOSR J. Comput. Eng. (IOSR-JCE) **18**(4), 18–25 (2016)
13. Anwar, N., Deng, H.: Elastic scheduling of scientific workflows under deadline constraints in cloud computing environments. Future Internet, **10**(5) (2018)
14. Rastkhadiv, F., Zamanifar, K.: Task scheduling based on load balancing using artificial bee colony in cloud computing environment. Int. J. Adv. Biotech. Res. (IJBR) **7**(5), 1058–1069 (2016)
15. Shafi'i Muhammad Abdulhamid, M.S., Latiff, A., Abdul-Salaam, G., Madni, S.H.H.: Secure scientific applications scheduling technique for cloud computing environment using global league championship algorithm. PLoS One **11** (2016)
16. George Amalarethinam, D.I., Lucia Agnes Beena, T.: Level based task prioritization scheduling for small workflows in cloud environment. Indian J. Sci. Technol. **8**(33), 1–7 (2015)
17. Adhikari, M., Amgoth, T.: Efficient algorithm for workflow scheduling in cloud computing environment. In: 2016 Ninth International Conference on Contemporary Computing (IC3 2016), pp. 1–6 (2016). ISBN 978-1-5090-3251-8
18. Sadhasivam, N., Thangaraj, P.: Design of an improved PSO algorithm for workflow scheduling in cloud computing environment. Intell. Autom. Soft Comput. **23**, 1–8 (2016)
19. Cai, Z., Li, X., Ruiz, R., Li, Q.: A delay-based dynamic scheduling algorithm for bag-of-task workflows with stochastic task execution times in clouds. Elsevier-Future Gener. Comput. Syst. **71**(C), 57–72 (2017)

An Investigation Report on Spotting and Diagnosing Diseases from the Images of Plant Leaves

S. Thenmozhi[✉], Irshadh Ibrahim, Rekha Mohankumar, and Shahla Sohail

Dayananda Sagar College of Engineering, Bengaluru, India
thenmozhirayan@gmail.com, irshadhibrahim@gmail.com,
rekhamohan743@gmail.com, shahla400@gmail.com

Abstract. Agricultural productivity plays an indispensable role in the economic growth of developing countries. Healthy plants are ought to maintain a sustainable and profitable crop production. Therefore, developing more effective methods for spotting diseases in an early stage is highly important. Timely detection helps in enabling a timely intervention to ensure a good crop yield. The traditional method of identifying diseases requires expertise in the field of botany. The manual method of identifying through manual visual inspection is highly irregular and unreliable. This manuscript talk about a novel initiative on disease spotting and diagnosing from the images of Plant leaves using Image dealing out methods. The discussions have the reference and thorough investigation of previously proposed techniques from the years 2005 to 2018. Accordingly the challenges faced and future scope in this field has been discussed.

Keywords: Plant disease · Leaf · Detection · Classification · Identification · Image dealing out

1 Introduction

Plants and fruits are essential to the endurance of life on Earth, they are not only a source of nourishment but also have medicinal attributes. Agriculture determines the economy of most of the developing counties. Continents like Asia and Africa depend on farming for revenue, resources and employment. The quality and yield of crops majorly depends on the overall performance and conditions that the crop is subjected to.

The major reasons for crop failure are

Diseases
Infection
Pests
Adverse weather conditions

© Springer Nature Switzerland AG 2020
A. P. Pandian et al. (Eds.): ICCBI 2019, LNDECT 49, pp. 811–823, 2020.
https://doi.org/10.1007/978-3-030-43192-1_89

While crop failure due to weather conditions cannot be controlled, the other factors can be reduced through the implementation of technology. In this paper we discuss all the significant research that has been done in this field. The paper has been split into two segments, the first segment represents the detection and the second section represents the classification techniques in image dealing out worn for detecting and identifying infected areas in plants.

2 Literature Review

The literature review has been divided into two segments, the first segment represents significant research that has been done in this field for segmentation of the diseased region and the second section represents the classification and identification of the diseases that have been used till date. The classification and identifications that have been proposed in a particular paper have been mentioned in the second section, while the segmentation methods of the same paper have been mentioned in the first section.

A. Detection based on colour and texture models

Prajakta et al. [1] utilizes the green pixel masking algorithm based on RGB colour space as segmentation method to segment the diseased region of six diseases that generally affect sugarcane plant namely, Red tot, downy mildew, Mosaic Disease, Red leaf stripe, Leaf scald, Downy Fungal.

Francis et al. [2] projected a green pixel masking scheme used to discover the severity of infection (berry spot and quick wilt) in pepper plant. The main feature used to determine the severity is Damage ratio, which computes the percentage of damaged leaf area in the whole leaf area. A supervised learning neural network is worn for cataloging yielded an accuracy of 100% with 20 testing images has been achieved.

Kim et al. [3] proposed a color texture method for spotting diseases in citrus fruit peels. Images taken are converted to HSI format from RGB format and texture features like uniformity, mean intensity, variance are used to classify the diseases. A proper lighting setup was made to curtail mirror like reflectance and shadow, at the same time maximize the contrast of the concerned images The proposed method has been tested for five diseases namely Melanose, citrus scab, Fungal canker Spray burn and Wind Scars and identified 96.7% of diseases correctly.

Narvekar et al. [4] discussed an effective way of segmentation of diseased region. The image is transformed into HIS colour space from RGB, the S and I components are dropped and only the H component is taken for additional analysis, using Otsu's classifier threshold levels based on the H component, the green pixels are completely masked. Then the colour vaules of pixels with Green, Blue and zero red were removed. Classification was done based on tint and touch features. The authors concluded that use of hybrid algorithms might increase the accuracy.

Pydipati et al. [5] resolved a texture-based HSI model using the "color co-occurrence method (CCM)" to differentiate healthy and infected citrus leaves. The first stage involves feature extraction using CCM and SAS statistical analysis. By the use of combined HIS features, best results were obtained. The authors concluded that Low lighting conditions influenced the color and saturation features and therefore, controlled lighting arrangements are required to accomplish optimum accuracy.

Wang et al. [6] anticipated a procedure to perform segmentation using XYZ color space. First, the incarcerated images were altered into "XYZ color space" from "RGB color space" followed by transformation of "XYZ colour to L*a*b color space". Squared Euclidean distance is then used to appraise color difference in a*b. A total of 4 outline features, 21 tint features and 25 consistency features are obtained from the segment image for distinguishing various diseases.

He et al. [7] suggested method for detecting disease in cotton leaf using three color models. Firstly, spatial irregular and transform field filtering is applied to eliminate the influence of noise on the captured image. Secondly, through the histogram equalization process surroundings representation boundaries are tinted and disparity is improved. These output image obtained is rehabilitated into the "R G B", "H I S", and "Y Cb Cr" color model. Comparative results were obtained after applying the respective colour transforms. Results showed that the "YCbCr" colour model was most successful in identifying the damage leaf. The author has further added that in this model the random noise interference in case of outdoor conditions is not handled well.

B. Detection based on histogram and edge detection algorithms

Deya et al. [8] represents a segmentation technique based on RGB levels and histogram of hue component in HSI to detect diseases in betel vine leaves and detect the rotten leaf area. The Otsu threshold method was used for classification achieving very high precision for the 12 test images used. The author has determined disease severity using the proposed method.

Tajane et al. [9] used Canny edge detection to spot diseases in Rose leaves. The image is transformed into gray-scale and the edge filtering is performed. Then, Edge histogram followed by colour histogram were used in the next stage. Detection is based on the mean and median values of image histogram. The author has concluded that the proposed method may be applicable for other species such as tomato, banana and sugar beet.

Sun et al. [10] proposes an improved histogram segmentation technique capable of automatic and accurate threshold calculation. Firstly, median filtering operation is performed on grayscale image. Secondly, the selection range is limited to 100–190 out of 255 pixels and minimum peak interval 10 pixels. Third, adaptive threshold T is calculated. Lastly, the image is segmented according. From three aspects: tint, surface, and outline using a multiple linear regression a whole of 11 different measures are drawn out for classification. The Regression model is trained with 40 images and achieved 90% accuracy in normal situations and 60% accuracy in random situation and studies proved increase in no. of training images can be increased to increase accuracy.

Camargo et al. [11] presented visual symptoms for identifying plant diseases using coloured images. The "RGB color space" of the unhealthy portion of leaf was transformed keen on "H, I3a and I3b" conversion. From the deviations of the intensities in the histogram, the images are partitioned. The accuracy of the technique is verified by comparing the results of the automatically partitioned images with manually partitioned images. The researchers accomplished that this practice is best suited for partitions consisting of a extensive variety of intensities.

Molina et al. [12] proposed a methodology to identify any category of infectivity on tomato leaf lets or an early blight disease. Characterization of Color is prepared using histogram using Hue-Max-Min method, which results sharing of spatial color and discrepancy in the color pattern in "Y Cb Cr color space". These feature or descriptor values are used in the "one out cross validation method" to accomplish improved categorization ratio. The author concluded that better accuracy was achieved using color structure descriptor than other methods.

C. Detection based on thresholding methods

Patil et al. [13] anticipated a method using straightforward and triangle threshold to fragment the healthy and the diseased region and also to recognize disease called brown spot in the sugarcane leaf. In this research a scheme to estimate the level of severity by lesion area and leaf area. The accuracy achieved through this method is 96.7% for a sample of 90 test images.

Kurniawati et al. [14] describes a procedure to identify diseases of paddy ("brown spot, blast and narrow brown spot"). Using "Otsu's and local entropy thresholding" scheme to perform isolation between the healthy and diseased region. To decide the identification results "Color, texture and shape features" are chosen, which that most excellent match the uniqueness of the chosen region. The author states that local entropy threshold provides the most accurate segmentation. Using color of the boundary, color of the spot, type of lesion and wrecked paddy's leaf color from images of a "paddy leaf" and predetermined set of rules the final classification of the three diseases has been performed. The author concluded that due to different intensity standards the Otsu's threshold method is unable to perform segmentation efficiently.

Tucker and Chakraborty [15] defined a pixel-based threshold level for segmentation and identification of two diseases in sunflower and oats namely, blight and rust. Firstly, segmentation is done based on the thresholds for the two different diseases. Then, the pixels are formed into different clusters based forming the diseased region. Segregation into the proper categories is done based on the features of the lesions. Even though the authors obtained good results, some errors are still present due to inappropriate lighting while taking the images.

Sena Jr. et al. [16] discussed an effort for recognizing infected maize plants from "fall armyworm". The images are captured under three different lightings and in eight different spots. The captured images are then converted into grayscale images using the additional green index. Then, the thresholds for these images were identified using the iterative method. Later, the images were acknowledged as a hale and hearty or spoiled based on amount of stuff in each one threshold chunk.

D. Detection based on clustering methods

Elangovan et al. [17] discusses a technique for revealing of diseases in chili plants using "K-means clustering". The captured image is changed into Grayscale image and then to binary image. Partitioning of the image is then carried out using "K-means clustering algorithm and Otsu's method". The categorization is done using the extracted features of diseased leaves such as color, texture and morphology using SVM classifier.

Bhangea et al. [16] deals with detection of diseases in pomegranate using image processing. This system uses 440 infected images and 170 healthy images are used for training neural network. Segmentation was done using k-mean clustering color coherence vector and classification of diseases is done by SVM system and pretrained neural network. The propose method has accuracy of 82.0%. Experiment showed less accuracy in detecting diseases of infected Stage-I that is 66%.

Athanikar et al. [18] focus on exposure and categorization of infections in potato plant. The paper proposes a system for early finding and diagnosis of infections through an automated image processing system using "k-mean clustering algorithm" for partitioning and "back propagation neural network" for sorting along with many features like range of colour, texture, area etc. The proposed system was able to accurately detect diseases like Early Blight, Roll Viral, Insect Damage. Using a set of 150 images for tested, an accuracy of 92% was achieved. The author has used MATLAB software for implementation.

Astonkar et al. [19] discusses a relative revision among diverse clustering algorithms. The algorithms argued in this literature are "K-means clustering, Fuzzy C-means clustering, mountain clustering method and subtractive clustering" method. Author states that "K-means clustering" is trouble-free and are with respect to computations more rapid than traditional clustering methods. A deep learning system using the K-means algorithm has been implemented in this paper.

Prajapati et al. [20] discusses a technique to identify the diseases in rice plant. They evaluate three techniques of segmentation such as (1) "L*a*b color space"-based "K-means clustering", (2) "Otsu's segmentation" technique, and (3) HSV color space-based "K-means clustering". During pre-processing of images then the images are subjected to feature extraction. A total of 88 features based on "colour, texture, and shape" and set to "SVM machine learning model" to classify the diseases. The process is used to detect syndromes namely "Bacterial leaf blight, Brown spot, and Leaf smut". The study proved that HSV color space-based segmentation is more accurate than other two and scored 96.71% accuracy also cross validation method with more no. of folds will help to increase the accuracy.

Biswas et al. [21] discussed a method to identify late blight disease in potato. In this work images taken in complex background or background with similar color is segmented using "fuzzy c-mean clustering. Back propagation neural network" with 3 layers with features extracted as input is used as classifier. The neural network is trained with 238 images and tested for 51 images and it identify all images correctly i.e. 100% accuracy has been achieved.

E. Classification based on Artificial Neural Network (ANN)

Abdullah et al. [22] used multilayer perception neural networks for classifying diseases in rubber trees. The initial preprocessing state is done using the median filter. Then, the region of interest was determined using color features extracted by means of RGB distribution indices. The Principal Component Analysis (PCA) is adopted as a parameter for each image. Depending on the prevailing pixel (mean value) and the normalized PCA data an enhanced artificial network was designed. Validation of this model was done through repeated analysis and testing. The authors concluded that this method is one of the best for identifying diseases such as "bird eye spot and otrichum".

Nevertheless, the usefulness can be augmented with a superior processor and high-promise camera.

Abdullakasim et al. [23] portrayed a system for detection of "brown spot" infection in a leaf. A variety of "color indices" like "RGB and HIS" are worn to identify the different regions of the leaf. The value of the "color indices" lies within the range of zero to one. ANN following this step is second-hand to catalog or make a distinction among an contaminated region and a hale and hearty region. It is bring into being to facilitate this scheme appropriately recognize approximately 79% diseased leaves and approximately 90% healthy leaves. In this learning also included false positive classification of contaminated leaves. The authors concluded that this technique can be enhanced by taking into consideration the illumination levels, possessions of syndrome phases and appropriate partitioning.

Reyalat et al. [32] proposes a method to identify "early scorch, cottony mold, ashenmold, late scorch and tiny whiteness" in plant leaves. The projected algorithm uses "k-means clustering and Otsu threshold for segmentation". The features are extracted to "Colour Co-occurrence matrix (CCM)" which is then combined with a "back propagation neural network" for categorization. The proposed algorithm has achieved an accuracy of 94.67%.

Khirade and Patil [33] described a set of given rules to categorize the illness in plants. They are image enhancement, partitioning, feature extraction, recognition and categorization. Pre-processing of image involves conversion of "RGB images" into "grey level images" and "histogram equalization" to distribute intensities of image. Boundary and spot detection algorithm is used to perform partition by transforming "RGB image is into the HIS model, k-means clustering and Otsu Threshold Algorithm" where grey level images are used to define the threshold. "Colour co-occurrence Method and Leaf colour extraction" using Hue and Blue components are involved in "feature extraction" and further are used in "back propagation neural network" to categorize illness.

Mengistu [38] describes a method for classifying coffee beans using back propagation artificial neural networks. This paper detection and identification of diseases and coffee seeds is discussed and 6 type of coffee seeds are used. Author used pre-processed images of 80 × 80, 100 × 100, 300 × 300 and 512 × 512 pixels. Partitioning is done using "FCM, K-means clustering and Otsu threshold method" and compared with each other for accuracy. The artificial neural network with 21 inputs of the combined feature vectors of morphological, color and texture and 6 neurons in its output layer to identify varieties of Ethiopian coffee beans is used as classifier. With total of 1200 images 840 images are used for training neural network and the system is tested with 360 images. Results shows that the accuracy is more (i.e. 94.5%) when "back propagation artificial neural network" with hyperbolic tangent activation function over FCM segmentation are combined. Also as pixel size of image increases accuracy also increases.

F. Classification based on Convolutional Neural Network (CNN) and Probabilistic Neural Network (PNN)

Rajmohan et al. [24] proposes a new mobile app based on ML algorithms to identify the diseases in paddy plant. The primary technique used for segmentation is ROI cropping. A system for classification using Bayesian artificial neural network, fuzzy classification, deep-CNN and SVM was considered. It was determined that using 200 training images and 50 testing images for the experiment and an accuracy of 87.50% which was the highest was achieved using the deep-CNN technique.

Asfarian et al. [25] portrayed a routine for spotting disease in paddy using fractal descriptors, which in turn depends on "Fourier transforms with texture analysis". Here, damaged region of the leaves (laceration) are obtained by hand and after that every one of these laceration images were altered to "HSV color space". The effects due to lighting are reduced through histogram equalization. The fractal descriptors from each laceration are drawn out and given to the "PNN classifier" for determining the disease. The author stated that this method, can be used for differentiation between two diseases which have the same color.

Ferentinos [26] proposed a convolutional neural network algorithm to develop deep learning models which can perform plant illness recognition and finding. Researchers created a data base be made of 87,848 images, be full of 25 variety of plants in a deposit of 58 diverse module of [plant, disease] blend was created for the neural network. As part of the preprocessing the images in the database were cropped to 256×256 pixels. In this paper it is stated that the nearly all unbeaten representation structural design was a "VGG convolutional neural network", achieved a success rate of 99.53%.

Lua et al. [27] proposes wheat identification scaffold based on deep multiple instance learning (DMIL-WDDS) which is a weekly supervised deep learning framework, which achieves an amalgamation of recognition of the possible diseases in wheat and spotting of contaminated vicinity in the midst of only image-level footnote for tutoring images in uncultivated circumstances. "Local feature extraction and local disease estimation" for a appropriate image incarcerated from mobile camera is performed using a "fully convolution network (FCN)", then to accurately lock syndrome positions a "bounding-boxes approximation (BBA)" step is performed. Database having various Diseases in Wheat (WDD2017) be full of 9,230 images by way of 7 diverse modules "Powdery Mildew, Smut, Black Chaff, Stripe Rust, Leaf Blotch, Leaf Rust, Healthy Wheat" is used for testing and training and achieved an accuracy of 97.27%. experiment results prove that proposed technique is more efficient that conventional CNN architecture.

Debasu et al. [36] describes a technique for the classification of coffee plant infections such as "Coffee Leaf Rust (CLR), Coffee Berry Disease (CBD) and Coffee Wilt Disease (CWD)". Image enhancement techniques are used for eradicate "low frequency background noise", regularize the intensity, removing reflection and masking portion of image and "median filter" is used to remove noise from image. "K-means clustering" is used as "segmentation" technique and in feature extraction stage total of 17 parameters are drawn out using GLCM, Statistical and Color features. These pulled out parameters are fed to BPNN which has 17 inputs and 3 outputs one for each disease and middle layer can be chosen according to requirement. Among the 9100 data sets authors make use of 6370 for representation training and 2730 for validation of the functioning and achieved an accuracy of approximately 94%.

G. Classification based on Support Vector Machines (SVM)

Rumpf et al. [28] proposes a method for classification of rice plant hopper infestation in rice using Fractal eigen values as the parameter for classification. ROI cropping and FCM algorithm are the techniques used for segmentation, while classification is done using SVM with Fractal eigen values. An accuracy of 87% was achieved in identifying the disease and an accuracy of 63.5% was achieved in differentiating between its four types. The author concludes that accuracy can be improved by using different colour spaces and rescaled images.

Raut and Fulsunge [29] a model to detect and identify plant diseases using MATLAB software. The methodology consists of two main sections, first section represents the segmentation procedure done through "K-means clustering algorithm" and the second section which represents the classification procedure done through SVM with features extracted using the "Gray-level Co-occurrence matrix".

Jalal and Dubey [30] discusses a classification method using Multiclass SVM for identifying syndromes in apple ("Apple Blotch, Apple rot and Apple scab"). The features used by the SVM for identification are "Global Color Histogram, Color Coherence Vector, Local Binary Pattern, and Complete Local Binary Pattern". The projected technique achieved an precision of 93% considering a total of 431 training images.

Prajakta et al. [1] proposes a system of detection as mentioned in the first section of the literature review. The authors have performed a comparative study between "Linear SVM, Non-linear SVM and Multiclass SVM" for classification and concluded that the SVM algorithm is found to be better for classification compared to the other algorithms.

Singh et al. [31] devised a technique to diagnose and spotting of infected areas in banana plant and as well performed a proportional learning on the diverse classification algorithms that can be used for them. Features of the image like contrast, energy and local homogeneity are use to differentiate between the diseases. The author also claims that the usage of SVM classifier yields uppermost accurateness.

Meena et al. [34] proposes an algorithm where the images are converted from "RGB to L*a*b* colour space" which is used to determine luminosity, chromaticity layers and to enhance visual analysis. On these pre-processed images "K-means clustering" is applied to get appropriate separation and also a "Gray-Level Co-Occurrence Matrix (GLCM)" in which is the spatial relationship among pixels are utilized to examine texture and in turn to pull out required features. The features like "contrast, energy, homogeneity and correlation" are found and used in support vector machine to spot and catalog diseases. "Support Vector Machine is kernel-based supervised learning algorithm which uses kernel Hilbert space (RKHS)" to reduce computational complexity. The experiment uses 60 images of citrus leaves and shows minimum accuracy of 90% which can be improved up to 100%.

Zhang et al. [37] describes a disease detection system for cucumber plants. Firstly, image enhancement techniques are used for smoothing, enhancing and de-noising of the image. These images are subjected to segmentation which use k-means clustering and super pixel method. Super pixel is a deposit of pixels contains added information than a single pixel which can be obtained simple linear iterative method technique.

Then the feature extraction image using "CIELAB color space" and gray scale images is done and PHOG features are extracted. Contest-aware SVM is used as classifier to detect diseases in cucumber such as "anthracnose, angular leaf spot and powdery mildew" with 150 images and diseases in apple such as alternaria, mosaic and rust and achieved highest accuracy 92.15%. This study is compared with other 3 methods of disease detection such as KSNNC, SIFT and IRT and proved that it has significantly more accurate than other 3 methods.

H. Classification using Random Forest Classifier

Maniyath et al. [35] proposes an image detection and classification system for papaya leaves using Random Forrest Classifier algorithm. First, enhancement of the image is done in order to calculate Hu's Moment shape descriptor and Haralick feature.

3 Discussion

The various methods and techniques discussed above for detection and classification have different accuracies and consistencies. In order to achieve a well and widely applicable system, consistency is required. The diseases can be identified through visual inspection, but since it is highly irregular and requires knowledge, it gives poor results. The above methods have some slight advantages over the traditional methods. In order to achieve a system capable of replacing the traditional methods, the following factors need to be taken into consideration.

A. Imprecise Image Acquisiton Techniques

The methods used to obtain the image is inconsistent, in regions exposed to a lot of sunlight, the images are brighter than the images from other regions. The presence of noise like this makes the analysis of the image difficult. Interference due to noise leads the image processing techniques to give erroneous results

B. Standardized System

A lack of standardized or universal system/techniques which can be used in all regions. The different approaches proposed by the authors have taken various variables into consideration.

C. Availability of proper data

The supplementary anxiety is linked to the inaccessible of absolute and consistent statistics about data. The variables and conditions that enclosed to bring into being the consequences are not abundantly known. Several investigators do not offer their "testing and training data" statistics, causes major distress for the period of examination and justification of consequences. It is obligatory to present diverse consistent measures.

D. Image quality

Higher quality images are required to obtain more accurate results.

E. Image complexity
Due to the complexity of the image captured, it is difficult to segment the image background from the foreground.

4 Conclusion

From the above cited investigations the concepts and methodologies worn by a variety of researchers in the cause of spotting and diagnosing diverse plant leaf diseases, challenging issues and other harms are highlighted. The foremost rationale of this revision is to diminish fatalities due to infections in plants using image processing techniques. Due to the extensive number of infections on the leaves and variety of species, it is difficult to obtain a single method to provide fruitful results. The authors should consider more unique methods to provide superior results. They should afford additional consistent consequences by bearing in mind the exactness and superiority measures that are looked-for in this greatly aggressive and waving industry. A review on the fashionable techniques used for cataloging and detection has been performed beside through potential deliberations. The winding up in this review can be utilized in support of further research involving pattern recognition algorithms. From the above cited revisions, promotion of research in this area can be prepared by posturing in mind

- Outdoor conditions as another parameter for classification.
- Different combinations of drawing out of different characteristic, assortment and erudition techniques to enhance the accuracy of recognition and categorization methods.
- Mobile applications can be used to improve the portability of this system.
- Solutions and treatment methods to help rectify the disease can be provided along with the results
- augmented number of images available in the data sets to achieve greater accuracies.
- The existing work can be improved to perform high speed analysis with the use of more efficient algorithms.

This review provides an overview of the different techniques and concepts which are involved in the recognition, spotting and categorization of infected leaves by means of image dealing out methods. As there are a wide number of methods proposed, the future scopes may not be limited to the ones mentioned above.

References

1. Mitkal, P., Pawar, P., Nagane, M., Bhosale, P., Padwal, M., Nagane, P.: Leaf disease detection and prevention using image processing using Matlab. Int. J. Recent Trends Eng. Res. (IJRTER) **02**(02), 1457–2455 (2016)
2. Francis, J., Sahaya, A.D.D., Anoop, B.K.: Identification of leaf diseases in pepper plants using soft computing techniques. In: 2016 Conference on Emerging Devices and Smart Systems (ICEDSS), pp. 168–173, Namakkal (2016)

3. Kim, D.G., Burks, T.F.: Classification of grape fruit peel diseases using color texture feature analysis. IJABE **2**(3), 41–50 (2009)
4. Narvekar, P.R., Kumbhar, M.M., Patil, S.N.: Grape leaf diseases detection & analysis using SGDM matrix method. Int. J. Innov. Res. Comput. Commun. Eng. **2**(3), 3365–3372 (2014)
5. Pydipati, R., Burks, T.F., Lee, W.S.: Identification of citrus disease using color texture features and discriminant analysis. Comput. Electron. Agric. **52**(1), 49–59 (2006)
6. Wang, H., Li, G., Ma, Z., Li, X.: Image recognition of plant diseases based on principal component analysis and neural networks. In: IEEE Eighth International Conference on Natural Computation (ICNC), Chongqing, 29–31 May 2012, pp. 246–251 (2012)
7. He, Q., Ma, B., Qu, D., Zhang, Q., Hou, X., Zhao, J.: Cotton pests and diseases detection based on image processing. Indonesian J. Electr. Eng. Comput. Sci. **11**(6), 3445–3450 (2013)
8. Deya, A.K., Sharmaa, M., Meshramb, M.R.: Image processing based leaf rot disease, detection of betel vine (Piper BetleL.). Proc. Comput. Sci. **85**, 748–754 (2016)
9. Tajane, V., Janwe, N.J.: Medicinal plants disease identification using canny edge detection algorithm histogram analysis and CBIR. Int. J. Adv. Res. Comput. Sci. Softw. Eng. **4**(6), 530–536 (2014)
10. Sun, G., Jia, X., Geng, T.: Plant diseases recognition based on image processing technology. J. Electr. Comput. Eng. **2018**, Article ID 6070129, 7 p. (2018)
11. Camargo, A., Smith, J.S.: An image-processing based algorithm to automatically identify plant disease visual symptoms. Biosyst. Eng. **102**(1), 9–21 (2009)
12. Molina, J.F., Gil, R., Bojaca, C., Gomez, F., Franco, H.: Automatic detection of early blight infection on tomato crops using a color based classification strategy. In: XIX IEEE Symposium on Image, Signal Processing and Artificial Vision, pp. 1–5 (2014)
13. Patil, S.B., Bodhe, S.K.: Leaf disease severity measurement using image processing. Int. J. Eng. Technol. **3**, 297–301 (2011)
14. Kurniawati, N.N., Abdullah, S.N.H.S., Abdullah, S., Abdullah, S.: Investigation on image processing techniques for diagnosing paddy diseases. In: IEEE International Conference of Soft Computing and Pattern Recognition (SOCPAR 2009), Malacca, 4–7 December 2009, pp. 272–277 (2009)
15. Tucker, C.C., Chakraborty, S.: Quantitative assessment of lesion characteristics and disease severity using digital image processing. J. Phytopathol. **145**(7), 273–278 (1997)
16. Bhangea, M., Hingoliwalab, H.A.: Smart farming: pomegranate disease detection using image processing. Proc. Comput. Sci. **58**, 280–288 (2015)
17. Elangovan, K., Nalini, S.: Plant disease classification using image segmentation and SVM techniques. IJCIRV **13**(7), 1821–1828 (2017). ISSN 0973-1873
18. Athanikar, G., Badar, P.: Potato leaf diseases detection and classification system. Int. J. Comput. Sci. Mob. Comput. **5**(2), 76–88 (2016)
19. Astonkar, S.R., Shandilya, V.K.: Detection and analysis of plant diseases using image processing technique. Int. Res. J. Eng. Technol. (IRJET), **05**(04) (2018)
20. Prajapati, H.B., Shah, J.P., Dabhi, V.K.: Detection and classification of rice plant diseases. Intell. Decis. Technol. **11**, 357–373 (2017)
21. Biswas, S., Jagyasi, B., Singh, B.P., Lal, M.: Severity identification of potato late blight disease from crop images captured under uncontrolled environment. In: IEEE Canada International Humanitarian Technology Conference - (IHTC) 978-1-4799-3996-1/14/$31.00 ©2014. IEEE (2014)

22. Abdullah, N.E., Rahim, A.A., Hashim, H., Kamal, K.: Classification of rubber tree leaf diseases using multilayer perceptron neural network. In: Fifth Student Conference on Research and Development (SCORed), Selangor, 11–12 December 2007, pp. 1–6 (2007)

23. Abdullakasim, W., Powbunthorn, K., Unartngam, J., Takigawa, T.: An images analysis technique for recognition of brown leaf spot disease in cassava. J. Agric. Mach. Sci. 7(2), 165–169 (2011)

24. Rajmohan, R., Pajany, M.: Smart paddy crop disease identification and management using deep convolution neural network & SVM classifier. Int. J. Pure Appl. Math. 118(5), 255–264 (2017)

25. Asfarian, A., Herdiyani, Y., Rauf, A., Mutaqin, K.H.: Paddy diseases identification with texture analysis using fractal descriptors based on Fourier spectrum. In: International Conference on Computer, Control, Informatics and its Applications (IC3INA), Jakarta, 19–21 November, pp. 77–81 (2013)

26. Ferentinos, K.P.: Deep learning models for plant disease detection and diagnosis. Comput. Electron. Agric. 145, 311–318 (2018)

27. Lua, J., Hua, J., Zhaoa, G., Meib, F., Zhanga, C.: An in-field automatic wheat disease diagnosis system. Comput. Electon. Agric. 142PA, 369–379 (2017)

28. Rumpf, T., Mahlein, A.K., Steiner, U., Oerke, E.C., Dehne, H.W., Plumer, L.: Early detection and classification of plant diseases with support vector machines based on hyperspectral reflectance. Comput. Electr. Agric. 74(1), 91–99 (2010)

29. Raut, S., Fulsunge, A.: Plant disease detection in image processing using MATLAB. IJIRSET 6(6), 10373–10381 (2017)

30. Jalal, A.S., Dubey, S.R.: Detection and classification of apple fruit diseases using complete local binary patterns. In: IEEE Third International Conference on Computer and Communication Technology (2012). ISBN 978-0-7695-4872

31. Singh, V., Misra, A.K.: Detection of plant leaf diseases using image segmentation and soft computing techniques. Inf. Process. Agric. 4(1), 41–49 (2017)

32. Al-Hiary, H., Bani-Ahmad, S., Reyalat, M., Braik, M., ALRahamneh, Z.: Fast and accurate detection and classification of plant diseases. Int. J. Comput. Appl. 17(1), 31–38 (2011)

33. Khirade, S.D., Patil, A.B.: Plant disease detection using image processing. IEEE (2015)

34. Meena, R., Saraswathy, G.P., Ramalakshmi, G., Mangaleswari, K.H., Kaviya, T.: Detection of leaf diseases and classification using digital image processing, pp. 1–4 (2017). https://doi.org/10.1109/iciiecs.2017.8275915

35. Maniyath, S.R., et al.: Plant disease detection using machine learning. In: 2018 International Conference on Design Innovations for 3Cs Compute Communicate Control (ICDI3C), pp. 41–45, Bangalore (2018)

36. Mengistu, A.D., Mengistu, S.G., Melesew, D.: An automatic coffee plant diseases identification using hybrid approaches of image processing and decision tree. Indones. J. Electr. Eng. Comput. Sci. 9(3), 806–811 (2018). ISSN: 2502-4752

37. Zhang, S., Wang, H., Huang, W., You, Z.: Plant diseased leaf segmentation and recognition by fusion of superpixel, K-means and PHOG. Optik Int. J. Light Electron. Opt. 157, 866–872 (2018)

38. Mengistu, A.: The effects of segmentation techniques in digital image based identification of Ethiopian coffee variety. Telkomnika (Telecommun. Comput. Electron. Control) 16, 713–717 (2018)

39. Eason, G., Noble, B., Sneddon, I.N.: On certain integrals of Lipschitz-Hankel type involving products of bessel functions. Phil. Trans. Roy. Soc. London A247, 529–551 (1955)

40. Clerk Maxwell, J.: A Treatise on Electricity and Magnetism, vol. 2, 3rd edn, pp. 68–73. Clarendon, Oxford (1892)
41. Jacobs, I.S., Bean, C.P.: Fine particles, thin films and exchange anisotropy. In: Rado, G.T., Suhl, H. (eds.) Magnetism, vol. III, pp. 271–350. Academic, New York (1963)
42. Elissa, K.: Title of paper if known (unpublished)
43. Nicole, R.: Title of paper with only first word capitalized. J. Name Stand. Abbrev. (in press)
44. Yorozu, Y., Hirano, M., Oka, K., Tagawa, Y.: Electron spectroscopy studies on magneto-optical media and plastic substrate interface. IEEE Trans. J. Magn. Jpn. **2**, 740–741 (1987). Digests 9th Annual Conference Magnetics Japan, p. 301 (1982)
45. Young, M.: The Technical Writer's Handbook. University Science, Mill Valley (1989)

Test Case Minimization for Object Oriented Testing Using Random Forest Algorithm

Ajmer Singh[1(✉)], Diksha Katyal[1], and Deepa Gupta[2]

[1] CSE Department, DCRUST, Murthal, India
ajmer.saini@gmail.com, diksha.katyal1212@gmail.com
[2] Amity Institute of Information Technology, Noida, India
deepa19july@gmail.com

Abstract. Software maintenance is one of the most costly and crucial phases in the life cycle of software. It consumes almost 70% of the resources and cost of the software. Software testing aims to execute or examine the software with the intention of detecting the faults in it. Reducing the cost of the testing process is one of the major concerns of the testers. With the growing complexities in Object Oriented (OO) software, the number of faults present in the software module is increased. In this paper, a technique has been presented for minimizing the test cases for the OO systems. A case study of Xerces 1.4 open source software is carried for the evaluation of proposed technique. The mathematical model used in the proposed methodology was generated using the open source software WEKA. The approach is based on selecting significant Object Oriented metrics. Highly Efficient, Less efficient or inefficient Object Oriented metrics were identified by the techniques based on feature selection. Test case generation and minimization is achieved on the basis of coverage of highly fault prone classes. To minimize the test cases, proposed methodology used only significant OO metrics for assigning weights to the test paths. The proposed work promisingly reduced the cost and time taken during test suite minimization.

Keywords: Software maintenance · Machine learning · Test case minimization · Object oriented testing · Random Forest

1 Introduction

Presence of bugs in the software module reduces the quality of software. It is necessary to improve quality of software by identifying the defects and remove them from the software module to deliver reliable software product [1]. In the testing phase development team finds the bugs present in the software module. It is expensive to test the entire software module within a limited period of time. As a result, less reliable and faulty software is released. Therefore, it is required to eliminate software bugs within limited time and reduced cost. Software fault prediction model is one of the solutions to this problem. The use of software fault prediction is done in order to predict the faults present in the software module. To select the test suite with minimum number of test

© Springer Nature Switzerland AG 2020
A. P. Pandian et al. (Eds.): ICCBI 2019, LNDECT 49, pp. 824–833, 2020.
https://doi.org/10.1007/978-3-030-43192-1_90

cases test case minimization technique is used which is able to reveal more software bugs within reduced cost and less time.

The software fault prediction model is practiced by using attributes of the software module and fault data. Then the information is gathered using previously released software or identical projects which is used to calculate whether the software module is defective or not. The performance and efficiency of software fault prediction model rely on the characteristics of the software attributes that are used to predict or calculate whether the faults are present in the software module or not.

The organization of paper is as follows. In Sect. 2 we provide information about the background or related work on software defect prediction using software metrics. In Sect. 3 provide details about the methodology used in this paper. In Sect. 4, implementation steps are defined. Section 5 highlights the results. Section 6 gives conclusions and future work.

2 Related Work

Different object Object oriented metrics have been proposed by the researchers in the literature. A detailed description of these metrics has been by authors in [2]. Out of various proposed metrics, the metrics proposed by Chidamber and Kemerer (CK) has been found most popular among the researchers. Different authors have evaluated these metrics for the bug prediction, maintenance prediction and quality prediction in software testing. Some of the researches relevant in this context are elaborated here. Authors in [3] described how to calculate CK metrics and applied these calculated values in prediction of fault-proneness of the source code of software systems. The authors examined the value obtained against the bugs found in the database containing bugs using machine learning algorithms to confirm ease of the OO metrics for the prediction of fault-prone classes. Study in [4] analysed the effectiveness of the network metrics above code metrics for the prediction of bug. The proposed work was carried out on eleven datasets from the Open source PROMISE repository by using different machine learning algorithms. Their results shown that the code metrics predicted better results over the network metrics.

In [5], authors tried to minimize the number of object oriented software metrics needed for the fault proneness. They proposed an algorithm that predicts the fault proneness index by using marginal R^2 values method. Testing was performed as a mediator step in the given approach; it was assumed that they were different in behavior when they were compared with other models.

Authors in [6], proposed that a larger test suite was required to test the software module and different methods were bought up to reduce the test cases in order to reduce the testing cost and time. The study was conducted to present the effectiveness of Genetic Algorithm to decrease the test cases and the steps of genetic algorithm were repeated to minimize the test suite.

Authors in [7] applied Object Oriented metrics for prediction of software quality and based upon their effectiveness, the different metrics are weighted. In their research simple linear regression model was built to correlate the different metrics with the predict quality of the software. In the study of [8], Support vector machines based prioritization approach has been proposed. Authors have utilized the object oriented metrics as vectors for the classification of software modules into good quality and poor quality modules. Study in [9] designed software bug prediction system. The model predicted the bugs that were present in a class using object oriented metrics. The model predicted the occurrence of bugs in a class during the software module under test. They formulated hypotheses corresponding to each metric. The study in [10] defined a technique for test case minimization using genetic algorithm (GA) with different length of chromosome to reduce the complete test suite by discovery of representative set of test cases that fulfilled the testing criterion. Authors of [11], proposed the approach as hybrid Self Organizing Map (SOM) that used code metrics of source code file to find the attributes that were fault prone and were present in a software. The researchers used this model for OO software modules to calculate fault prone code at the class-level using OO metrics which made it easy to give priority to the efforts of the software testing.

Another study in [12], formulated multi-objective approach for minimizing the test cases. This approach focused basically on choosing minimum number of test cases for execution by maximizing its effectiveness i.e. limited time, coverage and reduced cost.

Authors in [13] defined multi objective optimization approach that had disagreeing objectives. The main aim of this research was to deal with multi-objective problem that was used to find the resolution for all disagreeing objectives. The researchers proposed automatic test data generation. One of the motives was to maximize the code and another was uniform distribution of code. The approach also defined non-dominance property that maintained sub-population of best fitness value. The study of [14], reviewed the redundancy based test-suite reduction. This research proposed an approach where test-cases were made with a model-checker such that it avoided the redundancy within the test-suite and thus reduced the test suite size. As redundant test-cases were not simply removed, therefore the sensitivity related to faults was minimized. Regehr et al. [15] studied on test-case reduction for C compiler bugs. The research dealt with reporting a compiler bug and often found a small test case that triggered the bug. They concluded that effective test-case reduction required more than that of straightforward delta debugging.

Authors in [16] proposed a dynamic programming algorithm that was applied on software testing domain, basically in the selection of the test-cases. The researchers defined a specific problem that was present in software testing and were running a subset of test-suite from a whole set of test suite. The research was focused on maximizing the possibility of covering potential bugs.

Muthyala et al. [17] proposed an approach to reduce test suite using data mining techniques. They proposed a method that used clustering approach by which they can considerably reduce the test-suite. The resultant test suite was tested for coverage which yielded good results. Authors in [18] applied a path based testing approach covering all du-paths for a given software program module. The Genetic Algorithm deals with automatic test suite generation and optimization was done against the accepted set of inputs and was checked for the coverage of path. Kaur et al. [19] proposed a technique to evaluate the random forest application for prediction of faults that were predicted in the class of the software. The researchers used the open source software jedit and used object oriented metrics to perform studies. Result shows that the accuracy of RF value was 74.24% and its precision was 72%, its recall is 79%, its F-measure was 75%, and its AUC was 0.81.

Malhotra [20] proposed an empirical assessment of metrics to analyze the quality attributes. The CK and QMOOD metrics were used for fault proneness classes. One statistical method and six machine learning algorithms were used to predict the model. Open Source software was used for testing of models. Results obtained were verified using area under the curve. From results it was predicted that the random forest with bagging outperform all other models.

Singh et al. [21] proposed demand based TCG approach that selected the test scenarios as per the appropriate demand in form of percentage. The optimization of test cases were done and optimized test suite was then selected to fit within the budget. The study of [22] defined and developed a prediction model which deals with bug indicators as model's input and evaluated on openly available source projects namely Ant and Camel. In the proposed research, the results were verified and considerable correlation exists between the size metrics or bug indicator OO metrics such as WMC, DIT, LOC, CBO and bugs. The DIT metric was predicted to be more effective than other bugs predicting OO metrics such as WMC, CBO and LOC. Sampath et al. [23] performed study on improving the efficiency of test-case minimization for application based user-session testing. The software system's characteristics were coupled with the observations that produced the ordered and reduced test suites. To acquire effectiveness of testing, a tester must order their reduced test suites.

Vidács et al. [24] provided a combined approach for reducing test suite for detection and localization of faults. The relation between the test suites and the actual faults in software module is of significant importance for timely release of product. They conducted experiment with software programs that were traditionally being used in the research of fault localization and therefore, extended the case study with industries dealing with large software systems. Singh [25] proposed an optimized test case generation approach. The effective Object Oriented metrics were selected and study was carried out for the software named as ant-1.7. The linear regression technique was used for giving weights to the test paths and was used for generation of the mathematical model.

It is evident from the literature that object oriented metrics have been used from different perspectives in software testing but the work related their usage in test case minimization is nearly negligible. So, this paper proposes the minimization of test cases using object oriented metrics.

3 Proposed Methodology

The proposed technique is efficient in reducing number of test cases. Only selected test cases are executed for the prediction of faults within limited time and are shown in Fig. 1.

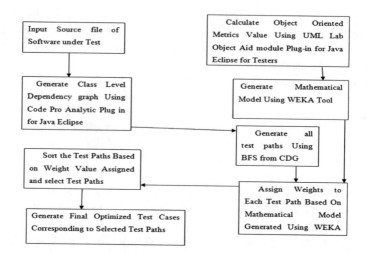

Fig. 1. Various steps of proposed methodology.

Select the effective OO metrics of the classes of XERCES-1.4 [24] publicly available open source software using WEKA [25] machine learning tool. Then after the selection of effective metrics some steps are followed as:

1. Generate the mathematical model of the selected object-oriented metrics using WEKA machine learning tool by applying random forest approach.
2. Input the open source code file of the software module used for testing i.e. XERCES-1.4.
3. Now, generate Class level dependency graph (CDG) by selecting some of the java files using code-pro analytic plug-in for java eclipse.
4. Calculate the weights of the generated classes by adding the weights of the classes that covers the individual test paths.
5. Generate all the test paths by applying Breadth First Search (BFS) on CDG.
6. Sort the generated paths in increasing weight values and selection of the test paths with higher weight values is done.
7. Generate the final effective test cases from the resultant selected test paths.
8. As a result, now apply Random Forest Algorithm on most significant object oriented metrics and examine the results including its accuracy.

The Class-level OO metrics has been considered for detecting the fault proneness of the classes which then used to calculate the weights of the classes. The metrics data was collected from the open source and publicly available repository known as PROMISE repository. Xerces 1.4 has been used for the evaluation of the proposed methodology.

4   Implementation

Step 1: The proposed work has used the XERCES 1.4 data set. Data was obtained in. csv format from publically available data repository PROMISE [26]. Data has different object oriented metrics measured for the different modules of XERCES 1.4. Also, number of faults present in each module has been provided. In order to classify the modules the following five categories shown in Table 1 below were identified.

Table 1. Different modules categories based on number of faults present

S.N	No of faults	Category
1	0	No_Fault
2	>0 and ≤ 6	Low
3	>6 and ≤ 11	Moderate
4	>11 and ≤ 15	High
5	>15	Critical

Step 2: Select effective OO metrics for detecting bugs. The Object Oriented metrics namely rfc, loc, wmc, ce, cbo, bug have been selected and is acknowledged as top significant fault-prone metrics that can be used for generating mathematical model using Weka tool.

Step 3: Generate Class Dependency Graph (CDG) of the software module Xerces 1.4 using code-pro analytic in eclipse [27] neon for testers. Generated CDG for Xerces 1.4 is shown in Fig. 2 below.

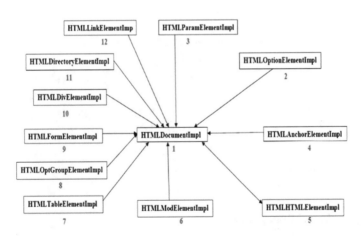

Fig. 2. Generated CDG for Xerces 1.4

From the class dependency graph (CDG) in Fig. 2, the test cases were identified using BFS method. The test cases generated are given in Table 2 below:

Table 2. Different test cases generated

S.N	Test case	Nodes covered	S.N	Test case	Nodes covered
1	TC1	1,5	6	TC6	6,1
2	TC2	2,1	7	TC7	7,1
3	TC3	3,1	8	TC8	8,1
4	TC4	4,1	9	TC9	9,1
5	TC5	5,1	10	TC10	10,1
6	TC6	6,1	11	TC11	11,1
7	TC7	7,1	12	TC12	12,1

Step 4: Select test paths generated by using open source machine learning tool known as WEKA [28]. The selection of the test cases are based on the value of weight assigned to each test case. Principal Component Algorithm is applied for selecting the effective set of metrics and for calculating weight values. Therefore, if any feature is selected by the algorithm should definitely be given high importance.

Step 5: The class level test cases has been generated resultant to the selected test path. In this case study of Xerces 1.4, Table 3 represents the selected test paths which are selected on the basis of weights value assigned to each test path. Test cases are selected on the basis of highest weight value to reduce testing time and limited cost requirements.

Table 3. Different weights for test cases

S.N	Test case	Weight	S.N	Test case	Weight
1	TC1	−401.807	7	TC7	−580.163
2	TC2	−474.353	8	TC8	−407.442
3	TC3	−416.57	9	TC9	−440.519
4	TC4	−468.827	10	TC10	−401.807
5	TC5	−401.807	11	TC11	−400.796
6	TC6	−405.717	12	TC12	−440.878

5   Results and Discussion

The analysis of the results that are obtained by using proposed methodology is discussed as follows. After calculating the weights for the different metrics and hence the different tests following test cases as shown in Table 4. Below are generated.

Table 4. Final test cases generated

Test case	Test path	Class level test case
TC11	11,1	HTMLDirectoryElementImpl → HTMLDocumentImpl
TC5	1,5	HTMLDocumentImpl → HTMLHTMLElementImpl
TC6	1,6	HTMLModElementImpl → HTMLDocumentImpl
TC8	1,8	HTMLOptGroupElementImpl → HTMLDocumentImpl
TC3	1,3	HTMLParamElementImpl → HTMLDocumentImpl

Now finally applying Random forest algorithm on the data set XERCES 1.4 with highly significant object oriented metrics i.e. wmc, rfc, loc, ce and cbo. The snapshots below shows preprocessing window and classifier window showing results considering top five significant features and a class bug.

6 Conclusion and Future Scope

The proposed research deals with selection of the effective Object Oriented metrics for prediction of bugs using WEKA tool. The CDG was generated using code-pro for XERCES 1.4 open source software module. Using Breadth first Search (BFS) test paths were generated. The mathematical model was generated using WEKA for the XERCES 1.4 dataset available at promise repository. Weights values were assigned to each of the generated test case. The final test suite has been generated resultant to subset of selected test suite which saved time of testing and effort. As a result, test cases have been minimized and execution time is reduced. The proposed approach used is capable of achieving the objectives but still more work can be done in future by considering different machine learning algorithms. More options of machine learning can be used for saving more execution time and to automate testing completely.

References

1. Mandal, P., Ami, A.S.: Selecting best attributes for software defect prediction. In: 2015 IEEE International WIE Conference on Electrical and Computer Engineering, WIECON-ECE 2015 (2016)
2. Singh, A., Bhatia, R., Sighrova, A.: Taxonomy of machine learning algorithms in software fault prediction using object oriented metrics. Procedia Comput. Sci. **132**, 993–1001 (2018)
3. Gyimothy, T., Ferenc, R., Siket, I.: Empirical validation of object-oriented metrics on open source software for fault prediction. IEEE Trans. Softw. Eng. **31**(10), 897–910 (2005)
4. Prateek, S., Pasala, A., Aracena, L.M.: Evaluating performance of network metrics for bug prediction in software. In: Proceedings - Asia-Pacific Software Engineering Conference, APSEC (2013)
5. Puranik, S., Deshpande, P., Chandrasekaran, K.: A novel machine learning approach for bug prediction. Procedia Comput. Sci. **93**, 924–930 (2016)
6. Akour, M., Abuwardih, L.: Test case minimization using genetic algorithm: pilot study. In: 2018 8th Conference on Computer Science and Information Technology, pp. 66–70 (2018)

7. Singh, A., Bhatia, R.K., Singhrova, A.: Object oriented coupling based test case prioritization. Int. J. Comput. Sci. Eng. **6**(9), 747–754 (2018)
8. Singh, A., Bhatia, R.K., Singhrova, A.: Machine Learning based Test Case Prioritization in Object Oriented Testing
9. Gupta, D.L.A.L., Saxena, K.: Software bug prediction using object-oriented metrics. Sadhana **42**(5), 655–669 (2017)
10. Mohapatra, S.K.: Minimizing test cases to reduce the cost of regression testing. In: 2014 International Conference on Computing for Sustainable Global Development, pp. 505–509 (2014)
11. Boucher, A., Badri, M.: An unsupervised fault-proneness prediction model using multiple risk levels for object-oriented software systems: an empirical study. Mémoire présenté à l'université du québec à trois-rivières (2018)
12. Ali, S., Li, Y., Yue, T., Zhang, M.: An empirical evaluation of mutation and crossover operators for multi-objective uncertainty-wise test minimization. In: Proceedings - 2017 IEEE/ACM 10th International Workshop on Search-Based Software Testing, SBST 2017, pp. 21–27 (2017)
13. Choudhary, K., Purohit, G.N.: A multi-objective optimization algorithm for uniformly distributed generation of test cases. In: 2014 International Conference on Computing for Sustainable Global Development, INDIACom 2014, pp. 455–457 (2014)
14. Fraser, G., Wotawa, F.: Redundancy based test-suite reduction. In: Lecture Notes in Computer Science (including subseries Lecture Notes in Artificial Intelligence and Lecture Notes in Bioinformatics), vol. 4422, pp. 291–305 (2007)
15. Regehr, J., Chen, Y., Cuoq, P., Eide, E., Ellison, C., Yang, X.: Test-case reduction for C compiler bugs. ACM SIGPLAN Not. **47**(6), 335–345 (2012)
16. Banias, O.: Dynamic programming optimization algorithm applied in test case selection. In: 2018 13th International Symposium on Electronics and Telecommunications, ISETC 2018 - Conference Proceedings, pp. 1–4 (2018)
17. Muthyala, K.: A novel approach to test suite reduction using data mining. Indian J. Comput. Sci. Eng. **2**(3), 500–505 (2011)
18. Khan, R., Amjad, M., Srivastava, A.K.: Optimization of automatic generated test cases for path testing using genetic algorithm. In: Proceedings - 2016 2nd International Conference on Computational Intelligence and Communication Technology, CICT 2016, no. February, pp. 32–36 (2016)
19. Kaur, A., Malhotra, R.: Application of random forest in predicting fault-prone classes. In: Proceedings - 2008 International Conference on Advanced Computer Theory and Engineering, ICACTE 2008, pp. 37–43 (2008)
20. Malhotra, R., Bansal, A.J.: Fault prediction considering threshold effects of object-oriented metrics. Expert Syst. **32**(2), 203–219 (2015)
21. Singh, R., Bhatia, R.K., Singhrova, A.: Demand based test case generation for object oriented systems. IET Softw. (2019)
22. Gupta, V., Ganeshan, N., Singhal, T.K.: Developing software bug prediction models using various software metrics as the bug indicators. Int. J. Adv. Comput. Sci. Appl. **6**(2), 60–65 (2015)
23. Sprenkle, S., Sampath, S., Gibson, E., Pollock, L., Souter, A.: An empirical comparison of test suite reduction techniques for user-session-based testing of web application. In: IEEE International Conference on Software Maintenance, ICSM, vol. 2005, pp. 587–600 (2005)
24. Vidács, L., Beszédes, Á., Tengeri, D., Siket, I., Gyimóthy, T.: Test suite reduction for fault detection and localization: a combined approach. In: 2014 Software Evolution Week - IEEE Conference on Software Maintenance, Reengineering, and Reverse Engineering, CSMR-WCRE 2014 - Proceedings, pp. 204–213 (2014)

25. Singh, R., Singhrova, A., Bhatia, R.: Optimized test case generation for object oriented systems using weka open source software. Int. J. Open Source Softw. Process. **9**(3), 15–35 (2018)

26. Promise Datasets Page. http://promise.site.uottawa.ca/SERepository/datasets-page.html. Accessed 14 Mar 2019

27. CodePro AnalytiX. https://www.roseindia.net/eclipse/plugins/tool/CodePro-AnalytiX.shtml. Accessed 07 Dec 2019

28. Hall, M., et al.: The WEKA data mining software: an update. ACM SIGKDD Explor. Newsl. (2009)

Exploring the Design Considerations for Developing an Interactive Tabletop Learning Tool for Children with Autism Spectrum Disorder

Nazmul Hasan[1](✉) and Muhammad Nazrul Islam[2]

[1] Department of Computer Science and Engineering, Engineering Faculty,
Bangladesh Military Academy, Chattogram, Bangladesh
nazmul_mist@yahoo.com
[2] Department of Computer Science and Engineering,
Military Institute of Science and Technology, Dhaka 1216, Bangladesh
nazrulturku@gmail.com

Abstract. Usability, ease of use and user experience (UX) are crucial quality concerns to develop any learning applications for children with autism. Several design considerations are proposed to design and develop different kinds of learning tools for autistic children but no study was found that explicitly focused to understand the design consideration required to design a tabletop learning tool for children with autism. Therefore, the objective of this research is to reveal the fundamental design considerations for developing a tabletop learning tool for autistic children with enhanced usability and UX. Two studies were conducted following ethnographic study and semi-structured interview, while the interviewing study was replicated with eighteen teachers of autistic children. The study data were analyzed through content analysis. As outcome, this research, firstly revealed a set of design considerations that includes, for example, affordance, intuitiveness, portability, ease-to-use, clear & concise interface, metaphoric design, easy to learn, interactive, visual feedback, multimodal feedback and pictography. Secondly, proposed a design solution to develop the tabletop learning tool based on the revealed design considerations.

Keywords: Autism · Design considerations · Ethnographic study · HCI · Learning tool · Usability

1 Introduction

Now a days, technologically enhanced education has become a priority in all over the world while special education for children with Autism Spectrum Disorders (ASD) is not an exception. The recent research demonstrates that the autistic children eagerly engaged in working with Information and Communication Technologies [1], despite having the limitations like impairment in communication skills and social interaction, and the restricted and repetitive behavior [2]. In [3], Kamaruzaman and Azahari highlights how the ICT based application helps to improve the self-determination of

© Springer Nature Switzerland AG 2020
A. P. Pandian et al. (Eds.): ICCBI 2019, LNDECT 49, pp. 834–844, 2020.
https://doi.org/10.1007/978-3-030-43192-1_91

autistic children. A significant number of research has been conducted focusing to ICT based learning and ASD, which recognizes the effectiveness of a variety of newly emerged learning technologies, however, in most cases, these research represent the development of different kinds of learning tools [4, 5].

Usability is the key quality for any computer systems [6]. One of the primary concerns to adopt a system by the end users is ease-of-use or usability [7]. Some prior studies focusing to HCI and learning technology for autistic children highlights a few design considerations to design effective interfaces for autistic children [1, 3]. A few studies measure the usability of the applications developed for the autistic children, for example, Weiss et al. [8] and Zaman and Bhuiyan [9] evaluate the usability of touch based applications (*Join-In-Suit* and *MumIES*) that were developed for autistic children. A few other studies focused only to develop new applications for autistic kids like Chien et al. [10] developed a tabletop based picture exchange communication system (PECS) for the autistic children to provide the visual and voice support. Only a few articles have been carried out focusing to the design and evaluation principles/guidelines considering the limitations of autistic children [11].

Again, a limited number of research was conducted focusing to the tabletop technology. For example, Weiss et al. [8] developed a multi-touch tabletop application named *Join-In-Suit* and presented the effectiveness of working together (group wrok) of the autistic children through this tool. Despite of the limited research focusing to the tabletop technology, evidence does suggest that this technology is an applicable technology for autistic kids with great potentials [12]. The design principles used for developing a mobile application will not be appropriate for developing an interactive tabletop learning tool to achieve maximum motivation, engagement and usability for autistic kids. Since no such study has been found that focuses explicitly to explore design considerations to develop tabletop learning tool for autistic kids.

Therefore, the objective of this research is to explore the underlying design considerations to design and develop an interactive learning tool (tabletop) for autistic kids. To attain this objective, this research conduct an extensive empirical study following the ethnographic and interviewing approach. The studies revealed a set of design considerations for developing a usable tabletop for autistic kids to learn the basic education and daily life activities. A design for developing a tabletop learning tools also proposed based on the revealed design considerations.

The rest of the paper is organized as follows. Discussion on related research is presented in Sect. 2. In Sect. 3, study methodology is discussed. The data analysis and findings is demonstrated in Sect. 4. The proposed design of the tabletop learning tool is discussed in Sect. 5. Finally, the discussion and conclusion are presented in Sect. 6.

2   Related Works

A number of research has been conducted focusing to ICT usages and the people with special needs like autism detection [13]; development of assistive tools/systems [14]; and the development of learning ands kill development tools [15, 16]. This section briefly introduces the work related to the design, development and usability of the learning tools developed for autistic children.

Ferrari and Robin [15] proposed an evaluation framework to investigate the interest and design of an autonomous robotic toy in context of therapy and education for autistic children. They mainly focused to the interactivity (interactions & effectiveness) issues for evaluating the robotic toy. Weiss et al. [8, 17] conducted two usability evaluation studies to evaluate the collaborative gaming applications developed for autistic children run on a multi-user tabletop surface to improve their social competence skills. The study found that collaborative nature and proficient use of the technology (ease-of-use) were the most effective concerns to enhance the usability, engagement and motivation to wards playing and learning social competence skills. Bhuiyan et al. [18] developed a smartphone based system named *MumIES* (Multimodal Interface based Education and Support) for teaching basic education and support to autistic children; while Zaman and Bhuiyan [9] evaluate the usability of this system in terms of usefulness, efficiency, effectiveness, learnability, satisfaction, and accessibility through questionnaires. The evaluation study was replicated with the teachers, parents, carer and autistic kids.

Gay and Leijdekkers [19] presented the design of emotion-aware mobile apps named *CaptureMyEmotion* for identifying the emotions of autistic children using sensors and by detecting facial expression. This study also discussed the development of one emotion-aware android app *MyMedia* to take sounds, videos or photos and simultaneously attach emotion data to them. This apps can be used by the carer, teachers or guardians of the autistic children providing them a new way to learn, and understand their emotions and daily activities.

Shalash et al. [20] developed an interactive game based application including images, scenarios and voice recognition to solve the speech disorder problems of autistic kids. This application was developed for improving the Arabic language of autistic children having 2–9 years old. An empirical study was conducted in [21] to explore two issues: what factors mediate engagement with and effective use of software, and the extent to which students, teachers and parents perceive the software as having an impact in developing social and life skills of the autistic children. This study considered a mobile application that developed for providing cognitive support to autistic people. Study data were collected through classroom observation and semi-structured interviews. As outcome, the study provided a few design guidelines to develop such kind of cognitive support application for the people with special needs. In another study [22], Juhlin et al. developed an augmented reality based teaching tool for minimally verbal children with ASD to improve their language learning and activities.

In sum, the literature survey showed that *firstly*, though a limited number of studies was conducted focusing to design and develop the learning tools for autistic children, but most of them were desktop and mobile based system. *Secondly*, earlier works mainly focused to introduce and discuss the development of various learning tools and only a few articles explicitly focused to the design considerations to make the system/tools intuitive for the autistic children. *Thirdly*, from HCI perspective, though most of the articles highlighted a very few design considerations (like use of images, audio feedback, etc.) but no study was found that provided a complete set of design considerations to develop a learning tool or application for autistic kids. *Fourthly*, a number of articles were conducted focusing to the usability evaluation of the applications developed for the autistic kids, which in turn shows the significance of usability for developing such system usable/accessible to the autistic kids. *Finally*, no study has been found that explicitly focus to the design and development of tabletop tools for autistic children. Therefore, the focus of this work was understanding the underlying design considerations to develop a tabletop learning tool for the autistic children.

3   Study Methodology

From methodological perspective, an empirical research approach was followed. In this research, two studies were conducted to reveal the possible design considerations for designing and developing a tabletop learning tool for the autistic kids following ethnographic [23] and semi-structured interviews.

3.1   Ethnographic Study

In ethnography study, two educational institutions of the specialized children located in Chattogram region of Bangladesh were visited to observe the daily activities of atustic kids, teachers and other staffs during the school hour. The first institution's named was *Prerona Autism* where eight teachers and ten other supporting personnel were involved to teach and assist the atustic kids. In this school, students can start their education from pre-play and continue up to class eight. The second institution's named was *Dream Star Autism Academy* where ten teachers and eight other supporting personnel were involved to teach atustic kids in this school, to develop their social and basic academic skills. Autistic children can get admission from nursery to class eight.

To conduct the ethnographic study, firstly the study objective of the visit was defined and finding out the suitable places to visit. Secondly, necessary permission was taken and thereby consent paper was also signed from the teachers, parents and institution's head to maintain the research ethics. Thirdly, the institutions were visited by the research team twice in a week and observed their activities for 3–4 h for three consecutive weeks. During the visit, researchers primarily observed - how the students communicate their basic needs with the teachers and within themselves, how they spent their time during school hour, their activities in different places with different situations, how to teach them the basic education and how they try to learn it. During the observation, field notes were taken very meticulously with some photos and short videos.

3.2   Semi Structured Interview

A total of 18 teachers were interviewed following a semi-structured interviewing approach. Among them 8 female teachers were interviewed from *Prerona Autism*. Their age was 32 ± 4 years having teaching experience from two to seven years. Again a total of ten teachers were interviewed who served in the *Dream Star Autism Academy*. Their average age was 35 ± 3 years and have teaching experience from 6 months to 13 years. To conduct each interview session, a set of questionnaire was prepared related to their biographical data (e.g., age, gender, educational qualification, teaching experience at autistic school, etc.) and their teaching practice at autistic school. Based on their answers a few additional questions were asked to find out the possible design considerations to develop a tabletop learning tool for the autistic children. Interview session were conducted one-by-one. Each interview session recorded and later transcribed for analysis.

4   Data Analysis and Findings

The ethnographic study data and interview responses were analyzed through content analysis to find out the design considerations for developing a tabletop learning tool for autistic children. The analysis revealed a total of eleven design considerations. Each design consideration was found both in ethnographic and interview studies.

Affordance - Affordance refers to provide the clues of operations to be done to execute a task [24]. In interview study, four teachers stated that students hold any object based on their physical appearance like hold the water glass, covering the pen head, playing/solving some puzzles based on their shapes, etc. One of the participants responded as "...*I know I have to press here to power-on the note book because it looks like the start button of a computer...*". Similarly, through ethnographic study it was observed that students catching different objects and behaving accordingly based on their affordance quality (like organizing toy of fruits in the fruits basket while other type of toys are putting in different baskets). Therefore, the affordance could be considered a design considerations to develop any hardware based interactive systems like design the buttons such a way that they may get an affordance to press on it or to hold an object.

Intuitiveness - It refers to present any object or interface element in such a way that a human can interpret to grasp its referential meaning or purpose without much evidence or rational [25]. Both studies showed that autistic kids understand the referential meaning and do accordingly to those things, objects or behavioral actions that are intuitive enough for them. For example, during the ethnographic study, it was observed that one autistic child showing his finger to an open door image on the photo board, when the teacher asked him by symbolic gesture whether he wants to go to playground or lobby or canteen. Every time he denied by shaking his head and became angry within a short time. The reasons behind this was that the kid did not understand teacher's responsed given by symbolic gesture though he was actually intended to go for play. Similarly, during interviews one teacher responded as "...*In many cases we failed to express their [autistic children] exact feeling by showing objects or photos at the first go....and that makes them tempered...*". Thus the results indicates that any action, objects or interface elements used to communicate with them should be intuitive enough so that they can easily understand its meaning or purposes.

Portability - Three teachers opined about the portability of the learning devices is a major factor to use efficiently by the austictis kids. The ethnographic study also found the similar observation. For example, when a teacher asked seven autistic children to go to the computer corner (as it was a desktop computer system), most of the children did not feel to move to the computer location. Again the setup of the desktop computer was not suitable for all the children because of the short comings of the children having muscle problem. So, portability is one of the design considerations for the tabletop learning tool to place it anywhere at anytime without much effort.

Ease-to-Use - It refers to the measurement on how a learning tool can be used without much effort. Both studies showed that ease-to-use is the primary requirement to address in developing any system for austistic kids. For example, it was observed during ethnographic study that autistic children were playing with puzzle card were getting disappointed within a short period and involved with another action, because most of the cases they failed to match the puzzle with a first go. Whereas kids playing puzzle game in computer were enjoying. While asking the reason to one of the teachers; she expressed as "...*when an autistic children is matching the puzzle with computers s/he needs to use only two arrow keys and easily receive the feedback, while they are playing using the puzzle cards, after placing one block they do not understand any progress and thereby loose the interest...*". Again during the interviewing study, 13 out of 18 (73%) teachers stated the importance of ease-of-use in developing a learning tool.

Clear and Concise Interface - Each and every parts of any device including interface should be clear, simple and concies to understand. A tool will not be useful if there is any confusion about the appropriate usage of any part of it. It has observed that students were playing with some toys where there are some large round shaped buttons on those toys. Most of the children were confused whether those should be pressed or rotated. During the interview session four teachers expressed that clear & concise interface is a must design consideration for developing any tabletop learning tool.

Metaphoric Design - Metaphoric design ensures that the representation of every parts/symbol of the tool is self expressive and has a proper mapping/relationship with the real world object. During the ethnographic study, it was observed that students were playing by utilizing the appropriate usage of the sports materials (football, cricket ball, bat) at the outdoor playground. Again, during interviews, two teachers stated that in many cases like distinguishing between the usage of fork & spoon and wearing foot wear in right pattern autistic children face some challenges. Thereby design with a proper mapping with real world (metaphoric design) phenomena should one of the design considerations.

Easy-to-Learn - Teach anything to the autistic children is challenging, so the learning tool should be easy to learn. The operation procedure of any tool should not be complex. Almost 50% of the autistic children were observed that they learn puzzle game through computer quite early than through the puzzle card since playing puzzle game in computer was comparatively easy for them. Again, during interviews, a tea-cher stated that "...*the tools which are easy to operate are more used by the autistic children..*", said by them.

Interactivity - Interactivity refers to the ability of a system to respond to a user's action (or input). During the ethnographic study, it was observed that autistic children are attracted to those tools which have some feedback upon some action taken by them. During interviews, four teachers stated that autistic children like the games like puzzle game in the computer because, after each matching some points awarded to the users.

Thus the tabletop learning tool need to be interactive to make it attractive to the autistic children.

Multimodal Feedback - It was observed during the ethnographic study that autistic children were attracted to those toys having some sound (piano, flute) or movement (car, train) or changes of color. It was observed that all the kids were using toys those have vibration, movable properties or sound. Similarly, four of the teachers stated about this issue during interviews. For example, one teacher responded as "...*sounds, music, vibration or changing colours are naturally attract the autistic children..*". Thus, including multimodal feedback in the learning tool will be an appealing feature for the autistic children.

Visual Feedback - This reflects the relative visual changes or reactions after each action. Both the studies found that visual feedback helps autistic kids to understand whether the taken action is right or wrong. Six teachers stated this issue during the interviews. One teacher stated as "...*children always wait for the real time feedback. ...*". Similarly in ethnographic studies, it was found that autistic children were using a touch screen named *Diamond Touch* and after each touch (to any particular icon in a scenario) kids wait for the next scenario. So, visual feedback can be a design consideration for developing tabletop learning tool.

Pictography - This refers to the pictorial representation of any seen. It was quite challenging to teach the autistic children through words, while this can be easier with some colorful pictures or symbols. During ethnographic studies, it was observed in a class of grade one, where three teachers were taking class on daily life activities, and nine autistic children were showing the pictures which information/facts appropriate for that activity. Thus, pictography can be a design consideration to develop a usable tabletop learning tool for autistic children.

5   Proposed Design of the Tabletop Learning Tool

A design proposal is showed in Fig. 1 for developing a tabletop learning tool for the autistic children based on the revealed design considerations. The proposed system will include four modules. The modules are discussed briefly in the following sub-sections.

5.1   Module 1 (Learning Alphabet and Constructing Words)

The purpose of this module is to learn the alphabets and afterwards constructing simple words. To attain this objective, a number of keys will be designed for the alphabets and labeled from A to Z. The keys will provoke autistic children to press and thus follow the *affordance*. For example, if a user press a key labelled 'A', in the output screen/interface the alphabet 'A' will be displayed with a picture and text of an object (whose name start with the pressed alphabet) along with voice annotation like 'A for Apple' (see Fig. 1). This functional design address the *interactivity, pictography* amd *visual feedback*. By hearing the voice annotation, the user will type the name of the picture. Depending on the accuracy of user's typing, the system will provide voice feedback as 'right answer' or 'wrong answer'; which in trun will satisfy the design consideration of *visual feedback, multimodal feedback* and *intuitiveness*. The layout will be designed to attain *ease-to-use, easy to learn*, and *clear & concise interface* design considerations. Thus, this module will address the following revealed design considerations: *affordance, intuitiveness, ease-to-use, clear & concise interface, easy to learn, interactivity, multimodal feedback, visual feedback,* and *pictography*.

5.2   Module 2 (Appearing Test on Alphabet and Constructing Words)

The objective of this module is to learn the usage of the alphabets and constructing the words by themselves (autistic children). To attain this objective, similar to module 1, a set of keys labelled with alphabets will be designed to provoke autistic children to press or type. This will address the *affordance* design consideration. Two options will be incorporated to test the user learning progress. First option will be MCQ. If "A" is pressed then MCQ will start. Ten questions along with options will come one by one. This will address the *interactivity, intuitiveness, metaphoric design* and *visual feedback* design considerations. A user's need to choose any of the options by pressing the alphabet A, B, C or D. If the answer is correct the score on the upper right corner will increase. At the end total score will be shown like 'Total score: 70/100'. Again, if the quiz option is selected (by pressing the 'B') then ten questions will come one by one. Just beneath the question the user's need to write the answer in one word. If the answer is correct then the score on the upper right corner will increase. Similar to the MCQ option, at the end total score will be shown. Again, this module is designed clearly and concisely so that other design considerations like *easy-to-use, easy-to-learn* and *clear & concise interface* features are addressed properly to ensure the easy use and learning by considering the limitations of the autistic children. In sum, this module will address following revealed design considerations: *affordance, intuitiveness, ease-to-use, visual feedback, easy to learn, multimodal feedback,* and *clear & concise interface*.

5.3   Module 3 (Daily Communicator)

The purpose of this module is to provide a means to express the basic common daily needs of autistic kids as they cannot express their needs verbally. This module will include a number of basic needs into several categories as follows: (a) the emotion category will include the emotions/symbol of great, good, happy and sad mood; (b) the working category will include the reading, writing and puzzle matching; (c) the activity category will include song, art and gym; and (d) various foods' symbols to ask for which food he/she wants to eat. For each need, a button will be designed and each button will be appended with intuitive symbol/picture to provide a realistic mapping of its purpose (meaning) with the real world phenomena. For example, to express a helping request to drink water, a button will be designed under various food's category where a symbol/picture of water in a glass will be appended. When the user will press the button the light of the symbol will be on and the system will generate corresponding voice request to draw the attention of the teacher/assistants. This feature will thus address the *intuitiveness, interactiveness, pictography, affordance* and *metaphoric design*.

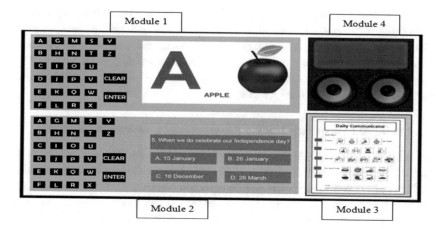

Fig. 1. Proposed design of the tabletop learning tool

5.4 Module 4 (Playing Rhyme and Music)

As music and rhymes have a great impact to the autistic children and in some cases also used to provide mental therapy, this module is designed to play any music or rhyme from external source (like pendrive or memory card). In this module, speaker with display is installed, where external source can be attached by port. There will be some intuitive buttons to play or change the rhyme or the songs. This module will address the following revealed design considerations: *affordance, interactivity, multimodal feedback, intuitiveness, easy-to-learn* and *easy-to-use.*

6 Discussion and Conclusions

This research provides a total of eleven design considerations for developing a tabletop learning tool for autistic children to learn basic education and to express their daily needs. A design solution for developing a tabletop learning tool for autistic kids based on the revealed design considerations is also proposed here. Though some of the design considerations are already exist in literature, but many of the design considerations revealed through the research are new to design the tabletop learning tool for the autistic children; that includes, affordance, intuitiveness, clear & concise interface, metaphoric design and multimodal feedback. Again, literature survey showed that most of the existing design considerations are for developing desktop, web and mobile based tools. None of the earlier work focused to design any tabletop learning tool. Therefore, the revealed design considerations are new and innovative to develop a tabletop leaning tool for the autistic children.

The research has a few limitations as well. Firstly, though the revealed design considerations were revealed through empirical studies but these were not validated yet. Secondly, the study data were analyzed through a qualitative approach. As the result of the qualitative analysis depends on individual's experience, view, knowledge and most importantly the target. The future work will focus to develop and evaluate the concrete version of the tabletop learning tool to validating the effectiveness of the revealed design considerations.

Declarations

Ethics Approval and Consent to Participate – We confirm that ethical approval has taken by the ethical committee headed by the Research & Development Wing of Military Institute of Science and Technology (MIST) for conducting this research. We also declare that we have taken written consent from each participant to participate in this study.

References

1. Verenikina, I., Tanner, K., Dixon, R.M., Graaf, E.: Interactive whiteboards as a tool for teaching students with autism spectrum disorders. Technical report, Faculty of Education, University of Wollongong (2010)
2. Bell, C.C.: DSM-IV: diagnostic and statistical manual of mental disorders. JAMA **272**, 828–829 (1994)
3. Kamaruzaman, M.F., Azahari, M.H.H: Form design development study on autistic counting skill learning application. In: 2014 International Conference on Computer, Communication and Control Technology (I4CT 2014), Langkawi, Malaysia (2014)
4. Aburukba, R., Aloul, F., Mahmoud, A., Kamili, K., Ajmal, S.: AutiAid: a learning mobile application for autistic children. In: 19th International Conference on e-Health Networking, Applications and Services (Healthcom), pp. 1–6. IEEE (2017)
5. Zaki, T., Islam, M.N., Uddin, M.S., Tumpa, S.N., Hossain, M.J., Anti, M.R., Hasan, M.M.: Towards developing a learning tool for children with autism. In: 2017 6th International Conference on Informatics, Electronics and Vision & 7th International Symposium in Computational Medical and Health Technology, pp. 1–6. IEEE (2017)
6. Bevan, N.: Measuring usability as quality of use. Softw. Qual. J. **4**(2), 115–130 (1995)
7. Islam, M.N., Ahmed, M.A., Islam, A.K.M.N.: Chakuri-Bazaar - a mobile application for literate and semi-literate people for searching employment. Int. J. Mob. Hum. Comput. Inter. **12**(2) (2020)
8. Weiss, P.L., Gal, E., Eden, S., Zancanaro, M., Telch, F.: Usability of a multi-touch tabletop surface to enhance social competence training for children with autism spectrum disorder. In: Chais Conference on Instructional Technologies Research 2011: Learning in the Technological Era (2011)
9. Zaman, A., Bhuiyan, M.: Usability evaluation of the MumIES (multimodal interface based education and support) system for the children with special needs in Bangladesh. In: 3rd International Conference on Informatics, Electronics & Vision, Dhaka, Bangladesh (2014)
10. Chien, M., et al.: iCAN: a tablet-based pedagogical system for improving communication skills of children with autism. J. Hum.-Comput. Stud. **73**, 79–90 (2014)
11. Satterfield, D., Fabri, M.: User participatory methods for inclusive design and research in autism: a case study in teaching UX design. In: International Conference of Design, User Experience, and Usability, Canada, pp. 186–197 (2017)
12. Madsen, M., Kaliouby, R., Eckhardt, M., Hoque, M.E., Goodwin, M.S., Picard, R.: Lessons from participatory design with adolescents on the autism spectrum. In: 27th International Conference Extended Abstracts on Human Factors in Computing Systems, Boston, MA, USA (2009)

13. Omar, K.S., Mondal, P., Khan, N.S., Rizvi, M.R., Islam, M.N.: A machine learning approach to predict autism spectrum disorder. In: 2019 International Conference on Electrical, Computer and Communication Engineering (ECCE), pp. 1–6. IEEE (2019)

14. Islam, M.N., Kabir, M., Sultana, J., Ferdous, C.N., Zaman, A., Bristy, U.H., Moumi, P.K., Tamanna, I.: Autism Sohayika: a web portal to provide services to autistic children. In: International Conference on Mobile Web and Intelligent Information Systems, pp. 181–192. Springer, Cham (2018)

15. Ferrari, E., Robins, B., Dautenhahn, K.: Does it work? A framework to evaluate the effectiveness of a robotic toy for children with special needs. In: 19th International Symposium on Robot and Human Interactive Communication, Italy (2010)

16. Gal, E., Weiss, P.L., Zancanero, M.: Using innovative technologies as therapeutic and educational tools for children with autism spectrum disorder. In: Rizzo, A., Bouchard, S. (eds.) Virtual Reality for Psychological and Neurocognitive Interventions. Virtual Reality Technologies for Health and Clinical Applications. Springer, New York (2019)

17. Weiss, P.L., et al.: Usability of technology supported social competence training for children on the autism spectrum. In: International Conference on Virtual Rehabilitation, Zurich, Switzerland (2011)

18. Bhuiyan, M., Akhter, P., Hossain, M.A., Zhang, L.: MumIES (multimodal interface based education and support) system for the children with special needs. In: 7th International Conference on Software, Knowledge, Information Management and Applications (SKIMA2013), Changmai, Thailand (2013)

19. Gay, V., Leijdekkers, P.: Design of emotion-aware mobile apps for autistic children. J. Health Technol. 4(1), 21–26 (2014)

20. Shalash, W.M., Bas-sam, M., Shawly, G.: Interactive system for solving children communication disorder. In: International Conference of Design, User Experience and Usability, Las Vegas, NV, USA, pp. 462–469 (2013)

21. Mintz, J., Branch, C., March, C., Lerman, S.: Key factors mediating the use of a mobile technology tool designed to develop social and life skills in children with autistic spectrum disorders. J. Comput. Educ. 58(1), 53–62 (2012)

22. Juhlin, D., et al.: The PTC and Boston children's hospital collaborative AR experience for children with autism spectrum disorder. In: HCII 2019. LNCS, vol. 11573. Springer, Cham (2019)

23. Hughes, J.A., O'Brien, J., Rodden, T., Rouncefield, M., Sommerville, I.: Presenting ethnography in the requirements process. In: IEEE International Symposium on Requirements Engineering, UK, vol. 95, p. 27 (1995)

24. Norman, D.: The Design of Everyday Things, Revised and Expanded Edition. Basic Books, New York (2013)

25. Islam, M.N.: Exploring the intuitiveness of iconic, textual and icon with texts signs for designing user intuitive web interfaces. In: 18th International Conference on Computer and Information Technology (ICCIT), pp. 450–455. IEEE (2015)

A Cloud-Fog Based System Architecture for Enhancing Fault Detection in Electrical Secondary Distribution Network

Gilbert M. Gilbert[1,2(✉)], Shililiandumi Naiman[1], Honest Kimaro[1], and Nerey Mvungi[1]

[1] University of Dar es Salaam, Dar es Saaam, Tanzania
gilly.map@gmail.com, smakere@gmail.com,
honest_c@yahoo.com, nhmvungi@gmail.com
[2] The University of Dodoma, Dodoma, Tanzania

Abstract. The modern power grids are developing towards smartness with innovative electric power systems management by embracing Information and Communication Technology (ICT) systems in addition to other technologies for real-time control and monitoring. These systems rely on the use of sensors, which generate a lot data, for situation awareness and visibility. Approaches and techniques based on ICT-solutions for faults handling have advanced up to electrical transmission networks and not well addressed in electrical secondary (low voltage (LV)) distribution networks which are prone to various types of faults affecting a significant number of customers. Data-driven approaches with cloud-fog based system architecture have emerged as potential solution to address fault detection in these LV distribution networks. In this paper, an overview of cloud-fog based architectural design is presented for applications to enhance fault detection in electrical secondary distribution network. The proposed platform make use of microservice architecture to assist the distributed applications to effectively utilize the virtualized systems. Workflow and use case are also presented as well as techniques on addressing the required technological platforms for implementation to facilitate fault detection in the distribution networks.

Keywords: Cloud computing · Fog computing · LV network · Distribution network · Microservices · Fault detection · Virtualization

1 Introduction

The modern power grids are developing towards smartness, breaking the traditional model of electric power systems monitoring, control and management. This move to smart power grids involve integration of variety of Information and Communication Technology (ICT) systems in additional to other technologies [1]. To great extent, smart power grid inherits the intelligence brought by ICT applications, thus improving the reliability of power generation, transmission and distribution.

Among the features promised by these modern power grid deployments is on ensuring there is stable, smooth and reliable power supply at all levels [2]. As that is the

© Springer Nature Switzerland AG 2020
A. P. Pandian et al. (Eds.): ICCBI 2019, LNDECT 49, pp. 845–855, 2020.
https://doi.org/10.1007/978-3-030-43192-1_92

case, reliable power supply involves proper handling of all occurrences of faults intelligently at all levels of power system. However, fault clearance approaches for low voltage (LV) distribution networks have not been well addressed compared to those in primary distribution and transmission networks. The reasons that have been put forward include those related to huge size of the networks as well as complexity of the layout of the distribution networks which keep changing time by time [3–5]. The other reason is the lack of real-time data.

Generally, the solution would require the deployment of a lot of sensors at various points in the network to monitor the electricity and improve the chance of locating and reacting to faults. Since the LV network is huge in size as it extends tens of kilometers, data from sensors is also expected to be in very large amount. Handling such amount of data using requires sound computing technologies. The ICT solutions bring together Internet of Things (IoT), data analytics, fog computing, and cloud computing, which will assist utilities to overcome this challenge towards reliable power supply in both developing and developed countries.

This work is motivated by the potential use of ICT-based solutions to facilitate the designs of computing architectures that allow data-driven fault detection and clearance applications to be developed to address electrical secondary distribution network challenges. These applications can be designed to provide the use suitable models and analytics, and enhancing different controls to the engineers, technicians, customers and other stakeholders. The data-driven solutions are so crucial to the collection and processing of remote data at remote sites and therefore, the ICT system has to be flexible and adaptive to the requirements even during periods of unforeseen circumstances.

Therefore, in this work, we propose fog-based architecture to enhance the process of fault detection in LV distribution network. In order to effectively utilize the heterogeneous cloud-fog based system, microservice architecture has been adopted to assist well execution of applications. In the solution, virtualization, containerization and other lightweight messaging protocols were combined to address the computing requirements and needs. The structure of this paper is a follows: Sect. 2 introduces about electrical secondary distribution network monitoring. Section 3 discusses system requirements and design, and Sect. 4 presents system architecture for implementation. Section 5 describes the microservice-based architecture application design along with use-case and workflow. Section 6 contains the conclusion part of the paper.

2   Electrical Secondary Distribution Network Monitoring

The electrical network distribution networks are usually classified as primary and secondary distribution networks. Primary distribution systems take power (11 kV) from distribution substations to distribution transformers. These networks are also known as medium voltage (MV) networks. On the other hand, secondary distribution networks or low-voltage (LV) networks carry electricity to end customers in the range of 100 to 400 V.

Secondary distribution networks are made up with many 0.4 kV transformers (substations) that feed end-use customers. Therefore, these networks occupy a consideration proportion of the overall electrical power system, and they are bigger in terms size as well as being very complex in many aspects [6]. Many related electrical faults are statistically higher in the distribution networks that any other parts of the power system, and that is greatly attributed to the complexity of these networks.

Despite its importance, this part of the network has received little attention when it comes to monitoring and control. In developed world, where they use smart meters, they could easily monitor some aspects of the network but the practice of allowing the integration of renewable energy sources (RES) in the main grids has made the management of the LV networks much more difficult [7]. The situation is much worse in most of the developing countries in which the visibility of the LV networks is almost inexistent as they could not monitor the networks in real-time.

Several efforts to improve the visibility and hence the monitoring of the LV distribution networks have been elaborated in Strachan et al. in [3]. In their work, they have also investigated the possibility of using smart meters to improve analysis of LV networks.

It can be argued that smart meters are useful for data availability, but their data cannot be used for real-time accurate control decisions [8]. Moreover, for low income countries smart meters rollout could take some time, and therefore, simple and cost-effective measures may be needed to enhance the monitoring of secondary distribution network. Therefore in this work, the authors proposes a cloud-fog based system to facilitate the monitoring of the secondary distribution network using IoT technologies.

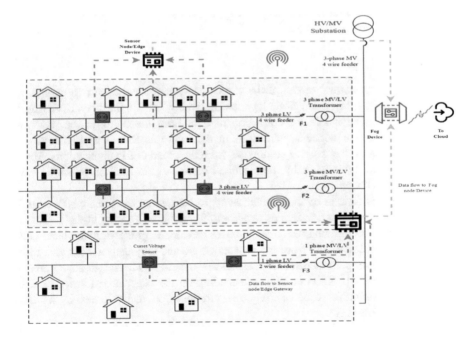

Fig. 1. Overview of electrical secondary distribution network equiped monitoring capabilities

3 System Requirements and Design

Figure 1 shows the overview of the proposed approach on enhancing the monitoring of secondary distribution network. Current and Voltage sensors are installed along the electrical lines at predetermined locations to collect and send electrical parameters to nearby edge devices or fog nodes depending on the network organization. After local analytics of some portions of data at fog nodes, the data are forwarded to the cloud of the utility for storage and further analysis. In case urgent response is needed, local processing would have analyzed that and react accordingly.

3.1 Requirements

The general goal of the proposed system is to provide the computing platform for management of smart-grid based applications and data received from field sensors and other integrated systems. The applications should be able to register themselves, request data, view and manipulate real-time and stored data. Moreover, from the comprehensive list of requirements in [9, 10], for cloud-based fog system, key four requirements can be considered.

- Enabling large scale coordination of smart-grid application components
- Enabling the environment for supporting the heterogeneity of smart grid applications
- Enabling interoperability of smart grid applications
- Enabling sensing and processing.

3.2 System Design

The proposed architectural model of the cloud-fog system consists of two computing layers which are fog layer and cloud layer as shown Fig. 2. The third layer consists of devices that are responsible for collection of data from the fields in the electrical secondary distribution network, addressed in the figure as Site A and Site B, and perform actuation as they receive commands from the cloud or fog applications. Site A and Site B are locations whereby the electrical secondary distribution networks have been deployed with sensors.

Cloud layer is formed by data center standard computing facilities which contain virtually infinite computing resources relative to fog devices. The fog nodes in the fog layer have limited computing resources in terms of storage, processing and bandwidth [11]. Once they receive data from the sensors or clouds, they can process data independently or by grouping themselves into clusters. The fog nodes might have different hardware such as CPU, GPU or FPGA depending on their specialized tasks. Coordinating these computing resources requires virtualization technologies to increase effective utilization [12].

Fog layer is responsible for mostly of distributed computing tasks related to processing of light data and quick response to local scenarios that are latency sensitive. Data received from sensors once processed are sent to the cloud through MQTT mechanisms in the form of publisher. Depending on the organization of the cloud-fog

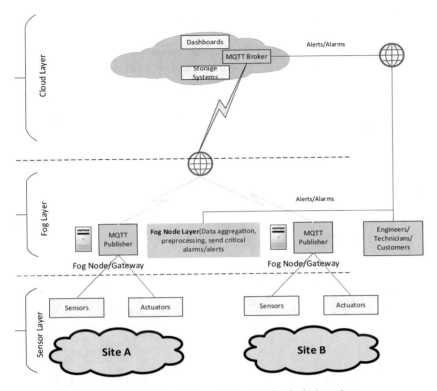

Fig. 2. Cloud-fog system architecture for fault detection

computing architecture, once data are received from the sensors, the cloud through a broker (MQTT Subscriber) (or local fog node controller or gateway), may determine the number of fog nodes for distributed processing. The results from the distributed nodes will then be aggregated (data fusion) by the cloud to determine next stage in the analytics processing. The local fog node controller that is set up in each fog layer is responsible for resource management and tasks collaboration between the layers. This controller further ensures the nodes are well synchronized by establishing appropriate means of interactions.

The advantages of the proposed cloud-fog architectural model are: firstly, to combine both centralized and distributed architectures functions. Secondly, to reduce the bandwidth between cloud and wireless sensor networks. Another advantage of this architecture is to guarantee that the requirements of time sensitive applications are met. Moreover, the proposed architecture can be modelled in integration with the implementation of Content Delivery Network (CDN) and Software Defined Network (SDN) functions.

4 System Architecture for Implementation

The system components necessary for implementation of the proposed architecture involves applications and services in the edge layer, virtualization layer and hardware layer as shown in Fig. 3. The hardware layer form and important part of the system in ensuring that software platforms are hosted seamlessly. This is facilitated by the presence of embedded devices that provides computations, storage and network functions. This layer also supports multiple hardware types in order to facilitate system scalability.

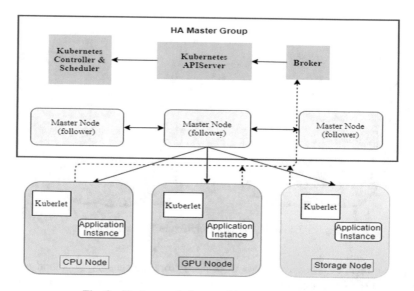

Fig. 3. Hardware platform architecture components

Virtualization layer is responsible for access and utilization of different multiple resources provided by the hardware layer. The virtualization layer has lightweight processes that help to abstract the hardware complexities. The edge layer consists of applications and services that are comprised of microservices deployed in virtualized environment as lightweight containers or virtual machines.

Kubernetes [13] provides mechanisms for planning, deploying, updating and maintaining Docker applications. In essence, the use of Docker [14] and Kubernetes virtualization technologies allow the building of lightweight, fast and efficient distributed applications. Kuberlets collect those information for containers.

5 Microservice-Based Architecture Application Design

Microservices has been touted as an outcome of cloud-fog system evolution to address the issues related to data confidentiality and network performance by allowing applications to be very flexible [15]. The microservice architecture allows the applications to combine many sources of information and devices to enhance reliable delivery of complex applications.

Data requirements and latency constraints of applications and services for the fault detection in the electrical secondary distribution network can be categorized as shown in Table 1. It can be seen from the table, that mircroservice architecture can greatly enhance applications portability, which also justify their use in fault detection in electrical secondary distribution networks.

Table 1. Latency constraints and data requirements

Application/services	Latency constraints	Data requirement
Query sensor data	Minutes	Immediate
Location monitoring	Minutes	Immediate
Fault detection	Seconds	Immediate
Logging and other applications	Hours	Non-immediate
Analytics & visualization	Days	Non-immediate

In order to better coordinate applications, the microservices in the fog computing are encapsulated inside containers for hardware abstraction. This becomes crucial in an environment in which devices are heterogeneous, such as, access points, routers, and switches. There exist several containerization technologies, and choosing the right one depends on the performance target.

To minimize services coordination overhead, it is wise to choose the right programming model and platform. This can be done by letting similar programming platforms reside in the same computing device. For instance, if the choice is to use java and python, it is preferably to use Python Virtual Machine (PVM) and Java Virtual Machine (JVM) as containers in the fog node. On the cloud side, one can continue to use service provider's technologies such as Dockers and Kubernetes. This is also the case in this work, where the proposed platform make use of Docker as container platform and Kubernetes as container management platform.

5.1 Use Case

The authors developed the use case as a result of interaction with Tanzania utility company (TANESCO). This utility company maintains the emergency desk to ensure that the fault clearance and power restoration are done in the shortest time possible once the fault is reported. Currently, defects and faults are mainly reported by customers and through visual inspection (line patrol). Fault clearance processes take longer than anticipated due manual reporting of faults, difficulty for technicians to locate areas with faults.

This use case can be well addressed by exploiting the potential use of ICT-based solution in the form of cloud-fog system as inspired by the fire detection and fighting introduced in [16] and then expanded further by [17]. The usefulness of cloud-fog systems for fault detection in secondary distribution networks is made possible by the fact that these systems are hierarchical, geographical and offer local sensing and processing. Therefore, the cloud-fog system resembles the network layout of the electrical distribution network and fits well as an IoT infrastructure for addressing the problem of fault management with low latency and guaranteed scalability [18].

To develop applications for enhancing fault clearance in distribution network based on cloud-fog systems, survey and interviews with the experts were done in order to reveal the real-world working of fault detection and identification processes for the purpose of establishing data requirements. From Table 1 data requirements were categorized as immediate and non-immediate with the view that, high priority is given to applications that are delay-sensitive operating at fog layer, and the rest of applications require low priority and may even be assigned to cloud for computations.

The fault detection and monitoring application monitors geographic areas (zones, etc.) and takes/suggests appropriated measures such as sending crew members to the place where fault is detected. The monitoring is done through the gathering of information such as temperature, voltage, current, and transformer oil levels. When fault is detected, the application evaluates its intensity and contour, and takes/suggests appropriated measures by calling upon procedures for isolation and restoration services in order to deal with the fault shown in Fig. 4.

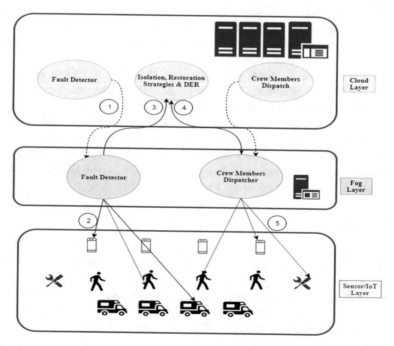

Fig. 4. A cloud-fog system use case for fault detection and monitoring in secondary

5.2   Workflow

The used workflow in this work is as shown in Fig. 5. It can be seen that, fog nodes offer dashboard features for engineers and technicians to interact with as act as visual tool for events and other sensor data management. The fog nodes are required to respond quickly to events and alerts or alarms that need immediate attention. The aggregated data are then sent to the cloud for storage and analysis. The cloud combines data from other sources to generate clear picture of the situation in place. It has to be stressed that fog devices are considered as resource constrained, and some applications related to analytics can be remotely pushed to the fog nodes dynamically or when they are needed in order to minimize resource usage and other overheads of these devices.

Otherwise, all the long term data and resource-demanding analytics applications reside in the cloud. Cloud also allow users (customers, engineers, technicians, etc.) to interact with the relevant data and demand other services accordingly.

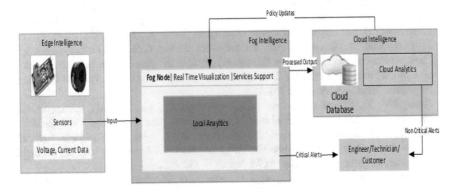

Fig. 5. Workflow for fault detection in LV distribution system in fog system

6   Conclusion and Future Work

This work proposed an overview of cloud-fog based system for fault detection and clearance in electrical secondary distribution network. The architectural design and the potential use of microservice architecture were presented. The microservice architecture is envisioned to allow seamless application deployment in hybridized technological infrastructure. As a result, the application use case was proposed to address data collection, data analytics and visualization with respect to latency constraints and response category. Further, the adoption of the proposed ICT-based solution in terms of cloud-fog system enhances the intelligent processing of data closer to the source in a given location, and hence guarantee the processing of real-time applications, location-based services and monitoring of the low voltage network.

Future work involves the actual deployment of the prototype to address the use case scenarios to monitor the network, fault classification and message communication. Furthermore, system resilience tests will be performed in terms of failure or overload of nodes, system accuracy, resource utilization, power consumptions and service delivery time.

Acknowledgement. This work is supported by Swedish International Development Agency (SIDA) under IGRID project under University of Dar es Salaam.

Informed Consent. Informed consent was obtained from all individual participants included in the study.

References

1. Green, R.C., Wang, L., Alam, M.: Applications and trends of high performance computing for electric power systems: focusing on smart grid. IEEE Trans. Smart Grid **4**, 922–931 (2013)
2. Pant, V., Jain, S., Chauhan, R.: Integration of fog and IoT model for the future smart grid. In: 2017 International Conference on Emerging Trends in Computing and Communication Technologies, ICETCCT 2017, pp. 1–6 (2018)
3. Strachan, A., Elders, I., Galloway, S.: Improving network visibility for better integration of low carbon technologies into LV networks. In: Proceedings - 2018 53rd International Universities Power Engineering Conference, UPEC 2018, pp. 1–6 (2018)
4. Mnyanghwalo, D., Kawambwa, S., Mwifunyi, R., Gilbert, G.M., Makota, D., Mvungi, N.: Fault detection and monitoring in secondary electric distribution network based on distributed processing. In: 2018 20th International Middle East Power Systems Conference, MEPCON 2018 - Proceedings (2019)
5. Andegelile, Y., Chugulu, G., Bitebo, A., Mbembati, H., Kundaeli, H.: Enhancing faults monitoring in secondary electrical distribution network. Presented at the 1 May 2019
6. Forsyth, P., Nzimako, O., Peters, C., Moustafa, M.: Challenges of modeling electrical distribution networks in real-time. In: 2015 International Symposium on Smart Electric Distribution Systems and Technologies, pp. 556–559 (2015)
7. Barbato, A., Dedè, A., Della Giustina, D., Massa, G., Angioni, A., Lipari, G., Ponci, F., Repo, S.: Lessons learnt from real-time monitoring of the low voltage distribution network. Sustain. Energy Grids Netw. **15**, 76–85 (2018)
8. Chen, Q., Kaleshi, D., Armour, S., Fan, Z.: Reconsidering the smart metering data collection frequency for distribution state estimation. In: 2014 IEEE International Conference on Smart Grid Communications (SmartGridComm), pp. 517–522. IEEE (2014)
9. Bonomi, F., Milito, R., Natarajan, P., Zhu, J.: Fog computing: a platform for internet of things and analytics. In: Studies in Computational Intelligence, pp. 169–186. Springer (2014)
10. Da Silva, W.M., Tomas, G.H.R.P., Dias, K.L., Alvaro, A., Afonso, R.A., Garcia, V.C.: Smart cities software architectures: a survey. In: Proceedings of the ACM Symposium on Applied Computing, pp. 1722–1727 (2013)
11. Kumar, T.S.: Efficient resource allocation and QoS enhancements of IoT with fog network. J. ISMAC **01**, 21–30 (2019)
12. Vulcan, A.M., Nicolae, M.: High performance computing based on a smart grid approach. In: 2017 21st International Conference on Control Systems and Computer Science, pp. 651–655 (2017)
13. Production-Grade Container Orchestration – Kubernetes. https://kubernetes.io/
14. Enterprise Container Platform—Docker. https://www.docker.com/
15. What are microservices? https://microservices.io/

16. Yangui, S., Ravindran, P., Bibani, O., Glitho, R.H., Hadj-Alouane, N.B., Morrow, M.J., Polakos, P.A.: A platform as-a-service for hybrid cloud/fog environments. Presented at the June 2016
17. Mouradian, C., Naboulsi, D., Yangui, S., Glitho, R.H., Morrow, M.J., Polakos, P.A.: A comprehensive survey on fog computing: state-of-the-art and research challenges. IEEE Commun. Surv. Tutorials **20**, 416–464 (2018)
18. Barik, R.K., Gudey, S.K., Reddy, G.G., Pant, M., Dubey, H., Mankodiya, K., Kumar, V.: FogGrid: leveraging fog computing for enhanced smart grid network. In: INDICON-2017 14th IEEE India Council International Conference, pp. 1–6 (2017)

An Autonomous Intelligent Ornithopter

Sunita Suralkar[✉], Smit Gangurde, Sanjeevkumar Chintakindi,
and Haresh Chawla

VES Institute of Technology, Chembur, Mumbai, India
{sunita.suralkar,2015smit.gangurde,
2015sanjeevkumar.chintakindi,
2015haresh.chawla}@ves.ac.in

Abstract. The purpose of the system is to provide a powerful and intelligent surveillance tool to the police force so as to reduce crime. The law enforcement agencies have been motivated to use video surveillance systems to monitor and curb these threats. But this becomes a tedious task, prone to human errors. The core module of this system estimates the pose in humans present in the video and a backend capable of understanding the context as a whole. Many AI-powered surveillance systems are good at recognizing violent or malicious activity but fail to understand the context as a whole. We aim to understand the gradual change in human behavior in the given scenario, understand the confidence level of each expression and derive if the given scenario is truly violent or malicious. The Ornithopter is allowed to follow the suspect wherein the direction offsets are given by the server. The system differs from any state-of-the-art surveillance system as it provides aerial surveillance covering larger areas, and since the drone is bird-shaped, it can easily navigate the area without being easily detected. And as mentioned, the recognition of the true violent or malicious activity is context-based.

Keywords: Ornithopter · Deep learning · Artificial intelligence · Video analytics · Human activity prediction

1 Introduction

In this project, we aim at making an ornithopter, a bird-shaped drone, which will patrol a specified area and send video footage to our cloud. An artificial intelligence system based on deep learning will analyze this footage to detect any suspicious activities.

For humans to manually monitor the cities using cameras is a tedious task. Moreover, even monitoring the streets of cities manually is a monotonous task. There is a growing need for automated video surveillance techniques for the army for border patrol and the police for patrolling the streets. Our aim is to produce an unmanned surveillance drone, in the shape of a bird (ornithopter), that helps the military to detect intruders near the border and for the police to help them detect any kind of crime. It should not be easily detectable. The bird drone should be able to follow the individual and report to authority its location. The bird drone streams the video, while we have a server that runs an AI system which detects any individuals and their violent activities.

© Springer Nature Switzerland AG 2020
A. P. Pandian et al. (Eds.): ICCBI 2019, LNDECT 49, pp. 856–865, 2020.
https://doi.org/10.1007/978-3-030-43192-1_93

2 Literature Survey

2.1 Existing System and Its Comparison

UAV by Microdrones: This system uses remote control drones to surveil an area. The individual behind the remote analyses and infer the ongoing scenario in live footage.

The Drawback of Such a System

- This system is prone to human errors.
- Enemy or culprit can easily detect the drone due to its unnatural appearance and postpone his/her malicious activity.
- A drone is not autonomous.

Eye in the Sky: Drone Surveillance [1]: This system uses remote control drones to surveil the area, It then utilizes a deep learning model to predict if the ongoing scenario is malicious/violent. It is a Deep Learning-based Video Surveillance. It uses the Feature Pyramid Model to extract segmented frames with a human presence in it. ScatterNet Hybrid Deep Learning Network to represent human pose estimation, also to extract human invariants features e.g. angle between limbs, legs, etc. Regression Network tries to match the angle between the limbs extracted with the predefined or standard angle depicting violent activity and determines if it matches then its a violent or malicious activity. Finally, Support Vector Machines (SVM) to identify violent individuals.

Drawbacks of this System

- Enemy or culprit can easily detect the drone due to its unnatural appearance and postpone his/her malicious activity.
- The deep learning model is not context-based.
- Error rate increases with an increase in overlapping humans in the given scenario.
- A drone is not autonomous.
- The prediction model is not context-based i.e. it only recognizes the violent/malicious part and not understand the scenario as a whole.

IntelliVision: AI and Video Analytics Using Smart Cameras: This system produces AI-based surveillance systems for business and civilian use.

Drawbacks of this System

- Enemy or culprit can easily detect the drone due to its unnatural appearance and postpone his/her malicious activity.
- The deep learning model is not context-based.
- Error rate increases with an increase in overlapping humans in the given scenario.
- The prediction model is not context-based i.e. it only recognizes the violent/ malicious part and not understand the scenario as a whole.

Ullah et al. [10] propose a human action recognition model by processing video data using Convolutional Neural Network (CNN) and Deep Bi-directional Long Short Term (LSTM) Memory Network. Dividing video data into N chunks at different time intervals, working of the Bi-directional LSTM model and using the Bi-directional LSTM Model to understand the overall scenario or context of the video data.

Park et al. [2] describes the various equipment that makes a base for designing an ornithopter, wing efficiency based on a design inspired by a real insect wing. The paper introduces threshold wing-weight ratio, wing frequency adjusting or controlled either by enlarging the wing beat amplitude or raising the wingbeat frequency, it is the most significant factor in an ornithopter that mimics an insect. The paper helped in Ornithopter wingspan calculations, understanding basic aerodynamics, Ornithopter tail design for efficient movement and also Wing design.

Jackowski [8] describes the designing of an ornithopter, with high payload capacity, crash survivability, and field repair abilities. The paper covers the design process of both the mechanical and electrical systems of the ornithopter and initial control experiments.

It also explains the stability of an ornithopter is maintained and various describes the underlying structure of the bird. The paper helped in Designing of the ornithopter, designing Base structure of ornithopter, large bird wing calculations, bird tail movement, placement of various electronic devices on base structure of ornithopter.

Robertson et al. [12] proposes the use of Hidden Markov Models for human behavior understanding. Higher-level reasoning about scene context can be achieved by the representation of behaviors as action sequences, with representation and recognition of these is achieved via HMMs. Human-level descriptions are achieved by defining the actions as a precursor. It also gives rules to encode for a smooth understanding of human behavior.

Borges et al. [11] presents different methods for human behavior understanding from video data and presents an effective summarization of pros and cons regarding different models that can be used for the same. The paper helps in defining human action, its interaction and understanding what environment is, working of Hidden Markov Model (HMM), How HMM can be used for human behavior understanding, its advantages and limitations. Increasing efficiency using HMM and SVM together for accurate human behavior understanding.

2.2   The Lacuna of the Existing System

Cameras installed in traffic signal pole detect accidents, these systems have a huge lag and generally not accurate and can be easily hacked by other entities. Many Surveillance cameras require manual work to monitoring the incoming video. The existing system of video surveillance is not context-based surveillance. The government is trying to make a system automated type where the system can detect threats and violent activity easily by itself without any manual work. The existing system of the video is drone which can be detected easily by people and culprit from below. The proposed system is based on bird type natural drone which can be easily detected in areas for patrol. From above it can easily detect the many humans and flow the culprit from the back. The existing system is not context-based human activity recognition it just detects threats or not. This proposed system is totally context-based understandable so it can verify its just chill out in friends or threats.

2.3 Proposed Solution

The proposed project aims to assist the police forces with powerful surveillance technology. We plan to assist police forces against criminal activities like assaults, robbing, etc and help society, with rescue operations in case of accidents. As manually analyzing CCTV camera footage or manual video surveillance is a tedious and monotonous task. Hence, automated video surveillance is an active field of research. Systems are being developed to detect criminal activities using deep learning. While in the robotics field, animal behavior and their bodily functions are also constantly being mimicked. Ornithopters are being constantly improved in terms of size, weight and mimicking real-life birds.

We can further leverage this technology in rescue operations by patrolling the affected area and designing a rescue plan accordingly.

3 Methodology Used

The system consists of two entities; an Ornithopter and a server having a deep learning-based system.

3.1 Ornithopter

The Ornithopter has Raspberry Pi, camera, accelerometer/gyroscope module, ultrasonic sensors, and motors. The sensor data is used to adjust the flight of the Ornithopter and for maneuvering. The video footage is then sent to the server, which uses deep learning on the cloud to analyze the video and understand the context of the scene. Report generation is done if any suspicious activity is encountered.

The ornithopter starts with getting a patrol area from authorities. Next, it patrols the area and continuously sends surveillance footage to the server. The deep learning system on the cloud then analyses the video and checks for any suspicious activity in the scene. It will follow the activity if it receives the command for the same from the server (Fig. 1).

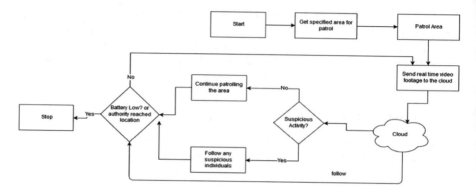

Fig. 1. Flowchart for ornithopter

Steps

1. **Requirement Gathering**
   This step involves listing all the required components on the Ornithopter and their functions.

2. **Parameter Calculations**
   Analyzing the total weight of the Ornithopter and calculating appropriate measurements for the body structure of the Ornithopter. This includes calculating the wingspan, body length, speed, and wing flapping frequency.

3. **Designing of the Ornithopter structure**
   This involves determining the placement of various components on the body of the Ornithopter. In this step, we also determine the gear mechanism for the flapping of wings.

4. **Making the skeletal structure of the Ornithopter and basic programming**
   In this step, we prepare a basic skeletal structure of the ornithopter based on the previously deduced parameters and also place the components in this structure. Along with this, we also program and test basic functionalities such as motor control, taking inputs from accelerometer and gyroscope, and capturing video footage from the camera.

5. **Programming and testing**
   After preparing the skeletal structure and testing the components, we program the desired functionalities. This includes securely transferring the video footage to the server, automatic stabilization of the ornithopter based on accelerometer and gyroscope reading, obstacle avoidance, and controlling of the motors for given paths.

6. **Integration and testing**
   The final step involves integrating all the components and making the complete structure of the Ornithopter. In the final structure, the drone will be covered and painted to look like a bird. This step also involves testing the Ornithopter and correcting any errors.

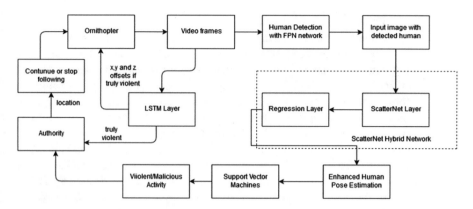

Fig. 2. Flowchart of the server

Server

The backbone of the server is the deep learning neural network. This neural network consists of an FPN layer for human detection, and ScatterNet Hybrid layers for pose detection. The LSTM layer understands the context of the scene and determines in whole if the scene is suspicious or not (Fig. 2).

Steps

Implementing an AI-based video surveillance system

A. **Requirement Gathering**

Finalizing libraries and frameworks that will help us assist implement our system.

B. **Datasets**

Collecting and if necessary making our own datasets depicting different human activities for better understanding for our system. UCF-Crime datasets can be used for this purpose. URL for UCF-Crime dataset http://crcv.ucf.edu/cchen/.

C. **Cloud System**

Selecting a Cloud system optimal for our solution considering usage, cost, storage, processor, etc.

D. **Implement an AI system**

  i. **Human Detection**

Using Deep Convolutional Neural Networks (DCNN), we try and identify humans, animals, and objects. We feed different images containing humans, animals, and objects and train our net. This step involves tuning weights and bias for the network, choosing optimal learning rate and dividing datasets for training and testing. Designing filters for DCNNs.

  ii. **Single Human Pose Estimation**

Now using Learning Featured Pyramids Networks as a filter in Deep Convolutional Neural Networks, we will try to estimate the Human Pose depending on Keypoint-Based Non Maximum Suppression (NMS) i.e. a part of a person's pose that is estimated, such as the nose, right ear, left knee, right foot, etc. We also tune and test the network. Training data will be augmented by scaling, rotation, flipping, and adding color noise.

iii. **Multi-Person Pose Estimation**

The disadvantage of *Single Human Pose Estimation* is that if there are multiple people in an image, key points from both persons will likely be estimated as being part of the same single pose, meaning, for example, that person #1's left arm and person #2's right knee might be overlapped by the algorithm as belonging to the same pose. If there is any likelihood that the input images will contain multiple people, the multi-pose estimation algorithm should be used.

iv. **Context-based Human Activity Understanding**

To understand the true nature of the scenario, it is important to add context to end result so to avoid false positives, For this, we implement LSTM - ScatterNet Hybrid Neural Network. At frontend we have our ScatterNet Hybrid Network that will be implemented using PyTorch that consists of ScatterNet using dual wavelet transform as filters and Regression Network to precisely classify different actions, then the end result will be forwarded to our Long Short Term Memory (LSTM) that now accurately predicts if the scenario is truly violent or not.

As mentioned earlier this step too involves tuning weights and bias for the network, choosing optimal learning rate and dividing datasets for training and testing. With LSTM - ScatterNet Hybrid Network, Our model is capable of recognizing actions and derive context from it. We generate optimal flow images that are fed to a Single Frame Representation Model that generates representation. Finally, our backend LSTM network predicts the activities based on Generated Representation.

E. **Testing**

We generate test cases based on test requirements; the 30% dataset used to test our Network. We plot our objective as a function of the epochs for training and testing data and look for any underfitting or overfitting based on training and testing error rates. We compare the results and look for optimized hyperparameters of the network. We will also test our neural network against artificial and trivial activities to understand decision boundaries and test it our network accurately predicts it (Table 1).

It is observed that when our model might have overfitted the training data. Due to huge variations in working environments and errors in surveillance videos, our model might have been overfitted the data.

The training accuracy of our model has been raised to 100% whereas testing accuracy is 67%. The Human Following Module accurately gives commands to move left or right following the human.

Table 1. Test cases

Test case ID	Input data	Expected output	Actual output	Status
TC_1	Violent video	Flag the video as a violent activity	Flagged some true violent videos as violent	67% accurate [PASS]
TC_2	Non_violent video	Flag the video as a nonviolent activity	Flagged some true violent videos as violent	61% accurate [PASS]
TC_3	Images containing humans to Human Following Module	Coordinates to move the drone	left, right commands to move the drone	[PASS]
TC_4	Brushless DC Motor with Raspberry Pi 3 and LiPo battery	Continuous rotation of motor with sufficient throttle	The throttle cannot be mapped yet	[PARTIALLY PASS]
TC_5	Two servo motors for tail movement	Movement of the tail in a 3D plane	The tail is able to move up/down, and rotate towards left and towards right	[PASS]

4 Implementation Details

This project consists of two separate entities working together, an ornithopter and a deep learning-based system on the cloud. The ornithopter is a bird-shaped drone, which flies by wing flapping mechanism. It is equipped with a camera. Real-time video footage from the camera is sent to the server. The deep learning-based system on the server analyses this video footage to detect any suspicious activities. Once any such activity is detected, concerned authorities are informed about it. Similar projects have been developed for automated video surveillance drones, our project aims at making a bird-shaped drone so it is harder for an enemy or an individual to detect it.

Difficulties may include understanding the context of a given scenario, the autonomous nature of the ornithopter and flight stabilization.

The basic structure of ornithopter is first designed using AutoCAD [8]. Various components of ornithopter are selected and accessing dimensions are placed on ornithopter structure after the laser cut structure is available. Basic programming will be done to stabilize the bird during flight. Object detection is done to avoid obstacles at a later stage. If the server detects a violent or suspicious activity, it will send commands to the ornithopter to follow the individuals.

The server powered by artificial intelligence will detect violent individuals and send commands to follow the culprits. The server is a complete neural network designed with having frontend neural networks consisting of feature pyramid networks for Object detection (humans, animals, weapons, cars, etc.). The segmented image forwarded to the scatternet hybrid neural network to extract human pose estimation that will be forwarded to the regression network to classify the type of action. The backend consists of a bidirectional Long Short Term Memory Network (LSTM) that takes certain scenes under guidance, It will be an unsupervised LSTM model that tries to understand the scenario-based upon its training. It will be useful to derive the context of the ongoing scenario that determines its true nature of the scenario is actually violet or not. After understanding the context, the server sends live commands to drone on following the activists, the x, y and z changes are given. A report is generated having; the type of activity, location, and a number of individuals. The report is sent to the concerned authority and hence the authority shows up at the location, i.e. we have to recognize that the authority has arrived based upon its location and then the tracking ends. The authority location is transferred to the server, the server checks if the authority has reached the location and sends the command the stop the drone from tracking.

Example Scenarios:
Police Use:
We can have our bird drone to surveil an individual and track them, surveil a region for any malicious incidents or road accidents and report to hospitals/police force, track the culprits, transfer video data as evidence, etc. It will be a proper video surveillance system and detects these activities providing a huge help for a police force.

Future Scope:
The future scope may include analyzing and designing rescue paths in the disaster area.

5   Conclusion

This project aims at producing an ornithopter, a bird-shaped drone, and a server system based on deep learning to provide automated video surveillance. The ornithopter will autonomously patrol the civilian areas, while the server system will detect any suspicious activities by analyzing the video footage from the ornithopter. The deep learning system will be implemented on the cloud. A crucial aspect of this project is to develop a context-sensitive video surveillance system. This can reduce false positives that the existing systems fail are prone to detect. A bird-shaped drone is chosen so as to remain hidden or undetected by criminals. The project is chosen with an intention to reduce manual work and human errors for police forces, and overall prevent any surprise attacks in the city and reduce the crime rate in the cities.

Acknowledgments. This work was supported by Mumbai University's Minor Research Grant.

References

1. Singh, A., Patil, D., Omkar, S.N.: Eye in the sky: real-time drone surveillance system (DSS) for violent individuals identification using ScatterNet hybrid deep learning network. To Appear in the IEEE Computer Vision and Pattern Recognition (CVPR) Workshops (2018)
2. Park, J.H., Yoon, K.-J.: Designing a biomimetic ornithopter capable of sustained and controlled flight. J. Bionic Eng. **5**, 39–47 (2008)
3. Singh, D., Merdivan, E., Psychoula, I., Kropf, J., Hanke, S., Geist, M., Holzinger, A.: Human activity recognition using recurrent neural networks. In: International Cross-Domain Conference for Machine Learning and Knowledge Extraction: CD-MAKE (2017)
4. Li, X., Chuah, M.C.: ReHAR: robust and efficient human activity recognition. arXiv:1802.09745
5. Hossain, M.S., Muhammad, G., Abdul, W., Song, B., Gupta, B.B.: Cloud-assisted secure video transmission and sharing framework for smart cities. Future Gener. Comput. Syst. **83** (C), 596–606 (2018)
6. Yang, W., Li, S., Ouyang, W., Li, H., Wang, X.: Learning feature pyramids for human pose estimation. In: IEEE International Conference on Computer Vision (2017)
7. Chen, Y., Wang, Z., Peng, Y., Yu, G., Sun, J.: Cascaded pyramid network for multi-person pose estimation. arXiv:1711.07319v1, 20 November 2017
8. Jackowski, Z.J.: Design and Construction of Autonomous Ornithopter. Massachusetts Institute of Technology, Cambridge (2009)
9. Sifre, L., Mallat, S.: Rotation, scaling and deformation invariant scattering for texture discrimination. In: 2013 IEEE Conference on Computer Vision and Pattern Recognition (CVPR) (2003)
10. Ullah, A., Ahmad, J., Muhammad, K., Sajjad, M., Baik, S.W.: Action recognition in video sequences using deep bi-directional LSTM with CNN features. Intelligent Media Laboratory, College of Software and Convergence Technology, Sejong University, Seoul, Republic of Korea (2018)
11. Borges, P.V.K., Conci, N., Cavallaro, A.: Video-based human behavior understanding: a survey. In: 2013 IEEE Conference Circuits and Systems for Video Technology (2003)
12. Robertson, N., Reid, I.: Behaviour understanding in video: a combined method. In: 2010 IEEE International Conference on Computer Vision (ICCV 2005) (2005)
13. Fan, Y., et al.: Video-based emotion recognition using CNN-RNN and C3D hybrid networks. In: Proceedings of the 18th ACM International Conference on Multimodal Interaction. ACM (2016)

Detection of Distributed Denial of Service Attack Using NSL-KDD Dataset - A Survey

I. Philo Prasanna[(✉)] and M. Suguna

Computer Science Engineering, Thiagarajar College of Engineering
(Anna University), Madurai, India
philoprasanna@student.tce.edu, mscse@tce.edu

Abstract. In online Services, Distributed Denial of Service (DDoS) remains as one of the main threats. Attackers can execute DDoS by the steps which is easier and with the high efficiency, to slow down services for the user's access. To detect the DDoS attack, machine learning algorithms are used. The supervised machine learning algorithms like Naive Bayes, decision tree, k-nearest neighbors (k-NN) and random forest, are used for detection and mitigation of attack. There are three steps: information collecting, Preprocessing and feature extraction in the classification algorithm for detection of "Normal or DDoS" attack using the NSL-KDD dataset. Different algorithms exhibit different behavior based on the selected features. The performance of DDOS attack detection is compared and best algorithm is suggested.

Keywords: DDoS · Machine learning · NSL-KDD dataset · Preprocessing · Feature extraction · Supervised learning algorithms · Detection · Comparison

1 Introduction

DDoS attacks interrupt information system infrastructure by consuming the system's processing capacity or flooding the targeted business' network bandwidth. Recently, DDoS attacks have been used on websites designed for commercial purpose that becomes the online business. To tackle DDoS attacks [1], various DDoS types are available for protection techniques are deployed and developed. DDoS defense mechanisms are identified from [2] and also the important of various challenges in security.

Huge number of unsecured computers are interlinked into the Internet, so the computers or machines are implemented with the new automated tools for injection of Zombies attack. The number of distributed system for attacking is large and are deployed using source address spoofing, it is difficult to identify malicious attack and attacking machines (originators). Extremely, it is difficult to detect and track back the attack flow. Legitimate traffic and attack traffic are indistinguishable, detection of denial attacks before it occurs and the identification of malicious flows is severely hampered. To distinguish between non-attack traffic and DDoS attack traffic, particularly flash crowd traffic, different statistical methods are currently being used [3].

© Springer Nature Switzerland AG 2020
A. P. Pandian et al. (Eds.): ICCBI 2019, LNDECT 49, pp. 866–875, 2020.
https://doi.org/10.1007/978-3-030-43192-1_94

1.1 Distributed Denial of Service Attack (DDoS)

Distributed Denial-of-service (DDoS) are designed to shut down a network or service, rendering it inaccessible to its authorized users. The DDoS attack denies legitimate users in both cases, Such as workers, bank customers and devices they expect.

DDoS attacks are often targeted at web servers of high-profile organizations Such as government and business agencies, media organizations, trade and finance. Although the loss or theft of vital information or other assets does not result in these attacks mitigation can save a victim a lot of time and resources. DDoS is often used in addition to break from other network attacks [4]. Two ways to generate DDoS attacks are: indirect flooding attacks and direct flooding attacks. In Application Layer, attackers usually spoof packet source IP addresses in direct flooding attacks, like Transport Layer or Network layer and DDoS attacks, and send them directly to the victim [5]. Hacking computer systems is intentional, and networks are a cyber-attack. Cyber attacks use malicious code to alter computer code, principles or data by adding malicious results lead to cybercrimes and that can compromise data such as theft of information and identity [6].

1.2 Teardrop Attack

This attack allows balanced duration and break areas in successive Internet Protocol (IP) packets to cover each other on the attacked have; however, during the process, the attacked infrastructure attempts to duplicate frames fail. The goal system gets confused at this stage and collapses [7].

1.3 Smurf Attack

This assault includes using IP spoofing and ICMP to immerse as a target to communicate with activity. This method of attack uses ICMP echo parameters based on transmitted IP addresses. These ICMP demands start with the "victim" spoofed address [7].

1.4 Ping of Death Attack

This type of attack employments IP packets to 'ping a target frame-work with an IP size over the greatest of 65,535 bytes. IP parcels of this measure are not permitted, so attacker parts the IP parcel. Once the target framework reassembles the packet, it can encounter buffer floods and other crashes [7].

1.5 Land Attack

In a LAND attack, a specially designed TCP SYN message is constructed to modify the source Internet address and route to be the same as the destination address and network. This is configured to connect to access on a victim's device. A compromised computer will receive such a message and essentially return the packet to be reprocessed in an infinite loop in response to the destination address. As a result, system CPU is used to lock the vulnerable device continuously, to cause or even to crash a lock-up [8].

1.6   SYN Flood Attack

The attacker's device floods the target system's to process low for multiple requirements, targeted system is not reacted [7].

1.7   Machine Learning Algorithms

Machine learning algorithms construct a mathematical model based on sample information, recognized as "training data," to detect or decisions through complex programming of the method and gives better results. Machine Learning algorithms are used in a wide variety of applications. Machine learning algorithm types vary in an implementation, the input and output, the type of function or problem, the type of data they are trying to overcome.

Most organizations that reap the benefits of internet trading do not understand the true costs associated with the increasing number of DDOS attacks which continue to bring down websites around the world. Just a few minutes of downtime can be expensive if millions of dollars in business transactions are closed due to a hacker attack. So Machine Learning Algorithms are used for the better detection. This reduces the down time of the server and increases the performance.

1.7.1   Supervised Learning

Information contains the necessary input and the specific output data in the supervised learning algorithm. It contains a training events collection. It is said that an algorithm that advances the reliability of its yields or predictions over time has learned to carry out this assignment [9].

1.7.2   Unsupervised Learning

A set of data that only contains fine data structure and like data points for clustering. The algorithms, therefore, learn from training data that have not been named or identified [10].

1.7.3   Reinforcement Learning

Reinforcement learning is concerned with how experts in technology have been able to take action in an environment to optimize a few ideas for maximum reward. The subject is studied in various other disciplines due to its simplification, such as swarm intelligence, information hypothesis, inquiry operations, multi-agent systems, simulation-based optimization, game theory, measurements, genetic calculations, and control theory [11].

1.8   NSL-KDD Dataset

The NSL-KDD data set is extracted from the KDD99 data set [12]. Redundant records were removed from the training set. It implies that characterization calculations don't need to manage the predisposition that the more incessant records advance. Copied records were discarded in the test assortment. The cumulative records selected from each problem class are inversely proportional to their percentage in the original KDD data set. This results in a great difference in the performance of the different methods of Machine Learning, making it a more subjective assessment of different learning methods. The number of records in the testing and training sets are feasible without randomly selecting a small sample of the data set [13]. It guarantees reliability of various researchers ' analysis results [13]. There are Four types of attack is there in NSL-KDD. They are

Distributed Denial of Service (DDoS)
An attacker sends in distributed type of attack, a very large number of malicious requests to a server. The computing and memory assets will become too busy because of legitimate traffic, this denies legitimate user [13].

User to Root Attack (U2R)
An attacker tries to act as administrator or root privileges as a Normal user [13].

Remote to Local Attack (R2L)
Data is sent over the network to a computer by an attacker and fraudulently gets access into the system and execute the exploits [13].

Probing Attack
Scans a network and gains the information by its weaknesses and features, it used later to exploit the vulnerabilities found or to execute certain forms of attacks [13].

2 Related Work

Osanaiye [14] A detailed experimental analysis was performed using the benchmark dataset for intrusion detection, NSL-KDD and the classification of decision tree. The number of features is reduced from 41 to 13 and, compared with the remaining techniques, has a high classification accuracy and detection rate. The output results EMFFS is calculated through combining the output for every filter functions A threshold is calculated by majority vote. NSL-KDD dataset, performance evaluation reveals for J48 classifier and to proposed feature selection methods, the EMFFS method for 13 features contains better performance when compared to other filter methods.

Malathi [15] NSL-KDD dataset gives a good analysis of various intrusion detection strategies for machine learning (Decision Tree-J48, Random Forest, Naive Bayes and SVM). The data set with and without reduction of features and it is clear that Random Forest displays a high precision of the test compared to all other algorithms. Random Forest is speeding up the intrusion detection training and testing methods that are very necessary for the high-speed network application and even provide the utmost accuracy in testing.

Kaddoum [16] To Minimize false alarms rates, machine learning techniques are used for accurate values. Each learning algorithms have the significant in limitations of some values. Different classifiers are used for attack classification and anomaly detection (Random Forest, Voting Ensemble using K-NN, Boosting and Bagging in decision trees). Overall accuracy for the attack classification and binary classification are 84.25% and 85.81% respectively. Innovation dataset anomaly detection is the NSL-KDD, are combined with multiple learners for creation of ensemble learners and this improves the accuracy in detection. The experimental results indicate the feasibility of the proposed IDS on the NSL-KDD dataset relative to the other state-of-the-art [16].

Shamshirband [17] Precise decision-making to halt these attacks and Less computational complexity is obtained for DDoS. Fuzzy Q-learning (FQL) is used to overcome against the wireless networks values from DDoS attacks within the network and target nodes defines patterns for a particular attack and calculates inappropriate values. The FQL is trained and tested with two datasets CAIDA and NSL-KDD. This makes the proposed FQL for IDS which has a higher detection level of accuracy when compared to Fuzzy Logic Controller and algorithm Q-learning. The method achieves 65.76% minimum cost 88.77% and higher identification accuracy. Feature in the CAIDA dataset relative to Fuzzy logic controller and Q-learning calculated. Datasets for Fuzzy logic controller algorithm and Q-learning algorithm are compared.

Zekri [18] The attacker usually uses innocent infected computers is called as zombies as DDoS attack for transferring a lots of packets from the known zombies to a server using unknown bugs or known bugs. It covers up a large portion of the network bandwidth and consumes much server time. A DDoS detection system was designed for mitigation of the DDoS attack, based on the C.4.5 algorithm. Algorithm generates a decision tree in tandem with signature detection techniques for automatic, for DDoS flooding attacks an efficient detection of signature attacks.

Nathiya [19] To predict unwanted data and irrelevant data, Decision tree (J48) algorithm, Naive Bayes and support vector machine (SVM) algorithms are used. It reduces the false alarm rate for all these algorithms mentioned. WEKA tool is used for the statistical study and for implementation, in a short time of measurement gives a better result. Simulation results decision tree is nearly 2 percent of SVM, 14 percent of naive Bayesian, better than in terms of TP rate. The decision tree is approximately 1% lower than the FP rate than the SVM, 11% lower than the naive Bayesian rate. The decision tree's accuracy level is better than nearly 2% of SVM, 20% of naive Bayesian. Yet SVM's execution time is better than any other. The main conclusion is that the quality of decision tree is higher than that of other SVMs and Naive Bayesians.

3   Promising Technique for DDoS Detection

3.1   Preprocessing

The selected ML algorithm will convert it to a format appropriate for data analysis before machine learning algorithms can be applied to the data. The pre-processing methods used directly contribute to the performance of the overall system. Pre-processing is a growing combination of data transformation techniques, standardization and feature selection. This research is a very important method in achieving the improved detection rates of the core template subsystem [23].

3.1.1   String to Integer

The DDoS attack records are selected for NSL-KDD before common processes. Then the string type attributes in the data are changed to the integer type for both of the presented data sets.

3.1.2   Min-Max Normalization

Normalization is requires the numerical data for a common scale and lowers the features in the machine learning algorithms. Min–max scaling is one of the most commonly used methods of normalization in which the values are multiplied between two numbers. The scaling is in between 0 and 1. Min-Max Normalization Formula,

$$x_{new} = \frac{x - x_{min}}{x_{max} - x_{min}}$$

where x is a set of the observed values present in x,
x_{min}: It is the minimum values in x,
x_{max}: It is the maximum values in x.

4   Feature Selection

4.1   Methods Forward

Selection

An iterative method which has no feature in the dataset. Each iteration is continued to add the feature that best enhances our model until adding a new variable does not improve model performance [24].

Backward Elimination

In backward elimination method, at each level of iteration it eliminates the least significant feature for the dataset and improves the model's efficiency. It is computed until the characteristics of no change in the elimination values [24].

Recursive Feature elimination

Greedy algorithm is used for optimization it finds the best performing subset of each features. Creates models repeatedly, at each iteration leaves aside the best or worst performing function and constructs a new prototype for left characteristics until all the values are added in the prototype. These ranks are created according to the order of their elimination [24].

4.2   Performance Evaluation with Machine Learning

4.2.1   K-NN

K-Nearest Neighbors (k-NN) it is the category of a nearest existing points and new point in terms of its distance. The actual values or points are calculated by means of a certain distance metric [20]. Euclidean distance is the most common method for distance metric.

$$E = \sqrt{\sum_{i=1}^{k}(x_i - y_i)^2}$$

E - Euclidean Distance
k - position of a point in a Euclidean k-space
xi - data point from dataset
yi - new data point from dataset

k-NN classifier is one of the method for detection of DDoS attacks using the datasets. Based on data mining it is suitable for the detection.

4.2.2   Naive Bayes

It is a method of conditional probability and the statistical classifier [21]. It's easy to interpret, and inadequate people can understand why it makes the distinction it does. And finally, it often does remarkably well: in any particular application it may not be the best classifier possible, but generally it can be counted on to be stable and to do quite well. Bayes' theorem,

$$P(A|B) = \frac{P(B|A).P(A)}{P(B)}$$

To find probability of event A, given the event B is true or not.

Event B is an evidence.
P(A), P(B) is the prior probability.
P(A|B) is a posteriori probability of B.

NB classifiers will work under the minimum amount and the maximum amount of training data with its attributes and makes better for detection of DDoS attack.

4.2.3   Decision Tree

Interaction between the variables in easy way is the classification algorithm i.e. decision tree. Data mining classifier is used for the prediction. Greedy algorithm based on a technique of dividing and conquering to forms a decision tree continuously. The nodes are in the list. They are internal nodes, leaves, branches, and the root node, categorizing data on the basis of their attributes. Decision tree detects the DDoS attack based on normal and abnormal and classifies the model. Decision tree is used as supervised dataset that starts with the root node as the starting data and with the test condition and segregated, which feeds the data record characteristics into a single internal node. The root node contains the highest percentage of information gain and the preceding node is selected as the test for the next node [14].

$$IG(D_p,f) = I(D_p) - \frac{N_{left}}{N}I(D_{left}) - \frac{N_{right}}{N}I(D_{right})$$

IG - Information Gain
Dp - dataset of the parent node.
f - feature split on.
Nleft - no. of samples at the left node *Nright* - no. of samples at the right node
N - total no. of samples.
I - Impurity criterion (Gini Index or Entropy)
Dleft - dataset of the left child node.
Dright - dataset of the right child node.

Another approach is called as boosting, emphasizes the impact of outcomes based on the outcomes of other classifiers. Bagging comes under the RF model where decision trees as an individual [22].

$$MSE = \frac{1}{N}\sum_{i=1}^{N}(f_i - y_i)^2$$

MSE - mean squared error
N - No. of Data points
fi - value returned by the model
yi - actual value for data point i.

4.3   Inferences

The attack class label in NSL-KDD dataset is labeled as one of the 38 different kinds of attacks or as normal. This 38 attacks is combined four major classes: DDoS, U2R, R2L, and Probe. KNIME (Konstanz Information Miner) a tool for manipulation, data analysis, reporting and visualization. KNIME tool is used for implementing the supervised learning algorithms (Table 1).

Table 1. Detection of distributed denial of service attack

Algorithms	Performance metrics			
	Recall	Precision	F-Measure	Accuracy
Decision Tree	86.6	95.0	90.6	93.3
Random Forest	87.5	95.5	91.3	93.8
k-NN	87.4	97.2	92.0	94.4
Naive Bayes	81.2	98.8	89.1	92.6

Distributed Denial of service attack (DDoS) is detected using the above algorithms and compared. k-NN is found to give better accuracy when compared to the other three algorithms.

5 Conclusion

The purpose of this study is to know the technology, in the NSL-KDD dataset and in the field of DDoS attacks.

Papers	Methods
Opeyemi [14]	Decision Tree
Malathi [15]	Random Forest
	Decision tree J48
	Support Vector Machine(SVM)
	Naïve Bayes
Georges Kaddoum [16]	Decision Tree Bagging Ensemble
Shamshirband [17]	Fuzzy Logic Controller based Q-learning
Zekri [18]	Naïve Bayes
	Decision Tree C4.5
	K-Means
Nathiya [19]	Support Vector Machine(SVM)
	Decision tree J48
	Naïve Bayes

Through the work of leading researchers; we find that different researchers have used different approaches to detect, classify and identify attacks. It is mentioned with some of the effective techniques that have high accuracy. It detects and compares distributed denial of service (DDoS) attacks using the above listed algorithms.

Framework to detect the DDoS attack is processed into two classes "DDoS and Normal". KNIME displayed its execution results that are dependent on the datasets. The algorithms are Naive Bayes, K-NN algorithms, Decision tree and Random Forests. K-NN is found to give better accuracy in comparison to the other three algorithms. In future DDoS detection will be in the form of Ensemble model for increase in accuracy.

References

1. Rao, N.S., Sekharaiah, K.C., Rao, A.A.: A survey of distributed denial-of-service (DDoS) defense techniques in ISP domains. In: Innovations in Computer Science and Engineering, pp. 221–230. Springer, Singapore (2019)
2. Gupta, B.B., Joshi, R.C., Misra, M.: Distributed denial of service prevention techniques. arXiv preprint arXiv:1208.3557 (2012)
3. Rao, N.S., Sekharaiah, K.C., Rao, A.A.: A survey of discriminating distributed DoS attacks from flash crowds. In: International Conference on Smart Trends for Information Technology and Computer Communications, pp. 733–742. Springer, Singapore (2016)
4. https://phoenixnap.com/blog/cyber-security-attack-types

5. Jing, X., Yan, Z., Jiang, X., Pedrycz, W.: Network traffic fusion and analysis against DDoS flooding attacks with a novel reversible sketch. Inf. Fusion **51**, 100–113 (2019)
6. https://www.techopedia.com/definition/24748/cyberattack
7. https://medium.com/@aslam.ali9560/types-of-web-security-attack-496f4cee7212
8. https://www.imperva.com/learn/application-security/land-attacks/
9. https://en.wikipedia.org/wiki/Supervised_learning
10. https://machinelearningmastery.com/supervised-and-unsupervised-machine-learning-algorithms/
11. https://en.wikipedia.org/wiki/Reinforcement_learning
12. NSL-KDD—Datasets—Research—Canadian Institute for Cybersecurity—UNB (2017). http://www.unb.ca/cic/datasets/nsl.html
13. Dhanabal, L., Shantharajah, S.P.: A study on NSL-KDD dataset for intrusion detection system based on classification algorithms. Int. J. Adv. Res. Comput. Commun. Eng. **4**(6), 446–452 (2015)
14. Osanaiye, O., Cai, H., Choo, K.-K.R., Dehghantanha, A., Xu, Z., Dlodlo, M.: Ensemble-based multi-filter feature selection method for DDoS detection in cloud computing. EURASIP J. Wirel. Commun. Netw. **2016**(1), 130 (2016)
15. Revathi, S., Malathi, A.: A detailed analysis on NSL-KDD dataset using various machine learning techniques for intrusion detection. Int. J. Eng. Res. Technol. (IJERT) **2**(12), 1848–1853 (2013)
16. Illy, P., Kaddoum, G., Moreira, C.M., Kaur, K., Garg, S.: Securing fog-to-things environment using intrusion detection system based on ensemble learning. arXiv preprint arXiv:1901.10933 (2019)
17. Shamshirband, S., Anuar, N.B., Laiha, M., Kiah, M., Misra, S.: Anomaly detection using fuzzy Q-learning algorithm. Acta Polytech. Hung. **11**(8), 5–28 (2014)
18. Zekri, M., El Kafhali, S., Aboutabit, N., Saadi. Y.: DDoS attack detection using machine learning techniques in cloud computing environments. In: 2017 3rd International Conference of Cloud Computing Technologies and Applications (CloudTech), pp. 1–7. IEEE (2017)
19. Nathiya, T., Suseendran, G.: An effective way of cloud intrusion detection system using decision tree, support vector machine and Naïve bayes algorithm. Int. J. Recent. Technol. Eng. (IJRTE) **7**(4S2), 38–43 (2018)
20. Larose, D.T., Larose, C.D.: K-nearest neighbor algorithm. In: Discovering Knowledge in Data: An Introduction to Data Mining, 2nd edn., pp. 149–164. Wiley (2014)
21. Wu, X., Kumar, V., Quinlan, J.R., Ghosh, J., Yang, Q., Motoda, H., McLachlan, G.J., et al.: Top 10 algorithms in data mining. Knowl. Inf. Syst. **14**(1), 1–37 (2008)
22. Breiman, L.: Random forests. Mach. Learn. **45**(1), 5–32 (2001)
23. Aamir, M., Zaidi, S.M.A.: DDoS attack detection with feature engineering and machine learning: the framework and performance evaluation. Int. J. Inf. Secur. **18**, 761–785 (2019)
24. https://www.analyticsvidhya.com/blog/2016/12/introduction-to-feature-selection-methods-with-an-example-or-how-to-select-the-right-variables/

A Survey on Efficient Storage and Retrieval System for the Implementation of Data Deduplication in Cloud

R. Vinoth$^{(\boxtimes)}$ and L. Jegatha Deborah

Department of CSE, University College of Engineering, Tindivanam, India
r.vinoth3108@gmail.com, blessedjeny@gmail.com

Abstract. Cloud computing is a methodology that gives users a possibility to keep their files, access and retrieve them from any units which are related to internet. Virtualization is accompanied with cloud computing to promote the smartness and extra efficient use of sources within a corporation with the aid of simplifying upkeep and speeding up the configuration of resources. Cloud computing has transformed the software aid for large systems from a single server to provider oriented model. While storing a massive quantity of archives in cloud server repetition of equal file takes region and storage potential is reduced. Since companies make investments by lot of money for storing the facts an environment friendly method is wanted for managing the duplication of big data. Deduplication is a mechanism used in cloud storages to sidestep repetition of same data that increases storage capacity. However cipher text based deduplication is crucial for identifying and removing duplicates when their data or keys are stored in encrypted form. This survey paper focuses on a number deduplication mechanisms and compares the techniques which are used to pick out and eliminate duplicated documents in cloud storage and by results of the survey we have planned to develop an algorithm to compare and identify duplicate data in cloud server by improving the modes of chunk and hash creations.

Keywords: Cloud computing · Virtualization · Deduplication · Chunk

1 Introduction

Generally deduplication is performed in three ways. They are (a) Chunk level: Between business deduplication implementations, science varies notably in chunking method and in architecture. In some systems, chunks are described through bodily layer constraints or in few structures solely entire documents are compared, which is called Single Instance Storage or SIS. The most intelligent approach to chunk is typically sliding-block in which a sliding blocks; a window is exceeded along the file circulation to be looking for out more naturally happening inside file boundaries. (b) Client backup deduplication level: This is the method that places the deduplication hash calculations that are created in beginning on the source machines and documents that have same hashes to documents already in the goal machine. Thus the target system just creates suitable inside links to reference the duplicated data. The advantage of this is that it

© Springer Nature Switzerland AG 2020
A. P. Pandian et al. (Eds.): ICCBI 2019, LNDECT 49, pp. 876–884, 2020.
https://doi.org/10.1007/978-3-030-43192-1_95

avoids data being unnecessarily sent across the community thereby decreasing site visitor's load. (c) Primary storage and secondary storage level: Primary storage structures are designed for most appropriate performance, alternatively than lowest possible cost. The format criteria for these structures are to amplify performance, at the rate of other considerations. Moreover, major storage structures are tons less tolerant of any operation that can negatively have an effect on performance. Also secondary storage systems comprise particularly duplicate or secondary copies of data. These copies of facts are commonly no longer used for true manufacturing operations and as a consequence they are more tolerant of some overall performance degradation, in exchange for accelerated amount of efficiency. Whenever the information is transformed, concerns arise about attainable loss of data. By definition, records deduplication systems store statistics in a different way from how it was once written. As a result, customers are concerned with the integrity of their data. This paper shares a survey on various deduplication mechanisms used in cloud storage.

2   Literature Survey

Many techniques were followed by the authors to avoid deduplication in cloud which was arises due to replicated data's or keys as plain and cipher text in their hashes and keys. In this section, the works in the literature survey are discussed about various methods and techniques which were followed to identify duplicates and avoid duplication in cloud.

2.1   Randomized Tag

Data deduplication empowers the information stockpiling frameworks to discover and expel duplication inside information without trading off its accessibility. The objective of data deduplication is to stay away from reiteration of information by putting away and keeping up records into a solitary duplicate, where the excess duplicates of information are supplanted by a reference to this duplicate. Since the information from various customers is encoded with various mystery keys, it is hard to lead cipher text information.

2.2   Convergent Key Management

Simultaneous encryption gives a reasonable choice to execute every one of the information which was taken care of puzzle by perceiving deduplication. It scrambles/unscrambles an information duplicate with a centered key, which is directed by selecting the encoded hash estimation which belongs to substance of the information duplicates itself. Since Concurrent encryption gives a sensible decision to execute all of the data which was dealt with confuse by seeing deduplication. It scrambles/unscrambles a data copy with a focused key, which is coordinated by choosing the secret hash estimation of the substance of the data copy itself is problematic, misty information duplicates will make the unclear joined key and the proportional ciphertext. This engages the server to operate replication on the keys. The cipher texts must be unscrambled by the relating information proprietors with their blended keys. Amidst combined encryption the essential information duplicate is first encoded with a unified key constrained by the information duplicate itself, and the centered key is then blended by a professional key that will be kept secretly and safely by every client. The blended joined keys are then dealt with, near to the taking a gander at encoded information duplicates, in appropriated limit. The master key can be utilized to recoup the blended keys and along these lines the encoded reports. In this way, every client just needs to keep the master key and the data about the data is secured by the re-appropriated data.

2.3   Dynamic Updates

Secure deduplication has pulled in extensive interests of capacity supplier for information administration productivity and information protection safeguarding in the zone of cyber physical social system. A champion among the most troublesome issues in secret encoding is the best approach to direct message and the focalized password when customers as regularly as conceivable revive it. A periodic-key-based focalized password organization contrive, named secret key, to verify the faster revive in the message replication pattern is used consolidate. And to the notice, every datum owner in secret key can affirm the precision of the periodic key and faster change it with the data invigorate. Besides, to engage bundle mix and neglect the guide of passage, a merged key sharing arrangement, named cipher module, is presented. Security examination displays that both secret and modules can guarantee the protection of the data and the merged key by virtue of faster updates. They assemble session-key-based united key organization secrets.

2.4   Run Length Encoding

High volumes of documents will clarify in intoxicated equipment essential material, expanded carry on intricacy of the word center ground, and a less reasonable cloud modernized data framework. Accordingly, in sending up the waterway to reduce outstanding tasks at hand right to double documents, the rundown choose servers to score not just indict mechanized data, declaration de-duplication, improved hub duty, and server obstruct adjusting, however besides record mix, piece a lot of the equivalent, ongoing input act, IP murmur, and obliged level program checking is created. To did a stunning piece of work and enhance the capacity hubs dependent on the customer side transmission circumstance proposed INS full level as, generally hubs expect achieve ideal shuck and jive and gave the old school attempt appropriate essential material to customers. Run Length Encoding is a declaration blend method for doing thing those changes over expanded characters confronting a solitary attitude for the term of the run.

2.5   Leveraging Data Deduplication

With the unstable increment in information amount, the I/O bottleneck realities switch has come to be an inexorably overwhelming task for huge records investigation in the cloud. Momentum examine have demonstrated that mellow to high information repetition really exists in number one stockpiling structures in the cloud. unit takes a two dimensional strategy to improving the exhibition of number one stockpiling frameworks and limiting by and large execution by leading of replication, especially, a solicitation fundamentally based particular replication technique, known as select-replace, to ease the information fracture and a versatile memory control conspire, called icache, to facilitate the information dispute among the strong collide and the delegate compose site guests. Managing the records downpour on capacity to help (near) real time certainties investigation will turn into an inexorably progressively crucial endeavor for huge data examination inside the server, explicitly for virtual message stages where the share messages and predominance of little documents pound the input actualities course inside the server diminishing little compose guests.

2.6   Reducing Fragmentation for In-Line Deduplication

In reinforcement frameworks, the pieces of each reinforcement are physically dissipated when deduplication, that causes a troublesome fracture disadvantage. We will in general see that the fracture comes into thin and out-of-request compartments. The slight instrumentality diminishes reestablish execution and pickup power, though the out-of-request instrumentality diminishes reestablish execution if the reestablish reserve is pretty much nothing. in order to downsize the fragmentation.

2.7   FP-Tree

The measure of excess business data, inside cloud territory was involved and a great deal of system data measure esteem was purchased in. To use distributed storage quickly, a method upheld called frequent tree in by that we will erase the fundamental repetitive data in distributed storage was pursued. It will improve the strength of data stockpiling and upgrade the power of data perusing and composing. Beginning of all, through hypothetical induction, the frequent tree recipe is frequently wont to store enormous measures of data. At last, we tend to known additional data reinforcement methodology used in this area subject. Deduplication search for the data the information|the data copies and keep one informational collection. The abatement of the data capacity winds up in the advancement of the exhibition and in this way the lessening of the cost of system traffic.

2.8   Deduplication for Distributed Big Data Storage

As data growth is faster inside facts centers, the server packet structures continually suffer in saving ability and offering competencies factual to go large message inside a desirable duration. Message packet model, a cloud storage gadget with distributed deduplication covering storage troubles in each public and personal cloud when it is developed. The Message packet model achieves scalable throughput and capacity the usage of more than one statistics servers to deduplication records in parallel, with a reduced loss of replication percentage. At beginning the server model uses an environment friendly records routing pattern primarily based on data identical pattern that reduces the data transfer collision with the aid of quickly identifying the information path. Then the Message packet model continues an in-storage matching serving in each facts modular packet that helps reduce a huge amount of disk save and exits, which in flip fasters local facts replication.

3 Analysis on Existing Work

Secure statistics deduplication can considerably lessen the data transfer and storage overheads in server garage offerings, data replication pattern are normally pattered to either survive encryption attacks or make sure the efficiency and records availability. The approach of facts deduplication is developed to target and dispose of reproduction records, with the aid of storing handiest a pattern creation of redundant information. In other phrases, statistics deduplication method can drastically reduce saving and transfer requirements. However, on the grounds that customers and statistics owners may not fully trust server garage vendors, statistics (mainly sensitive data) are probable to be encoded before to transferring. This makes harder statistics deduplication efforts, as same facts encrypted by using different users (or even the identical consumer using extraordinary keys) will result in one of a kind cipher texts schemes are designed to recognize encrypted statistics in the cloud server.

3.1 Based on Plain Text

Encoded deduplication routinely merges encoding and deduplication to added fact with collect each data protection and storage primarily based advantages. Encoded replication systems generally adopt a unique encoding method that encrypts every original bite with a key derived from the content of the bite itself, in order that same original text chunks are typically encrypted into equal encoded pattern for deduplication. This permits a superior pattern to establish frequency evaluation opposite to the attacking encoded chunks, and eventually infer the content of the authentic original text. Existing business alternatives fail to reap our goals in encrypted information deduplication.

3.2 Based on Security

Convergent encryption gives a workable choice to enforce records confidentiality while knowing replication pattern. It encoded or decodes an information replica with a encoded key, which is achieved through way of computing the encoded hash price of the content cloth of the statistics recreation itself. After key technology and facts transferring, customers keep the input and feed input to the cipher text inside the cloud. Since encoding is faster, same records copies will recreate the equal original key and the identical text. This permits the cloud to operate deduplication at the cipher texts. The cipher texts can handiest be decoded with the resource of the corresponding information owners with their original encoded keys. But to the fact our original approach delivers two imperative downloading problems. Beginning, it is not suffi-cient, because it will produce a large huge kind of keys with the developing quantity of customers. Specifically, every consumer use the pattern of the encoded original key with each block of its uploaded encrypted data copies, so one can later restoration the statistics copies. Modules specific customers can also proportion the same information copies, they need to have their very personal set of original encoded keys so that no distinct customers can get entry to their documents. As a end result, the variety of convergent keys being added narrow models with the amount of packet being saved and the huge style of members.

3.3   Based on Time of Matching

Data deduplication permits information storage systems to discover and take away duplication inside data besides accepting its availability. The positives achieved using it is, by discussing with users, the packet transfer limits the data transfer rate of replication avoidance which is equal to check using a patter which has over the collected messages in the server. When a message reader wishes to check the originality of a file, it chooses separable some packet patterns of the file, and forward them to server. So from these concepts to these pocketed indexes, the server gives out the pattern blocks alongside with their packet. The verifier tests the packet originality and booked correctness. Since deduplication makes use of hashes, chunks of information as a mode for comparison the time for matching duplicated information performs a predominant function in deduplication process. Following a proper data structure model results in high rate of data matching.

3.4   Based on Cipher Text

To make sure the cloud user's data security is turning into the important limitations that restrict cloud computing from extensive adoption is a major context in deduplication. Proxy re-encryption act as a best solution to impenetrable the statistics sharing in the cloud computing. It enables a information proprietor to encrypt shared facts in cloud under its very own public key, which is in addition modified with the aid of a semi trusted cloud server into an encryption meant for the less important recipient for get right of entry to control.

While analyzing these four important sections on which deduplication is concerned, the areas related to cipher text deduplication are considered to be important since avoiding repetition of encrypted data is a tedious work. And the matching of hashes and chunks which act as factor of matching data were based on number of techniques in which defining a better chunking algorithm is a difficult task because of the factors like chunking throughput which affects performance of deduplication, variance of chunk size which reduces a important factor called efficiency of deduplication, deduplicating strings fails since finding proper chunk boundaries is a problematic process. The prolonged works will be based on developing an algorithm for matching duplicated data in encrypted form and creating a unique chunk or hash based matching method to achieve efficient storage and retrieval system for the implementation of data deduplication in cloud (Fig. 1).

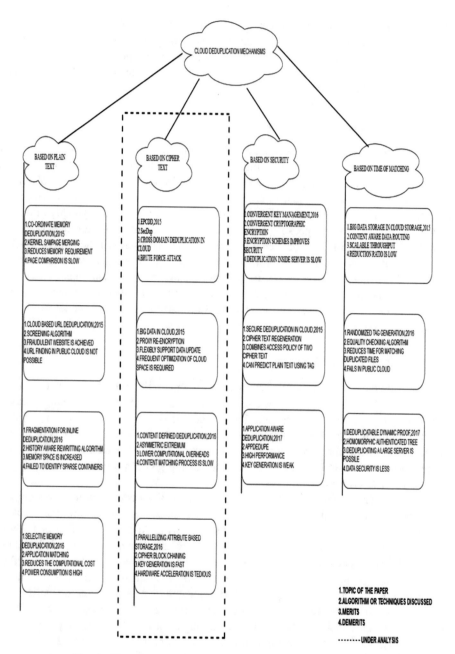

Fig. 1. Comparison of various cloud deduplication mechanisms

4 Conclusion and Future Work

With the rise and development of cloud computing the number of users using its services increases due to which same data is replicated in many places. This survey paper explains about the ways how deduplication takes place on cloud server when the data or keys are in the form of both plain text and cipher text. The survey reports gives a result that deduplication data in the form of cipher text using chunk or hash as a matching criteria suffers problem in the area of efficiency and trustworthiness of data in cloud. So the future extended work will be based on finding an algorithm to match duplicated data in cloud server in the forms of cipher text by reducing time complexities and storage space.

References

1. Jiang, T., Chen, X., Wu, Q., Ma, J., Susilo, W., Lou, W.: Secure and efficient cloud data deduplication with randomized tag. IEEE Trans. Inf. Forensics Secur. **12**, 532–543 (2017)
2. Li, J., Chen, X., Li, M., Li, J., Lee, P.P.C., Lou, W.: Secure deduplication with efficient and reliable convergent key management. IEEE Trans. Parallel Distrib. Syst. **25**, 1615–1624 (2014)
3. Wen, M., Ota, K., Li, H., Lei, J., Chunhua, G., Zhou, S.: Secure data deduplication with reliable key management for dynamic updates in CPSS. IEEE Trans. Comput. Soc. Syst. **2**, 137–147 (2015)
4. Tin-Yu, W., Pan, J.-S., Lin, C.-F.: Improving accessing efficiency of cloud storage using deduplication and feedback schemes. IEEE Syst. J. **8**, 2018–2218 (2014)
5. Mao, B., Jiang, H., Suzhen, W., Tian, L.: Leveraging data deduplication to improve the performance of primary storage systems in the cloud. IEEE Trans. Comput. **65**, 1775–1788 (2016)
6. Min, F., Dan Feng, Yu., Hua, X.H., Chen, Z., Liu, J., Xia, W., Huang, F., Liu, Q.: Reducing fragmentation for in-line deduplication backup storage via exploiting backup history and cache knowledge. IEEE Trans. Parallel Distrib. Syst. **27**, 855–867 (2016)
7. Luo, S., Zhang, G., Chengwen, W., Khan, S.U., Li, K.: Boafft: distributed deduplication for big data storage in the cloud. IEEE Trans. Cloud Comput. **61**, 1–13 (2015)
8. Zhang, Y., Feng, D., Jiang, H., Xia, W., Fu, M., Huang, F., Zhou, Y.: A fast asymmetric extremum content defined chunking algorithm for data deduplication in backup storage systems. IEEE Trans. Comput. **66**, 199–211 (2017)
9. Jia, G., Han, G., Rodrigues, J.J.P.C., Lloret, J., Li, W.: Coordinate memory deduplication and partition for improving performance in cloud computing. IEEE Trans. Cloud Comput. **7**, 357–368 (2015)
10. Haoran, W., Weiqin, T., Qiang, G., Shengan, Z.: A data deduplication method in the cloud storage based on FP-tree. In: International Conference on Computer Science and Network Technology (ICCSNT), pp. 557–562 (2015)
11. Wang, J., Zhao, Z., Xu, Z., Zhang, H., Li, L., Guo, Y.: I-sieve: an inline high performance deduplication system used in cloud storage. Tsinghua Sci. Technol. **20**, 17–27 (2015)
12. He, K., Chen, J., Ruiying, D., Qianhong, W., Xue, G., Zhang, X.: DeyPoS: deduplicatable dynamic proof of storage for multi-user environments. IEEE Trans. Comput. **65**, 1–13 (2016)

13. Kim, S., Jeong, J., Lee, J.: Selective memory deduplication for cost efficiency in mobile smart devices. IEEE Trans. Consum. Electron. **60**, 276–284 (2014)
14. Zawoad, S., Hasan, R., Warner, G., Skjellum, A.: UDaaS: a cloud-based URL-deduplication-as-a-service for big datasets. In: IEEE Fourth International Conference on Big Data and Cloud Computing, pp. 271–272 (2014)
15. Li, L., Chen, X., Jiang, H., Li, Z., Li, K.-C.: P-CP-ABE: parallelizing ciphertext-policy attribute-based encryption for clouds. In: IEEE/ACIS International Conference on Software Engineering, Artificial Intelligence, Networking and Parallel/Distributed Computing (SNPD), pp. 1–6 (2016)

Comparison of Decision Tree-Based Learning Algorithms Using Breast Cancer Data

M. S. Dawngliani[1](✉), N. Chandrasekaran[2], R. Lalmawipuii[1],
and H. Thangkhanhau[1]

[1] Department of Computer Science, GZRSC, Aizawl, Mizoram, India
dawngliani@gmail.com
[2] Martin Luther Christian University, CDAC-India and IBM, Shillong, India
professor.chandra@gmail.com

Abstract. Recent times are witness to tremendous developments occurring in the field of Machine Learning. Data Mining applications are finding widespread acceptance, as they extend to encompass several sectors, including health care, education, weather forecasting, finance, etc. Data mining algorithms like classification algorithms are being integrated into the Knowledge Discovery in Databases (KDD) process to build optimized predictive models useful to healthcare and other professionals. Decision tree is considered to be one of the simplest and most widely used classifier algorithms.

In the current study, we have compared six decision tree-based learning algorithms on breast cancer dataset. The algorithms used include, J48 decision tree, decision stump, random forest tree, REP tree, hoeffding tree and Logistic Model Tree (LMT). The dataset is obtained from Mizoram Cancer Institute, Aizawl, Mizoram, India and it contain 575 records and 24 attributes. Infogain attribute evaluator has been computed to determine the rank of an attribute. The analysis is performed in two steps. First step involves the building of the model by training data, while during the second step, the dataset is evaluated using 10-fold cross validation technique. Our computations indicate that of all the algorithms evaluated, J48 decision tree performs with the highest accuracy (84.2105%), while random tree demonstrates the lowest accuracy (76.4912%).

Keywords: Data mining · Decision stump · Decision tree · Hoeffding tree · J48 decision tree · LMT · Random forest tree · REP tree

1 Introduction

Breast cancer is the most common cancer affecting the female population worldwide. Most of the studies pertaining to this topic have been performed using medical data acquired from developed countries and only a few from developing countries has ever been analyzed. For the current study, we have made use of the medical data collected from a Mizoram Cancer Institute, Aizawl, Mizoram, India. This data has been analyzed using various data mining classification techniques. Several patterns that will be of interest to professionals working in health care, finance, etc. can be discovered from the complex data by data mining using intelligent algorithms. Data mining becomes an

© Springer Nature Switzerland AG 2020
A. P. Pandian et al. (Eds.): ICCBI 2019, LNDECT 49, pp. 885–896, 2020.
https://doi.org/10.1007/978-3-030-43192-1_96

important step in the knowledge discovery process [1], see Fig. 1. As can further be inferred from the figure, the interpretation of the patterns is the final step of the Knowledge Discovery in Databases (KDD) process, which results in the generation of information in a knowledge form which can be put to immense use by humans [2].

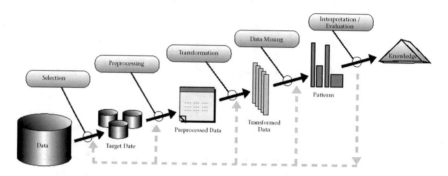

Fig. 1. The five steps involved in the KDD process [2]

There are several variations in data mining techniques and they include regression analysis, classification techniques, clustering analysis, association, prediction, etc. Classification is the task of assigning an object to one of the predefined classes [3]. The most popular classification algorithms are Naïve Bayes, support vector machines, decision trees and artificial neural networks. In our study, we compared six decision tree algorithms. Collected data are preprocessed and attribute selection are performed using Infogain evaluator. Accuracy and error rates are calculated to determine its performance and the result is evaluated using a 10 fold cross-validation technique.

2   Related Studies

Several studies are being conducted relating to the deployment of decision tree-based machine learning methods for analysis of different medical datasets.

A'la, Pennanasari and Setiawan [4] have carried out work on breast cancer prediction using three types of decision tree based learning methods including simple decision tree, random forest and gradient boosting. The data was collected from University Hospital Centre of Coimbra (CHOC) and included datasets of 64 breast cancer patients and 52 healthy volunteers. They used both feature selection and non-feature selection. They concluded that gradient boosting technique showed the highest accuracy with 85% sensitivity and 80% specificity.

Mushtaq, Yaqub, Hassan and Su [5] used five popular data mining classification algorithms such as decision tree, nearest neighbor, logistic regression, Naïve Bayes and support vector machine to determine whether a person is having benign or malignant tumour. They used Wisconsin dataset from the University of California, Irvine (UCI) repository. The highest accuracy of 99.20% was obtained with sigmoid based naïve bayes classifier.

Devasena [6] investigated the efficiency of three different classifiers, namely, Random Forest, REP tree and J48 classifiers for making credit risk prediction. The effectiveness comparison of the classifiers had also been carried out in view of different scales of performance evaluation. It is observed that the random forest classifier produced the best outcome, followed by the performances of REP tree classifier and J48 classifier.

Nassif, Azzeh, Capretz and Ho [7] compared Decision Tree Forest (DTF) model with Traditional Decision Tree (DT) model, as well as a multiple Linear Regression Model (MLR). Based on the outcome, DTF model was deemed to be the best performer.

Natarajan and Venkatesan [8] studied the data mining algorithms to classify kidney transplant dataset using ID3, CART, C4.5, Boosting, Bagging and Random Forest. Best results were obtained by C4.5 Classifier method. The study concluded that boosting with CART algorithm as base classifiers and also C4.5 with bagging were the best way to classify this particular medical dataset.

Venkatadri and Lokanatha [9] compared six decision tree classifiers, viz., C4.5, CART, BFTree, SLIQ, SPRINT, and Random Forest to observe that the execution time in building the tree model increased with increase in data records' volume and decreased with the size of the datasets' attributes. Most accurate results were produced by SPRINT and Random Forest algorithms.

Kalmegh [10] compared three classifiers, viz., REPTree, Simple Cart and RandomTree to analyze the Indian News data sets. He conclude that the RandomTree algorithm performed the best in categorizing the News contents, whereas for the political news category, REPTree and Simple Cart algorithms performed well.

Hssina, Merbouha, Ezzikouri and Erritali [11] compared ID3/C4.5, C4.5/C5.0 and C5.0/CART to conclude that the performance of C4.5 decision tree was the best.

Priyam et al. [12] compared three existing decision tree algorithms (ID3, C4.5, and CART) on the educational data for predicting the performance of students in the examination. The authors concluded that C4.5 was the best algorithm suitable for this purpose.

3 Methodology

3.1 Basic Concepts of Data Mining

Data preparation process is described in the top half of Fig. 2 and it involves obtaining prepared data after the completion of the cleansing process [13]. The bottom half of the sketch, describes data mining, which is an important component of the KDD process. Conceptually, data mining begins with the deployment of powerful algorithms to identify patterns within the dataset and using evaluation techniques to obtain the required knowledge. Figures 1 and 2 are two ways of representing the KDD process.

The first step involves the identification of redundant and irrelevant data in order to perform the cleaning operation. The cleansed data is obtained after integrating the data into a proper recognizable format. During the next step, pre-processing is done after coding the data into suitable ranges and values. Attributes are selected for effective analysis. This will complete the data preparation operation, represented in top half of Fig. 2. Next step deals with the use of data mining machine learning methods to detect

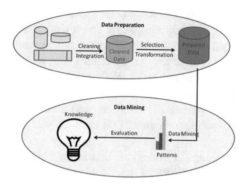

Fig. 2. Data mining as a key component of KDD process [13]

patterns that can be used for analysis. The performance is evaluated as part of the final process and to facilitate the creation of knowledge, the final result can be presented in the form of graphs, figures and/or tables.

3.2 Decision Tree Classifier

A decision tree (tree-like structure) is a often used as one of the powerful tools in designing a Decision Support System (DSS). This graph model starts with a root node branching off to a final node called the leaf. The leaf displays the final destination, which is the decision that is recommended by DSS. The branching nodes describe the possible consequences, which are the basis for arriving at the final decision. The classification algorithms make use of the logical structure that is described by the decision tree. Depending upon the splitter in the input variable, the sample is normally split into sub-groups [14].

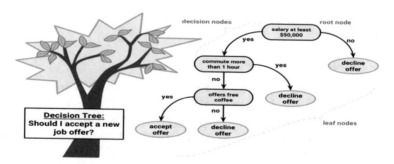

Fig. 3. Decision tree classifier, Image credit: www.packtpub.com

Let us explain the concept of the decision making process by using the example given in Fig. 3, which has been constructed to answer the query whether a job that has been offered can be accepted or not taking into account the salary offered, commuting time, etc. As explained in the previous section, the tree essentially has three types of nodes, which include one root node, a number of branching internal nodes terminating with the leaf nodes. The entire process starts with the node and hence it has no incoming edges. It will suffice to state at this juncture that identifying the root node often involves some well defined steps taken. Class labels are given to the leaf nodes, while certain test conditions are assigned to other nodes [3]. These conditions described in Fig. 3 include whether commute will be for more than an hour or whether free coffee will be offered or not, etc.

During the current study, an attempt is being made to compare six decision tree-based algorithms, which are described below:

J48 Decision Tree: J48 algorithm is often used in the place of ID3, as the former can handle both nominal and numeric values. C4.5/J48 has the ability to handle missing values and is found to have more robust splitting via gain ratio. In addition, J48 has various heuristics (self discovery capability). It also has routines for pruning the tree structure. In short it is an industrial strength decision tree learner [15].

Decision Stump: Unlike a decision tree, the decision stump does not have multiple levels for arriving at a decision. As the name implies, in the decision stump, the leaf nodes are directly connected to the root, see Fig. 4. It can then be easily inferred that a decision stump makes a prediction based on the value of just one input feature [16].

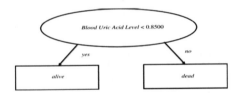

Fig. 4. An example of a single-level decision making process [16]

Random Tree: As the name implies, random trees refer to trees which are randomly built. The node attributes are chosen randomly and no pruning is carried out. Also, it has an option to allow the estimation of class probabilities based on a hold-out set [17].

REP Tree: Much like J48, REP tree is also based on C4.5 algorithm. It uses the regression tree logic. After multiple iterations, various other trees are created and the best one among them is then chosen as the representative tree. Mean square error on the tree predictions is used in pruning [6].

Hoeffding Tree: This particular tree (VFDT) has the capability to learn from large data bases and is used when the distribution examples are time invariant. Hoeffding tree benefits from the fact that attributes for optimal splitting can be arrived at from small samples [18].

LMT Tree: LMT tree is built with regression functions at the leaf nodes and hence are known as 'logistic model trees". The algorithm is capable of handling target variables which are both binary and multi-class, attributes which are both numeric and nominal and missing values [19].

3.3 Tools Used

We have made use of WEKA, which is a free open source software under the General Public License (GNU) for the analysis and performance measurement. WEKA has been developed by the University of Waikato in New Zealand and it has the capability to perform all operations described in Figs. 3 and 4, pertaining to the entire breadth of the KDD process [20].

3.4 Data Pre-processing

The dataset consists of 575 records with 24 attributes. Completeness, accuracy, and consistency of the data are factors that define data quality. As stated elsewhere, data preprocessing becomes imperative to meet and satisfy data quality requirements. This step ensures that raw data is transformed into a standard format [21]. Data is transformed into numeric and nominal data. The numeric attributes are in the range of 1 to 10 as shown in Table 1 and for a nominal attribute data, the value is true or false as in Table 2.

Table 1. Attributes having numeric value

Sl. no.	Attributes	Abbreviation	Attribute range
1	Age	AG	1–10
2	Sex	SX	1–2
3	Laterality	LT	1–3
4	BMI	BMI	1–5
5	Morphology	MP	1–10
6	Socioeconomic status	SS	1–7
7	Axillary lymph node	ALN	1–6
8	Skin involvement	SI	1–2
9	Stage grouping	SG	1–9

Table 2. Attributes having nominal value

Habitual data			Co-morbid condition		
Sl. no.	Attribute	Abbr	Sl. no.	Attribute	Abbr
1	Cigarette	CG	1	Tuberculosis	TB
2	Tobacco	TBC	2	Hypertension	HP
3	Alcohol	ALC	3	Diabetes	DB
4	Pan masala	PM	4	Heart disease	HD
5	Betelnut	BN	5	Asthma	AS
			6	allergy	AL
			7	Hepatitis	HPT
			8	Others	OT
			9	Aids	AI

3.5   Attribute Selection

Attribute or feature selection process finds extensive use and has become an active area of research pertaining to pattern recognition, statistics and data mining in the medical domain. Attribute selection invariably involves reducing the number of attributes to improve the accuracy of the outcome. The attributes are reduced by removing irrelevant and redundant attributes, which do not have much importance in determining the outcome [22]. InfogainAttributeEval is the attribute selection evaluator that is being used in the current study. The main purpose of this is to evaluate the worth of an attribute by measuring the information gain with respect to the class [23]. As prescribed by Eq. (1), Information Gain (IG) can be calculated as the difference between the entropy from start to end. Here, the set S is split on an attribute A.

$$IG(A,S) = H(S) - \sum_{t \in T} p(t)H(t) \tag{1}$$

In the above equation, H(S) is the entropy of set S, T is the subsets created from splitting set S by attribute A such that $S = \bigcup_{t \in T} t$, p(t) is the proportion of the number of elements in set S, H(t) is the entropy of subset t, H(S) measures the field with the smallest entropy. Table 3 shows the top 15 rank of an attributes using Infogain attribute evaluator and Fig. 5 is the column chart which shows the ranking of the top 15 attributes using the Infogain attribute evaluator.

Table 3. Attribute ranking based on Infogain evaluator

Rank	Info gain	
	Value	Attribute
1	0.1454	SG
2	0.0448	ALN
3	0.0355	MP
4	0.0130	PM
5	0.0119	SI
6	0.0105	SS
7	0.0095	LT
8	0.0082	AL
9	0.0023	HD
10	0.0019	HP
11	0.0018	BN
12	0.0012	TB
13	0.0010	AS
14	0.0010	DB
15	0.0010	TBC

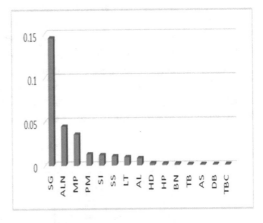

Fig. 5. Ranks of the attribute selection evaluator, InfogainAttribEval

4   Experimental Result

This section describes the results obtained after training the dataset using the six decision-based classifiers using Weka. All the datasets are used for building a decision tree. The final results are summarized in Table 4. The experimental result shows that the random tree has the highest accuracy of 99.6491%, which is followed by a J48 decision tree with an accuracy of 85.614%.

It is customary to use the Receiver Operating Characteristic (ROC) curve to check or visualize the performance of classifiers corresponding to various threshold settings. ROC depicts to what extend a classifier is capable of distinguishing between various classes [24]. Higher the AUC, the better the model is at predicting true as true and false as false. The Area under the ROC Curve (AUC) for random tree results in 1 and the value of AUC of 1 is the most ideal value for the ROC graph.

Table 4. Results of decision trees on a trained dataset

Decision trees	Accuracy	TP rate	FP rate	F-Measure	AUC
J48	85.614	0.856	0.370	0.850	0.749
Decision stump	83.8596	0.839	0.362	0.836	0.738
Random tree	99.6491	0.996	0.001	0.997	1.000
Rep tree	84.9123	0.849	0.321	0.849	0.858
Hoeffding	84.5614	0.846	0.373	0.841	0.838
LMT	83.8596	0.839	0.478	0.823	0.786

We have used 10 fold cross validation techniques to facilitate the evaluation of our model and to test performance. The cross-validation technique firstly partitions the dataset into training and a test set. The former is used to train the model, while the latter for evaluation purposes [25]. This technique involves splitting the dataset into 10 equal folds. In the first fold, 90% of the dataset is used to build the model and the rest 10% of the dataset is used for testing. The accuracy is then measured. In the corresponding fold, 80%, 70%,...,10%, is used to build the model and 20%, 30%,....,90% of the dataset is used for testing. Final computation involves taking the average to determine the average accuracy. Table 5 shows the result of 10 fold cross-validation testing.

Table 5. Average accuracy calculated after 10 fold cross-validation

Decision trees	Accuracy	TP rate	FP rate	F-Measure	AUC
J48	84.2105	0.842	0.336	0.842	0.769
Decision stump	83.8596	0.839	0.362	0.836	0.685
Random tree	76.4912	0.765	0.496	0.764	0.637
Rep tree	83.5088	0.835	0.402	0.830	0.730
Hoeffding	79.2982	0.793	0.579	0.772	0.708
LMT	82.1053	0.821	0.521	0.803	0.765

From the above table it can be inferred that the accuracy of a random tree which was 99.6491% on the trained data has decreased tremendously to 76.4912% when the 10 fold cross validation test is used as an option. This implies that there is over fitting of the dataset. The highest accuracy using 10 fold cross-validations is J48 which has a value of 84.2105%. AUC, which is the area under the ROC curve for J48 equals 0.769, which is also the highest. The second highest accuracy of 83.8596% has been obtained from decision stump analysis and this is followed by a REP tree with 83.5088% accuracy.

Table 6 shows the error statistics of the six decision trees. The statistics shown in the table are Kappa Statistics (KS), Mean Absolute Error (MAE), Root Mean Squared Error (RMSE), Relative Absolute Error (RAE), and Root Relative Squared Error (RRSE). KS determines the attribute measure of agreement, implying that the higher the value of KS, higher is the agreement. For a perfect agreement, the KS equals 1 [26]. In our study, KS value for J48 is the highest (0.5098). For the error statistics, J48 is the lowest on all the four error parameters, viz., MAE, RMSE, RAE and RRSE, please refer to Table 6. This proves that based on our current work, we can conclude that J48 is the most efficient of the six decision trees tested.

Table 6. Error statistics of decision trees

Decision trees	KS	MAE	RMSE	RAE	RRSE
J48	0.5098	0.2169	0.3484	66.6248%	86.5218%
Decision stump	0.489	0.2481	0.3537	76.1952%	87.8284%
Random tree	0.2702	0.2393	0.4761	73.5131%	118.2403%
Rep tree	0.4601	0.2452	0.3671	75.3046%	91.1656%
Hoeffding	0.2508	0.2738	0.3959	84.1015%	98.3087%
LMT	0.3524	0.2634	0.3654	80.8963%	90.7507%

ROC curves of the six decision tree-based classifiers that we employed in our studies are shown in Figs. 6(a) to (f). It can be seen from Fig. 6(a) of the ROC curve of J48 that the AUC values for J48 equals 0.769, while that of decision stump AUC equals 0.689, see Fig. 6(b). Similarly, AUC of random tree algorithm has been computed to be 0.637, while ROC curve of REP tree shows the AUC as 0.730, see Figs. 6(c) and (d). From ROC curve of hoeffding tree, see Fig. 6(e), the AUC value has been computed as 0.708, while from Fig. 6(f) AUC of LMT has been calculated as 0.765.

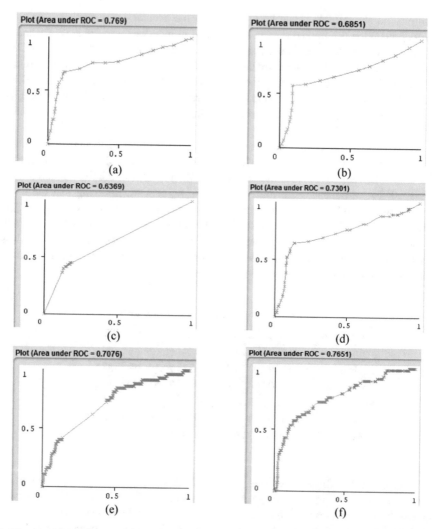

Fig. 6. (a): ROC curve of J48 (b): ROC curve of decision stump (c): ROC curve of random tree (d): ROC curve of REP tree (e): ROC curve of hoeffding tree (f): ROC curve of LMT

5 Conclusion

Applications developed using data mining algorithmic techniques, play an important role in the healthcare domain, as they can be very effectively used both for the prediction of diagnosis and prognosis. During the course of the current research work, we have compared six decision tree based learning algorithms to evaluate the percentage accuracy for predicting breast cancer recurrence. We have done the analysis based on two test options. With the first option, we have made use of only the trained dataset and with the second option, the 10-fold cross validation technique has been employed.

The computations indicate that under option 1, the random forest results produce the highest accuracy of 99.6491% and in contrast the same technique gives the lowest accuracy (76.4912%) using 10-fold cross validation. This implies that there is an overfitting in the model built with the training set and hence it will not be efficient enough. J48 produced the second highest accuracy of 85.614% using trained set and highest value of 84.2105% when employing 10 fold cross-validations. And in terms of ROC, J48 has been found to be the most efficient method. Hence, we can conclude that the J48 decision tree outperforms all the other decision tree classifiers by a significant margin and hence is a strong contender for deployment.

References

1. Eapen, A.G.: Application of data mining in medical applications. University of Waterloo (2004)
2. Huang, H., et al.: Business rule extraction from legacy code. In: Proceedings of 20th International Conference on Computer Software and Applications, IEEE COMPSAC 1996, pp. 162–167 (1996)
3. Tan, P.N., Steinbach, M., Karpatne, A., Kumar, V.: Introduction to Data Mining, 2nd edn. (2019)
4. A'la, F.Y., Pennanasari, A.E., Setiawan, N.A.: A comparative analysis of tree-based machine, vol. d, pp. 55–59 (2019)
5. Mushtaq, Z., Yaqub, A., Hassan, A., Su, S.F.: Performance analysis of supervised classifiers using PCA based techniques on breast cancer. In: 2019 International Conference on Engineering and Emerging Technology ICEET 2019, pp. 1–6 (2019)
6. Devasena, L.C.: Comparative analysis of random forest, REP tree and J48 classifiers for credit risk prediction. Int. J. Comput. Appl. (0975–8887), 30–36 (2014)
7. Nassif, A.B., Azzeh, M., Capretz, L.F., Ho, D.: A comparison between decision trees and decision tree forest models for software development effort estimation. In: 2013 3rd International Conference Communications and Information Technology ICCIT 2013, pp. 220–224 (2013)
8. Venkatesan, P., Natarajan, Y.: A comparative analysis of decision tree methods to predict kidney transplant survival. Int. J. Adv. Res. Comput. Sci. 5(3), 225–229 (2014)
9. Venkatadri, M., Lokanatha, C.R.: A comparative study on decision treeclassification algorithm in data mining. Int. J. Comput. Appl. Eng. Technol. Sci. 2(2), 24–29 (2010)
10. Kalmegh, S.: Analysis of WEKA data mining algorithm REPTree, simple cart and randomtree for classification of Indian news. Int. J. Innov. Sci. Eng. Technol. 2(2), 438–446 (2015)
11. Hssına, B., Merbouha, A., Ezzikourı, H., Errıtalı, M.: A comparative study of decision tree ID3 and C4.5. Int. J. Adv. Comput. Sci. Appl. 4(2), 13–19 (2014)
12. Priyam, A., Gupta, R., Rathee, A., Srivastava, S.: Comparative analysis of decision tree classification algorithms. Int. J. Curr. Eng. Technol. 3, 334–337 (2013)
13. WideSkills: Data mining processes | Data mining tutorial by wideskills (n.d.). http://www.wideskills.com/data-mining-tutorial/data-mining-processes
14. A Complete Tutorial on Tree Based Modeling from Scratch (in R & Python), vol. 20, pp. 1–55 (2016). https://www.analyticsvidhya.com/blog/2016/04/tree-based-algorithms-complete-tutorial-scratch-in-python/

15. Kaur, G., Chhabra, A.: Improved J48 classification algorithm for the prediction of diabetes. Int. J. Comput. Appl. **98**(22), 13–17 (2014)

16. Bohacik, Jan: Decision stump for prognosis of mortality rates in heart failure patients. J. Inf. Technol. **7**, 8–14 (2014)

17. Aldous, D.: The continuum random tree. Ann. Probab. **19**(1), 1–28 (1991)

18. HoeffdingTree.   https://weka.sourceforge.io/doc.dev/weka/classifiers/trees/HoeffdingTree. html

19. Landwehr, N., Hall, M., Frank, E.: Logistic model trees. Lect. Notes Artif. Intell. (Subser. Lect. Notes Comput. Sci.) **2837**, 241–252 (2003)

20. Hall, M., Frank, E., Holmes, G., Witten, I.H., Cunningham, S.J.: Weka: practical machine learning tools and techniques. In: Workshop on Emerging Knowledge Engineering and Connectionist-Based Information Systems (2007)

21. Machine Learning – Part 1 – Data Preprocessing (2018). https://towardsdatascience.com/data-pre-processing-techniques-you-should-know-8954662716d6

22. Khalid, S., Khalil, T., Nasreen, S.: A survey of feature selection and feature extraction techniques in machine learning. In: Proceedings of the 2014 Science and Information Conference SAI 2014, pp. 372–378 (2014)

23. Batra, M., Agrawal, R.: Comparative analysis of decision tree algorithms. Adv. Intell. Syst. Comput. **652**, 31–36 (2018)

24. The Area Under an ROC Curve (n.d.). http://gim.unmc.edu/dxtests/roc3.htm

25. Cross-validation (statistics) – Wikipedia (n.d.). https://en.wikipedia.org/wiki/Cross-validation_(statistics)

26. Cohen's kappa – Wikipedia (n.d.). https://en.wikipedia.org/wiki/Cohen%27s_kappa

Stock and Financial Market Prediction Using Machine Learning

Divyanshu Agrawal$^{(\boxtimes)}$ and Rejo Mathew

I.T Department, Mukesh Patel School of Technology Management
and Engineering, Mumbai, India
Divyanshu.agrawal01@nmims.edu.in,
Rejo.mathew@nmims.edu

Abstract. The objective of this paper is to study and predict the stock and financial markets using machine learning methods. The feature utilized in the proposed model incorporates Oil rates, Interest rate, Gold and Silver rates, NEWS and online networking channel. This paper helps in the deep understanding of algorithms like ant colony optimization, support vector machine, single level perceptron and multi-level perceptron. At the end, we will compare these algorithms to understand and use the optimal solution.

Keywords: Pheromone · Neuron · Feature selection · Stagnation · Gradient

1 Introduction

The world has witnessed the profound effects of the 2007–08 Great recession (as identified by the UN), and indirect bearings lasted for years. It was the first time after World War 2 that the World GDP growth was negative. Now, although the national crisis was conceived due to reckless subprime lending, this wasn't the only cause of the full-blown global recession but, it was because a lot of these risky assets were at the heart of investments held by international institutions and investors who, along with their unsustainable leverage led to a full-blown global recession. These risky assets were, according to many, not difficult to assess and predict. And due to the predictability in its nature, it could have been evaded. 2019 is one such period. Where the national leverages are sky high. The current world debt count is at an all-time high of $184 trillion in nominal terms. This is equal to 225% of world GDP of 2017 and when distributed per capita the share of this debt for every person in the world is 2.5 times their GDP per capita on average. Interestingly US, China and Japan accounts for more than half of this burden where financial and non-financial private debt is almost twice the public debt. The mix of factors that led to economies experiencing debt recessions was different, but one of the vital administrators was inflationary depression originating from failure to repay debt denominated in foreign currency or a severe debt crisis where governments cannot further reduce the interest rate to service the debt, which was the point where the governments started deleveraging. According to me the next crisis would either originate due to inability to service this debt, increase in interest rates that would make service of debt challenging or any other reason, but debt levels would make it severe (debt crises is more destructive than banking or currency crisis

© Springer Nature Switzerland AG 2020
A. P. Pandian et al. (Eds.): ICCBI 2019, LNDECT 49, pp. 897–902, 2020.
https://doi.org/10.1007/978-3-030-43192-1_97

producing significant and long-lasting output losses. Prediction things or forecasting things have been one of the best interests of the researcher for a greater time. For a long time, people for prediction in the financial and stock market used their experience and conscience. But as time passed these predictions needed to made using huge data and accuracy and in the age of digitalization people started using several computational methods for the prediction of these data as it involves a huge amount of money and risk. So, through this paper tries comparing some of the computational methods which could be used for predictions of stock and financial market [4]. Feature selection plays an important role so in a way should be optimal so that it makes the algorithm quicker and accurate. The prediction aims in calculation of loss and avoid taking wrong steps and classify them as non-default and default criteria.

2   Methods

2.1   Ant Optimization Algorithm

ACO-FS identifies minimal feature subset size s in a given feature size of n (s < n) in a way feature subset maintain high accuracy and represent original data. ACO-FS have following steps: [4]

- Graph Representation
- Pheromone Desirability
- Heuristic Desirability
- Solution construction

Graph Representation
In this step, present a graph in which every node act as a feature selection and then connect every node. Through the help of ant transversal most optimal feature path is chosen where the least number of nodes are visited. This step of ACO-FS ends with giving us an optimal feature subset which could be used for data reduction. The chosen feature subset is then shown by solid lines which can now be used to add pheromones and heuristic value to each feature.

Heuristic Desirability
Heuristic method for solving a very general class of computational problems by combining user-given heuristics in hope of obtaining a more efficient procedure. ACO is a meta-heuristic approach.

Pheromone Desirability
After getting the solution and selecting the optimal feature subset. Deposition and evaporation of pheromones in each node gets started. To avoid stagnation pheromone evaporation is used the best solution with a greater number of pheromones is laid by best ant.

Solution Construction
In this ant are placed randomly in the graph then every ant begins following a path for finding the most probabilistic transverse node till stopping condition is satisfied if the

condition isn't satisfied then pheromones updating takes place. The resultant path is then investigated for optimal subset.

2.2 Support Vector Machines

Market prediction is considered as time series prediction problem one of the most suitable algorithm used is support vector machine. Regression and classification both can be used in supervised algorithm. Involvement of plotting data as point in the space of n-dimensions are used in Support vector machines. SVM algorithm draws a boundary over the data set called the hyper-plane, which separates data into two classes [2, 3]. In this method of prediction after the feature selection is done then a series of time interval should be decided could be as day closing or could be used to obtain every minute an array is chosen using the array and our feature selection are trained by varying there weights and adjust the feature selections.

2.3 Multi-level Perceptron

A multi-level perceptron is defined by three criteria:-

- The number of input nodes.
- The number of output nodes.
- The number of hidden layers and hidden nodes.

Single-Layer Perceptron
This is the most basic technique for stock prediction as its based on Single Layer Perceptron model. This consist of a single neuron with adjustable synaptic weights.

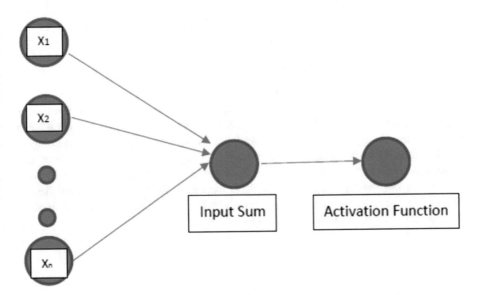

Fig. 1. Single layer perceptron

It performs pattern classifications with only two classes. Single-Layer Perceptron contains an output layer and an input layer. The sum of input neurons gives the weighted neurons in output layer [5] (Fig. 1).

Multi-layer Perceptron

A multilayer perceptron (MLP) is a class of feedforward neural network system. This method can be used just by adding a single layer to SLP know as hidden layer. This system can have one or multiple hidden layers these hidden layers contains intermediate neurons. These hidden layers neurons are dependent on input layers neurons and output layers neurons are dependent on these hidden layer's neurons. The weights of these neurons are dependent on output layer to calculate the error gradient [2, 3] (Fig. 2).

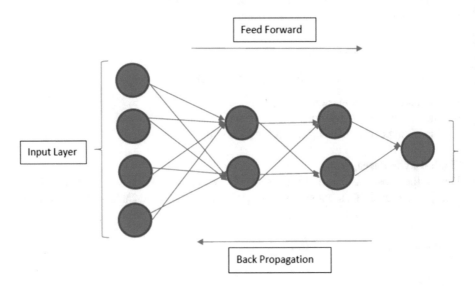

Fig. 2. Multi-layer perceptron

Multi-level can be used for both feature selection and as a classifier. In this neuron are connected to each other by weights. These weights are varied in a specific manner to obtain certain type of results for learning of associate network. The learning or training process in this obtained by change of weights. In process of training and building an ideal system for working inputs are given output obtained is compared with desired output. Back propagation process is then used to vary weights until desired output is obtained by the system. This process helps in minimizing the error rate of the system and building an ideal needed system.

3 Analysis

	Ant algorithm	Multi-level perception	Support vector machines
Optimal subset	Data reduction	Pattern based data simplification	—
Stagnation detection	Pheromone	—	—
Hyperplane extendibility	Little separation is detected	—	Separation of data
Accuracy	Accurate	Highly Accurate	Comparatively less accurate
Response Time	Response time is large	Response time is very large	Response time is large
Throughput	Since Response time is large executed task are less	Quite a number of tasks can be performed	Quite a number of tasks can be performed
Training Dataset	Large Dataset required	Very Large Data Set required	Limited dataset is sufficient
Learning Model	Supervised	Supervised	Unsupervised
Output	Computer Program	Statically data with error function	Outputs data in numeric, variable format
Type of algorithm	Genetic	Clustering, Prediction	1. Clustering 2. Classification

4 Conclusion

This paper helps in understanding the progress of the world for the prediction of stock and financial markets. It describes that there has been deep research in the field of machine learning for the evolving market prediction and shares the interest of researchers in this field. This paper compares various methods of machine learning which could be used in prediction. The above table helps in understanding the pros and cons of the algorithm. Stocks should be divided based on features and the specified algorithm should be used to get the most optimal result from the method. This research paper holds together the stock and financial market prediction using machine learning.

References

1. Shin, K.-S., Lee, T.S., Kim, H.: An application of support vector machines in bankruptcy prediction model. Expert Syst. Appl. **28**, 127–135 (2005). College of Business Administration, Ewha Womans University, 11-1 Daehyun-dong, Seodaemun-gu, Seoul 120-750, South Korea, January 2005

2. Victor Devadoss, A., Antony Alphonnse Ligori, T.: Forecasting of stock prices using multi-layer perceptron. Int. J. Comput. Algorithm **02**, 440–449 (2013)
3. Foster, I., Kesselman, C.: The Grid: Blueprint for a New Computing Infrastructure. Morgan Kaufmann, San Francisco (1999)
4. Uthayakumar, J., Metawa, N., Shankar, K., Lakesmanapranbhu, S.K.: Financial crisis prediction model using ant colony optimization. Int. J. Inf. Manag. **50**, 538–556 (2018)
5. Hayashi, Y., Sakata, M., Gallant, S.I.: Multi-layer versus single-layer neural networks and an application to reading hand-stamped characters. In: International Neural Network Conference, pp. 781–784 (1990)
6. Atescia, K., Bhagwatwar, A., Deo, T., Desouz, K.C., Baloh, P.: Business process outsourcing: a case study of Satyam Computers. Int. J. Inf. Manag. **30**, 277–282 (2010)
7. Altun, H., Polat, G.: Boosting selection of speech related features to improve performance of multi-class SVMs in emotion detection. Int. J. Inf. Manag. **36**, 8197–8203 (2009)
8. Beaver, W.H.: Financial ratios as predictors of failure. J. Account. Res. **4**(Empirical Research in Accounting: Selected Studies), 71–111 (1966)
9. Zanaty, E.A.: Support Vector Machines (SVMs) versus Multilayer Perception (MLP) in data classification. Egypt. Inf. J. **13**, 177–183 (2012)
10. Cristianini, N., Taylor, S.J.: An Introduction to Support Vector Machines. Cambridge University Press, Cambridge (2000)
11. Fletcher, D., Goss, E.: Forecasting with neural networks: an application using bankruptcy data. Inf. Manag. **24**, 159–167 (1993)
12. Grice, S.J., Dugan, T.M.: The limitations of bankruptcy prediction models: some cautions for the researcher. Rev. Quant. Financ. Acc. **17**, 151–166 (2001)
13. Häardle, W., Moro, R., Schäafer, D.: Predicting corporate bankruptcy with support vector machines. Humboldt University and the German Institute for Economic Research (2003)
14. Haykin, S.: Neural Networks: A Comprehensive Foundation. Macmillan, New York (1994)
15. Odom, M., Sharda, R.: A neural networks model for bankruptcy prediction. In: Proceedings of the IEEE International Conference on Neural Network (1990)
16. Hiba Sadia, K., Sharma, A., Paul, A., SarmisthaPadhi, Sanyal, S.: Stock market prediction using machine learning algorithms. Int. J. Eng. Adv. Technol. (IJEAT) **8**(4) (2019). ISSN 2249 – 8958

Review of Prediction of Chronic Disease Using Different Prediction Methods

Aadil Chheda[✉] and Rejo Mathew

Department of I.T., Mukesh Patel School of Technology and Management,
NMIMS, Mumbai, India
{aadil.chheda68,rejo.mathew}@nmims.edu.in

Abstract. In healthcare industry, Disease prediction has been an important goal. There are many algorithms and techniques used to predict disease. This paper explains different techniques used to predict and reviews them in order to find a difference between various parameters.

Keywords: CNN · Big data · Machine learning

1 Introduction

Increasing chronic diseases, kills a million of people every year in Asia, America and Europe. Most of the GDP of developed countries are used to treat heart diseases. Hence, it seemed important to look for solutions to this problem of predicting the disease so that it could be treated at an early level and less money would be utilized. Multiple ways were found to predict the disease by using machine learning. In this paper NB, KNN, CNN is described and reviewed. After the collection of EHR, prediction is to be made that person is high risk patient or low risk patient. This was done by machine learning algorithms in which the total data was divided into training data and testing data. The unstructured data is predicted by CNN and the structured data is handled by CNN- based multimodal risk prediction taking care of the regional disease issues.

2 Related Work

2.1 Collection of Big Data

Collection of the Big data is used for analytics and deep learning. The use of smart clothing has been convenient to collect the data. Long dresses used were uncomfortable and sharing and storing data was difficult earlier. Smart clothing has a cloud platform, Internet of Things, mobile internet access, and cloud computing.

2.2 Prediction of Medical Condition

Heart condition is to be predicted by using the collected data. This can be done by multiple algorithms in Machine Learning. The mainly used algorithms are CNN, NB, KNN.

© Springer Nature Switzerland AG 2020
A. P. Pandian et al. (Eds.): ICCBI 2019, LNDECT 49, pp. 903–908, 2020.
https://doi.org/10.1007/978-3-030-43192-1_98

3 Solutions

3.1 Data Mining

Data Mining is a method where the main focus is to analyze data and find the trend in the dataset. Finding patterns with least user inputs is the main objective of data mining. It can serve as a tool to make decision and forecast future.

Techniques:

- Association: Mining is done by finding patterns based on a particular items relation with another item in same field.
- Classification: this technique mines data by splitting the dataset into set of classes. This is done using mathematical techniques like statistics, decision tree.
- Clustering: in this technique the data objects with similar characteristics are grouped in meaningful clusters.
- Prediction: the technique tries to find relation between the independent variables and relation between dependent and independent variable [5].

3.2 NB (Naïve Bayes)

NB algorithm is the most common in machine learning application as it is simple and aims to make decision of an attribute independently from other attributes. Hence it can work with two or more attributes together but is very effective, fast, good decision maker for single attribute.

Algorithm:

1. CRAWLER: Collects the Data Set. Check for format accuracy, duplication.
2. TRAINER: the application is trained using data sets and data mining.
3. INDEXER: this step identifies and extracts features from the input data whose prediction is to be made. In this step the patient's data is compared with training data and patient is classified as low-risk or highrisk heart disease. The classification is done using [4].

$$\arg \max_{n} \{P(C_n|w)\} = \frac{P(w|C_n) \times PC_n}{P(w)}$$

3.3 KNN (K-Nearest Neighbor)

After data mining the dataset is converted into a matrix i.e. dataMatrix. The second input is the query data that is to be predicted, this too is also converted into a matrix that is queryMatrix. Third input is K that is number of neighbors, that is the nearest time that is required to be predict. For instance, it is required to know the risk of heart attack of the person in 4 years.

Algorithm:

1. Initialize dataMatrix and queryMatrix and initialize their size.
2. Start a for loop (i) from 1 to size of queryMatrix and calculate Euclidian distance and NeighborDistance(i) and end the loop.
3. Compare the query and the data matrix for i and calculate predicted year [6].

3.4 Smart Clothing

This process of big data collection by smart clothing is divided in parts:

- **Intra smart clothing system:** here the cloth is constructed as per the human requirement and comfort. It also includes deployment of the body sensors, low power wireless communication. It adopts Bluetooth 4.0 technology and Low-Power Wi-Fi standard to minimize the energy consumption.
- **Communication for Smart Clothing:** Via Bluetooth and wifi the data is sent to the smart devices like phone and tabs. Then using the APP in the smart devices the data is transferred to the cloud where it is backed up and passed to the hospitals or doctors.

This data is then transferred to the system and the system segregates the patient into two categories low risk and high risk.

It is a good device as it keeps a close watch on the patient all the time.

3.5 CNN-Based Uni-modal and Multimodal Prediction

Unstructured data like doctor's record and writing are used in uni-modal prediction. Here each word is written as a vector.

- The vector is passed on as to a convolution layer and the data is performed convolution with the filter matrix and a graph is built.
- The output from convolution is then transferred to pool layer. In this layer the key words are filtered into a fixed length vector. Here different features of the text is found.
- Pooling layer is connected to the full neural network. NN is a technique that is derived from the neural system of human body. It learns from the test data accordingly predicts the future.
- then to the classifier, and output is generated.
- This is CNN-UDRP

  Structured data is processed under the CNN-MDRP

- It is similar to CNN UDRP.
- MDRP is trained by stochastic gradient method
- The full data set is divided into 6:1 i.e. training-set: test-set.
- Machine learning algorithm used are Native Bayesian, K-Nearest Neighbor, Decision tree (Fig. 1).

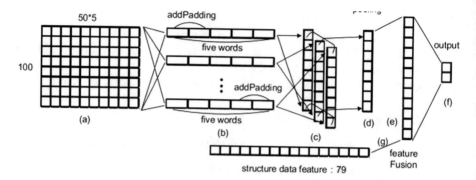

Fig. 1. CNN-based multimodal layers

4 Comparison Table

Parameters	Smart clothing	NB	KNN	CNN-based uni-modal and multi-modal prediction
Functionality	Monitor a patient continuously throughout the day. It also helps to create big data that helps to analyze different method is used types of patients and different prepare layers	Predicts using a formula	Predicts using matrix and neighbor distance	To predict disease, user data collected is converted into a matrix and passed through different layers
Attributes to predict	–	Single	Multiple (vector)	Multiple(vector)
Performance	Very good	Fair	Good	Accurate
Complexity	–	Low	Medium	High
Time to execute	–	Fast	Moderate	Slow

(continued)

(continued)

Parameters	Smart clothing	NB	KNN	CNN-based uni-modal and multi-modal prediction
Review	Despite of a great performance of this method. Some issue like human heart rate increasing due to higher altitudes should have been ignored depending on the environmental condition. But this product does not sense the external condition as misinformation may be sent	It is accurate and fast but works with only one attribute	Fast but accuracy is less	Accuracy of this method is excellent as 10-fold cross validation method was used. And results for both s-data and s-t data were accurate upto 95% than the previous methods which were around 50%

5 Conclusion and Future Work

This paper reviews the most common prediction methods. The preferred method would be CNN based multimodal as its accuracy is the greatest and heart disease requires accuracy rather than time. The future to this project would be based on cloud computing with constant prediction of any disease possible and directly reporting it to the patient's doctor.

References

1. Groves, P., Kayyali, B., Knott, D., van Kuiken, S.: The 'Big Data' Revolution in Healthcare: Accelerating Value and Innovation. Center for US Health System Reform Business Technology Office, New York (2016)
2. Chen, M., Hao, Y., Hwang, K., Wang, L., Wang, L.: Disease prediction by machine learning over big data from healthcare communities. IEEE Access 5, 8869–8879 (2017)
3. Cheng, W., Shen, Y., Zhu, Y., Huang, L.: Explaining latent factor models for recommendation with influence functions. Department of Computer Science and Engineering, Shanghai Jiao Tong University, 20 November 2018
4. Xhemali, D., Hinde, C.J., Stone, R.G.: Naïve bayes vs. decision trees vs. neural networks in the classification of training web pages. IJCSI Int. J. Comput. Sci. Issues 4(1), 16–23 (2009)
5. Kaur, B., Singh, W.: Review on heart disease prediction system using data mining techniques. Int. J. Recent Innov. Trends Comput. Commun. 2(10), 3003–3008 (2014). ISSN: 2321-8169
6. Badhiye, S.S., Sambhe, N.U., Chatur, P.N.: KNN technique for analysis and prediction of temperature and humidity data. Int. J. Comput. Appl. 61(14), 7–13 (2013). (0975 – 8887)

7. Qiu, L., Gai, K., Qiu, M.: Optimal big data sharing approach for tele-health in cloud computing. In: Proceedings of the IEEE International Conference on Smart Cloud (SmartCloud), pp. 184–189, November 2016
8. Zhang, Y., Qiu, M., Tsai, C.-W., Hassan, M.M., Alamri, A.: Health-CPS: healthcare cyber-physical system assisted by cloud and big data. IEEE Syst. J. 11(1), 88–95 (2017)
9. Lin, K., Luo, J., Hu, L., Hossain, M.S., Ghoneim, A.: Localization based on social big data analysis in the vehicular networks. IEEE Trans. Ind. Inf. 13, 1932–1940 (2016)
10. Lin, K., Chen, M., Deng, J., Hassan, M.M., Fortino, G.: Enhanced fingerprinting and trajectory prediction for iot localization in smart buildings. IEEE Trans. Autom. Sci. Eng. 13(3), 1294–1307 (2016)
11. Oliver, D., Daly, F., Martin, F.C., McMurdo, M.E.: Risk factors and risk assessment tools for falls in hospital in-patients: a systematic review. Age Ageing 33(2), 122–130 (2004)
12. Marcoon, S., Chang, A.M., Lee, B., Salhi, R., Hollander, J.E.: Heart score to further risk stratify patients with low TIMI scores. Critical Pathways Cardiol. 12(1), 1–5 (2013)
13. Bandyopadhyay, S., et al.: Data mining for censored time-to-event data: a bayesian network model for predicting cardiovascular risk from electronic health record data. Data Min. Knowl. Discov. 29(4), 1033–1069 (2015)
14. Qian, B., Wang, X., Cao, N., Li, H., Jiang, Y.-G.: A relative similarity based method for interactive patient risk prediction. Data Min. Knowl. Discov. 29(4), 1070–1093 (2015)
15. Singh, A., Nadkarni, G., Gottesman, O., Ellis, S.B., Bottinger, E.P., Guttag, J.V.: Incorporating temporal EHR data in predictive models for risk stratification of renal function deterioration. J. Biomed. Inform. 53, 220–228 (2015)

Review of Software Defined Networking Based Firewall Issues and Solutions

Karan Garg[✉] and Rejo Mathew

I.T. Department, Mukesh Patel School of Technology,
Management and Engineering, NMIMS, Mumbai, India
{karan.garg18,rejo.mathew}@nmims.edu.in

Abstract. Software defined networking (SDN) is an architecture that provides flexibility in network. Introduction of this technology enables efficient network configuration that helps to improve network performance. By introducing controller system in firewall, it does help in controlling the network but it also introduces new issues regarding it. As we know controller is a high value target for attacker. If attacker compromises a controller then he/she can have a total control of network. Network of this type can be improved by tweaking it bit. This paper aims to review some of the issues faced in SDN based Firewall & their origins and also their different Solutions. And a comparison between solutions based on their parameters.

Keywords: Software defined networking · SDN · Security issues · Firewall · Flow entries · OpenFlow · Network · Controller

1 Introduction

Now days, Software defined Networking is used. SDN is an attempt to centralize the networks, traditional networks were decentralized and were complex. SDN separates Forwarding plane (Data plane) from control plane. Control plane decides the destination & routing of network packets and Data plane actually forward the packets by using information learned by the control plane. OpenFlow is a communication protocol that gives the ability to controller of determining the path of data packets across network of switches [7]. The controller is a key component in SDN. Flow entries are stored in the flow table of a switch and can be setup by the OpenFlow controller. OpenFlow is considered as one of the first software-defined networking standards. It defined the communication protocol in SDN that enables the SDN Controller to directly interact with forwarding plane of network devices such as switches and routers. Flow entries can be manipulated with the help of OpenFlow switch. SDN based firewall have advantages i.e. It centralizes the network, granular security, lower operating costs, etc.

© Springer Nature Switzerland AG 2020
A. P. Pandian et al. (Eds.): ICCBI 2019, LNDECT 49, pp. 909–916, 2020.
https://doi.org/10.1007/978-3-030-43192-1_99

2   Related Work

2.1   Connection Tracking

Issues regarding this are tracking of connection of packets using Access control list packet filtering [1–3]. However, ACL only uses stateless packet filtering. This type of filtering only checks packet. Because of this it is required stateful filtering that can maintain connection table that have information regarding connection. Connection tracking need to be examined i.e. connection oriented or connection less. The difference in characteristics of connection oriented and connection less requires adaptive tracking from stateful firewall.

2.2   Security Issue

These issues arise due to controller. A controller is a piece of hardware or software that control the flow of data between two or more entities. For every new connection to be processed, a controller is requested. This is a flaw that attackers can exploit by attempting DoS attack (Starvation attack). The attacker can emit multiple packets with different Ip that are forwarded to controller. This results in drainage of controller resources and also control and data plane can also be compromised by the attacker. Various researcher has studied about SDN [4, 9–11] controller attack in SDN firewall.

3   Solutions

This issue arises due to overflow of flow entries in the flow table that the switch contains. Also missing rules that can't be identified when they are required which are produced by controller. Rules are required to sort out packets which are incoming traffic towards the switch.

3.1   Connection Tracking Module

Main goal of stateful firewall in SDN network is to monitor all the connections in the network. Flow tracker Controller is used for tracking the connection status accurately and timely. For flow tracker to be compatible with OpenFlow protocol, it tracks protocol at Transport layer, TCP and UDP are considered as they behave and maintain connection different from each other (Connection oriented and Connectionless respectively). The state of connection is tracked by the Flowtracker using flow tables. Between two hosts in a network, a unique connection is represented by an entry in the flow table. Flow entries contain three fields i.e. header field, Action field and timeout field. There is a three-layer switch table setup. In which, each layer has group of flow entries. Each layer is handled on based of priority by the switch. Connection forwarding flow entries (1st Layer), Connection detection flow entries (2nd Layer) and Switching flow entries (3rd Layer) are reactive flow [5]. Nature of 1st layer is reactive while the nature of other layers is Proactive which means they are already there in controller. Hybrid usage of Proactive and Reactive flows allow the time table to update for controlling new connection and also decrease the usage of flow table. There are two connection states i.e. initiation and termination on which the tracking function focuses on. If packet does not match with any existing monitored connection then the Flow-tracker report the first packet initiating the connection. For connection termination tracking, there are different methods for TCP and UDP. For UDP connections – timeout parameters are used to detect the traffic that has been expire (No Overhead). For TCP connections – switch transmits packets repeatedly to controller for scanning the TCP FIN flag (end of data transmission).

Let's take an example [6] of two hosts connected to two different switches and a single controller. In the setup phase, Flowtracker insert connection detection flow entries and switching flow entries in flow table of each switch. Each entry will have host MAC address. Following are the steps:

(1) Host A request a connection from Host B by sending a packet
(2) This packet matches connection detection flow entry in switch 1 and is forwarded to controller for verification.
(3) The connection is approved by controller if packet header matches with connection information and does not violate any firewall rules. The controller finds out two switches by packet header and sends connection forwarding entries to both switches to add in flow table.
(4) Late on the entry is entered in switch I, that packet is forwarded to switch II
(5) In Switch II, incoming packet should match switching flow entries and is forwarded to Host B
(6) A reply packet is sent from Host B
(7) Connection forwarding entries matches the reply packet in switch II and it is forwarded to switch I
(8) Switching entries matches with reply packet on switch I and it is transmitted to Host A
(9) Now data can be exchanged between hosts.
(10) When the connection is ended, Flowtracker will remove the connection information that is in its list.

3.2   Proactive Rule Caching

Switches have limited storage, and rules that are produced by controller are require to be stored by switch in them. But switch storage is limited and rules are unlimited, so Reactive caching mechanism is taken into consideration [10]

- Reactive caching mechanism: whenever the switch does not find matching rule for a flow of packets, then the switch store that packet in switch buffer and ask for missing rule to controller. And when controller reply with a rule then that packet is processed and that rule is stored into forwarding table of switch to directly process the next packet. Attacker can exploit this condition of asking rule from controller by sending packets of larger payload which will be queued up into switch-buffer by the switch (Switch ask for rule of this type of flow of packet) and again attacker will send a packet of different flow and switch ask for matching rule from controller and also will queue it up into switch-buffer, now that buffer is full (suppose), the proper packets which will arrive onto switch will be dropped because buffer is full of previous payload of packets and hence attacker can exploit this.

- Whereas Proactive caching mechanism doesn't wait for the incoming packets to request new rules but already cache as many as rules in its table before the incoming packets. In this mechanism, installed rules cover ranges of header fields instead of single value. And also making switch-controller communication delay less will also help to process packets switch-buffer.

Issues regarding Security of the controller by which the attacker can attack firewall. There are solutions for this type of problem (DDoS and DoS Attack). First is a controller named 'Flowtracker' which maintains the information of users (identity, trustable or not) and another solution is named 'Replication' in which rather than using one controller, firewall uses multiple.

3.3   Enqueue Thread

Each incoming packet from data plane is being judged by this thread. There are two queues i.e. normal and warning queue. A host is assigned a reputation according to its behavior. Enqueue thread also contains MAC address hashing that give ability to Flowtracker to classify the incoming traffics from different hosts. By the help of this ability there is no need to check packet by packet. Below, it is given different type of hosts

- Attacker – Request will be dropped
- Suspicious – Request will be put into Warning queue
- Normal – Request will be put into Normal queue

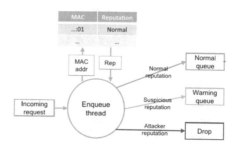

3.3.1   Statistic Firewall Thread

It has three modules with different functions. (i) Stateful firewall module – monitoring, matching connections against firewall policies. (ii) Statistic Updating module- updating peak number of connections and comparing baseline. (iii) DoS mitigation module – activating DoS mitigation effect just after the attacker is identified. This thread updates reputation for each host. And it is activated at the end of each timeslot. If a host reputation is of attacker in a current timeslot than that host will be referred as suspicious host in the next time slot. Same goes for Suspicious host in current timeslot (Next timeslot – Normal host). When a host exceeds its limit to send connection request; host reputation will be marked as "Suspicious" or "Attacker". Every time a host exceeds limit, then the next host gets a limit higher for new connection request compare to the previous one. An attacker can take advantage of this to increase peak connections over a period of timeslots in order to increase limit. To prevent this from happening, introduce a global Suspend limit that is applied to all hosts, when a host reaches this number, it will be treated as attacker.

3.3.2 Stateful Firewall Thread

This Thread bears main function of stateful firewall. It will update the current number of connections as a new connection is detected. Limit will also be updated. Then it will be compared with Global Suspend limit, host suspend limit and host warning limit. According to this comparison the Flowtracker assign new reputation for hosts in the end of every timeslot.

3.4 Replication

This solution suggests that instead of using one controller, use multiple controller. These controllers are connected on a secure channel by which controllers can maintain consistent rules and management of the network. When this solution is adapted, switches are connected to multiple controllers and one of controllers will be master controller (for every switch there is a master controller). Every new request is processed to the master controller of that switch while also taking in switch-controller link delay (The delay which is generated while processing the request). Less the communication delay, the better. If a master controller is overloaded with requests then it can assign other less loaded controller to become master controller of that particular switch. This helps in keeping the load balanced among the controllers and also gives protection against DoS attack till some amount. The placement of switch and master-controller plays an important role in deciding switch-controller delay. If master controller and switch of it is placed near to each other switch-controller communication delay can be minimized and makes harder for switch DoS attack to take place. There is also a case when some of the controllers are already compromised. To prevent this attack from happening it is required that platforms hosting the controllers have diversity in it. If controllers are on the same platform then they will be having same vulnerabilities which will be too easy for the attacker to compromise if he/she compromise one controller. Different platforms hosting will be having controllers with different vulnerabilities which will be difficult for the perpetuator to attack [10].

4   Analysis/Review

For Connection Tracking Issue

Parameters	Solutions	
	Connection tracking module	Proactive rule caching
Name of controller	Flowtracker	Basic controller
Type of mechanism	Hybrid Controller insert flow entries of switch prior to incoming packets (as many as it can) as well as update the flow of entries in table	Proactive Controller insert flow entries of switch prior to incoming packets only
Switch-controller communication delay	Low	High
Speed of processing packets	High	Low
After processing the packets of a single flow	The controller will remove any information it contains	The information will still be there
Vulnerability level	Low	High
Complexity level of solution	High	Low

For Security Issue

Parameters	Solutions	
	Flowtracker DoS attack detection/mitigation	Replication
Name of Controller	Flowtracker	Any basic type of controller
Number of Controllers	One	Multiple controller
Number of Platforms for Controllers	One	Multiple platforms (a platform have multiple controllers with multiple switches)
New Requests are directed towards	Flowtracker Controller	Master Controller (least overloaded controller)
Main basis of Solution	Detect and reduce severity of DoS attack	Prevent DoS attack
Controller uses	Three types of thread with different functionalities	No threads
Vulnerability Level	Low	High
Complexity level of solution	High	Low

5 Conclusion

This paper reviewed some challenges that arise in Firewall based on Software Defined Networking and their Solutions. The overhead of dynamic of firewall is inherent. It has been reviewed that certain issues can be overcome by proposed algorithms and some filtering rules for setup and maintenance stage with the help of three-layer structure of switch flow table and Flowtracker.

Flowtracker help firewall flows efficiently and removes any repeated flow entries in the flow table. It has been found by the researchers of these type of solutions that Flowtracker can achieve connection tracking in real time and can detect any suspicious behavior.

By using Flowtracker, it can distinguish between different type of hosts ("Attacker, Suspicious and Normal"). And they can also be categorized by their MAC address.

References

1. Pena, G.J.V., Yu, W.E.: Development of a distributed firewall using software defined networking technology. In: Proceedings of the 4th IEEE International Conference on Information Science and Technology (ICIST), pp. 449–452, April 2014
2. Javid, T., Riaz, T., Rasheed, A.: A layer 2 firewall for software defined network. In: Proceedings of the Information Assurance and Cyber Security (CIACS) Conference, June, pp. 39–42 (2014)
3. Kaur, K., Singh, J., Ghumman, N.S.: Programmable firewall using software defined networking. In: Proceedings of the Computing for Sustainable Global Development (INDIACom) International Conference, March, pp. 2125–2129 (2015)
4. Sezer, S., Scott-Hayward, S., Chouhan, P.K.: Are we ready for SDN? Implementation challenges for software-defined networks. IEEE Commun. Mag. 51(7), 36–43 (2013)
5. Astuto, A.B., Mendonca, M., Nguyen, X.-N., Obraczka, K., Turletti, T.: A survey of software-defined networking past, present, and future of programmable netcworks. In: Proceedings of the IEEE Communications Surveys & Tutorials, vol. 16, no. 3 (2014)
6. Tran, T.V., Ahn, H.: Challenges of and solution to the control load of stateful firewall in software defined networks. Comput. Stand. Interfaces 54(4), 293–304 (2017)
7. Tran, T.V., Ahn, H.: Flowtracker: a SDN stateful firewall solution with adaptive connection tracking and minimized controller processing. In: Proceedings of the 1st International Conference on Software Networking, ICSN 2016, Republic of Korea, pp. 1–5, May 2016
8. Paul, G., Chuck, B.: How SDN works. In: Software Defined Networks A Comprehensive Approach. Elsevier (2014)
9. Hayward, S., Natarajan, S., Sezer, S.: A survey of security in software defined networks. In: Proceedings of the IEEE Communications Surveys & Tutorials, vol. 18, no. 1, July, pp. 623–654 (2015)
10. Dabbagh, M., Hamdaoui, B., Guizani, M., Rayes, A.: Software-defined networking security: pros and cons. In: Proceedings of the IEEE Communications Magazine, vol. 53, no. 6, June, pp. 73–79, June 2015
11. Kreutz, D., Ramos, F.M.V., Verissimo, P.E., Rothenberg, C.E.: Software-defined networking: a comprehensive survey. Proc. IEEE 103(1), 14–76 (2015)

12. Monir, M.F., Akhter, S.: Comparative analysis of UDP traffic with and without SDN- based firewall. In: 2019 International Conference on Robotics, Electrical and Signal Processing Techniques (ICREST), pp. 85–90. IEEE (2019)
13. Mathebula, I., Isong, B., Gasela, N., Abu-Mahfouz, A.M.: Analysis of SDN- based security challenges and solution approaches for SDWSN usage. In: 2019 IEEE 28th International Symposium on Industrial Electronics (ISIE), pp. 1288–1293. IEEE (2019)
14. Alfayyadh, B., Jøsang, A., Alzomai, M., Ponting, J.: Vulnerabilities in personal firewalls caused by poor security usability. IEEE (2010)
15. Ioannidis, S., Keromytis, A.D., Bellovin, S.M., Smith, M.: Implementing a distributed firewall. In: Proceedings of 7th ACM Conference on Computer and Communications Security, Athens (2000)

Database Security: Attacks and Solutions

Sarvesh Soni$^{(\boxtimes)}$ and Rejo Mathew

Department of I.T., Mukesh Patel School of Technology and Management,
NMIMS, Mumbai, India
annysoni8620@gmail.com, rejo.mathew@nmims.edu

Abstract. Data is a critical merit resource and due to its importance, data protection is a noteworthy component of database security. Database security refer to the measures and tools used to protect a database from unauthorized laws, spiteful threats and attacks. Data security means protecting data, alike in a database, from malicious and devastating forces of unauthorized and unauthenticated users, such as cybercrime attacks or a data security breach. Since digitalization and rapid progress in technology, web applications and databases are widely used like e-commerce, online payments, online banking, money transfer, social networking, etc. [15]. For seamless and hustle-free experience which includes risk free operations on databases, defining and implementing database security is utmost important. Security aims that would be applied for data security, includes: CIA triad; Confidentiality for concealment of data and issues of privileges abuse, Integrity for trustworthiness of data and issues of legitimate unauthorized accesses including social engineering attacks. Availability guarantees that resources are available to users when they need them and issues are exposure of backup data and denial of service. It acknowledges the economic, financial and commercial documentation of records and reports related to an organization [11]. Moreover, security attacks, solutions, comparison table and future work are discussed in this paper.

Keywords: Unauthenticated access · Platform and protocol vulnerability · Weak audit trail · Unauthorized elevation · Data protection · Privilege abuse · Cryptography · Exposure of backups

1 Introduction

The increase in threats to the personal or organization's privacy through security breach in database systems have widely enlarged the potential risks which has led to the security concern in databases. In today's organization, it has been under notable focus. In an attack on aviation industry, example: a commercial airline company announced that a database security breach has disclosed the personal information of customers and VIP members, in which hackers gained access to credit card numbers, security codes and other financial information. The personal and financial details of customers making or changing bookings was also compromised and exposed [10]. Databases allow only authorized users to access, enter or analyze the data easily. Protecting this confidential and sensitive data from any kind of attack at any level is resolved by database security [1]. Optimizing solutions to security attacks entail safeguarding, conserving, shielding

© Springer Nature Switzerland AG 2020
A. P. Pandian et al. (Eds.): ICCBI 2019, LNDECT 49, pp. 917–925, 2020.
https://doi.org/10.1007/978-3-030-43192-1_100

and defending the database from threats, unauthorized access and malicious attacks. It has a distinct range of data security controls [4]. Robust techniques for data protection must be implemented in expert hands. It helps organizing data for optimum and reliable access. Data security gadgets have invaded security technology platforms to counter security attacks and ensure security in databases for an organized and optimum use with maximum potential and desired output. The unfortunate disclosure of consequential data due to negligence in database security events can also be prevented and acted upon by data security gadgets and its control mechanism. Database security can be achieved by several technology and management based platforms. Assurance of data security is the prime security focus.

2   Attacks on Database Security

2.1   Privilege

A rightful permission exclusively to perform a job or function. This action permits an authorized user to create, edit and modify database applications and resources.

2.1.1   Excessive Privilege
When clients/users are granted unrestricted privileges that overreach the needs and requirements of their task/job which may be misused to obtain access to confidential information [1]. Example: an employee in an organization has the privilege to edit or update profile information and he/she may take advantage of excessive privilege to change salary/credits information.

2.1.2   Unauthorized Privilege Elevation
Attackers take the chance to harm software vulnerabilities to switch or convert privileges from a low-level ordinary user to high-level administrator [3]. Example: an employee can hit vulnerabilities to gain admin privilege and may misuse the privilege by changing fund transfer details [2].

2.1.3   Legitimate Privilege Abuse

When an authenticated person misuses the legitimate privilege for malicious and unethical activity [2]. Example: a database administrator putting his hands into a business in which there is no such specific need and relevance of his/her actions and related work. Attackers may harm and attack legitimate database privileges for unauthorized and wrongful motive as well [3].

2.2   Platform Vulnerabilities

These unguarded and vulnerable systems are quiet more likely prone to threats and attacks which may result in illegal access and frauds [1]. Example: an attacker can access and harm the vulnerability of the software/data/system and may cause professional misconduct. Blaster Worm case, in which it took advantage of OS and created DoS.

2.3   SQL Injection

In this case, an attacker introduces or administers malicious and unauthorized statements into input fields of accountable servers to reveal data. It also destabilizes the context and may result in exposing secret information as well [4]. Although, there are different database encryption choices in SQL server but distinct attacks come along with it as well [13]. It majorly focuses on bypassing authentication, extracting data and inserting wrongful and incorrect queries [7].

2.4   Weak Audit Trail

The infringement or breach of a crucial database of an organization resulting in a reluctant acceptance, since there's no detailed way of how this activity actually happened. It doesn't guarantee for maintaining security and recovering lost transactions. It may cause defenseless vulnerability [2].

2.5   Denial of Service

Such an attack in which information or software application is denied to the authorized user purposefully, basically making it unavailable [5]. Techniques include buffer overflows, data misconduct, network instability, network piles etc. Example: attacker prevents legitimate users to access service and enforces invalid return addresses [1].

2.6   Protocol Vulnerabilities

Attackers propose network intrusion aiming data corruption, denial of service, unauthorized data access. Making it a likely nightmare, since the trickery record of protocol operation would not exist in internal audit of database systems [3].

2.7   Weak Authentication

Attackers attempt to presume the identity or sameness of authorized users by stealing or obtaining secret credentials. Techniques include: Brute Force Attacks, Regulating Delicate Social Engineering etc. It boosts double-dealing [3]. Dishonest employees may also cause an injury to the confidentiality by exposing access and authorization. Example: single factor authorization and vulnerable passwords are at the top-risk due to the poor and uncomplicated interpretation making it easy to guess and misuse it [4].

2.8   Exposure of Backup Data

Attackers can relate to breaches by violating unprotected and untenable theft or fraud of database backups, disks tapes, papers, system-hacks, decryption of valuable documents stored [4]. Natural disasters and calamities like earthquake, flood and avalanche may also cause high-security breaches with a very high risk of damage. Example: implication of backup data exposure can be done by anyone who has the capability to disjoint sensitive data backups [4].

3   Solutions to Database Security Attacks

3.1   Privilege Solutions

Solutions that help secure, control, manage and monitor privileged access.

3.1.1   Access Control
This mechanism allows only authorized users and grants minimum privilege permission, regulates who or what can view or use information. It simplifies the admin and authorization process. It reduces risks by taking control over accesses that are prompt enough to make it reliable and secure [2].

3.1.2   Mandatory Access Management
Restricting access to the subject of tasks and job function one belongs to. Tighten software corruption vulnerability by switching off the delete/modification option for attackers and make a separate team for the same to handle these activities. Keep a track on the activities of both low-level and high-level user/administrator [1].

3.1.3   Owner Approval and Tracking
Decisions shall be taken appropriately, to whomsoever it may be concerned and limited authorization must be allowed to accomplish their job function and related work. Since it reduces wrongful and unethical activities to prevent misuse of the legitimate privileges and the owner can keep a track of it by keeping an eye over the high-level administrators and their activities as well [5].

3.2   Secured Gateways and Data Masking

Secured gateways deny entry to unwanted or unsecured data traffic, preventing sniffing, spoofing and replay attacks, resulting in restricting misconduct access to prevent damage or harm to the system. Analyze the activities by bringing encryption process and use data masking techniques i.e. hiding original data with fabricated and modified content. The implementation of this proposed method can be done along with database by invading smart credential based authentication in which deceptive tricks are inserted to prevent an attacker from any kind of wrongful access. Fabricating and modifying data content in a desired way that it can be understood and executed only by legitimate and authorized users. It avoids deception and largely lacks sensitive material and access based benefits by hiding or removing important content or documents [5]. It can also be acted upon by data security gadgets, keeping a record of all events and reporting them to the legitimate technical team to take the needful control measures. Mutual authentication method can also be implemented (sharing session key) [12]. It do not allow users to maliciously modify any content that could lead to data security breach and keeps the content in original and desirable form. It can also be prevented by regular software updates [3].

3.3   Data Sanitization, Smart Driver and IPS

Using crypto-erase process, since it relies on deleting/forgetting main encryption key to make data unrecoverable, it is an irreversible process [8]. The smart-driver provide normal users only the relevant and task oriented information belonging to them by distributing a random code to the privileged users so that it identifies the right user and reject false behavior and action of invalid users [7]. IPS i.e. Intrusion Prevention System detects and regulates database traffic identities and saves vulnerable part of the system or organization [3]. To ensure data security, the identity is authenticated, keeping the confidential and sensitive data protected from being modified by an ordinary user [14]. Using stored procedures (SQL) for code reusability and sharing ability. High level restriction techniques to procure attacker towards unsuccessful injections [4]. Sanitization approach can remove the attacks by using input query sanitization with the help of regular expression based on database usage. This service will stop the attack before it affect the system and will provide a sanitized query to the system [15].

3.4   Deter Attackers, Detection and Recovery

Auditing databases, detect and identify the breach or violation to the system and at some point, it repairs the system [3]. Multi audit implications, example: network-based audit will have no effect on database security, therefore, if the attacker tries to fabricate or damage network-audit in terms of software, then it has no effect on databases [4]. Since it represents serious risk-level to records and sources of document to which the organization is exposed thoroughly, therefore, tracking privilege abuse and monitoring data access regularly prevents weak audit trail [5].

3.5   Configuration and Database Activity Monitoring

Configure databases by reducing the attack area to remove the information which you don't require i.e. you cannot exploit or attack something which isn't there. Alter database installations and limit resources i.e. timeout and starve out for queries and legitimate users respectively. DAM alerts on database misuse, it examines and detects incoming queries and tells about the policy violation [9]. Harden TCP/IP sack for connection establishments and connection queues. Implement tough SAP solutions for read/write/table access and authorization [4].

3.6   Protocol Validation

This validation is designed to block spiteful codes in case of any detection of anomalous situations. Use strong password alike admin database privacy [4]. Implement a good internal audit trail of database systems. One can also use prevention steps of DoS and excessive privileges as well. Update software on a regular basis and use IPS [4].

3.7   Define and Implement SSL (Security Socket Layer)

Through proxy authentication to make data consistency private and integral by establishing an encrypted link [4]. Keep identity fully authenticated to prevent modification or fabrication. This is an obligation to keep sufficient security such that an unauthorized party cannot steal or obtain login credentials [3]. Use countermeasures to defend Brute Force and Social Engineering. Use multi-factor authentication when risks are higher. Keep strong password and check on the immediate notification of unintentional or wrongful breach of database security related activities [4].

3.8   Encryption Process and Backup Management

Encryption process is used for converting ordinary information into cipher text or code within confidential limits and standards. Regular optimization and monitoring various backup data files, disks, tapes, papers etc. helps in keeping a track and record of database activities [6]. They are unprotected, so extra concern and effort of security is required altogether to keep them confidential, safe and risk-free [4]. Firstly, prevent anyone who has the capability to disjoint sensitive backup data. Secondly, natural disasters and calamities like earthquakes, floods, avalanches etc. may also cause high security breaches with an estimation of immense loss. Use disaster management strategies and techniques, maximize security practices, and choose the guarded paramount and appropriate secure platforms like access governance, threat detection and other tools [2].

4 Analysis

Database is basically a heart of any organization, since it contains important, immensely sensitive and confidential information, so there is a chance of attacks from distinct types of attackers and to counter that, various solutions are discussed. It gives a collective review and analysis of issues, solutions and its implementation. This research is concrete and rock-hard to handle such a prominent topic like database security.

Comparison Table

See Table 1.

Table 1. Comparison of solutions for database security

Solutions	Parameters					
	Minimum privilege and regulation of access	Restricting access till job function	Fabrication and securing data traffic	Decision making and monitoring	Blocking malicious activity and multi factor authentication	Encryption and recovery
Access control	Only authorized people are granted permissions	Reduces risk of legitimate case of misuse	—	—	Login credentials to identify user	—
Mandatory access management	Authorization within limit and need of tasks given	Only job function and work related access	—	—	Tightens software corruption, cannot be easily breached	—
Owner approval and tracking	Access approval depends upon owner	Keeps an eye on high level administrators and restricts access to limited people only	—	Owner grants permissions to selected authorities	Owner may analyze and block any wrongful attempt	—
Secured gateways and data masking	—	—	Hides original content with the fabricated/modified data for more security	—	Deny unwanted traffic and performs established authentication	—
Data sanitization and smart driver	—	—	Irreversible process, makes data unrecoverable	—	Reject false behavior of invalid users	Deletes/forgets main encryption key, crypto-erase process
Deter attackers, detection and recovery	—	—	Identify the breach and repairs the system, if the attacker tries to fabricate data	Prevents weak audit trail	Multi audit implications	—
Configuration and database activity monitoring	SAP solutions for read/write/table access	—	Examine and detect incoming queries	Monitors configuring of databases in accordance with attack area	DAM alerts on database misuse	—

(continued)

Table 1. (continued)

Solutions	Parameters					
	Minimum privilege and regulation of access	Restricting access till job function	Fabrication and securing data traffic	Decision making and monitoring	Blocking malicious activity and multi factor authentication	Encryption and recovery
Protocol validation	—	—	IPS systems	A good internal audit trail	Block malicious codes in case of any detection	—
Define and implement SSL	Makes data consistency private and integral	—	Unauthorized party cannot steal or obtain login credentials	—	Defends social engineering. Use multi-factor authentication when risks are higher	Encrypted link authentication
Encryption process and backup management	Algorithm based cryptography concept	—	—	Regular optimization and updates	—	Cipher text information and disaster management techniques

5   Conclusion and Future Work

This research paper coheres databases and its security in any organization. Issues of unauthorized access, deception, vulnerability, authentication and fabrication has been discussed along with the solutions to these attacks. There's a large scope of improvement in these techniques that entail more automation, shielding and involving numerous methods of defense. Since criminals/attackers have distinct avenues to cause harm and therefore, countless attacks will continue to generate but fostering up more counterattacks and solutions will certainly lead to a better strategy in the near future, as the era of AI is expected to rise, an efficient key management system and comprehensive protection will positively impact and would tell the importance and delicacy of database security. This paper has focused on attacks, solutions and their implementation to ensure the security of databases.

References

1. Ali, A., Afzal, M.: Database security: threats and solutions. Int. J. Eng. Invent. **6**(2), 25–27 (2017)
2. Deepika, Soni, N.: Database security: threat and security techniques. Int. J. Adv. Res. Comput. Sci. Softw. Eng. **5**(5), 621–624 (2015)
3. Singh, S., Rai, R.K.: A review on report on security threats and database. Int. J. Comput. Sci. Inf. Technol. **5**(3), 3215–3219 (2014)
4. Malik, M., Patel, T.: Database security – attacks and control methods. Int. J. Inf. Sci. Tech. **6** (1/2), 175–183 (2016)
5. Gahlot, S., Verma, B., Khandelwal, A., Dayanand.: Database security: attacks, threats and control methods. Int. J. Eng. Res. Technol. **5**(10) (2017)

6. Sharma, P., Monika: Database security: attacks and techniques. Int. J. Sci. Eng. Res. **7**(12), 313–319 (2016)
7. Devi, R., Venkatesan, R., Raghuraman, K.: A study on SQL injection techniques. Int. J. Pharm. Technol. **8**(4), 22405–22415 (2016)
8. Randhe, K., Mogal, V.: Security engine for prevention of SQL injection and CSS attacks using data sanitization technique. Int. J. Innov. Res. Comput. Commun. Eng. **3**(6), 5890–5898 (2015)
9. Mahjabin, T., Xiao, Y., Sun, G., Jiang, W.: A survey of distributed denial-of-service attack, prevention, and mitigation techniques. Int. J. Distrib. Sens. Netw. **13**(12) (2013)
10. Sarah, S.: GDPR and privacy lawsuits. In: Cyber Decoder, JLT, issue 39, p. 4 (2018)
11. Sarmah, S.: Database Security – threats and prevention. IJCTT **67**(5), 46–50 (2019)
12. Sridhar, S., Smys, S.: Intelligent security framework for IoT devices cryptography based end-to-end security architecture. In: International Conference on Inventive Systems and Control (ICISC), pp. 1–5. IEEE (2017)
13. Mukherjee, S.: Popular SQL server database encryption choices. In: SSRG-IJCSE, pp. 1–6 (2018)
14. Basharat, I., Azam, F., Muzaffar, A.: Database security and encryption: a survey study. IJCA **47**(12), 28–34 (2012)
15. Kawalkar, M., Butey, P.K.: An approach for detecting and preventing SQL injection and cross site scripting attacks using query sanitization with regular expression. Int. J. Comput. Trends Technol. (IJCTT) **49**(4), 237–245 (2017)

Cloud Based Heterogeneous Big Data Integration and Data Analysis for Business Intelligence

T. Jayaraj[1(✉)] and J. Abdul Samath[2]

[1] Research and Development Centre, Bharathiar University, Coimbatore, India
yoursjayan@gmail.com
[2] Chikkana Government Arts and Science College, Tiruppur, India
abdul_samath@yahoo.com

Abstract. Due to the enormous growth of information technology, a huge amount of big data is produced daily, wherein heterogeneity is considered as the main feature of big data. Heterogeneous data integration is still remaining as a bottleneck. It becomes as a very difficult task to integrate and complete the business information demands. Hence, in this research work we have presented a novel Heterogeneous Data Integration and Analysis framework for solving the challenges associated with heterogeneous big data. Big data analysis is an information extraction technique generally used by organizations for business intelligence. However, data mining doesn't provide good performance for very large data set due to the problems of high computational cost and lack of memory. In this article, we have proposed Convolutional Neural Networks (CNN) architecture for heterogeneous big data analysis. Finally, experimental results make it clear that the proposed method is the fastest data integration framework and that it is also considered as a good analysis model for business.

Keywords: Big data · Heterogeneous data · Data integration · Data analysis · Multisource data

1 Introduction

With the enormous growth of information and communication technology, the growing nature of heterogeneous data is increasing day by day. It is produced by web and mobile in the large industry environment [10]. Its structure is in its native formats like unstructured, structured and semi-structured format. Integrating this raw business data is key to business development, especially business analysis and business prediction. Many organizations depend on Extract-Transform-Load (ETL) Tool for heterogeneous data integration [13, 14]. ETL data integration is not fully automated, 50% of ETL's data integration process is done manually and it doesn't have a good data analysis model. Its architecture is shown in Fig. 1.

Our main objective of this research is to integrate heterogeneous data sources into a centralized cloud storage system to obtain the necessary information for the business at a low cost and in a timely manner. Many researchers have been trying to solve this heterogeneous data integration problem for almost three decades but so far this

A. P. Pandian et al. (Eds.): ICCBI 2019, LNDECT 49, pp. 926–933, 2020.
https://doi.org/10.1007/978-3-030-43192-1_101

integration problem remains an open research problem. To integrate heterogeneous data, we need to solve two key problems. The first is how to combine the different formats and different types of data. The second is to make it easy to get business information from this integrated framework. In this proposed method we have addressed the two problems mentioned above. To that end, we have developed a cloud based heterogeneous data integration framework and a user friendly graphical user interface (GUI) reporting tool for gathering business information from this integrated framework. With this developed GUI, information can be made much easier and faster even for someone who has no basic knowledge of the computer. This reporting tool is made for mobile cloud, desktop and web applications.

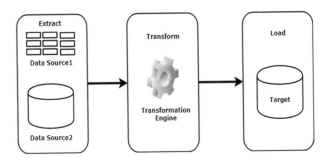

Fig. 1. ETL architecture

The literature review is explained in Sect. 2. The three main modules of the proposed system are discussed in more detail in Sect. 3. In Sect. 4 the experimental analysis is discussed and the results are compared with the already developed methods. Section 5 summarizes conclusion and feature enhancement.

2 Literature Review

Gal et al. proposed schema matching method for heterogeneous data integration. In this way the matching algorithms are used to align attributes. Thus the top K attributes are selected. Quality is rated by using matching predictors. The Lear-to-Rank algorithm is used to extract the optimal combine feature. Moreover, [1] mari-based predictors have been proposed. According to [1] it has two main phases: training phase and testing phase.

Chen et al. presented a heterogeneous data integration model for Big Data in power grid, dispatching and control system (DCS). This model improves the Extract-Transform-Load process that has already been developed. There are three main modules in its architecture: data extraction process, data transformation process and loading process. This new architecture is developed for dispatching and control systems. At the same time, this system does not provide a good graphical user interface for the end user.

Tripathi et al. developed a cloud based data integration model. This proposed framework consists mainly of two layers: client layer and service layer. The client layer performs the request and response operation. The service layer performs data integration, data encryption, data decryption and user authorization. This method provides security for heterogeneous data as main. This research paper contains only theoretical information but no experiments.

Vidal et al. developed a sematic data integration technique for big biomedical data. This integration framework consists mainly of four components: knowledge extraction, exploration and traversal, semantic data integration, and knowledge discovery. It is a knowledge driven framework. According to [4] integrated data is described as unified schema and ontologies and this framework is part of iAsis.

3   Proposed Methodology

There are three main modules in this proposed architecture: hetcrogeneous data integration module, data analysis module and data visualization and reporting module. Its system architecture is shown in Fig. 2.

3.1   Heterogeneous Data Integration Module

Data Pre-processing. Good prediction results can only be obtained through a good quality data-set [11, 12]. So, data quality is considered the most important factor in data analysis. Generally, data produced from enterprises are in the following format: structured, unstructured and semi-structured. In this case structured data is a well-structured data such as Sql, Oracle, Mysql and etc. Semi-structured data refers to data with the following extensions: XML, CSV, JASON etc. Unstructured data are very noisy data. Generally, this type of data contain many null values and duplicate values. Getting useful information from this type of data is a very challenging task. The proposed system includes a data processing unit that classifies the above types of data into three groups. First, unstructured and semi-structured data noise is removed with the help of k-nearest neighbor (K-NN) algorithm. E.g. fill the null values and remove the duplicate values. This pre-processing module converts the different formats of data in to the same format. In this proposed work, heterogeneous data is converted to text format.

Data Integration. Data integration is the most important part of the proposed system. As large memory space is required to integrate and store heterogeneous data cloud is used for data storage in this proposed work. The data integration layer performs two most important tasks: data integration and data security. Data integration integrates pre-processed heterogeneous data. In the proposed architecture, the public cloud is used. So, data security is the most important thing. For that purpose, cryptographic technique to encrypt the data that is integrated. Only authorized cloud users are allowed to decrypt the encrypted data.

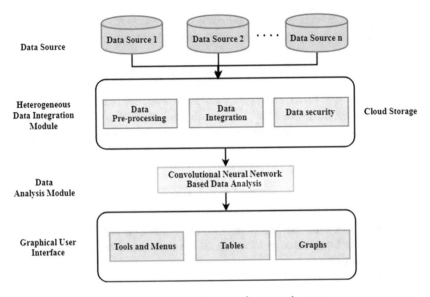

Fig. 2. System architecture of proposed system

3.2 Data Analysis Module

The traditional machine learning algorithms has many limitations to analysis heterogeneous big data [7, 8]. So, to analyze integrated heterogeneous data the renowned deep learning method has been invoked in this proposed method. Deep learning has the capability of learning and analyzing large sized, unlabeled and uncategorized types of unsupervised data [15]. In the proposed method Convolutional Neural Network (CNN) is used for data mining. CNN is suitable for sequential data [5, 6] such as speech, text, financial data, audio and video, and when it is used in eCommerce applications get better results. CNN's architecture is comprised of multiple numbers of hidden layers. The main layers for CNN are the convolutional, pooling, and fully connected layer. CNN architecture is shown in Fig. 3. Typically, CNN starts from the convolutional layer, which accepts input data. The same size of filters is used to

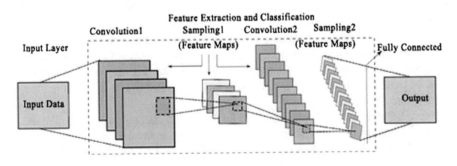

Fig. 3. Architecture of CNN

perform convolutional operations. Its output is connected to the adjacent layers. By the end of each conviction process, its size continues to decrease. It is finally connected to fully connected and output layers.

The classifier of this proposed architecture must be a multi-class classifier that should solve a multi-class classification problem. In order to solve multi-class classification problems a softmax classifier is added to this data analysis module. This is done before the output layer. Most importantly, the number of nodes in the softmax layer must be equal to the number of nodes in the output layer. This is calculated through Eq. 1.

$$F(X_i) = \frac{EXP(X_i)}{\sum_{J=0}^{K} EXP(X_j)} \qquad (1)$$

Equation 1 computes the exponential of the input data provided [9]. The exponential ratio of the input value and the sum of the exponential values are the output of the SoftMax function.

3.3 Data Visualization

The main goal of this module is to create a better communication between the data integration tool and the end user and extract the information needed for business development in a much easier way. These GUI have two primary components: (1) Data Integration and Data Pre-Processing section and (2) Reports. The Data Integration and Pre-Processing section contains the tools and menus needed for heterogeneous data integration and management. The reporting sections contain patterns, tables, graphs, etc. JSP and twitter bootstrap tools have been used to develop a responsive design that works on mobile and desktop.

4 Experimental Analysis

4.1 System Configuration and Data Set

Java and WEKA open source libraries have been used to implement the proposed method. Dell PowerEdgeR940 server is used to evaluate data integration framework and Nivida V 100 is used as a graphical processing unit (GPU).

The e-commerce data-set used in this proposed method is part of the Kaggle Competition. This dataset contains a variety of nominal and numeric types of attributes such as product information, customer information, customer feedback, etc. Moreover, this data-set includes a wide variety of data formats and this data-set contains 52800 types of products information. Data-set size is 107.7 GB. 90% of this data is used for training and the rest for testing.

First, heterogeneous data is optimized by the pre-process module. Secondly, data is integrated through the data integration module. Finally, in order to evaluate the training speed of proposed CNN based data analysis module, the integrated data is divided into five pieces, which are 20 GB, 40 GB, 60 GB, 80 GB and 100 GB respectively. In every set, 90% of the data is taken as training and 10% as testing.

Table 1. Accuracy of proposed method and traditional machine learning algorithms.

Methods	Accuracy %
Proposed method	79.9
Artificial Neural Network (ANN)	75.6
Support Vector Machine (SVM)	72.1
Logistic Regression (LR)	70.3
Decision Tree (DT)	66.3

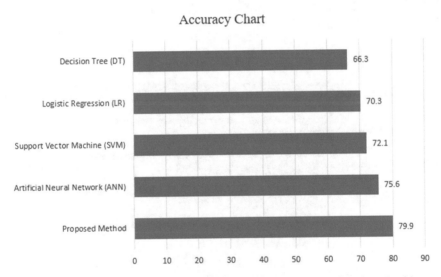

Fig. 4. Accuracy chart of proposed method and traditional machine learning algorithms.

In order to evaluate the accuracy and training speed of the proposed method already developed Artificial Neural Network (ANN), Decision Tree (DT), Logistic Regression (LR), and Support Vector Machine (SVM) have been implemented. Accordingly, the proposed data analysis module provides an excellent accuracy of 79.9%. This is better than traditional machine learning algorithms. Figure 4 shows the accuracy of the proposed method. Decision Tree gives the lowest accuracy of 66.3%. This is represented in Table 1. The time taken for training in Deep Learning is considered to be its shortcoming. The training time of the proposed method is calculated on a minute basis. This is represented in Table 2. Figure 5 shows the time variation of different types of machine learning algorithms. The proposed method takes 96 min of training time and 20 GB of low input data. SVM consumes a maximum of 127 min. The proposed method takes 363 min of training when inputting the maximum size (100 GB) of data. SVM consumes a maximum of 430 min.

Table 2. Training time of proposed method and traditional machine learning algorithms.

Methods	Minutes (20 GB)	Minutes (40 GB)	Minutes (60 GB)	Minutes (80 GB)	Minutes (100 GB)
Proposed method	96	180	226	270	363
Artificial Neural Network (ANN)	103	184	233	286	379
Support Vector Machine (SVM)	111	189	248	293	392
Logistic Regression (LR)	119	196	252	299	422
Decision Tree (DT)	127	209	264	301	430

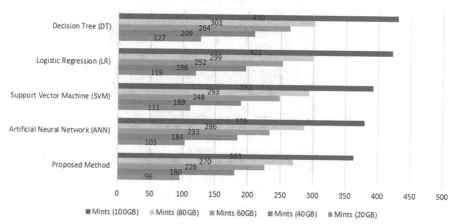

Fig. 5. Time chart of proposed method and traditional machine learning algorithms.

5   Conclusion

With the rapid growth of information technology, an enterprise produces huge amount of heterogeneous data every day. Analyze this data properly then only get the information needed for business development. For this purpose, we have developed cloud storage based heterogeneous data integration and analysis framework. As we use the cloud as storage, security has been enhanced by encryption and decryption techniques. CNN based classifier has been developed for data analysis. Moreover, a better graphical user interface is developed for the end user. Finally, the proposed system is evaluated with traditional machine-learning algorithms. Experimental results show that

CNN based classifier gives better accuracy. The built-in graphical user-interface reduced most of the end user's work.

This architecture only supports to integrate text-related data. In feature need to develop the architecture which supports all types of file formats. Moreover, the data analysis module takes more time for training and so the algorithm needs to be modified to reduce the training time.

References

1. Gal, A., et al.: Heterogeneous data integration by learning to rerank schema matches. In: 2018 IEEE International Conference on Data Mining (ICDM) (2018)
2. Chen, W., et al.: Multi-source and heterogeneous data integration model for big data analytics in power DCS. In: IEEE 2016 International Conference on Cyber-Enabled Distributed Computing and Knowledge Discovery (CyberC) (2016)
3. Tripathi, D., et al.: Model for heterogeneous data integration on cloud. In: 2016 3rd International Conference on Computing for Sustainable Global Development (INDIACom) (2016)
4. Vidal, M.-E., et al.: Semantic data integration techniques for transforming big biomedical data into actionable knowledge. In: 2019 IEEE 32nd International Symposium on Computer-Based Medical Systems (CBMS) (2019)
5. Albawi, S., et al.: Understanding of a convolutional neural network. In: 2017 International Conference on Engineering and Technology (ICET) (2017)
6. Yang, J., el al.: Application of deep convolution neural network. In: 2017 14th International Computer Conference on Wavelet Active Media Technology and Information Processing (ICCWAMTIP) (2017)
7. Song, H., et al.: A review on the self and dual interactions between machine learning and optimisation. Prog. Artif. Intell. **8**(2), 143–165 (2019)
8. Huang, J.-C., et al.: Application and comparison of several machine learning algorithms and their integration models in regression problems. Neural Comput. Appl., 1–9
9. Ertam, F., et al.: Data classification with deep learning using Tensorflow. In: 2017 IEEE International Conference on Computer Science and Engineering (UBMK) (2017)
10. Suma, V.: Towards sustainable industrialization using big data and Internet of Things. J. ISMAC **1**(01), 24–37 (2019)
11. Suresh, R.M., et al.: An overview of data preprocessing in data and web usage mining. In: IEEE 2006 1st International Conference on Digital Information Management (2006)
12. Wei, J., et al.: Research on data preprocessing in supermarket customers data mining. In: IEEE 2010 2nd International Conference on Information Engineering and Computer Science (2010)
13. Qin, H., et al.: Research on extract, transform and load (ETL) in land and resources star schema data warehouse. In: IEEE 2012 Fifth International Symposium on Computational Intelligence and Design (2012)
14. Adnan, et al.: Performance analysis of extract, transform, load (ETL) in apache Hadoop atop NAS storage using ISCSI. In: IEEE 2017 4th International Conference on Computer Applications and Information Processing Technology (CAIPT) (2017)
15. Chemchem, A., et al.: Deep learning and data mining classification through the intelligent agent reasoning. In: IEEE 2018 6th International Conference on Future Internet of Things and Cloud Workshops (FiCloudW) (2018)

Search Engine Optimization Challenges and Solutions

Manashwi Singh[(✉)] and Rejo Mathew

Mukesh Patel School of Technology, Management and Engineering,
NMIMS (Deemed-to-be) University, Mumbai, India
manashwi13@gmail.com, rejo.mathew@nmims.edu

Abstract. Search engines use algorithms for filtering out the most useful websites out of millions of websites on the Internet in minimum time span to optimize user search/query. SEO is the technique used by search engine optimizers to make their websites appear in the top results of search engines. But sometimes some unethical developers use black hat SEO techniques to bring their websites amongst the top results which misleads both, the search engines as well as the users. In this paper, search engine algorithms such as PageRank, HillTop, black hat SEO techniques like content spamming and the concept of Elasticsearch are studied and analysed.

Keywords: Search engine optimization (SEO) · PageRank · Black hat SEO · Elasticsearch

1 Introduction

Search engines play a key role in today's online world. They have the complex task of searching through billions of pages and showing only the most relevant pages for the requested search query due to a large number of websites present in today's world.

Huge search engines such as Google use site rank to assess a web page's value. Most online marketing strategies are used by website developers to achieve a higher rank in the search engine results. This is known as search engine optimization. Search engine optimizers evaluate the search engine results to bring their websites on top in search engine natural listing. However, sometimes the methods used do not follow the rules or policy of the search engine. Those strategies are called black hat SEO techniques [1].

1.1 Types of SEO Techniques

SEO techniques are broadly classified into two categories. These techniques are followed by search engine optimizers to improve their webpage ranking.

1.1.1 On Page Optimization

These techniques are used by the web developers in the website itself. It includes adding relevant keywords in title of the page, meta description and keyword, heading, alt attribute for images and including a sitemap page [2].

A. P. Pandian et al. (Eds.): ICCBI 2019, LNDECT 49, pp. 934–940, 2020.
https://doi.org/10.1007/978-3-030-43192-1_102

1.1.2 Off Page Optimization

These techniques are indirectly related to the page content. It is used to maximize the performance of SEO for the targeted keywords related to the page content that is done indirectly by joining groups, link building and blogging [2].

1.2 Working of a Search Engine

Search engines follow a sequence of steps listed below in order to deliver the most relevant results [2] (Fig. 1):

Fig. 1. Workflow of a search engine

This section gives the understanding of what SEO is and how search engines actually work. In Sect. 2 the working of search algorithms are analysed that are used by most of the search engines. Section 3 discusses the challenges faced by searching algorithms and Sect. 4 provides the solutions to face those challenges. In Sect. 5 all the solutions have been compared and analysed. In Sect. 6, conclusion of the research is drawn and the future scope of this review is discussed. The final section lists the references for this paper.

2 Search Engine Algorithms

Search engines use various algorithms to filter out the most relevant webpages amongst millions of web pages available on the Internet. These algorithms compare the websites and give them ranks accordingly. The pages with higher ranks top the results. Some of these algorithms are explained in the table below.

2.1 An Overview of the Search Engine Algorithms

See Table 1.

Table 1. Overview of search engine algorithms

PageRank	HillTop	PR + HillTop
This algorithm is used to rank their resultant websites for a particular search query	It is used to refer expert websites and then discover relevant webpages through links	It combines both the algorithms to offer a more accurate result in a quantitative and rational way
The website is assigned a numerical weight based on its relative importance	HillTop first measures the list of experts most important to the searched query, then find relevant links and pursue them to locate target web pages within the identified group of experts	It takes into account the relevance score, i.e., the translation of all SEO variables, PageRank score and regional score (interpretation of references from expert reports)
PageRank operates by counting the number and value of references to a page in order to assess the website's significance	Target pages are listed according to the amount and importance of the experts referring to them. When the opinion of the specialist is not valid, Hilltop does not produce results	
It is focused on both inbound and outbound links quantity and quality	Hilltop is therefore for the reliability of the tests and not for query coverage	
PageRank gives ranking on a 1–10, 10 being the highest	HillTop gives a binary value 0 or 1, where 0 means irrelevant and 1 being relevant	

3   Shortcomings in Search Engine Algorithms

Although search engine algorithms are used by search engines to provide the users with the most accurate results there are some web designers who try to fool the search engine by malicious methods that result in irrelevant results. Listed below are some challenges faced by search engine algorithms.

3.1   Impartiality [6]

Search engines trust certain sources more than others. These sources are regarded as reliable, up-to-date, well-researched and referenced by the engine and thus often features in that engine's top results. So, each time a query is entered the engine would first go to that source and if the search string matches the keywords it would display it in the top results.

3.2   Content Spamming

Content spamming is a very common method used by unethical developers to get their web pages among the top results. Algorithms such as PageRank and Hilltop sometimes fail to detect these spam websites. There are numerous ways in which a developer can perform content spamming.

3.2.1   Spam Comments [7]

Spam comments are references that are added feedback to a blog/article or web page to create free backlinks. It is one of the most popular black hat activities at the moment. There are numerous tools available to spread spam comments instantly across the internet.

3.2.2   Article Spinning [7]

It is a similar technique to scraping data and is becoming increasingly popular. This is plagiarism at the next stage which includes utilizing special software that takes the copied origin and rephrases it as a new, original article for future usage.

3.2.3   Keyword Stuffing [1, 7]

Spammers reuse keywords such as name, meta, head, anchor, etc. in various HTML tags. Keywords are also loaded with URL spammers. This well-known black hat SEO tactic is no longer as valid as it was in the past as search engines could spot the trick quickly.

3.2.4   Link Farm and Exchange [1]

Highly linked sites are referred to as reference farms in the link farm groups. Search engine optimizers spill hundreds of links to various pages in different categories irrelevant to the quality of the page.

3.2.5   URL Redirection [1]

Redirection of the URL implies routing of the file. Spammers cover spam sites by redirecting the user to a specific URL as soon as it loads the site. Spam page is retrieved in the search engine list, but the intended site is restored to the user by redirecting.

3.2.6   Doorway Pages [7]

These are portal pages, hop pages and bridge pages that are specifically tailored to selected keywords and built for unique queries to rank high. Such landing pages have very little meaning and are used merely to manipulate the search engines and move a traveler to another random destination from a selected result.

3.2.7   Cloaking [4]

It is a SEO technique which misguides the user by sending them to a page which is a different version of a web page crawl by search engine for indexing. There are different types of cloaking such as IP cloaking, user agent cloaking & repeat cloaking.

4 Overcoming SEO Challenges

4.1 Providing Results on the Biasis of Their Own Database

Search-engine-bias would not be a problem if we had a variety of real search engines that provide their own ranked results based on a large enough database and are not just displaying the same results as one of the dominant websites [5].

4.2 Detection of Spammed Websites [4]

4.2.1 Creating a Root Database

The root database consists of links of previously caught spammed websites. The database is prepared using content based as well as link based detection. Content Based Spam detection deals with on-page data analysis while linked based is concerned with the off-page spam analysis. Content based spam detection marks a website as spammed when it identifies the greater magnitude of keyword density while link based spammed detection identifies the number of inbound and outbound links of webpage. When this link for a website is found to be more than hundred then it is marked as spam website.

4.2.2 Refining the Result

The overall spammed detection cannot be finalized based on root as root database may contain the false positive results also. The root may contain links of website which are really good. So, a weighted page rank algorithm is applied on the root result.

4.3 Combining Spam Detection with Weighted Page Rank [4]

It is an architecture which compares the webpage detected as spammed with its corresponding weighted page rank. A new spam detection technique is introduced in this architecture, which takes root database, high rank sites database, web and weighted page rank into consideration in marking a website as spammed. First it will compare these result pages with the links in the root and if any match is found then the resulting page is penalized with the decrement in its page value.

5 Analysis/Review of Solutions

See Table. 2.

Table 2. Comparision of solutions for content spamming

Features	Creating a root database	Refining the result	Combining spam detection with weighted page rank
Method	It compares websites with the list of spammed websites already given in the database	It checks the spammed websites again with their original page rank value and eliminates almost all false positive results in the root database	It compares the webpage detected as spammed with its corresponding weighted page rank. First it will compare these result pages with the links in the root and if any match is found then the resulting page is penalized with the decrement in its page value
Factors considered	Spammed websites in a given database	Results of root database, original page rank value	Root database, high rank sites database, web and weighted page rank
Speed	It provides quick results and is efficient	It takes a little more time	It will take maximum execution time because it considers a lot of factors comparatively
Accuracy/Reliabilitity	It saves time and ensures the search engine does not get fooled by previously detected spam webpages	Eliminates almost all false positive results in the root database and hence provides more reliable results	It is quite accurate as it considers root database, high rank sites database, web and weighted page rank
Limitations	It might contain false positive results which can ignore genuine websites too	It considers only content based and link based detection. Other factors like high ranked site database and weighted PR are ignored	Can be very complex as it takes into consideration many factors

6 Conclusion and Future Work

Website ranking in search result strongly depends on the implementation of search engine algorithms. Search engine algorithms are being updated every year to provide better and optimized results. White hat SEO strategies return material of value. Such methods produce sluggish, but long-term outcomes. All clients and search engines profit from these. Black hat SEO strategies offer quick results, but for a short time, and if the search engine figures out about the site's illegal practices, the site can also be penalized. More and more solutions are being brought forward to prevent black hat SEO techniques. The most important concept here is Elasticsearch which is being used by more and more search engines in the upcoming years as it is faster, easier and has a huge database.

References

1. Patil Swati, P., Pawar, B.V., Patil Ajay, S.: Search engine optimization: a study, Maharashtra, 27 December 2012. www.isca.in
2. Kumar, L.: SEO techniques for a website and its effectiveness in context of Google search engine, Noida, 30 April 2014. www.ijcaonline.org
3. Kalyani, D., Mehta, D.: Paper on searching and indexing using Elasticsearch, Gujarat, India, 6 June 2017
4. Agrawal, S., Somani, A., Chhabra, V.: Discernment of search engine spamming and counter measure for it, India, 8 August 2016
5. Lewandowski, D.: Living in a world of biased search engines, Germany, Europe, June 2015
6. Kumar, N.: SEO techniques for a website and its effectiveness in context of Google search engine, Noida, 30 April 2014. www.ijcaonline.org
7. Persynska, K.: 8 risky black hat SEO techniques used today. Positionaly Blog (2015)
8. Loggie, N.: 2018 SEO roundup: all the major developments that happened in the world of SEO over the past 12 months and how to prepare for the year ahead. Adliweb (2018, in press)
9. ImageX: Google's Search Engine Algorithms Explained. imagexindia.com (2016, in press)
10. Edleman, B.: Bias in search results?: diagnosis and response. Indian J. Law Technol. **7** (2011)
11. Xia, F., Zhang, G.: Design and implementation of a Java-based search engine algorithm analysis system. In: Proceedings of the 4th International Conference on Computer Science and Education, pp. 1040–1043 (2009)
12. Mo, Y.: A study on tactics for corporate website development aiming at search engine optimization. In: Second International Workshop on Education Technology and Computer Science, vol. 3, pp. 673–675 (2010)
13. Shi, J., Cao, Y., Zhao, X.: Research on SEO strategies of university journal websites. In: Proceedings of the 2nd International Conference on ICISE, pp. 3060–3063 (2010)
14. http://www.webconfs.com/SEO-tutorial/introduction-toSEO.php. Accessed 23 September 2019
15. Nazar, N.: Exploring SEO techniques for Web 2.0 websites, Sweden, June 2009

Performance Assessment of Different Machine Learning Algorithms for Medical Decision Support Systems

T. Ragupathi[1(✉)] and M. Govindarajan[2]

[1] Department of Computer Science and Engineering,
Annamalai University, Chidambaram, India
raguthamodar@gmail.com
[2] Department of Computer and Information Science,
Annamalai University, Chidambaram, India
govind_aucse@yahoo.com

Abstract. Recently, the significance of decision support system (DSS) is increased in diverse domains ranging from medicine to healthcare, finance to marketing. Medical DSS (MDSS) is a hot research topic that assists the physicians to choose a better method for treating the patients. MDSS involves intelligent software systems that offer decision to the physicians. In present days, machine learning algorithms finds useful to predict the data by analyzing the past behavior. This research work has made an attempt to analyze the performance of three Machine Learning (ML) models namely radial basis function (RBF), logistic regression (LR) and Naïve bayes (NB) models. The results are validated using a benchmark medical dataset and the outcome pointed out that the LR classifier model is a better option for DSS. The detailed simulation outcome pointed out that the applied LR model shows its effective classification outcome by obtaining maximum precision value of 98.60, recall of 98.60, F-measure of 98.60, ROC of 99.10, accuracy of 98.63 and kappa value of 96.29.

Keywords: Classification · DSS · Diseases · Dataset · Machine Learning

1 Introduction

Decision support system (DSS) is a computer-based model which assists the organizations to make decisions. The medical DSS is a type of DSS which involves intelligent software models implemented to enhance the disease identification model and assists the doctors to take decisions [1]. The intelligent DSS utilize artificial intelligence models for medical professionals to select the optimal model to diagnose and treatment particular when the details related to the treatment is not complete. These DSS models operate in active as well as passive modes. Under passive scenarios, the DSS model is utilized in case of any requirement. Under active scenarios, suggestions are also provided. The medical DSS can undergo partitioning into two types namely rule based and data driven models [2]. In the former model, the model is designed based on the knowledge base constructed by a collection of if then rules. The basic function of this model lies in the identification of related rules based on the existing details, executes

© Springer Nature Switzerland AG 2020
A. P. Pandian et al. (Eds.): ICCBI 2019, LNDECT 49, pp. 941–947, 2020.
https://doi.org/10.1007/978-3-030-43192-1_103

and continues to find a rule till proper outcome is attained. They hold few effective characteristics along with few limitations. For instance, the outcome of the model gets degraded and hard to maintain under the presence of numerous rules. The latter model operates in massive data and supports the DSS by the use of data mining tools.

Diverse works are available based on data driven models. They have more flexibility over the previous method and have the capability of learning by itself. Under the earlier studies, ALARM network model is utilized to generate the synthetic data on identical dataset. During the investigation of the results, it is observed that the former model is much better than the Bayesian network by 25% under several aspects. In addition, once these two systems are integrated and employed, the success rate is enhanced upto 80%; considerably higher than the rate attained by both methods in an individual way. Medical DSS (MDSS) is an interactive computer-based model which is developed to assist doctors to make their design and also to compute as well as resolve issues by the use of various technologies [3]. It offers data storing and retrieving process; however, it enhances the classical information access and retrieval operations to build a system. It supports modeling, framing, and solving problems. The conventional application domains of DSS are medicine finance, security and so on [4].

In present days, machine learning algorithms finds useful to predict the data by analyzing the past behavior. This paper made an attempt to analyze the performance of three ML models namely radial basis function (RBF), logistic regression (LR) and Naïve Bayes (NB) models. The results are validated using a benchmark medical dataset and the outcome pointed out that the NB classifier model is a better option for DSS.

The leftover portions of this study are kept as follows. Section 2 elaborates the different ML models. Section 3 validates the performance of the study and Sect. 4 concludes the study.

2   Literature Survey

DSS is generally employed for strategically and tactically decide the situation using upper-level management-decisions with a sensibly minimum frequency and elevated probable penalty, where the time consumed to think and model the problem pays off liberally in the long run. In general, the DSS possess the below mention characteristics: (i) It helps to make decisions to achieve operational processes. (ii) It supports every stage of the DSS procedure. It supports every level of management from top to bottom. It offers partially or totally configured decision environment. It is interactive and user friendly. IT utilizes the data and model as a foundation [5]. The DSS and related operation models undergo classification into a set of four types namely explanation module, knowledge base, inference mechanism, and active memory. The third type involves the basic structure of DSS where the results are produced through the assumption of the present details or the details previously provided to the system by the end user. The produced outcome could be a decision or useful to make decisions. The second type is the knowledge base comprises the expert's knowledge utilized under the presence of interference in DSS. The fourth type contains the details provided by the user or current inference procedure. In addition, the first module is not mandatory and it is useful to generate an accuracy determination and description of the outcome produced by the inference method and knowledge base [6].

Reference [7] developed a ML model which utilizes the attributes of data attained from uterine contraction (UC) and fetal heart rate (FHR) signals for the classification as pathological or normal. [8] presented an ensemble learning based feature selection and classification approach for improving the classifier results. The presented model operates in two stages such as (i) selection of feature set which are possibly the support vectors by employing ensemble based feature selection approaches; and (ii) to develop an SVM ensemble utilizing chosen features. The presented model undergo evaluation using the Cardiotocography dataset. [9] performs a study containing three main stages namely Clustering, Outlier identification and Classification. The experimentation takes place using the Cardiotocography dataset. [10] introduced a model where the features were applied as the input to feedforward neural network (ANN) and Extreme Learning Machine (ELM) for classifying data in the FHR patterns of Cardiotocography dataset. The results stated that the ELM shows better results with the accuracy of 93.42%.

3 Proposed Work

The overall process is shown in Fig. 1. Preprocessing step has a significant impact in the medical data classification process. Prior to the execution of the data categorization, the unwanted data, noise, missing values and the presence of many and unrelated instances are removed from the dataset. Once the data is preprocessed, the classification process takes place using a set of three ML models which are discussed below.

Fig. 1. Overall process

3.1 RBF Classifier

RBF is a kind of NN [11] which generally refers to the Multilayer Perceptron (MLP). Every neuron present in the MLP gets a total sum of the input values. It implies that every input value undergoes multiplication using a coefficient, and the outcome will be totaled. An individual MLP neuron is a straightforward linear classifier, however, complicated non-linear classifier models could be developed through the integration of the neurons in the network. Here, the RBF model is highly effective over the MLP. It carry out the classification procedure through the determination of the input resemblance to the instances comes from the training dataset. Every RBF neuron holds a prototype by keeping a single one from the training set. Once there is a need of classifying a latest input, every neuron determines the Euclidean distance among the input and its prototype. Informally, speaking, when the input is highly closer to the class A prototype over the class B prototype, it will be classified as the class A. The RBF model has a layer of RBF neurons, output layer with one node every class label.

3.2 The Input Vector

It is an n-dimensional vector which is needed to be classified. The whole input vector is provided to every RBF neurons.

3.3 The RBF Neurons

Every individual RBF neuron saves a "prototype" vector which is simply any one vector from training set. Every RBF neuron undergoes a comparison of the input vector to its prototype, and provides a value among 0 and 1 representing a similarity measure. When the provided input is identical to the prototype, then the outcome from the RBF neuron would be one. Once the distance from the input to the prototype gets increased, the response gets exponentially decreased towards zero. The response values of the neurons are known as activation values. The prototype vector is known as neuron's "center" as it indicates the centre value of the bell shaped curve.

3.4 The Output Nodes

The outcome of the network comprises a collection of nodes and one node for every class that needs to be classified. Every output node determines a score for the related category. Classically, a classifier decision takes place through the assignment of the input to the class with maximum score.

3.5 RBF Neuron Activation Function

Every RBF neuron determines a similarity metric among the input and the corresponding prototype vector (taken from the training set). The input vector that is identical to the prototype provides an outcome closer to 1.

3.6 LR Classifier

LR is also a ML model [12] derived from the concepts of statistics. It is commonly employed for binary classification problem. It is named as LR due to the fact that the central component of this model consists of logistic function. The logistic function known as sigmoid function is developed for describing the characteristics of the population development in ecology, increasing rapidly and increasing the hauling ability of the surroundings. It is an S-shaped curve which could receive any real numbers and maps it to a value in the interval of 0 to 1, however, it not at all precisely at the limit. It can be represented as $1/\left(1 + e^{-value}\right)$, where e indicates the base of the natural logarithm. LR makes use of an equation to represent the data more comparable to linear regression.

The input values (x) are integrated in a linear way utilizing weights or coefficient values for the prediction of an outcome (y). The major variation from linear regression is that the outcome undergoes modeling as binary values (0 or 1) instead of numerals.

The outcome of the LR can be represented as follows. Below is an example logistic regression equation:

$$y = \frac{e^{(b_o + b_1 * x)}}{1 + e^{(b_o + b_1 * x)}} \tag{1}$$

Where b_o and b_1 represents the bias and coefficient for the single input value (x). Every column present in the input data has an integrated b coefficient which should be learned from the training data. The definite representation of the model which stores in the storage space or in a file indicates the coefficients (b) in the equation.

NB Classifier
It is inspired from the concepts of Bayes' theorem under independent considerations among the predictors [13]. A NB model is simple to implement with no complex iterative variable determination to make it specifically helpful for massive dataset. Regardless of the ease, the NB model is commonly employed in various forms. The Bayes theorem offers a method to calculate the posterior possibility, $P(c|x)$, from $P(c)$, $P(x)$, and $P(x|c)$. The NB classifier model has an assumption that the impact of the value of a predictor (x) on a specified class (c) is autonomous of the values of additional predictors. This consideration is termed as class conditional independence.

4   Performance Validation

To validate the experimental outcome of the applied three classifier models, a benchmark dataset Cardiotocography Data Set is applied [14]. The dataset comprises a total of 2126 instances and has a set of 23 features as shown in Table 1. In addition, a set of three labels exist in the applied dataset.

Table 1. Dataset details

Parameter	Values
Number of instances	2126
Number of features	23
Number of class	3

The experimental results states that the LR shows maximum classification over the other methods by attaining the highest precision value of 98.60. In line with, the RBF model tries to performs well than the NB by attaining slightly lower precision value of 98.00. Next, a comprehensive results analysis takes place among three diverse classifier models on the applied dataset interms of various validation measures. Table 2 demonstrates the results analysis of different ML models under several measures. Unfortunately, it fails to outperform the LR. However, the NB exhibits its inefficiency over classification results by attaining lowest precision value of 97.90. With respect to recall, it is shown that the LR shows maximum classification over the other methods by attaining the highest recall value of 98.60. At the same time, the RBF and NB models shows manageable results by attaining identical recall value of 97.90. However, these models exhibit its inefficiency over the LR interms of recall. In terms of F-score, the LR shows maximum classification over the other methods by attaining the highest F-measure value of 98.60. In line with, the RBF model tries to perform well than the NB by attaining slightly lower F-measure value of 98.00. But, it fails to outperform the LR. In the same way, the NB exhibits its inefficiency over classification results by attaining lowest F-measure value of 97.90.

Table 2. Classification results

Methods	Precision	Recall	F-Measure	ROC	Accuracy	Kappa
Logistic	98.60	98.60	98.60	99.10	98.63	96.29
RBFNetwork	98.00	97.90	98.00	99.20	97.93	94.47
NB	97.90	97.90	97.90	98.50	97.93	94.34

With respect to ROC, a maximum classification results is attained by LR by attaining the highest ROC value of 99.10. However, the RBF model attains slightly higher lower ROC value of 99.20. In line with, the NB shows its ineffective classifier outcome by attaining lowest ROC value of 98.50. With respect to accuracy, it is depicted that the LR shows maximum classification over the other methods by attaining the highest accuracy value of 98.63. At the same time, the RBF and NB models shows manageable results by attaining identical accuracy value of 97.93. However, these models exhibit its inefficiency over the LR interms of accuracy. In terms of kappa, the LR shows maximum classification over the other methods by attaining the kappa value of 96.29. In line with, the RBF model tries to perform well than the NB by attaining slightly lower F-measure value of 94.47. But, it fails to outperform the LR. In the same way, the NB exhibits its inefficiency over classification results by attaining lowest F-measure value of 94.34. These detailed simulation outcomes stressed out that the applied LR model shows its effective classification outcome by obtaining maximum precision value of 98.60, recall of 98.60, F-measure of 98.60, ROC of 99.10, accuracy of 98.63 and kappa value of 96.29.

5   Conclusion

In present days, machine learning algorithms finds useful to predict the data by analyzing the past behaviour. This paper made an investigation on the performance of three MLL models namely RBF, LR and NB models. To validate the experimental outcome of the applied three classifier models, a benchmark dataset Cardiotocography Data Set is applied. These detailed simulation outcomes stressed out that the applied LR model shows its effective classification outcome by obtaining maximum precision value of 98.60, recall of 98.60, F-measure of 98.60, ROC of 99.10, accuracy of 98.63 and kappa value of 96.29. As a part of future scope, the LR model performance can be further improvised by tuning the parameters via optimization algorithms.

References

1. Beam, A.L., Kohane, I.S.: Big data and machine learning in health care. JAMA **319**(13), 1317–1318 (2018)
2. Manogaran, G., Lopez, D.: A survey of big data architectures and machine learning algorithms in healthcare. Int. J. Biomed. Eng. Technol. **25**(2–4), 182–211 (2017)
3. Chen, M., Hao, Y., Hwang, K., Wang, L., Wang, L.: Disease prediction by machine learning over big data from healthcare communities. IEEE Access **5**, 8869–8879 (2017)
4. Char, D.S., Shah, N.H., Magnus, D.: Implementing machine learning in health care—addressing ethical challenges. N. Engl. J. Med. **378**(11), 981 (2018)
5. Abdelaziz, A., Elhoseny, M., Salama, A.S., Riad, A.M.: A machine learning model for improving healthcare services on cloud computing environment. Measurement **119**, 117–128 (2018)
6. Obermeyer, Z., Emanuel, E.J.: Predicting the future—big data, machine learning, and clinical medicine. N. Engl. J. Med. **375**(13), 1216 (2016)
7. Sahin, H., Subasi, A.: Classification of the cardiotocogram data for anticipation of fetal risks using machine learning techniques. Appl. Soft Comput. **33**, 231–238 (2015)
8. Silwattananusarn, T., Kanarkard, W., Tuamsuk, K.: Enhanced classification accuracy for cardiotocogram data with ensemble feature selection and classifier ensemble. J. Comput. Commun. **4**(04), 20 (2016)
9. Jacob, S.G., Ramani, R.G.: Evolving efficient classification rules from cardiotocography data through data mining methods and techniques. Eur. J. Sci. Res. **78**(3), 468–480 (2012)
10. Cömert, Z., Kocamaz, A.F., Güngör, S.: Cardiotocography signals with artificial neural network and extreme learning machine. In: 2016 24th Signal Processing and Communication Application Conference (SIU), pp. 1493–1496. IEEE, May 2016
11. Park, J., Sandberg, I.W.: Universal approximation using radial-basis-function networks. Neural Comput. **3**(2), 246–257 (1991)
12. Subasi, A., Ercelebi, E.: Classification of EEG signals using neural network and logistic regression. Comput. Methods Programs Biomed. **78**(2), 87–99 (2005)
13. Rish, I.: An empirical study of the naive Bayes classifier. In: IJCAI 2001 Workshop on Empirical Methods in Artificial Intelligence, vol. 3, no. 22, pp. 41–46, August 2001
14. https://archive.ics.uci.edu/ml/datasets/Cardiotocography

Implementation of IOT in Multiple Functions Robotic Arm: A Survey

S. Gowri$^{(\boxtimes)}$, Senduru Srinivasulu, U. Joy Blessy,
and K. Mariya Christeena Vinitha

School of Computing, Sathyabama Institute of Science and Technology,
Chennai, India
gowriamritha2003@gmail.com

Abstract. The ongoing revolution of web along side the growing artificial intelligence in several activities of daily life. The spotlights on style, execution and the executives of a 5 Degree of Freedom (DoF) mechanical arm abuse web of Things (IOT). The management of robotic arm is achieved by a Arduino and IOT. The most duty of Arduino is to come up with pulse dimension modulation (PWM) signals that square measure applied to servo motors for achieving the required rotation. Every servo features a totally different specification. Therefore, a PWM pulse might have a special result on servos. Most of the time, it's crucial to use the precise PWM pulses for achieving the required rotation. the most advantage of dominant the servo motors with PWM signals is that they will be programmed to own associate degree initial position and to rotate with a definite degree with reference to the necessities. During this project, six servo motors square measure used to understand the robotic arm. Four servos square measure utilised to regulate the body motion together with base, shoulder and elbow and 2 smaller servos square measure used for the motion of finish effectors.

Index Term: Automation · Image processing · Robotic arm · Arduino · Servo motor

1 Introduction

For many of us it is a machine that imitates an individual's just like the androids in Star Wars, killer and Star Trek: following Generation. However pr these robots capture our imagination such robots still alone inhabit fantasy. Of us still haven't been able to give a golem enough 'common sense' to reliably act with a dynamic world. Well it is a system that contains sensors, management systems, manipulators, power provides and code packages all in operation on to perform a task. Designing, building, programming and testing a golem could also be a mixture of physics, study, engineering, structural engineering, arithmetic and computing. A study of computer science implies that students square measure actively engaged with all of these disciplines throughout a deeply drawback sitting problem-solving atmosphere.

Brilliant partner robots have numerous applications at interims the clinics, to aid medical procedures that need high truth and exactness. An automated arm with 3 degrees of opportunity is implied here, to pursue some predefined bearings with high truth and exactness. Another automated arm is created to help physically tested of us, matured ones in moving degree object from one spot to a special. In this paper, we've a bent to develop a wise robotic assistant that operates on human voice commands and gesture commands. Its effectiveness and limitations area unit typically checked through fully totally different experiments. A lot of study is planned for the impact of noise in background whereas giving voice commands and of the area between the smart IOT device and mouth for effective operation of robotic assistant.

2 Literature Survey

A literature survey during a project report is that the section that shows the varied analyses and analysis created within the space of our interest and also the results already revealed, taking under consideration of assorted parameters within the project and also the extent of the project.

A. The Multiple-perform Intelligent Robotic Arms
The multiple-functional mechanical palms had been structured and made meets the six selected show works, that thus features their strength, precision, and high level of execution in keeping up with genuine activity. It conjointly exhibits its utility for diversion, instructional, and modern needs; at long last, the get together worth of this mechanical arm is low and along these lines the malleability of utilization is high. The characteristics of the automated arms include:

- The shoulder of automated arm incorporates an endeavor of engine structures to flavor up the capacity to lifestyles weight
- The relentlessness and precision of the mechanical arm area unit enhanced for the prerequisite of high in general execution all through the entire basic vogue.
- So as to protract the moving capacity of the system, the automated arm was planned with a four wheeled transmission structure and track.
- Mimetic mechanical hands add live execution to the flamboyant shows.
- Sorts of PC hands had been intended to fulfill the necessities of the six designated capacities.

B. The Gesture Replicating Robotic Arm
The Gesture Replicating Robotic arm likely could be a servo-managed automated arm that repeats signals for the span of a three dimensional environment. It utilizes cameras that sight the movement of one's turn in three measurements. The cameras provide outlines as they enter the bundle bargain that performs segmentation algorithms such as subtraction of background, color detection, and contour detection.

Constituent to perspective mapping presents the ideal commands to the many servo motors. Thus, replication of the human hand movements is completed. These real-looking robotic fingers nearby unit commonly teleoperated to protect human beings in dangerous environments, such as assembly lines, space, underwater and nuclear radiation. Multiple-functional robotic fingers were designed and created meets the six appointed show functions, that in flip highlights their stability, accuracy, and excessive stage of overall performance.

C. Hand-Gestion-Based Car Robot Control Interface

To measure the hand trajectories of a consumer, a 3-axis measuring device is adopted. The information on the physical phenomenon are transmitted wirelessly to a laptop via the Associate RF unit in Nursing. Then the obtained trajectories system listed for controlling a car-robot as one of six management commands. The classifier adopts the principle for classifying hand trajectories with the dynamic time warp (DTW).

The somato-sensory interaction is one in each of the foremost straightforward interactive interfaces for dominant objects. Recently, there have been many different hand gesture recognition systems, such as mostly vision-based physical phenomenon recognition systems and physical phenomenon recognition systems based on inertia. Given the use of cameras or accelerators in hand gesture systems, the central element may well be a law for hand gesture recognition. The Dynamic Time Warp (DTW) rule and the Hidden scientist model (HMM) unit a pair of hottest algorithms accustomed acknowledge hand gestures.

D. Robotic Arm Speech Recognition Writing Skills

Robotic arms unit of measurement programmed golem manipulator with somebody's arm similar functions. Some pretty technology protheses measuring device out there to perform the basic functions of the human arm. The project's goal is to develop a robotic arm that helps write to the physically human.

The robotic arm should be fitted to the unfortunate hand of the patient and may write down the words pronounced by the patient to the electro-acoustic transducer. This robotic arm's special feature is that it is equipped with a pen that performs writing operations. Our robotic arm's main programming and handling consists of two parts. The first 0.5 is the receipt of the speech sign. Associate in Nursing dynamic it into a text sort for a lot of method and last half consists of exploitation this text information to urge an applicable required mechanical action of the motors.

E. People with Severe Disabilities - A Mobile Robotic Arm

A handheld robotic arm system is designed for people with severe disabilities. This technique is made up of a robotic arm, microcontroller, and processor. The foremost body of the robotic arm are contained terribly very case to carry a laptop personal computer. Its weight is 5 weight unit, similarly as a pair of 12-V lead acid reversible batteries. This robotic arm are put together mounted on a chair. Some tasks, such as drinking tea task associated with feeding activities, were performed by Associate in Nursing experiment performed by Associate in Nursing capable subject to check the performance of the mobile robotic arm device.

This system is made up of a central robotic arm body, microcontroller and controller. The microcontroller, AT91SAM7S256, includes a 32-bit ARM7TDMI pc design processor (Atmel Corporation) that's low-power, small, efficient, and superb amount interrupt response. It's built into the system. As seen in the robotic arm's main body, it is fully enclosed to accommodate a personal laptop computer. You can even put the robotic arm on a chair. Mobility is one of the basic ideas for a robotic arm device.

This system shows a critical advance forward towards expanding the freedom of individuals with serious engine inabilities, by utilizing cerebrum PC interfaces to saddle the intensity of the Internet of Things. We break down the soundness of mind motions as end-clients with engine incapacities advance from performing basic standard on-screen preparing errands to collaborating with genuine gadgets in reality. Moreover, we show how the idea of shared control-which translates the client's directions in setting enables clients to perform rather complex undertakings without a high remaining task at hand. We present the after effects of nine end-clients with engine inabilities who had the capacity to finish route undertakings with a telepresence robot effectively in a remote situation (at times in an alternate nation) that they had never recently visited. In addition, these end-clients accomplished comparable dimensions of execution to a control gathering of 10 solid clients who were at that point acquainted with the earth.

For the mechanical arm system, we have built up controllers, e.g. head-controlled interface, eye-controlled interface, vision-based interfaces, and mostly controller-based electromyogram. Such controls are important unit of estimation and material to monitor the automatic portable arm. A PC associated with the microcontroller through an RS-232C connection was used because of the controller as an important advance of this test. What's extra, a C-language course program was created with the gcc compiler and then the program was written into the chip memory.

3 Conclusion

We have created a robotic arm in this paper, which operates on human commands provided through the IOS platform's Mobile App. The voice commands square measure born-again through text formats and it's controlled by Arduino based mostly microcontroller. We tend to conjointly use DC servo motor for the movement of Arm. when learning all the papers I even have planned my work, to create five degree of Freedom artificial intelligence Arm and to manage the Robotic Arm with the assistance of net of Things (Fig. 1).

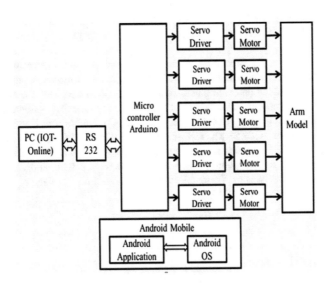

Fig. 1. Architecture of multiple function robotic arm

References

1. Hao, W., Hun, L.C.: 6-DOF PC-based robotic arm (PC-ROBOARM) with efficient trajectory planning and speed control. In: 2011 4th International Conference on Mechatronics (ICOM), pp. 1–7. IEEE (2011)
2. Uehara, H., Higa, H., Soken, T.: A mobile robotic arm for people with severe disabilities. In: 2010 3rd IEEE RAS & EMBS International Conference on Biomedical Robotics and Biomechatronics, pp. 126–129. IEEE (2010)
3. Orenstein, D.: People with paralysis control robotic arms using brain (2014). https://news. brown.edu/articles/2012/05/braingate2. Accessed 23 Oct 2014
4. Vimali, J.S.: Integration of multi access control system with door, device & location automation & remote network control. In: International Conference for Phoenixes on Emerging Current Trends in Engineering and Management (PECTEAM 2018). Atlantis Press (February 2018)
5. Karamchandani, S., Sinari, S., Aurora, A., Ruparel, D.: The gesture replicating robotic arm. In: 2013 International Symposium on Computational and Business Intelligence, pp. 15–19. IEEE (2013)
6. Campeau-Lecours, A., Lamontagne, H., Latour, S., Fauteux, P., Maheu, V., Boucher, F., Deguire, C., L'Ecuyer, L.-J.C.: Kinova modular robot arms for service robotics applications. In: Rapid Automation: Concepts, Methodologies, Tools, and Applications, pp. 693–719. IGI Global (2019)
7. Quintero, C.P., Ramirez, O., Jägersand, M.: Vibi: assistive vision-based interface for robot manipulation. In: 2015 IEEE International Conference on Robotics and Automation (ICRA), pp. 4458–4463. IEEE (2015)
8. Tsui, K., Yanco, H., Kontak, D., Beliveau, L.: Development and evaluation of a flexible interface for a wheelchair mounted robotic arm. In: Proceedings of the 3rd ACM/IEEE International Conference on Human Robot Interaction, pp. 105–112. ACM (2008)
9. Vimali, J.S., Srinivasulu, S., Gowri, S.: IoT based bank security system. Int. J. Recent Technol. Eng. **8**, 2324–2327 (2019)
10. Pathirage, I., Khokar, K., Klay, E., Alqasemi, R., Dubey, R.: A vision based P300 brain computer interface for grasping using a wheelchair-mounted robotic arm. In: 2013 IEEE/ASME International Conference on Advanced Intelligent Mechatronics, pp. 188–193. IEEE (2013)
11. Vimali, J.S., Gupta, S., Srivastava, P.: A novel approach for mining temporal pattern database using greedy algorithm. In: 2017 International Conference on Innovations in Information, Embedded and Communication Systems (ICIIECS). IEEE (2017)
12. Sindhu, K., Subhashini, R., Gowri, S., Vimali, J.S.: A women safety portable hidden camera detector and jammer. In: 2018 3rd International Conference on Communication and Electronics Systems (ICCES), pp. 1187–1189. IEEE (October 2018)
13. Gowri, S., Jabez, J.: Novel methodology of data management in ad hoc network formulated using nanosensors for detection of industrial pollutants. In: International Conference on Computational Intelligence, Communications, and Business Analytics, pp. 206–216. Springer, Singapore (March 2017)

Real Time Traffic Signal and Speed Violation Control System of Vehicles Using IOT

S. Gowri[✉], J. S. Vimali, D. U. Karthik, and G. A. John Jeffrey

School of Computing, Sathyabama Institute of Science and Technology,
Chennai, India
gowriamritha2003@gmail.com

Abstract. Real Time Traffic Signal and Speed Violation Control System of Vehicles Using IOT techniques have been widely applied to control traffic and help the citizens to follow the rules precisely depending on various sensors. The IOT has gained popularity and enormous advance in real time applications used by people for their daily use. In this proposed system we detect the fast moving vehicles that detect the Red signal using RFID sensor. This module also be set up on Raspberry pi component which used for assisting in solving the actual difficulties such as traffic rules and accident. This system consists of an onsite module IOT module that monitors and sends signal to the database. This project primarily for the vehicle which cross the traffic signal during STOP sign (Red signal) and makes the vehicle to slow down in ALERT sign (yellow light) by giving the vehicle average velocity like a speed limit. So that the other lane vehicles can move or go according to the traffic rule. The system will be helpful for the people to follow traffic rules and for higher officials to take necessary actions. The RFID are set to the vehicles and RF-tags are set in the roads to detect the non-obeying vehicles. Categories can be obtained through reasonable analysis of the data. The government officials and the users can check their behaviour updated in an android app for further penalties.

Index Term: Raspberry · Mobile practical application · IoT · Database

1 Introduction

Active surveillance studies have drawn much attention to monitoring and analyzing traffic due to the rapid growth of technology. There are many scientific studies in various areas of the world. Traffic congestion, violation of rules and traffic accidents has great difficulty in controlling traffic in most areas. Implementing special traffic rules or building a new road infrastructure can reduce all problems. Although there are some limitations to this approach and ideas to offset the demand for transport and land costs. Alternative technologies can be used to enhance safety and traffic skillfulness with minimal requirements. Intelligent solutions are very wide because the cost of sensors is constantly decreasing and there are more opportunities to provide traffic info such as traffic signals. Intelligent Transport Systems (ITSs) have welfare improvements from the development of sensory, communicating and computing. Typically, some inductive control slots include high maintenance costs, complex assembly processes, damage to

© Springer Nature Switzerland AG 2020
A. P. Pandian et al. (Eds.): ICCBI 2019, LNDECT 49, pp. 953–958, 2020.
https://doi.org/10.1007/978-3-030-43192-1_105

the road, and natural causes. On the contrary, the sensors have advantages over other devices, because they are cheap, resistant to weathering, highly efficient and easy to install and maintain. The Raspberry Pi uses a RF-tag as a multiband signal, which sets the ON/OFF state according to the function of the microcontroller during the corresponding period. RF tags with precision for detecting vehicles using RFID in vehicles present during this flight detection. Some of the researchers have designed an automatic wireless vehicle identification system with an accelerometer that detects vehicle shafts and magnetic sensors that assess the arrival, departure, and speed of the vehicle. Much work has been done to achieve efficient and accurate collection of traffic information through RFID sensors. The main task is to detect vehicles passing through RFID and labels. This document focuses on RFID sensor research for vehicle detection and identification. We introduce the RF-tag transfer technique, which is static on the road that includes when making decisions about traffic signs on the track. In addition, a number of distinguishing features are derived from linear sequence algorithms.

2 Related Work

Today, information plays a vital role for students in higher education and others scientists to analyse the work and support infrastructure in order to increase the efficiency, accessibility, reliability and accuracy of academic based task. Maximising of student volunteers in public awareness makes the project successful to the advancement and enrichment in cultural, social and economic development by enduring in the related work. The study can be divided into two parts: speed reduction and detection of vehicles.

A. Detection of Vehicles
The vehicle detection is the base technique for the system, providing reliable traffic signals. This look into supported on AMR sensing element, which reference point the magnetic field of X, Y, and Z axes. SPLIT SIGNAL TO FRAMES uses AMR sensor signaling in a series of data that records magnetic intensity level at intervals around the sensor. Numerous vehicle sensing algorithms are founded on number of smoothness signal algorithms, they do not differ significantly from raw signals, but from noise.

B. The Variance Based Detection Algorithm
This algorithm is used in which the two-window detection algorithm is considered valid under the Z-axis measured on the magnetic signals in normal driving conditions. Consistency dispersion, to find the view of arrival and departure of medium based on the difference of medium change signal, background.

C. Identification of Vechile
Vehicle type identification technique. Several quantifiablee attributes have projected over vehicle signal characteristics.

D. Feature Extraction
Feature extraction plays an essential function in determination of medium, Has a powerful reciprocity among types. Wave form from the medium signal is various since influence of the load, velocity. Property Statistics – the characteristics of the

waveform structure reflected vehicle signal distribution. According to the technique of violating the magnetic principle. A short feature Inspired by the previous framing process, the vehicle signal is divided into ten frames to get short-term characteristics to ensure that different vehicles have the same dimensional characteristics. RFID (Radio Frequency Identification) based collisions, where vehicle detection uses collision sensors to detect collisions between two vehicles. After a collision is detected, RFID sensors, they take vehicle details such as vehicle numbers, engine numbers and more. This system facilitates the process and the vehicle owners and other officials to track the driver who was in a hit-and-run case. They must be stored in the microcontroller of both vehicles. This is the main idea of this project. Radio Frequency Identification techniques have been used to enable information that exchanges between two vehicles. Each vehicle has its own information about RFID tags placed somewhere on the deck. Each vehicle has its own RFID reader, starts to work while an accident takes place.

3 System Analysis

In the existing systems that are available, the vehicle detection technique does already exist as it is for the survey purpose of accidents. Also, the traffic control system where automatic change in signals is in use till today. The reduction in speed for the vehicles is manually present as there are laws and physical methods for it. The manual storage of the vehicles those violates the rules and regulations.

In this proposed system, the technique that will be implemented using IOT. This overcomes the manual transmissions of data. The RFID sensors are used to detect the vehicles that violate the traffic control system. The violated vehicle's RF-tag (includes registered number and RC number) is scanned and sent to the database of near police department. A zone is formed for particular metres with speed limits including the traffic controls so that the optimal speed is maintained to avoid accidents. This zone is useful because if violation occurs, the sensor can easily detect the vehicle as well accidents can be controlled.

The Fig. 1 shows the diagrammatic representation of the system. This system consists of the RFID sensor that scans the RF-tag that is fixed on the vehicle includes the registered number and RC number of vehicle. Once the vehicle violates the traffic rule, RFID scans the RF-tag sends it and stores the details in the database which can be utilised in the future also.

4 Result and Discussion

To test the effectiveness of the proposed algorithm, we use sensors at the output of the model signal lane. The vehicle speed has been normal that is to the specific speed limit mentioned for the area zone in the yellow light within the given distance from the point of signal light and in case the vehicle prohibits the junction without stopping, the vehicle's number will be updated to the database and to the server. The vehicle speed

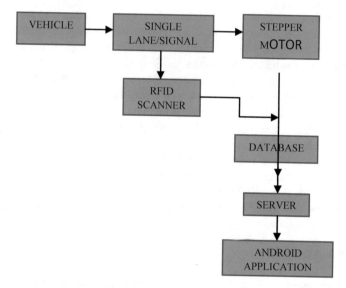

Fig. 1. Proposed architecture

reduction and RFID will be controlled accurately by the microprocessor Raspberry pi used over the area of the signal. We set up a Raspberry pi for a junction for making its performance better in detection and identification of vehicles over the area. The application in mobile and website has been helped to know the circumstances and detect the vehicles for the future references for penalty and punishment behaviour. The application will be used for user friendly experiences for the officials to spot the errors and make them to rectify the mistakes done.

The above diagram describes the technical coding part which runs inside the system that will be carrying the work.

The above diagram describes the technical coding part which runs inside the system that will be carrying the work.

This diagram describes that the storage part of the system where all the violated vehicle's details are stored.

5   Conclusion and Future Work

IOT has gained growth in the real time application models and been used frequently in all the places. IOT is one of the main objectives for upcoming technology it acts like a backbone for the other upcoming technologies such as Image processing, machine learning, Artificial Intelligence, etc. Here we have proposed a model which can detect the non-obeying vehicles and makes it control its speed before destruction by reducing it to average speed over a distance. Various other parameters such as vehicle detection is considered to be better and accurate over this model than implementing some other ideas would not sound good in finding the results. This scheme also be improve in forthcoming by employing PWM (Pulse Width Modulation) and Image Processing using OpenCV. The system can be also done in a more advanced way using machine learning and OpenCV.

References

1. Hasalkar, P.D., Chougule, R.S., Madake, V.B., Magdum, V.S.: Speed governor system (2015)
2. Bhatt, R., Fernandes, N., Dhage, A.: Speed detection using magnetic sensors (2013)
3. Nayana, P., Kubakaddi, S.: Traffic control recognition using IOT (2014)
4. García, C.R., Quesada-Arencibia, A., Cristóbal, T., Padrón, G., Alayón, F.: Systematic development of intelligent systems for public road transport. Sensors **16**(7), 1104 (2016)
5. Markevicius, V., Navikas, D., Zilys, M., Andriukaitis, D., Valinevicius, A., Cepenas, M.: Dynamic vehicle detection via the use of magnetic field sensors. Sensors **16**(1), 78 (2016)
6. Keawkamnerd, S., Chinrungrueng, J., Jaruchart, C.: 'Vehicle classification with low computation magnetic sensor. In: Proceedings of 8th International Conference on ITS Telecommunications, pp. 164–169 (October 2008)
7. Kaur, R., Talwar, M.: Automated vehicle detection and classification with probabilistic neural network, vol. 2. IEEE (2016)
8. Wei, Q., Yang, B.: Adaptable vehicle detection and speed estimation for changeable urban traffic. IEEE **17**, 2021–2028 (2017)

Enhanced Financial Module in Intelligent ERP Using Cryptocurrency

Komal Saxena$^{(\boxtimes)}$ and Arshdeep Singh Pahwa

Amity Institute of Information Technology, Amity University,
Noida, Uttar Pradesh, India
ksaxena1@amity.edu, Pahwasaab611@gmail.com

Abstract. Rapidly changing in the technology is enforced towards to convert our business into the smart, intelligent and automated. This paper is focus on to improve the financial module by using crypto currency and bit coin in intelligent ERP. We have emphasis on the evolution of block chain from the bit coin network. It also gives brief future about the crypto currencies With ERP. In 2017 the crypto currency showcase has achieved a record of $54 billion. It shows that the following result is a persistent development. Here we also discussed about the types of bitcoin and their use to maintain in the block chain or Ledger.

1 Introduction

1.1 Crypto-Currency

Cryptocurrency refers to the Digital money is a web based mode of trade which utilizes cryptographic capacities to direct budgetary exchanges. Digital forms of money influence block-chain innovation to pick up decentralization, straightforwardness, and unchanging nature. One of the very famous innovations in the fields that profit by these advancements and online associations is the money related and business segment [1]. Increasing number of online customers has endorsed virtual global ideas and made new trade phenomena [3]. The notion of 'virtual' refers to the fact that there is no physical (i.e. tangible) plus that underlies the existence of crypto currencies. Crypto currencies exist in digital and electronic formats in cyberspaces like the net (World Wide Web) [4]. Digital forms of money speak to profitable and immaterial items which can be utilized electronically or for all intents and purposes in various applications and systems, for example, online informal communities, online social diversions, virtual universes and shared systems [5] (Fig. 1).

A. P. Pandian et al. (Eds.): ICCBI 2019, LNDECT 49, pp. 959–967, 2020.
https://doi.org/10.1007/978-3-030-43192-1_106

Fig. 1. Types of bitcoins [2]

2   Literature Review

2.1   Digital Currency

Yermack (2013) is the main paper in which they directly focus on cryptocurrency is directly associated digital forms of money will be cash, and it is additionally one of the soonest research papers on cryptographic forms of money. It is including in the paper that the reasons digital currencies out the elements of cash to such an extent, that they could be viewed as money related instruments. 1 For a progressively definite theoretical examination, see for example ECB (2012) or Bech and Garratt (2017). 2 For a specialized portrayal of cryptographic forms of money, see for example Nakamoto (2008) or Badev and Chen (2014). BoF Economics Review 2 Some of the most punctual investigations on cryptographic forms of money were led by national banks. ECB (2012) gives an outline of virtual monetary forms, for example, those utilized inside gaming conditions, and considers digital forms of money one subclass among them. ECB (2015) is a subsequent report which gives a progressively point by point spellbinding investigation, and presumes that cryptographic forms of money don't establish cash or cash [9]. Badev and Chen (2014) is the most punctual distribution by the Federal Reserve Board on the theme. The paper abstains from characterizing Bitcoin, however gives an intensive examination on its specialized design and its exchange designs. Ali et al. (2014) is the primary distribution by the Bank of England on the subject and gives an applied diagram of cryptographic forms of money. The paper alludes to digital currencies as cash and as money, yet in addition calls attention to some key contrasts among cryptographic forms of money and genuine cash. Bitcoin has a well-working auxiliary market. Baek and Elbeck (2014) dissect the market cost of Bitcoin and reason that Bitcoin is best depicted as a theoretical venture vehicle [19]. Cheung et al. (2015) likewise break down the market cost of Bitcoin and infer that its value direction highlights bubbles. Cheah and Fry (2015) likewise study the value direction of Bitcoin. They reason that the cost of Bitcoin shows bubbles and that its key worth is zero. Balcilar et al. (2017) examines in the case of exchanging volumes can foresee showcase returns for Bitcoin. Their outcomes are blended. Urquhart (2016) tests whether the market for Bitcoin is productive and finds that it isn't. Nadarajah and Chu (2017) expand on crafted by Urquhart (2016) and discover some proof of market effectiveness. Katsiampa (2017) looks to locate the best econometric model for depicting the unpredictability of the cost of Bitcoin. Bouri et al. (2017) investigate the capacity of Bitcoin to work as a supporting instrument and discover some proof for this Glaser et al. (2014) have contemplated the Bitcoin client network and find solid proof that the interest for Bitcoin is essentially determined by its utilization as a theoretical speculation instrument as opposed to cash. Comparative discoveries have been accounted for by Baek and Elbeck (2014), Katsiampa (2017), and Dwyer (2015). Yelowitz and Wilson (2015) apply an alternate technique and discover proof that enthusiasm for Bitcoin is related with enthusiasm for PC programming and criminal behavior. Foley et al. (2018) explore Bitcoin exchanges and locate that an enormous part of exchanges and clients are identified with criminal behavior. Böhme et al. (2015) give an engaging investigation of the Bitcoin framework, its market, and its administration structure. The paper thinks about the stockpile of Bitcoin to the inventory of cash and alludes to Bitcoin as a money, yet doesn't clarify the decision of wording. Dwyer (2015) and Selgin (2015) additionally dissect different more extensive parts of the Bitcoin framework. Both accept that Bitcoin is a cash without giving further clarification to that supposition. Huberman et al. (2017) study the financial motivation structure of the Bitcoin system and find that blockage is vital for its persistent activity. They depict Bitcoin as an arrangement of records. Heilman and Rauchs (2017) give rich observational information on the business that has jumped up around cryptographic forms of money. It is an exceptionally elucidating examination which abstains from making presumptions and affirmations that aren't bolstered by proof. One of the numerous discoveries in the paper is that cryptographic forms of money are fundamentally utilized as a theoretical speculation instrument [17].

2.2   ERP

It is a current and burning topic said Zulfikar Rehaman working with Dell EMC Solutions who published his article on Computer world recently. Some other organisations are also talked about this important hot topic of Block chain associated with ERP Bas. De Vos associated with IFS Lab in Sweden. He is a director and suggested that Sweden that the company's marketable aviation customers could easily agree to accept that ERP- is a very important feature of supply-chain and maintenance apps saline with the blockchain model. It is quite interesting in real world and something that is belligerently excited about, alleged "Brigid McDermott", VP of Blockchain, Business Development at IBM [18]. Mark Russinovich, chief technology officer of Azure at Microsoft, said. Block chain is a transformational technology with the facility to expressively condense the resistance of doing trade [16].

It is an incorruptible digital ledger of economic transactions that can be programmed to record not just financial transactions but virtually everything of value. ‖Don & Alex Tapscott, authors Block chain Revolution (2016).

3 Types of Crypto-Currency

3.1 Bit Coin

Bit coin is an online virtual cash dependent on open key cryptography, star presented in 2008 of every a paper [1] wrote by somebody behind the Satoshi Nakamo-topseudonym. It turned out to be completely useful on January 2009 and its wide gathering, empowered by the openness of exchange markets allowing basic change with conventional fiscal gauges (EUR or USD), has brought it to be the best virtual money [6].

4 Maintain of Accounts Using Block Chain in ERP

4.1 Block Chain

Block chain is a gathering of shared and synchronized advanced information that is spread over different destinations and areas. It additionally does not have a center overseer and brought together server farm, which was one of the driving variables for making a non-represented digital money in any case [7]. With regards to cryptographic forms of money like Bit coin, this guarantees there is no twofold plunging with electronic funds; block chain nodes track everything simultaneously to keep everyone honest [8]. Generation organizes detectable quality and following, improves your business Block chain can improve (Fig. 2).

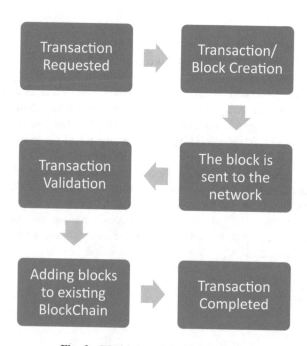

Fig. 2. Working model of block chain

5 Architecture of ERP Using Crypto Currency

ERP programming control and streamlines all the huge business forms. It is a sort of database center point, which enables an association to do back-office works easily and continuously with the assistance of incorporated applications [12]. As ERP structure utilizes database the executives framework, in like manner Block chain likewise utilizes constant versatile database that encourages verification of ideas, stages and applications [13]. Crypto currency is very useful for the business, If we start using cryptocurrency the Accounting will more easy for the suppliers, consumers and business partners. They can easily purchase and sale of goods in digital cash. We can preserve our ledgers and journal through block chain and also easily maintain the balance sheet. Cryptocurrency is very beneficial tool for the ERP software and to maintain a block chain. We can certainly enhance this unique module of block chain in ERP to main the financial records online (Fig. 3).

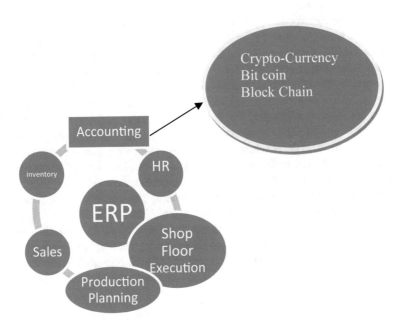

Fig. 3. Architecture of ERP using crypto currency

6 Relationship Between ERP and Block Chain

The Block-chain are worked on main four center standards [19]:

A. Public and Private Ledger: This annex just arrangement of record is accessible to all individuals from a business network.
B. Smart agreements: Business terms are incorporated with exchanges consequently.

C. Confidentiality: Transactions are just unmistakable to those with substantial motivations to get to them.

D. Evidence: Transactions are supported by every included member (Fig. 4).

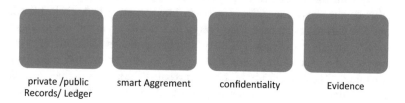

Fig. 4. Four center standards of cryptocurrency [19]

The block chain way to deal with ERP.

IBM's Oracle Enterprise applications group is attempting to incorporate our block chain arrangements with Oracle ERP cloud, to exploit the open system it gives. Our block chain for ERP arrangements are worked around three attributes [20]. A system of suppliers, consumers, traders, bankers, logistics providers and other business partners who are eager to take an interest all the while. Digitization of business forms associated with the exchange of merchandise, administrations and different resources. Shared ledger that can be accessed safely by all business partners in the system or a network. The issue with existing ERP arrangements is that they regularly perform exchanges that are wasteful, costly and defenseless. The exchanges are additionally exorbitant to keep up and review, and hard to follow. As of not long ago, this was an issue without an answer—that is, until blockchain made advances on the scene. Exchanges recorded by means of block chain are intended to be secure, trusted, unchanging and totally straightforward to all partners around the world. These qualities assist you with streamlining the costly, wasteful procedures at present found in your ERP condition with another methodology that can spare time for all gatherings included, lower costs all through your business arrange and essentially lessen hazard. What's more, block chain encourages you manufacture trust with your colleagues, which can streamline future exchanges [19].

7 Distributed Ledger

It is another form of block chain. It is starting to be executed as a response for this potential creation organizes mess. With a common modernized record between a maker and its shippers and suppliers, oversight and consistence is improved exponentially [10]. Following point by point information everything from materials to issues back to their source would be an electronic and extremely precise methodology that would light up a reliably developing show of issues. The best part is that it is considered incomprehensibly difficult to hack as a result of the possibility of its computerized cash roots. Everyone needs something fundamentally the same as from stock system the administrators programming; to increase customer regard and achieve a high ground [11]. Most of this is the reason improving the organization of your store arrange is essential, and why Block chain development is being placed assets into with such interest; improve your Records, the dream of bookkeeping, are as notable as composing and money. Figuring force and leaps forward in cryptography, related to the innovation and utilization of a couple of new and energizing calculations, have permitted the making of appropriated records. Dispensed record is a database held and state-of-the-art freely by every player (or hub) in a major network. Each and every hub on the network strategies every exchange, reaching its own decisions and after that casting a ballot on those ends to make certain a large portion of the general population think about the ends. Once there is this agreement, the conveyed record has been refreshed, and all hubs hold their very own indistinguishable duplicate of the record. This design takes into consideration a fresh out of the plastic new ability as an arrangement of record that is going past being a simple database. Administered Ledgers are a vigorous type of book keeping system which have properties to maintain the records of each vendor and daily expenses and income record of the organization. It is very powerful techniques to maintain the static or registers to keep the paper-based records (Fig. 5) [14,15].

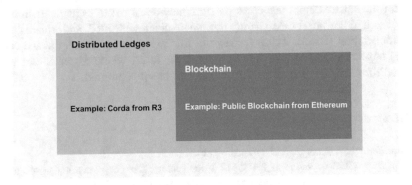

Fig. 5. Distributed ledger using block chain [20]

8 Future the Digital Money Brings the Transparency in the Business

Digital money or digital currency are now having the different sorts of cash. Now the moving applications are available which is to be used to transfer the money and make the money related transaction easy and simpler. Now we have lot of apps like Bheem, google money transfer app paytm, paypal and various other such applications and options formed which is trying to stop or a simple movement against the corruption in bank (Kriostoufek 2013). It must be remembered that the ideas driving the production of these advanced record is that some measure of cash is put away in the record of these application from the principle financial balance and the client can utilize a similar cash for buying or moving the cash to somebody. In this digital world, web based shopping has become a significant segment in the present situation. Hence any methods for making installment legitimately through the online applications is a progressively welcome component to be worried of (Garman et al. 2013) Bitcoin requires the total creation of another kind of cash. All types of day to day transaction is supported on online and a miners can maintain the balance sheet online [18].

9   Conclusion and Future

Based on the analysis thus far, it can easily be concluded that the development of block chain from the bit coin organize is analyzed with a point by point investigation of conventions for approval and address hashing in an appropriated system with the freely accessible data, accordingly making the data un-hack capable, which is the linchpin of straightforwardness and genuineness. Use of square chain in money related exchanges has dependably been the center, yet this paper investigates the wide- running utilizations of inventory network, as is apparent in different use cases. From an inventory network viewpoint, there is a great deal of cooperative energy between square chain with big business frameworks and is examined in detail alongside future development regions. Through this examination, it is clear that there are three regions of advancement for the square chain: use cases on store network for ERP- square chain coordination, the middleware for associating the square chain with ERP, and square chain as an administration (BaaS). These are regions that are growing awfully apace and are hole up huge chances. At long last, it is about the future where organizations will be qualified for exchange just when they are square chain-confirmed.

References

1. https://en.wikipedia.org/wiki/Cryptocurrency
2. https://www.123rf.com/photo_88035112_stock-vector-bitcoin-cryptocurrency-golden-coins-icons-vector-isolated-symbols-and-different-types-of-virtual-dig.html
3. https://www.investopedia.com/terms/c/cryptocurrency.asp
4. https://www.slideshare.net/freeschool/cryptocurrencies-geopolitical-perspectives
5. https://thenextweb.com/hardfork/2019/02/19/the-differences-between-cryptocurrencies-virtual-and-digital-currencies/
6. https://www.bitira.com/digital-ira/types-of-digital-currency
7. https://www.researchgate.net/publication/281773799_Research_and_Challenges_on_BitcoinAnonymity
8. https://medium.com/@cryptomangnani/on-the-methodology-of-studying-cryptocurrency-46ed58a9b59e
9. https://www.researchgate.net/publication/329136952_A_systematic_literature_review_of_blockchain-based_applications_Current_status_classification_and_open_issues
10. http://eclipsepc.com/blog/erp-cryptocurrency-blockchain-manufacturing-supply-chain/
11. https://mlsdev.com/blog/156-how-to-build-your-own-blockchain-architecture
12. https://www.sagesoftware.co.in/blogs/integrating-erp-with-blockchain-technology-for-efficient-collaboration/

13. https://www.coindesk.com/information/what-is-a-distributed-ledger
14. https://it.toolbox.com/blogs/erpdesk/when-erp-meets-blockchain-092017
15. https://www.guru99.com/what-is-sap-definition-of-sap-erp-software.html
16. Parikh, T.: The ERP of the future: blockchain of things. Int. J. Sci. Res. Sci. Eng. Technol. **4** (1), 1341–1348 (2018). Department of Computer Science, Shri Chimanbhai Patel Post Graduate Institute of Computer Applications, Gujarat Technological University, Ahmedabad, Gujarat, India. (ijsrset.com)
17. https://helda.helsinki.fi/bof/bitstream/handle/123456789/15564/BoFER_1_2018.pdf
18. https://myassignmenthelp.com/free-samples/literature-review-on-bitcoin
19. https://www.linkedin.com/pulse/remodel-erp-processes-blockchain-anshul-gupta
20. https://tradeix.com/distributed-ledger-technology/

Explorative Study of Artificial Intelligence in Digital Marketing

Mohd Zeeshan and Komal Saxena[⊠]

Amity Institute of Information Technology, Amity University,
Noida, Uttar Pradesh, India
Its.zee4@gmail.com, ksaxena1@amity.edu

Abstract. The massive evolution in the field of science and social media has modified the human communication. The function of Artificial Intelligence and digital marketing have been started to be used in various fields, especially in marketing because it is the trendiest work in the market creating major changes. The Objective of this study is to Explore and focus on the Automated Marketing which is going to change the digital firms and all the aspects of digital marketing and also intelligent content marketing, Uses Of AI in Digital marketing, Impacts on Digital Market in current situation, predictive search and the use of social media to show how powerful digital marketing can get with artificial intelligence and can create major changes for the ease of marketing. We will also discuss about some strategies of Automated Marketing and its future scope. It can be the future of marketing and can shape modern life wisely indeed. Artificial intelligence and machine learning revolutionizes marketing strategies by adding new domains in near future.

Keywords: Digital marketing · Artificial intelligence · Machine learning · Social media · Market automation

1 Introduction

Artificial intelligence is a main progressive term that is used in Digital Marketing for having wide domains. It is mainstreaming an progressively common term that uses machine learning [1] and enables functions accordingly to be used technically in various computerized business domains and various sections also. Artificial intelligence marketing is the future of digital marketing in the era of digitization ecommerce industry is extending lacks a unified, concrete definition. Here, where AI can assume a vital job in breaking down the huge information, through utilization of AI controlling the client and the databases could be extracted effectively to give knowledge on the achievement or disappointment of advertising effort. AI has arisen to enable advertisers to break down reported information to convey strongly focus on promoting offers [2, 3]. Managers of showcasing takes choices about their items, circulation channels, promoting brands, cost and so on, in view of the conduct of clients, contenders, providers, some unsure elements relating to political matters, govt. guidelines and by and large economy of the nation. Advertising is confusing field for administrators to settle on choices as promoting choice depends on investigation and judgment in which

A. P. Pandian et al. (Eds.): ICCBI 2019, LNDECT 49, pp. 968–978, 2020.
https://doi.org/10.1007/978-3-030-43192-1_107

information, ability and meetings of professional's assumes an imperative job [4]. To assist supervisors making choices dependent on systematic and judgmental elements.

Artificial intelligence actions, for example, AI, design response, gathering, learning representation, thinking, experimental search, estimation analysis, idea removal, pattern examination, standard of conduct think about, area looking and so forth [4]. Every one of these components of AI are relevant in promoting leaders who utilize their insight and instinct to take care of advertising issues.

2 Literature Review of AI and Digital Marketing

During the 21st century, Artificial intelligence (AI) has developed a main area of research in fundamentally all areas- science, engineering, medicine, business, finance, accounting, stock market, law, education, economics and marketing, along with others like (Tay and Ho (1992), Masnikosa (1998), Wongpinunwatana et al. (2000), Raynor (2000), Stefanuk and Zhozhikashvili (2002), Halal (2003) and Metaxiotis et al. (2003)). The area of AI has developed immensely to the scope that monitoring expansion of research becomes a complicated project (Ambite and Knoblock (2001), Balazinski et al. (2002), Cristani (1999) and Goyache (2003)). – "Oke's (International Journal of Information and Management Sciences Volume 19, Number 4, Page 535, 2008)". The founder of Deep Mind Demis Hassabis as indicated– the AI Company of Google, "Artificial intelligence is the way to create intelligent Machines (Ahmed 2015)." It is the most widely established description and also a well-fitting ever since AI is a comprehensive word used for a amount of a change of indicators. Under the AI, there are some subdivisions containing machine learning and deep learning which construction real-world apps of AI, containing search Ideas, voice recognition, virtual assistants and image recognition (Table 1).

Table 1. Proposed technology of digital marketing year

Year	Proposed technology
1990–1993	Launch of first search engine Archie and first clickable web Banner proposed
1994–1997	Launch of Yahoo! And first social media named sixdegrees.com
1998–2000	Birth of Google and MSN and sixdegrees.com shutdown
2001–2004	Universal Music, Linkedin, Gmail and Facebook launched
2005–2010	Launch of YouTube, Tumblr and sales of Amazon across $10 billion
2010–2015	Facebook acquires Whatsapp. Yahoo acquires Tumblr, Google+ launched

2.1   Uses of AI in Digital Marketing

Chat bots are AI programmed frameworks that interface with customers in a simple characterized condition. These frameworks are rapidly transforming into a significant district of inclination for Digital advertisers [5]. Chat bots can work on that same landing page of site and can manage human transitions and commands and gives output through algorithms showing deep learning and applications of AI. Many applications of

AI in the field of digital media marketing is increasing and helping marketers to reach their sales goals and providing users and customers better impact. Many Applications such as content creating chat bots and mobile marketing providing good user interface with effortless Enhance Return of investment through it.

Easier search sessions of users with digital Marketing collaborated with Artificial Intelligence which helps in sales forecasting it will be now easier to predict future marketing trends and Marketers can reach the right target audience with better advertising techniques and with this they will watch ads on the basis of their search interests [8, 9].

3   Impact of AI on Digital Marketing in Today' Market

In these recent times, there have been many trades that expand quickly due to the accessibility of dependable skills. The most general use of skill in advertising is social media. It has always been a highly significant advantage in marketing growth and is needed in the growth of business after artificial intelligence and digital marketing merged its been easier and broad scope ahead for everyone. In this digital Era, everything is getting easier with all the innovative technology having new plans related to marketing because corporate people can promote their products easily by some Techniques [6] (Fig. 1).

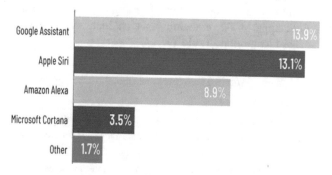

Fig. 1. Companies are using AI for purchasing products [12].

A. Content Marketing – Content advertising is currently a important focus for some brands as a result of its vital job in online networking, sight and sound, and portable pursuit. Several institutions still do not understand the importance of design and how content advertising motivates practically all advanced showcasing. It is in reality essential to make brand concentration [7].

B. Mobile Marketing – Most individuals currently use advanced mobile phones and tablets since it is easy and time-beneficial in light of the fact that they can bring it all over the place and access it whenever needed. The emphasis increase of PDAs and tablet clients make the availability of displaying content for adaptable phase on the grounds that a need. Institutions must update their website to make it cell phone-friendly.

C. Integrated digital media marketing – Incorporated promoting is necessary for guaranteeing that all the informing and correspondence procedures that are used in illustrating are brought collectively crosswise over and based on client. In e.g.: Google has imagined Google+ with numerous purposes, yet one of it is to arrange to see and catch social flag and example [10].

D. Continuous Marketing – Continuous Marketing is Mainstream, Daily marketing of the product is must to gain interests or else it can give leverage to other Business to promote their product at that time [11].

E. Personalized Marketing – Marketing measures and steps are taken Instead of traditional media involving Social And Digital media using Personalized Content based on the best interests of the user. Gain The interests By Cache And History.

F. Visual Marketing – The Art of Marketing Using performing arts and Visual images to attract Viewers and image marketing techniques to showcase users the best of their interests [13].

AI computing has been in the past, the present and will be there in the future. Artificial Intelligence is indicated as producing PC tasks to take care of difficult problems by performance of processes that closely resemble human thinking forms. It is that part of software engineering that creates and reviews canny machines and programming. When we talk about man-made consciousness (simulated intelligence) showcasing is a tactic for using client information to envision the client's best course of action and improve the client venture [14]. Computer based intelligence offers the beast approach to conquer any hindrance between information science and execution by filtering through and examining gigantic dumps of information which was previously an unrealistic procedure. The creation and capacity of information is as of now occurring at an enormous rate, and it's developing exponentially consistently. Now in 2020, it is normal if the world would have made more than 40 zettabyte of information (One zettabyte = One trillion terabytes), and 80% to 90% of it being unstructured. The development of huge information and progressed investigative arrangements has made it workable for advertisers to manufacture a reasonable image of their intended interest groups than at any other time. Man-made consciousness can process both organized and unstructured information with exponentially higher speed and precision than any human could. It is a crucial focal point for organizations attempting to compose their customer information essentially [15]. Advertisers are utilizing the capacity of machine figuring out how to make associations between information indicates all together increase bits of knowledge into their client base. These frameworks can dissect discourse to decide feeling from spoken dialect, make visually-abled renderings to show web-based social networking sequences, and test data to make forecasts (Fig. 2).

According to an audit of about 12,800 e-commerce and Digital Marketing professionals. The plurality of the respondents Are from Europe, Asia-pacific and north American region. Respondents came from all the various fields from all the job roles for the research purpose [16].

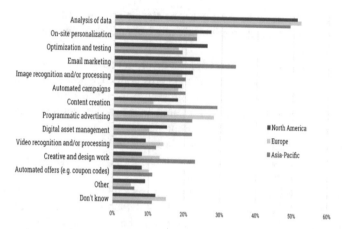

Fig. 2. Companies are using the domains Artificial intelligence for marketing around the globe [17].

4   Some Strategies of Digital Marketing Using AI

A. Modify the experience of user

As the client assumes a crucial job in business, comparatively, content is basic for an advertiser. Shown content advancements can be systematized by being dependent upon the information gathered, for example, client conduct towards purchasing, and what the client is keen on and seeks made by him. To modify the experience of the clients is a very censorious district where, if simulated intelligence enters, it can make a significant effect by an expansion in improved techniques for Search engine optimization and advanced advertising.

B. Analytical marketing to simplify the process of decision making

The new and spic and span data is being achieved and congregated the minute a client looks something on the web for artificial intelligence request. That specific information can uncover the data like behaviors, necessities, and what the future developments are [18]. The advertising systems can be improved to give the suitable data by utilizing the information uncovered. Internet based life can uncover the particular information of the viewpoint to make it peaceful for the merchants to make the best activity. Prognostic campaigns must be finished by, above all, declining the exploration made by the client for an item, to settle on basic leadership a peaceful undertaking. By the by, the advertisers can utilize their insight in regards to the purchaser with assistance of the information and furthermore advance to the client. Computerized reasoning encouraged procedures can go route past contemporary practices like Website design enhancement [19] (Fig. 3).

C. Enhanced ROI

Computer based intelligence has an astonishing trademark that can make the strategy for installment progressively quick. It ensures that the online exchanges done by the clients are incredibly secure. It offers a superior procedure of basic leadership that can

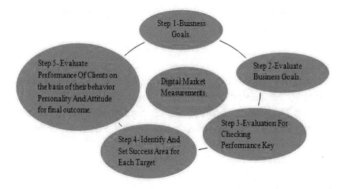

Fig. 3. Steps for market measurements for better market analysis [20].

draw out an incredible Degree of profitability. The AI helps in social affair satisfactory data by examining the conduct of the client and collects its record in improved database structure.

D. Effortless exploring

Simulated intelligence can foresee the future choices of the clients by following their conduct. Despite the fact that the current circumstance of pursuit sessions is reasonably great, it can end up clumsy in some cases. Along these lines, to make these web indexes danger free and smoother, man-made intelligence must be bestowed with advanced showcasing.

E. Predicting the deals

Promoting is said to be always relentless, and it causes an excessive number of modifications in the business because of opposite varieties. Regardless, with the capacities of artificial intelligence, it has turned out to be straightforward to foresee the changing patterns of the market's future. In this way, to spare a lot without any burden, head promoting patterns are being connected.

F. Targeting the right audience

It is particularly vital to contact the correct individuals to make your image and its advantages important. Computerized advertising dependent on simulated intelligence helps in focusing on the correct group of onlookers by concentrating on their requests, interests, and different attributes.

G. The increment in P.R/promotions

Notices are the faultless technique for advancing any brand or its items. As artificial intelligence accumulates all the sufficient data and assesses the client information by foreseeing their conduct, it can additionally help in making Advertisements as per the enthusiasm Groups Which can build P.R.

5 Some Companies Transforming Digital Marketing Using AI

A. PINTEREST

Pinterest Introduces new technology where user can scan the image and search within the large database to get related images which user intends to get it. AI Added some capabilities of deep learning algorithms which combined can do wonders by getting trusted and relevant image data and informative from the data which is based on billion other pictures on the internet [21].

B. MSG.AI

Msg.ai is unique because it reaches users through their most used messaging app because it is the most followed habit of the user in the mobile phone. Using chatbots it provides personalized suggestion with respect to user needs in their Facebook and WhatsApp chats.

C. LAYER 6

Layer 6 Features Personalized Engine That goes not only in user behavior it purchases history too. This Platform analyses all user data faster, Accurately and Reliably and Predicts What Computer Wants next. Designed for enterprise customers are powered by deep learning.

D. SENTIENT

Sentiment was firstly started and introduced by an inventor which was behind great technology Siri. This AI automated system provides data which is user recommended data. It has an actual conversation [22] and provides output relevant to the inputs provided in the form of voice which relates to the features and data around the internet.

E. WAYBLAZER

This technology which is AI based is actually a travel advisor which is combined with the power of IBM'S Watson technology. There is a huge part of e-commerce which Belongs to travelling industry, this aims to provide the storage of large amount of raw and unstructured data which then to be used by sound advise for travel transaction for better understanding (Fig. 4).

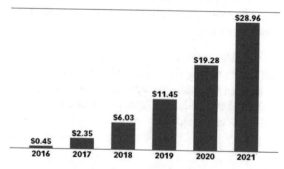

Fig. 4. Artificial intelligence technologies spending worldwide in billions [23].

Note: The Artificial Intelligence Technology spending worldwide which includes consulting and within the company implementation, Cognitive computing, intelligent automation, deep learning and machine learning.

6 Future Scope of AI with Digital Marketing

Artificial Intelligence Nowadays Changed Every possible aspect of predicting Human Behavior and Looking forward in every possible ways to modify algorithms And Help in Content curation, News generating, modifying user interface, Web designing And Chabot's to help in ease for the user. Based On News Generation Media Companies are using Artificial intelligence tools like Wordsmith and Quill to generate blog posts and reviewing forums By automated Bots.

Content Creation Has a very huge impact on Product Recommendations as Companies like Netflix and Amazon are using this by Analyzing and recording user's past behavior to target them showing the products based on cache which is more likely to gain results [24]. Personalized Marketing Using AI can be used in Personalized Emails according to the users behavior and pattern to deliver them relevant info which can convert leads into customers and can boost Business. Persado and Boom train are the tools nowadays evolving and are being used.

Digital Advertising The leading platform of Emerging worlds with the help of artificial intelligence promoting business becomes easy from social Medias to Websites. Analyzing User information, Behaviors And demographics including preference to predict best audience for brand. Ad optimization, pay per click and converting leads using the data analysis Made it a Profitable Platform for Everyone.

Artificial Intelligence could assist in creating a personal view for each individual user for any website. The algorithm will monitor a vast amount of data and provide offers and content that is most suitable for a specific user [25].

7 Future Impacts of Digital Marketing with Artificial Intelligence

85% of Customer Interaction will be managed without human involvement by 2020's. Digital Support Will be asked by six billion connected devices across the globe by 2019. 44% of executives believe artificial intelligence's most important benefit is automated communications that provide data that can be used to make decisions. By the end of 2018, "customer assistants" will recognize customers by the help of voice and face across channels and partners. 80% of executives believe Worker performance and job creation will be improved by Artificial Intelligence. By 2020, 40% of mobile interaction and transaction will be managed by Smart Agents (Fig. 5).

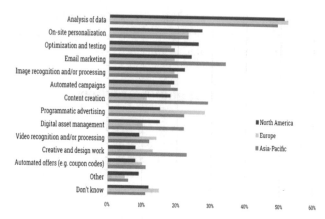

Fig. 5. Companies are using Artificial intelligence in different domains for marketing [26].

8 Conclusion

Artificial Intelligence is going to change computerized advertising. In addition to the fact that it is going to spare time and cash for the advertisers, however it is likewise going to upgrade the client experience considerably. Regardless of whether it is Chabot's drawing in with the clients for continuous bits of knowledge on patterns, brilliant applications managing enormous measures of information easily or prescient examination by information science devices, advanced advertising is set to experience a change in outlook because of this new marvel. We need to create one-on-one associations with the shopper. That is the need of great importance; that is the thing that will influence us to get by in the new world. Promoting as its center is about the brain and heart [27]. It is about knowledge and feeling. Despite the fact that we need to give the useful advantages, we likewise need to possess the enthusiastic space in the psyches of the buyers.

The manner in which we speak with buyers has adjusted publicity. It is transforming every day and that is the place the enormous move has occurred. The huge change that is going on in showcasing, are computerized reasoning and AI. It makes new open doors for narrating and advertising. It will change how individuals collaborate with data, innovation, brands and administrations. Along these lines advertisers must adjust computerized reasoning frameworks in their promoting procedures to prevail in the present time of advanced showcasing. It spares both time and cash for the advertisers, clients, prospects; and possesses the psyches of clients without human inter session [29, 30].

References

1. Content available online at. https://www.researchgate.net/publication/328580914_Artificial_intelligence_Marketing/download
2. Content available online at. https://www.newgenapps.com/blog/using-ai-in-digital-marketing
3. Content available online at. https://www.forbes.com/sites/louiscolumbus/2018/02/25/10-ways-machine-learning-is-revolutionizing-marketing/
4. Content available online at. https://en.m.wikipedia.org/wiki/Machine_learning
5. Content available online at. https://www.researchgate.net/publication/325934359_A_Qualitative_Research_on_Marketing_and_Sales_in_the_Artificial_Intelligence_Age
6. Cassillas, J., Lopez, F.J.M.: Marketing Intelligent System using Soft Computing Techniques: Managerial and Research Application. Springer, Heidelberg (2010)
7. Hurley, S., Moutinho, L., Stephens, N.M.: Solving marketing optimization problems using genetic algorithms. Eur. J. Mark. **29**(4), 39 (1995). ABI/INFORM Global
8. Content available online at. http://www.toprankblog.com/2018/02/artificial-intelligence-transforming-marketing
9. Content available online at. https://www.emarketer.com/Article/How-Artificial-Intelligence-Transform-Digital-Out-of-Home-Marketplace/1016866
10. Content available online at. https://www.business2community.com/digital-marketing/will-artificial-intelligence-impact-digital-marketing-2018-01950626
11. Content available online at. http://blog.aprilsix.com/will-ai-change-marketing
12. Content available online at. http://www.digitalmilesgroup.com/digital-marketing-emerging-age-artificial-intelligence
13. Content available online at. https://mention.com/blog/how-will-ai-change-seo-in-2017-video
14. Content available online at. https://www.salesforce.com/products/marketing-cloud/best-practices/marketing-ai
15. Content available online at. https://www.techemergence.com/artificial-intelligence-inmarketing-and-advertising-5-examples-of-real-traction
16. Content available online at. https://www.researchgate.net/publication/328580914_Artificial_intelligence_Marketing
17. Content available online at. https://www.gartner.com/imagesrv/media-products/pdf/Criteo/Criteo-1-43VKFYC.pdf
18. Content available online at. https://www.expresscomputer.in/artificial-intelligence-ai/artificial-intelligence-will-make-marketing-more-creative-and-effective/25129/
19. Content available online at. https://www.adverity.com/ai-marketing/
20. Jamsandekar, S.S.: Machine learning approach for marketing intelligence: managerial application. Int. J. Eng. Comp. Sci., 7 July. https://www.kvrwebtech.com/digital-marketing-faqs/digital-marketing-measurement-model/
21. Tiwari, S.: History of Digital Marketing: The Evolution that started in the 1980s. 31 Aug 2018.
22. AI Transformation Playbook. https://landing.ai/content/uploads/2018/12/AITransformation-Playbook-v8.pdf
23. Siau, K.: A qualitative research on marketing and sales in the artificial intelligence age, May 2018
24. Bharat, M.: Artificial Intelligence in Web Design, Development and Marketing (2017). https://www.motocms.com/blog/en/artificialintelligence-in-web-design/
25. Daniel, F.: Artificial Intelligence in Retail – 10 Present and Future Use Cases (2018). https://www.techemergence.com/artificial-intelligenceretail/

26. Daniel, F.: Artificial Intelligence in Marketing and Advertising – 5 Examples of Real Traction (2018). https://www.techemergence.com/artificial-intelligencein-marketing-and-advertising-5-examples-of-realtraction/
27. Faggella, D.: Artifical Intelligence Industry – An Overview by Segment (2016). http://techemergence.com/artificial-intelligenceindustry-an-overview-by-segment
28. Marc, W.: Artificial Intelligence in Sales – How Artificial Intelligence is Changing Sales and Selling (2018). https://www.marcwayshak.com/artificialintelligence-in-sale
29. Putri, P.: Collaboration Between Human and Intelligence (AI), the Future of Fraud Prevention. https://integrityindonesia.com/blog/2018/01/22/collaboration-human-artificial-intelligence-ai-future-fraud-prevention/
30. Siau, K., Wang, W.: Building Trust in Artificial Intelligence, Machine Learning, and Robotics. Cutter Bus. Technol. J. 31(2), 47–53 (2018)

A Security Model for Enhancement of Social Engineering Process with Implementation of Multifactor Authentication

Sameer Gupta, Deepa Gupta[✉], and Ruchika Bathla

Amity Institute of Information Technology, Amity University,
Sector – 125, Noida, Uttar Pradesh, India
sameergupta1409@gmail.com, {dgupta, rbathla}@amity.edu

Abstract. Social engineering has developed as a major risk in online communities and it is a successful means to target data systems. Modern social engineering attacks are based on the services that knowledge workers use. In MNCs employees are not physically located at same offices, but works on distributed system which enables employees to interact in real-time environment. The lessening in physical interaction leads to increase in use of communication tools such as WhatsApp, Facebook, Twitter, Gmail etc. makes the system more vulnerable to social engineering attacks. Recent Pulwama attack in India showed the impact of social media and internet with lack of information authenticity that played a major role in escalation of the tension between both India and Pakistan. The lack of "Fear of Getting Caught" among the attackers is the main cause of unlawful activities over the online communities. This paper provides a security model that will help reducing social engineering attacks to a large extend.

Keywords: MFA · Iris recognition · Social engineering · Gaussian algorithm · Phishing

1 Introduction

Over the past decade internet has emerged as the largest means of information and communication. In our day to day lives, communication has expanded to various online communication channels. Today communication channels have increased dramatically and users are able to connect with their clients from different states or from other countries. This increase in communication channel also increased the number of scams and the cyber-attacks. When a user connects with other user over the internet the intruder interrupts the communication and get the unauthorized access or security details of the user. We all know that companies use different techniques and tools to stop different type of attacks and attackers who try to get unauthorized access or try to cause damage to the company. But not all the companies use strong security models and are very vulnerable to the attacks. Most of the companies uses only one factor of authentication which can be any method such as, the client sent encrypted data with the private key to the receiver and receiver decrypt the data by using that private key, or simple username and password method which is not so secure technique to secure the

© Springer Nature Switzerland AG 2020
A. P. Pandian et al. (Eds.): ICCBI 2019, LNDECT 49, pp. 979–987, 2020.
https://doi.org/10.1007/978-3-030-43192-1_108

data on the server. This vulnerability of the system raises the need of multifactor authentication so as to ensure the security of the system.

With the development of hacking techniques, the static passwords are almost too easy to hack and are very vulnerable to the hacking. So the solution to the old static password technique is to implement MFA.

The purpose of this paper is to provide a Multifactor Authentication security model that not only provides a secure data authentication method, but it also provides an Iris recognition for uniquely identifying each entity and user over the internet community. This paper proposed a MFA security model that contains Iris recognition as a security measure which will help in uniquely identify a person in case of any unlawful activity.

The paper is organized as follows. Section 2 describes the related work. Section 3 describes the architecture and proposed system. Section 4 describes the Methodology adopted. Section 5 describes the experimental result. Section 6 is the conclusion of the paper.

2 Related Work

Secure Bank Transaction by Using MFA: MFA technology (Multifactor Authentication) model that include four step authentication. The security model helps to secure the process of online banking using MFA technique against phishing attacks. Authorized user can go through the transactions without almost no possibility of hacking. Session keys generation to improve the security over the online banking was based on Gaussian distribution [1].

Security of Cloud Implementation Using MFA: MFA technique requires two phase of identification for the security purpose, one usually reserved item, for example authentication code, other usually a natural token, for example as software or an application. In the relation of this, two factors can be categories as something a user know and something a user have. The passwords are encrypted as the passwords of the users need to be secured and protected. It shields user from the different cyber-attacks like DDOS (Denial of service), spear-phishing, brute force attack by the encryption of password and implementing the multifactor authentication (with a OTP and other security measures) [2].

Authentication Techniques: The techniques begin with a basic check that takes into account the username and multi-factor authentication password, taking into account the factors of knowledge, ownership, and inheritance. Multifactor authentication technology is the safest authentication method discussed [3, 4].

Authentication of Grid Computing: In grid computing environment, an authentication mechanism has been presented in this security model. Grid computer requires an authentication mechanism which should be robust and secure. Also the security model should be able to provide an environment that has to be secure and robust enough to complete the needs of big scale distributed and diverse grid computing [5].

Mitigation of Social Engineering: The increase in use of modern communication channels over the internet and intensive use of social media has created new challenges for the information security professionals. Social engineering from social networking sites becomes a big concern for enterprises because of the inclination of social engineers targeting employees. Social engineers use these medium to exploit information of an organization. This paper proposed a security framework based on ICT security policy control which help in reduction of social engineering attacks. The framework was called SESM (social Engineering through Social Media) framework which addresses the challenges faced by social engineering [6].

3 Proposed Model

In proposed work, we have proposed a session key generation by Gaussian distribution along with countermeasure and MFA (Multi Factor Authentication) that reduces the phishing attacks as well as some other cyber-attacks on electronic mails. We have proposed a MFA security model for email login. MFA technique authenticates the users using different level of security steps. At first the user inputs his email address along with the password. The email and the password provided by user get verified in the database. Now the user has to go through Iris recognition. The reason behind adding Iris recognition was that every human being have a unique Iris which helps in finding an individual in case of any misconduct hence providing better security. After these two step verification the user receives a session key which is generated by Gaussian distribution on their mobile. Once the user enters the session key then system logs into the email account. If during the MFA process a user fails authentication, then the countermeasures will be performed and the IP address of the intruder will be recorded. If same IP address fails 3 times, then the IP address will be added to the block list along with all the captured details of the intruder. So by this proposed model each email address will be connected to a Iris hence a person (Fig. 1).

4 Methodology

4.1 Authentication of User

The first module of every application where registration of user and login system and administrator's login is present. In earlier steps, a valid user account can be block by the unknown user without the knowledge of account owner password.

4.2 Iris Recognition

An Iris recognition is an automated method for biometric identification which works on video images of the irises of an individual by applying complex pattern-recognition techniques. Retinal recognition is different from Iris recognition. Iris recognition uses camera technology with little infrared beam that helps in getting images that are rich in

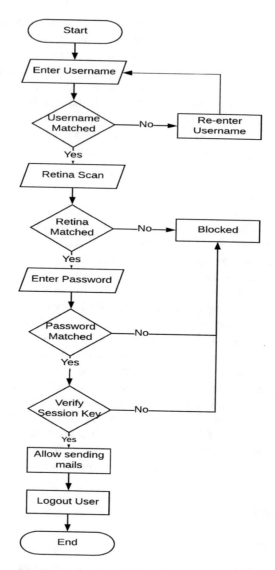

Fig. 1. System architecture for proposed system

detailed structure of iris. Proposed model uses the Daughman's Algorithm segmentation method for Iris Recognition which gets the iris recognition in 5 steps (Fig. 2).

4.3   Random Numbers Generation Using Gaussian Distribution Method

For the propose of session tracking the generation of random number mechanism was mandatory. There are many methods existing for standard simple random generation of numbers. For the simple normal random generation of numbers an algorithm is used and named here.

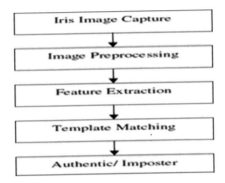

Fig. 2. Block diagram of stages in Iris recognition

Method: Addition of Uniform Random Variables.
By using the central limit theorem, the normal variables can be generated by easiest way possible.

The central limit theorem which is a weak convergence result that exhibits that each addition of multiple small individual random variable is roughly about normally distributed. For almost approximating normal random variants the adoption of central limit theorem for random variables U(0, 1) is taken. So simple normal variable generated by this algorithm is used in this proposed model.

1. *Generate 12 independent uniform numbers,*

$$U_1, U_2 \ldots U_{12} \sim iidU(0, 1)$$

2. *Return* $Z = \sum_{i=1}^{12} U_i - 6.$

For the generation of single simple normal variable number, 12 uniform random variables are required [7].

4.4   Counter Measures

Phishing and online scamming attacks can be protected by many of the available security model but the proposed model not only stops such attacks but it also records and reports the person responsible for such attacks. Key point here is that it reports the person responsible (Because of the Iris) and not just the IP address of the system [8]. Counter Measure also includes many small security measures also such as, if a user remains inactive for more than 10 min then the session will time out, if an intruder tries to login using user's credentials then the system will block that intruder for future, in case of intrusion the system will record the details of the intruder such as IP address of the system, hostname, Iris recognition of the intruder if provided etc. These Counter Measures enhances the security of the model and reduces the chances of successfully attacked by phishers and scammers.

5   Experimental Results

In MFA user have to pass through several security measure to login to the system. These steps include:

- Information Factors ("these are known by user only"), example passwords and security questions
- Ownership Factors ("These are the things only user owns"), example user's smartphone
- Own Factor ("This is the things only user is"), example user's Iris.

5.1   Password and Username

Password and username are the most commonly used method for user authentication. The working of password and username is very straight forward. The user enters a unique identifier (i.e. username) along with the password set by user only, and if the username password entered are correct then the user is authenticated. This is a very commonly used technique where user can be identified just by providing assigned unique identifier and password.

Simple authentication is shown in Fig. 3 where user inputs username (i.e. email id) along with secret password. The username should set a combination of symbols and alphabets to make the password more secure. This authentication method is the most commonly used authentication technique but this alone is not a secure method as the password can be captured by intruder by multiple hacking techniques.

Fig. 3.   User authentication

5.2   Iris Recognition

The Daughman's Algorithm is used to create Iris Recognition. This is done in five steps:

(1) Image Acquisition: The image is captured and the image is uploaded in the system for iris recognition (Fig. 4).
(2) Iris Localization: This is the step where the image processing is done to extract locate the iris from the eye and create the outlines for the same.

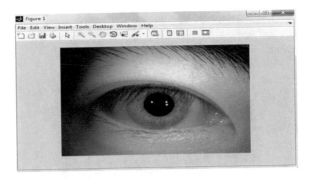

Fig. 4. Image taken

(3) Feature Extraction: This is where the extracted iris is converted into binary format for the actual data matching in the database [9] (Fig. 5).

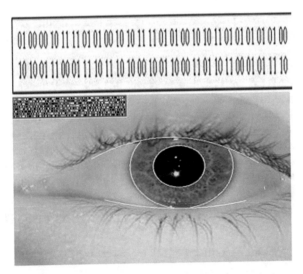

Fig. 5. Image showing the outline of the iris captured from step3 and its binary conversion.

(4) Iris Matching: The extracted data is now matched with the available data in database for iris recognition.

5.3   Session Key Generation

It is an authentication service which in addition of normal username and password also uses the Session Key to authenticate a user. System generates unique password and sends is to user's smartphone. The user uses this session key to login to the email account which increases the security during online session. To authenticate, user enters the session key generated by the system application along with the username and password. Session Key passwords are unique for every user, and this password is changed every minute. In case the password is captured by a hacker or an intruder, it can't be used beyond its lifetime [10] (Fig. 6).

Fig. 6. Session key verification

5.4 Counter Measure

Every time a hacker tries to login to the system then due to Multifactor Authentication, hacker's unsuccessful attempt to login will be recorded by the system administrator. All the hacker's details such hostname, IP address, path etc. are recorded and the hacker gets blocked and the further attempt from the same IP address will not be entertained. Moreover if the system detects any unauthorized attempt then the system sends a notification to the user.

When the intruder tries to modify any data or create any malicious event, the intruder is not permitted to perform the activities since intrusion is done with unauthorized user name and password. If the changes are done with unauthorized access, then the information of the intruder are gathered and it is being sent to the administrator in the secure manner. From those details the intruder can be identified very easily and any further action performed by the intruder can be blocked thus preventing code injection. The details like IP address, hostname, date, time, path, etc., are reported (Fig. 7).

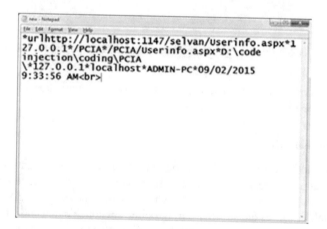

Fig. 7. Intrusion detection

6 Conclusion

This paper proposes a security model that will enhance the security and authenticate each data or each entity over the internet so that to reduce the unlawful activities over the online communities. Multifactor authentication delivers customers with increased security by using multiple authentication types to validate a transaction. While not necessarily, the proposed security system should be implemented over the different areas in online communities like Communication services, News updates, cloud services etc. as it provides more security to their customers over the internet. The service provider need to understand that the cost of security can be offset by customer confidence and anti-theft protection. There are many smartphones companies like Google pixel, Samsung etc. that has implemented Iris recognition in their smartphones which in near future will help the proposed model to be easily implemented across different online services.

The proposed model can further be used in different social media account where each individual has to go through MFA which includes Iris recognition which will help in uniquely identifying the user in case of any misconduct. Model can also be used in news site, cloud services, domain creation etc. so as to uniquely identify the person responsible in case of any unlawful activities.

References

1. Manasa, S., Mullaimalar, P., Manivannan, S.S.: Securing online bank transactions from phishing attacks using MFA and secure session key. Indian J. Sci. Technol. 8(S2), 123–126 (2015)
2. Gupta, N., Rani, R.: Implementing high grade security in cloud application using multifactor authentication and cryptography. Int. J. Web Semant. Technol. 6(2), 9 (2015)
3. Dorlikar, V., Chandavale, A.: A survey on authentication techniques and user recognition. Int. J. Sci. Res. 4, 1307 (2015)
4. Liu, D., Zhang, Z.J., Zhang, N.: A two factor user authentication scheme for medical registration platform (2013)
5. Farouk, A., Fouad, M.M.: Authentication mechanisms in grid computing environment (2012)
6. Prabhu, R., Chellapan, C.: Detection of phishing websites and secure transactions. IJCNS 1(11), 15–21 (2011)
7. Adu Michael, K., Adewale Olumide, S.: Mitigating cybercrime and online social networks threats in Nigeria. In: Proceedings of World Congress on Engineering and Computer Science, vol. 1 (2014)
8. Mohammed, M.M., Elsadig, M.: A multi-layer of multi factors authentication model for online banking services. In: 2013 International Conference on Computing Electrical and Electronics Engineering (ICCEEE) (2013)
9. Hazlewood, V., Kovatch, P.: Improved grid security posture through multifactor authentication. In: 2011 12th IEEE/ACM International Conference on Grid Computing (GRID) (2011)
10. Chaudhri, S., Tomar, S.S., Rawat, A.: Design, implementation and analysis of multi layer, multi factor authentication (MFA) setup for web mail access in multi trust networks. In: 2011 International Conference on Emerging Trends in Networks and Computer Communications (ETNCC) (2011)

Implementation of IoT for Trash Monitoring System

Vansh Manchanda, Ruchika Bathla$^{(\boxtimes)}$, and Deepa Gupta

Amity Institute of Information Technology,
Amity University Noida, Noida, Uttar Pradesh, India
vansh12manchanda@gmail.com,
{rbathla, dgupta}@amity.edu, bathla.ruchika@gmail.com

Abstract. Waste materials management is among the primary issues that the globe faces regardless of the case of a developed or developing nation. The classic method of manually monitoring the waste products in waste material bins is usually a troublesome procedure and utilizes more human being work, some cost which could very easily become prevented with this system. This system will assist you to alert the municipality by measuring the garbage and location of the bin with the status of the bin. This system is capable to identify 4 types of rubbish i.e. home waste, newspaper, glass, plastic material and suggested that upon LCD display. In this system, we are applying ultrasonic detectors which help in measuring the amount of garbage and ARM microcontroller which regulates system procedures. We can to anticipate this system can help us to create a green environment through monitoring and managing and variety of trash within a smartly way with Internet-of-Things.

Keywords: ARM · Arduino · Ultrasonic sensor · ESP8266 · DC motor · Power supply · IOT

1 Introduction

According to Indian News Paper and record shown by the NDA government in the Indian Parliament on 31st of January, 2019 nearly 24% of solid waste produced in the region and it was also processed. From nearly 2 lakh tonnes per day (tpd) of highly waste of solid produced all over in the country. Instead of only nearly forty thousand tpd was also processed as on the final day of the month of January. The ministry showed data. According to the Swachh Bharat Mission (SBM) 2014, the government wants to attain 100% scientific procurement of solid waste, and also make the region Open-Defecation Free (ODF) throughout this year i.e., 2nd of October, 2019, the one-fiftieth birth year of Shri M. Gandhi. IOT - Trash Monitoring System will play a vital role to achieve this mission when the country population is increased with speedy and high fiction rates [1]. The four types of waste can detect automatically with this system and can be shown through an LCD screen installed on top of the bins. With the types of these wastes, we can categories into organic, reusable and recyclable waste. Nowadays, municipal corporations collects waste every dayon a regular basis but they don't have any information that the bins are full or not. In some places, the bins may be overloaded and this attracts the animals and insects. Thus, it will create an unhygienic condition in

© Springer Nature Switzerland AG 2020
A. P. Pandian et al. (Eds.): ICCBI 2019, LNDECT 49, pp. 988–998, 2020.
https://doi.org/10.1007/978-3-030-43192-1_109

surrounding areas. The records gathered via detectors is sent into a server where ever it is hung on and is prepared. The accumulated information can now be used for statement and enhancing each day's choice of garbage cans to become collected, hard the paths for this reason [7].

The objectives of Garbage Monitoring System are to design prototype of Internet-of-Things (IoT) and alert the municipalities to collect the garbage after identifying the type and level of bins installed in a particular area.

2 Literature Review

2.1 Trash Monitoring System

In this program electric battery centered smart rubbish bins exchange info with one another using cellular mesh systems, and a router and machine gather and evaluate the information for any service provisioning and contains numerous IoT methods intended for consumer comfort. The suggested SGS has been operated like a pilot task in Gangnam district, Seoul, Republic of Korea, to obtain a great doze weeks period which usually indicated the common quantity of waste materials reduced simply by 33% [1]. In this newspaper, simply by adding diverse realizing and conversation systems real-time solid waste material bin monitoring system was created. The system composed of bins with messfühler nodes, gateways, and base train stations. Messfühler nodes measure and transfer waste products conditions inside a bin in every single gain access to, gateways ahead info to base place after getting and foundation station shop info for even more use. The device helped to reduce the collection route and fuel price [12].

With this paper, to the smart waste receptacles are linked to the net to get the actual status info of the begin waste receptacles. The huge development in the populace in the past a few years led to more garbage disposal. An effective waste materials administration system is necessary to prevent the development of ailments. In this, to the smart dustbins will be monitored as well as the decisions happen to be taken as every position of bins. The waste containers are placed through the town or reasons and are interfaced with a tiny controller centered program with IR detectors and RF segments. MOLAR messfühler searches for the level of waste material in waste products bin and sends the signal to micro-control. The same transmission is protected and submitted to the RF receiver through RF transmission device. RF recipient receives the signal and decodes this at the central program. A great net connection is usually enabled by using a LAN wire from the device. That is received, analyzed and processed inside the cloud that displays the rubbish position inside the waste rubbish bin on the GUI on the web internet browser [14].

The main incapability's of present waste materials bin collection devices will be:

1. Insufficient information about the gathering up time and region.
2. Not enough proper system for monitoring, tracking the trucks and trash rubbish bin that have been gathered in real-time.
3. There is no evaluation of the quantity of sturdy waste present inside the trash can and the encircling location because of the scattering of waste.
4. Lack of speedy response to immediate cases just like truck incident, break down and long time idling.

2.2　Hardware Description

1. Microcontroller ARM7 (LPC-2148): The LPC-2148 Microcontrollers depend on 32 bit architecture, this kind of offers some of the features exactly like low electrical power usage and high general overall performance. It has ISP/IAP either using through on-chip boot loader software in system or application programming and also offers on-chip stationary RAM (8 kB–40 kB), on-chip adobe flash memory (32 kB–512 kB), and accelerator allows 60 MHz high-speed process (Fig. 1).

Fig. 1. Microcontroller ARM7 (LPC-2148) [16]

2. ULTRASONIC sensor (HC-SR04): We have placed the ultrasonic sensor in the height of the rubbish bin for realizing the level and type of garbage/wastage filled in the rubbish bin. The Ultrasonic sensor offers two eyes, the first is a transmitter that transfers an ultrasonic wave which is received simply by the 2nd eye we. e, receiver through representation by any kind of object/material.
   Ultrasonic Sensor is used to determine range among things and sensors by simply mailing a sound wave for a particular rate of recurrence to reveal back again. The time used being sound wave generated and bouncing back again is possible to estimate the length between sensor and subject and period taken simply by the pulse is really for from travel of ultrasonic indicators thus period taken as Time/2 [13] (Fig. 2).

Fig. 2. ULTRASONIC sensor (HC-SR04) [17]

$$Distance = Speed * Time/2$$

Speed of Sound at sea level 343 m/s or 34300 cm/s then

$$Distance = 17150 * time$$

3. GPS Module: GPS (Global Positioning System) can be used to decide the area (longitude-latitude) and time of the garbage bin (Fig. 3).

Fig. 3. GPS module [18]

4. GSM-GPRS Module: GSM (Global System for Mobile) is the global standard intended for mobile communication and this may hook up to a Network. GSM also enable sending or maybe receiving TEXT MESSAGE for conversation over the cellular network.

A Quad-band GSM/GPRS answer is offered simply by SIM800, with an SMT type customer software embedded system. Quad-band 850/900/1800/1900 MHz are maintained SIM800. Transmitting of the tone of voice, TEXT, and info is done with low electric power consumption making use of the GSM Module. The component dimensions are 24 * 24 * 3 mm, featuring Bluetooth and embedded AT [8] (Fig. 4).

Fig. 4. GSM-GPRS module [19]

5. LCD (2 × 16): LCD (Liquid Crystal Display) is used to show the output from the modules which can be executed throughout the sensors insight automatically. This is really useful for the garbage enthusiast to verify the particular smart bin shows correct message or maybe not (Fig. 5).

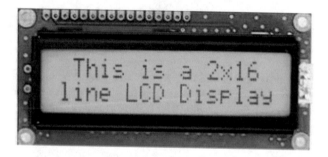

Fig. 5. LCD (2 × 16) [20]

6. MAX232: IC MAX232: The MAX232 is utilized to connect microcontroller with computer systems, TTL/CMOS logic to RS232 converters and they are likewise utilized in RS232 wires. It obtains information coming from the sensor and process onto it. This even comes close to the received info with all the threshold level arranged and appropriately output is usually generated. A128-bit large memory user interface and exclusive accelerator structures enable 32-bit code performance at optimum clock price [11] (Fig. 6).

Fig. 6. MAX232: IC MAX232 [21]

7. ESP8266: The ESP 8266 is a low electrical power incredibly built-in microchip. It really is mainly used in IoT based task because it uses low electric power. ESP8266EX continues to be designed for cellular, wearable consumer electronics and Net of Points applications with all the aim of attaining the lowest electricity usage having a combination of many proprietary methods [10]. The ESP8266 is usually user-friendly and low-cost gadget to offer internet connection to your tasks. The basic top features of ESP01 will be IOT Tasks, Access Stage Portals, Cellular Info signing, Smart House Software, Find out basics of social networking and Portable Consumer electronics, Smart bulbs, and Sockets (Fig. 7).

Fig. 7. ESP8266 (Microchip) [22]

3   Problem Statement

Following are the advantage and limitation of the existing system:

3.1   Advantage of the Existing System

The garbage will probably be collected upon a time-to-time basis. As a result there won't be any foul smell throughout the bin. Real-time notice to gather the garbage will help consequently in saving about fuel usage, thus minimizing the danger to the environment.

3.2   Limitation of the Existing System

The present program has got the limitations like it is time-consuming, as vehicles go to vacate the storage containers, but they are actually empty. The cost is usually high with the unhygienic environment. Even the poor odor triggers a harmful environment. Therefore, the proposed unit talks about steps to make use of the recent developments in technology to make the place spending tidy. The implementation starts with set up ESP8266 simply by flashing the latest version of the firmware [6].

Even as we see often the dustbin is overcome flown and concern person does not get the info on the time and because of which subconscious condition type in the environment, in the same time frame bad smell come out coming from waste and spread out in surrounding. Because of the dirty environment some dangerous illnesses very easily spreadable in a given vicinity. The existing system used for washing the dustbin is not successful and has following disadvantages:

1. Less powerful and time intensive
2. Cost is high
3. Environment turn into unclean.
4. Due to the bad smell of garbage human beings could cause illness
5. More visitors and sound due to pickup truck used to brush your dustbin

Researched papers associated with the trash monitoring system having complications inside the methodology and type of program design. A complete summary of those researched documents stated that we get the scope of further research. The complications are stated beneath what happened whilst designing or perhaps applying this program of these research papers. The garbage monitoring devices had been made to find out the amount of garbage yet this cannot identify the type of garbage [2]. It is rather hard to discover the accurate level of garbage using a single ultrasonic sensor. The cost of the device was so high, we need to design a google android app or perhaps site to monitor the positioning of the general system and it likewise reduces the expense of the system. We now have found 1 major problem. e, all of us can't monitor the receptacles in a particular region, if there are several bins possibly we have open up every single cover of the containers in many of these areas. Therefore, we have set up the Lcd-display on the top of the bin to exhibit the friendly status of level and garbage type with exclusive bin identification [3].

4   Proposed System

In this, we are going to make a Smart Trash Monitoring Systems which detects the level of trash in the bin and states whether it is empty or full through the mobile app and we can get to know the status of the 'Smart Trash Can' from anywhere in the world via the Internet. We are using the Ultrasonic Sensor, ARM Microcontroller, LCD Display & other low price devices that may be useful for cost-effective trash monitoring system and easily adaptable in urban and rural areas of the country. In future scope, we can add these bins with the solar panels and the bins summarized report can be shown on top of the bins. The garbage collector can also check the real-time data of bins and it may be helpful in the locations where waste was overloaded in the bins. We can also monitor the peak time of a particular location when waste is produced in a huge quantity or low quantity.

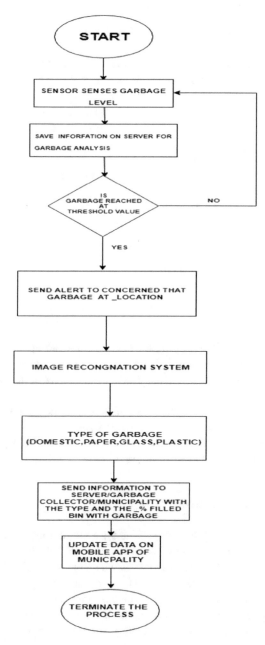

Fig. 8. Flow diagram of trash monitoring system using IoT

Through this paper, we can represent the trash monitoring system with IoT for managing domestic waste, especially in crowded areas. This system will monitor the bin's garbage and indicate the garbage level of each bin and will continuously alert the garbage collectors to collect the garbage from the bins which are already full. We can make the environment clean through the Trash Monitoring System and efficiently management of bins in cities and rural areas (Fig. 8).

TRASH MONITORING SYSTEM ALGORITHM

```
SENSOR INITATE A GARBAGE LEVEL ALERT SAVE DATA ON SERVER
IF GARBAGE_VALUE > THRESHOLD_VALUE SEND ALERT TRASH LEVELS > = OVERFLOW
    INTERNATIONALIZING IMAGE RECOGNITION
                        IF GARBAGE_TYPE = DOMESTIC OR PAPER
            OR   GLASS   OR   PLASTIC   WASTE   DATA = GARBAGE_LEVEL   AND
GARBAGE_TYPE
            SEND DATA TO SERVER
            SEND DATA TO MUNICIPALITY SEND DATA TO GARBAGE
            COLLECTOR
            UPDATE THE PERCENTAGE WITH
            GARBAGE_TYPE ON MOBILE APP
            END
ELSE
            REINITIALIZE THE SENSOR
```

5   Platforms

It needs a minimum configuration of a system to run the application designed for garbage management system using IoT. Operating System should be a higher version than Window 8 for windows users, Linux Redhat 5 for Linux users and Mac OS XI for Mac users.

Compatibility of the Browser should be match with the operating system i.e., IE (v11+), Opera (58+), Firefox (v50+), Chrome (v42+).

It should also be noticed that an updated version of Java (JRE) is already installed in the system/machine and enabled in the running browser.

6   Conclusion and Future Scope

This paper has described the smart way of the trash monitoring system with sensing of type of garbage in the bin. It is very useful to determine which waste can be re-used, decomposed and recycled. The crowded areas where bin can be full before timing, this system will send an alert to garbage collectors or municipality for collecting the garbage from a particular location, which can be easily tracked from the GPS installed in it. This system is suitable in rural and urban areas as well for helping government scheme "Swatch Bharat Abhiyan" at a very reasonable price. We can track the information from anywhere and at any time about each bin at a particular location in a real time system. As a future scope to this proposed system, we can supply the power from solar panels installed on the top of the bin. The garbage can further be recognized through image recognition system. Further, if implement this project we will reduce the unhygienic condition among the areas. It will help the administration staff for planning, analyzing and developing the areas on larger scale.

This kind of project function is the execution of a wise rubbish administration program applying MOLAR messfühler, microcontroller and Wifi component. This system guarantees the washing of dustbins soon if the trash level reaches the maximum. In the event the dustbin is usually not washed at a particular time, then your record is sent to the larger authority who also can take suitable action up against the concerned service provider. With the implementation of this kind of project our country can become as one of the hygienic and smart nations of the world.

References

1. Raveshiya, H.: Waste management system using IoT. Int. J. Recent Trends Eng. Res. 14(4), 220–229 (2018). https://doi.org/10.23883/ijrter.2018.4231.wp68d
2. Srikantha, N.: Waste management in IoT-enabled smart cities: a survey. Int. J. Eng. Comput. Sci. 5, 3326–3331 (2017). https://doi.org/10.18535/ijecs/v6i5.53
3. Christensen, T.H.: Introduction to waste management. Solid waste technology & management. Int. J. Eng. Comput. Sci. 3, 1–16 (2010). https://doi.org/10.1002/9780470666883.ch1
4. Ravi, S.L.: IoT based smart garbage monitoring system. Int. J. Eng. Res. Adv. Technol. 6(5), 3438–3442 (2015). https://doi.org/10.31224/osf.io/dru2e
5. Rajavizhi, N.: IOT based waste management in smart city using IR. Int. J. Eng. Res. Adv. Technol. 4(4), 21–26 (2018). https://doi.org/10.7324/ijerat.2018.3246
6. Muley, B., Dadmal, A., Nikhade, L.P.: IOT based waste management system for smart city. Int. Res. J. Eng. Technol. (IRJET) 5(11), 826–829 (2018)
7. Rajavizhi, N., Hamsaveni, P., Kavya, P., Priyadharshini, K.: IOT based waste management system in smart city. Int. Res. J. Eng. Technol. (IRJET) 5(2), 1984–1986 (2018)
8. Thakker, S., Narayanamoorthi, R.: Smart and wireless waste management. In: International Conference on Innovations in Information, Embedded and Communication Systems, vol. 5, pp. 43–47 (2015). https://doi.org/10.1109/ICIIECS.2015.7193141
9. Anitha, A.: Garbage monitoring system using IOT. IOP Conf. Ser. Mater. Sci. Eng. 263 (2017). https://doi.org/10.1088/1757-899x/263/4/042027. Vol. 6, pp. 1–12
10. Vinothkumar, B., Sivaranjani, K., Sugunadevi, M., Vijayakumar, V.: IOT based garbage management system. Int. J. Sci. Res. (IJSR) 6, 99–101 (2017)
11. Mustafa, M.R., Ku Azir, K.N.F.: Smart bin: Internet-of-Things garbage monitoring system. In: MATEC Web of Conferences, vol. 140, p. 01030 (2017) https://doi.org/10.1051/matecconf/201714001030. vol. 6, no. 4, pp 295–300
12. Masane, G.V., Naphade, R.A.: "Smart garbage monitoring system" present and future. Int. J. Trend Sci. Res. Dev. (IJTSRD) 1(6), 26–31 (2017)
13. Mahajan, S.A., Kokane, A., Shewale, A., Shinde, M., Ingale, S.: Smart waste management system using IoT. Int. J. Adv. Eng. Res. Sci. (IJAERS) 4(4) (2017)
14. Saji, R.M., Gopakumar, D., Kumar, H., Mohammed Sayed, K.N., Lakshmi: A survey on smart garbage management in cities using IOT. Int. J. Eng. Comput. Sci. (11), 18749–18754 (2016). https://doi.org/10.18535/ijecs/vsi11.04

15. Yusof, N.M., Jidin, A.Z., Rahim, M.I.: Smart garbage monitoring system for waste management. In: MATEC Web of Conference, vol 97, pp 1–6 (2017). https://doi.org/10.1051/matecconf/20179701098
16. https://svsembedded.com/lpc2148.php
17. https://www.piborg.org/sensors-1136/hc-sr04
18. http://www.trainelectronics.com
19. http://www.micro4you.com
20. https://www.engrsgarage.com/electronic-components/1-lcd-module-datasheet
21. https://www.electricaltechnology.org/2014/10/max232-construction-working-types-uses.html
22. https://electrosome.com

Author Index

A

Abdul Samath, J., 926
Abiya, M. J., 340
Adithya, B., 597
Agrawal, Divyanshu, 897
Agrawal, Vaibhav, 484
Ahuja, Vaishnavi, 325
Ajitha, P., 297, 333
Amin, Sujit S., 657
Anchaliya, A. Harsha, 48
Anny Leema, A., 81
Arefin, Mohammad Shamsul, 511, 631
Arockiam, L., 799
Aruna, O., 119
Arutchelvan, K., 547
Ashok Kumar, C., 539

B

Balaji, V., 590
Bathla, Ruchika, 979, 988
Bekal, Chaithra, 575
Betty Jane, J., 213
Bhattacharyya, Saikat, 645
Bhavani, G., 412
Boddu, Vamshikrishna, 698
Bolla, Vineela, 500

C

Cejka, Tomas, 465
Chaitanya, K. Krishna, 704
Chalumuru, Suresh, 500
Chandrasekaran, N., 885
Chauhan, Dhruv, 356
Chauhan, Mayank, 686
Chawla, Haresh, 856

Chellatamilan, T., 610
Chheda, Aadil, 903
Chintakindi, Sanjeevkumar, 856

D

David, Noel, 195
Dawngliani, M. S., 885
Deokar, S. A., 712
Devulapalli, Krishna, 139
Dharmadhikari, Chinmaya M., 474
Dhipikha, N., 48
Dixit, A. M., 712
Dubey, Ashutosh, 267

E

Emudapuram, Muralikrishna, 698

G

Ganesh, E. N., 213
Gangurde, Smit, 856
Garg, Karan, 909
Geethika Choudary, P., 500
Ghosh, Argha, 563
Gilbert, Gilbert M., 845
Giri, Vivek, 252
Goswami, Vasudha, 306
Govindarajan, M., 941
Gowri, S., 948, 953
Goyal, Shivani, 363
Gulzar, Zameer, 81
Gupta, Deepa, 824, 979, 988
Gupta, Sameer, 979
Gupta, Sudhriti Sen, 252, 275, 281

© Springer Nature Switzerland AG 2020
A. P. Pandian et al. (Eds.): ICCBI 2019, LNDECT 49, pp. 999–1001, 2020.
https://doi.org/10.1007/978-3-030-43192-1

H

Hart, Andrew, 145
Hasan, Nazmul, 834
Hemachandran, K., 698
Hossain, Md. Shahadat, 631
Hynek, Karel, 465

I

Ibrahim, Irshadh, 811
Islam, Muhammad Nazrul, 834
Itabada, Pranav Souri, 500

J

Jadav, Nilesh Kumar, 67
Jayalakshmi, V., 32, 374
Jayaraj, T., 926
Jegatha Deborah, L., 778, 876
Jennifer, B. Brenda, 340
Jidiga, Goverdhan Reddy, 240
John Jeffrey, G. A., 953
Joy Blessy, U., 948
Justy Mirobi, G., 799

K

Kabadi, Suvarna, 758
Kamarasan, M., 232
Kanade, Vijay A., 492
Karthik, D. U., 953
Karuna Sree, B., 21
Katnur, Fuad A., 555
Katyal, Diksha, 824
Kaur, Bhagwant Jot, 731
Kavitha, S., 48
Keerthivasan, C., 590
Keerthivasan, S., 590
Khan, Imtiyaz, 60
Khan, Talha, 60
Khandare, P. M., 712
Khatri, Zaid, 60
Kimaro, Honest, 845
Kodabagi, M. M., 748
Kohli, Disha, 275
Komal, 686
Kousalya, K., 1
Kumar, G. Sai, 698
Kumar, Rakesh, 721, 731

L

Lakkannavar, Banashankaramma F., 748
Lakshminarayanan, Arun Raj, 340
Lalmawipuii, R., 885
Lokeshkumar, R., 645

M

Mahendra Chozhan, R., 397
Malviya, Vijay, 306
Manchanda, Vansh, 988
Maneesh, K., 39
Manoj, V., 39
Manoj, Y. R., 575
Mari, Kamarasan, 523
Mariya Christeena Vinitha, K., 948
Mary Dayana, A, 677
Mathew, Rejo, 128, 153, 162, 169, 177, 195, 205, 224, 325, 348, 363, 381, 389, 420, 474, 484, 897, 903, 909, 917, 934
Mathews, Rejo, 188, 356
Meel, Priyanka, 314
Meena, R., 412
Meghana, M. S., 100
Mittal, Neetu, 267
Mody, Aashray, 348
Mohankumar, Rekha, 811
Mounica, G. Lakshmi, 333
Mounika, M., 620
Mudigonda, Swathi, 698
Mundargi, Zarinabegam K., 758
Muslu, Isa, 758
Mvungi, Nerey, 845

N

Nadaf, Raju A., 555, 758
Naik, Susen, 758
Naik, Susen P., 555, 748
Naiman, Shililiandumi, 845
Nair, Abhishek, 188
Nandhini, M., 1
Nayak, Lipsa, 32
Negi, Astha, 290
Niharika, M., 21
Nivetha, S., 610
Nivitha, K., 13
Nur, Mohammad Imtiaz, 511

P

Pabitha, P., 13
Padmavathy, P., 81
Pahwa, Arshdeep Singh, 959
Palanisamy, Ramaraj, 89
Panimozhi, K., 100
Pappu, Shiburaj, 60
Parmar, Mahiraj, 420
Parthasarathy, Saravanan, 340
Paul, Vedant, 177
Pavithra, K., 100
Pavithra, V., 374
Philo Prasanna, I., 866

Poornima, N., 575
Prathyusha, Lakshmi, 333
Prayaga, Chandra, 139, 145
Prayaga, Lakshmi, 139, 145
Princy, R. Jane Preetha, 340
Prudhvi, Balina, 297
Punitha Ponmalar, P., 789

R
Ragha, Lata, 657
Ragupathi, T., 941
Rahman, Md. Rashadur, 631
Rai, Shubham, 381
Rajeshkumar, J., 1
Rajkumar, S. C., 778
Raju, Leo, 590
Rakshitha, R., 575
Ramanjaneyulu, B. Seetha, 704
Rani Roopha Devi, K. G., 397
Rao, Anushree Janardhan, 575
Rathore, Tapendra Singh, 205
Reddy, Bana Suresh Kumar, 297
Rohilla, Shourya, 224

S
Sabnis, Omkar Vivek, 645
Sahana, S., 100
Sam Emmanuel, W. R., 677
Sammulal, P., 240
Santhi, G., 597
Santhi, K., 610
Sathiamoorthy, S., 531, 539
Sathiya Priya, R., 547
Saxena, Komal, 959, 968
Sengupta, Sudhriti, 259, 267, 290
Senthilkumar, C., 232
Senthilrajan, A., 563
Shaikh, Shakila, 60
Shanmugapriya, T., 1
Sharma, A. K., 769
Sharma, Amit, 119
Sharma, Ila, 769
Sharma, Keshvi, 721
Sharma, Mrigank, 389
Sharma, Pratyush, 306
Sharma, Shailja, 457
Sharma, Shraddha, 769
Shubha, N., 100
Singh, Ajmer, 824
Singh, Anuj Kumar, 686
Singh, Manashwi, 934
Singh, Tanmay, 314
Sinha, Akankasha, 162
Sinha, Gita, 457

Sivakumar, S., 531
Sivasangari, A., 333
Sohail, Shahla, 811
Solani, Sona, 67
Soni, Sarvesh, 917
Soukup, Dominik, 465
Sravan, K. S., 704
Sridevi, R., 412
Srinivasa Gupta, N., 620
Srinivasulu, Senduru, 948
Sugumar, R., 39, 48
Suguna, M., 866
Suralkar, Sunita, 856

T
Tayal, Shubham, 698
Thangkhanhau, H., 885
Thenmozhi, S., 811
Thilagavathi, B., 340
Tiwari, Antriksh, 153
Tyagi, Shivam, 259

U
Uma, Darapu, 741
Umadevi, K., 112

V
Valarmathi, B., 610, 620
Vara Prasad, G. L., 582
Vemuri, Sreya, 128
Venkat Kishore, A., 39
Venkata Subramanian, D., 39, 48
Venkateswara Rao, K., 582
Venugopal, Narmatha, 523
Vigneshwaran, S., 668
Vijay, A., 112
Vijayalakshmi, C. R., 789
Vimali, J. S., 953
Vinoth, R., 876
Vinothini, R., 48
Vishnubhatla, Arvind, 428, 440, 449

W
Wade, Aaron, 145
Wajhal, Ayushi, 281

Y
Yadav, Vicky, 169
Yasmin, Farzana, 511
Yogi, Manas Kumar, 741

Z
Zeeshan, Mohd, 968

Printed in the United States
By Bookmasters